Multi-Agent Systems

Simulation and Applications

Computational Analysis, Synthesis, and Design of Dynamic Models Series

Series Editor

Pieter Mosterman
*The MathWorks
Natick, Massachusetts*

Discrete-Event Modeling and Simulations: A Practitioner's Approach, *Gabriel A. Wainer*

Discrete-Event Modeling and Simulations: Theory and Applications, *Gabriel A. Wainer and Pieter J. Mosterman*

Model-Based Design for Embedded Systems, *Gabriela Nicolescu*

Multi-Agent Systems: Simulation & Applications, *edited by Adelinde M. Uhrmacher and Danny Weyns*

Computational Analysis, Synthesis, and Design of Dynamic Models Series

Multi-Agent Systems
Simulation and Applications

Edited by
Adelinde M. Uhrmacher
Danny Weyns

CRC Press
Taylor & Francis Group
Boca Raton London New York

CRC Press is an imprint of the
Taylor & Francis Group, an **informa** business

CRC Press
Taylor & Francis Group
6000 Broken Sound Parkway NW, Suite 300
Boca Raton, FL 33487-2742

© 2009 by Taylor and Francis Group, LLC
CRC Press is an imprint of Taylor & Francis Group, an Informa business

No claim to original U.S. Government works

Printed in the United States of America on acid-free paper
10 9 8 7 6 5 4 3 2 1

International Standard Book Number: 978-1-4200-7023-1 (Hardback)

This book contains information obtained from authentic and highly regarded sources. Reasonable efforts have been made to publish reliable data and information, but the author and publisher cannot assume responsibility for the validity of all materials or the consequences of their use. The authors and publishers have attempted to trace the copyright holders of all material reproduced in this publication and apologize to copyright holders if permission to publish in this form has not been obtained. If any copyright material has not been acknowledged please write and let us know so we may rectify in any future reprint.

Except as permitted under U.S. Copyright Law, no part of this book may be reprinted, reproduced, transmitted, or utilized in any form by any electronic, mechanical, or other means, now known or hereafter invented, including photocopying, microfilming, and recording, or in any information storage or retrieval system, without written permission from the publishers.

For permission to photocopy or use material electronically from this work, please access www.copyright.com (http://www.copyright.com/) or contact the Copyright Clearance Center, Inc. (CCC), 222 Rosewood Drive, Danvers, MA 01923, 978-750-8400. CCC is a not-for-profit organization that provides licenses and registration for a variety of users. For organizations that have been granted a photocopy license by the CCC, a separate system of payment has been arranged.

Trademark Notice: Product or corporate names may be trademarks or registered trademarks, and are used only for identification and explanation without intent to infringe.

Library of Congress Cataloging-in-Publication Data

Multi-agent systems : simulation and applications / Adelinde M. Uhrmacher and Danny Weyns.
    p. cm. -- (Computational analysis, synthesis, and design of dynamic models series ; 4)
  Includes bibliographical references and index.
  ISBN 978-1-4200-7023-1 (alk. paper)
  1. Intelligent agents (Computer software) 2. Distributed artificial intelligence. 3. Computer simulation. 4. Application software--Development. I. Uhrmacher, Adelinde. II. Weyns, Danny. III. Title. IV. Series.

QA76.76.I58M864 2009
006.3--dc22   2009016309

Visit the Taylor & Francis Web site at
http://www.taylorandfrancis.com

and the CRC Press Web site at
http://www.crcpress.com

Contents

Preface vii

Acknowledgments ix

About the Editors xi

Contributors xiii

Part I: Background

1 Multi-Agent Systems and Simulation: A Survey from the Agent Community's Perspective
   Fabien Michel, Jacques Ferber, and Alexis Drogoul 3

2 Multi-Agent Systems and Simulation: A Survey from an Application Perspective
   Klaus G. Troitzsch 53

3 Simulation Engines for Multi-Agent Systems
   Georgios K. Theodoropoulos, Rob Minson, Roland Ewald, and Michael Lees 77

Part II: Simulation for MAS

4 Polyagents: Simulation for Supporting Agents' Decision Making
   H. Van Dyke Parunak and Sven A. Brueckner 109

5 Combining Simulation and Formal Tools for Developing Self-Organizing MAS
   Luca Gardelli, Mirko Viroli, and Andrea Omicini 133

6 On the Role of Software Architecture for Simulating Multi-Agent Systems
   Alexander Helleboogh, Danny Weyns, and Tom Holvoet 167

7 Replicator Dynamics in Discrete and Continuous Strategy Spaces
   Karl Tuyls and Ronald Westra 215

8 Stigmergic Cues and Their Uses in Coordination: An Evolutionary Approach
   Luca Tummolini, Marco Mirolli, and Cristiano Castelfranchi 243

Part III: MAS for Simulation

9 Challenges of Country Modeling with Databases, Newsfeeds, and Expert Surveys
   Barry G. Silverman, Gnana K. Bharathy, and G. Jiyun Kim 271

10 Crowd Behavior Modeling: From Cellular Automata to Multi-Agent Systems
   Stefania Bandini, Sara Manzoni, and Giuseppe Vizzari 301

11 Agents for Traffic Simulation
   Arne Kesting, Martin Treiber, and Dirk Helbing 325
12 An Agent-Based Generic Framework for Symbiotic Simulation Systems
   Heiko Aydt, Stephen John Turner, Wentong Cai, and Malcolm Yoke Hean Low 357
13 Agent-Based Modeling of Stem Cells
   Mark d'Inverno, Paul Howells, Sara Montagna, Ingo Roeder, and Rob Saunders 389

Part IV: Tools

14 RoboCup Rescue: Challenges and Lessons Learned
   Tomoichi Takahashi 423
15 Agent-Based Simulation Using BDI Programming in **Jason**
   Rafael H. Bordini and Jomi F. Hübner 451
16 SeSAm: Visual Programming and Participatory Simulation for Agent-Based Models
   Franziska Klügl 477
17 JAMES II - Experiences and Interpretations
   Jan Himmelspach and Mathias Röhl 509

Glossary 535

Index 543

Preface

Motivation

Multi-agent systems (MAS) consist of multiple entities called agents that interact in a shared environment aiming to achieve some individual or collective objective. Simulation studies the modeling of the operation of a physical or conceptual system over time. For more than two decades, the field of MAS and the field of simulation have been combined in very active strands of research. Where, on the one hand, agents have been used extensively as a tool for designing modeling and simulation problems, on the other hand, simulation has often been used for the design of MAS in a variety of application domains. Bringing these research endeavors together promises valuable insight and benefits between the two strands of research. This book aims to integrate and consolidate the acquired knowledge and experiences, and outline promising directions for future research in this field.

Scope

The focus of this book is on the intersection of MAS, simulation, and application domains. The book starts from the observation that simulation is being used for agents and agents are being used for simulation in a variety of application domains.

Simulation is being used for agents. MAS are particularly useful in application domains that are characterized by an inherent distribution of resources and highly dynamic operating conditions. Examples are open electronic markets, traffic and transportation systems, and supply chain management. Because of the highly dynamic conditions in the environment, testing by simulation becomes imperative in the software development process of MAS. Simulation, however, is also exploited for agents' decision making. For example, software agents in military systems are equipped with simulation capabilities to reason about the future effects of their decisions enabling them to select the most suitable actions.

Agents are being used for simulation. A typical example where MAS models are used in simulation is for simulating traffic situations. In microscopic simulation of traffic situations, drivers' behavior can naturally be modeled as agents. Traffic simulations are used to predict traffic flow, to study the effects of control measures, etc. Another example is the domain of online simulations that are used to optimize industrial processes. Such systems apply MAS to exploit their inherent adaptability to cope with the fast evolving distributed business environment. A simulator gathers data from sensors that are connected to the real world; it simulates possible future states, and adapts the system based on the outcome of the simulations. Such systems have been built, for example, for online optimization of semiconductor manufacturing where quick decision making is required in order to improve the operational performance of a semiconductor wafer fabrication plant.

Key Aims

The objective of the book is twofold. The first aim is to provide a comprehensive overview of the research in the intersection of MAS, simulation, and application domains, thereby consolidating knowledge and experiences. The study of various aspects such as modeling abstractions, methodological approaches, formal foundations, and platforms in different domains will yield insights to the underlying principles of MAS and simulation. The second aim of the book is to initiate avenues for future research. In particular, the aim is to outline how the obtained insights provide opportunities and enable future research in each of the related areas: MAS, simulation, and different application domains.

Audience

This book serves different communities. The first audience is the multi-agent system community, including students in computer science and artificial intelligence, and researchers and engineers in these fields. The book provides an overview of experiences with MAS simulation and tools for exploiting simulation in MAS software development, and outlines promising venues for using simulation in MAS engineering. The second audience is the modeling and simulation community, including students in computer science, simulation, mathematics, economics, etc., and researchers and engineers in these fields. The book presents modeling and simulation techniques and tools for applying MAS in simulation systems, and identifies opportunities for future research in this area. Finally, the third audience includes professionals from various application domains such as civil engineering, biological systems, transportation and traffic, and factory scheduling and control. The book depicts experiences with MAS simulation in these various application domains; it offers a repertoire of concepts and tools to enrich modeling and simulation practice.

Structure of the Book

The book is structured in four parts. Each part consists of a coherent set of chapters that discuss aspects of MAS and simulation from a particular viewpoint and outline promising directions for future research. Part I systematically introduces background and basic principles. It gives an overview of the intersection and co-evolution of MAS and simulation from two different perspectives, i.e., the agent community and social science. To illuminate the richness of methodological developments, the area of simulation engines for MAS is explored.

Part II discusses the use of simulation in MAS. It explains simulation support for agent decision making, the use of simulation for the design of self-organizing systems, the role of software architecture in simulating MAS, and the use of simulation for studying learning and stigmergic interaction in MAS. Part III zooms in on MAS for simulation. Discussed subjects are an agent-based framework for symbiotic simulation, the use of country databases and expert systems for agent-based modeling of social systems, crowd-behavior modeling, agent-based modeling and simulation of adult stem cells, and agents for traffic simulation. Finally, Part IV presents a number of state-of-the-art platforms and tools for MAS and simulation, including *Jason*, JAMES II, SeSAm, and RoboCup Rescue.

Acknowledgments

This book has been a joint effort and many people contributed to it. We would like to take this opportunity to express our thanks.

- For making the book possible: all contributors of this book.
- For enthusiastic and lively discussions about the structure and direction of this book: all contributors, but in particular Alexander Helleboogh and Jan Himmelspach.
- For their work in compiling this book, which revealed the non-cooperativity of diverse LaTeX packages: Nadja Schlungbaum and Jan Himmelspach
- For constructive and detailed reviews on the chapters: all contributors and Bikramjit Banerjee, Peter Vrancx, Andreas Tolk, Levent Yilmaz, Michael North, David Hales, and Bruce Edmonds.
- For assistance and support: the editor of the series "Computational Analysis, Synthesis, and Design of Dynamic Models," Pieter Mosterman, and Taylor & Francis Group employees Nora Konopka and Amy Blalock.

Adelinde M. Uhrmacher and Danny Weyns
Rostock and Leuven

About the Editors

Adelinde M. Uhrmacher is head of the modeling and simulation group at the Institute of Computer Science at the University of Rostock. She completed her Ph.D. in Computer Science at the University of Koblenz in 1992. Afterward she did her postdoctoral research at the Arizona Center for Integrative Modeling and Simulation at the University of Arizona in Tucson. She was senior researcher at the Artificial Intelligence Lab at the University of Ulm from 1994 until 2000, when she joined the University of Rostock. Her research is aimed at methodological developments referring to modeling formalisms, simulation methods, tools, and applications. To the former belong extensions of the DEVS formalism which address the agent's need for dynamic interaction and composition patterns, and parallel discrete event approaches for simulating multi-agent systems. Application areas span from agent-based modeling of social, economic and cell biological systems to exploiting simulation as a means in agent-oriented software engineering. The diverse needs of multi-agents applications led to developing the plug-in based simulation system JAMES II (a JAva based Multipurpose Environment for Simulation, aka a Java-based Agent Modeling Environment for Simulation). Adelinde Uhrmacher has served on program committees and in the organizations of various international conferences. She has been member of the editorial board of several journals and editor-in-chief of the *Simulation - Transactions of the SCS* from 2000 until 2006.

Danny Weyns received a Master's degree in Electronics from the Hoger Instituut Kempen in 1980. He worked as a lector at the Hogeschool voor Wetenschap en Kunst in Brussels. In 2001, Danny received a Master's degree in Informatics from the Katholieke Universiteit Leuven; and in 2006, he received a Ph.D. in Computer Science from the same university for work on multi-agent systems and software architecture. From 2004–2007, Danny participated in a joint IWT project with Egemin on the development of a decentralized control architecture for automated guided vehicles. The essential role of simulation in the architectural design of this real-world multi-agent application is described in one of the chapters in this book. After finishing his Ph.D. Danny started working as a senior researcher in DistriNet and is now funded by the Research Foundation Flanders. His main research interests are in software architecture, self-managing systems, multi-agent systems, and middleware for decentralized systems. Danny has published over 80 reviewed articles in these research areas. He is co-editor of four books and several special issues in journals. Danny co-organized several international workshops and special sessions at various international conferences. He served on the program committee of various international conferences, and he has performed review work for several journals.

Contributors

Heiko Aydt
Parallel and Distributed Computing Center
Nanyang Technological University
Singapore

Rafael H. Bordini
Department of Computer Science
University of Durham
UK

Cristiano Castelfranchi
Institute of Cognitive Sciences and Technologies
National Research Council (CNR)
Italy

Jacques Ferber
LIRMM
Université Montpellier II, CNRS
France

Alexander Helleboogh
Department of Computer Science
Katholieke Universiteit Leuven
Belgium

Paul Howells
Cavendish School of Computer Science
University of Westminster
London, UK

Arne Kesting
Faculty of Transportation and Traffic Sciences "Friedrich List"
Dresden University of Technology
Germany

Stefania Bandini
Complex Systems and Artificial Intelligence Research Center
University of Milan-Bicocca
Italy

Sven A. Brueckner
NewVectors: Division of TechTeam Government Solutions, Inc.
USA

Alexis Drogoul
Research Institute for Development (IRD)
Paris, France

Luca Gardelli
Department of Electronics, Computer Sciences and Systems
Studiorum-Università di Bologna
Italy

Jan Himmelspach
Institute of Computer Science
University of Rostock
Germany

Jomi F. Hübner
Multi-Agent System group, G2I ENS Mines
Saint-Etienne
France

G. Jiyun Kim
Department of Electrical and Systems Engineering
University of Pennsylvania
USA

Gnana K. Bharathy
Department of Electrical and Systems Engineering
University of Pennsylvania
USA

Wentong Cai
Division of Computer Science
Nanyang Technological University
Singapore

Roland Ewald
Institute of Computer Science
University of Rostock
Germany

Dirk Helbing
Department of Humanities and Social Sciences
ETH Zürich, Switzerland

Tom Holvoet
Department of Computer Science
Katholieke Universiteit Leuven
Belgium

Mark d'Inverno
Department of Computing
University of London
UK

Franziska Klügl
Department of Artificial Intelligence and Applied Computer Science
University of Würzburg
Germany

Michael Lees
School of Computer Science and Information Technology
University of Nottingham
UK

Fabien Michel
LIRMM
Universitè Montpellier II, CNRS
France

Sara Montagna
Department of Electronics, Computer Sciences and Systems
Università di Bologna
Italy

Ingo Roeder
Institute for Medical Informatics, Statistics and Epidemiology
University of Leipzig
Germany

Barry G. Silverman
Department of Electrical and Systems Engineering
University of Pennsylvania
USA

Martin Treiber
Faculty of Transportation and Traffic Sciences
"Friedrich List" Dresden University of Technology
Germany

Stephen John Turner
Parallel and Distributed Computing Center
Nanyang Technological University
Singapore

Giuseppe Vizzari
Complex Systems and Artificial Intelligence Research Center
University of Milan-Bicocca
Italy

Malcolm Yoke Hean Low
School of Computer Engineering
Nanyang Technological University
Singapore

Rob Minson
School of Computer Science
University of Birmingham
UK

Andrea Omicini
Department of Electronics Computer Sciences and Systems
Università di Bologna
Italy

Mathias Röhl
Institute of Computer Science
University of Rostock
Germany

Tomoichi Takahashi
Department of Information Science
Mejo University
Japan

Klaus G. Troitzsch
Computer Science Faculty
University of Koblenz
Germany

Karl Tuyls
Faculty of Industrial Design
Eindhoven University of Technology
Netherlands

Ronald Westra
Department of Mathematics
Maastricht University
Netherlands

Sara Manzoni
Complex Systems and Artificial Intelligence Research Center
University of Milan-Bicocca
Italy

Marco Mirolli
Institute of Cognitive Sciences and Technologies
National Research Council (CNR)
Italy

H. Van Dyke Parunak
NewVectors: Division of TechTeam Government Solutions, Inc.
USA

Rob Saunders
Key Center of Design Computing and Cognition
University of Sydney
Australia

Georgios K. Theodoropoulos
School of Computer Science
University of Birmingham
UK

Luca Tummolini
Institute of Cognitive Sciences and Technologies
National Research Council (CNR)
Italy

Mirko Viroli
Department of Electronics Computer Sciences and Systems
Università di Bologna
Italy

Danny Weyns
Department of Computer Science
Katholieke Universiteit Leuven
Belgium

I

Background

1 **Multi-Agent Systems and Simulation: A Survey from the Agent Community's Perspective** 3
   Fabien Michel, Jacques Ferber, and Alexis Drogoul
   Introduction • M&S for MAS: The DAI Case • MAS for M&S: Building Artificial Laboratories • Simulating MAS: Basic Principles • The Zeigler's Framework for Modeling and Simulation • Studying MAS Simulations Using the Framework for M&S • Conclusion

2 **Multi-Agent Systems and Simulation: A Survey from an Application Perspective** 53
   Klaus G. Troitzsch
   Simulation in the Sciences of Complex Systems • Predecessors and Alternatives • Unfolding, Nesting, Coping with Complexity • Issues for Future Research: The Emergence of Communication

3 **Simulation Engines for Multi-Agent Systems** 77
   Georgios K. Theodoropoulos, Rob Minson, Roland Ewald, and Michael Lees
   Introduction • Multi-Agent System Architectures • Discrete Event Simulation Engines for MAS • Parallel Simulation Engines for MAS • Issues for Future Research

Simulation studies have accompanied the development of multi-agent systems from the beginning. Simulation has been used to understand the interaction among agents and between agents and their dynamic environment. The focus has been on test beds, and the description and integration of agents in dynamic virtual environments. Micro and individual-based simulation approaches also became aware of the new possibilities that the agent metaphor and the corresponding methods offer. The area of social simulation played a key role, being enriched and equally challenged by more detailed models of individuals and contributing itself to a better understanding of the effects of cooperation and coordination strategies in multi-agent environments. During the first decade the focus of research has been on the modeling layer. Only gradually, the need to take a closer look at simulation took hold, e.g., how to ensure efficient and repeatable simulation runs.

The chapter by Fabien Michel, Jacque Ferber, and Alexis Drougol on "Multi-agent systems and simulation: A survey from the agent community's perspective" gives a historical overview of the methodological developments at the interface between multi-agent systems and simulation from an agent's perspective. The role of test-beds in understanding and analyzing multi-agent systems in the 1980s, the development of abstract agent models, the role of social simulation in promoting research in multi-agent systems and simulation, and the challenges of describing agents and their interactions shape the first decade of research. With the environment of agents becoming an active player, the questions about timing move into focus and with them traditional problems of simulator design, e.g., how to handle concurrent events. For in-depth analysis simulation questions like validity of models, design and evaluation of (stochastic) simulation experiments need to be answered, but also new one emerge in the context of virtual, augmented environments.

The significant impact of social science on multi-agents research is reflected in the realm of simulation. In the chapter "Multi-Agent Systems and Simulation: A Survey from an Application Perspective," Klaus Troitzsch traces the first simple agent-based models back to the 1960s. Particularly, analyzing the micro and macro link of social systems, i.e., the process of human actions being (co-) determined by their social environment and at the same time influencing this social environment, permeates agent-based simulation approaches from the beginning, despite the diversity of approaches which manifests itself in varying level of details, number of agents, interaction patterns (e.g., direct or in-direct via the environment), and simulation approach. The aim of these simulation studies is to support or falsify theories about social systems. However, in doing so, they also reveal mechanisms that help to ensure certain desirable properties in a community of autonomous interacting entities and as such can be exploited for the design of software agent communities as proposed by the "socionics" initiative.

A long neglected area of research has been the question of how to execute multi-agent models in an efficient and correct manner. This question is addressed in the chapter by Georgios Theodoropolous, Rob Minson, Roland Ewald, and Michael Lees on "Simulation Engines for Multi-Agent Systems". Often agent implementations were translated into discrete stepwise "simulation" with no explicit notion of simulation time. However, the need to associate arbitrary time with the behavior of agents and synchronize the behavior of agents with the dynamics of the environment led to discrete event simulation approaches. As the simulation of multiple heavy weight agents require significant computation effort, sequential discrete event simulators are complemented by parallel discrete ones and help an efficient simulation of multi-agent systems. Interestingly, in the opposite direction we find the agent approach exploited to support the distributed simulation of latency simulation systems. Simulation systems are interpreted as agents and the problem of interoperability and synchronization of these simulation systems is translated into terms of communication and coordination.

1

Multi-Agent Systems and Simulation: A Survey from the Agent Community's Perspective

Fabien Michel
CReSTIC - Université de Reims

Jacques Ferber
LIRMM - Université Montpellier II

Alexis Drogoul
IRD - Paris

1.1	Introduction...	3
1.2	M&S for MAS: The DAI Case	5
	The CNET Simulator • The DVMT Project • MACE: Toward Modern Generic MAS Platform	
1.3	MAS for M&S: Building Artificial Laboratories ..	7
	The Need for Individual-Based Modeling • The Microsimulation Approach: The Individual-Based Modeling Forerunner • The Agent-Based Modeling Approach • Agent-Based Social Simulation: Simulating Human-Inspired Behaviors • Flocks and Ants: Simulating Artificial Animats	
1.4	Simulating MAS: Basic Principles	13
	Agent • Environment • Interactions • Modeling Time • Simulating MAS as Three Correlated Modeling Activities • A Still-Incomplete Picture	
1.5	The Zeigler's Framework for Modeling and Simulation...	29
	Source System • Experimental Frame • Model • Simulator • Modeling Relation: Validity • Simulation Relation: Simulator Correctness • Deriving Three Fundamental Questions	
1.6	Studying MAS Simulations Using the Framework for M&S	32
	Does the Model Accurately Represent the Source System? • Does the Model Accommodate the Experimental Frame? • Is the Simulator Correct?	
1.7	Conclusion	41
	References	42

1.1 Introduction

This chapter discusses the intersection between two research fields: (1) *Multi-Agent Systems* (MAS) and (2) *computer simulation*.

On the one hand, MAS refer to a computer research domain that addresses systems which are composed of micro level entities -agents-, which have an autonomous and proactive behavior and interact through an environment, thus producing the overall system behavior which is observed at the macro level. As such, MAS could be used in numerous research and application domains. Indeed, MAS are today considered as an interesting and convenient way of understanding, modeling, designing and implementing different kind of (distributed) systems. Firstly, MAS could be used as a programing paradigm to develop operational software systems. MAS are particularly suited to deploy distributed software systems that run in computational contexts wherein a global control is hard or not possible to achieve, as broadly discussed in [Zambonelli and Parunak, 2002]. At the same time, MAS also represent a very interesting modeling alternative, compared to equation based modeling, for representing and simulating real-world or virtual systems which could be decomposed in interacting individuals [Parunak et al., 1998; Klügl et al., 2002].

On the other hand, computer simulation is a unique way of designing, testing and studying both (1) theories and (2) real (computer) systems, for various purposes. For instance, according to Shannon, simulation is defined as [Shannon, 1975]:

> *"The process of designing a model of a real system and conducting experiments with this model for the purpose either of understanding the behavior of the system and/or of evaluating various strategies (within the limits imposed by a criterion or a set of criteria) for the operation of the system."*

With respect to this definition, simulation could be thus considered as a *computational tool* used to achieve two major motivations which are not mutually exclusive:

- The understanding of a real system;
- The development of an operational real system.

So, the opportunities of using both MAS and simulation are numerous, precisely because they can be applied and/or coupled in a wide range of application domains, and for very different purposes. In fact, the number of research works and software applications that belong to the intersection between MAS and simulation is simply huge. To have an idea of how close MAS and simulation are today, one can consider that there are about 800 instances of the word *simulation*, distributed among more than 35% of the 273 papers published the 2007 agent community's most known conference: AAMAS'07 [Durfee et al., 2007]. Moreover, considering this already very high percentage, one has to take also into account that, in this conference, there was no session directly related to simulation at all.

So, numerous works belong to the intersection between MAS and simulation. In this book, this intersection is considered according to two main perspectives:

1. Modeling and Simulation (M&S) for MAS;
2. MAS for M&S.

Roughly, the first case refers to projects wherein computer simulation is used as a **means** for designing, experimenting, studying, and/or running a MAS architecture, whatever the objectives. Especially, simulation could be used to ease the development of MAS-based software, by following a *software-in-the-loop* approach (e.g., [Riley and Riley, 2003]): Simulation allows one to design, study and experiment with a MAS in a controlled and cost-efficient way, using simulated running contexts in place of the real running context (e.g., the Internet). Examples of related application domains are Supply Chain Management (SCM), Collective Robotics, self-organized systems, and Distributed Artificial Intelligence (DAI) to cite just a few of them.

The second case is related to simulation experiments that use MAS as modeling paradigm to build *artificial laboratories*. Well-known examples are the simulation of virtual insect colonies (e.g., [Drogoul and Ferber, 1992]), artificial societies (e.g., [Epstein and Axtell,

1996]), social systems (e.g., urban phenomena [Bretagnolle et al., 2006]), etc. Today, using artificial laboratories represents a unique way of testing theories and models for many application domains.

The first part of this chapter provides the reader with an overview of some historical motivations belonging to each perspective. As it would be endless to enumerate them all, the chapter only focuses respectively on (1) DAI aspects for illustrating the M&S for MAS perspective, and on (2) some relevant works done in the scope of Artificial Life (AL) for illustrating the MAS for M&S perspective.

Beyond the previous distinction, there is of course only a thin line between concerns belonging to each category as they both rely on simulating MAS. So, the second part presents and studies some basic concepts of MAS in the scope of simulation. However, although this study highlights some general concerns related to MAS simulations, in the third part we will argue on the idea that it does not represent the full picture of the intersection between MAS and simulation. In fact, understanding this intersection requires us to shift our point of view from the MAS field to the simulation field. Indeed, simulation is not only a computational tool, but a real scientific discipline. As such, M&S already defines some general issues, whatever the application domain. That is why one has also to consider these issues to apprehend some of the challenges that the MAS community is facing regarding simulation.

So, the next part of the chapter studies MAS simulation works according to a pure M&S perspective, derived from the Zeigler's *framework for M&S* proposed in the early seventies [Zeigler, 1972]. Doing so, our goal is twofold: (1) put in the light the relevance of studying MAS simulation according to this perspective, and (2) highlight some major challenges and issues for future research.

1.2 M&S for MAS: The DAI Case

In the literature, MAS are often related to the history of DAI [Ferber, 1999]. Not surprisingly, most of the first MAS researches wherein simulated systems are involved belong to the DAI field. To understand why there is a need for simulation to engineer MAS, let us here consider three historical researches which are often cited as forerunner examples of MAS: (1) the Contract Net Protocol [Smith, 1980], (2) the Distributed Monitoring Vehicle Testbed (DVMT)[Lesser and Corkill, 1983], and (3) the MACE platform [Gasser et al., 1987].

1.2.1 The CNET Simulator

Proposed by Smith, the Contract Net Protocol [Smith, 1980] specifies problem-solving communication and control for task allocation (or *task sharing*) over a set of distributed nodes. To experiment with this negotiation protocol, Smith needed some instances of relevant distributed problems. To this end, Smith developed a simulated system called CNET, the purpose of which was to simulate such instances. For instance, CNET was used to simulate a distributed sensing system (DSS): A network of sensor and processor nodes distributed in a virtual geographic area. In this experiment, the agent was in charge of constructing and maintaining a dynamic map of vehicle traffic in the area.

With these experiments, Smith was able to test and refine his protocol. Obviously, it would have been very hard to achieve such experiments without the use of a simulated environment. The reason is twofold: (1) The deployment of the system in a real running context would have been costly and (2) real-world experiments cannot be entirely controlled (e.g., hardware reliability) so that they do not ease the development process, as they include irrelevant noise in it. So, CNET simulations were used to provide the suitable **testbed** which was required by Smith to experiment with the Contract Net Protocol.

The point here is that the CNET simulator has played a fundamental role in this research: CNET was **the** means by which the Contract Net Protocol has been evaluated and validated. Therefore the CNET settings were a critical parameter doing this research and this was already noticed by Smith in the original paper. This historical research is a forerunner example of the importance that simulation has always taken in DAI research, and also MAS engineering. This is even more clear with the second example.

1.2.2 The DVMT Project

Initiated in 1981 by Lesser and Corkill and continued through 1991, the Distributed Monitoring Vehicle Testbed (DVMT) project [Lesser and Corkill, 1983] has been some of the most influential research for the MAS field. The purpose of the DVMT project was clear: Provide a generic software architecture, a so-called **testbed**, for experimenting with cooperative distributed problem solving networks composed of *semi-autonomous* processing nodes, working together to solve a problem (an abstract version of a vehicle monitoring task in this project). In this perspective, the DVMT was clearly presented as a simulation tool, the purpose of which was to enable researchers to explore design issues in distributed problem solving systems.

The interesting thing is that this research emphasized the need of a real DAI tool rather than on a particular DAI problem to solve. Indeed, considering the DAI research experiments which were previously conducted up until that time, such as the well known Hearsay II project [Erman et al., 1980], it was clear that further investigations were required to understand and explore all the issues raised by the use of these *Functionally Accurate, Cooperative Distributed Systems* (FA/C) [Lesser and Corkill, 1981], especially regarding the size of the network and the communication topology of the nodes. Questions like, how to select an appropriate network configuration with respect to the selected task characteristics, was at the heart of such research, and this did require extensive experimentation. Lesser and Corkill pointed out the difficulties of doing such experiments because of the inflexibilities in the design of the existing systems: The design of these systems were focused on the problem to solve rather than on the corresponding testbed. Thus, experimenting with the existing systems was time consuming or even not feasible.

Therefore, designing the DVMT, the approach of Lesser and Corkill was to [Lesser and Corkill, 1983]:

- Abstract a realistic distributed solving task to make it more generic and parameterizable.
- Develop the related distributed problem solving system using a software architecture as flexible as possible.
- Build a **simulation system** able to run this system under different environmental scenarios, node and communication topologies, and task data.

Motivating this approach, Lesser and Corkill said [Lesser and Corkill, 1983]:

> *"We feel this approach is the only viable way to gain extensive empirical experience with the important issues in the design of distributed problem solving systems. In short, distributed problem solving networks are highly complex. They are difficult to analyze formally and can be expensive to construct, to run, and to modify for empirical evaluation... Thus, it is difficult and expensive to gain these experiences by developing a "real" distributed problem solving application in all its detail."*

Obviously, Lesser and Corkill were right as they were generalizing approaches like Smith's. But, more than a pragmatic method of doing this kind of research at that time, the approach followed by Lesser and Corkill clearly highlights (1) that modeling and simulation is a key for most of the DAI research and thus (2) why the simulation aspects on which this research relies cannot be bypassed or underestimated.

Here, considering the scope of this chapter, one question is: Is it possible to generalize the importance of modeling and simulation aspects to MAS engineering in general? The answer is a strong yes. Indeed, in the preceding citation, it is possible to replace the words *distributed problem solving systems* by *MAS* to have a statement that still holds today in most cases when engineering multi-agent-based software.

Still, DVMT was only a first step toward a more general way of engineering agent-based systems as it was focused on the simulation of a particular domain, considering a specific paradigm (FA/C), and motivated by pure DAI purposes. The next step relied on proposing a more generic testbed which would not have been built with a dedicated domain in mind. This step was achieved few years later by another historical pioneer work that has also deeply influenced MAS research: the MACE platform [Gasser et al., 1987].

1.2.3 MACE: Toward Modern Generic MAS Platform

The *Multi-Agent Computing Environment*, MACE, is the result of a work initiated by Gasser and his colleagues in the mid-eighties [Gasser et al., 1987]. At that time, thanks to the previous works done on DAI testbeds, there was a general agreement in the DAI community that having a framework embedding simulation tools was essential for exploring DAI issues (see [Decker, 1996] for a detailed discussion on historical DAI testbeds and their essential features). However, MACE represented a historical shift for several reasons:

- MACE was not related to a particular domain nor based on a particular DAI architecture;
- MACE was completely focused on providing DAI system development facilities able to model various kinds of DAI systems, and considering different levels of granularity for the agents;
- MACE proposed both (1) modeling and simulation tools to experiment DAI systems and (2) a physically distributed software environment able to deploy the corresponding systems over a real network;
- the MACE software was itself designed as a *community of system agents* that defined the concrete means for building agent-based software systems with MACE: MACE was agentified.

So, even though MACE was tagged as a DAI tool, it could be considered as the ancestor of many generic MAS platforms that were designed in the years that followed (e.g., MADKIT [Gutknecht et al., 2001], JAMES [Himmelspach et al., 2003], Spades [Riley and Riley, 2003], and the new version of MACE, MACE3J [Gasser and Kakugawa, 2002]). Indeed, most of the existing MAS testbeds and platforms (DAI related or not) provide simulation tools, particularly to enable a software-in-the-loop approach for designing MAS architectures.

1.3 MAS for M&S: Building Artificial Laboratories

1.3.1 The Need for Individual-Based Modeling

The interest of using MAS in the scope of M&S mainly appears when it comes to the simulation of complex systems (i.e. systems which are composed of many interacting entities).

The modeling of complex systems has always been a motivation for scientific researchers. One historical example is the continuous deterministic model which has been proposed by Volterra to represent the population dynamic of two animal species (predators and prey) [Volterra, 1926].

By mathematically relating the population dynamics of two species using a differential equation system (DES), the purpose of this model was to explore the mechanism of interaction between predator and prey (sharks and sardines in the original Volterra's study). This model was found appropriate because the solution of this DES shows an intuitively sound result which is that both populations should oscillate: When the prey density is high, the predator multiply until there is not enough prey, which leads in turn to the regeneration of the prey, thanks to the lethal competition which occurs between the predators.

However, despite the apparent appropriateness of this model, some researches gave results which did not exhibit oscillatory behavior with experimental predator/prey populations [Gause, 1934]. Therefore, many variations of the original model were done in order to reconcile the discrepancy between observed experimental results and the model: Adding more complex dynamics (e.g., variation of the predator's voracity over time), new constraints (e.g., a maximum number of animals in the environment), and/or new parameters (e.g., gestation time of prey and predator) to the system [Rosenzweig, 1971].

There were also early attempts to formulate a stochastic version of the system, but it was only in the sixties that, thanks to the advent of computer simulations, it has been possible to experiment with such versions of the model (e.g., [Bartlett, 1960]). At that time, these researches highlighted that one major drawback of deterministic models precisely relies on the fact that they are unable to take into account the apparent not deterministic nature of real life complex systems such as a predator/prey ecosystem. Indeed, real life interaction situations seem to always involve randomness because of the actual complexity of real world processes.

However, the counterpart of the stochastic versions of the original equations was that they exhibit more complex behaviors and great variability, even when used with the same parameters [Buettner and Siler, 1976]. This raises the question of the relevance of the parameters with respect to the targeted system: Are they really capturing something from reality? Such models finally seem to be more related to a mathematical exercise than a modeling that helps to understand the targeted system. So, although this model has some interesting behavior with respect to real population dynamics, it is criticized for the lack of insights it gives about the dynamics of the true components of the systems: The prey and predators.

So, several problems remain with these approaches considering the modeling of complex systems which involve individual entities [Ferber, 1999]:

- Only a global perspective is possible
- Equation parameters hardly take into account the complexity of micro-level interactions
- The modeling of individual actions is not possible
- Integrating qualitative aspects is hard

As we have seen in the previous section, MAS platforms, relying on the modeling and simulation of autonomous proactive entities, did appear practically at the time the agent paradigm was defined, in the early eighties. However, the notion of agent itself, as a fundamental unit of modeling could be found in models and experiments which took place much earlier, at least in the late fifties.

Indeed, the first attempts that tried to solve the previously cited problems, by integrating several levels of analysis in the modeling, were proposed by the microsimulation approach.

1.3.2 The Microsimulation Approach: The Individual-Based Modeling Forerunner

In the scope of social sciences, the notion of agent has always been essential. So, one of the first approaches that has been proposed as an alternative to mathematical models comes from social science and is named microsimulation [Orcutt, 1957].

In [Orcutt, 1957], Orcutt pointed out that the macroeconomic models which were proposed until that time failed to provide relevant information about the influence of governmental policies on the evolution of micro level entities (households and firms). Therefore, Orcutt's point was to emphasize the need for integrating micro level entities in the development of simulation models, making their results more relevant and useful for the study of social systems' dynamics. To this end, the basic principle of microsimulation is to concretely integrate the micro level by means of rules, either deterministic or stochastic, that apply on the attributes of individual units, leading to the modeling of changes in their state and behavior.

Mainly used for population, traffic, and firm based models, microsimulation (or microanalytic simulation) is a modeling approach which is still intensively used today, as the activity of the International Microsimulation Association (IMA) shows*. The IMA website provides many references to actual works in the microsimulation field. Other introductions to this field could be found in books such as [Conte et al., 1997; Harding, 1996].

So, although microsimulation has its roots within the scope of social sciences, it can be considered as the original starting point of Agent-Based Modeling (ABM) approaches (sometimes called Individual-Based Modeling, e.g., in the modeling of ecological systems [Grimm and Railsback, 2005]).

1.3.3 The Agent-Based Modeling Approach

Considering the integration of the micro level within the modeling, an ABM approach goes a bit further than microsimulation. ABM suggests that the model not only integrates the individuals and their behaviors, but also focus on concretely modeling the actions and interactions that take place between the entities, through the environment. Similarities and differences between ABM and previous approaches are discussed in more details in this book, Chapter 2. So, with respect to Equation-Based Modeling (EBM), using the MAS paradigm to model a system provides a completely different perspective which represents an attractive alternative regarding the problems raised previously: Contrary to EBM, wherein the system global dynamics are defined a priori using mathematical relations between global system properties (e.g., the total number of prey), ABM relies on the explicit modeling of micro level entities and dynamics (e.g., individual characteristics and behaviors, actions and interactions between the entities and the environment, etc.). The observed global behavior of the system being thus considered as the result of these micro level dynamics. For instance, considering a predator/prey system, each individual has to be modeled (cf. Figure 1.1).

More generally, quoting [Parunak et al., 1998], the two main differences between EBM and ABM rely on (1) the way they model relations between entities and (2) the level at which

---

*http://www.microsimulation.org. Accessed June 2008.

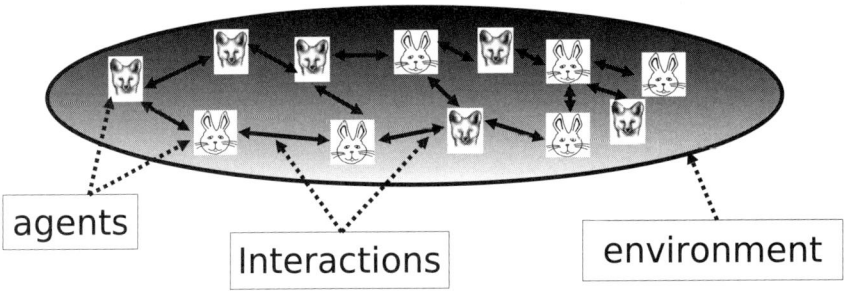

FIGURE 1.1 An agent-based model: The micro level entities, their actions and interactions, and the environment.

they focus their attention. EBM model relations between system observables (i.e. measurable characteristics of interest) while ABM represent individuals that evolve by interacting with one another and the environment.

The interest of MAS relies on four main concepts:

1. Autonomous activity of an agent, i.e. its ability to carry out an action on its own initiative, (pro-activity), by controlling its behavior in order to increase its satisfaction and by deciding to help or to prevent others from satisfying their goals. As it could be noticed, the previous definition of agent stresses the autonomy of decision, which results from the independence with which an agent tries to satisfy its objectives (in the broad sense of the term), by using its competences and its own resources, or by asking for help from others.

2. The sociability of the agents, i.e. their ability to act with other agents from a social point of view. An agent in a MAS is not an isolated entity, but an element of a society. A society emerges from the interactions which take place between the agents, and conversely, the organization of a society constrains the behavior of the agents by ascribing them roles which will restrain their action potentialities.

3. Interaction is what connects the two preceding concepts. This interleaving of actions, where each action is decided inside an agent's mind, produces organized patterns of activities, emerging social forms, which, in return, force and constrain the behavior of agents. So it is through these interactions that forms of interaction emerge, such as cooperation, conflict or competition. This, in return, produces more or less stable organizational patterns which structure the individual action of each agent.

4. The situatedness of the agents, i.e. the fact that the agents are placed into an environment which defines the conditions in which the agents exist, act, and interact. The environment is the glue that connects the agents together, enabling interaction between the agents so that they are able to achieve their goals.

All the power of MAS comes from this cyclic process: Agents act in an autonomous way in a space constrained by the structure of the society in which they act, this structure resulting itself from the behaviors of these agents. There is a dependency loop between agents and societies, between the micro and the macro level, between individual and collective, which is finally at the core of complex system issues.

It is thus not by chance if MAS seem to be a major tool to model complex systems, as societies of agents. They propose much more than a simple technique of modeling. They are not simple abstract tools making it possible to numerically characterize the evolution of a system from its parameters. Being societies by themselves, and being built on the same basis as any complex systems, MAS prove to be "artificial micro-worlds", of which it is possible to control all characteristics and reproduce series of experiments as in a laboratory. Compared to animal and human societies, and to complex systems in general, MAS may be seen as "microcosms", small-scale models of real systems, like reduced size models of boats and buildings, while having the same dynamics and following the same principles as the real social systems.

Moreover, one of the main qualities of multi-agent modeling stands in their capacity of integration and in their flexibility. It is possible to put together quantitative variables, differential equations and behaviors based on symbolic rules systems, all in the same model. It is also very easy to incorporate modifications in the behavior of the individuals, by adding behavioral rules which act at the individual level. It is also possible to add new agents with their own behavioral model, which interact with the already defined agents. For example, in a forest management model, it is possible to introduce new animals, new plant species, new farmers, and to analyze their interactions with already modeled agents.

Because of emergent processes due to interactions, MAS makes it possible to represent complex situations where global structures result from interactions between individuals, i.e. to spring up structures of the macro level from behaviors defined at the micro level, thus breaking the barrier between levels.

Establishing an exhaustive list of all the simulations which rely on an ABM approach would be endless. Therefore, we will only focus on some representative and forerunner examples in the following sections.

1.3.4 Agent-Based Social Simulation: Simulating Human-Inspired Behaviors

Naturally, inspired by the forerunner works which have been done in the scope of microsimulation, social science was one of the first research fields wherein an ABM approach has been applied. In this perspective, the seminal work of Schelling on residential segregation [Schelling, 1971] has deeply inspired the field. Schelling's model was one of the first to show clearly how global properties (segregation among the agents in this case) may emerge from local interactions.

Following this trend of research, Agent-Based Social Simulation (ABSS) became widely used only in the mid-nineties (a number of examples can be found in [Conte et al., 1997]). These works model artificial (human-inspired) societies, considering various levels of abstraction as discussed in [Gilbert, 2005].

To depict the approach underlying these models, Epstein and Axtell coined the notion of *generative social science* in the book *Growing Artificial Societies* [Epstein and Axtell, 1996]. As discussed in more details by Epstein in [Epstein, 2007], this notion enables one to highlight the differences between ABM and both inductive and deductive social science. More on the different motivations of using agent-based models for social science could be found in papers such as [Conte et al., 1998; Axtell, 2000; Goldspink, 2002] and in the second edition of the book of Gilbert and Troitzsch which gives a comprehensive view of the field [Gilbert and Troitzsch, 2005].

ABSS concrete examples are the modeling of urban phenomena (e.g., [Vanbergue et al., 2000; Bretagnolle et al., 2006]), works done in game theory on the iterated prisoner dilemma (e.g., [Beaufils et al., 1998]), and the modeling of opinion dynamics [Deffuant et al., 2002].

One can find many other examples in the electronic journal JASSS*, the *Journal of Artificial Societies and Social Simulation*.

Some generic ABSS-related platforms have also been proposed (e.g., ASCAPE [Parker, 2001], MODULECO [Phan, 2004]). Moreover, a modeling language relying on rule-based agents has been proposed in [Moss et al., 1998], namely SDML. Today, the RePast toolkit [North et al., 2006] is a representative example of platform which is used for ABSS.

1.3.5 Flocks and Ants: Simulating Artificial Animats

The Reynolds's Boids

Considering the use of the agent paradigm for M&S purposes, a forerunner work was done by Reynolds on flocks [Reynolds, 1987]. In the scope of computer graphic animation, the goal of Reynolds was to achieve a believable animation of a flock of artificial birds, namely *boids*. Reynolds remarked that it was not possible to used a scripted flock motion to achieve a realistic animation of group motion. So, Reynolds was inspired by two stream of research: (1) particle systems [Reeves, 1983] and (2) the Actor paradigm [Agha, 1986].

Particle systems were already used for the animation of complex phenomena, such as clouds or fire, which were modeled as collections of individual particles having their own behavior and state. However, particles did not interact as their behavior only relied on their own internal state (position, velocity, lifetime, etc.) and potentially on some global parameters that could represent phenomena such as gravity.

The idea of Reynolds was that boids have to be influenced by the others to flock in a coherent manner: "*Boid behavior is dependent not only on internal state but also on external state*". So, Reynolds used the actor abstraction to define the boids behavior so that they do interact to flock. Boids change their directions according to others' to stay at the center of the local flock they perceive.

The results obtained were very compelling and the impact of the Reynolds's boids on the community of reactive agents can still be perceived today as it was one of the first works to be *bio-inspired*.

Indeed, beyond graphic animation, boids-inspired behaviors can be found in domains such as mobile robotics (e.g., flocks of unmanned aerial vehicles (UAV) [Nowak et al., 2007]), and human crowd simulation, e.g., like in [Musse and Thalmann, 2000] or [Pelechano et al., 2007]. In this last reference, the research is also inspired by the pioneer work of Helbing and Péter Molnár on human crowd which uses the concept of *social forces* to model how pedestrian are *motivated to act* with respect to external environmental conditions (e.g., others pedestrians or borders) [Helbing and Molnár, 1995]. A famous example of an application derived from these different technologies is the use of the Weta Digital's $MASSIVE^{TM}$ software for special visual effects in movies such as the *Lord of the Rings*, in which there are huge battle scenes.

Although the boids paradigm is well suited for modeling collective moves, it is limited by the fact that it only relies on the direct perception of the others: The environment does not play any role in the interaction between the agents. On the contrary, the environment plays an essential role for the modeling of ant colonies.

---

*http://jasss.soc.surrey.ac.uk

Ant Colonies

Regarding bio-inspired MAS related works, ant colonies probably represent the natural system which has inspired the most. Indeed, first, ant colonies exhibit many features which are expected from MAS such as self-organization, self-adaptation, robustness, emergent properties, and so on. Secondly, the ability of ants in solving complex problems in an efficient and constant way is fascinating considering their limited individual characteristics. The most well known example of that being how ants use pheromones to find the shortest path from the nest to food sources [Beckers et al., 1992].

So, as an ant colony was quickly recognized as a fascinating model of MAS, many MAS simulations done in the early nineties rely on simulating and studying characteristics of ant-based systems (e.g., [Collins and Jefferson, 1992]). Among these forerunner works on ants, one of the most ambitious has been done in the scope of the MANTA (Modeling an ANTnest Activity) project [Drogoul and Ferber, 1992]. Firstly, MANTA relied on an innovative modeling and simulation framework called EMF (EthoModeling Framework) which was sufficiently simple to be used by non computer scientists to implement multi-agent simulations. Secondly, the purpose of MANTA was not only to simulate some characteristics of ants but to translate all the parameters of a real biological study into a virtual ant farm. So, MANTA simulations were composed of all the creatures that can be found inside an ant nest (ants, larvae, cocoons, eggs) and also took into account time and some environmental factors such as light and humidity.

It is worth noting that, at the same time, Dorigo also took inspiration from ants to define a new kind of optimization algorithm, namely the *Ant system* heuristic [Dorigo, 1992]. Especially, he derived from this approach a distributed algorithm called *Ant Colony System* (ACS) which was applied to the traveling salesman problem [Dorigo and Gambardella, 1997]. Today, *Ant Colony Optimization* (ACO) [Dorigo and Stützle, 2004] represents a whole trend of research in the domain of swarm intelligence.

The common point between all these researches is that they all identified that one major characteristics of ants is their ability to use pheronomes (evaporative scent markers) to indirectly communicate and coordinate through the environment. Indeed, pheromones enable the ants to achieve complex tasks thanks to a simple environmental mechanism: Pheromones evaporate, so that the system can forget obsolete information which is the fundamental feature of such systems [Parunak, 1997]. For instance, paths leading to depleted food sources progressively disappear with time. This powerful mechanism has been successfully translated into computer programs and simulations, thus defining the concept of digital pheromones which has been used in numerous works. Additionally, using a pheromone-based approach of course does not entail one to only consider ant-like agents. One example can be found in this book: *Polyagents* [Parunak and Brueckner, 2008].

1.4 Simulating MAS: Basic Principles

MAS have been developed around a set of principles and concepts. These concepts are Agents, Environment, Interaction and Organizations, as presented in the Vowels approach [Demazeau, 1997]. We will give a brief overview of the first three aspects in this chapter. Then, we will discuss issues related to the modeling of time in MAS simulations.

1.4.1 Agent

Many definitions of agency have been proposed in the field of MAS, each one being more adapted to a specific flow of research (see [Woolridge, 2002]). The following one is adapted from [Ferber, 1999]: An *agent* is a software or hardware entity (a process) situated in a virtual or a real environment:

1. Which is capable of acting in an environment
2. Which is driven by a set of tendencies (individual objectives, goals, drives, satisfaction/survival function)
3. Which possesses resources of its own
4. Which has only a partial representation of this environment
5. Which can directly or indirectly communicate with other agents
6. Which may be able to reproduce itself
7. Whose autonomous behavior is the consequence of its perceptions, representations and interactions with the world and other agents (cf. Figure 1.2).

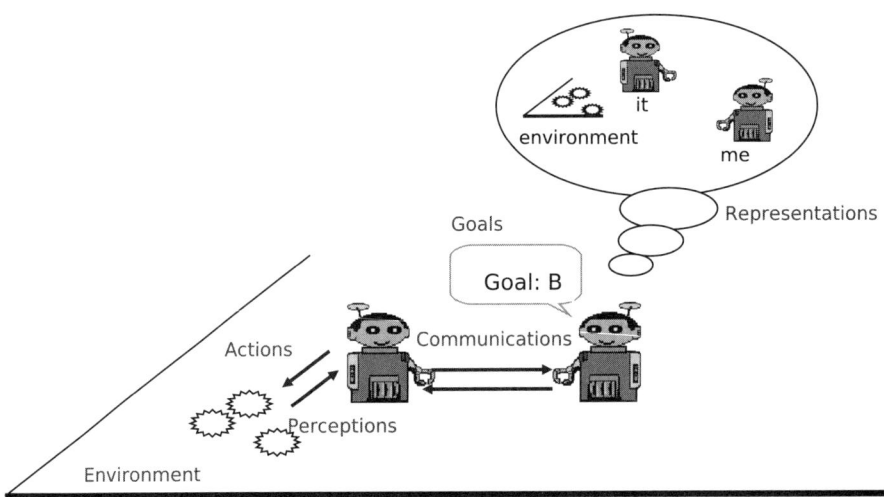

FIGURE 1.2 A multi-agent world.

Agent Architectures

The term generally used to describe the internal organization of an agent is that of *architecture*, by analogy with the structure of computers.

It is generally considered that there are two main approaches for analyzing agent architectures: (1) The reactive approach in which we only consider perception-action (or stimuli-response) architectures and (2) the cognitivist approach which relies on mental issues, such as the explicit representation of the environment (and other agents) by agents. A third approach, called hybrid, consists in trying to get together the first two.

Reactive Architectures. A reactive agent does not have an explicit representation of its environment nor of other agents. Its behavior is entirely described in terms of stimuli-response loops which represent simple connections between what they perceive and the set of available operations that may be performed. The most well-known architectures in this domain are the subsumption architecture in which tasks in competition are arbitrated along predefined priorities [Brooks and Connell, 1986], the competitive task architecture, where concurrent tasks have their weight modified through a reinforcement learning process [Drogoul and Ferber, 1992], and connectionist architectures which are based on neural nets. Some approaches combine these different structures within an integrated architecture, like the one of Tyrell for instance, which combines control by priorities and neurons, within a hierarchical structure [Tyrrell, 1993]. There are also architectures that combine behaviors, each behavior being represented as a vector of actions. The *Satisfaction-Altruism* function allows one to combine behaviors centered on the desires of agents with cooperative behaviors centered on the needs of others [Simonin and Ferber, 2000].

Cognitive Architectures. Cognitive architectures are founded on the computational metaphor which considers that agents reason from knowledge described with a symbolic formalism. This knowledge explicitly represents their environment (states, properties, dynamics of objects in the environment) and the other agents. The most well-known architecture of this type is the BDI (Belief-Desire-Intention) which postulates that an agent is characterized by its beliefs, its goals (desires) and intentions [Rao and Georgeff, 1992]. It is assumed that cognitive agents are intentional, i.e. that they intend to perform their action, and that these actions will allow them to satisfy their goals. In other words, a BDI agent acts rationally from its beliefs about world states, its knowledge (and those of others), its intentions (and those of others) to achieve its goals. A BDI agent is assumed to possess a library of plans, each plan being designed like a recipe making it possible to achieve a particular goal. An intention is set when an agent makes a commitment about achieving a specific goal by using a particular plan. The management and update of beliefs, goals and intentions are carried out by the BDI engine which selects the plans and the actions to be undertaken.

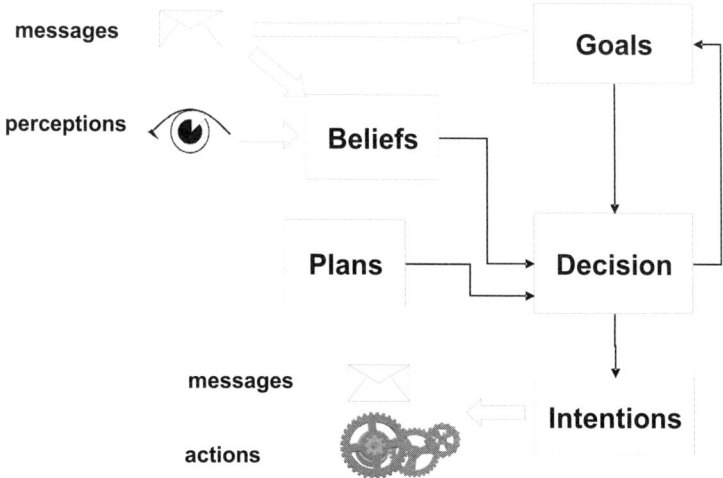

FIGURE 1.3 A BDI architecture with its different modules.

Figure 1.3 illustrates the functional representation of a BDI architecture. External perceptions and messages are used to form beliefs. It is supposed that an agent may initially have several goals. These initial goals, added with the goals resulting from requests of other agents, are analyzed and selected in the decision component to trigger the set of plans compatible with the agent's beliefs. If there are no plans that can be found, a problem solver (not shown here) has the responsibility to decompose the initial problem into sub-problems, by producing sub-goals that the agent will have to satisfy. When an agent chooses to execute a plan, the actions of this plan are transformed into intentions which will produce environmental actions and communications with other agents. The main quality of BDI architectures is to create a behavior which mimics that of a rational human being. As an example, The Jason platform, which is described in this book [Bordini, 2008], enables to design Agent-Based Simulation Using BDI Programming.

Hybrid Architectures. The two main types of agents, cognitive and reactive, propose solutions apparently diametrically opposite, but in fact, they may be viewed as complementary. In order to build best suited architecture to solve a problem (in terms of response time, precision or efficiency), it is possible to create hybrid architectures which combine the two types of approaches, and then build more flexible agent architectures.

In such architectures, agents are composed of modules which deal independently with the reflex (reactive) and reflexive (cognitive) aspect of the agent behavior. The main problem is then to find the ideal control mechanism ensuring a good balance and a good coordination between these modules. Let us cite the Touring Machine [Arango et al., 1993] and InteRRap [Müller and Pischel, 1993], the most well know examples of hybrid architectures.

Developing MAS with a specific architecture in mind (cognitive, reactive and even hybrid approaches), may be a disadvantage when developing open MAS: some developers think in terms of reactive agents whereas others prefer to use a cognitive approach. In an open system, all these agents must live together, conform to a framework of execution and of behavior, and thus must be able to interact within the same interaction space.

Modeling the Behavior of Agents

In order to simplify the model of agents, the process which takes place between the perception of inputs and the production of outputs can be considered as the *deliberation function* of the agent. So, an agent is a cyclic three phase process: (1) perception, (2) deliberation, and then (3) action (cf. Figure 1.4).

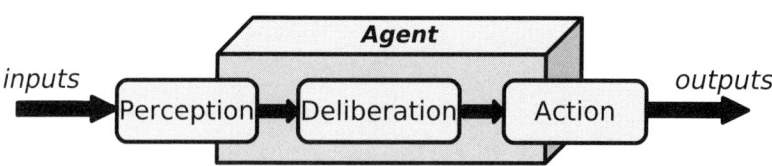

FIGURE 1.4 An agent as a three phases process.

As the purpose of this section is not to explain the modeling details of existing agent architectures, we here just present a formalism which is inspired by the work of Genesereth and Nilsson [Genesereth and Nilsson, 1987] and focuses on representing the cyclic nature of the behavior of an agent. Let $\sigma \in \Sigma$ be the actual state of the world, the behavior cycle (perception / deliberation / action) of an agent a is represented using a function $Behavior_a : \Sigma \mapsto A_a$, A_a representing the set of possible actions that the agent a can take:

- $Perception_a : \Sigma \mapsto P_a$, that computes a percept $p_a \in P_a$ from the system's state Σ.
- $Deliberation_a : P_a \times S_a \mapsto S_a$, that computes the new internal state s_a of the agent.
- $Action_a : P_a \times S_a \mapsto A_a$, that produces the action of a.

So, firstly an agent obtains a percept p_a which is computed by the *Perception* function using the current state of the environment σ. Such a function may simply return some raw data (quantitative variables) used to defined the state of the environment (e.g., coordinates of objects, the current temperature in celsius degrees, etc.), but it may also represent more complex processes that transform raw data to high level percepts representing qualitative aspects of the environment (e.g., near or far from an object, the temperature is hot or cold, etc.), thus easing the deliberation of an agent as discussed in [Chang et al., 2005].

Secondly, the *Deliberation* function (*Memorization* in [Genesereth and Nilsson, 1987]) defines how the agent uses p_a to make its internals evolve according to p_a, updating its own representation of the world for instance. The deliberation process of an agent defines the core part of its behavior and characterizes its architecture (reactive or cognitive). As such, it is obviously the part which has been studied the most. Still, it is worth noting that this function is skipped for tropistic agents (without memory) as they directly match percepts to actions.

Finally, the *Action* function represents how an agent makes its decision, based on its new internal state and current percept, and thus chooses the action to take. Most of the time, this action is directly concretized by the modification of the environment which is supposed to succeed the action (e.g., $\sigma = \{door(closed)\} \mapsto \sigma = \{door(open)\}$). In other words, the direct modification of the environment is the means by which is the result of the action of an agent is computed.

1.4.2 Environment

An Essential Compound of MAS

In the beginning of MAS, in the old days of DAI, the concept of environment had not been given an important consideration. But the development of MABS has shown the importance of the environment because, in MABS models, agents are situated in a concrete environment: the simulated environment.

More recently, the environment has also been pointed out as an essential compound of MAS as it in fact broadly defines all the perceptions and actions that an agent may have or take: The environment defines the conditions in which the agents exist in the MAS [Odell et al., 2002]. Especially, the term environment could also refer to the infrastructure in which agents are deployed and thus be studied as a first order abstraction from an agent-oriented software engineering perspective, as thoroughly discussed by Weyns et al. [Weyns et al., 2005b].

To distinguish the different concerns which could be related with the concept of environment in MAS, Valckenaers et al. have proposed a structured view on environment-centric MAS applications, thus identifying three base configurations which one is simulation [Valckenaers et al., 2007]. So, in the scope of a simulation configuration, according to this proposition the environment simply refers to the part of the simulation that models the relevant portion of the real world (either it exists, needs to be realized or no longer exists) or an imaginary world. From now on, we will only consider this aspect of the environment, i.e. the modeling part that represents the world in which the simulated agents evolve.

Modeling the Environment

The inputs an agent receives come from the environment it is situated in (and of which the other agents are part). Similarly, the outputs an agent produces go in the environment.

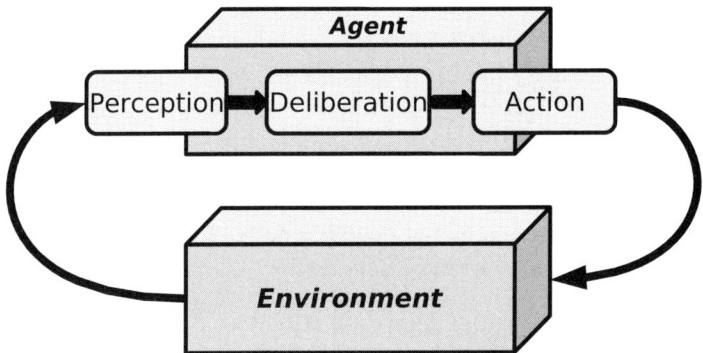

FIGURE 1.5 An agent receives inputs from the environment and produces outputs on it.

From a M&S perspective, the environment is another dynamic system (ABM relies on a *multi-model* approach). However, contrary to an agent model, the dynamic of the environment does not represent any autonomous behavior. Regarding the characteristics and dynamics of the environment, Russell and Norvig propose to consider several properties [Russell and Norvig, 2003]:

- *Accessible* vs. *inaccessible*. An environment is defined as accessible if its complete state can be can perceived by an agent.
- *Deterministic* vs. *nondeterministic*. An environment is deterministic if its next state is entirely determined by its current state and the actions selected by the agents.
- *Episodic* vs. *non-episodic*. In an episodic environment, an episode (an agent perceiving and then acting) does not depend on the actions selected in previous episodes.
- *static* vs. *dynamic*. In a static environment, changes only occur under the influences of the agents, contrary to a dynamic environment which posses an endogenous evolution.
- *Discrete* vs. *continuous*. In a discrete environment, the number of possible perceptions and actions is limited and clearly defined.

The modeling of the environment usually embeds the representation of some physical places wherein the agents evolve. As the environment defines the perceptions and actions of the agents, two main approaches can be distinguished with respect to the granularity of these perceptions and actions:

1. Discretized: The environment is discretized in bounded areas that define space units for the perception/action of the agents (environment-centered)

2. Continuous: The range of each perception/action depends on the acting agent and the nature of the perception/action (agent-centered)

The first approach consists in modeling the environment as a collection of connected areas, which thus defines the topology of the environment. Figure 1.6 shows three examples for such an approach: The rooms of a house thus defining non-uniform cells (a), a physical space divided in regular zones (uniform cells) (b) and the nodes of a network (c). With such a modeling, the idea is to use the environmental characteristics to define the range of the perceptions and actions (e.g., an entire room, a cell, or a network node).

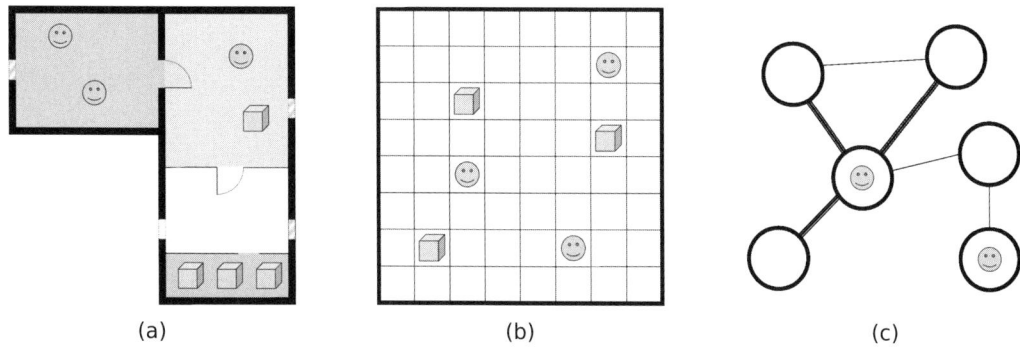

FIGURE 1.6 Examples of discretized environments.

The most usual examples of this kind of approach are those wherein the environment is discretized in a regular grid of cells (or patches) (cf. Figure 1.6 (b)). Platforms allowing such a modeling of the environment are numerous (e.g., STARLOGO [Resnick, 1994], TURTLEKIT [Michel et al., 2005]).

Widely used for their simplicity of implementation, grid-based environment models have also the advantage of easing the modeling of environmental dynamics such as the diffusion and evaporation of digital pheromones. As major drawback, a grid-based model raises the problem of the granularity of the perceptions/actions which an agent can make. Indeed, whatever the perceptions/actions of an agent, which can be very heterogeneous, their range will always be the same (modulo a factor): The cell.

On the contrary, the continuous approach considers each agent as the reference point from which the range of perceptions/actions is computed (cf. Figure 1.7. This kind of modeling is required when accurateness is needed. For instance, this modeling is usually used to simulate soccer robots (e.g., in RoboCup soccer simulators [Kitano et al., 1997]): The granularity of the movements of both the agents and the ball has to be accurately modeled.

Except from the fact that such a modeling could be required, the main advantage is that the continuous approach is far more flexible than the former considering the integration of heterogeneous perceptions/actions within the modeling of the agents. The counterpart is that it is of course more difficult to model and implement such an approach.

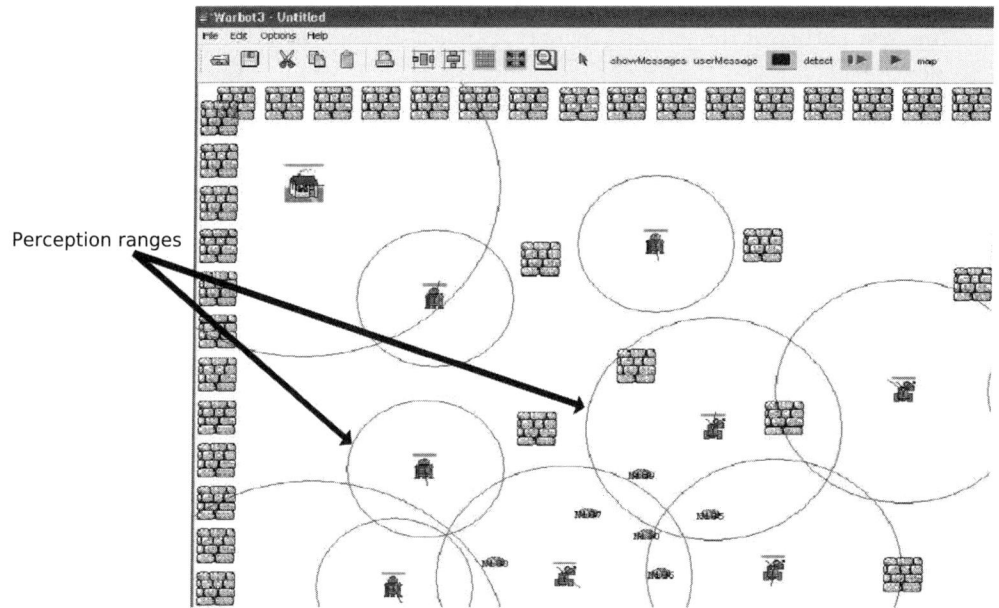

FIGURE 1.7 Continuous approach for the perception of the agents.

1.4.3 Interactions

Communications between agents are at the core of interactions and social organizations. Without communicating, an agent is totally isolated, deaf and dumb, closed on its perception-deliberation-action loop. Because they communicate, agents can cooperate and coordinate their actions, and become true social beings. Communication is expressed through language items (messages, signals) which, once interpreted, will produce an effect on agents. There are a great number of approaches to communication: the social sciences, linguistics, and philosophy of language each developed a whole set of concepts, that deal with what are known as speech acts. Biology and ethology have also produced theories on communication through signals.

Communication by Message Passing

The traditional model of communication between agents relies on message passing. A transmitter sends a message to a recipient knowing directly (or indirectly via a directory) its address. This simple model is used in most of the MAS. It has been extended by the speech act theory, which gives precise semantics to the interaction. First appearing in the work of Austin [Austin, 1962] and Searle [Searle, 1969], the concepts of speech act theory have been simplified with the development of the KQML [Finin et al., 1994] and ACL [for Intelligent Physical Agents] languages. Speech act theory considers that communications may be interpreted as mental acts. For example, if an agent A, with the goal of having the world in state S sends a request to an agent B to perform the action a, it supposes that A believes that the world is not in state S, that a has not been performed, and that the achievement of a would result in the state S, and that A believes that B is able to do a. It is the same if A informs B that a proposition p is true. Its meaning resides in mental issues: A believes that p is true, that B has no beliefs about p, and the expected result by A is that finally B eventually believes that p is true.

Multi-Agent Systems and Simulation

Research works have also been conducted to define interaction protocols, i.e. to define the sequence of messages which characterize an interaction situation. For example, if A requests B to perform a, it waits for an answer from B saying if B agrees to accomplish the job. If B does agree, it will inform B when the task a will be achieved. This means that it is possible to represent interactions between agents with communication diagrams. Proposed by Odell, Bauer and Müller, AUML [Bauer et al., 2001] is a MAS description language which extends UML sequence diagrams by adding specific extensions and by reducing some of the obviously too specifically "object-oriented" aspects.

Interactions Using Signals

Besides speech acts and communication standards, agents may communicate by using signals. This type of communication is based on biology and ethology, where animals tend to behave collectively by using signals. In the MAS domain, two main kinds of signals are used: marks and fields. Marks are traces that agents make while moving. Agents may drop marks on their way (pheromones, tracks, pebbles, objects of any sort,..) which could then be interpreted by other agents. This mechanism produces the famous ant lines. The tracks, in the form of pheromones dropped by the ants while returning to their nest, are used as interaction medium to signify to other ants where food can be found (Figure 1.8).

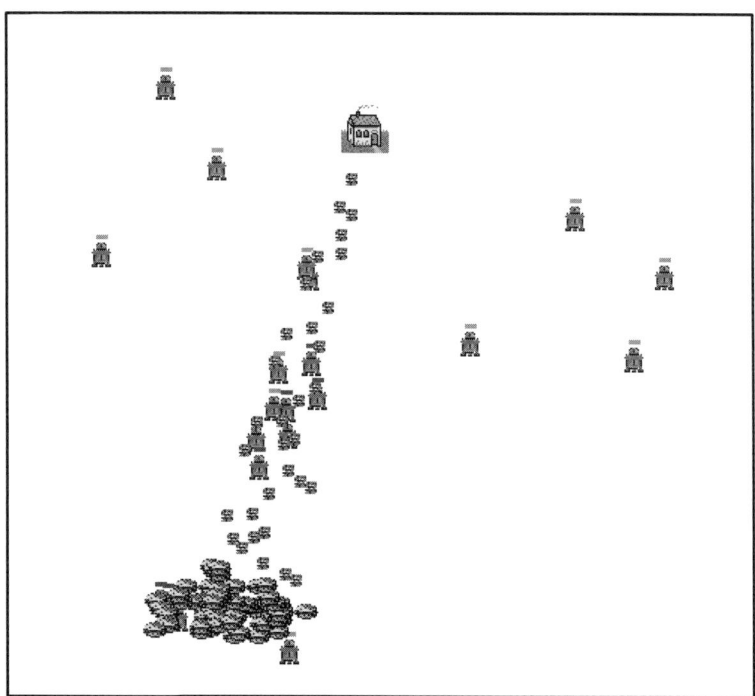

FIGURE 1.8 Ant agents, represented as robots, carry resources back to their base while dropping pheromones which are used as marks.

Signals may be spread in the environment. And in this case, they may be used to indicate, remotely, the presence of obstacles, desirable objects, agents to help or to avoid. These signals are used to coordinate a set of agents acting in a common environment. Let us consider the intensity of signals relative to the distance between the source and a location in the environment. If these signals are propagated in a uniform and homogeneous way, they form potential fields. It is then possible to define attractive and repulsive forces from the gradient of these fields [Arkin, 1989]. The goals are then represented as attractive fields and obstacles as repulsive fields. The movement is obtained by a combination of attractive and repulsive fields, and an agent has only to follow the greatest slope direction (Figure 1.9). When obstacles can move, i.e. when there are other agents, it is necessary to add avoidance behaviors [Zeghal and Ferber, 1993; Simonin and Ferber, 2000; Chapelle et al., 2002].

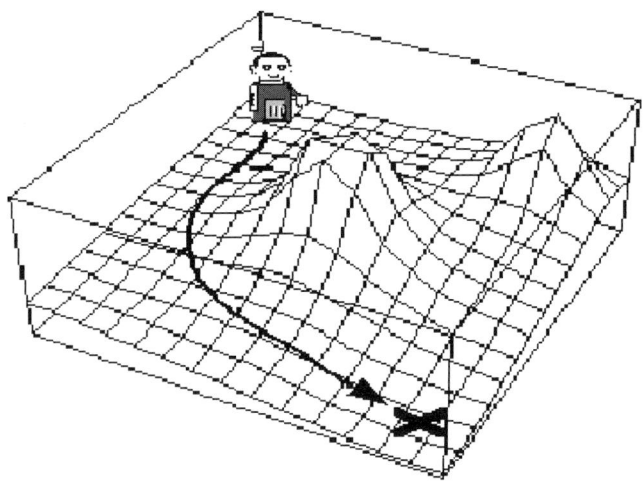

FIGURE 1.9 An agent moves in a field by following its gradient.

One has to notice that these signals may either be real (and part of the environment) or virtual, i.e. rebuilt by agents as simple means to coordinate their movements in the environment, like it is done in [Mazouzi et al., 2007] in the scope of image segmentation: Simulated agents create potential fields, without affecting the pixels, to attract others in areas of interest (object edges).

1.4.4 Modeling Time

Like the simulation of other kinds of system, the evolution of time should be modeled when simulating MAS, especially if all the agents are supposed to act and interact concurrently with respect to the principle of causality [Fyanio et al., 1998].

Time could be modeled using three main approaches: (1) continuous time (by means of time functions which can compute the system state for any time stamp, e.g., DES), (2) discrete time (time evolves discretely with respect to constant time intervals), and (3) discrete event-based (time evolves discretely from one event to the next considering a continuous time line, i.e. the time interval between two events could be any real number).

In this respect, modeling time for a MAS needs to consider at least three aspects:

1. The modeling of the behavior time of the agents
2. The modeling of the endogenous dynamics of the environment
3. The evolution of the environment with respect to the agent actions.

Modeling the Behavior Time of an Agent

The models which are used to represent how an agent reacts to external events and deliberates are inherently discrete. Indeed, although an agent could be put in an environment which dynamics are modeled using a continuous approach, the deliberative process of an agent is usually modeled as a process that makes its internal state variables change discretely, i.e. instantaneously. Moreover, in almost every agent-based model, the perception/deliberation/action cycle is associated to a single instant t for simplicity reasons.

However, in some application domains, the temporal thickness of the behavior of an agent has to be explicitly taken into account, and therefore modeled. For instance, in the scope of DAI, the evaluation of the deliberation efficiency of an agent could be the aim of the study. For example, this efficiency could be modeled as a function of the actual computation time which is used by a deliberating agent, like in the Phoenix simulator [Cohen et al., 1989]. The goal is to be able to measure and compare the efficiency of different agent architectures.

For example, the Multi-Agent System Simulator (MASS), which has been designed to study the efficiency of agents' deliberation in multi-agent coordination/negotiation processes, explicitly integrates time models of the mechanisms involved in the deliberation processes (method invocations, access to resources, etc.) [Vincent et al., 2001]. A similar approach has been used in the Sensible Agents platform [Barber et al., 2001] to experiment with agent-based systems in dynamic and uncertain environments. In another example, Uhrmacher and Schattenberg have proposed to model the overall behavior of an agent as a discrete event system using the DEVS (Discret Event Systems Specification) [Uhrmacher, 2001]. Doing so, the authors are able to model not only the deliberation time of an agent, but also the reaction time of an agent to external events, thus explicitly distinguishing, within the modeling of time, between the reactive and the proactive processes of an agent. The JAMES simulation platform relies on this approach [Schattenberg and Uhrmacher, 2001]. In this book, the reader will find a chapter about the new version of this platform, namely JAMES II [Himmelspach and Röhl, 2008].

Modeling the Temporal Evolution of the Environment

Depending on the application domain, modeling the temporal evolution of the environment could be crucial. For instance, simulating agents in a network, it could be interesting to model different lag times to evaluate the impact of network congestion, node failure, and so on. Moreover, the environment may not only react to the agents inputs but also evolve according to its own dynamic, namely its *endogenous* dynamic. For instance, in a robocup simulation, a rolling ball continues to move even when the agents do not perform any actions.

As the environment could represent very different kind of systems, continuous or discrete time modeling could be considered depending on the experiment requirements. However, due to the fact that the agents have their perception and produce their actions in a discrete manner, the environment temporal dynamics generally embeds some event-based mechanisms to ease the coupling between the model of the environment and the agent models.

Coupling the Agents and the Environment: Scheduling the MAS

Once the agent behaviors and environment dynamics are defined, they have to be coupled to finally model the targeted MAS, so that a MAS could be considered as three-tuple: $MAS = <Agents, Environment, Coupling>$ [Parunak, 1997]. Figure 1.10 gives the overall picture of this coupling.

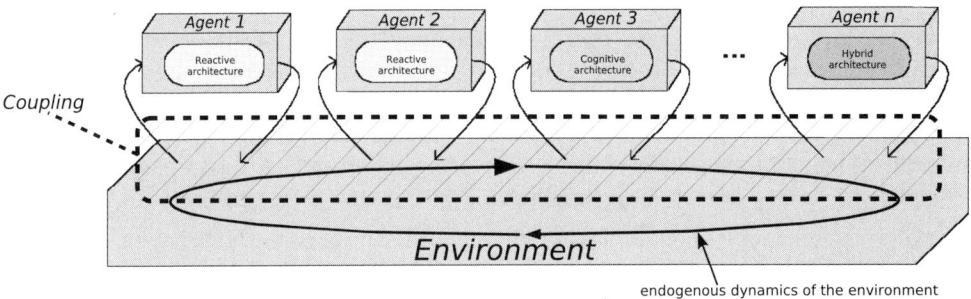

FIGURE 1.10 Coupling the agents and the environment.

The main problem of the coupling is to make it coherent with respect to time [Parunak, 1997; Fyanio et al., 1998]. In other words, achieving this coupling relies on defining a function *Evolution* such that the evolution of the MAS from one moment t to the next $t+dt$ results from the combination of the agent actions, $A_1(t), A_2(t)...A_n(t)$ with the dynamics produced by the natural evolution of the environment, $E_n(t)$, at t:

$$sigma(t+dt) = Evolution(\uplus(A_n(t), E_n(t)), \sigma(t)) . \qquad (1.1)$$

The symbol $\uplus$ is used here to denote the action composition operator. It defines how the actions produced at the instant t must be composed in order to calculate their consequences on the previous world state $\sigma(t)$. Intuitively, this coupling does not seem to be more than scheduling the activation of each model considering the modeling of time which is chosen (i.e. discrete time or event-based). However, several difficulties exist in the scope of ABM as we will now briefly see.

Due to the simplicity of its implementation, discrete time simulation (i.e. dt is constant) is the most used technique for simulating MAS. Indeed, in a discrete time simulation, all the models are sequentially activated for a time t, and then the global clock of the system is increased of one time unit. So, the implementation can be as simple as the following loop:

```
while(globalVirtualTime != endOfSimulation){
        for (SimulatedAgent agent : AllTheAgents)
                agent.act(); //perception, deliberation, action
        virtualEnvironment.evolve(); // for a dynamic environment
        globalVirtualTime++;
}
```

However, there is a major issue with such a scheduling technique: The order in which each model is activated may change the result obtained for the system state since the environment evolves for each action. This in turn may lead to very different system dynamics with the same behaviors [Michel et al., 2004]. Figure 1.11 illustrates this issue on a prey/predator problem: the prey (the triangle) could be dead or alive depending on the activation list.

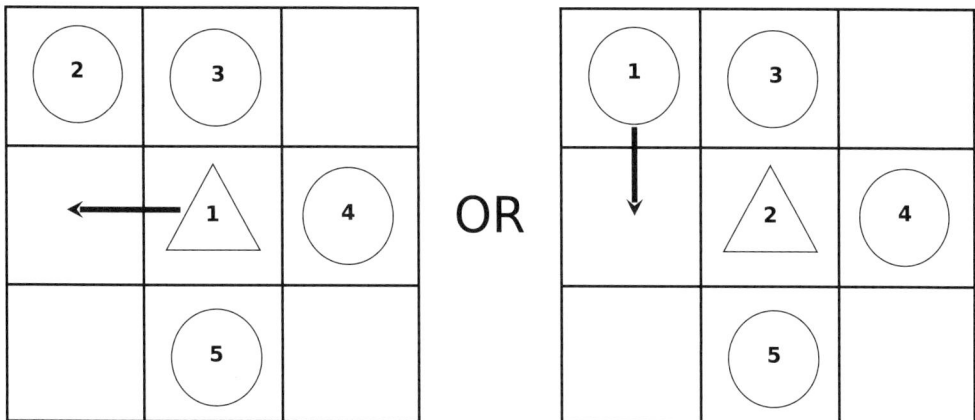

FIGURE 1.11   The prey could be dead or alive depending on its rank in the activation list.

Several variations of the discrete time approach have been proposed to solve this problem. One solution is to systematically shuffle the activation list at each time step in order to smooth the problem (e.g., [Epstein and Axtell, 1996]). Another solution is to simulate concurrency by having all the agents operate on temporary variables, so that the perceived environment is the same for all the agents. Once done, the next system state is computed. So, the activation list has no effect on the results. However, the new problem is that one has to solve the conflicts which may exist between two or more actions that produce different states for the system (e.g., the same ball at two different places). Solutions to such conflicts could be hard to find and lead to complex solving algorithms, as discussed in [Dumont and Hill, 2001] for instance.

More generally, whatever the time management (discrete time or event-absed), modeling interaction (concurrent actions) in MAS is hard to achieve using action representations such as the one presented in the end of Section 1.4.1 because they directly match the decision of an agent to a modification of the environment: One action gives one result. Quoting Parunak [Parunak, 1997]: *This leads to an (unrealistic) identity between an agent's actions and the resulting change in the environment, which in turn contributes to classical AI conundrums such as the Frame Problem*, as also further discussed in [Ferber and Müller, 1996]. Therefore, the actions (Equation 1.1) are not composed to produce a result but rather produce results one by one, which indeed leads to conflicting, sometimes unrealistic, situations. So, implementing simultaneity using such procedures may still be done but it takes complex codes which are more related to programming tricks than to an efficient and generic simultaneity modeling solution [Ferber, 1999]. In Section 1.6.2, we will see that several works have addressed this particular problem by relying on the Influence / Reaction model [Ferber and Müller, 1996].

Besides concurrency, another issue which is raised by a discrete time approach is the temporal granularity of actions. Indeed, in the same way the discretisation of the environment raises problems about the modeling of heterogeneous perceptions/actions, the discretisation of time implies taking care of the temporal granularity of actions. Figure 1.12 illustrates this problem: If an agent is able to move across more than one spatial unit in one time step, then several agents may have crossing paths without generating any collision, whatever the discrete time algorithm (i.e. there is no conflict here). This is an important issue in MABS involving mobile agents like traffic simulations (e.g., [Dresner and Stone, 2004]).

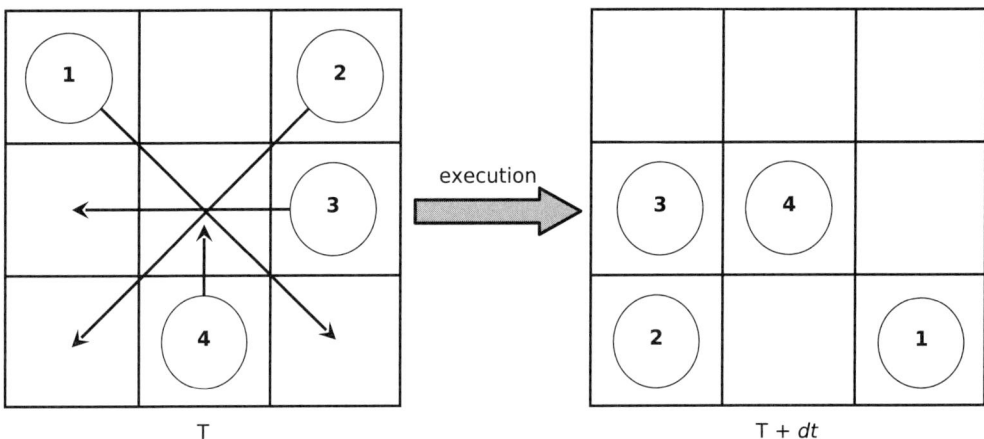

FIGURE 1.12 The problem of the temporal granularity of actions.

So, depending on the application domain, an event-based approach could be required as it represents a far more flexible approach when it comes to the modeling of heterogeneous actions because it eases the integration of actions having different granularities. Moreover event-based simulations are often considered as more representative of the reality that we observe in the everyday life. Indeed, it is difficult to think about reality as a system in which all the entities would be "updated" simultaneously [Huberman and Glance, 1993].

Like for the discrete approach, there are various way of modeling and implementing an event-based approach, depending on how the events are generated and handled (classical event-based implementations are presented in [Balci, 1988]). The event list could be determined using a predefined order, like in the SWARM platform [Minar et al., 1996], or generated dynamically during the simulation, like in JAMES which uses DEVS [Himmelspach et al., 2003], or in DIMA which relies on the activity notion [Guessoum, 2000].

1.4.5 Simulating MAS as Three Correlated Modeling Activities

To sum up this section, we propose here a high level view of agent-based simulation models. To this end, we consider that simulating a MAS relies on three correlated modeling activities: (1) the modeling the agents, (2) the modeling of the environment, and (3) the modeling of the coupling between the agents and environment models. Figure 1.13 illustrates this view:

- The first module is related to the modeling of the agent behaviors, namely the *behavior module*. Considering this module especially involves choosing an agent architecture and a behavior time model.
- The second module, the *environment module*, is related to the definition of the virtual place wherein the agents evolve and interact. Modeling the environment

notably supposes to decide if the environment should be (1) discrete or continuous, and (2) static or dynamic. If the environment has an endogenous dynamic, then a time model of this dynamic should also be defined. Additionally, as agent interactions are mediated by the environment, how the agents interact is mainly defined through the modeling of the environment, by defining (1) how the actions of the agents affect the environment state and (2) how these actions are perceived by the agents (messages and/or signals).

- Finally, the *scheduling module* defines how all the models, environment and agents, are coupled and managed with respect to time. Besides choosing a particular time management, considering this module implies defining how the environment evolution, the agent perceptions, actions and interactions are done with respect to one another, thus defining the MAS causality model. It is worth noting that this fundamental part of the modeling is not always specified nor even considered during the elaboration of agent-based models as it is often regarded as something for which the simulation platform should be responsible, thus ignoring a major part of the model dynamics*. In Section 1.6.3, we further discuss why defining this module should be considered as a primary modeling activity.

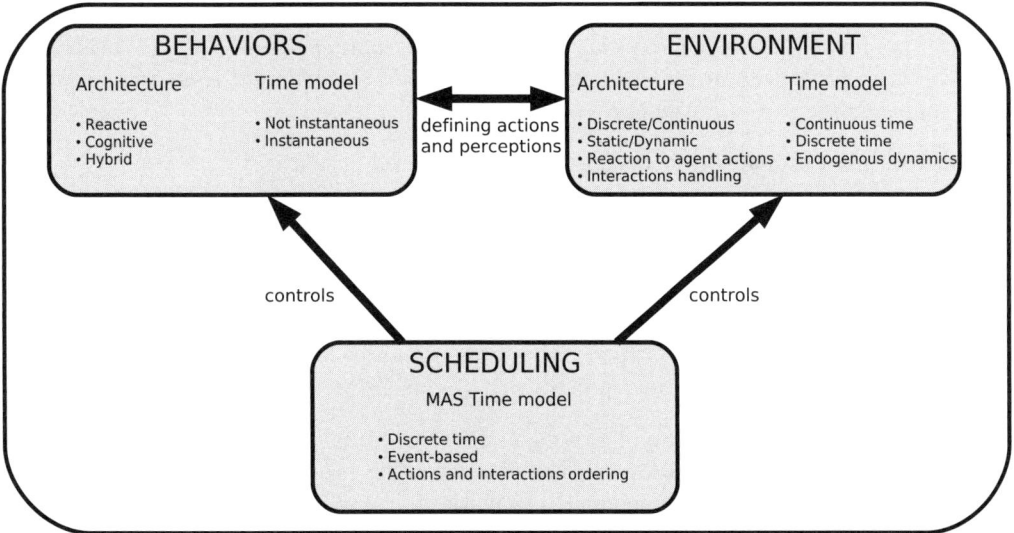

FIGURE 1.13 Simulating a MAS as three modeling modules.

This tripartite perspective also defines how the modules are related to each other. The scheduling module controls how the agents and the environment are executed considering the representation of time (e.g., actions and interactions management and ordering). The relation between the agent module and the environment module concerns the definition of

---

*Other researchers usually have a different perspective and consider scheduling as an issue at the level of the simulation framework, rather than at the level of modeling.

all the possible agent actions and perceptions with respect to the environment. This relation is bidirectional since adding new perceptions or actions requires modifying both modules.

1.4.6 A Still-Incomplete Picture

This overall picture proposed in Figure 1.13 highlights the different modeling activities which have to be considered when designing a MAS simulation model. Each of these activities raises very different issues according to the considered application domain. Most of them are discussed in detail throughout this book. However, it is difficult to identify what are the actual challenges for these different modules without going into the details of some application domains. However, in the scope of this chapter, we want to provide a more global perspective of the intersection between MAS and simulation that will help to identify some fundamental questions and general concerns for the next years.

Going toward a global perspective of the intersection between MAS and simulation, it is necessary to understand that the module view is rather incomplete. Indeed, it is only a "MAS-side" view of the problem, in the sense that it focuses on MAS design issues. The other side of the coin is simulation itself. Indeed, simulating a system, whatever it is, also relies on considering some general issues about the design process of the experiment. To make this point clear, let us consider the Fishwick's definition for simulation [Fishwick, 1997]:

> *"Computer simulation is the discipline of designing a model of an actual or theoretical physical system, executing the model on a digital computer, and analyzing the execution output."*

Fishwick thus distinguishes between three fundamental activities:

1. Model design: The simulation is expressed in a model that specifies the characteristics of the simulated system.
2. Model execution: The model specifications are implemented in concrete computational structures that constitute the *simulator* of the experiment.
3. Execution analysis: The outputs of the simulation are checked according to some validation rules and then interpreted.

By defining simulation as a discipline composed of three fundamental activities, Fishwick's idea is to highlight that simulation is a concrete scientific process, which could be studied from a general perspective, independently from the purpose or the reason why it is used. In other words, simulation is more than just a tool. Simulation is a discipline that has its own characteristics and requirements (terminology, methodology, etc.) which should be considered whatever its context of use.

Therefore, understanding the different issues which exist considering both MAS and simulation, it is also necessary to have some insights about the general issues which are related to M&S. To this end, the next section briefly presents the *Framework for M&S* which was originally proposed by Zeigler in the beginning of the seventies [Zeigler, 1972; Zeigler et al., 2000]. In the scope of this chapter, the interest of the Framework for M&S is twofold: (1) Put in the light the relevance of studying MAS simulation according to this framework, and (2) highlight some major challenges and issues for future research in the MAS community.

1.5 The Zeigler's Framework for Modeling and Simulation

In [Zeigler et al., 2000]*, Zeigler proposed the *Framework for M&S* which relies on defining the *entities* of the framework and their fundamental *relationships*. This section briefly presents these entities, namely the *source system*, the *experimental frame*, the *model*, and the *simulator*, and these relationships, namely the *modeling relation* and the *simulation relation* (see Figure 1.14).

1.5.1 Source System

As shown by Fishwick's definition, the terms *real system* used in Shannon's definition do not necessarily refer to a real-world system, but more generally to the system which is the object of the current study, i.e. the phenomenon to simulate. This is what is called the *source system* in the Framework for M&S: The real or virtual system which is targeted by the simulationist. Thus, a source system may define a theoretical phenomenon which first exists only in the mind of somebody, and which does not exist before it is simulated. So, this terminology enables one to distinguish the real or virtual targeted system from the system which is finally obtained by executing the implementation of the model on a simulator.

If the source system does exist in the real world, it should be considered as the source of observable data that defines what is called the *behavior database* of the system (i.e. a relation between inputs and outputs). For instance, if the purpose of the simulation is to study a real ant farm, it is possible to collect data about the ants (e.g., lifetime, population dynamics, etc.).

1.5.2 Experimental Frame

Since a source system could be modeled considering very different purposes and objectives, the definition of a source system should be always done in the scope of an associated *experimental frame*. Formally, an experimental frame is a specification of the conditions under which the system is observed or experimented with. An experimental frame is the operational formulation of the objectives that motivate a modeling and simulation project.

Stating the experimental frame is crucial. Firstly, it deeply influences how the modeling, validation and verification processes are done with respect to (1) the information available about the source system and (2) the level of abstraction at which the source system is considered with respect to the resources and objectives of the experiment*. For instance, for the study of a real ant colony, taking into account the velocity of the ants within the modeling could be of interest, or even crucial in some cases, while not relevant in others. Therefore, in a more general perspective, the experimental frame could be considered as the definition of the experimental context. Especially, taken in a broad sense, this includes the used modeling paradigm and its associated constraints as discussed later on.

1.5.3 Model

Model is an overloaded term. However, in the scope of the Framework for M&S, a computer simulation model has a precise meaning. A model refers to all the computational

---

*The first edition of this book, *Theory of Modeling and Simulation*, was published in 1976.
*See [Boero and Squazzoni, 2005] for a focused discussion on these issues in the scope of ABSS.

specifications (instructions, rules, equations, and so on) that entirely define how all the possible state transitions of a system operate. In other words, a model is the specification of the *Evolution* function (equation 1.1). This specification could be written using different simulation formalisms or languages.

1.5.4 Simulator

A simulator is any software architecture on which the model could be executed to generate its behavior. In other words the simulator is responsible for computing the *Evolution* function and producing the outputs of the model with respect to acceptable inputs. A simulator could also embed some verification and validation tools. Depending on its general purpose, a simulator could be designed for only one experiment or a wide class of models.

1.5.5 Modeling Relation: Validity

The modeling relation concerns the study of the relation between a model and a source system in the scope of a particular experimental frame. This relation thus globally raises issues about the validity of the experiment by comparing the model (and its behavior) with the source system (and its behavior database if it exists). Here, the main question is to know if the modeling which has been done could be validated as an acceptable abstraction of the source system, considering the chosen qualitative criteria and experiment objectives. To answer this question, Zeigler proposes three main levels of validity:

1. *Replicative validity* is achieved if the behavior of the model which is observed matches the behavior of the source system considering an acceptability threshold.
2. *Predictive validity* implies replicative validity but also requires that the model is able to predict unseen source system behavior, which requires that the state transition function does reflect how the source system actually reacts to any acceptable inputs.
3. *Structural validity* implies predictive validity but also requires that the model mimics the way in which all the components of the source system do their state transitions.

1.5.6 Simulation Relation: Simulator Correctness

The *simulation relation* concerns the verification of the simulator with respect to the model. The question is to ensure that the simulator does generate correctly the model behavior. In other words, the simulation relation holds if the simulator guarantees that the consecutive states of the model are correctly reproduced. Conversely, it is interesting to note that this implies that the model specifications should entirely define the state transitions in both a (1) computerizable and (2) unambiguous way, otherwise the simulation relation cannot hold as it is possible to implement different interpretations of the model, thus implementing simulators relying on different state transition functions, which in turn produce different results for the same model.

Additionally, as the model specifications must be independent from the way they are executed by the simulator, the simulation relation also implies that the evolution of a simulated system must not depend on any simulator or hardware-related issues, enabling the replication of the experiment, which is fundamental from a M&S perspective. For instance, let us considered a simulated MAS executed on a Java Virtual Machine (JVM). If the evolution of any part of the system does depend on how the JVM manages the Java threads, replication is not achievable: the system is not simulated, but more simply executed.

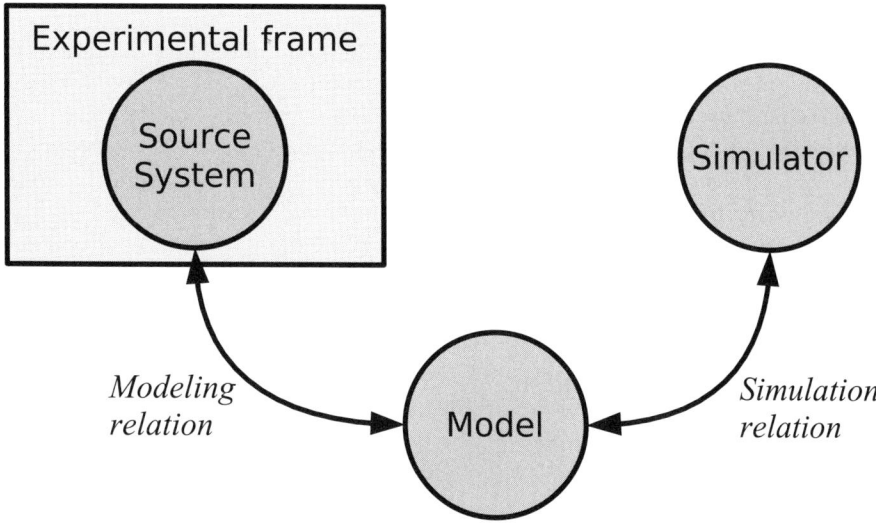

FIGURE 1.14 The Zeigler's Framework for M&S.

1.5.7 Deriving Three Fundamental Questions

Considering the use of simulation, the Framework for M&S has several advantages. Firstly, it gives a precise meaning to the different keywords which are related to simulation. Secondly, by relating these different key words to each other, thanks to the modeling and simulation relations, the Framework for M&S highlights the basic issues and problems which are related to any M&S experiment. Especially, it gives some clear guidelines to study the problem of the validation and verification (V&V). Although the purpose of this chapter is not to directly discuss V&V issues in the scope of MAS simulations, it is interesting to study MAS simulation works with respect to the basic relations of the Framework for M&S. Especially, we propose to do this study according to three questions derived from the Framework for M&S:

1. Does the model accurately represent the source system? (modeling relation: model vs. source system)
2. Does the model accommodate the experimental frame? (modeling relation: model vs. experimental frame)
3. Is the simulator correct? (simulation relation: model vs. simulator)

So, studying how these questions could be answered in the scope of MAS simulations, the main goal of the next section is to extract and discuss some important issues which represent relevant directions for future research within the MAS community.

1.6 Studying MAS Simulations Using the Framework for M&S

1.6.1 Does the Model Accurately Represent the Source System?

Identifying the Nature of the Source System

One can easily see that the issues related to the study of the modeling relation (validity) strongly depend on the source system being actual or theoretical. Indeed, in the case where the source system is purely theoretical, structural validity is a given property since the model does represent the actual state transitions of the system: They do not exist elsewhere.

On the contrary, if the source system is actual, then the modeling should be done with great care according to the behavior of the source system in order to, at least, achieve replicative validity. For instance, in a DAI platform that simulates the environment as a network, like in the DVMT experiment, the behavior of the simulated network has to match with real-world conditions (messaging time, resource access time, node failure, etc.) to some extent to be valid. So, doing a simulated MAS, one has to clearly state the nature of the source system which is considered: Actual or theoretical.

One could argue that it could be subtle to decide if a source system is actual or theoretical, which is true. For instance, although ABSS are inspired by real-world human societies, the level of abstraction used is sometimes so high that the model is considered as purely theoretical, while being inspired by real-world dynamics (e.g., SugarScape [Epstein and Axtell, 1996]). Validation issues related with an actual vs. theoretical perspective are discussed in papers such as [Marietto et al., 2004; Boero and Squazzoni, 2005] where continuum classifications of ABSS models are proposed according to different goals, levels of abstraction, and empirical data availability and gathering. The former proposes a general layered classification of paradigmatic models in ABSS with respect to their targets. The latter gives a focused discussion on related methodological issues and proposes a model maker perspective taxonomy of ABSS models (from case-based models to theoretical abstractions), suggesting various validation strategies according to the characteristics of the *empirical target matter*. So, by characterizing and layering the level of abstraction of ABSS models, these approaches help to decide how the modeling should be validated and such approaches could be useful in other domains.

Nonetheless, defining the level of abstraction of a model is a subjective task to some extent, and it would probably be possible to define hundreds of model classes if all application domains would be considered at once. In the scope of this chapter, we need a more general criterion to decide if a source system should be rather stated as actual or theoretical. In this respect, it is important to understand that such a criterion should not rely on the level of abstraction which is used in the modeling. And, in fact, it is possible to solve this problem in a general way with respect to the existence of the behavior database. Indeed, if a behavior database could be established, then the results could be checked according to it, thus the source system has to be considered as actual, whatever the level of abstraction.

Paradoxically, this also holds when it is not possible to establish the behavior database before the experiment. For instance, a large number of DAI operational systems, such as DVMT, are first design using simulation models before being deployed in real-world computational structures (when they finally are). Therefore, no behavior database is available a priori. However, if the system could be deployed, a behavior database could be established and used to check replicative validity for the model. So, deciding if a source system is theoretical or actual, one has to check if it could be possible to establish a behavior database for the system. If it is not the case, the source system is theoretical.

Different Modules, Different Issues

Still, the previous analysis is not complete and should go further: In the scope of MAS, raising the question about the nature of the source system, it is important to make a distinction between the behavior module and the environment module. Indeed, it is possible to define a system with theoretical behaviors embedded in a very realistic environment. So, it is possible to distinguish between four situations as illustrated in Figure 1.15.

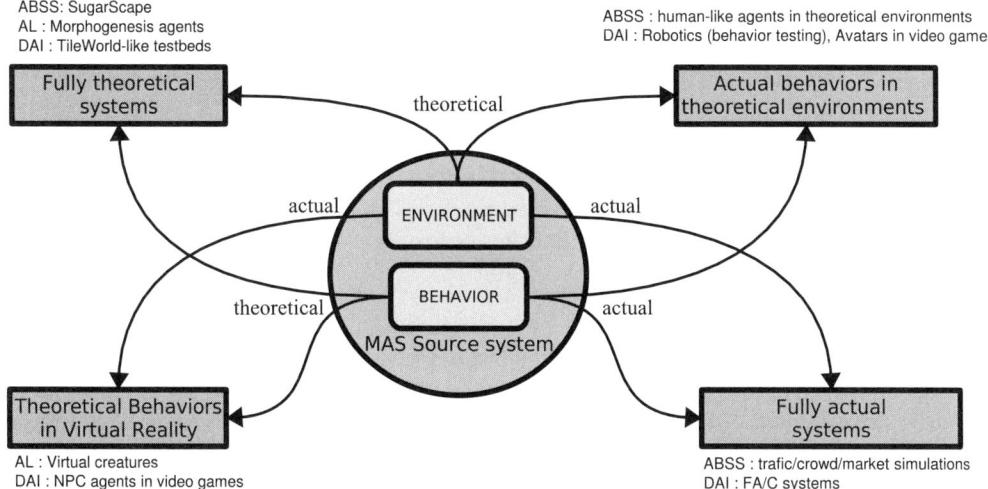

FIGURE 1.15 Simulation and MAS with respect to the source systems of the behavior and environment modules.

Each situation requires a different approach for validating the modeling: From fully theoretical systems for which structural validity is always true to fully actual systems for which both the behavior module and the environment module have to validated (at least with respect to replicative validity) according to the behavior of their targeted source systems. Thus each situation raises different issues when considering the modeling validation. Let us now discuss some of these issues according to the classification proposed in 1.15.

Environment Module Issues: The Need for Virtual Reality

From a validation point of view, the simplest case is to have a fully theoretical model. For instance, in [Beurier et al., 2006], a morphogenesis model for multi-agent embryogeny is proposed. Although this model is inspired by real biological processes, the purpose of both the behavior and environment modules is not to mimic any real system at all, but only to define a system which is able to produce emergent phenomena having interesting properties such as self-organization and the like. Therefore, there is no need to check the validity of the model with respect to the source system: The model is the source system itself, and therefore structurally valid. As we will see later on, that does not mean that the overall modeling could be directly considered as correct, but it clearly remains the less problematic case.

Next in difficulty are models where theoretical behaviors operate in an actual environment, namely in virtual reality (VR). For instance, in the Sims's work [Sims, 1994], virtual creatures evolve in a 3D world. As the Sims's goal is to study the evolution of the virtual creatures in a realistic world, the environment model embeds complex dynamics such as gravity, collisions, friction and so on. In such a case, structural validity is true for the behavior module as the agents are purely virtual (i.e. the source system of the behavior module is theoretical), while the model of the environment has to be validated according to real world dynamics (i.e. the source system of the environment module is actual). Therefore, from a validation point of view, the environment module is more important than the behavior module considering Simulated MAS applications that use VR. For instance, designing the agents AI in realistic video games, it is crucial that, like for the virtual creatures of [Sims, 1994], *"the environment restricts the actions to physically plausible behaviors"*: Observing and studying the behavior of the agents is not relevant if the modeling of the environment is not valid.

If having a realistic model of the environment is a strong requirement in the scope of VR-based applications, it is very interesting to see that the need for realism has also been pointed out as a primary concern in the DAI field relatively early, in the late eighties. For instance, studying teamwork and coordination of fire fighting units, a primary goal of the Phoenix testbed was to simulate an environment as realistic as possible* [Cohen et al., 1989]:

> *"Our position is that most AI systems have been built for trivial environments which offer few constraints on their design and, thus, few opportunities to learn how environments constrain and inform system design. To afford ourselves this opportunity, we began the Phoenix project by designing a real-time, spatially distributed, multi-agent, dynamic, ongoing, unpredictable environment."*

Later, during the DAI testbeds explosion of the nineties, Hanks revealed the *"danger of experimenting in the small"* in [Hanks et al., 1993]. Hanks's point was that DAI testbeds such as the TileWorld do not really fulfill their objectives, i.e. making progress in both the engineering and the understanding of (efficient) DAI systems, because most of the time they consider small, controlled, and deterministic world:

> *"The ultimate danger of experimentation in the small is that it entices us into solving problems that we understand rather than ones that are interesting. At best it gives the mistaken impression that we are making progress toward our real goal. At worst, over time it confounds us to the point that we believe that our real goal is the solution of the small, controlled problems."*

Today, it is clear that experimentation in the small is still a reality, not only in the DAI field. Of course, from a pragmatic point of view, we always have to experiment in the small to some extent: We just cannot simulate all the details of reality and should focus on the aspects of the environment which are the most relevant according to the experiment objectives. However, one has to bear in mind that, with respect to the experiment objectives, the modeling of the environment could be crucial when evaluating the appropriateness of the overall modeling. In this respect, one major challenge for DAI researches in the years to come

---

*Studying forest fires management in the Yellowstone National Park, Phoenix defined a highly dynamic environment thanks to a complex simulation integrating several realistic parameters such as weather conditions, physical features of the park, and burn rates of different land types.

is to move from theoretical environments to actual ones: Putting DAI systems into VR, and thus *experimenting in the large*. Such a point of view also nicely bridges the gap between concerns from Artificial Life and DAI, highlighting the importance of the environment module for both [Luck and Aylett, 2000]. Considering such an issue, the RoboCup [Kitano et al., 1997] is a really interesting step in the history of DAI testbeds because it represents a good deal between realism and DAI testing. Additionally, one can find in this book a chapter dealing with the fundamental role of the environment for software-in-the-loop simulations [Helleboogh et al., 2008].

Behavior Module Issues: Toward the Participatory Design of Simulation

Although checking the validity of an actual environment module could be a hard and time consuming task, the general approach itself is relatively straightforward, because of our everyday life experience of the causality principle. Indeed, as we are living inside this real environment, we can directly observe it so that we are able to deeply analyze its functioning. So it is easy (relatively of course) to see that an environment module does not agree with real world dynamics, thanks to our own experience. Objects are supposed to be under the influence of the gravity force for instance. Comparatively, it is much more difficult to decide about the appropriateness of the behavior module when facing the modeling of real-world behaviors.

In fact, considering the modeling of actual behaviors, we have to distinguish between two different cases: (1) behaviors that refer to real-world computational structures (robots, software agents, etc.) and (2) behaviors of real-world living entities (insects, animals or human).

In the first case, checking the validity of the behavior module is not more difficult than for the environment because we deeply know the real structure of the related behaviors as we created them. For instance specifying the behavior module of a real robot simply consists in modeling how the behavior program of the robot runs on some concrete hardware in real conditions (software-in-the-loop). The same goes for an e-commerce agent as well.

However, when it comes to the modeling of living entities, we do not have such information. In contrast with the environment, we cannot observe the internal mechanisms that effectively conduct living entities' abilities to make decisions: A living creature's behavior is a black box whose internal mechanisms could be only modeled using very high level abstractions.

Consequently, it could be difficult to decide about the validity of a behavior module with respect to the experiment objectives as we are, once again, always experimenting in the small to some extent: Validating a behavior module is a highly subjective task in such cases. Moreover, evaluating a human-like behavior module could be by a huge order of magnitude more problematic than one representing an ant-like behavior. It is obviously a reason why there is a growing trend toward a participatory design of simulation: We need to "put the human in the loop".

In [Drogoul et al., 2002], Drogoul et al. highlight the huge gap which usually exists between the behaviors of the domain model (the behavior module) and the computational structures which are finally used to implement them:

> "*This means that the resulting 'computational agents' (if we can still call them this way) do not possess any of the properties generally assigned to the agents used in MAS or DAI: they do not have any structural nor decisional autonomy, are not proactive, and cannot, for instance, modify their knowledge and behaviors through learning or adaptation.*"

The authors thus proposed a MABS methodological process that more concretely involves real stakeholders in the modeling activity, thus advocating the idea of the participatory design of simulations. Agent-Based Participatory Simulations could be considered as a more strong merging between MAS and Role Playing Game (RPG) [Guyot and Honiden, 2006]. RPG was used in [Barreteau et al., 2001] to work on both the design and the validation of social MABS models. Still, in such works, the participants of the RPG do not directly control an agent of the simulation and the agent's behaviors are still encoded as computational structures. So, the main idea of agent-based participatory simulations is to have real humans playing the role and taking control of simulated agents to design and then refine the behavior module through an iterative modeling process (with steps with or without real human agents) [Nguyen-Duc and Drogoul, 2007; Guyot and Honiden, 2006]. The main advantage is that, as all the events could be recorded, the behavior of the participants could be analyzed in details during and, more importantly, after the simulation runs using the trace of the simulation execution. This helps to understand why and how the participants make decisions with respect to the information they gather during the simulation. So, the participatory design of simulation helps for the modeling of more accurate artificial behaviors, thanks to the refining of the perception/deliberation/action processes which is done at each iteration of the modeling loop proposed by this approach.

Although motivations for the design participatory simulation are numerous, in the scope of this chapter, we can summarize the major idea of such approaches as follows. Modeling the behavior of a real living entity is hard and structural validity could not be an objective at present time considering the behavior module. We need high level abstractions. Building these abstractions, rather than trying to model existing behaviors from scratch using complex deliberation models, and thus searching for an unreachable structural validity, we should focus on studying what we can observe and analyze from real living entities: The actions they produce with respect to their perceptions. In other words, rather than focusing on the structural aspect of the behavior module, we should first focus on its ability to achieve replicative validity, thus relating perceptions with actions, i.e. inputs with outputs.

If the relevance of participatory simulations is obvious considering social simulations, it could also be very interesting to use a similar approach for the modeling of non human species. The interest of the idea still holds: It could help for the design of the behavior of artificial entities as it is of course possible for a human to play the role of a predator agent for instance. Studying what makes a human decide what are the relevant perceptions/actions may significantly help for refining both the behavior module (modeling deliberation) and the environment module (adding perceptions/actions).

Merging Participatory Design and VR: A Promising Future for MABS

In the previous sections, we have discussed only few issues which are related to the study of the modeling relation in SMAS. However we have highlighted two very interesting tracks of research, each of which being related to either the environment module or the behavior module: That is respectively VR and the participatory design of simulation. In this respect, it is worth noting that very recent works integrate both in a very innovative and unexpected manner.

In the pioneer work proposed in [Ishida et al., 2007], Ishida introduces the notion of *augmented experiment*, in the scope of real-world experiments on disaster evacuation. The idea of a multi-agent-based augmented experiment is to do real-world experiments, with real-world subjects evolving in an *augmented reality*: That is, in a real-world environment where actions of simulated entities, evolving in a corresponding VR, could be perceived by real-world subjects thanks to the use of suitable sensor devices. In this perspective, the related design process of the experiment is particularly interesting as it involves a participatory design loop for modeling VR, which is in turn used to do augmented experiments. Doing so, such a work establishes a link between reality and VR as well as between humans and virtual entities.

We are at the frontier of what we can do with our technologies today, but augmented experiments raises a number of technical problems [Ishida et al., 2007]. However, this is obviously a forerunner example of how we could finally experiment in the large in a near future. In such a perspective, VR and participatory simulation are key concepts.

1.6.2 Does the Model Accommodate the Experimental Frame?

MABS are Supposed to Model MAS

One could think that considering a theoretical model implies that the overall modeling could be directly validated. Especially, as discussed by Troitzsch in the scope of ABSS, a simulation program can be seen as a full model of a theory as it can be considered as a *structuralist reconstruction* of this theory as discussed [Troitzsch, 2004].

However, beyond this legitimate consideration, it is also crucial to not forget that the modeling relation is tripartite, implying that the source system and the model should always be considered in the scope of a particular experimental frame, which defines modeling constraints both explicitly and implicitly. Therefore these constraints must be taken into account in the validation process, whatever the nature of the source system. Notably, this is also true for fictitious worlds because they do also define environmental laws that impose consistency and coherence, as does the real world [Valckenaers et al., 2007]. For instance, if a theoretical source system is supposed to embed usual physical laws, then the modeling should not define state transitions which contradict these laws: It is not a valid modeling otherwise as it is an implicit constraint of the experimental frame.

So, as pointed out by Zeigler, it is crucial for the validity of a model that it at least coherently accommodates the experimental frame, thus verifying the *accommodation relation* [Zeigler et al., 2000]. Considering this relation in the scope of MAS is more challenging that it seems to be at first sight.

Verifying the accommodation relation, it is obvious that both the behavior module and the environment module must follow some modeling guidelines so that they do not integrate state transitions contradicting the experimental frame. For instance, it is obvious that an agent must not be able to perform an action that makes it go through a wall. Therefore, there is not really a need to put forward the importance of such concerns here.

Nonetheless, all MABS rely on an implicit, but strong, hypothesis: They use MAS as modeling paradigm. As such, their modeling should accommodate the properties of the agent paradigm, whatever the objective of the simulation study. Verifying this particular accommodation is related to the study of what we define as *paradigmatic validity/coherency*[*], i.e. the validity of the model with respect to the multi-agent paradigm itself. In the following section, we only focus on two consequences underlying the consideration of such a level of validity.

---

[*]The word *validity* is intentionally used to make a parallel between paradigmatic validity and the other forms of validity defined in the Framework for M&S.

Modeling Simultaneous Actions and the Autonomy Feature

Firstly, that means we should be able to represent simultaneous actions as agents are supposed to act in a concurrent manner on the environment. As Ferber has highlighted, implementing simultaneity with usual formalisms is not trivial [Ferber, 1999].

Secondly, one has to bear in mind that agents are supposed to be autonomous entities. As pointed out in [Michel et al., 2004], it is easy to model systems wherein agents are finally not autonomous, since some of their behaviors are directly controlled by other agents.

Without going into the details of the problems which are related to each case, it is interesting to see that the Influence/Reaction model has been considered as a solution for both. Indeed, the Influence/Reaction model [Ferber and Müller, 1996] relies on modeling the agent actions as influences that do not modify the environment (and the other agents as well), but which are treated as a whole to compute the reaction of the environment to these influences, and thus the next system state. So, it is possible to easily model simultaneous actions by combining the influences to compute the actual result on the environment. Works such as [Helleboogh et al., 2007; Michel, 2007; Weyns and Holvoet, 2004; Dàvila and Tucci, 2000] show the interest of such an approach for modeling simultaneous actions in different contexts. Moreover, as the agents do not directly modify the others' internal variables, it is easy to check that the autonomy feature is taken into account [Michel et al., 2004].

Toward Paradigmatic Validity: The Need of Linking the Micro and Macro Levels

In the previous section, we discussed the opportunity of using the Influence/Reaction model for simulating MAS. From a more general point of view, this shows that we need approaches and formalisms specifically dedicated to the ABM approach in order to accommodate the experimental frame. All this is about studying issues related to paradigmatic validity.

Considering this study, one difficulty relies on the fact that MAS define at least two levels of dynamics: the micro level and the macro level. Most approaches only consider the micro level specifications and do not evaluate their impact on the macro level dynamics (e.g., the possibility of representing simultaneous actions for instance). So, most of the time, there is a discrepancy between the system we want to model (i.e. the source system) and the model we finally produced. In this respect, [David et al., 2002] shows the interest of specifying both the micro and macro levels for ABM. Notably, [David et al., 2002] proposes modeling specifications at different levels of granularity: *atomic agentified entities (AE)* are specified at the micro level, and then aggregates of AEs (macro-AE) are specified at different levels of aggregation (first order macro level, second order, and so on). The Influence/Reaction model also makes a difference between the micro level (influences) and the macro level (the reaction of the environment to the influences) within the modeling.

Nonetheless, it is clear that many efforts have still to be done in such directions. One major challenge for ABM approaches is indeed to reduce the gap between what we are supposed to model (i.e. MAS) and the modeling abstractions and formalisms we use to do it. In the scope of paradigmatic validity, this is a necessary requirement if we want to have a clear understanding of the models we build, knowing they are becoming more complex everyday. In this perspective, the problem is not only to represent simultaneous actions or autonomy, but also to be able to express complex concepts such as stigmergy, self-organization, emergence and the like, in the model specifications. To this end, making an explicit link between micro level and macro level specifications is definitively an avenue for future research. As we will see now, such considerations also impact the study of the simulation relation.

1.6.3 Is the Simulator Correct?

We now switch our focus from the modeling relation to the simulation relation, that is, to the study of concerns related to the simulator correctness.

A New Day, a New Simulator

From a MAS perspective, it possible to classify MAS simulators according to three categories according to their capability and focus:

1. limited to a single class of model: *One shot*;
2. able to accept a whole class of models related to a particular domain: *Domain related* (e.g., Ecology: Cormas [Bousquet et al., 1998], Echo [Hraber et al., 1997]);
3. designed for implementing MAS simulations using a particular agent software architecture, independently of a particular domain: *Generic* (e.g., Swarm [Minar et al., 1996], MadKit [Gutknecht et al., 2001], Jason [Bordini, 2008], SeSam [Klügl, 2008]).

It is worth noting that there are far more One-shot MAS simulators than the two other kinds. This clearly shows that there still no consensus on the way agent-based systems have to be modeled and implemented. Indeed, one question is: Why is a new simulator required almost every time? One could say that it is due to the ability of MAS to address numerous application domains, each having their own modeling requirements. This is probably true to some extent. However, another answer is possible if we consider the problems raised in the previous section: The gap between the MAS paradigm and the modeling tools we used. This gap does not allow a clear understanding of MAS simulation frameworks. Consequently, it is not surprising that researchers prefer to define their own simulation tools in order to (1) fulfill their requirements and (2) control entirely the simulation process.

Having a majority of one shot simulators raises several problems for the community. Firstly, it severely reduces the possibility of reusing previous works, which is an undisputed drawback from a software engineering point of view. It is a sure bet to claim that similar things (agent and environment models, simulation engines, etc.) have been coded hundreds of times: The community hardly capitalizes experiences. Of course, it also increases the possibility of facing bugged simulators. So, one major challenge for the community is to find ways of doing MABS so that we can reuse and capitalize on experiences in a more efficient manner.

Secondly, it raises the problem of reusing the simulation results themselves. Indeed, the simulation relation is rarely studied and few MAS works do consider the simulator as a first order entity of the experiment. Most of the time, the implementation of the model is considered as a step of the experiment which correctness would be true a priori, which is obviously a mistake. Moreover, this decreases the confidence we have in published simulation results. Therefore, not considering the simulation relation is obviously a weak point that should be addressed in the years to come because it dramatically limits how simulation experiments can be evaluated: Verifying the simulation relation is not less important than validating the modeling relation in the simulation process. To emphasize more on that point, let us now consider a fundamental consequence of this state of affairs: The difficulty of replicating existing models.

The Fundamental Problem of Replication

In Section 1.4, we have shown that ABM implementations could raise many problems, in many different ways. So, it is clear that going from model specifications to the implementation is not a trivial task in the scope of MAS. Not surprisingly, simulation results are highly sensitive to the way the model is defined and implemented (see [Merlone et al., 2008] for a recent and very interesting study of that issue). Works such as [Axelrod, 1997; Lawson and Park, 2000; Rouchier, 2003; Edmonds and Hales, 2003] also clearly address this issue by showing all the difficulties we face when trying to replicate published results. Especially, they show that, since simulation results are usually published without detailing enough the model and its implementation, replications have a high probability of producing results which are (very) different from the originals.

This problem is identified as the *engineering divergence phenomenon* in [Michel et al., 2004], i.e., the possibility of implementing different simulations, considering a unique model. Such a problem contradicts at the heart a basic hypothesis of the simulation relation of the Framework for M&S : A model should be defined so that the produced results do not depend on how it is implemented (cf. Section 1.5.6). In the previous section we claim that there is a gap between the MAS paradigm and the modeling tools/formalisms we used. We can make a similar statement here: There is a gap between the agent-based models which are defined and the computational structures we used to implement them. In fact, the complexity of the implementation, which is dramatically real, is usually simply bypassed when agent-based models are designed.

The scheduling module is particularly neglected during the modeling phase because it is often considered as only related to the implementation. This is a mistake: The scheduling module is fundamentally part of the modeling. Indeed, if this part is fuzzily or not defined at all, thus the model specifications are ambiguous and could be implemented following different interpretations, which in turn may lead to huge differences in the simulation results.

So, as highlighted by Axelrod in the scope of ABSS [Axelrod, 1997], one major challenge for the community is to find ways of sharing modeling specifications in an efficient way, enabling the replication, and thus the study, of published experiments. To this end, it is clear that making an explicit link between computational structures and models is the key. In this respect, the Swarm platform design is of great interest [Minar et al., 1996]. Indeed, establishing a Swarm-based model explicitly relies on defining, not only the behavior and the environment modules, but also the order in which they are finally executed by the simulator. In other words, a swarm-based model is directly related to its implementation, so that it is easy to replicate it.

Generalizing such an approach is crucial for the years to come: Enabling replication is what the community should go for. To this end, we claim that two things should be placed on the agenda of multi-agent simulationists: (1) Promoting the reuse of existing works as far as possible and (2) defining unambiguous models. The former will enable the community to go toward the elaboration of a real simulator software engineering process, ultimately making the reuse of existing works possible. The latter will help to really study the simulation relation in ABM by making an explicit distinction between the model and its implementation, which in turn will help to better understand the true dynamics of MAS as they will be explicitly modeled. Only such approaches will enable us to ultimately answer the fundamental question of this section: Is the Simulator Correct?

1.7 Conclusion

MAS and simulation are two research fields having a close relationship. Studying the intersection between these two fields, it is important to keep in mind that simulation is not only a tool, but a real scientific discipline which already defines some design guidelines. This is the reason why we have argued on the interest of considering MAS simulation works in the scope of a pure M&S perspective such as Zeigler's. Indeed, the Framework for M&S permitted us to study MAS simulation works according to the modeling and simulation relations, which are essential for any M&S experiment.

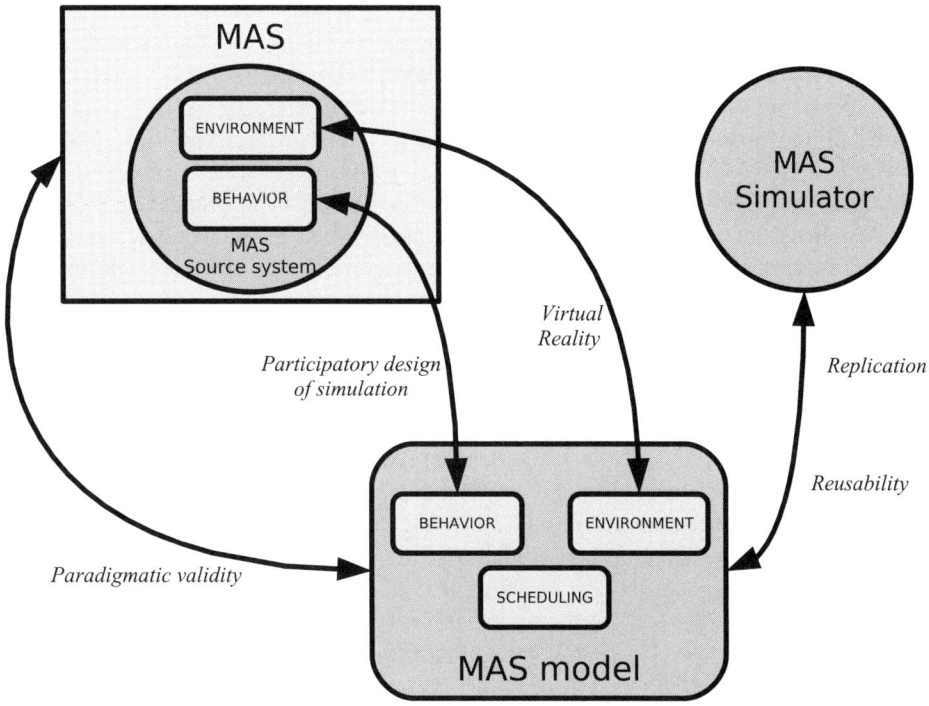

FIGURE 1.16 Defining MAS simulation issues using the Framework for M&S.

Figure 1.16 illustrates the different challenges we have highlighted by studying MAS simulation according to Zeigler's Framework for M&S.

First of all, we have seen that VR is a concept of primary importance with respect to the environment module. Experimentation in the small has been useful for exploring some characteristics of the MAS paradigm in the past two decades. Today, the available technologies urge us to make the step toward VR-based models of the world to prove the usefulness of the existing DAI architectures and develop new ones, more adapted to real-world applications. Additionally, considering software-in-the-loop approaches, VR is the only way to go: Without an accurate model of the deployment environment, the system cannot be developed in an efficient way. The reader will find more about the crucial importance of the modeling of the environment in this book [Helleboogh et al., 2008]. For the MAS community, VR is the key for experimenting in the large.

About the behavior module, we have emphasized the difficulty of building representative models of real living entity behaviors, whether they are human or animal beings. Indeed, contrary to software agents (e.g., robots), the real structure of the decision process cannot be observed with our technologies today, so that we use theoretical approaches to model them. Therefore we have to find ways to reduce, as far as possible, the inescapable gap which exists between what we want to model (i.e., real decision processes) and what we could model using the theories and computational modeling abstractions available today. In this respect, we have discussed the relevance of the participatory design of simulation. There is a major idea in this trend of research: Rather than model real behaviors so that they somehow fit our computational structures, we should make the computational structures evolve so that they can reflect how a real living entity is motivated to act (which perceptions for which actions). In other words, we should not directly use computational structures to model real behaviors, but use real behaviors to model new computational structures, which will in turn be more suited to our needs. Doing this shift of perspective is a major challenge for the MAS community, considering behavioral modeling.

The two previous directions of research represent paths toward the accommodation of future MAS models with the MAS experimental frame. Indeed, they force us to make progress toward simulation formalisms that take into account the specificities and features of the MAS paradigm, which in turn enables the study of issues related with paradigmatic validity. For instance, the Influence/Reaction formalism permits one to model MAS basic concepts such as action simultaneity and autonomy. In a more general perspective, a major feature of ABM relies on the fact that it requires at least two fundamental levels of modeling: The micro (agent) and the macro (multi-agent) levels. Most of the approaches do only consider the modeling of the micro level entities and do not clearly specify how their dynamics should be composed, leaving the developer free to choose during the implementation phase. Providing abstractions for the macro level dynamics, such as simultaneous actions, is a first step toward a concrete linking between these two fundamental levels.

Ultimately, studying the simulation relation, we have emphasized on a crucial issue for the MAS community: Capitalizing on passed experiences. This is necessary if we want to speed up the research on MAS simulation in general, whatever the objectives. In this respect, it is obvious that the software engineering perspective is neglected by most of the MAS simulation works as the emphasis is always put on modeling issues. The MAS community has to find the way of concretely sharing implementation experiences, promoting reusability as a must. Without such concerns in mind, reinventing the wheel will continue to be a reality for the community. More than that, it is the only way through which we will be able to really do verification on MAS simulation experiments as it will ease replication, which is fundamental from a M&S perspective. Trusting simulator implementations a priori is not a viable option.

References

G. Agha. *Actors: a model of concurrent computation in distributed systems*. MIT Press, Cambridge, MA, USA, 1986. ISBN 0-262-01092-5.

M. Arango, L. Bahler, P. Bates, M. Cochinwala, D. Cohrs, R. Fish, G. Gopal, N. Griffeth, G. E. Herman, T. Hickey, K. C. Lee, W. E. Leland, C. Lowery, V. Mak, J. Patterson, L. Ruston, M. Segal, R. C. Sekar, M. P. Vecchi, A. Weinrib, and S.-Y. Wuu. The touring machine system. *Communications of the ACM*, 36(1):69–77, 1993. ISSN 0001-0782. doi: http://doi.acm.org/10.1145/151233.151239.

R. C. Arkin. Motor schema-based mobile robot navigation. *International Journal of Robotics Research*, 8(4):92–112, 1989.

J. L. Austin. *How to Do Things With Words*. Oxford University Press, 1962.

R. Axelrod. *Advancing the art of simulation in the social sciences*, pages 21–40. Volume 456 of Conte et al. [1997], 1997. ISBN 3-540-24575-8.

R. L. Axtell. Why agents? on the varied motivations for agent computing in the social sciences, csed working paper no. 17. Technical report, Center on Social and Economic Dynamics, The Brookings Institution, November 2000.

O. Balci. The implementation of four conceptual frameworks for simulation modeling in high-level languages. In *Proceedings of the 20th conference on Winter simulation*, pages 287–295. ACM Press, 1988. ISBN 0-911801-42-1. doi: http://doi.acm.org/10.1145/318123.318204.

K. S. Barber, R. McKay, M. MacMahon, C. E. Martin, D. N. Lam, A. Goel, D. C. Han, and J. Kim. Sensible agents: an implemented multi-agent system and testbed. In *Proceedings of the fifth international conference on Autonomous agents*, pages 92–99. ACM Press, 2001. ISBN 1-58113-326-X. doi: http://doi.acm.org/10.1145/375735.376007.

O. Barreteau, F. Bousquet, and J.-M. Attonaty. Role-playing games for opening the black box of multi-agent systems: method and lessons of its application to senegal river valley irrigated systems. *JASSS, The Journal of Artificial Societies and Social Simulation*, 4(2), 2001.

M. S. Bartlett. *Stochastic Population Models in Ecology and Epidemiology*. Wiley, 1960.

B. Bauer, J. P. Müller, and J. Odell. Agent uml: A formalism for specifying multiagent software systems. In P. Ciancarini and M. Wooldridge, editors, *Agent-Oriented Software Engineering, First International Workshop, AOSE 2000, Limerick, Ireland, June 10, 2000, Revised Papers*, volume 1957 of *Lecture Notes in Computer Science*, pages 91–104. Springer, 2001. ISBN 3-540-41594-7.

B. Beaufils, J.-P. Delahaye, and P. Mathieu. Complete classes of strategies for the classical iterated prisoner's dilemma. In *Proceedings of the 7th International Conference on Evolutionary Programming VII*, pages 33–41. Springer-Verlag, 1998. ISBN 3-540-64891-7.

R. Beckers, J. Deneubourg, and S. Goss. Trails and u-turns in the selection of a path by the ant lasius niger. *Journal of Theoretical Biology*, 159:397–415, 1992.

G. Beurier, F. Michel, and J. Ferber. A morphogenesis model for multiagent embryogeny. In L. M. Rocha, L. S. Yaeger, M. A. Bedau, D. Floreano, R. L. Goldstone, and A. Vespignani, editors, *Proceedings of the Tenth International Conference on the Simulation and Synthesis of Living Systems*, pages 84–90, Cambridge, MA, USA, August 2006. MIT Press. ISBN 0-262-68162-5.

R. Boero and F. Squazzoni. Does empirical embeddedness matter? methodological issues on agent-based models for analytical social science. *Journal of Artificial Societies and Social Simulation*, 8(4):6, 2005. ISSN 1460-7425. URL http://jasss.soc.surrey.ac.uk/8/4/6.html.

R. H. Bordini. Agent-based simulation using bdi programming in Jason. In Uhrmacher and Weyns [2008], chapter 15.

F. Bousquet, I. Bakam, H. Proton, and C. L. Page. Cormas: Common-pool resources and multi-agent systems. In *Proceedings of the 11th International Conference on Industrial and Engineering Applications of Artificial Intelligence and Expert Systems*, pages 826–837. Springer-Verlag, 1998. ISBN 3-540-64574-8.

A. Bretagnolle, E. Daudé, and D. Pumain. From theory to modelling : urban systems as complex systems. *CyberGeo: European Journal of Geography*, (335):1–17, March 2006.

R. A. Brooks and J. H. Connell. Asynchronous distributed control system for a mobile robot. In W. Wolfe and N. Marquina, editors, *SPIE's Cambridge Symposium on Optical and Opto-Elecronic Engineering*, volume 727, pages 77–84, October 1986.

G. M. Buettner and W. Siler. Variability in predator-prey experiments: simulation using a stochastic model. In *Proceedings of the 14th annual Southeast regional conference*, pages 194–201. ACM Press, 1976. doi: http://doi.acm.org/10.1145/503561.503602.

P. H. Chang, K.-T. Chen, Y.-H. Chien, E. Kao, and V.-W. Soo. From reality to mind: A cognitive middle layer of environment concepts for believable agents. In Weyns et al. [2005a], pages 57–73. ISBN 3-540-24575-8.

J. Chapelle, O. Simonin, and J. Ferber. How situated agents can learn to cooperate by monitoring their neighbors' satisfaction. In *Proceedings of the 15th European Conference on Artificial Intelligence ECAI'2002*, July 21-26 2002.

P. R. Cohen, M. L. Greenberg, D. M. Hart, and A. E. Howe. Trial by fire: understanding the design requirements for agents in complex environments. *AI Magazine*, 10(3): 32–48, 1989. ISSN 0738-4602.

R. J. Collins and D. R. Jefferson. Antfarm: Towards simulated evolution. In C. G. Langton, C. Taylor, J. D. Farmer, and S. Rasmussen, editors, *Artificial Life II*, pages 579–601. Addison-Wesley, Redwood City, CA, 1992. URL citeseer.ist.psu.edu/collins91antfarm.html.

R. Conte, R. Hegselmann, and P. Terna, editors. *Simulating Social Phenomena*, volume 456 of *Lecture Notes in Economics and Mathematical Systems*, Berlin, 1997. Springer-Verlag. ISBN 3-540-24575-8.

R. Conte, N. Gilbert, and J. S. Sichman. MAS and social simulation: A suitable commitment. In J. S. Sichman, R. Conte, and N. Gilbert, editors, *Multi-Agent Systems and Agent-Based Simulation*, volume 1544 of *LNCS*, pages 1–9. Springer-Verlag, 1998. ISBN 3-540-65476-3.

N. David, J. S. Sichman, and H. Coelho. Towards an emergence-driven software process for agent-based simulation. In J. S. Sichman, F. Bousquet, and P. Davidsson, editors, *Multi-Agent-Based Simulation II, Proceedings of MABS 2002, Third International Worshop*, volume 2581 of *LNAI*, pages 89–104. Springer-Verlag 2003, July 2002. URL citeseer.nj.nec.com/540531.html.

J. Dàvila and K. Tucci. Towards a logic-based, multi-agent simulation theory. In *International Conference on Modeling, Simulation and Neural Networks MSNN'2000*, October 22-24 2000.

K. S. Decker. Distributed artificial intelligence testbeds. In G. M. P. O'Hare and N. R. Jennings, editors, *Foundations of distributed artificial intelligence*, chapter 3, pages 119–138. John Wiley & Sons, Inc., 1996. ISBN 0-471-006750.

G. Deffuant, F. Amblard, G. Weisbuch, and T. Faure. How can extremism prevail? a study based on the relative agreement interaction model. *The Journal of Artificial Societies and Social Simulation, JASSS*, 5(4), January 2002. URL http://www.soc.surrey.ac.uk/JASSS/5/4/1.html.

Y. Demazeau. Steps towards multi-agent oriented programming. slides Workshop, 1st International Workshop on Multi-Agent Systems, IWMAS '97, October 1997.

M. Dorigo. *Optimization, learning and natural algorithms*. PhD thesis, Politecnico di Milano, Italy (in Italian), 1992.

M. Dorigo and L. M. Gambardella. Ant Colony System: A cooperative learning approach to the traveling salesman problem. *IEEE Transactions on Evolutionary Computation*, 1(1):53–66, 1997. doi: http://dx.doi.org/10.1109/4235.585892.

M. Dorigo and T. Stützle. *Ant Colony Optimization*. MIT Press, Cambridge, MA, 2004.

K. Dresner and P. Stone. Multiagent traffic management: A reservation-based intersection control mechanism. In N. R. Jennings, C. Sierra, L. Sonenberg, and M. Tambe, editors, *Proceedings of The Third International Joint Conference on Autonomous Agents & Multi Agent Systems (AAMAS'2004)*, volume 2, pages 530–537. ACM Press, July 19-23 2004. URL http://www.cs.utexas.edu/users/kdresner/papers/2004aamas.

A. Drogoul and J. Ferber. Multi-agent simulation as a tool for modeling societies: Application to social differentiation in ant colonies. In C. Castelfranchi and E. Werner, editors, *Artificial Social Systems*, volume 830 of *Lecture Notes in Computer Science*, pages 3–23. Springer-Verlag, 1992. ISBN 3-540-58266-5.

A. Drogoul, D. Vanbergue, and T. Meurisse. Multi-agent based simulation: Where are the agents? In J. S. Sichman, F. Bousquet, and P. Davidsson, editors, *Multi-Agent-Based Simulation II, Proceedings of MABS 2002, Third International Worshop*, volume 2581 of *LNAI*, pages 89–104. Springer-Verlag, July 2002. URL citeseer.nj.nec.com/540531.html.

B. Dumont and D. R. Hill. Multi-agent simulation of group foraging in sheep: effects of spatial memory, conspecific attraction and plot size. *Ecological Modelling*, 141: 201–215, July 1 2001.

E. H. Durfee, M. Yokoo, M. N. Huhns, and O. Shehory, editors. *Sixth International Joint Conference on Autonomous Agents and Multiagent Systems (AAMAS 2007), Honolulu, Hawaii, USA, May 14-18, 2007*, New York, NY, USA, 2007. ACM. ISBN 978-81-904262-7-5.

B. Edmonds and D. Hales. Replication, replication and replication: Some hard lessons from model alignment. *The Journal of Artificial Societies and Social Simulation, JASSS*, 6(4), October 2003.

J. M. Epstein. Agent-based computational models and generative social science. In *Generative Social Science Studies in Agent-Based Computational Modeling*, Princeton Studies in Complexity, pages 4–46. Princeton University Press, January 2007. ISBN 0691125473. URL http://www.amazon.ca/exec/obidos/redirect?tag=citeulike09-20\&path=ASIN/0691125473.

J. M. Epstein and R. L. Axtell. *Growing Artificial Societies*. Brookings Institution Press, Washington D.C., 1996.

L. D. Erman, F. Hayes-Roth, V. R. Lesser, and D. R. Reddy. The hearsay-II speech-understanding system: Integrating knowledge to resolve uncertainty. *ACM Computing Surveys*, 12(2):213–253, 1980. ISSN 0360-0300. doi: http://doi.acm.org/10.1145/356810.356816.

J. Ferber. *Multi-Agent Systems: An Introduction to Distributed Artificial Intelligence*. Addison-Wesley Longman Publishing Co., Inc., 1999. ISBN 0201360489.

J. Ferber and J.-P. Müller. Influences and reaction : a model of situated multi-agent systems. In M. Tokoro, editor, *Proceedings of the 2nd International Conference on Multi-agent Systems (ICMAS-96)*, pages 72–79. The AAAI Press, December 10-13 1996. ISBN 0-1-57735-013-8.

T. Finin, R. Fritzson, D. McKay, and R. McEntire. KQML as an agent communication language. In *CIKM '94: Proceedings of the third international conference on Information and knowledge management*, pages 456–463, New York, NY, USA, 1994. ACM. ISBN 0-89791-674-3. doi: http://doi.acm.org/10.1145/191246.

191322.

P. A. Fishwick. Computer simulation: growth through extension. *Transactions of the Society for Computer Simulation International*, 14(1):13–23, 1997. ISSN 0740-6797.

F. for Intelligent Physical Agents. http://www.fipa.org. Accessed June 2008. FIPA Communicative Act Library Specification.

E. Fyanio, J.-P. Treuil, E. Perrier, and Y. Demazeau. Multi-agent architecture integrating heterogeneous models of dynamical processes : the representation of time. In J. S. Schiman, R. Conte, and N. Gilbert, editors, *Multi-Agent Systems and Agent-Based Simulation*, volume 1534 of *LNCS*, pages 226–236. Springer-Verlag, 1998. ISBN 3540654763.

L. Gasser and K. Kakugawa. MACE3J: fast flexible distributed simulation of large, large-grain multi-agent systems. In *Proceedings of the first international joint conference on Autonomous agents and multiagent systems*, pages 745–752. ACM Press, 2002. ISBN 1-58113-480-0. doi: http://doi.acm.org/10.1145/544862.544918.

L. Gasser, C. Braganza, and N. Herman. Mace: A flexible testbed for distributed ai research. *Distributed Artificial Intelligence*, pages 119–152, 1987. doi: http://doi.acm.org/10.1145/67386.67415.

G. F. Gause. *The Struggle for Existence*. Williams and Wilkins, Baltimore, 1934.

M. R. Genesereth and N. J. Nilsson. *Logical foundations of artificial intelligence*. Morgan Kaufmann Publishers Inc., 1987. ISBN 0-934613-31-1.

N. Gilbert. Agent-based social simulation: Dealing with complexity. http://www.soc.surrey.ac.uk/staff/ngilbert/ngpub/paper165_NG.pdf. Accessed June 2008, 2005.

N. Gilbert and K. G. Troitzsch. *Simulation for the Social Scientist*. Open University Press, 2nd edition, February 2005. ISBN 0-335-19744-2.

C. Goldspink. Methodological implications of complex systems approaches to sociality: Simulation as a foundation for knowledge. *The Journal of Artificial Societies and Social Simulation, JASSS*, 5(1), January 2002. URL http://jasss.soc.surrey.ac.uk/5/1/3.html.

V. Grimm and S. F. Railsback. *Individual-based Modeling and Ecology (Princeton Series in Theoretical and Computational Biology)*. Princeton Series in Theoretical and Computational Biology. Princeton University Press, July 2005. ISBN 069109666X. URL http://www.amazon.ca/exec/obidos/redirect?tag=citeulike09-20{\&}path=ASIN/069109666X.

Z. Guessoum. A multi-agent simulation framework. *Transactions of the Society for Computer Simulation International*, 17(1):2–11, 2000. ISSN 0740-6797.

O. Gutknecht, J. Ferber, and F. Michel. Integrating tools and infrastructures for generic multi-agent systems. In *Proceedings of the fifth international conference on Autonomous agents, AA 2001*, pages 441–448. ACM Press, 2001. ISBN 1-58113-326-X. doi: http://doi.acm.org/10.1145/375735.376410.

P. Guyot and S. Honiden. Agent-based participatory simulations: Merging multi-agent systems and role-playing games. *JASSS, The Journal of Artificial Societies and Social Simulation*, 9(4), 2006.

S. Hanks, M. E. Pollack, and P. R. Cohen. Benchmarks, test beds, controlled experimentation, and the design of agent architectures. *Artificial Intelligence Magazine*, 14(4):17–42, 1993. ISSN 0738-4602.

A. Harding. *Microsimulation and Public Policy*. North Holland Elsevier, Amsterdam, 1996.

D. Helbing and P. Molnár. Social force model for pedestrian dynamics. *Physical Review E*, 51(5):4282–4286, May 1995. doi: 10.1103/PhysRevE.51.4282. URL http:

//dx.doi.org/10.1103/PhysRevE.51.4282.

A. Helleboogh, G. Vizzari, A. Uhrmacher, and F. Michel. Modeling dynamic environments in multi-agent simulation. *Autonomous Agents and Multi-Agent Systems*, 14(1):87–116, 2007.

A. Helleboogh, T. Holvoet, and D. Weyns. The role of the environment in simulating multi-agent systems. In Uhrmacher and Weyns [2008], chapter 6.

J. Himmelspach and M. Röhl. JAMES II - experiences and interpretations. In Uhrmacher and Weyns [2008], chapter 17.

J. Himmelspach, M. Röhl, and A. M. Uhrmacher. Simulation for testing software agents - an exploration based on James. In S. Chick, P. J. Sánchez, D. Ferrin, and D. J. Morrice, editors, *WSC'03: Proceedings of the 35th Winter Simulation Conference*, volume 1, pages 799–807, Piscataway, NJ, USA, 2003. IEEE press. ISBN 0-7803-8132-7.

P. T. Hraber, T. Jones, and S. Forrest. The ecology of Echo. *Artificial Life*, 3(3): 165–190, 1997. ISSN 1064-5462.

B. A. Huberman and N. S. Glance. Evolutionary games and computer simulations. *Proceedings of the National Academy of Science USA*, 90(16):7716–7718, August 1993.

T. Ishida, Y. Nakajima, Y. Murakami, and H. Nakanishi. Augmented experiment: Participatory design with multiagent simulation. In M. M. Veloso, editor, *IJCAI 2007, Proceedings of the 20th International Joint Conference on Artificial Intelligence*, pages 1341–1346, 2007.

H. Kitano, M. Asada, Y. Kuniyoshi, I. Noda, and E. Osawa. Robocup: The robot world cup initiative. In W. L. Johnson, editor, *AGENTS '97: Proceedings of the first international conference on Autonomous Agents*, pages 340–347, New York, NY, USA, 1997. ACM Press. ISBN 0-89791-877-0. doi: http://doi.acm.org/10.1145/267658.267738.

F. Klügl. SeSAm: Visual programming and participatory simulation for agent-based models. In Uhrmacher and Weyns [2008], chapter 16.

F. Klügl, C. Oechslein, F. Puppe, and A. Dornhaus. Multi-agent modelling in comparison to standard modelling. In F. J. Barros and N. Giambasi, editors, *Proceedings of AIS 2002: Artificial Intelligence, Simulation and Planning in High Autonomous Systems*, pages 105–110, San Diego, CA, April 7-10 2002. SCS Publishing House.

B. G. Lawson and S. Park. Asynchronous time evolution in an artificial society mode. *The Journal of Artificial Societies and Social Simulation, JASSS*, 3(1), January 2000. URL http://www.soc.surrey.ac.uk/JASSS/3/1/2.html.

V. R. Lesser and D. D. Corkill. Functionally accurate cooperative distributed systems. *IEEE transactions on systems, machines and cybernetics*, SMC-11(1):81–96, 1981. URL citeseer.ist.psu.edu/lesser81functionally.html.

V. R. Lesser and D. D. Corkill. The distributed vehicle monitoring testbed: A tool for investigating distributed problem solving networks. *AI Magazine*, 4(3):15–33, 1983. URL http://mas.cs.umass.edu/paper/276.

M. Luck and R. Aylett. Applying artificial intelligence to virtual reality: Intelligent virtual environments. *Applied Artificial Intelligence*, 14(1):3–32, 2000.

M. B. Marietto, N. David, J. S. Sichman, and H. Coelho. A classification of paradigmatic models for agent-based social simulation. In D. Hales, B. Edmonds, E. Norling, and J. Rouchier, editors, *Multi-Agent-Based Simulation III*, volume 2927 of *LNCS*, pages 193–208. Springer-Verlag, February 2004. ISBN 3-540-20736-8.

S. Mazouzi, Z. Guessoum, F. Michel, and M. Batouche. A multi-agent approach for range image segmentation with bayesian edge regularization. In J. Blanc-Talon, W. Philips, D. Popescu, and P. Scheunders, editors, *Advanced Concepts for Intelligent Vision Systems, 9th International Conference, ACIVS 2007, Delft, The Netherlands, August 28-31, 2007*, pages 449–460, 2007.

U. Merlone, M. Sonnessa, and P. Terna. Horizontal and vertical multiple implementations in a model of industrial districts. *Journal of Artificial Societies and Social Simulation*, 11(2):5, 2008. ISSN 1460-7425. URL http://jasss.soc.surrey.ac.uk/11/2/5.html.

F. Michel. The IRM4S model: the influence/reaction principle for multiagent based simulation. In Durfee et al. [2007], pages 903–905. ISBN 978-81-904262-7-5.

F. Michel, A. Gouaïch, and J. Ferber. Weak interaction and strong interaction in agent based simulations. In D. Hales, B. Edmonds, E. Norling, and J. Rouchier, editors, *Multi-Agent-Based Simulation III*, volume 2927 of *LNCS*, pages 43–56. Springer-Verlag, February 2004. ISBN 3-540-20736-8.

F. Michel, G. Beurier, and J. Ferber. The TurtleKit simulation platform: Application to complex systems. In A. Akono, E. Tonyé, A. Dipanda, and K. Yétongnon, editors, *Workshops Sessions, First International Conference on Signal & Image Technology and Internet-Based Systems SITIS' 05*, pages 122–128. IEEE, november 2005.

N. Minar, R. Burkhart, C. Langton, and M. Askenazi. The Swarm simulation system: A toolkit for building multi-agent simulations. Technical Report 96-06-042, Santa Fe Institute, Santa Fe, NM, USA, 1996. URL http://www.santafe.edu/projects/swarm/overview.ps.

S. Moss, H. Gaylard, S. Wallis, and B. Edmonds. SDML: A multi-agent language for organizational modelling. *Computational & Mathematical Organization Theory*, 4(1):43–69, 1998. ISSN 1381-298X. doi: http://dx.doi.org/10.1023/A:1009600530279.

J. P. Müller and M. Pischel. The agent architecture inteRRaP: Concept and application. Technical Report RR 93-26, German Research Center for Artificial Intelligence, 1993. URL http://jmvidal.cse.sc.edu/library/muller93a.pdf.

S. R. Musse and D. Thalmann. From one virtual actor to virtual crowds: requirements and constraints. In *AGENTS'00: Proceedings of the fourth international conference on Autonomous Agents*, pages 52–53, New York, NY, USA, 2000. ACM press. ISBN 1-58113-230-1. doi: http://doi.acm.org/10.1145/336595.336975.

M. Nguyen-Duc and A. Drogoul. Using computational agents to design participatory social simulations. *JASSS, The Journal of Artificial Societies and Social Simulation*, 10(4/5), 2007.

M. North, N. Collier, and J. Vos. Experiences creating three implementations of the repast agent modeling toolkit. *ACM Transactions on Modeling and Computer Simulation*, 16(1):1–25, January 2006.

D. J. Nowak, I. Price, and G. B. Lamont. Self organized UAV swarm planning optimization for search and destroy using swarmfare simulation. In *WSC '07: Proceedings of the 39th Winter Simulation Conference*, pages 1315–1323, Piscataway, NJ, USA, 2007. IEEE Press. ISBN 1-4244-1306-0.

J. Odell, H. V. D. Parunak, M. Fleischer, and S. Breuckner. Modeling agents and their environment. In F. Giunchiglia, J. Odell, and G. Weiss, editors, *Agent-Oriented Software Engineering (AOSE) III*, volume 2585 of *Lecture Notes on Computer Science*, pages 16–31, Berlin, 2002. Springer-Verlag.

G. H. Orcutt. A new type of socio-economic systems. *The Review of Economics and Statistics*, 58:773–797, 1957.

M. T. Parker. What is ascape and why should you care? *The Journal of Artificial Societies and Social Simulation, JASSS*, 4(1), January 2001. URL `http://www.soc.surrey.ac.uk/JASSS/4/1/5.html`.

H. V. D. Parunak. "go to the ant": Engineering principles from natural multi-agent systems. *Annals of Operations Research*, 75(0):69–101, January 1997. doi: 10.1023/A:1018980001403. URL `http://dx.doi.org/10.1023/A:1018980001403`.

H. V. D. Parunak and S. A. Brueckner. Polyagents: Simulation for supporting agents' decision making. In Uhrmacher and Weyns [2008], chapter 4.

H. V. D. Parunak, R. Savit, and R. L. Riolo. Agent-based modeling vs. equation-based modeling: A case study and users' guide. In J. S. Schiman, R. Conte, and N. Gilbert, editors, *Proceedings of the 1st Workshop on Modelling Agent Based Systems, MABS'98*, volume 1534 of *Lecture Notes in Artificial Intelligence*. Springer-Verlag, Berlin, July 4-6 1998. ISBN 3540654763.

N. Pelechano, J. M. Allbeck, and N. I. Badler. Controlling individual agents in high-density crowd simulation. In *SCA '07: Proceedings of the 2007 ACM SIGGRAPH/Eurographics symposium on Computer animation*, pages 99–108, Aire-la-Ville, Switzerland, 2007. Eurographics Association. ISBN 978-1-59593-624-4.

D. Phan. From agent-based computational economics towards cognitive economics. In P. Bourgine and J.-P. Nadal, editors, *Cognitive Economics*, pages 371–395. Springer Verlag, 2004. ISBN 978-3-540-40468-2.

A. S. Rao and M. P. Georgeff. An abstract architecture for rational agents. In B. Nebel, C. Rich, and W. R. Swartout, editors, *Proceedings of the 3rd International Conference on Principles of Knowledge Representation and Reasoning (KR'92)*, pages 439–449. M. Kaufmann Publishers, October 25-29 1992. ISBN 1-55860-262-3.

W. T. Reeves. Particle systems – a technique for modeling a class of fuzzy objects. *ACM Transactions on Graphics*, 2(2):91–108, 1983. ISSN 0730-0301. doi: http://doi.acm.org/10.1145/357318.357320.

M. Resnick. *Turtles, termites, and traffic jams: explorations in massively parallel microworlds*. MIT Press, Cambridge, MA, USA, 1994. ISBN 0-262-18162-2.

C. W. Reynolds. Flocks, herds and schools: A distributed behavioral model. In *SIGGRAPH '87: Proceedings of the 14th annual conference on Computer graphics and interactive techniques*, pages 25–34, New York, NY, USA, 1987. ACM. ISBN 0-89791-227-6. doi: http://doi.acm.org/10.1145/37401.37406.

P. Riley and G. Riley. SPADES — a distributed agent simulation environment with software-in-the-loop execution. In S. Chick, P. J. Sánchez, D. Ferrin, and D. J. Morrice, editors, *WSC'03: Proceedings of the 35th Winter Simulation Conference*, volume 1, pages 817–825, Piscataway, NJ, USA, 2003. IEEE press. ISBN 0-7803-8132-7.

M. L. Rosenzweig. Paradox of enrichment: Destabilization of exploitation ecosystems in ecological time. *Science*, 171(3969):385–387, January 1971.

J. Rouchier. Re-implementation of a multi-agent model aimed at sustaining experimental economic research: The case of simulations with emerging speculation. *The Journal of Artificial Societies and Social Simulation, JASSS*, 6(4), October 2003.

S. J. Russell and P. Norvig. *Artificial Intelligence: A Modern Approach*. Pearson Education, Upper Saddle River, NJ, 2nd edition, 2003. ISBN 0137903952.

B. Schattenberg and A. M. Uhrmacher. Planning agents in james. *Proceedings of the IEEE*, 89(2):158–173, February 2001.

T. Schelling. Dynamic models of segregation. *Journal of Mathematical Sociology*, 1: 143–186, 1971.

J. R. Searle. *Speech Acts*. Cambridge University Press, Cambridge, MA, 1969.

R. E. Shannon. *Systems Simulation: the art and science*. Prentice-Hall, Upper Saddle River, NJ, 1975.

O. Simonin and J. Ferber. Modeling Self Satisfaction and Altruism to handle Action Selection and Reactive Cooperation. In J. A. Meyer, A. Berthoz, D. Floreano, H. Roitblat, and S. W. Wilson, editors, *From Animals to Animats 6, the sixth International Conference on Simulation of Adaptive Behavior SAB'00*, pages 314–323, Cambridge, Massachussets, USA, 11-16 septembre 2000. MIT Press.

K. Sims. Evolving 3d morphology and behavior by competition. *Artificial Life*, 1(4): 353–372, 1994. ISSN 1064-5462.

R. G. Smith. The contract net protocol: High-level communication and control in a distributed problem solver. *IEEE Transactions on Computers*, C29(12):1104–1113, 1980.

K. G. Troitzsch. Validating simulation models. In G. Horton, editor, *18th European Simulation Multiconference. Networked Simulations and Simulation Networks*, pages 265–270. The Society for Modeling and Simulation International, SCS Publishing House, 2004. URL http://www.uni-koblenz.de/~kgt/EPOS/RevisedAbstracts/Troitzsch.pdf.

T. Tyrrell. The use of hierarchies for action selection. *Adaptive Behavior*, 1(4):387–420, 1993. ISSN 1059-7123. doi: http://dx.doi.org/10.1177/105971239300100401.

A. M. Uhrmacher. Dynamic structures in modeling and simulation: a reflective approach. *ACM Transactions on Modeling and Computer Simulation (TOMACS)*, 11(2): 206–232, 2001. ISSN 1049-3301. doi: http://doi.acm.org/10.1145/384169.384173.

A. M. Uhrmacher and D. Weyns, editors. *Agents, Simulation and Applications*. Taylor and Francis, Boca Raton, FL, 2008.

P. Valckenaers, J. A. Sauter, C. Sierra, and J. A. Rodríguez-Aguilar. Applications and environments for multi-agent systems. *Autonomous Agents and Multi-Agent Systems*, 14(1):61–85, 2007.

D. Vanbergue, J.-P. Treuil, and A. Drogoul. Modelling urban phenomena with cellular automata. *Advances in Complex Systems*, 3, 1-4:127–140, 2000.

R. Vincent, B. Horling, and V. R. Lesser. An agent infrastructure to build and evaluate multi-agent systems: The java agent framework and multi-agent system simulator. *Infrastructure for Agents, Multi-Agent Systems, and Scalable Multi-Agent Systems*, 1887:102–127, January 2001. URL http://mas.cs.umass.edu/paper/200.

V. Volterra. Variations and fluctuations of the number of individuals in animal species living together. In R. Chapman, editor, *Animal Ecology*, pages 409–448. McGraw-Hill, New York, 1926.

D. Weyns and T. Holvoet. Formal model for situated multi-agent systems. *Fundamenta Informaticae*, 63:1–34, 2004.

D. Weyns, H. V. D. Parunak, and F. Michel, editors. *Environments for Multi-Agent Systems, First International Workshop, E4MAS 2004, New York, NY, USA, July 19, 2004, Revised Selected Papers*, volume 3374 of *LNAI*, 2005a. Springer. ISBN 3-540-24575-8.

D. Weyns, H. V. D. Parunak, F. Michel, T. Holvoet, and J. Ferber. Environments for multiagent systems: State-of-the-art and research challenges. In Weyns et al. [2005a], pages 1–47. ISBN 3-540-24575-8.

M. Woolridge. *Introduction to Multiagent Systems*. John Wiley & Sons, Inc., New York, NY, USA, 2002. ISBN 047149691X.

F. Zambonelli and H. V. D. Parunak. From design to intention: signs of a revolution. In

Proceedings of the first international joint conference on Autonomous agents and multiagent systems, pages 455–456. ACM Press, 2002. ISBN 1-58113-480-0. doi: http://doi.acm.org/10.1145/544741.544846.

K. Zeghal and J. Ferber. Craash: A coordinated collision avoidance system. In *European Simulation Multiconference*, 1993.

B. P. Zeigler. Toward a formal theory of modeling and simulation: Structure preserving morphisms. *Journal of the ACM (JACM)*, 19(4):742–764, 1972. ISSN 0004-5411. doi: http://doi.acm.org/10.1145/321724.321737.

B. P. Zeigler, T. G. Kim, and H. Praehofer. *Theory of Modeling and Simulation*. Academic Press, Inc., Salt-Lake, UT, 2nd edition, 2000. ISBN 0127784551.

2

Multi-Agent Systems and Simulation: A Survey from an Application Perspective

Klaus G. Troitzsch
Universität Koblenz-Landau

2.1 Simulation in the Sciences of Complex Systems .. 53
2.2 Predecessors and Alternatives 56
   Simulation of Voting Behavior in the Early Sixties •
   Dynamic Microsimulation to Predict Demographic
   Processes • Cellular Automata: Simple Agents Moving
   on a Topography • Discrete Event Simulation •
   Similarities and Differences Among These Approaches
2.3 Unfolding, Nesting, Coping with Complexity 64
   Agents with Different Roles in Different Environments
   • The Role of Interactions • The Role of the
   Environment
2.4 Issues for Future Research: The Emergence of
   Communication 69
   Agent Communication • Concluding Remarks
References 71

2.1 Simulation in the Sciences of Complex Systems

As social and economic systems are among the most complex systems in our world, the chapter will mainly deal with applications of simulation in general and agent-based simulation in particular in economics and the social sciences. Thus it will start with a discussion of the predecessors and origins of agent-based simulation mainly, but not only, in these sciences from the time when the first simulation models were created that used, or rather should have used, multi-agent systems. If one accepts that multi-agent systems have object-oriented languages as their prerequisites, one has also to accept that multi-agent systems proper could only be implemented after the early 1980s, but much earlier, namely in the 1960s the first simulations, for instance in political science, were built that can be described as forerunners of multi-agent systems.

At the same time, ingredients were developed that nowadays are a defining part of the agents in multi-agent systems, such as fact and rule bases [Abelson and Carroll, 1965] in which early "agents" stored information that they communicated among each other, although they lacked the defining feature of autonomy. But for a long time, simulation approaches prevailed that did not address the fact that in social and economic systems there are actors who are endowed with a very high degree of autonomy and with the capability to deliberate. Although not for all purposes of the sciences dealing with these systems, autonomy and deliberation are necessary ingredients of theory and models, one would not content oneself with humans being modeled as deterministic or stochastic automata but prefer models that reflect some typically human capability. And one would not content oneself with models that deal only with the macro level of a society. As early as in the nineteenth century, Emile Durckheim [Durckheim, 1895] proposed "sociological phenomena [that] penetrate into us by force or at the very least by bearing down more or less heavily upon us", thus anticipating what Coleman [Coleman, 1990, pg. 10] introduced as the well-known "Coleman boat" (see Figure 2.1), a representation describing the process of human actions (co-) determined by their social environment and at the same time changing this environment, such that social change is not just a change of the macro state of a society (or organization, or group) but at the same time always a change in the micro state of most or all of the individual beings.

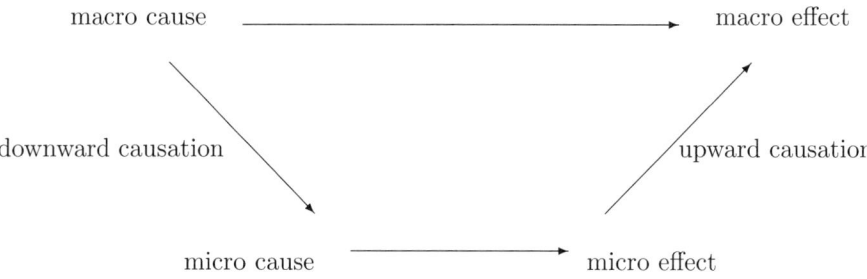

FIGURE 2.1 The Coleman boat describing downward and upward causation and the link between the micro and the macro level.

Coleman's idea of upward and downward causation coincides with the concept of exactly two societal levels (micro and macro) — which for many socio-economic models might seem too simple (cf. [Tilly, August 1997]), which makes it necessary to elaborate on the concept of levels and to discuss alternatives.

Thus a concept of levels will always be a part of any modeling of complex systems (but see [Ryan, 2006]). With "level" a set of things of the same "natural kind" [Bunge, 1979] is understood, and two subsequent levels $L_i < L_j$ are sets of things for which the following holds:

$$L_i < L_j =_{df} (\forall x)[x \in L_j \Rightarrow (\exists y)(y \in L_i \wedge y \in \mathcal{C}_{L_j}(x))] \tag{2.1}$$

where $\mathcal{C}_{L_i}(x)$ means the composition of system x. In plain words this means that the micro level L_i consists of entities which we call y, and these compose the entities x of the macro level L_j.

Again this refers to Bunge's definition in [Bunge, 1979] according to which a system σ is defined as a triple consisting of its composition $\mathcal{C}_{L_i}(\sigma)$, its environment $\mathcal{E}_{L_i}(\sigma)$ and its structure $\mathcal{S}(\sigma)$, where both $\mathcal{C}_{L_i}(\sigma)$ and $\mathcal{E}_{L_i}(\sigma)$ are subsets of the same set of things belonging to the same "natural kind" ($\mathcal{C}_{L_i}(\sigma), \mathcal{E}_{L_i}(\sigma) \subseteq L_i$) and where $\mathcal{S}(\sigma)$ is a set of relations, among them "bonding relations", defined on L_i ("bonding relations" are those due to which one element $y_1 \in L_i$ acts on another element $y_2 \in L_j$ — $y_1 \triangleright y_2$ — in a way that "it modifies the latter's behavior line, or trajectory or history" [Bunge, 1979, pg. 6]). $\mathcal{S}(\sigma)$, in turn, is a subset of the set of all relations that can be defined on L_i, where different kinds of systems may have different structures (see below 2.3 on p. 65).

Using L_i and L_j as abbreviations for two natural kinds, for instance L_i the set of all human beings and L_j the set of all groups of human beings, equation 2.1 can be interpreted as follows: "the micro level L_i *precedes* the macro level L_j," which means by definition that for every group on this macro level there exist individuals belonging to the micro level, and these are components of their group. Ryan [Ryan, 2006] debunks the idea of levels and recommends replacing them with scopes and resolutions although he concedes that Bunge's concept of levels is sounder than the usual talk of levels, but does not explain why he finds this concept "still not explanatory". The idea of resolutions instead of levels nevertheless allows for a more appropriate modeling than the one with levels, as unlike levels, resolutions are not defined as sets of entities of the same natural kind, but as spatial or temporal distinctions, such that they can encompass entities belonging to different natural kinds. On the other hand, Ryan's concept of scope and resolution is perhaps less appropriate for socio-economic systems, as will be seen later in this chapter. For now it might be sufficient to mention that Ryan, when he addresses socio-economic phenomena, restricts himself to interesting, but not very complex models such as the tragedy of the commons and the prisoners' dilemma [Ryan, 2006, pg. 17], where he distinguishes between "local and global structure", but with only two levels, resolutions, scopes or scales (to add still another related concept also mentioned by Ryan) it seems not really necessary to make a difference between these four concepts. Only when we have to deal with a deeper nesting does difference seem to matter.

Whatever concept will be agreed upon among complexity researchers, reducing the analysis of complex systems to only one level (or scope or resolution) will never be satisfactory. This is why this chapter touches the so-called system dynamics approach to analyzing complex systems only superficially — although the system dynamics community also claims to deal with complex systems (see, for instance, [Schwaninger, 2006]), but their interpretation of "complex system" is obviously different from the one adopted by the multi-agent systems community, as their understanding of complexity refers to the interactions between the attributes or variables of the one and only object of a system dynamics model.

The role simulation plays in the context of complex systems (as these are viewed from the perspective of the multi-agent systems community) can be described as the role of a method that helps to construct "would-be worlds" [Casti, 1996] and artificial systems with the help of which the behavior of real systems can be understood better. When Casti states that and "how simulation is changing the frontiers of science" he obviously has in mind that simulation is a tool to develop "a proper theory of complex systems that will be the capstone to this transition from the material to the informational." [Casti, 1996, pg. 215] His idea that "the components of almost all complex systems" are "a medium-sized number of intelligent, adaptive agents interacting on the basis of local information" necessitates a new formalism that currently cannot be provided by mathematics. Mathematics is capable of dealing with a large number of not very intelligent agents interacting on the basis of global information — as the synergetics approach [Haken, 1978; Weidlich and Haag, 1983] and sociophysics [Helbing, 1994a,b] have impressively shown. But in the case of medium-sized numbers of agents, the approximations and simplifications used to find a closed solution (for instance, of the master equation) will not always be appropriate.

Although Casti's "intelligent, adaptive agents" might also move in a "social field" [Lewin, 1951] and be driven by "social forces" [Helbing and Johansson, 2007], both concepts cannot capture the cognitive abilities of human beings, as these, unlike particles in physics, move autonomously in a social field and can evade a social force. Thus, for instance in a situation of pedestrians in a shopping center (as it is modeled in [Helbing and Johansson, 2007, pg. 631]) humans decide autonomously whether they obey the "social force" exerted on them by the crowd, or the "social force" exerted by their children or by the shop windows. Admittedly, often "people are confronted with standard situations and react 'automatically' rather than taking complicated decisions, e.g., if they have to evade others" [Helbing and Johansson, 2007, pg. 625], but more often than not they do make complicated decisions. Another assumption that will not always hold is "the vectorial additivity of the separate force terms reflecting different environmental influences" — again, this assumption is doubtful and might be replaced with the assumption that only the strongest "force" is selectively perceived and obeyed by a pedestrian (to keep to the example). Again, Casti's "intelligent, adaptive agents" (the term not only refers to human beings, but to other living things as well) are different from physical particles in so far as they are selective. And this results in the consequence that emergence in complex physical systems (such as lasers) is quite different from emergence in living and, particularly, cognitive systems. This describes the role of simulation in the context of the complex systems in Casti's sense as quite different from the role it plays in general complex systems.

Definitions of "complexity" and "complex system" are manifold and they refer to different aspects of what these terms might mean. For the purpose given here, the following characteristics may suffice: Complexity deals with a dynamic network of many agents acting and reacting to what other agents are doing. The number of different kinds of interactions may be arbitrarily high (not restricted to the forces of physics), and the state space of agents may have a large number of dimensions.

2.2 Predecessors and Alternatives

This section deals with the similarities and differences between agent-based simulation in multi-agent systems and all the earlier approaches to simulation: continuous and discrete, event-oriented and oriented at equidistant time steps, microsimulation and system dynamics, cellular automata, genetic algorithms and learning algorithms. Many of these earlier approaches were developed for applications in biology, ecology, production planning and other management issues as well as in economics and the social sciences, and at the same time physicists exported some of their mathematical and computer based methods to disciplines such as economics and sociology, forming interdisciplinary approaches nowadays known as econophysics and sociophysics, and — as already discussed in Section 2.1 — although they call their simulations often agent-based, these agents are often not much more than interacting particles whose interactions are based on "forces" that are described in the same manner as gravitational or electrostatic and electromagnetic forces. Agent-based computational demography is a child of both classical microsimulation and multi-agent systems, and the respective ancestry also holds true for many other interdisciplinary approaches, such as systems biology, evolutionary economics and evolutionary game theory, to name just a few. Socionics is another field where multi-agent systems, simulation and classical theory building methodologies of an empirical science come together.

System Dynamics, already mentioned before, also deals with complex systems, but as it is a one-of-a-kind method of modeling systems in so far as it models exactly one object with a potentially large number of attributes (state variables) interacting via a large number of equations (mostly difference or differential equations), this approach has very little to do with multi-agent systems — but a system dynamics model can serve as one agent in a multi-agent system, usually describing part of the environment of many interacting agents [Möhring and Troitzsch, 2001].

2.2.1 Simulation of Voting Behavior in the Early Sixties

Among the early forerunners of multi-agent systems in the social sciences at least two can be named where processes of voter attitude changes are modeled and simulated. Although the poor computer languages of the late 1950s and early 1960s did not allow for agents in the sense of our days, any re-implementation would nowadays be a multi-agent system with several classes of agents (representing voters, candidates, media channels, as in [Abelson and Bernstein, 1963]) or in the Simulmatics project supporting John F. Kennedy's election campaign [Sola Pool and Abelson, 1962] as they dealt with the communications among citizens, between citizens and candidates as well as between citizens and media channels and modeled their behavior and actions in a rule-based manner.

This is especially true for the referendum campaign simulation presented by Abelson and Bernstein as early as in 1963 [Abelson and Bernstein, 1963]. Their simulated individuals were exposed to information spread by different communication channels about the topic of the referendum (fluoridation of drinking water, an issue in the United States at that time), and they exchanged information among themselves. Abelson and Bernstein defined some 50 rules determining how the individuals dealt with the incoming information. As an example, rule A1 says that the probability that an individual will be exposed to a particular source of information in a particular channel is a direct function of this individual's attraction toward this channel. To continue, rules A2 and B1 say that the exposure probability to each source in any channel is also a direct function of the individual's interest in the issue, and that the probability that this individual talks to another individual about the issue is direct function of the former's interest in the issue. Another rule (B26) says that interest in the issue increases as a direct function of the number of both conversational and channel exposures; if there are no exposures, interest decreases. There is a total of 22 rules (or propositions, as Abelson and Bernstein call them) about the influence of sources in channels and another 27 about the influences the individuals exert on to each other, as well as two additional rules determining whether an individual goes to the polling station or not and, if it does, which vote it will cast in the end. The simulation ran over ten or more simulated "weeks", and every "week" the simulated individuals were exposed to information from channels and peers, changed their interest in and their attitudes toward the issue. The information is coded in terms of "assertions" (which are either pro or con with respect to the fluoridation issue); individuals keep track of all the assertions they are exposed to and change their own attitude as a consequence of the evaluation of the sources of the assertions. Several rules deal with the acceptability of an assertion (ideological match between individual and source, consistency between the assertion and the individual's predisposition). Most of these rules are probabilistic rules (compare rules A1, A2 and B1 mentioned above), many rules, however, are formulated such as "An assertion is more apt to become accepted by [individual] i if he has not previously encountered it than if he has previously disagreed with it." (A8) It is not entirely clear from the paper how exactly this was formalized (but this "more apt" is likely to have been formalized with the help of some transition probability).

2.2.2 Dynamic Microsimulation to Predict Demographic Processes

Microsimulation [Orcutt et al., 1961] is another early forerunner of agent-based systems, as here, too, agents had to be modeled that changed their attributes according to certain stochastic rules — although up to now most microsimulation models do not include interaction between agents (except perhaps for some kind of marriage market), but agent-based computational demography [Billari and Prskawetz, 2003] makes heavy use of inter-agent processes, for instance as in some recent papers the propensity to bear children is modeled as dependent on the perceived attitudes of a friendship network [Aparicio Diaz and Fent, 2007] or as dependent on the biography of an individual [Leim, 2007] (whereas in the mainstream microsimulation only the current state of a simulated individual is used for calculating state changes).

The typical case of dynamic microsimulation uses a large database for initializing agents and households (the latter form an intermediate level between the individual and the aggregate level) which is simulated on a yearly base. Each member participates in a number of subprocesses — aging, death, child-bearing, marriage, to name the demographic subprocesses first, changes in educational and employment status, and perhaps other changes (of income, for instance). Death, child-bearing, marriage and divorce make a reorganization of households necessary, child-bearing, marriage and divorce change the networks to which the individuals belong. Thus the microsimulation keeps track of the changing structure of kinship networks and of the aggregate as a whole. In a way, demographic reproduction in a wider sense is reduced to biographies [Birg et al., 1991], and the macro level is linked with the micro level. Usually this link is only unidirectional as there is no feedback from the macro level to the micro level.

2.2.3 Cellular Automata: Simple Agents Moving on a Topography

Many modern tools for multi-agent simulation use the technique of cellular automata [Farmer et al., 1984] to give agents an environment with a topography that is sufficiently similar to the environment the simulated animals [Drogoul and Ferber, 1994; Drogoul et al., 1995] or humans [Schelling, 1971, 1978] live in, an approach that was extended into a full grown multi-agent model which reconstructs "social science from the bottom up" [Epstein and Axtell, 1996]. Other early approaches to reconstructing the emergence of complex systems used continuous spaces [Doran et al., 1994; Doran and Palmer, 1995], as is also the case in the ambitious NEW TIES project [Gilbert et al., 2006]. In discussing these bottom-up approaches — and most simulation approaches to complex systems are of the bottom-up type — one has to take into account that the bottom-up approach is not always useful, see the discussion in [Senn, a,b], where the argumentation is that a bottom-up approach that meticulously mimics the movement of each single molecule would have been misleading to explain the flow of heat in a gas. But this argumentation neglects that the averaging of the impulses of molecules could only be successful as all these molecules obeyed the same laws and were by no means selective with respect to several different forces by which they were driven — as there was only one force. Biological and social systems underlie the effects of several forces and are often not only reactive but proactive, have goals, sometimes conflicting, such that the mathematical reduction often proposed by physicists would not lead to success in cases where the interactions between the micro level entities are manifold.

Whereas the early usage of the concept of a two-dimensional cellular automaton such as Conway's game of life [Berlekamp et al., 1982; Gardener, 1970] was restricted to very simple agents who could just be born and die, other models — such as Schelling's segregation model [Schelling, 1971] or Hegselmann's migration model [Hegselmann, 1998] — allowed their agents to move around in order to look for cells that made them "happier". Both Conway's and Schelling's models have been programmed over and over again as demo models in nearly all simulation toolboxes. Nevertheless, a short discussion of Schelling's findings may still be in order: In his model, many, but not all of the cells of the cellular automaton are occupied by agents which come in exactly two versions, say red and green agents, and the distribution over the cells is entirely random in the beginning. The agents analyze their Moore neighborhood and count how many of their neighbors are of the same color. If this number is above a certain threshold, they are "happy", otherwise they try to find an unoccupied cell where the number of neighbors of the same color is greater and move there. Usually after several time steps, agents stop moving as all of them are "happy" or cannot find cells that make them "happier" than they currently are on their cells (the latter is a rather rare effect). And one finds that in the end the agents are by no means distributed randomly over the cells, instead clusters have formed, either red or green, and these are separated by empty stripes. If one compares the capabilities of these agents to one of the standard agent definitions [Wooldridge and Jennings, 1995] one finds that Schelling's agents are autonomous, reactive, perhaps even proactive, but their social ability is rather restricted: although they interact in a way with their neighbors, this interaction is not mediated by any kind of language or message passing. These agents have only very simple beliefs (their perception of the neighboring eight cells) and only one intention, namely to stay in a cell where the number of neighbors of the same color exceeds an exogenously given threshold. In spite of the extreme simplicity (which Schelling's model shares with the game of life), the model displays some emergent behavior (which is also true for Conway's game of life with its gliders, oscillators and other amazingly complicated and long-lived patterns). But these emergent phenomena are of a relatively simple kind — if one accepts that "temperature is an emergent property of the motion of atoms" [Gilbert and Troitzsch, 2005, pg. 11], then, of course, the game-of-life patterns and the ghettos in Schelling's model are emergent phenomena as well. Without formalizing the notion of an emergent phenomenon in this chapter, it can be described as a phenomenon that "requires new categories to describe it which are not required to describe the the behavior of the underlying components." [Gilbert and Troitzsch, 2005, pg. 11] In this sense, the property of gliders and oscillators to be able to glide or to oscillate obviously is an emergent property, as the ability to move is not a property of the game-of-life cells at all.

Similar behavior of simulation models can be observed, for instance in MANTA, a model that simulates the behavior of ants [Drogoul and Ferber, 1994; Drogoul et al., 1995] or in models simulating insect behavior on continuous surfaces [Calenbuhr and Deneubourg, 1991]. Of course, insect societies can be rather complex systems, but one could argue that human societies are much more complex, such that simple models do not do justice to the complexity of human societies.

A next step in doing more justice to human society by modeling them with the help of cellular automata is, for instance, Hegselmann's [Hegselmann, 1998] approach to analyzing social dynamics in the evolution of support networks. His agents can also move on a two-dimensional grid, but they play a two-person support game with selected neighbors. The agents in this model have several attributes, mainly a (constant) probability of becoming needy and the capability to select another cell according to the outcome of past games. The game they play can be described as follows: both players have a certain probability of becoming needy, i.e. falling into a river, in the illustrative example. If both are good swimmers, both move on, if both are bad swimmers, both drown; if only one of the two players is in need of help the other has to decide whether it helps or not. The payoffs are M for moving on, D for drowning, S for being saved and H for helping, where $S > D$ and $M > H$. Under certain conditions, a propensity to mutually support can be profitable to both players, such that co-operation can occur. All agents know the "risk class" (the probability of becoming needy) of all their neighbors (these probabilities are .1, .2, ...9), they know the payoffs M, D, S and H, and they can decide whether they rescue needy neighbors or not. Moreover, they can migrate to other position on the surface; this might be worthwhile, as there are good and bad "social positions": the best social position is being surrounded by four members (von-Neumann neighborhood) belonging to "the best risk class willing to engage in support relationships with them", and the worst social position is being surrounded "by empty cells only or only by those individuals with whom support relations are not possible" [Hegselmann, 1998, pg. 51]. Compared to Schelling's model, much more deliberation is necessary for the agents to make their decisions, but again there is little social ability, as there is no communication — agents look at each other and know what their potential partners' attributes are like. The outcome of this model is a stable state in which agents with the smallest probability of becoming needy support each other and members of the next higher risk class. The former sit in the middle of clusters, surrounded by members of the next higher risk classes, such that the risk classes form more or less concentric clusters, at the borders of which one can find the poor guys of the highest risk class, but these, too, are often involved in stable support relations.

Another approach is the one developed by Bibb Latané and Andrzej Nowak [Latané and Nowak, 1997; Latané, 1981, 1996; Latané and Nowak, 1994; Nowak and Lewenstein, 1996; Nowak and Latané, 1993]. Their cellular automata are populated by immobile agents, and these are characterized by mainly two attributes: a binary attitude and a persuasive strength. According to Latané's dynamic social impact theory (which he sees as "as a 'field theory' in which individuals are affected not only by their own personal experience, but by the strength, immediacy and number of their communication partners" [Latané, 1996, pg. 361], individuals flip their binary attitude when the sum of the products of strength and immediacy of those people with the opposite attitude outweighs the respective product sum of the people with the same attitude — immediacy being calculated as the inverse of the Euclidean distance between persons. The simulation model (SITSIM, programmed by Nigel Gilbert, can be found at http://ccl.northwestern.edu/netlogo/models/community/Sitsim and run with NetLogo [Wilensky, 1999]) displays clustering after some time, but unlike the Schelling and Hegselmann models this effect is not due to migration but to near-neighbor persuasion: cluster of people of the same attitude are formed around the proponents of each attitude with the highest persuasive strength (see Figure 2.2).

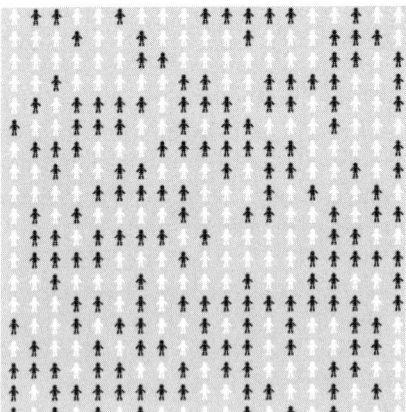 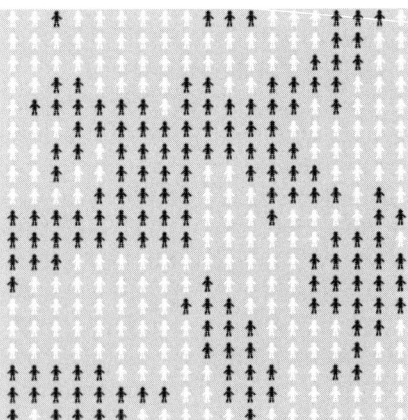

FIGURE 2.2 Start and final state of a SITSIM run.

Although Latané described his model as "a multi-agent 'bottom-up' conception of a complex self-organizing social system", again the SITSIM agents lack at least some of the characteristics ascribed to agents by Wooldridge and Jennings [Wooldridge and Jennings, 1995]. They are in a way autonomous and reactive, but their proactivity is not very convincing (at least less than in the Schelling and Hegselmann models), and their social ability is also deficient (they can persuade and be persuaded, but not by means of message passing but by directly looking into each other's minds, so to speak, or as [Hutchins and Hazlehurst, 1995] put it, by telepathy). Nevertheless, it is a "bottom-up" conception, much like Epstein's and Axtell's approach at "growing artificial societies" [Epstein and Axtell, 1996] which was a large step forward in simulating complex systems.

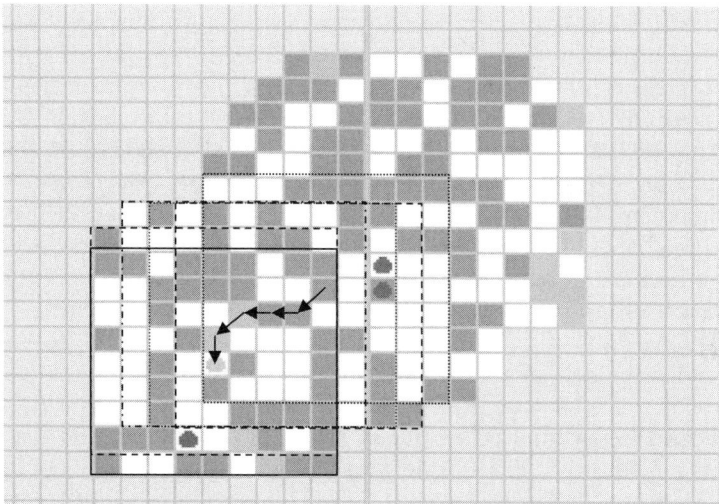

FIGURE 2.3 The memory of an agent in [König et al., 2002]. The chain of arrows shows the yellow agent's path, the quadratic frames show the information available at the past six points of time. If the agent had been a co-ordinator, the memory could have been augmented by information from followers. The blue ones are other agents, the color of the cells depends on the amount of resources available there.

Their artificial world called sugarscape is populated with agents who make their living on two complementary types of products (sugar and spice), both of which they need for surviving, but at individually different proportions. Both products grow on the surface of a cellular automaton and can be harvested, stored, traded and consumed. In the different versions of the model reported in [Epstein and Axtell, 1996] and several extensions published by other authors (see for example [König et al., 2002; Epstein et al., 2006; Flentge et al., 2000, 2001; Zaft, 2001]), agents are not only autonomous and reactive, but also proactive as they have goals they try to achieve, and they have some social abilities as they exchange goods and information. In [König et al., 2002] they explore their environment actively (moving around satisfies their curiosity) and store information about the cells they saw within their range of vision, about the resources and other agents in all these cells (see Figure 2.3); part of this information is forgotten after some time, whereas older information tends to become invalid after some time as resources may have grown or taken away and agents will not have stayed in the positions where they had been seen. The agents even form primitively structured groups whose chieftains collect this information and redistribute it to their tribe, and in [Flentge et al., 2000, 2001] agents can mark a cell that they currently occupy and sanction agents trespassing a marked cell — from which a possession norm emerges. The latter paper also extends the sugarscape concept of tags (which allows some cultural transmission [Epstein and Axtell, 1996, pg. 77] resembling the spread of an infectious disease) in a way that the poorer agent adopts a tag (meme) of the richer agent whom it meets in its neighborhood. The two tags (memes) the agents have play a particular role: only those having the possession meme mark cells and never trespass marked cells except when they do not find any exploitable cell in their range of vision, those having both the possession and the sanction meme additionally punish trespassers (taking away some of their wealth and consuming also some of their own, as a cost of sanctioning). Agent societies in [Flentge et al., 2001] without the possession meme were less sustainable than those with the possession meme, and, similarly, in [König et al., 2002] agent societies where chieftains redistribute information among their tribes are more sustainable than agent societies without the exchange of information.

Cultural transmission, absent in Schelling's and Hegselmann's segregation and migration models, seems only to be possible if agents have at least some memory (as in the tags in [Epstein and Axtell, 1996], the memes in [Flentge et al., 2000, 2001] and the memory in [König et al., 2002]). Thus cultural transmission necessitates the storage of information about the environment, i.e. agents must be able to develop models of their environment in their memories, and they must be able to pass part of these models to other agents. "No social mind can become appropriately organized except via interaction with the products of the organization of other minds, and the shared physical environment" [Hutchins and Hazlehurst, 1995, pg. 160]

One of the oldest models of this kind is the EOS testbed [Doran et al., 1994; Doran and Palmer, 1995] which reconstructs Upper Paleolithic social change. Its agents have a comparatively complex internal structure, as they "have a production system in the AI sense" [Doran et al., 1994, pg. 203] — rule base that works on a social model and a resource model. The latter stores (as in [König et al., 2002]) information about resources and their locations whereas the former stores an agent's belief about itself and other agents. A central concept of the EOS model is the request for co-operation that one agent can send another agent when the former is not able to collect a certain resource in reach individually — which can happen when the cost of exploiting a resource is higher than this agent can afford (the background is that these resources represent prey animals instead of plants). Co-operation relations between agents can develop into leader-follower relations (in a more complicated way than in [König et al., 2002]) when an agents succeeds in recruiting several other agents which in turn succeed in exploiting the resource.

2.2.4 Discrete Event Simulation

Discrete-event simulation, too, has extended into the field of multi-agent systems, as in the traditional approach customers could never be modeled as human members of waiting queues often behave: They would move from one queue to look for another queue that seems to be served more quickly, or they leave the system before being served at all, they would negotiate with the server — behavioral features that are difficult to model with classical toolboxes and call for agent-based models.

One of the first object-oriented discrete-event simulation approaches was DEVS, developed by Bernhard Zeigler [Zeigler, 1985, 1990] as early as in the 1970s. The DEVS concept describes atomic models formally as tuples consisting of inputs, internal states, outputs, two transition functions (one responsible for state changes due to internal events, another for state changes due to external inputs), an output function and a time advance function. These atomic models can be coupled into coupled models. These, too, are formally described as tuples consisting of inputs and outputs, the component models, the coupling between the latter, and a set of rules which select the component model firing next. From this formalization it was only a small step to an agent-oriented DEVS, called AgedDEVS [Uhrmacher, 1996], an extension that adds "internal models" to the standard DEVS models (for other extensions of DEVS see also the contribution of Jan Himmelspach in this book).

Some modern toolboxes such as AnyLogic [Ltd] which in its release 6 is announced as "the first and only dynamic simulation tool that brings together System Dynamics, Process-centric (AKA Discrete Event), and Agent Based approaches within one modeling language and one model development environment" take a more informal way of integrating these approaches. The demos that can be downloaded from the provider's web site include Schelling's model, a sophisticated version of a model similar to Latané's ("Social Response Dynamics"), and also a microsimulation model ("Urban Dynamics") with monthly update which additionally has a topography which in turn is used to model individual and public transport and the emission of CO_2 in the city of Urupinsk.

Features of discrete event simulation are often used in agent-based models to get rid of a unified time scale according to which all agents of all kinds have to be updated in every time step even if nothing happens in and with them. This feature was also used for classical dynamic microsimulation for the same purpose in the DYNAMOD model [Antcliff, 1993]. Examples of multi-agent models using discrete event features are [Becker et al., 2006; Dubiel and Tsimhoni, 2005; Riley and Riley, 2003; Troitzsch, 2004; Zaft, 2001].

2.2.5 Similarities and Differences Among These Approaches

All these approaches have more or less different concepts of complexity. Table 2.1 tries to give a comparison of these approaches with respect to how they handle different aspects of complexity — as already mentioned above, the number of levels, of agent kinds, the dimensionality of agents' state space, the kinds of interactions and the structure of the time domain.

TABLE 2.1 Overview of different simulation approaches in the social sciences with respect to their treatment of complexity

Approach	Levels	Object classes	State space of agents (dimensions)	Interactions	Time
System dynamics	one	one	many	only between variables	equidistant
Microsimulation	two	one	few	few between objects (local), aggregation to upper level (global)	most often equidistant
Early agent models	two	few	few	both between levels (global) and objects (local), often by "telepathy"	equidistant
Discrete event	one or few	several	few	mainly between objects	event oriented
Cellular automata	two	few	few	local	equidistant
Current agent models	several	several	many	both between levels (global) and objects (local), often by message passing	most often equidistant

Although, as discussed above, the concept of levels is a little problematic it allows for a first structuring of complexities: From the point of view of the *kinds of entities and systems represented* in a model, there is a clear divide between system dynamics and all the rest, as the former approach only generates "one-of-a-kind" models: the world, an enterprise, any target system is seen as an undifferentiated whole whose components are hidden or neglected (which is then compensated by a large number of attributes describing this indivisible whole, such that the *complexity of the target system in terms of its properties* is represented by a complex system of differential or difference equations — note that in the latter case, the word "system" denotes a concept, not a real-world entity [Bunge, 1979, pg. 52]). All the other approaches "open the box" in so far as they make a difference between a target system and its components, where microsimulation and the early forerunners of agent models usually do not talk about subsystems in the sense that the components of these subsystems are of the same kind as the components of the embedding system. And this applies to cellular automata as well, even when they are populated with agents that move around as in Schelling's model and in Sugarscape. The discrete event case is a little different as its models usually consist of entities of very different kinds (customers, server, queues), but as the reference to compound models above shows, most discrete event simulation models are "systems of nested systems" [Bunge, 1979, pg. 12], and this especially true for agent-oriented discrete-event systems (AgedDEVS [Uhrmacher, 1996]). Even if microsimulation is usually called a two-level approach, the "upper level" is not an object or agent itself as it is only used to count or average or do other statistics on the elements of the "lower level". The other forerunners of agent-based models define only a few types of objects or agents as well, only in the case of contemporary agent-based models one finds a greater number of object classes as the architecture of an agent representing a human, for instance, consists of several parts (sensor, effectors, decision engines) which do not, of course, represent human beings, but instead parts of the these. And these models might also represent teams or groups explicitly, which are necessary to define different roles agents have in different contexts.

In terms of *time*, most multi-agent simulations in economics and the social sciences still use simple equidistant time step all over a model, neglecting the fact that real-world target systems often use different time scales in different subsystems. Event-oriented approaches overcome this simplicity, but not all tool boxes are prepared for this modeling style.

In terms of *interaction*, complexity is coped with by different means — system dynamics restricts itself to interactions between the properties of the one and only target system, microsimulation is mainly interested in the fate of the individual components of its models and statistics on these and uses interactions mainly where they are necessary to generate new individuals in the birth process. Early agent models like the ones referred to above modeled communication more or less explicitly [Abelson and Bernstein, 1963], but even today communication is often done just by telepathy, i.e. with `get` and `set` methods of modern object-oriented programming languages instead of an explicit message passing which in turn necessitates something like interpretation on the side of the receiving agent instead of the direct manipulation of agent attributes.

2.3 Unfolding, Nesting, Coping with Complexity

Multi-agent systems also lend themselves to coupling models of different types and unfolding models in a top-down way, starting with a macro model of the top level (see eq. 2.1) of the system and then breaking it off, replacing part of the rules of the macro system with autonomous software entities representing real-world elements of the modeled overall system, as for instance exemplified in [Brassel et al., 1999]. An approach like this allows researchers to start with a macro view on a complex real-world system — given that all interesting real-world systems are complex. Whereas the complexity of many models derived from some of the existing system theories is restricted to complexity of the interactions between state variables of the system as such [Forrester, 1980], we usually observe that systems are decomposable into interacting system elements which in turn might be systems of another "natural kind" [Bunge, 1979]. On all the levels of such a nested system, agents can be used for representation, although not on all levels the respective agents would need to have all the features that are commonly attributed to them [Wooldridge and Jennings, 1995] — autonomy, reactivity, proactivity, social ability. Moreover, in such a view, not only the complexity of the domain can be mapped into a simulation model, but also the complexity of time — different time scales for the different levels of a nested system — as for every kind of agents different mechanisms of representing time can be used.

In a way, "agents cover all the world" [Brassel et al., 1997] in that multi-agent systems can be used for all simulation purposes, as agents can always be programmed in a way that they behave as continuous or discrete models, can be activated according to event scheduling or synchronously or in a round-robin manner, can use rule bases as well as stochastic state transitions, and all these kinds of agents can even be nested into each other, thus supporting a wider range of applications than any of the classical simulation approaches. This leads to a third aspect of complexity (after the complexity of domains and the complexity of time): agent-based models can encompass several different approaches, both from a technical and implementation point of view, but also from the disciplines making use of simulation (for instance, disciplines such as ecology, economics, sociology and political science can combine their contributions into a deeply structured simulation model [Möhring and Troitzsch, 2001], and the same holds for neurophysiology, cognitive psychology, social psychology and sociology in other conceivable examples).

It goes too far to say that all this would not be possible in a non-agents world (as everything is programmable in Assembler), but examples make clear that models combining aspects from even neighboring disciplines as the ones enumerated in the paragraph above are only understandable and communicable when they come in a modular form that is typical for agent-based models — and the same applies to ecological models where disciplines from physics, via biochemistry to population biology would play their co-operative roles.

2.3.1 Agents with Different Roles in Different Environments

First of all, one has to observe that real world entities can be components of several different systems at the same time — perhaps a fourth type of complexity. This is most obvious for humans who typically belong to a family, a peer group, a school form, an enterprise department and a military unit at the same time. All these systems are of different "natural kinds", to keep to Bunge's system theory [Bunge, 1979], although the micro level is the same for all these kinds of systems:

$$\forall \sigma \in \{F, P, S, D, M\} \left[\mathcal{C}_H(\sigma) \subset H \wedge \mathcal{E}_H(\sigma) \subset H \right] \qquad (2.2)$$

where

- H is the set of humans,
- σ is an element of one of the different system kinds F being the set of all families, P being the set of all peer groups, ..., M being the set of all military units, and all instances of these kinds of systems have humans as their components and populating their environments,

the set of (bonding) relations is different between a family and a military unit: not all relations that hold in a family would also hold in a military unit:

$$\exists R_i \in \mathcal{S}_F \wedge R_i \notin \mathcal{S}_M \tag{2.3}$$

where F and M are again the natural kinds of families and military units, respectively, and R_i could, for instance, be the (bonding) relation of nursing ($\langle a, b \rangle \in R_i \Rightarrow a \triangleright b$, meaning a nurses b). To put it the other way round:

$$\sigma_1 \in F \wedge \sigma_2 \in M \Rightarrow \mathcal{S}(\sigma_1) \backslash \mathcal{S}(\sigma_2) \neq \varnothing \tag{2.4}$$

such that it is not very reasonable to think of the different kinds of social systems mentioned above (as forming a unified (meso) level of social subsystems between the individual (micro) level and the macro level of society): $F, P, S, D, M \subset L_j$ does not make much sense when obviously $\sigma_1 \in F$ and $\sigma_2 \in M$ do not belong to the same set of systems: their structures are different as the nursing relation does not belong to the structure of military units.

In the end this means that the concept of level defined in 2.1 is not very helpful, but for other reasons then those mentioned in [Ryan, 2006], and Ryan's scope and resolution would not help either to cope with the problem of individuals belonging to different kinds of systems at the same time. Agent-based simulations, however, can easily model all these relations.

Only very few papers on simulation models have ever made use of the versatility of the agent-based approach in a way that took into account that real-world humans can belong to several systems at the same time. In systems of different kinds, agents perform different roles, and even in the same system (e.g., a team [Schurr et al., 2006]), a member can perform different roles at different times. In [Doran and Palmer, 1995; Doran et al., 1994], agents can perform the role of a leader or a follower, in [König et al., 2002] they can additionally perform the role of a single.

[Schurr et al., 2006, pg. 313] discusses the "assignment of roles to team members" as "of critical importance to team success". Teams are here defined as in [Kang et al., 1998] as a special kind of "group whose members share a common goal and a common task ... The most critical distinction between teams and groups [sc. at large] is the degree of differentiation of roles relevant to the task and the degree of member interdependence." Thus for modeling and simulating teams, it is necessary to endow team members with a problem solving capacity, a symbol system and a knowledge base. [Schurr et al., 2006], extending Soar [Laird et al., 1987] to Team Soar, emphasize that "each member in Team-Soar develops two problem spaces: a *team problem space* and a *member problem space*", as each members tries to achieve both the common goals and the member goal (which they just describe as "make a good recommendation to the leader", but the individual goals could, of course, be manifold, and in modeling and simulating project teams, members could even work for different projects at the same time). For the evolution of such teams, see [Dal Forno and Merloine, 2008].

Geller and Moss, too, make use of the concept of roles when they [Geller and Moss, 2007, pg. 115–116] describe the "complex interpersonal network of political, social, economic, military and cultural relations" in Afghan society. This network, called a *qawm*, consists of a number of different actor types. In reality (though not in their model), individual actors may "incorporate a variety of roles", and, moreover, members of different *qawms* compete among each others (and perhaps there might be even individuals who belong to more than one *qawm* at a time). The knowledge that agents (as representatives of real-world individuals) have in [Geller and Moss, 2007] is packed in so-called endorsements. "Endorsements are symbolic representations of different items of evidence, the questions on which they bear, and the relations between them. Endorsements can operate on each other and hence lead to the retraction of conclusions previously reached, but since there is no formal accounting of final conclusions, the process is seen as a procedural implementation of non-monotonic patterns of reasoning rather than as a logic." [Shafer and Pearl, 1990, pg. 626], referring to [Sullivan and Cohen, 1985].

Another approach to modeling agents that "engage in several relations simultaneously" was recently published in [Antunes et al., 2007]. The authors here represent agents on different planes in a cube, where on each plane the inter-agent network is displayed. Technically speaking, the agents move between planes, and this movement between planes represents the change in focus an agent has on its respective relations. The idea behind is that "an agent can belong to social relations, but possibly not simultaneously" [Antunes et al., 2007, pg. 492]. Although this idea can be criticized from a real-world perspective (where, e.g., husband H's relation with his wife W is simultaneous with his relation to his boss B, and both B and W influence H at the same time when they ask him a favor each, and these two favors are conflicting), the authors' approach is a step forward, and they extend the original approach a few pages later where "agents will have to face several contexts constantly, either simultaneously, or at rhythms imposed by the different settings." [Antunes et al., 2007, pg. 494]. This concept might be combined with the endorsement concept used by [Geller and Moss, 2007], as — at least in the case of "rhythms", the agents would not deliberate with respect to the current state of their environment, but on what they remember about the different settings.

2.3.2 The Role of Interactions

Interaction between agents is usually modeled in different modes. On the one hand, and this is the simpler case, agents just read from other agents' memories. This is easily programmed, but not very naturalistic when agents represent human beings, as these communicate via messages, most of them verbal, but also using facial expression or gesticulation, but these messages need not necessarily express the real opinions or attitudes that these humans have in their minds. Thus software agents in simulation of socio-economic processes should also be able to exchange messages that hide or counterfeit their internal state. For the exchange of messages, an environment (see below) is most often necessary (at least in the real world, but for message passing in the simulated world one would have blackboards for one-to-all messages or mail systems for one-to-one or one-to-few messages), but what is even more necessary is that agents have something like a language or some other symbol system for communicating. Communication can be achieved with the help of pheromones as in ant colonies, and the pheromone metaphor [Drogoul et al., 1995] may work well with simple types of messages (such as "follow this way" or "avoid this place"), but communication is richer when it can use symbolic languages (see the special issue of *Autonomous Agents and Multi-Agent Systems*, vol. 14 no. 2, or for an early example KQML [Finin et al., 2007]). These symbol systems have to refer to the components of the agents' environments and to the actions agents can perform.

2.3.3 The Role of the Environment

Environment in multi-agent simulation plays a special role. As discussed in [Brassel et al., 1997, 1999; Möhring and Troitzsch, 2001] it is possible to represent the environment (or several distinguishable parts of the environment) with an agent and several different kinds of agents, respectively. Representatives of the non-human environment in social simulations would be relatively simple agents then, lacking the social and proactive capabilities. The representation of elements of the agents' environments by additional agents is partly justified by the fact that from the perspective of an individual agent, all other agents belong to its environment [Weyns et al., 2007, pg. 8]. On the other hand, as e.g., discussed in [Parunak and Weyns, 2007, pg. 2] as "it provides the conditions for agents to exist" and, as one could add, to communicate (as for the communication, see Section 2.4). But unlike real world agents, software agents can exist without an explicit representation of a real-world environment (they need, however, the environment of a computer, its operating system and its runtime environment, but these are not the correlate of any real-world environment). But only with a simulated environment they are able to interact in a realistic manner (reading other agents' memories directly does not seem very realistic, the *no telepathy)* assumption [Hutchins and Hazlehurst, 1995, pg. 160]. This environment allows them to communicate, by digital pheromones [Drogoul et al., 1995] or by abstract force fields (see the discussion in Section 2.1 on p. 55, but also — and this should be the typical case in social simulations — symbolically, as it may contain blackboards and other means for sending messages. But at the same time it is also necessary to allow agents to take actions other than those that directly affect other agents of the same kind (e.g., harvesting, as in [Epstein and Axtell, 1996]) and thus to affect other agents indirectly. And perhaps agents need an environment as an object to communicate about and as an object for representation. In an early description of the NEW TIES project [Gilbert et al., 2006], there is a generic description of what a minimal environment for multi-agent social simulation should consist of. Beside a topography, an interesting requirement is that the environment should provide agents with "tokens", "distinguishable, movable objects, some of which can be used as tools to speed up the production of food, but most of which have no intrinsic function, and can be employed by agents as location markers, symbols of value (e.g., 'money'), or for ritual purposes". This is quite similar to the role of the environment for communication and coordination, but goes beyond as application of this environment to the famous Kula Ring [Ziegler, 2007] model shows.

In many special applications, for instance in traffic simulation, the representation of the physical environment is inevitable, as it "embeds resources and services" and defines rules for the multi-agent system [Weyns et al., 2007, pg. 16–17] in so far as it allows and forbids the usage of particular routes at particular times. Thus traffic simulation systems typically consist of a topography and agents (including both stationary agents such as traffic lights and signs and dynamic agents such as cars, bikes and pedestrians) [Lotzmann, 2006]. In some interesting cases, the topography might not be static at all: evacuation scenarios need an environment that can rapidly change (outburst of fire, spreading of smoke, all of which have to be modeled in terms of their physical dynamics).

Ecological models, too, need an explicit modeling of the environment in which agents representing humans, enterprises and other social and economic entities act and interact with this environment. Here the description of the environment is usually even more complex than in traffic simulation, as one is particularly interested in the physical dynamics of the environment and its responses to social and economic activities. For a more detailed analysis of the role of the environment see the contribution of Hellebogh, Weynes and Holvoet in this volume, especially their Sections 2, 4 and 5.

2.4 Issues for Future Research: The Emergence of Communication

2.4.1 Agent Communication

Although agent communication languages (see the special issue of *Autonomous Agents and Multi-Agent Systems*, vol. 14 no. 2) such as KQML [Finin et al., 2007] have been developed for a long time (back to 1993), it is still an open question how agents in a simulation model could develop a communication means on their own and/or extend their communication tool to be able to refer to a changing environment (see the special issue of *Autonomous Agents and Multi-Agent Systems*, vol. 14 no. 1). In their introduction [Dignum and van Eijk, 2007, pg. 119] to the latter special issue, Dignum and van Eijk even state that there is "a wide gap between being able to parse and generate messages that conform to the FIPA ACL standard and being able to perform meaningful conversations." Malsch and his colleagues also complain that "communication plays only a minor role in most work on social simulation and artificial social systems" [Malsch et al., 2007].

The problem of inter-agent communication has at least two very different aspects — one is about agents that use a pre-defined language for their communication, while the other is about the evolution of language among agents that are not originally programmed to use some specific language.

As far as simulation is concerned, the first aspect deals with languages that are appropriate for the particular scenarios simulated. [Fornara et al., 2007] center this aspect around the concept of commitment, as agent communication in general (but also in the special case of simulating social systems) is most often used for commitments (and humans chatting just to kill time are seldom simulated).

The second aspect, evolution of language, starts from agents having no predefined language at all, but have goals that they cannot achieve by themselves, which makes it necessary for them to ask others for help. It is not entirely clear whether a grammar is actually necessary as a starting point of the evolution of language. What seems to be necessary is the capability of agents to draw conclusions from regularities in other agents' behaviors. One behavior b_1 regularly accompanying another behavior b_2 might lead to a rule in an observing agent enabling it to predict that another agent will soon display behavior b_2 after it used b_1 immediately before. One of the first example of this approach is the paper by Hutchins and Hazlehurst [Hutchins and Hazlehurst, 1995] who made a first step into the field of the emergence of a lexicon, but their agents were only able to agree on names of things (patterns — moon phases) they saw. In another paper they developed agents that were able to learn that moon phases were regularly connected to the turn of the tide [Hutchins and Hazlehurst, 1991]. This enabled the agents in this extended model to learn from others about the role of moon phases for the turn of the tide and endowed them even with a very simplistic grammar.

The NEW TIES project [Gilbert et al., 2006], ambitious as it is, aims at creating an artificial society that develops its own culture and will also need to define agent capabilities that allow them to develop something like a language although it is still questionable whether the experimenters will be able to understand what their artificial agents talk about. As compared to all earlier attempts at having artificial agents develop a language, the NEW TIES project is confronted with the problem of large numbers [Gilbert et al., 2006, 7.1], both of language learners and of language objects (many agents have to agree on names for many kinds of things and their properties).

A similar objective is aimed at in the current EMIL project (Emergence in the Loop [Andrighetto et al., 2007; Conte et al., 2007]) which attempts at creating an agent society in which norms emerge as agents observe each other and draw conclusions about which behavioral feature is desirable and which is a misdemeanor in the eyes of other agents — which, as in the case of language emergence, makes it necessary that agents can make abstractions and generalizations from what they observe in order that ambiguities are resolved. But even in this case it seems to be necessary to define which kinds of actions can be taken by agents in order that other agents can know what to evaluate as desirable or undesirable actions. The current prototypical implementations of EMIL models [Troitzsch, 2008] include an agent society whose members contribute to a large text corpus resembling a wikipedia. The language used is entirely fictitious, but has sufficient features to make it resemble natural language, and the actions include writing, copying, adding to existent texts, checking spelling and searching for plagiarisms. Without a definition of possible actions that agents can perform nothing will happen in these simulation models, and before any norms can emerge in such an agent society, at least some rules must exist according to which agents plan and perform their actions, but on the other hand, if all possible actions and their prerequisites were predefined, no emergence would be possible. The architecture of these agents [Andrighetto et al., 2007] contains a normative board which keeps track of all information relevant to norms and several engines to recognize a norm or not, to adopt it or not, to plan actions and to decide whether to abide by or violate a norm as well as to defend a norm by sanctions taken against others.

2.4.2 Concluding Remarks

Although the use of multi-agent systems for simulating social and economic phenomena is not much older than about 15 years (if one neglects the early ancestors of this kind of approach) it has made rapid progress during its short lifetime and used a wide range of methods and tools. Sociologists using multi-agent simulation for their purposes often even claim that they could contribute to the further development of computer science while developing simulations of socio-economic systems which in turn are self-adaptive. Thus there is a claim that the development of self-adapting software could use the insights of social science to construct something such as more co-operative, secure agent societies, for instance on the web. Socionics [Müller et al., 1998, section 1.1] "start[ed] a serious evaluation of sociological conceptions and theories for computer systems", thus "leaving the path of 'folks-sociology'" of which it was not clear whether its protagonists used notions such as agents forming "'societies', 'teams', 'groups', 'an organization'" and "behave 'socially', ... help each other, ... are selfish" only "for the limited purpose of simplifying the complex technical background of a distributed system" or whether they took these terms seriously. The founders of the socionics approach claimed that sociological paradigms such as "social roles, cultural values, norms and conventions, social movements and institutions, power and dominance distribution" should be useful paradigms to teach computer scientists to build "powerful technologies" endowed with "adaptability, robustness, scalability and reflexivity". David Hales, also mentions "socially-inspired computing", reasoning that human social systems have "the ability ... to preserve structures ... and adapt to changing environments and needs" — even to a higher degree than biological systems that have already been used as a template for "design patterns such as diffusion, replication, chemotaxis, stigmergy, and reaction-diffusion" in distributed systems engineering [Babaoglu et al., 2006]. But still there seems to be a long way to go until socially-inspired computing has advanced in a way that well-understood social processes can be used as design patterns in distributed systems engineering — anyway, it might be "an idea whose time has come" [Hales, 2007].

References

R. P. Abelson and A. Bernstein. A computer simulation of community referendum controversies. *Public Opinion Quarterly*, 27(1):93–122, 1963.

R. P. Abelson and J. D. Carroll. Computer simulation of individual belief systems. *American Behavioral Scientist*, 8:24–30, 1965.

G. Andrighetto, M. Campenni, R. Conte, and M. Paolucci. On the immergence of norms: a normative agent architecture. In *Proceedings of AAAI Symposium, Social and Organizational Aspects of Intelligence*, Washington DC, 2007.

S. Antcliff. An introduction to DYNAMOD: A dynamic microsimulation model. Technical Report 1, National Centre for Social and Economic Modelling (NATSEM), University of Canberra, Canberra, 1993.

L. Antunes, J. a. Balsa, P. Urbano, and H. Coelho. The challenge of context permeability in social simulation. In F. Amblard, editor, *Interdisciplinary Approaches to the Simulation of Social Phenomena. Proceedings of ESSA'07, the 4th Conference of the European Social Simulation Association*, pages 489–300, Toulouse, 2007. IRIT Editions.

B. Aparicio Diaz and T. Fent. An agent-based simulation model of age-at-marriage norms. In F. C. Billari, T. Fent, A. Prskawetz, and J. Scheffran, editors, *Agent-Based Computational Modelling*, pages 85–116. Physica, Heidelberg, 2007.

O. Babaoglu, G. Canright, A. Deutsch, G. A. Di Caro, F. Ducatelle, L. M. Gambardella, N. Ganguly, M. Jelasity, R. Montemanni, A. Montresor, and T. Urnes. Design patterns from biology for distributed computing. *ACM Transactions on Autonomous and Adaptive Systems*, 1(1):26–66, September 2006.

M. Becker, J. D. Gehrke, B.-L. Wenning, M. Lorenz, C. Görg, and O. Herzog. Agent-based and discrete event simulation of autonomous logistic processes. In W. Borutzky, A. Orsoni, and R. Zobel, editors, *Proceedings of the 20th European Conference on Modelling and Simulation, Bonn, Sankt Augustin, Germany*, pages 566–571. European Council for Modelling and Simulation, 2006.

E. Berlekamp, J. Conway, and R. Guy. *Winning Ways for Your Mathematical Plays, Vol. 2: Games in Particular*. Academic Press, London, 1982.

F. C. Billari and A. Prskawetz. *Agent-Based Computational Demography*. Physica, Heidelberg, 2003.

H. Birg, E. J. Flöthmann, I. Reiter, and F. X. Kaufmann. *Biographische Theorie der demographischen Reproduktion*. Campus, Frankfurt, 1991.

K. Brassel, O. Edenhofer, M. Möhring, and K. G. Troitzsch. Modeling greening investors. In R. Suleiman, K. G. Troitzsch, and N. Gilbert, editors, *Tools and Techniques for Social Science Simulation*. Physica, Heidelberg, 1999.

K. H. Brassel, M. Möhring, E. Schumacher, and K. G. Troitzsch. Can agents cover all the world? In R. Conte, R. Hegselmann, and P. Terna, editors, *Simulating Social Phenomena*, volume 456 of *Lecture Notes in Economics and Mathematical Systems*, pages 55–72. Springer-Verlag, Berlin, 1997.

M. Bunge. *Ontology II: A World of Systems. Treatise on Basic Philosophy, Vol. 4*. Reidel, Dordrecht, 1979.

V. Calenbuhr and J.-L. Deneubourg. Chemical communication and collective behaviour in social and gregarious insects. In W. Ebeling, M. Peschel, and W. Weidlich, editors, *Models of Selforganization in Complex Systems*. Akademie-Verlag, Berlin, 1991.

J. L. Casti. *Would-Be Worlds. How Simulation Is Changing the Frontiers of Science*. Wiley, New York, NY, 1996.

J. S. Coleman. *The Foundations of Social Theory*. Harvard University Press, Boston, MA, 1990.

R. Conte, G. Andrighetto, M. Campenni, and M. Paolucci. Emergent and immergent effects in complex social systems. In *Proceedings of AAAI Symposium, Social and Organizational Aspects of Intelligence*, Washington DC, 2007.

A. Dal Forno and U. Merloine. The evolution of coworker networks: An experimental and computational approach. In B. Edmonds, C. Hernández, and K. G. Troitzsch, editors, *Social Simulation. Technologies, Advances, and New Discoveries*, pages 280–293. Information Science Reference, Hershey, 2008.

F. Dignum and R. M. van Eijk. Agent communication and social concepts. *Autonomous Agents and Multi-Agent Systems*, 14:119–120, 2007.

J. E. Doran and M. Palmer. The EOS project: integrating two models of Palaeolithic social change. In N. Gilbert and R. Conte, editors, *Artificial Societies: The Computer Simulation of Social Life*, pages 103–125. UCL Press, London, 1995.

J. E. Doran, M. Palmer, N. Gilbert, and P. Mellars. The EOS project: modelling Upper Paleolithic social change. In N. Gilbert and J. E. Doran, editors, *Simulating Societies: The Computer Simulation of Social Phenomena*, pages 195–222. UCL Press, London, 1994.

A. Drogoul and J. Ferber. Multi-agent simulation as a tool for studying emergent processes in societies. In N. Gilbert and J. E. Doran, editors, *Simulating Societies*. UCL Press, London, 1994.

A. Drogoul, B. Corbara, and S. Lalande. Manta: new experimental results on the emergence of (artificial) ant societies. In N. Gilbert and R. Conte, editors, *Artificial Societies: The Computer Simulation of Social Life*. UCL Press, London, 1995.

B. Dubiel and O. Tsimhoni. Integrating agent based modeling into a discrete event simulation. In *WSC '05: Proceedings of the 37th conference on Winter simulation*, pages 1029–1037. Winter Simulation Conference, 2005. ISBN 0-7803-9519-0.

E. Durckheim. *The rules of the sociological method*. The Free Press, New York, 1895. Translated by W.D. Halls.

J. G. Epstein, M. Möhring, and K. G. Troitzsch. Fuzzy-logical rules in a multi-agent system. *Sotsial'no-ekonomicheskie yavleniya i protsessy*, 1(1-2):35–39, 2006.

J. M. Epstein and R. Axtell. *Growing Artificial Societies – Social Science from the Bottom Up*. MIT Press, Cambridge, MA, 1996.

D. Farmer, T. Toffoli, and S. Wolfram. *Cellular automata: proceedings of an interdisciplinary workshop, Los Alamos, New Mexico, March 7-11, 1983*. North-Holland, Amsterdam, 1984.

T. Finin, Y. Labrou, and J. Mayfield. KQML as an agent communication language. In *Software agents*, chapter 14, pages 291–316. MIT Press, Cambridge MA, 2007.

F. Flentge, D. Polani, and T. Uthmann. On the emergence of possession norms in agent societies. In *Proc. 7th Conference on Artificial Life*, Portland, 2000.

F. Flentge, D. Polani, and T. Uthmann. Modelling the emergence of possession norms using memes. *Journal of Artificial Societies and Social Simulation*, 4/4/3, 2001. http://jasss.soc.surrey.ac.uk/4/4/3.html.

N. Fornara, F. Viganò, and M. Colobetti. Agent communication and artificial institutions. *Autonomous Agents and Multi-Agent Systems*, 14:121–142, 2007.

J. W. Forrester. *Principles of Systems*. MIT Press, Cambridge, MA, 2nd preliminary edition, 1980. First published in 1968.

M. Gardener. The game of life. *Scientific American*, 223(4), April 1970.

A. Geller and S. Moss. Growing *qawms*: A case-based declarative model of afghan power structures. In F. Amblard, editor, *Interdisciplinary Approaches to the*

Simulation of Social Phenomena. Proceedings of ESSA'07, the 4th Conference of the European Social Simulation Association, pages 113–124, Toulouse, 2007. IRIT Editions.

N. Gilbert and K. G. Troitzsch. *Simulation for the Social Scientist.* Open University Press, Maidenhead, New York, 2nd edition, 2005.

N. Gilbert, M. den Besten, A. Bontovics, B. G. Craenen, F. Divina, A. Eiben, R. Griffioen, G. Hévízi, A. Lőrincz, B. Paechter, S. Schuster, M. C. Schut, C. Tzolov, P. Vogt, and L. Yang. Emerging artificial societies through learning. *Journal of Artificial Societies and Social Simulation*, 9/2/9, 2006. http://jasss.soc.surrey.ac.uk/9/2/9.html.

H. Haken. *Synergetics. An Introduction. Nonequilibrium Phase Transitions and Self-Organization in Physics, Chemistry and Biology.* Springer Series in Synergetics, Vol. 1. Springer-Verlag, Berlin, 2nd enlarged edition, 1978.

D. Hales. Social simulation for self-star systems: An idea whose time has come? In F. Amblard, editor, *Interdisciplinary Approaches to the Simulation of Social Phenomena. Proceedings of ESSA'07, the 4th Conference of the European Social Simulation Association*, page 13, Toulouse, 2007. IRIT Editions.

R. Hegselmann. Modeling social dynamics by cellular automata. In W. B. Liebrand, A. Nowak, and R. Hegselmann, editors, *Computer Modeling of Social Processes*, pages 37–64. Sage, London, 1998.

D. Helbing. *Quantitative Sociodynamics. Stochastic Methods and Models of Social Interaction Processes.* Kluwer, Dordrecht, 1994a.

D. Helbing. A mathematical model for the behavior of individuals in a social field. *Journal of Mathematical Sociology*, 19(3):189–219, 1994b.

D. Helbing and A. Johansson. Quantitative agent-based modeling of human interactions in space and time. In F. Amblard, editor, *Proceedings of the 4th Conference of the European Social Simulation Association (ESSA'07)*, pages 623–637, Toulouse, September 10–14 2007.

E. Hutchins and B. Hazlehurst. Learning in the cultural process. In C. G. Langton, C. Taylor, J. D. Farmer, and S. Rasmussen, editors, *Artificial Life II*, pages 689–706. Addison-Wesley, Redwood City CA, 1991.

E. Hutchins and B. Hazlehurst. How to invent a lexicon: the development of shared symbols in interaction. In N. Gilbert and R. Conte, editors, *Artificial Societies: The Computer Simulation of Social Life*, pages 157–189. UCL Press, London, 1995.

M. Kang, L. B. Waisel, and W. A. Wallace. Team soar. a model for team decision making. In M. J. Prietula, K. M. Carley, and L. Gasser, editors, *Simulating Organisations.* AAAI Press, MIT Press, Menlo Park CA, Cambridge MA, London, 1998.

A. König, M. Möhring, and K. G. Troitzsch. Agents, hierarchies and sustainability. In F. Billari and A. Prskawetz-Fürnkranz, editors, *Agent Based Computational Demography*, pages 197–210. Physica, Berlin/Heidelberg, 2002.

J. E. Laird, A. Newell, and P. S. Rosenbloom. SOAR: An architecture for general intelligence. *Artificial Intelligence*, 33:1–64, 1987.

B. Latané. The psychology of social impact. *American Psychologist*, 36:343–356, 1981.

B. Latané. Dynamic social impact. Robust predictions from simple theory. In R. Hegselmann, U. Mueller, and K. G. Troitzsch, editors, *Modelling and Simulation in the Social Sciences from a Philosophy of Science Point of View*, Theory and Decision Library, Series A: Philosophy and Methodology of the Social Sciences, pages 287–310. Kluwer, Dordrecht, 1996.

B. Latané and A. Nowak. Attitudes as catastrophes: From dimensions to categories with increasing involvement. In R. Vallacher and A. Nowak, editors, *Dynamical Systems in Social Psychology*, pages 219–250. Academic Press, San Diego, CA, 1994.

B. Latané and A. Nowak. Self-organizing social systems: necessary and sufficient conditions for the emergence of clustering, consolidation and continuing diversity. In G. A. Barnett and F. J. Boster, editors, *Progress in Communication Sciences: Persuasion*, volume 13, pages 43–74. Ablex, Norwood, NJ, 1997.

I. Leim. *Die Modellierung der Fertilitätsentwicklung als Folge komplexer individueller Entscheidungsprozesse mit Hilfe der Mikrosimulation*. PhD thesis, Universität Duisburg-Essen, Duisburg, 2007.

K. Lewin. *Field Theory in Social Science*. Harper & Brothers, New York, 1951.

U. Lotzmann. Design and implementation of a framework for the integrated simulation of traffic participants of all types. In *EMSS2006. 2nd European Modelling and Simulation Symposium, Barcelona, October 2–4, 2006*. SCS, 2006.

X. T. C. Ltd. xj technologies: Simulation software and services: Anylogic. http://www.xjtek.com/. Accessed June 11, 2008.

T. Malsch, C. Schlieder, P. Kiefer, M. Lübcke, R. Perschke, M. Schmitt, and K. Stein. Communication between process and structure: Modelling and simulating message reference networks with COM/TE. *Journal of Artificial Societies and Social Simulation*, 10(1), 2007. http://jasss.soc.surrey.ac.uk/10/1/9.html.

M. Möhring and K. G. Troitzsch. Lake anderson revisited. *Journal of Artificial Societies and Social Simulation*, 4/3/1, 2001. http://jasss.soc.surrey.ac.uk/4/3/1.html.

H. J. Müller, T. Malsch, and I. Schulz-Schaeffer. Socionics: Introduction and potential. *Journal of Artificial Societies and Social Simulation*, 1(3), 1998. http://www.soc.surrey.ac.uk/JASSS/1/3/5.html.

A. Nowak and B. Latané. Simulating the emergence of social order from individual behaviour. In N. Gilbert and J. E. Doran, editors, *Simulating Societies: The Computer Simulation of Social Phenomena*, pages 63–84. UCL Press, London, 1993.

A. Nowak and M. Lewenstein. Modeling social change with cellular automata. In R. Hegselmann, U. Mueller, and K. G. Troitzsch, editors, *Modelling and Simulation in the Social Sciences from a Philosophy of Science Point of View*, Theory and Decision Library, Series A: Philosophy and Methodology of the Social Sciences, pages 249–286. Kluwer, Dordrecht, 1996.

G. H. Orcutt, M. Greenberger, J. Korbel, and A. Rivlin. *Microanalysis of socioeconomic systems: a simulation study*. Harper and Row, New York, NY, 1961.

H. V. D. Parunak and D. Weyns. Guest editors' introduction, special issue on environments for multi-agent systems. *Autonomous Agents and Multi-Agent Systems*, 14(1):1–4, 2007. doi: http://dx.doi.org/10.1007/s10458-006-9003-4.

P. Riley and G. Riley. SPADES — a distributed agent simulation environment with software-in-the-loop execution. In S. Chick, P. J. Sánchez, D. Ferrin, and D. J. Morrice, editors, *Winter Simulation Conference Proceedings*, volume 1, pages 817–825, 2003.

A. Ryan. Emergence is coupled to scope, not level. *arXiv:nlin/0609011v1 [nlin.AO]*, 2006.

T. C. Schelling. Dynamic models of segregation. *Journal of Mathematical Sociology*, 1:143–186, 1971.

T. C. Schelling. *Micromotives and Macrobehavior*. Norton, New York, NY, 1978.

N. Schurr, S. Okamoto, R. T. Maheswaran, P. Scerri, and M. Tambe. Evolution of a teamwork model. In R. Sun, editor, *Cognition and Multi-Agent Interaction*. Cambridge University Press, Cambridge MA, New York etc., 2006.

M. Schwaninger. System dynamics and the evolution of the systems movement. *Systems Research and Behavioral Science*, 23:583–594, 2006.

W. Senn. Mathematisierung der Biologie: Mode oder Notwendigkeit?, a. Collegium Generale, Universität Bern, 6. Dezember 2006, Vortragsserie "Aktualität und Vergänglichkeit der Leitwissenschaften", http://www.cns.unibe.ch/publications/ftp/Mathematisierung_Biologie.pdf. Accessed June 11, 2008.

W. Senn. Das Hirn braucht Mathematik. ein Plädoyer für Top-down-Modelle in der Biologie und den Neurowissenschaften. Neue Zürcher Zeitung 22.08.07 Nr. 193 Seite 57, b.

G. Shafer and J. Pearl, editors. *Readings in Uncertain Reasoning*. Morgan Kaufman, San Francisco, 1990.

I. D. Sola Pool and R. P. Abelson. The simulmatics project. In H. Guetzkow, editor, *Simulation in Social Science: Readings*, pages 70–81. Prentice Hall, Englewood Cliffs, NJ, 1962. Originally in *Public Opinion Quarterly* 25, 1961, 167–183.

M. Sullivan and P. R. Cohen. An endorsement-based plan recognition program. In *IJCAI-85*, pages 475–479, 1985.

C. Tilly. Micro, macro, or megrim?, August 1997. http://www.asu.edu/clas/polisci/cqrm/papers/Tilly/TillyMicromacro.pdf. Accessed June 10, 2008.

K. G. Troitzsch. A multi-agent model of bilingualism in a small population. In H. Coelho, B. Espinasse, and M.-M. Seidel, editors, *5th Workshop on Agent-Based Simulation*, pages 38–43, Erlangen, San Diego, CA, 2004. SCS Publishing House.

K. G. Troitzsch. Simulating collaborative writing: software agents produce a wikipedia. In *Proceedings of the 5th Conference of the European Social Simulation Association (ESSA'2008)*, Brescia, Italy, 2008.

A. Uhrmacher. Object-oriented and agent-oriented simulation: implications for social science applications. In K. G. Troitzsch, U. Mueller, N. Gilbert, and J. E. Doran, editors, *Social Science Microsimulation*, pages 432–445. Springer-Verlag, Berlin, 1996.

W. Weidlich and G. Haag. *Concepts and Models of a Quantitative Sociology. The Dynamics of Interacting Populations*. Springer Series in Synergetics, Vol. 14. Springer-Verlag, Berlin, 1983.

D. Weyns, A. Omicini, and J. Odell. Environment as a first class abstraction in multi-agent systems. *Autonomous Agents and Multi-Agent Systems*, 14:5–30, 2007.

U. Wilensky. NetLogo. http://ccl.northwestern.edu/netlogo. Accessed June 11, 2008., 1999.

M. Wooldridge and N. R. Jennings. Intelligent agents: theory and practice. *Knowledge Engineering Review*, 10(2):115–152, 1995.

G. C. Zaft. Social science applications of discrete event simulation: A DEVS artificial society. Master's thesis, University of Arizona, 2001.

B. P. Zeigler. *Object-Oriented Simulation with Hierarchical, Modular Models — Intelligent Agents and Endomorphic Systems*. Academic Press, San Diego, 1990.

B. P. Zeigler. *Theory of Modelling and Simulation*. Krieger, Malabar, 1985. Reprint, first published in 1976, Wiley, New York, NY.

R. Ziegler. The Kula ring of Bronislaw Malinowski. a simulation model of the co-evolution of an economic and ceremonial exchange system. Bayerische Akademie der Wissenschaften, Philosophisch-Historische Klasse. Sitzungsberichte, München, 2007. Heft 1.

3

Simulation Engines for Multi-Agent Systems

Georgios K. Theodoropoulos
University of Birmingham, UK

Rob Minson
University of Birmingham, UK

Roland Ewald
University of Rostock, Germany

Michael Lees
University of Nottingham, UK

3.1	Introduction..	77
3.2	Multi-Agent System Architectures................	78
3.3	Discrete Event Simulation Engines for MAS	79
	The Discrete Event Simulation Paradigm • A Survey of MAS Simulation Toolkits • Taxonomy of Discrete Event Simulation Toolkits	
3.4	Parallel Simulation Engines for MAS...............	86
	Parallel Discrete Event Simulation • The High Level Architecture and Simulation Interoperability • A Survey of Parallel MAS Simulation Toolkits • Taxonomy of Parallel DES Toolkits for MAS	
3.5	Issues for Future Research	96
	Scalability of Parallel Engines • Other Areas	
References		99

3.1 Introduction

Multi-agent systems (MAS) are often extremely complex and it can be difficult to formally verify their properties. As a result, design and implementation remains largely experimental, and experimental approaches are likely to remain important for the foreseeable future. Simulation is therefore the only viable method to rigorously study their properties and analyze their emergent behavior.

Over the last two decades, a wide range of MAS toolkits and testbeds have been developed and applied in different domains. A number of surveys of these systems have appeared in the literature in recent years, e.g., [Serenko and Detlor, 2002; Gilbert and Bankes, 2002; Tobias and Hofmann, 2004; Mangina; Railsback et al., 2006]. However, these have tended to focus on the high level support that these toolkits offer for the process of agent specification and model creation. In contrast this chapter focuses on the simulation engines integrated in to these toolkits to facilitate the execution of the resulting MAS models. In this context a simulation engine is taken to be any virtual machine which provides controlled execution of the model.

This chapter is organized as follows: Section 3.2 first discusses what types of MAS model - in terms of the architecture of an individual agent and the mechanisms by which it interacts with its environment - are executed using these engines; section 3.3 then surveys the engines used by the currently available toolkits in the context of these observations about MAS model types; Sections 3.4 and 3.5 then explore recent and future research in to distributed simulation engines which aim to provide both scalability and interoperability to MAS simulation models.

3.2 Multi-Agent System Architectures

Toolkits for building MAS are generally targeted at one of three types of application:

1. **MAS for studying Complexity.** Examples are social models (e.g., Schelling's segregation model [Schelling, 1971], or Ormerod's low-dimensional paradigm [Ormerod, 2007]), artificial life (e.g., Axtell's Sugarscape [Epstein and Axtell, 1996] or Reynolds's Boids [Reynolds, 1987] models) or logistics (e.g., Traffic Simulations [Burmeister et al., Feb 1997]). Such models use very simple agents which engage in little, if any, planning or coordination. The models are interpreted usually via some quantitative (average life expectancy, average queuing time, etc.) or qualitative (emergent segregation patterns, emergent flocking, etc.) observation at the macro-level of the population itself.

2. **MAS for studying Distributed Intelligence.** Examples range from planning (e.g., Blocksworld [Fahlman, 1973], Tileworld [Pollack and Ringuette, 1990]), to more cognitively 'accurate' social simulations (as advocated by researchers such as Nigel Gilbert [Gilbert, 2005], John Doran [Doran, 2001] or Ron Sun [Sun, 2001]) all the way up to research in to human cognition itself (such as the work of the CoSY project [Hawes et al., 2007], or of researchers like Aaron Sloman [Sloman and Logan, 1999] or Mattias Scheutz [Scheutz and Logan, 2001]). Such models use internally complex, situated, communicating agents, and are often designed to study the behavior of one particular cognitive formalism such as SOAR [Wray and Jones, 2005], ACT-R [Taatgen et al., 2005] or BDI [Rao and Georgeoff, 1995].

3. **Development of Software MAS.** These toolkits provide support for building software agents such as those described by Wooldridge & Jennings [Jennings et al., 1998], Franklin & Gaesser [Franklin and Graesser, 1996] or as implied by the FIPA [Foundation for Intelligent Physical Agents FIPA] or KQML [Finin et al., 1994] standards. Typical applications are Semantic Web agents, Beliefs-Desires-Intentions (BDI) agents in expert systems, or agents for network meta-management (e.g., load-balancing or service discovery). Many of these toolkits include a pre-deployment environment for debugging or verification of the implemented MAS which may be considered equivalent to a simulation engine.

These different target applications have an obvious impact on the modeling facilities offered by a toolkit in order to *develop* MAS models. However, they also have implications for the type of simulation engine that will be used (and usually packaged along with the toolkit itself) to *execute* the models.

The first two types provide an execution environment which would be most widely recognizable as a simulation engine. That is, a virtual machine with a notion of logical time advancing in discrete steps. This reflects the fact that, with these applications parameterization, repeatability and introspection are key to understanding the model's behavior, whether this be at a micro- or macro-level.

In contrast the types of execution environment which support development platforms for software agents are primarily concerned with controlled emulation of a real execution environment (e.g., a computer network). This reflects the fact that such applications are primarily concerned with debugging or validating the system in some way, while repeatable or parameterizable behaviors are less of a concern.

On this basis, the remainder of this chapter will focus on the type of execution environments which genuinely constitute simulation engines. This being those found in toolkits supporting the experimental development of multi-agent models as in types 1 and 2 above.

3.3 Discrete Event Simulation Engines for MAS

MAS Simulation toolkits, in the interests of robustness, repeatability and micro-scale analysis of the model, generally conceive of the model as a discrete system, transitioning through time in discrete steps. Toolkits of this type are discussed in this Section, but first it is important to establish the common paradigm - independent from MAS modeling specifically - which is generally used to simulate discrete systems: Discrete Event Simulation.

3.3.1 The Discrete Event Simulation Paradigm

Discrete Event Simulation (DES) [Fishman, 1978] is a common paradigm for the simulation of discrete (or discretized) systems. DES is a special case of the more general approach of Discrete Time simulations (such as a Time-Stepped approach) in which state transitions occur instantaneously rather than continuously with time. Under the DES paradigm these transitions do not occur as a linearly spaced sequence of steps, but occur instead at instants in time at which a transition has semantic significance. This is a more flexible paradigm (which is, of course, capable of simulating a Time-Stepped approach via a linearly spaced sequence of events).

DES engines are generally implemented as a queue of events (commonly implemented as a heap) each having a logical timestamp. A loop iterates over the queue, at each iteration dequeuing and executing the event with the lowest timestamp. This event may cause some transition in the system and, crucially, may also insert more events in to the queue to be executed in the future. Generally events encapsulate a small amount of state and are passed to event *handlers* which can perform conditional branching over the model's state at the time of execution to implement complex dynamic behavior. This paradigm is more flexible than a standard Time-Stepped approach as the speed of system transition is entirely dictated by the system being modeled. This allows, for example, both long- and short-cycle phenomena to be modeled concurrently and efficiently within a single model.

An example of the mechanics of a standard sequential DES engine are given in Figure 3.1. In this example an event at time 2 is dequed which represents the arrival of a truck at a postal sorting depot. When the handler for the event is invoked, depending on the state of the sorting depot, an event which represents the beginning of the mail sorting process may begin, or may continue if it is already in progress. The delay between the two events (from time 2 to time 7) may represent the overhead in starting the sorting process from cold which the model wishes to represent.

3.3.2 A Survey of MAS Simulation Toolkits

This Section surveys the simulation engines integrated in to many of the most popular and influential toolkits for building MAS simulations. This survey is not exhaustive due to the unavailability of technical detail for some popular, closed-source toolkits (eg. NetLogo), however it does provide a good coverage of the various engineering options when implementing a DES engine for MAS models.

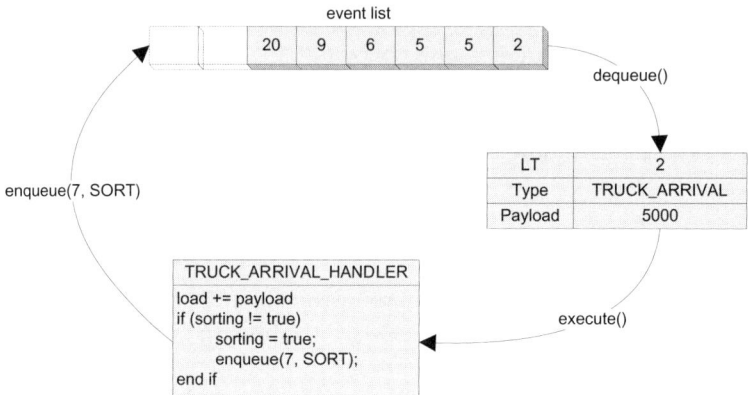

FIGURE 3.1 The operation of a standard sequential DES engine. The example application is one of a model of a postal sorting depot. Note that an event with time T may be scheduled regardless of the existence in the schedule of events with time $> T$, however an event may never be scheduled in the past (the time of the latest processed event).

LEE

The LEE (Latent Energy Environment) toolkit [Menczer and Belew, 1993] is implemented in C and targets research on evolutionary complex systems. Its underlying assumptions regarding the structure of agents, their actions, and their environment are therefore highly specific: Agents consist of a user-defined neural network that uses sensor information and the internal state to trigger actions. Agents may only interact with their environment, which is a two-dimensional toroidal grid. Each cell of the grid may contain consumable elements that provide the agents with energy. This model of a multi-agent system is combined with a genetic algorithm approach that allows individual agents to proliferate or die. A replenishment function is used to generate new consumable elements and place them on the grid, while a reaction table defines which elements react with each other, so that agents can combine elements to increase their energy consumption. The simulation is executed in a Time-Stepped manner. At every step, each agent has a certain chance of being executed. This 'virtual concurrency' mechanism was devised to model the parallel execution of agents in an unbiased serial way. The LEE has been successfully applied to model the co-evolution of motor-sensor systems [Dagorn et al., 2000]. Figure 3.2 shows a screenshot of the LEE graphical user interface (GUI).

JAMES II

JAMES II (Java-based Agent Modeling Environment for Simulation) [Himmelspach and Uhrmacher, 2007] has become a general-purpose modeling and simulation system over the years, but nevertheless provides several functions that are tailored to the needs of agent modeling and simulation. In JAMES II, agents need to be modeled in a formalism of the user's choice. The system provides various extensions of the DEVS [Zeigler et al., 2000] formalism (e.g., ml-DEVS [Uhrmacher et al., 2007]), to support multi-level modeling, dynamic structures, broadcasting, and the integration of external processes like planners or complete agents [Himmelspach and Uhrmacher, 2004]. Other supported formalisms are StateCharts, Petri Nets, Cellular Automata, and several process algebras, e.g., Space-π. The latter could in principle be used to model large sets of simple, reactive agents situated within a 3-dimensional environment (e.g., the 'swarm' of Euglena cells in [John et al., 2008]).

It is also possible to run a simulation in paced mode, i.e., in sync with the wallclock time.

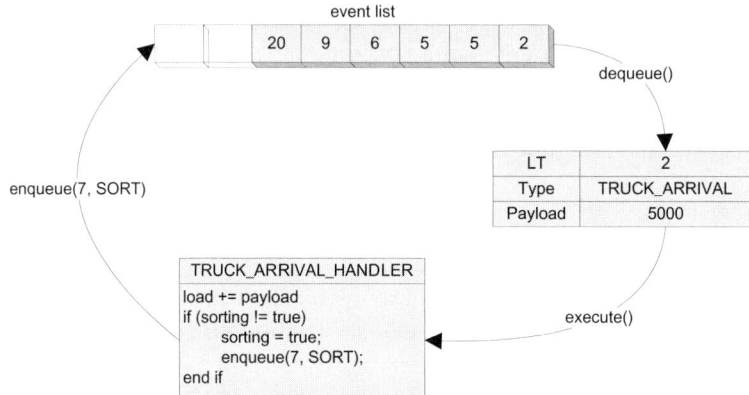

Simulation Engines for Multi-Agent Systems

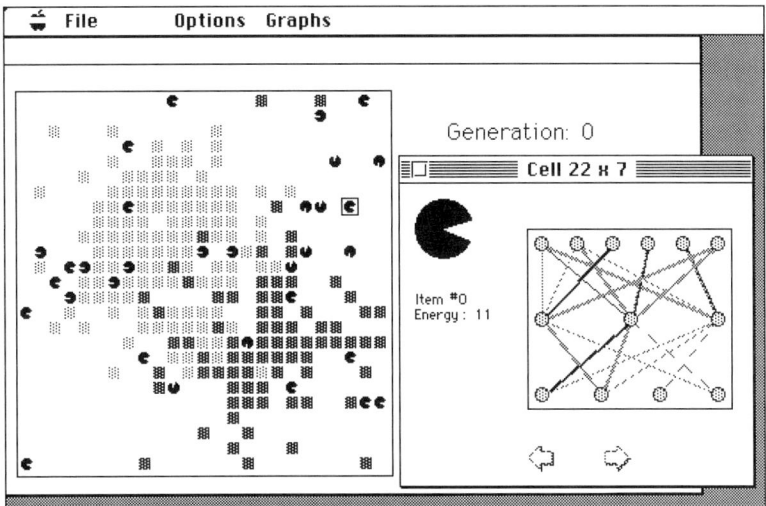

FIGURE 3.2 A screenshot of LEE (from [Menczer and Belew, 1993]).

This is useful for testing an existing agent software within a controlled environment [Gierke et al., 2006]. Under these circumstances, JAMES II could also be viewed as a real-time execution engine, consisting of the model simulated by JAMES II on the one hand, and some external processes on the other.

When using DEVS to model agents, however, there is no notion of space, i.e., situational agent models require an explicit, user-defined model of the environment. As DEVS is a discrete-event formalism, all DEVS simulation algorithms provided by the system are essentially discrete-event simulators, with varying performance profiles and capabilities, e.g., regarding the model's structure and the mode of execution (sequential/distributed). The scheduling of the agents depends on the model and has to be defined by the user. The models themselves are implemented as Java subclasses or as XML files containing Java code. More details on JAMES II can be found in Section 3.4.3. A screenshot of the JAMES II GUI showing a 2D visualization is shown in Figure 3.3.

SeSAm

SeSAm (Shell for Simulated Agent Systems) [Klügl and Puppe, 1998; Klügl et al., 2006] is a full-fledged simulation system for multi-agent systems, written in Java. It has a powerful graphical user interface that allows to model multi-agent systems without programming. In SeSAm, agents are embedded in a two- or three-dimensional grid. Agents are defined by sensors and effectors, as well as an internal function to select the appropriate action. The simulator proceeds in discrete time steps. There is, however, a plugin that enables the system to process models in a discrete-event manner. SeSAm has been widely used in research and teaching, often for modeling social systems like the behavior of honey bees or consumers in a supermarket. A screenshot of the SeSAm GUI is shown in Figure 3.4.

RePast

RePast (Recursive Porus Agent Simulation Toolkit) [Project; North et al., 2006] is a free, open-source toolkit with implementations in multiple languages, the most popular of these being the pure Java version. RePast borrows heavily from and is targeted at similar models to the Swarm framework (see below). RePast implements a full Discrete Event Scheduler (DES), allowing the scheduling of events in the form of BasicAction instances which can

82   Multi-Agent Systems: Simulation and Applications

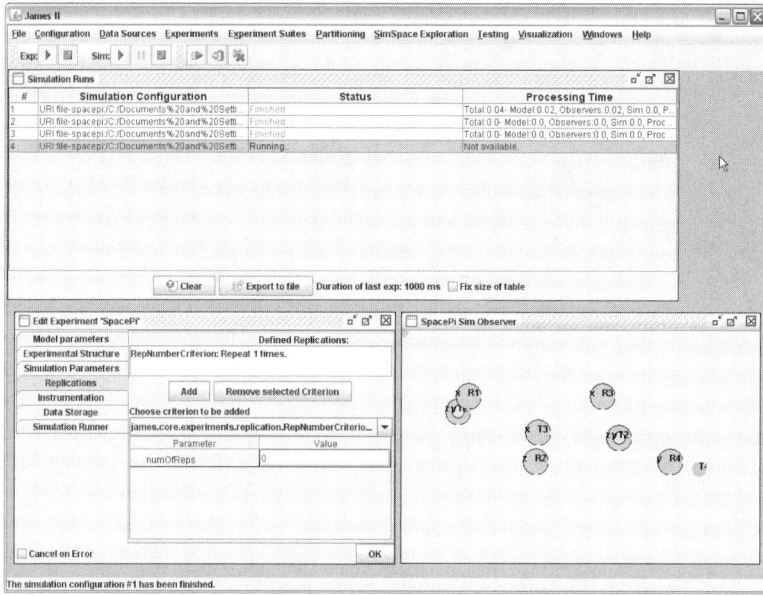

FIGURE 3.3 The JAMES II GUI is still work in progress. It contains a rudimentary experiment editor and supports to plug in arbitrary model editors and visualization components. Here, a two-dimensional visualization of molecular transport in Space-π is shown.

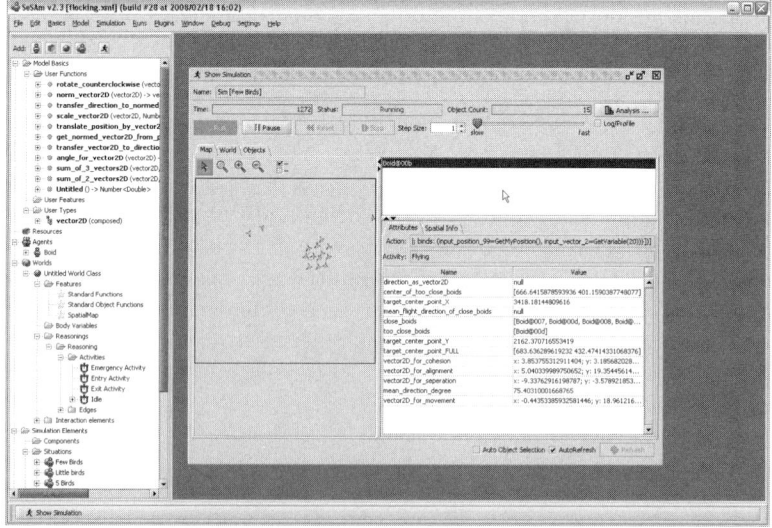

FIGURE 3.4 A screenshot of SeSAm simulating and animating a Boids [Reynolds, 1987] model. SeSAm's GUI is very convenient and well-designed. It provides a broad range of advanced features: debugging, model editing (including a Statecharts editor), output analysis, and experimentation. The user is also able to trace and introspect individual agents at runtime.

Simulation Engines for Multi-Agent Systems

execute arbitrary code, including the scheduling of future events in the schedule. RePast's particular schedule implementation includes facilities for automatically rescheduling events with some frequency and for the randomization of the order of execution of logically concurrent events, the latter facility being another example of virtual concurrency as mentioned above in the case of LEE. Although, from the modeling perspective, RePast provides several components (such as 2D or 3D grids, networks and GIS (Geographic Information System) environments) which are generally useful for modeling social phenomena as Multi-Agent systems, no assumptions are made inside the engine about the structure or behavior of the model. Figure 3.5 shows a screenshot of the RePast GUI controls being used to manage a GIS-based MAS simulation.

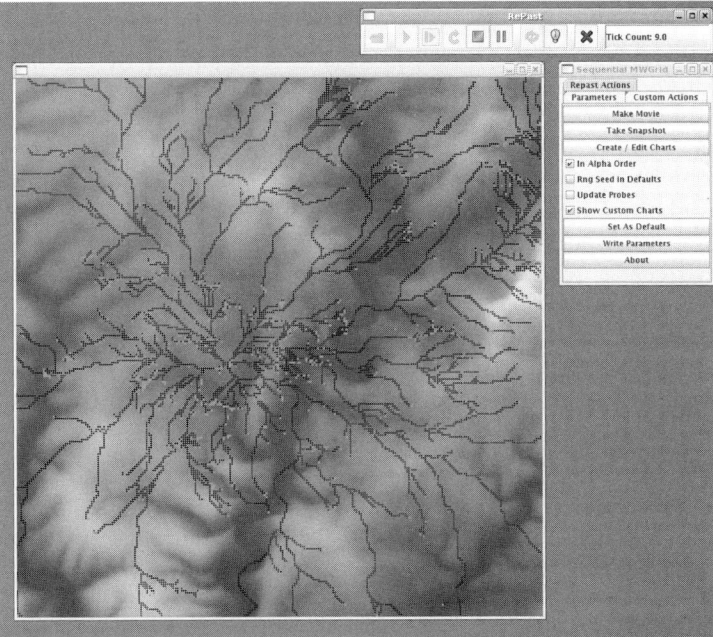

FIGURE 3.5 A screenshot of RePast in operation. The GUI offers methods to start, stop, pause and step iteratively through the simulation whilst the configuration panel provides model parameterisation, data collection and creation of video from running models.

Swarm

Swarm [The Swarm Development Group; Minar et al., 1996] is a free, open-source toolkit implemented in the Objective-C language. In terms of Swarm as an execution engine, it is very similar to RePast (see above) in that it implements a Discrete Event Scheduler which is populated explicitly by the model with events executing arbitrary code (including event (re-)scheduling). The DES extensions offered by RePast (automatic rescheduling, virtual concurrency, etc.) are also present in almost identical form here. Swarm encapsulates all this functionality, as well as data visualization, execution control and data analysis in the 'swarms' metaphor. The 'Model Swarm' represents the state and populates the schedule, whilst 'Observer Swarms' can insert or extract data from the model and perform analysis and visualization.

MAML

MAML (The Multi-Agent Modeling Language) [Gulyás et al., 1999; Gulyás] is a high-level modeling formalism for the creation of Multi-Agent models in the vein of the RePast and Swarm toolkits. MAML explicitly compiles in to Swarm code and shares many of the abstractions used by that toolkit. MAML adds a higher degree of structure to its models, particularly with regard to the planning and interaction of the agents in the model. This moves MAML slightly further away from the pure Discrete Event System of Swarm/RePast and toward a more Time-Stepped model in which agents run largely on a loop, though still with the ability to explicitly interact with the schedule itself in a standard Discrete Event manner.

SIM_AGENT

SIM_AGENT [Sloman and Poli, 1996] is an architecture-neutral agent toolkit developed as a series of libraries written in the Pop-11 programming language. It can be used to simulate many different types of MAS, such as software agents on the internet or physical agents in an environment. An agent is described in the toolkit by a series of rule-based modules which define its capabilities, e.g., sensing, planning etc. The execution mode of the modules can be varied, including options for firing sets of satisfied rules sequentially or concurrently and with specified resources. The system also provides facilities to populate the agent's environment with user-defined active and passive objects (and other agents). The execution model is Time-Stepped and each step operates in logical phases: sensing, internal processing and action execution, where the internal processing may include a variety of logically concurrent activities, e.g., perceptual processing, motive generation, planning, decision making, learning etc.

In the first phase an agent executes its sensing procedures and updates its internal memory to reflect changes in its environment and any messages it receives. The next phase involves decision making and action selection. The contents of the agent's database together with the new facts created in phase one are matched against the conditions of the condition-action rules which constitute the agent's rule system. Finally the agents send any messages and perform the actions queued in the previous phase. These external actions will typically cause the agent to enter a new state (e.g., change its location) or manipulate the environment and hence sense new data.

CHARON

CHARON [Alur, 2008] is a toolkit comprising a language, compiler and simulation engine designed to model continuous/discrete hybrid systems. These are systems which are modeled in a continuous manner, using a set of analog variables, but whose holistic behavior switches in a discrete manner between a set of 'modes'. CHARON uses the agent abstraction to describe such systems, generally modeling it as a hierarchy of interacting agents with shared analog variables triggering mode changes based on a set of invariants. CHARON itself is a language used to describe these modes, invariants and hierarchies. CHARON code can be compiled to standard Java code which is then executed by a simulation engine.

This engine progresses the system in a timestepped manner, with continuous steps being executed as timeslices using numerical integration, and discrete steps being executed as the tests of invariants in between continuous steps. In this sense the sequential CHARON engine can be seen as a simple timestepped engine which models continuous variables in a discrete manner. A parallel implementation has also been devised which is described in Section 3.4.3.

3.3.3 Taxonomy of Discrete Event Simulation Toolkits

Table 3.1 gives a summary of the surveyed discrete event toolkits. The toolkits are characterized by the execution engine (Discrete Event, Time-Stepped, etc.) which they employ and by the assumptions this engine makes about the structure of the model (if any).

TABLE 3.1 Engines found in MAS simulation toolkits.

Toolkit	Execution Model	Type of MAS Targeted	Type of Environment Supported
LEE	Time-Stepped (DES simulated with randomized agent selection)	Artificial Life (using Neural Networks)	Assumes 2D or 3D Grid
JAMES II	Full DES	Depending on formalism (planner integration, communication, etc. supported via plugins)	None prescribed
SeSAm	Time-Stepped (plugin for DES available)	Reactive Paradigm	Assumes 2D or 3D grid
RePast	Full DES	Complex Systems	None prescribed (library support for 2D/3D grids, GIS environments, etc.)
Swarm	Full DES	Complex Systems	None prescribed
MAML	Full DES (via Swarm)	Complex Systems (but larger emphasis on coordination/communication)	None prescribed
SIM_AGENT	Time-Stepped	Cognitive Agents (some prescribed external and internal structure)	None prescribed
CHARON	Time-Stepped	Discrete/Continuous Hybrid Systems (modeled as MAS)	None Prescribed

From this summary one interesting observation is that toolkits which seek to prescribe some structure to the MAS being modeled also tend to constrict execution to a Time-Stepped model. Though the phrase 'prescribe some structure' seems to imply a lack of flexibility it could equally be given a positive connotation. Toolkits such as SIM_AGENT provide a clear design template within which several cognitive architectures and agent designs can be implemented, without the need for re-implementation of 'boilerplate' code.

Indeed, it appears from this that there is a gap in the current design space for a toolkit which both provides/prescribes some structure for implementing agents but also provides a full Discrete Event scheduling implementation for the model's execution. It is possible that the two goals are not complementary in that prescribing structure to an agent also implies the prescription of how and how frequently an agent is called on to sense, think, act, communicate, etc.

3.4 Parallel Simulation Engines for MAS

As agent-based modeling is applied to ever more complex models, two problems of existing toolkits are emerging, hampering the development and deployment of complex multi-agent models: lack of scalability and lack of interoperability. Distributed simulation can offer a solution to these two problems and integrating this formalism with MAS simulation engines has received significant recent attention amongst Distributed Simulation and MAS Simulation researchers.

In this Section we review these projects, and discuss in Section 3.5 advanced issues of data management and load balancing which are raised.

3.4.1 Parallel Discrete Event Simulation

A parallelized form of Discrete Event Simulation (DES, discussed in Section 3.3.2 above) was formally introduced by Chandy and Misra in [Chandy and Misra, 1981] as the PDES paradigm. In common with general parallel execution algorithms, PDES was primarily concerned with making very large (in terms of required memory and required CPU-time) computations tractable, rather than concerns of interoperability between simulation models (see Section 3.4.2 below).

In PDES the simulation model is decomposed in to many *Logical Processes* (LPs) which each represent a closed subsystem. Each LP is internally driven by a standard DES scheduling algorithm, however it can also send and receive events to and from other LPs, this representing interactions between the subsystems. Incoming external events are integrated in to the local DES schedule and executed in the standard way, possibly scheduling future local or external events. This system is shown in Figure 3.6.

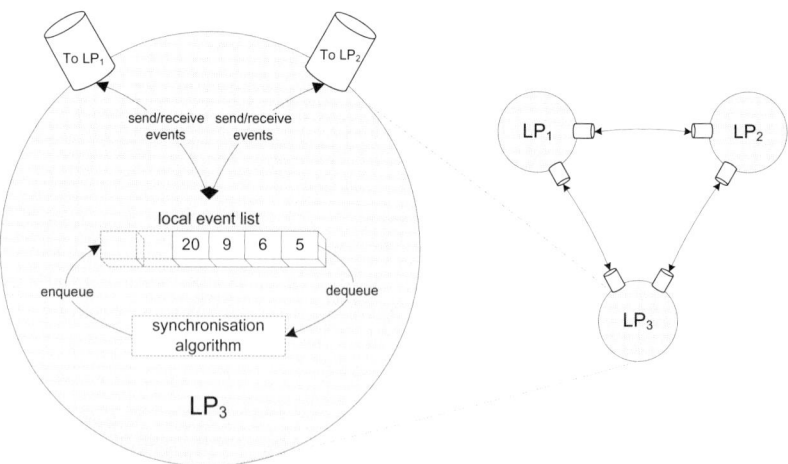

FIGURE 3.6 A system of three LPs with the internals of LP_3 shown in detail on the left.

In this context a synchronization algorithm is required to ensure that events are still processed in timestamp order, this condition is known as the *local causality constraint*. This can be trivially guaranteed in a sequential DES environment, but is far from trivial in a parallel environment. Several such algorithms have been proposed [Paul F. Reynolds, 1988] though the two main families remain those based on Chandy and Misra's original *conservative* scheme (in which an LP does not execute an event with timestamp t unless it can guarantee no external event will be received with timestamp $< t$) and the later *optimistic* scheme (in which an LP executes events in an unconstrained manner but also has the ability to *rollback* executed events with timestamp t' if an external event with timestamp t is received such that $t < t'$) derived from Jefferson's general-purpose *Timewarp* concurrency algorithm [Jefferson, 1985].

3.4.2 The High Level Architecture and Simulation Interoperability

The High Level Architecture (HLA) is a simulation interoperability standard and framework specification developed by the U.S. Department of Defense which aims to facilitate interoperation between independently developed simulation models in a single simulation [US Defence Modelling and Simulation Office, 1998; IEEE, 2000].

The HLA is a protocol based around the notion of the 'federation'. An individual federate in a given federation is an instance of some simulation executive, which is currently modeling a portion of the larger simulation (this is roughly analogous to an LP in the standard PDES model). The federates may be written in different languages and may run on different machines. The federates in a federation communicate through a central 'Runtime Infrastructure' (RTI) by issuing timestamped events which in some way modify the global state and synchronize their local schedules with the global schedule through one of the RTI's time management services. These services are instances of one of the standard PDES synchronization mechanisms described in Section 3.4.1 above.

Each federate shares in the global model through a common semantic understanding of the data delivered to it by the RTI. The structure of this data is defined in a 'Federation Object Model' (FOM), while the actual interpretation of this data is the responsibility of the federate itself. This semantic independence of data in the HLA provides the basis for model interoperation. Within a given FOM the classes of objects which are to be used in a specific federation are defined by a name (unique within the hierarchy) and a set of un-typed attributes.

At run-time the federate interfaces with the RTI through the use of an `RTIambassador` instance. This object provides access to the remote invocation services provided by the HLA specification. In a similar way, the federate itself must provide an implementation of the `FederateAmbassador` interface, which accepts callbacks from the RTI to notify of events pertinent to this federate. This coupling process is depicted in Figure 3.7.

3.4.3 A Survey of Parallel MAS Simulation Toolkits

Parallel and Distributed simulation technologies offer many important tools to MAS modeling, both in terms of increasing the complexity of feasible models and in terms of facilitating the controlled interoperation of existing models. However, both the tasks of decomposing a MAS model in to distinct computational units and of defining the semantics of interaction amongst different MAS models are non-trivial.

Several research projects have aimed to open up these important technologies to MAS models, exploring these challenges in the process. They are surveyed in the following Section.

Gensim and DGensim

DGensim (Distributed Gensim) [Anderson, 2000], developed in Allegro Common Lisp (ACL) under Linux, is an extension of the original Gensim and the authors state that the motivation for development was to increase the fidelity of the original Gensim. The authors

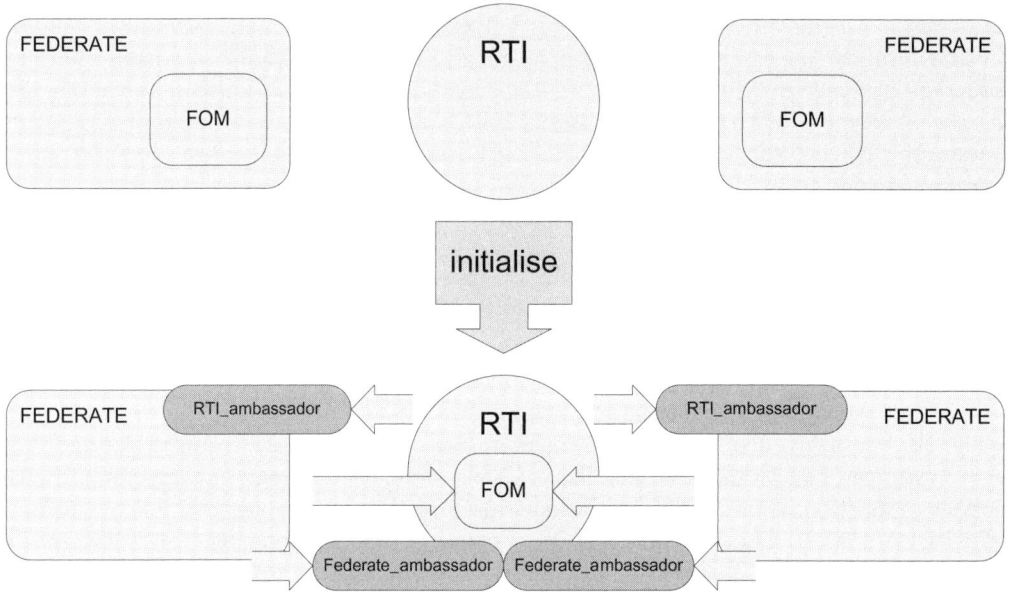

FIGURE 3.7 The 'Ambassador' abstraction used by the HLA.

also point out other motivations for distribution: the gain in computational power; greater experimental control; and reduced timesharing overhead. The major change in DGensim is concerned with physical organization and the execution layer. DGensim divides the simulation onto n node processors, $n-1$ of which execute internals of agents and an *agent manager*. The remaining processor executes the *environment manager*. When the simulation starts each agent process starts its agent manager which connects to the environment manager via a port. When the environment manager is started it contacts the agent managers and sends agent code to them. The agent manager is also sent limited environmental information from the environment manager.

In the original Gensim, changes made by agents are processed on an agent-by-agent basis cyclically. This has the undesired effect that the results of certain agents actions become apparent before others in any one cycle. Agents have local views of their environment which are maintained by *perception agents*. Each perception agent registers with the environment manager stating how frequently it requires sensory information. The local DGensim agent then senses its environment through its own perception agent. In DGensim agents send their decisions (which are timestamped) asynchronously to an action monitoring agent inside the environment manager. It is the job of the action monitoring agent to re-order the incoming decisions using the associated timestamps. Although agents make their decisions asynchronously in DGensim the environment manager is a time-driven simulation. While agents can make decisions at any point, the result of the decision will not be processed by the environment until the environment's simulation time reaches the timestamp associated with that particular decision.

The model in DGensim is susceptible to network delays, it is possible that an agent decision might be delayed enough to affect the sensory actions in the future, i.e., network delay could break the local causality constraint. Small delays are accounted for by a window of safety, which is the time step used by the environment manager. That is if an agent decision is processed and a slight delay occurs, as long as it is received within the same time step the action monitoring agent will rearrange the action into the correct order. If the delay is sufficiently long that the action is received in a future time step then the solution is slightly more complex. DGensim offers various alternatives in this case, firstly an action can be invalidated as though it never occurred. The second option is to process the event as though it had occurred when it was received rather than when it was sent. Although neither of these alternatives are ideal the authors prefer this option to using some method of rollback. The third and final option involves the agents regularly transmitting their actions and the environment processing them upon receiving a full set. DGensim also allows each action to have an associated time period, i.e., the number of units of simulation time an action takes to execute. This avoids agents executing too quickly internally and losing sensory information. The authors do point out that the mechanisms DGensim provides for dealing with delay are not ideal. For their own experiments they use a small dedicated network and hence delays are rare and so not an issue.

In DGensim perception occurs at a point between the agent decision making components and the environment. Most of the agents' perception is performed on the agent node by the agents' manager. The agent manager receives a description of objects when sensing. It is up to it to filter the relevant information. This setup is conceptually more appealing and also has the added benefit of spreading the computationally intensive task of perception across multiple machines. However, because the filtering is done on the receiving end the transmission overhead is large. To overcome this agent managers contain some basic views of objects. The environment then conveys information to the agents by sending information regarding particular changes to the state.

HLA_Agent

HLA_AGENT [Lees et al., 2002, 2003, 2004] is an extension of the SIM_AGENT [Sloman and Poli, 1996] toolkit developed at Birmingham University (see Section 3.3.2) which allows SIM_AGENT simulations to be distributed across a network using the HLA (see Section 3.4.2).

In HLA_AGENT each HLA federate corresponds to a single SIM_AGENT process and is responsible both for simulating the local objects forming its own part of the global simulation, and for maintaining *proxy* objects which represent objects of interest being simulated by other federates. The SIM_AGENT toolkit has been adapted in four different areas for use with HLA:

1. Extended SIM_AGENT to hold federate and federation information.
2. Object creation, deletion and attribute updates are transparently forwarded to the RTI.
3. Modified the scheduler so that only local (non-proxy) objects are executed on the local machine. The scheduler now also processes all callbacks created by the RTI calls made in step two (above), i.e., callbacks from object creation, deletion and attribute updates.
4. Added startup and synchronization code which is required to initialize the HLA_AGENT simulation.

HLA_AGENT uses external calls to C functions and a series of FIFO queues written in C to communicate with the RTI. This enables SIM_AGENT to request callbacks when it is able, removing the need to deal with asynchronous callbacks directly. All necessary RTI and federate ambassador calls are wrapped within a C style function, which in effect provides an implementation of the RTI in Pop-11.

SIM_AGENT works in time-stepped cycles, with each agent sensing, thinking and acting at each cycle. With the modifications to SIM_AGENT a cycle of HLA_AGENT now consists of

5 stages. The main simulation loop is outlined below:

1. Wait for synchronization with other federates.
2. For each object or agent in the scheduler list which is not a proxy:
   (a) Run the agent's sensors on each of the objects in the scheduler list. By convention, sensor procedures only access the publicly available data held in the slots of an object, updated in step 5.
   (b) Transfer messages from other agents from the input message buffer into the agent's database.
   (c) Run the agent's rule-system to update the agent's internal database and determine which actions the agent will perform at this cycle (if any). This may update the agent's internal database, e.g., with information about the state of the environment at this cycle or the currently selected action(s) etc.
3. Once all the agents have been run on this cycle, the scheduler processes the message and action queues for each agent, transfers outgoing messages to the input message buffers of the recipient(s) for processing at the next cycle, and runs the actions to update objects in the environment and/or the publicly visible attributes of the agent. This can trigger further calls to the RTI to propagate new values.
4. We then process the object discovery and deletion callbacks for this cycle. For all new objects created by other federates at this cycle we create a proxy. If other federates have deleted objects, we delete our local proxies.
5. Finally, we process the attribute update callbacks for this cycle, and use this information to update the slots of the local objects and proxies simulated at this federate. The updates performed at this stage are not forwarded to the RTI as these would otherwise trigger further callbacks.
6. Repeat.

HLA_AGENT has been tested using the SIM_TILEWORLD testbed on Linux clusters with varying numbers of nodes, a screenshot of a running instance of this system is shown in Figure 3.8. In the SIM_TILEWORLD federation, the environment is usually simulated by a single federate and the agents are distributed in one or more federates over the nodes of the cluster. In [Lees et al., 2004] results show the system obtaining performance increase for lightweight and heavyweight agents. The system achieves best speed up with the CPU intensive (heavyweight) agents. The results also show however that communication overhead becomes a dominating factor with relatively small numbers of nodes.

HLA_RePast

HLA_REPAST is another HLA enabled agent toolkit developed at the University of Birmingham [Minson and Theodoropoulos, 2004]. It is based on the RePast agent toolkit which is extended for HLA compliance. The key difference between HLA_REPAST and HLA_AGENT is that HLA_REPAST is based on a discrete event agent simulation whereas HLA_AGENT is time-stepped. RePast itself was developed for large-scale social simulations of agents and it provides a collection of tools and structures useful for agent simulation. More information on RePast can be found in Section 3.3.2.

HLA_REPAST is implemented as middleware between the sequential RePast executive (a standard heap-based DES engine - see Section 3.3) and the HLA's RTI executive. Each RePast instance now has two event queues, one populated with local events and one with

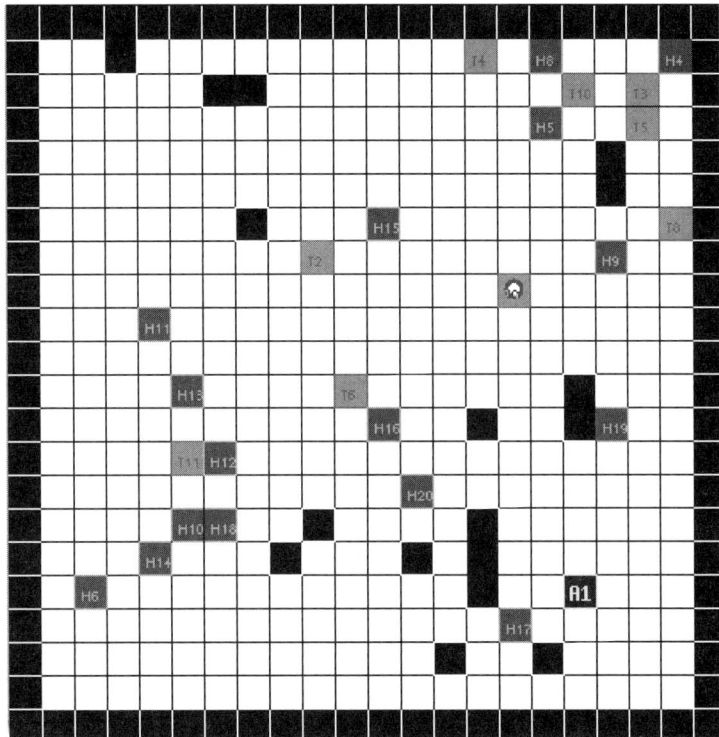

FIGURE 3.8 A screenshot of HLA_AGENT running an instance of the SIM_TILEWORLD testbed simulation.

external events. This is a simplified form of the LP abstraction discussed in Section 3.4. This design is depicted in Figure 3.9.

Complications to this design arise due to the slightly unusual structure of RePast as a DES platform.

RePast has no constraints on the number of state updates which can occur on a single event. It is possible to implement an entire simulation as a single event. While this offers great flexibility, it adds complexity to the HLA integration. The HLA_REPAST system therefore constrains the traditional RePast event system so that it is possible for the system to observe state transitions and propagate these throughout the HLA federation.

To pass state changes between RePast and the RTI HLA_REPAST defines a `PublicObject` and `PublicVariable` class. These classes and their children have wrapped primitives which forward appropriate calls to the RTI. On the receiving side HLA_REPAST processes the HLA callbacks by updating local proxies which model the equivalent variables and objects at the sender. Proxies are created upon receipt of object discovery notification and modeled using the HLA_REPastRemoteObject class. The `PublicVariable` and `PublicObject` classes are also used by HLA_REPAST to automatically generate an appropriate FOM which can be passed to the RTI at startup.

HLA_REPAST uses the notion of exclusive variables for conflict resolution. More specifically any variable defined as exclusive in HLA_REPAST can only be owned (in the HLA sense) by at most one federate per time stamp. The middleware achieves this by only granting one HLA ownership request per timestamp on exclusive variables. Deletion is performed by embedded code within the Java memory reclamation system. However, this relies on the

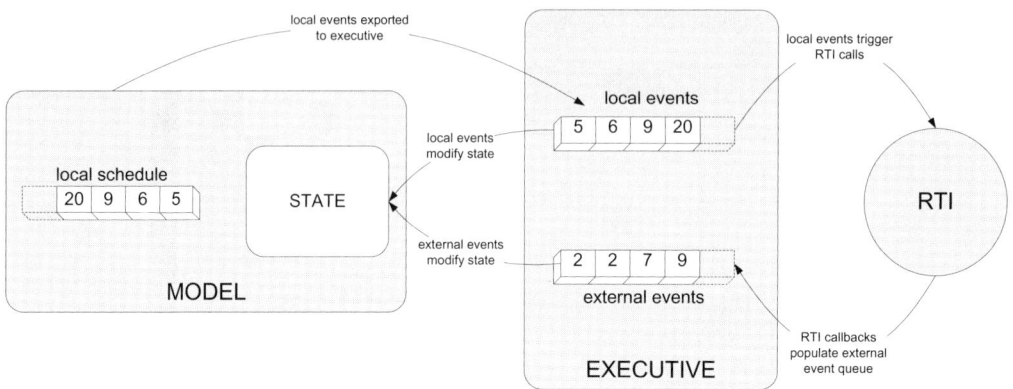

FIGURE 3.9 The structure of a single node in an HLA_RePast Federation.

assumption that the model and Java view of an object's life-cycle are consistent. Development of a more flexible method of deletion is listed as further work.

One of the key goals for HLA_RePast is that the integration be as transparent as possible, so it should be possible with little work to use existing RePast simulations with the HLA_RePast middleware. For this reason HLA_RePast uses a conservative synchronization algorithm as this prevents the need for implementation of rollback algorithms within the user model. To do this RePast implements a new class which inherits from the existing scheduler class, this new scheduler only executes events at time t if it is possible to ensure no future events will arrive with time $< t$. This guarantee is satisfied by using an HLA time advance to time $t - 1$ with a lookahead of one.

The results in [Minson and Theodoropoulos, 2004] are arrived at using a RePast implementation of the Tileworld testbed. The results found in this context are qualitatively similar to those of HLA_AGENT. Namely that, when decomposing the system around its agents, the best scalability can be achieved when an agent is computationally-bound (i.e., does a large amount of 'thinking' or other computation that produces only internal events, in comparison to a small amount of 'acting' or other actions which produce external events).

HLA_RePast has also been used as a testbed application for running distributed simulations in a Grid environment. The HLA_Grid_RePast project [Zhang et al., 2005; Theodoropoulos et al., 2006] ran a federated simulation between sites in Singapore and Birmingham, UK using grid services to provide RTI connectivity.

HLA JADE

In [Wang et al., 2003a] a system is presented which integrates HLA with the JADE (Java Agent DEvelopment) [Bellifemine et al., 1999] agent platform. JADE provides FIPA compatible middleware for agent development and testing. The system is composed of a series of JADE instances running on different machines. Each instance of JADE (which may have multiple agents) interfaces with the RTI through a *gateway federate*. The gateway federate is responsible for translating output from JADE into specific RTI calls. The gateway federate also handles any callbacks received from the RTI through the federate ambassador. Each gateway federate resides on the same JVM (Java Virtual Machine) as the JADE instance, this avoids any extra overhead due to Java RMI calls. Each federate is conservatively synchronized with two timesteps per simulation cycle. Various extensions have been applied to this system. These include methods for interest management[Wang et al., 2003b] to reduce bandwidth-utilization based on the DDM (Data Distribution Management) services of HLA. There are also schemes for conflict resolution[Wang et al., 2004b] and agent communication[Wang et al., 2004a].

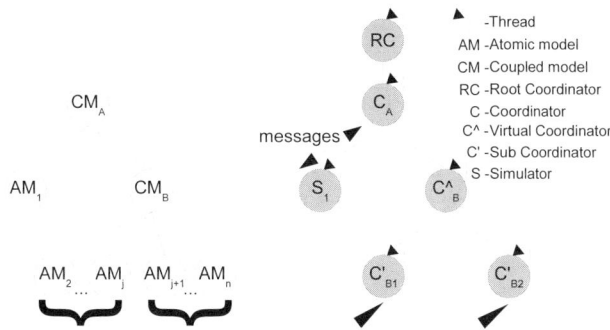

FIGURE 3.10 Mapping from a PDEVS-model structure to a parallel-sequential PDEVS simulator tree (from [Himmelspach et al., 2007]). Sub coordinators and virtual coordinators allow to manage subsets of a coupled model's sub-models within a single thread. While coupled models define the hierarchical structure of the model, agents would be represented by atomic models.

JAMES II

As already mentioned in Section 3.3.2, JAMES II is a Java-based general-purpose simulation system that provides several tools for MAS simulation, e.g., support for the parallel DEVS (PDEVS) modeling formalism. A distributed simulation of a PDEVS model in JAMES II begins with the distribution of models and associated simulators (atomic) or coordinators (coupled), which execute the model, across the nodes (Figure 3.10). Each node within the simulation executes a copy of the JAMES II program. A single node is defined as the master server, which manages all other nodes available for simulation. A client starts a simulation by sending the model to the master server.

The master server determines an appropriate partition of the simulation across the simulation servers [Ewald et al., 2006b]. Then it sends the model, the partition information, and the addresses of the simulation servers to a single simulation server. This single simulation server hosts the root coordinator and propagates sub-partitioning information to simulation servers hosting the corresponding sub-coordinators (or simulators), which in turn will continue to distribute it recursively. Upon completion of the distribution, each child node (starting with the leaves) informs its parent that the distribution is complete. Eventually this is propagated back through to the head node and then on to the master node, which then initiates the execution of the simulation.

The distribution of such PDEVS model trees implies a hierarchical communication scheme between nodes, which simplifies movement of agents between models. However, conservative PDEVS simulation only processes events in parallel if they occur at *exactly* the same time stamp, so usually only some parts of the model can be executed concurrently. Moreover, the synchronization protocol of PDEVS involves a considerable amount of communication, which may lead to a significant slow-down when simulating large sets of heavily interacting agents. To alleviate the former problem, a load balancing scheme that distributes PDEVS-models according to their inherent parallelism has been developed [Ewald et al., 2006c].

For efficient (conservative) simulation of PDEVS, JAMES II provides - among others - a parallel-sequential simulator that can be executed on several nodes while still using a single thread per node and thus saving threading overhead [Himmelspach et al., 2007].

Additional approaches of distributed simulation are available for other formalisms. For example, Beta-Binder models can be simulated in an optimistic manner by an algorithm that employs a coordinator/simulator scheme similar to PDEVS [Leye et al., 2007].

SPADES

SPADES [Riley and Riley, 2003; Riley, 2003]is a conservative parallel discrete event simulation engine designed for modeling Robocup teams. It uses a *Software-in-the-loop* methodology for modeling agent thinking time, which assumes a sense-think-act cycle for agents and that the time it takes an agent to think is non-negligible.

SPADES adopts the PDES paradigm for synchronization, though it does allow some out-of-order event processing. Agent communication in SPADES is done using a communication server on each processor which contains agents (see Figure 3.11). The agents communicate with the communication server via Unix pipes, allowing the agents to be written in any language which supports pipes. The world model is then created by the user code linking to the simulation engine library, resulting in the simulation engine and world model running in the same process. SPADES provides a series of C++ classes which world model objects inherit from to interact with the simulation engine.

From the agent's perspective the interaction with the communication server is a three stage process:

1. Receive sensation
2. Action selection (including sending actions to communication server)
3. Send completion message when all actions are sent

In SPADES agent sensing is done in a 'pushing' manner, in that all updates to the world model are sent via the communication server to all agents. This means an event created by agent A on machine M (e.g., moving an object) will have to be sent to all other communication servers and then on to all other agents regardless of whether the event is relevant. This type of scheme involves a large communication overhead, especially in highly dynamic simulations where the world model is constantly being updated. Agents are only able to act upon receipt of a message (this being analogous to the general assumptions in the original PDES literature in which LPs are considered as fairly deterministic systems which only act in response to some stimulus), therefore agents have a special action *request time notify* which can be sent asynchronously. If an agent wants to send an action independently of a stimulus message it can send a *request time notify* and respond to it appropriately.

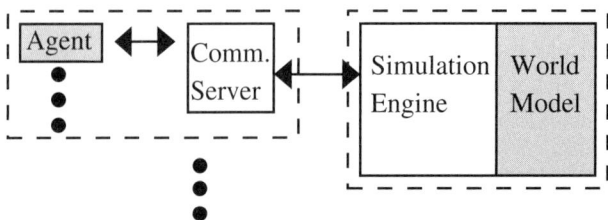

FIGURE 3.11 The basic SPADES architecture (from [Riley and Riley, 2003]). Agents interact with communication servers, which in turn interact with the simulation engine running on a remote machine.

For PDES synchronization, SPADES uses the *centralized event list* approach whereby when an event is scheduled it is sent to a master process. In this centralized approach the master process has global knowledge of the event queues and so can easily determine the minimum virtual time stamp of any future message. This approach has major drawbacks in terms of communication overhead and hence scalability. Having all events sent to a single master process will create a serious communication bottleneck as the number of agents and hence events increases. Riley and Riley do point out that for their purposes (modeling Robocup teams) they have kept the number of agents small and so they haven't noticed any severe overhead.

In experiments with the SPADES engine, an agent simulation is used where each agent is represented as a ball in a two dimensional world. Each sensation the agent receives contains the positions of all the other agents in the world. The only action event the agents can perform is to request a particular velocity vector. The simulation has two types of agent: the wanderers, who move randomly, and the chasers, who set their velocity vector toward nearest wanderer. Using the Ferrari cluster at Georgia Tech, the experiment was performed on 1 to 13 machines, with 2 to 26 agents. To emulate varied processing requirements in the different agents, an internal delay loop of 1.0ms (fast agents) and 9.0ms (slow agents) was used.

With the faster agents, the maximum speed up achieved is about a factor of 1.8 compared to a single machine. This is achieved with 14 agents and 9 machines. It seems with the faster agents the communication overhead starts to dominate at around 5 processors (for 26 agents). With the slower agents speed up is better, achieving a maximum speed up of 5.5 times faster than the single processor case. However, again the communication overhead becomes apparent. With the slower agents at around the 10 processor stage the speed up achieved by adding additional processors seems to tail off. Again these results suggest that simulations involving heavyweight agents which spend a long time 'thinking' compared to 'acting' scale better when parallelized.

Charon

CHARON [Alur, 2008] (see Section 3.3.2) is implemented in sequential form as a timestepped engine. [Hur and Lee, 2002] propose and evaluate two approaches to parallelization of this engine, one using conservative synchronization and one using optimistic. **CHARON** is already a highly modularized system and hence appropriate for parallelisation, with each processor modeling a set of modules. The main issue is in ensuring that the mode transitions within modules still operate with respect to the continuous shared variables and their invariants now that these are split across several processors and continuous steps executed in parallel.

In the conservative implementation, each processor completes a single timeslice in lockstep with the rest of the simulation. Within this slice, a complicated 7-phase barrier process ensures the triggering of discrete transitions is still correct with respect to the continuous behavior of each module. This implementation, the authors report, incurs a communication overhead far in excess of any speedup gained by parallel execution for most models.

In the optimistic implementation, each timeslice is again lock-stepped across all processors. However, in this case each processor computes the triggering of discrete mode transitions by remote modules by optimistically approximating their outputs. If these approximations are always correct the entire system can execute in parallel with only minor computational overhead and only a single barrier operation per-timestep. When an approximation is incorrect, the effected module will need to rollback its processing for that timestep after the barrier is complete and re-compute, possibly triggering a rollback for other connected modules. For the models tested, this approach offered far superior speedups than the conservative approach.

3.4.4 Taxonomy of Parallel DES Toolkits for MAS

Table 3.2 gives a summary of the surveyed parallel DES toolkits.

TABLE 3.2 A Summary of the surveyed PDES toolkits

Toolkit	Support for Conflict Resolution	Synchronization Technique	Automated Partitioning
DGensim	No	Real-time execution with logical time stamps and centralized timeslicing	Yes
HLA_AGENT	Yes	Conservative, via HLA RTI	No
HLA_RePast	Yes	Conservative, via HLA RTI	No
HLA_Jade	Yes	Conservative, via HLA RTI	No
JAMES II	No	Depending on formalism; Conservative for PDEVS	Yes
SPADES	N/A	Conservative, via Centralized Event List (no synchronization necessary)	No
CHARON	N/A	Lockstep, with Conservative and Optimistic implementations for intra-step consistency.	Yes

3.5 Issues for Future Research

3.5.1 Scalability of Parallel Engines

As discussed in Section 3.4, parallel approaches to DES decompose a simulation model in to disjoint subsystems, each of which is encapsulated by an LP: an air traffic control system is decomposed in to a set of airport LPs; a postal network is decomposed in to a set of sorting offices; and so on. Logical interactions between these subsystems are then represented by their LPs explicitly sending and receiving events.

Parallel implementations of MAS simulations are therefore unusual in that they generally partition the population of agents amongst the LPs leaving a large amount of *shared state* (shared resources, communication media, physical space, etc.) which agent LPs both observe (read) and modify (write) concurrently. When they are executed in a parallel environment, this property of MAS simulations leads to multiple engineering challenges which are the subject of current ongoing research.

Shared State and Data Distribution

When data items in a simulation need to be read and written by several LPs concurrently, the mechanisms for making these data available in a distributed setting can be non-trivial. This issue of data distribution - at which nodes are data stored and under what circumstances is communication between nodes necessary? - already has a rich research corpus in the fields of Distributed Memory models, Distributed Virtual Environments and elsewhere.

The parallel engines surveyed in Section 3.4.3 all took the approach of fully replicating all shared state at all simulation nodes. One node is generally responsible for continual maintenance of the environment (eg. for modeling state change over time) but all nodes replicate the current state of the shared data items. In this situation much of the communication spent updating data items is wasted if only a small proportion of the items are accessed by the simulation process at each LP.

Interest Management (IM) is a field of research associated with both Distributed Simulation [Morse, 2000; Tacic and Fujimoto, 1998; Rak et al., 1997] and with Distributed Virtual Environments (DVEs) [Abrams et al., 1998; Barrus et al., 1996; Morgan et al.]. IM seeks to distribute information about data items in a distributed system such that the locality inherent in the access patterns can be exploited to reduce the total communication and computation load of the system. Such an approach is particularly appropriate to MAS simulations in which agents are physically situated and therefore have a limited perceptual range.

The exploitation of this property using IM has been explored by the PDES-MAS project at the Universities of Birmingham and Nottingham, UK [Logan and Theodoropoulos, 2001; Lees et al., 2006; Ewald et al., 2006a; Oguara et al., 2005]. The PDES-MAS architecture assumes two classes of LP, an *Agent Logical Process* (ALP) which simulates the actions of an agent in the environment, and a *Communication Logical Process* (CLP) which maintains some set of the data items representing the shared state of the simulation. Events passed between ALPs and CLPs take the form of individual *reads* and *writes* of individual shared state variables. When a write occurs the value of the variable is updated at the CLP responsible for that variable. When a read occurs, the event travels to the relevant CLP, retrieves the value, and travels back to the issuing ALP.

This paradigm allows the PDES-MAS architecture to reduce irrelevant communication as a result of updates to shared state. However, the CLP now becomes a bottleneck as the single end point for all read and write operations. To ameliorate this problem, a CLP can subdivide assigning both its attached ALPs and the variables it currently maintains to the new child CLPs. The process of subdivision and reassignment is depicted in Figure 3.12.

This process of subdivision and reassignment creates a tree with ALP accesses emanating from the leaf ALPs and traversing the tree to the destination CLPs. In this scenario, the access patterns of an ALP and the distribution of variables to CLPs will drastically effect both the average latency for shared state access and the overall communication burden on the CLP network. The PDES-MAS framework uses heuristic methods for assignment and online migration of variables between CLPs to both achieve load balance across the network and affect interest management.

Similar approaches to online adaptation to access patterns for scalability have been taken in the work of Minson and Theodoropoulos [Minson and Theodoropoulos, 2005]. This framework uses heuristic techniques to choose between two forms of update processing: *push*-processing sends an update message when a variable is written (similar to most toolkits surveyed in Section 3.4.3); *pull*-processing sends a request message when a variable is read (similar to the PDES-MAS architecture). This technique is extended in [Minson and Theodoropoulos, 2007] to support range-query operations similar to those encountered in

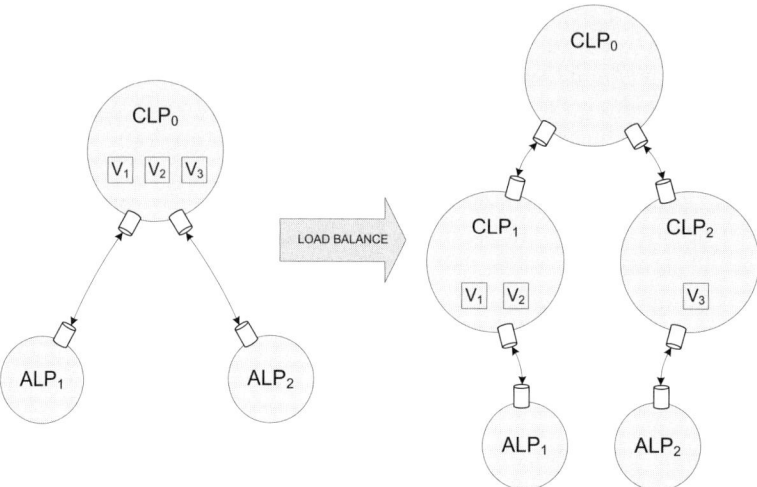

FIGURE 3.12 A CLP in the PDES-MAS architecture becoming overloaded and subdividing in to child CLPs. Both ALPs and the shared state variables it maintains are reassigned between the resulting child CLPs.

MAS simulations or DVEs.

This area of research in to adaptive, scalable data distribution techniques for parallel MAS simulations and similar application areas is the subject of considerable ongoing work.

Shared State and Conflict Resolution

When a simulation model, such as a MAS model, involves the concurrent modification of shared resources, it is necessary to define the semantics by which this concurrency is controlled to produce correct and repeatable results. For instance, in the TileWorld MAS testbed, Tile objects are a shared state resource which can be modified (picked up) concurrently by multiple agents. However, the actual effects carry model-specific semantics (a tile should only be successfully picked up by one of the many agents that attempt to do so).

Since different models may need to represent different semantics for different types of shared resource, it is not possible to define a single policy by which all concurrent modifications can be resolved. The HLA_AGENT project (see Section 3.4.3) defined a special type of *exclusive* variable which constrained the ability to modify a variable to a single LP per-time step. HLA_REPAST extended these semantics further by adding *cumulative* variables which could be concurrently modified by several LPs, but only using modifications relative to the current value, never absolute assignments.

Clearly each new model may require semantics which are not covered by those offered by a given library. Techniques for providing an extensible solution to this problem remain the subject of future research.

3.5.2 Other Areas

Instrumentation and visualization for MAS

To visualize data of a simulation run requires that the data is observed from the simulated model in the first place. Methods of model instrumentation, i.e., the association of observers with certain parts of the model, have already been proposed (e.g., [Dalle and Mrabet, 2007]), but in the context of MAS simulation the problem even aggravates: since agents may be created at any time during the simulation, a MAS model would have to be continuously re-instrumented after changing its state, and the user would need to specify the entities to be observed via generic rules. Furthermore, agents can have complex states, of which only a very small part may be of interest. The user should be able to choose which entities and their sub-states shall be observed. This would decrease the amount of unnecessarily stored result data and hence could speed up simulation significantly.

Symbiotic Simulation for MAS

Another important recent development in the area of simulation is Symbiotic Simulation, a class of (Dynamic Data Driven Application Systems (DDDAS) [Darema, 2005]).

This is a method where data from a system is absorbed into a simulation of the system in order to continually adapt the model to the reality, if necessary making changes to the assumptions on which it is based. The aim of this data-driven adaptation of the model is to gradually increase the reliability of its forecasts. At the same time the states predicted by the simulation can be potentially used to steer the observed system as well as the data selection process.

Symbiotic simulation is increasingly being deployed in simulations of social or socio-technical systems where the presence of MAS models present new challenges. Issues such as the assimilation of qualitative data for cognitively-rich agent models, automated consistency checking and semantic matching of MAS events and states with data are at the heart of research on DDDAS for social-technical simulations. This is the primary aim of the AIMSS project at Birmingham, UK*[Kennedy et al., 2007b,a; Kennedy and Theodoropoulos, 2006a,b, 2005; Darema, 2005].

Symbiotic simulation imposes new requirements on the underlying simulation engine. To support the data-driven adaptation of the model the simulation engine should incorporate mechanisms for the preemption of the simulation and the update of the simulation state and rules. This issue becomes more challenging in the case of parallel MAS simulation engines where model adaptation can lead to synchronization inconsistencies. For model-driven data selection, the simulation engine needs to provide support for the configuration of the data analysis and learning tools.

References

H. Abrams, K. Watsen, and M. Zyda. Three-tiered interest management for large-scale virtual environments. In *VRST '98: Proceedings of the ACM symposium on Virtual reality software and technology*, pages 125–129, New York, NY, USA, 1998. ACM Press. ISBN 1-58113-019-8.

R. Alur. CHARON homepage. http://rtg.cis.upenn.edu/mobies/charon/people.html. Accessed June, 2008., 2008.

J. Anderson. A generic distributed simulation system for intelligent agent design and evaluation. In *Proceedings of AI, Simulation and Planning In High Autonomy Systems*, Tucson, 2000.

J. Barrus, D. Waters, and R. Anderson. Locales and beacons: Efficient and precise support for large, multi-user environments. Technical Report TR-95-16a, MERL: A Mitsubishi Electric Research Laboratory, August 1996.

---

*http://cs.bham.ac.uk/research/projects/aimss

F. Bellifemine, A. Poggi, and G. Rimassa. JADE - a FIPA-compliant agent framework. In *Proceedings of the Practical Applications of Intelligent Agents*, 1999. URL http://jmvidal.cse.sc.edu/library/jade.pdf.

B. Burmeister, A. Haddadi, and G. Matylis. Application of multi-agent systems in traffic and transportation. *Software Engineering. IEE Proceedings*, 144(1):51–60, Feb 1997. ISSN 1364-5080.

K. M. Chandy and J. Misra. Asynchronous distributed simulation via a sequence of parallel computations. *Communications of the ACM*, 24(4):198–206, 1981. ISSN 0001-0782.

L. Dagorn, F. Menczer, P. Bach, and R. J. Olson. Co-evolution of movement behaviours by tropical pelagic predatory fishes in response to prey environment: a simulation model. *Ecological Modelling*, 134(2):325–341, October 2000. ISSN 0304-3800. doi: 10.1016/S0304-3800(00)00374-4. URL http://www.sciencedirect.com/science/article/B6VBS-41C2S1C-F/2/947295f956d36215b479497b6191d55d.

O. Dalle and C. Mrabet. An instrumentation framework for component-based simulations based on the separation of concerns paradigm. In *Proceedings of the 6th EUROSIM Congress on Modelling and Simulation (EUROSIM'2007)*, September 2007.

F. Darema. Grid computing and beyond: The context of dynamic data driven application systems. *IEEE Special Issue on Grid Computing*, pages 692–697, 2005.

J. Doran. *Cooperative Agents: Applications in the Social Sciences*, chapter Can Agent-Based Modelling Really be Useful?, pages 57–81. Springer, 2001.

J. M. Epstein and R. Axtell. *Growing artificial societies: social science from the bottom up*. The Brookings Institution, Washington, DC, USA, 1996. ISBN 0-262-55025-3.

R. Ewald, C. Dan, T. Oguara, G. Theodoropoulos, M. Lees, B. Logan, , T. Oguara, and A. M. Uhrmacher. Performance analysis of shared data access algorithms for distributed simulation of mas. In S. J. Turner and J. Lüthi, editors, *Proceedings of the Twentieth ACM/IEEE/SCS Workshop on Principles of Advanced and Distributed Simulation (PADS 2006)*, pages 29–36, Singapore, May 2006a. IEEE Press.

R. Ewald, J. Himmelspach, and A. M. Uhrmacher. A Non-Fragmenting Partitioning Algorithm for Hierarchical Models. In *Proc. of the 2006 Winter Simulation Conference*, pages 848–855, 2006b.

R. Ewald, J. Himmelspach, A. M. Uhrmacher, D. Chen, and G. K. Theodoropoulos. A Simulation Approach to Facilitate Parallel and Distributed Discrete-Event Simulator Development. In *Proceedings of the 10th IEEE International Symposium on Distributed Simulation and Real Time Applications, DS-RT 2006*, pages 209–218, 2006c.

S. E. Fahlman. A planning system for robot construction tasks. Technical report, Massachusetts Institute of Technology, Cambridge, MA, USA, 1973.

T. Finin, R. Fritzson, D. McKay, and R. McEntire. Kqml as an agent communication language. In *CIKM '94: Proceedings of the third international conference on Information and knowledge management*, pages 456–463, New York, NY, USA, 1994. ACM. ISBN 0-89791-674-3. doi: http://doi.acm.org/10.1145/191246.191322.

G. Fishman. *Principles of Discrete Event Simulation*. Wiley, New York, USA, 1978.

Foundation for Intelligent Physical Agents FIPA. FIPA standard status specifications. http://www.fipa.org/repository/standardspecs.html. Accessed December, 2007.

S. Franklin and A. Graesser. Is it an agent, or just a program?: A taxonomy for autonomous agents. In *Proceedings of the Third International Workshop on Agent Theories, Architectures, and Languages*. Springer-Verlag, 1996. Available at http://www.msci.memphis.edu/~franklin/AgentProg.html. Accessed December, 2007.

M. Gierke, J. Himmelspach, M. Röhl, and A. M. Uhrmacher. Modeling and simulation of tests for agents. In *Multi-Agent System Technologies (MATES'06)*, volume 4196/2006 of *Lecture Notes in Computer Science*, pages 49–60. Springer Berlin / Heidelberg, 2006. URL http://dx.doi.org/10.1007/11872283_5.

N. Gilbert. *Cognition and multi-agent interaction: From cognitive modeling to social simulation*, chapter When does Social Simulation need Cognitive Models, pages 428–432. Cambridge University Press, Cambridge, 2005.

N. Gilbert and S. Bankes. Platforms and methods for agent-based modeling. *Proceedings of the National Academy of Sciences of the USA*, 99(3):7197–7198, May 2002.

L. Gulyás. The MAML technical manual. http://www.maml.hu/maml/technical/index.html. Accessed December, 2007.

L. Gulyás, T. Kozsik, and J. B. Corliss. The multi-agent modelling language and the model design interface. *Journal of Artificial Societies and Social Simulation*, 2 (3), 1999.

N. Hawes, A. Sloman, J. Wyatt, M. Zillich, H. Jacobsson, G.-J. Kruijff, M. Brenner, G. Berginc, and D. Skočaj. Towards an integrated robot with multiple cognitive functions. In *Proceedings of the Conference of the Association for the Advancement of Artificial Intelligence*, pages 1548–1553. AAAI Press, 2007.

J. Himmelspach and A. M. Uhrmacher. A component-based simulation layer for JAMES. In *PADS '04: Proceedings of the eighteenth workshop on Parallel and distributed simulation*, pages 115–122, New York, NY, USA, 2004. ACM Press. doi: http://doi.acm.org/10.1145/1013329.1013349.

J. Himmelspach and A. M. Uhrmacher. Plug'n simulate. In *Proceedings of the 40th Annual Simulation Symposium*, pages 137–143. IEEE Computer Society, 2007. URL http://dx.doi.org/10.1109/ANSS.2007.34.

J. Himmelspach, R. Ewald, S. Leye, and A. M. Uhrmacher. Parallel and distributed simulation of parallel devs models. In *Proceedings of the SpringSim '07, DEVS Integrative M&S Symposium*, pages 249–256. SCS, 2007.

Y. Hur and I. Lee. Distributed simulation of multi-agent hybrid systems. *Object-Oriented Real-Time Distributed Computing, 2002. (ISORC 2002). Proceedings. Fifth IEEE International Symposium on*, pages 356–364, 2002. doi: 10.1109/ISORC.2002.1003781.

IEEE. IEEE 1516 (Standard for Modelling and Simulation High Level Architecture Framework and Rules), 2000.

D. R. Jefferson. Virtual time. *ACM Trans. Program. Lang. Syst.*, 7(3):404–425, 1985. ISSN 0164-0925. doi: http://doi.acm.org/10.1145/3916.3988.

N. R. Jennings, K. Sycara, and M. Wooldridge. A roadmap of agent research and development. *Autonomous Agents and Multi-Agent Systems*, 1(1):7–38, 1998. ISSN 1387-2532. doi: http://dx.doi.org/10.1023/A:1010090405266.

M. John, R. Ewald, and A. M. Uhrmacher. A spatial extension to the [pi] calculus. *Electronic Notes in Theoretical Computer Science*, 194(3):133–148, January 2008. doi: 10.1016/j.entcs.2007.12.010. URL http://portal.acm.org/citation.cfm?id=1342425.1342630.

C. Kennedy and G. Theodoropoulos. Towards intelligent data-driven simulation for policy decision support in the social sciences. Technical Report CSR-05-9, University of Birmingham, School of Computer Science, October 2005.

C. Kennedy and G. Theodoropoulos. Adaptive intelligent modelling for the social sciences: Towards a software architecture. Technical Report CSR-06-11, University of Birmingham, School of Computer Science, October 2006a.

C. Kennedy and G. Theodoropoulos. Intelligent management of data driven simulations to support model building in the social sciences. In *Proceedings of the ICCS, Part III*, number 3993 in Lecture Notes in Computer Science, pages 562–569, Berlin Heidelberg, 2006b. Springer.

C. Kennedy, G. Theodoropoulos, E. Ferrari, P. Lee, and C. Skelcher. Towards an automated approach to dynamic interpretation of simulations. In *Proceedings of the Asia Modelling Symposium 2007*, pages 589–594, March 2007a.

C. Kennedy, G. Theodoropoulos, V. Sorge, E. Ferrari, P. Lee, and C. Skelcher. Aimss: An architecture for data driven simulations in the social sciences. In *Proceedings of the ICCS, Part I*, number 4487 in Lecture Notes in Computer Science, pages 1098–1105, Berlin Heidelberg, 2007b. Springer.

F. Klügl and F. Puppe. The multi-agent simulation environment SeSAm. In *Proceedings of Workshop "Simulation in Knowledge-based Systems"*, April 1998. URL http://citeseer.ist.psu.edu/528152.html.

F. Klügl, R. Herrler, and M. Fehler. SeSAm: implementation of agent-based simulation using visual programming. In *AAMAS '06: Proceedings of the fifth international joint conference on Autonomous agents and multiagent systems*, pages 1439–1440, New York, NY, USA, 2006. ACM. ISBN 1595933034. doi: 10.1145/1160633.1160904. URL http://dx.doi.org/10.1145/1160633.1160904.

M. Lees, B. Logan, and G. Theodoropoulos. Simulating agent based systems with HLA: The case of SIM_AGENT. In *Proceedings of the European Simulation Interoperability Workshop (Euro-SIW'02)*, University of Westminster, United Kingdom, June 2002.

M. Lees, B. Logan, T. Oguara, and G. Theodoropoulos. Simulating agent-based systems with HLA: The case of SIM_AGENT – Part II. In *Proceedings of the 2003 European Simulation Interoperability Workshop*. European Office of Aerospace R&D, Simulation Interoperability Standards Organisation and Society for Computer Simulation International, June 2003.

M. Lees, B. Logan, T. Oguara, and G. Theodoropoulos. HLA_AGENT: Distributed simualtion of agent-based systems with HLA. In *Proceedings of the International Conference on Computational Science (ICCS'04)*, pages 907–915, Krakow, Poland, June 2004. Springer.

M. Lees, B. Logan, C. Dan, T. Oguara, and G. Theodoropoulos. Analysing the performance of optimistic synchronisation algorithms in simulations of multi-agent systems. In S. J. Turner and J. Lüthi, editors, *Proceedings of the Twentieth ACM/IEEE/SCS Workshop on Principles of Advanced and Distributed Simulation (PADS 2006)*, pages 37–44, Singapore, May 2006. IEEE Press.

S. Leye, C. Priami, and A. Uhrmacher. A parallel beta-binders simulator. Technical Report 17/2007, The Microsoft Research - University of Trento Centre for Computational and Systems Biology, 2007.

B. Logan and G. Theodoropoulos. The Distributed Simulation of Multi-Agent Systems. In *Proceedings of the IEEE - Special Issue on Agent-Oriented Software Approaches in Distributed Modelling and Simulation*, pages 174–185. IEEE, February 2001.

E. Mangina. Review of software products for multi-agent systems. http://www.agentlink.org/admin/docs/2002/2002-47.pdf. Accessed December, 2007.

F. Menczer and R. K. Belew. Latent energy environments. Technical Report CS93-301, University of California, San Diego, La Jolla, CA, 1993. URL http://portal.acm.org/citation.cfm?id=249721.

N. Minar, R. Burkhat, C. Langton, and M. Askenazi. The swarm simulation system: A toolkit for building multi-agent simulations. Technical Report 96-06-042, The Santa Fe Institute, Santa Fe, 1996.

R. Minson and G. Theodoropoulos. Distributing RePast agent based simulations with HLA. In *Proceedings of the 2004 European Simulation Interoperability Workshop*, Edinburgh, 2004.

R. Minson and G. Theodoropoulos. An adaptive interest management scheme for distributed virtual environments. In *Proceedings of the Workshop on Principles of Advanced and Distributed Simulation (PADS), 2005*, pages 273–281. IEEE, 2005.

R. Minson and G. Theodoropoulos. Adaptive support of range-queries via push-pull algorithms. In *Proceedings of the workshop on principles of advanced and distributed simulation (PADS)*, pages 53–60, 2007.

G. Morgan, K. Storey, and F. Lu. Expanding spheres: A collision detection algorithm for interest management in networked games. In *Proceedings of the Third International Conference on Entertainment Computing (ICEC '04)*.

K. L. Morse. *An Adaptive, Distributed Algorithm for Interest Management*. PhD thesis, University of California, Irvine, 2000.

M. North, N. Collier, and J. Vos. Experiences creating three implementations of the repast agent modeling toolkit. *ACM Transactions on Modelling and Computer Simulation*, 16(1):1–25, 2006.

T. Oguara, D. Chen, G. Theodoropoulos, B. Logan, and M. Lees. An adaptive load management mechanism for distributed simulation of multi-agent systems. In *Proceedings of the Ninth IEEE International Symposium on Distributed Simulation and Real Time Applications (DS-RT 2005)*, pages 179–186, Montreal, Quebec, Canada, October 2005.

P. Ormerod. Low dimensional models with low cognition agents: Understanding social and economic phenomena. Talk given to the Institute of Advanced Studies Workshop on Modelling Social Behaviour, University of Durham, November 2007.

J. Paul F. Reynolds. A spectrum of options for parallel simulation. In *WSC '88: Proceedings of the 20th conference on Winter simulation*, pages 325–332, New York, NY, USA, 1988. ACM. ISBN 0-911801-42-1. doi: http://doi.acm.org/10.1145/318123.318209.

M. E. Pollack and M. Ringuette. Introducing the tileworld: Experimentally evaluating agent architectures. In *National Conference on Artificial Intellegence*, pages 183–189, 1990.

T. R. Project. RePast: Recursive porous agent simulation toolkit. http://repast.sourceforge.net/index.htm. Accessed December, 2007.

S. F. Railsback, S. L. Lytinen, and S. K. Jackson. Agent-based simulation platforms: Review and development recommendations. *Simulation*, 82(9):609–623, 2006.

S. J. Rak, M. Salisbury, and R. S. MacDonald. HLA/RTI data distribution management in the synthetic theater of war. In *Proceedings of the Fall 1997 DIS Workshop on Simulation*, 1997.

A. Rao and M. Georgeoff. BDI agents: From theory to practice. In *Proceedings of the 1st International Conference on Multi-Agent Systems (ICMAS-95)*, pages 312–319, June 1995.

C. W. Reynolds. Flocks, herds and schools: A distributed behavioural model. In *Proceedings of the 14th Annual on Computer Graphics and Interactive Techniques, pages 25–34. ACM press, 1987.

P. Riley. MPADES: Middleware for parallel agent discrete event simulation. In G. A. Kaminka, P. U. Lima, and R. Rojas, editors, *RoboCup-2002: The Fifth RoboCup Competitions and Conferences*, number 2752 in Lecture Notes in Artificial Intelligence, pages 162–178. Springer, Berlin, 2003.

P. Riley and G. Riley. SPADES — a distributed agent simulation environment with software-in-the-loop execution. In S. Chick, P. J. Sánchez, D. Ferrin, and D. J. Morrice, editors, *Winter Simulation Conference Proceedings*, pages 817–825, 2003.

T. Schelling. Dynamic models of segregation. *Journal of Mathematical Sociology*, 1: 143–186, 1971.

M. Scheutz and B. Logan. Affective vs. deliberative agent control. In *Proceedings of the AISB '01 Symposium on Emotion, Cognition and Affective Computing*, pages 1–10. The Society for the Study of Artificial Intelligence and the Simulation of Behaviour, 2001.

A. Serenko and B. Detlor. *Agent Toolkits: A General Overview of the Market and an Assessment of Instructor Satisfaction with Utilizing Toolkits in the Classroom (Working Paper 455)*. McMaster University, 2002.

A. Sloman and B. Logan. Building cognitively rich agents using the sim_agent toolkit. *Communications of the ACM*, 42(3):71–ff., 1999. ISSN 0001-0782. doi: http://doi.acm.org/10.1145/295685.295704.

A. Sloman and R. Poli. SIM_AGENT: A toolkit for exploring agent designs. *Intelligent Agents Vol II (ATAL-95)*, pages 392–407, 1996.

R. Sun. Cognitive science meets multi-agent systems: A prolegomenon. *Philosophical Psychology*, 14(1):5–28, 2001.

N. Taatgen, C. Lebiere, and J. Anderson. Modeling paradigms in act-r. In R. Sun, editor, *Cognition and Multi-Agent Interaction: From Cognitive Modelling to Social Simulation*, pages 29–52. Cambridge University Press, 2005.

I. Tacic and R. M. Fujimoto. Synchronized data distribution management in distributed simulations. In *PADS '98: Proceedings of the twelfth workshop on Parallel and distributed simulation*, pages 108–115, Washington, DC, USA, 1998. IEEE Computer Society. ISBN 0-8186-8457-7.

The Swarm Development Group. The swarm platform. http://www.swarm.org Accessed December, 2007.

G. Theodoropoulos, Y. Zhang, D. Chen, R. Minson, S. J. Turner, W. Cai, Y. Xie, and B. Logan. Large scale distributed simulation on the grid. In *CCGRID 2006 Workshop: International Workshop on Distributed Simulation on the Grid*. IEEE, May 2006.

R. Tobias and C. Hofmann. Evaluation of free java-libraries for social-scientific agent based simulation. *Journal of Artificial Societies and Social Simulation*, 7(1), 2004.

A. M. Uhrmacher, R. Ewald, M. John, C. Maus, M. Jeschke, and S. Biermann. Combining micro and macro-modeling in devs for computational biology. In *Winter Simulation Conference, 2007*, pages 871–880, 2007. doi: 10.1109/WSC.2007.4419683. URL http://dx.doi.org/10.1109/WSC.2007.4419683.

US Defence Modelling and Simulation Office. HLA Interface Specification, version 1.3, 1998.

F. Wang, S. Turner, and L. Wang. Integrating agents into hla-based distributed virtual environments. In *4th Workshop on Agent-Based Simulation*, pages 9–14, Montpellier, France, 2003a.

F. Wang, S. J. Turner, and L. Wang. Agent communication in distributed simulations. In *In Proceedings of the Joint Workshop on Multi-Agent and Multi-Agent-Based Simulation*, pages 11–24, 2004a.

L. Wang, S. J. Turner, and F. Wang. Interest management in agent-based distributed simulations. In *Seventh IEEE International Symposium on Distributed Simulation and Real-Time Applications (DS-RT03)*, pages 20–27, Delft, Netherlands, 2003b.

L. Wang, S. J. Turner, and F. Wang. Resolving mutually exclusive interactions in agent based distributed simulations. In *Winter Simulation Conference*, pages 783–791, 2004b.

R. E. Wray and R. M. Jones. An introduction to soar as an agent architecture. In R. Sun, editor, *Cognition and Multi-Agent Interaction: From Cognitive Modelling to Social Simulation*, pages 53–78. Cambridge University Press, 2005.

B. P. Zeigler, T. G. Kim, and H. Praehofer. *Theory of Modeling and Simulation*. Academic Press, Inc., Orlando, FL, USA, 2000. ISBN 0127784551.

Y. Zhang, G. Thedoropoulos, R. Minson, S. J. Turner, W. Cai, Y. Xie, and B. Logan. Grid-aware large scale distributed simulation of agent-based systems. In *European Simulation Interoperability Workshop (EuroSIW 2005)*, June 2005.

II

Simulation for MAS

4 Polyagents: Simulation for Supporting Agents' Decision Making 109
H. Van Dyke Parunak and Sven A. Brueckner
Introduction • The Polyagent Model • Some Applications of Polyagents • Future Research • Conclusion

5 Combining Simulation and Formal Tools for Developing Self-Organizing MAS 133
Luca Gardelli, Mirko Viroli, and Andrea Omicini
Introduction • The A&A Meta-Model for Self-Organizing Systems • Methodological Issues Raised by Self-Organizing Systems • A Methodological Approach for Engineering Self-Organizing MAS • Using the PRISM Tool to Support the Method • Case Study: Plain Diffusion • Conclusion

6 On the Role of Software Architecture for Simulating Multi-Agent Systems 167
Alexander Helleboogh, Danny Weyns, and Tom Holvoet
Introduction • Background • AGV Transportation System • Modeling Multi-Agent Control Applications in Dynamic Environments • Architecture of the Simulation Platform • Evaluating the AGV Simulator • Related Work • Conclusions and Future Work

7 Replicator Dynamics in Discrete and Continuous Strategy Spaces 215
Karl Tuyls and Ronald Westra
Introduction • Elementary Concepts from Game Theory • Replicator Dynamics • Evolutionary Dynamics in Discrete Strategy Spaces • Evolutionary Dynamics in Continuous Strategy Spaces • Example of the Resulting Dynamics of the Continuous Replicator Equations • Conclusions

8 Stigmergic Cues and Their Uses in Coordination: An Evolutionary Approach 243
Luca Tummolini, Marco Mirolli, and Cristiano Castelfranchi
Introduction • Stigmergy: Widening the Notion but Not Too Much • Two Uses of Stigmergy in Coordination • Stigmergy in Cooperation and Competition • Why Pheromonal Communication Is Not Stigmergic • Understanding Stigmergy through Evolution • Future Work • Conclusion • Acknowledgments

A typical characteristic of MAS is an inherent distribution of resources. Agents have only partial access to the environment. System-wide properties result from the local actions of the agents and their interactions. Because of the distributed nature of MAS, testing by simulation becomes imperative in the software development process of MAS. Software agents can also use agent-based simulation to guide their decisions. This way, agents run a model of the world to see what might happen under various decision alternatives and make the decision that leads to the most desirable outcome. This part presents five chapters that show different uses of simulation for MAS.

In "Polyagents: Simulation for Supporting Agents' Decision Making", Parunak and Brueckner, introduce a construct for multi-agent modeling that encapsulates a technique of simulation-based decision-making. A polyagent represents an entity in the domain. The polyagent generates a stream of transient ghosts to explore various issues of interest to the entity. These ghosts can be applied to explore alternative futures, evaluate plan structures, compare the usefulness of alternative options, etc. Each ghost explores a possible trajectory for the entity, which then chooses its behavior based on the experiences of its ghosts.

In "Combining Simulation and Formal Tools for the Development of Self-Organizing Multi-Agent Systems", Luca Gardelli, Mirko Viroli, and Andrea Omicini present a systematic use of simulation and formal tools in the early design phase of the development of self-organizing systems. The authors propose an iterative process consisting of modeling, simulation, verification, and tuning. Due to the state explosion problem inherent to model checking, the approach is currently limited to verifying only small instances of systems. However, the authors show that use of model checking techniques is very useful in the first iterations on small instances of the problem.

Common knowledge in simulation platforms is typically reified in reusable code, libraries, and software frameworks. In "On the Role of Software Architecture for Simulating Multi-Agent Systems", Alexander Helleboogh, Tom Holvoet, and Danny Weyns put forward software architecture in addition to such code libraries and software frameworks. Software architecture captures the essence of a simulation platform in an artifact that amplifies reuse beyond traditional code libraries and software frameworks. It supports consolidating and sharing expertise in the domain of multi-agent simulation in a form that has proven its value for software development.

In "Replicator Dynamics in Discrete and Continuous Strategy Spaces", Karl Tuyls and Ronald Westra study multi-agent evolutionary dynamics from a game theoretic perspective. The authors simulate and analyze the properties and asymptotic behavior of multi-agent games with discrete and continuous strategy spaces. To this end they use existing models of the replicator dynamics and introduce a new replicator dynamics model for continuous strategy spaces. Experiments show that the new model outperforms existing models in a simple game.

Finally, in "Stigmergic Cues and Their Uses in Coordination: an Evolutionary Approach", Luca Tummolini, Marco Mirolli, and Cristiano Castelfranchi, explore the evolution of stigmergic behavior adopting a simulative approach. The authors simulate a population of artificial agents living in a virtual environment containing safe and poisonous fruits. The behavior of the agents is governed by artificial neural networks whose free parameters (i.e. the weights of the networks' connections) are encoded in the genome of the agents and evolve through a genetic algorithm. By making the transition from a MAS in which agents individually look for resources to one in which each agent indirectly coordinates with what the other agents do (stigmergic self-adjustment) and, finally, to a situation in which each agent sends a message about what kind of resources are available (stigmergic communication), the simulations offer a precise analysis of the difference between traces that are signs with a behavioral content and traces that are signals with a behavioral message.

4

Polyagents: Simulation for Supporting Agents' Decision Making

	4.1	Introduction	109
	4.2	The Polyagent Model	111
		A Challenge for Simulation-Based Decision Making • Two Big Ideas • The Architecture • The Environment • Related Work	
	4.3	Some Applications of Polyagents	118
		Factory Scheduling and Control • Vehicle Routing • Prediction	
	4.4	Future Research	124
		Theoretical Opportunities • New Applications	
	4.5	Conclusion	127
		References	128

H. Van Dyke Parunak
NewVectors, Division of TechTeam Government Solutions, Inc.

Sven A. Brueckner
NewVectors, Division of TechTeam Government Solutions, Inc.

4.1 Introduction

People often support decisions by thinking through the consequences of alternative courses of action-in effect, by simulating them mentally. In fact, one of the major uses of simulations is as decision-support aids.

Software agents can also use simulation to guide their decisions. That is, they may run a model of the world to see what might happen under various decision alternatives, then make the decision that leads to the most desirable outcome. Such simulation can be either equation-based or agent-based. Our focus here is on agent-based simulation,[*] in which software agents representing the domain entities are situated in a representation of the environment and interact with the environment and with one another, acting out a possible trajectory of the future. The agent responsible for making a decision activates an agent-based simulation of the domain, and chooses an action based on the evolution of that simulation.

This approach to decision-making is complementary to a number of other decision mechanisms. By way of contrast, consider several such mechanisms that are popular in research on AI and autonomous agents.

---

[*]Elsewhere [Parunak et al., 2006b] we discuss the benefits of agent-based modeling over simulation-based modeling; more recently, see [Shnerb et al., 2000].

Derivational methods, such as rule-based reasoning and theorem proving, derive consistent conclusions from axioms and inference rules such as modus ponens or modus tolens. These methods require translating the domain into a set of well-formed symbolic expressions with truth values that are amenable to formal manipulation. This translation process, an aspect of knowledge engineering, can be time-consuming and error-prone.

Constraint-based methods also work with sets of statements about the domain, but in this case the emphasis is on finding values for variables in these statements that satisfy certain conditions, rather than on deriving new statements from them. Constraint-based methods fall into two broad categories. Constraint satisfaction seeks a set of assignments to the variables that makes all of the statements true, while constraint optimization associates a cost function with conflicts among statements and seeks a set of assignments that minimizes the overall cost.

Both derivational and constraint-based methods favor a static world, in which the truth values of the statements on which they rely can be assumed to remain constant. By contrast, **game theoretic** methods explicitly model the interaction between two adversarial reasoners. They evaluate the distinct strategy choices on each side, but require enumeration in advance of all possible choices, and make two assumptions about the adversaries that may not be true: that the adversaries know the payoffs of the various options, and that they make their decisions rationally. Iterated game-theoretic methods can be viewed as a version of simulation, but one that focuses on the players' strategies and minimizes the impact of the environment.

These approaches are valuable tools in the decision-maker's kit. But they have certain fundamental limitations that agent-based simulation can avoid.

1. **Spatial irreducibility** recognizes the intractability of logical analysis of spatial constraints on a domain-independent, purely qualitative representation of space or shape [Forbus et al., 1987, 1991]. While sophisticated representations of spatio-temporal information are available (e.g., [Parent et al., 1999; Tryfona et al., 2003]), relevant reasoning about geometrically constrained problems still requires direct measurements on a spatial model. Such reasoning is much more natural using a simulation whose environment embeds the required topological structure.

2. **Process irreducibility** [Hopcroft and Ullman, 1979] means that for systems beyond a certain (very low [Wolfram, 2002]) level of complexity, direct emulation is the most efficient approach to predicting their evolution. Formal logical systems are rife with intractability barriers [Garey and Johnson, 1979] that require either restricting their application to small problems or weakening the expressiveness of the underlying logical formalism. Simulation works with an iconic representation of the domain rather than a symbolic one, and avoids the formal logical operations that lead to intractability.

3. **Dynamic uncertainty** means that nonlinear interactions among domain entities can generate uncertainty, even if the states and configuration of the entities are precisely known. Two successive runs of the same system may not yield the same outcome. The problem is not external noise or nondeterminism, but the sensitive dependence of an iterated nonlinear system on initial conditions. Classical reasoning methods are either deterministic or propagate the uncertainty in the inputs through to the outputs. Simulation can emulate the generation of dynamic uncertainty and estimate its impact, if it includes an explicit model of how the environment actively integrates the actions of various entities [Michel, 2004].

These issues make simulation a valuable tool for decision-making. The agent facing a decision runs a simulation of the domain, and consults its results to inform the choice that it makes.

This chapter describes a particular construct for multi-agent modeling, the polyagent, that encapsulates the technique of simulation-based decision-making. The construct is highly imitative, drawing from concepts from a number of disciplines, including computer science, biology, and physics. In the next section, we define the polyagent. In Section 4.3, we describe its use in several application domains. In Section 4.4, we outline directions for further research, and conclude in Section 4.5, summarizing the multidisciplinary inspirations for the polyagent.

4.2 The Polyagent Model

In spite of the benefits of simulation as a decision tool, it faces a significant challenge. This section articulates this challenge, motivates two basic principles involved in the construct, then outlines in detail the polyagent architecture and the environment in which it operates, and discusses related research.

4.2.1 A Challenge for Simulation-Based Decision Making

Simulation faces an important challenge compared with more traditional decision mechanisms. The logical formalisms on which those mechanisms are based impart a certain generality to their conclusions. A run of a simulation is just that, a single run, with no way to generalize it. Particularly if the simulation permits the generation of dynamic uncertainty, multiple runs must be made to sample the possible outcomes of the scenario. The necessary number of runs is much higher than one might initially think.

Imagine $n + 1$ entities in discrete time. At each step, each entity interacts with one of the other n. Thus at time t its interaction history $h(t)$ is a string in n^t. Its behavior is a function of $h(t)$. This toy model generalizes many domains, including predator-prey systems, combat, innovation, diffusion of ideas, and disease propagation.

It would be convenient if a few runs of such a system told us all we need to know, but this is not likely to be the case, for three reasons.

1. We may have imperfect knowledge of the agents' internal states or details of the environment (for example, in a predator-prey system, the carrying capacity of the environment). If we change our assumptions about these unknown details, we can expect the agents' behaviors to change.
2. The agents may behave non-deterministically, either because of noise in their perceptions, or because they use a stochastic decision algorithm.
3. Even if the agents' reasoning and interactions are deterministic and we have accurate knowledge of all state variables, nonlinearities in decision mechanisms or interactions can result in overall dynamics that are formally chaotic, so that tiny differences in individual state variables can lead to arbitrarily large divergences in agent behavior. A nonlinearity can be as simple as a predator's hunger threshold for eating a prey or a prey's energy threshold for mating. This process is responsible for the generation of dynamic uncertainty.

An equation-based model typically deals with aggregate observables across the population. In the predator-prey example, such observables might be predator population, prey population, average predator energy level, or average prey energy level, all as functions of time. No attempt is made to model the trajectory of an individual entity. This aggregation can leads to serious errors in the results of such a model [Shnerb et al., 2000].

An ABM explicitly describes the trajectory of each agent. In a given run of a predator-prey model (for example), depending on the random number generator, predator 23 and prey 14 may or may not meet at time 354. If they do meet and predator 23 eats prey 14, predator 52 cannot later encounter prey 14, but if they do not meet, predator 52 and prey 14 might meet later. If predator 23 happens to meet prey 21 immediately after eating prey 14, it will not be hungry, and so will not eat prey 21, but if it did not first encounter prey 14, it will consume prey 21. And so forth. A single run of the model can capture only one set of many possible interactions among the agents.

In our general model, during a run of length τ, each entity will experience one of n^τ possible histories. (This estimate is of course worst case, since domain constraints may make many of these histories inaccessible.) The population of $n + 1$ entities will sample $n + 1$ of these possible histories. It is often the case that the length of a run is orders of magnitude larger than the number of modeled entities ($\tau >> n$).

Multiple runs with different random seeds offer only a partial solution. Each run only samples one set of possible interactions. For large populations and scenarios that permit multiple interactions on the part of each agent, the number of runs needed to sample the possible alternative interactions thoroughly can quickly become prohibitive. In the application described in Section 4.3.3, $n \sim 50$ and $\tau \sim 10{,}000$, so the sample of the space of possible entity histories actually sampled by a single run is vanishingly small. We would need on the order of τ runs to generate a meaningful sample, and executing that many runs is out of the question.

We need a way to capture the outcome of multiple possible interactions among agents in a few runs of a system. Polyagents are one solution to this problem. In essence, each agent uses a lower-level multi-agent system representing itself to explore alternative futures in guiding its decisions.

4.2.2 Two Big Ideas

The next few sections explain the polyagent modeling construct. To help motivate it, we begin by introducing two concepts that it embodies: environmentally-mediated interactions, and continuous short-length prediction.

Environmentally-Mediated Interactions

Because of dynamic uncertainty, we need to explore many possible futures. Computational expense is a major obstacle to such exploration with conventional agent-based models, and much of this expense is due to the cost of computing direct agent-to-agent interactions.

Social insects achieve complex coordination tasks with very limited processing resources, by means of stigmergy [Grassé, 1959] (coordination through a shared environment rather than by direct interaction). Many insect species deposit and sense chemicals, known as pheromones, in their environment. A species might have a vocabulary of several dozen such chemicals. The strengths of the resulting fields reflect the frequency with which individuals that deposit them have been at the deposit location, and are sufficient to generate a wide range of cooperative behaviors among the organisms, including generation of minimal path networks and construction of complex three-dimensional nests [Parunak, 1997].

Insects construct their pheromone fields using chemicals in a physical environment. We use a digital pheromone field, consisting of scalar variables localized in a structured digital environment. This field reflects the likelihood that the class of agent that deposits that

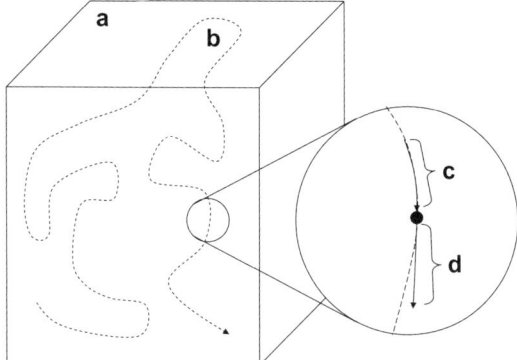

FIGURE 4.1 Tracking a Nonlinear Dynamical System. a = system state space; b = system trajectory over time; c = recent measurements of system state; d = short-range prediction.

particular flavor is present at a given space-time location. Each agent's field is generated by a swarm of representatives, or "ghosts," that build up the field by their movements and make their own decisions on the basis of the fields generated by other agents. Thus agents' decisions take into account a large number of possible interactions in a single run of the system.

Continuous Short-Range Prediction

Pierre-Simon Laplace was confident that to an observer with enough information about the present and sufficient computing capability, no detail of the future could remain hidden [Laplace, 1820]. His optimism foundered on the sensitivity of nonlinear processes to initial conditions. Nonlinearities in the dynamics of most realistic systems drive the exponential divergence of trajectories originating close to one another.

In many applications, we can replace a single long-range prediction with a series of short-range ones. The difference is between driving a car in the daytime and at night. In the daytime, one may be able to see several miles down the road. At night, one can only see as far as the headlamps shine, but since the headlamps move with the car, at any moment the driver has the information needed to make the next round of decisions.

In physical systems, one typically describes the systems with vector differential equations e.g.,

$$\frac{d\vec{x}}{dt} = f(\vec{x}).$$

At each moment, we fit a convenient functional form for f to the system's trajectory in the recent past, and then extrapolate this fit (Figure 4.1 [Kantz and Schreiber, 1997]). Constant repetition of this process provides a limited look-ahead into the future. The process can be applied in reverse as well, allowing us to project from a series of current observations into the past to recover likely historical antecedents of the current state. This program is straightforward with a system described numerically. The architecture described in the next few sections applies it to agent behavior.

4.2.3 The Architecture

Each polyagent represents a single domain entity. It consists of a single *avatar* that manages the correspondence between the domain and the polyagent, and a swarm of *ghosts* that explore alternative behaviors of the domain entity. Sometimes it is useful to consider the swarm of ghosts in their own right, as a delegate MAS. Let's discuss each of these concepts in more detail.

Avatar

The avatar corresponds to the agent representing an entity in a conventional multi-agent model of the domain. It persists as long as its entity is active, and maintains state information reflecting its entity's state. Its computational mechanisms may range from simple stigmergic coordination (coordination that is mediated by a shared environment) to sophisticated BDI reasoning. A typical polyagent model incorporates multiple polyagents, and thus multiple avatars.

The avatar observes its ghosts to decide on its own actions. Depending on the application, it may simply climb the aggregate pheromone gradient laid down by its ghosts, or evaluate individual ghost trajectories to select those that maximize some decision criterion. For example, in a military application, it may select the trajectory offering the least risk, or the greatest likelihood of success. (The two are often not the same!)

Avatars may also deposit digital pheromones directly. For example, in a military application, an avatar modeling a target will emit a target pheromone that attracts the ghosts of units seeking to attack that target, even if the target avatar does not use ghosts to plan its own movements (a situation that occurs when the target is stationary).

Ghosts

Each avatar generates a stream of ghost agents, or simply ghosts. Ghosts typically have limited lifetime, dying off after a fixed period of time or after some defined event to make room for more ghosts. The avatar controls the rate of generation of its ghosts, and typically has several ghosts concurrently active.

Ghosts explore alternative possible behaviors for their avatar and generate a digital pheromone field recording those possible behaviors for reference by other agents. The field is a function of both location and time, increasing as ghosts make their deposits and decreasing through a constant background evaporation that removes obsolete information. Each ghost chooses its actions stochastically based on a weighted (not necessarily linear) "combining function" of the strengths of the various pheromone flavors in its immediate vicinity, and deposits its own pheromone to record its presence. A ghost's "program" consists of its combining function, the vector of weights defining its sensitivity to various pheromone flavors, and any other parameters in its combining function. The ghost simply climbs the gradient defined by the output of the combining function.

Having multiple ghosts multiplies the number of interactions that a single run of the system can explore. Instead of one trajectory for each avatar, we now have one trajectory for each ghost. If each avatar has k concurrent ghosts, we explore k trajectories concurrently. But the multiplication is in fact greater than this.

The digital pheromone field supports three functions [Brueckner, 2000; Parunak, 2003]:

1. It *aggregates* deposits from individual agents, fusing information across multiple agents and through time. In some of our implementations of polyagents, avatars deposit pheromone; in other, ghosts do. Aggregation of pheromones enables a single ghost to interact with multiple other ghosts at the same time. It does not interact with them directly, but only with the pheromone field that they generate, which is a summary of their individual behaviors.

2. It *evaporates* pheromones over time. This dynamic is an innovative alternative to traditional truth maintenance in artificial intelligence. Traditionally, knowl-

edge bases remember everything they are told unless they have a reason to forget something, and expend large amounts of computation in the NP-complete problem of reviewing their holdings to detect inconsistencies that result from changes in the domain being modeled. Ants immediately begin to forget everything they learn, unless it is continually reinforced. Thus inconsistencies automatically remove themselves within a known period.

3. It propagates pheromones to nearby places, disseminating information.

This third dynamic (propagation) enables each ghost to sense multiple other agents. If n avatars deposit pheromones, each ghost's actions are influenced by up to n other agents (depending on the propagation radius), so that we are exploring in effect $n * k$ interactions for each entity, or $n^2 * k$ interactions overall. If individual ghosts deposit pheromones, the number of interactions being explored is even greater, on the order of k^n. Of course, the interactions are not played out in the detail they would be in a conventional multi-agent model. But our empirical experience is that they are reflected with a fidelity that is entirely adequate for the problems we have addressed.

Pheromone-based interaction not only multiplies the number of interactions that we are exploring, but also enables extremely efficient execution. In one application, we support 24,000 ghosts concurrently, faster than real time, on a 1 GHz Wintel laptop.

The avatar can do several things with its ghosts, depending on the application.

- It can activate its ghosts when it wants to explore alternative possible futures, modulating the rate at which it issues new ghosts to determine the number of alternatives it explores. It initializes the ghosts' weight vectors to define the breadth of alternatives it wishes to explore.

- It can evolve its ghosts to learn the best parameters for a given situation. It monitors the performance of past ghosts against some fitness parameter, and then breeds the most successful to determine the parameters of the next generation.

- It can review the behavior of its swarm of ghosts to produce a unified estimate of how its own behavior is likely to evolve and what the range of likely variability is.

Delegate MAS

The avatar-ghost relationship encapsulates the notion of an agent's using simulation to inform its decisions. An avatar's ghosts are in fact conducting a simulation of the domain to inform the avatar's next decision. The swarm of ghosts is a multi-agent system in its own right. Because it performs a service for its avatar, it can be described as a "delegate MAS" [Holvoet and Valckenaers, 2006a,b]. Figure 4.2 summarizes the unification of the polyagent and delegate MAS model, discussed in more detail elsewhere [Parunak et al., 2007c].

4.2.4 The Environment

The stigmergic coordination among ghosts requires an environment within which they have a well-defined location at any time. The importance of the environment is not limited to swarming agents. Even more complex agents are limited by the active role that the environment plays in integrating their actions, and inattention to this detail can lead to erroneous results [Michel, 2004]. In addition, the topology of the environment enables agents to limit their interactions to a subset of other agents, and thus to avoid the computationally explosive task of interacting with all other agents.

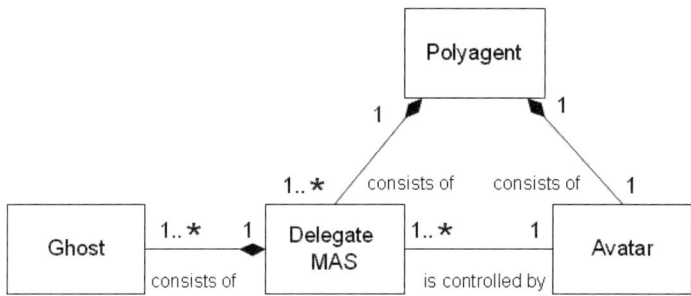

FIGURE 4.2 Each Polyagent is the combination of an Avatar and one or more delegate MAS. Each delegate MAS may render a specific service for the Avatar, and the Avatar may use a combination of delegate MAS to handle a single one of its concerns.

The simplest environment for stigmergic agents is a manifold, such as the surface of the earth. This is the environment in which the biological antecedents of stigmergy first developed. It offers a number of benefits, including uniformity in the structure of the topology seen locally by agents (typically, a lattice), and the existence of a well-defined metric.

Manifolds are natural environments for polyagents, but by no means the only feasible topologies. The factory scheduling applications discussed in Section 4.3.1 use a process graph, in which the nodes are process steps and the links indicate the order in which they may be performed. In such a structure, different nodes may have different numbers of successors, so that an agent's local view of the topology changes from one node to the next. In addition, there may be no well-defined metric on such a graph. In spite of these shortcomings, our experience shows that polyagents can function effectively on such a structure.

In our current implementation, the environment consists of a "book" of pheromone maps, one for each time step of interest. Thus the strength of the pheromone from an avatar's ghosts at a given location on a given page reflects the collective estimate of the avatar's ghosts that the avatar will be at that location at that time. In some applications, this period includes the recent past as well as the future through which prediction is desired. As ghosts move ahead in time, they advance from one page to the next. All agents share a single book, since the pheromone maps are the only means available for ghosts of different agents to interact with one another.

4.2.5 Related Work

Our polyagent bears comparison with several previous multi-agent paradigms, two varieties of enhanced simulation, and three other uses of the term "polyagent."

Polyagents and Multi-Agent Paradigms

Polyagents are distinct from the common use of agents to model different functions of a single domain entity. For example, in ARCHON [Wittig, 1992], the domain entity is an electrical power distribution system, and individual agents represent different functions or perspectives required to manage the system. In a polyagent, each ghost has the same function: to explore one possible behavior of the domain entity. The plurality of ghosts provides, not functional decomposition, but a range of estimates of alternative behaviors.

Many forms of evolutionary computation [Jacob, 2001] allow multiple representatives of a single entity to execute concurrently, to compare their fitness. In these systems, each agent samples only one possible series of interactions with other entities. Pheromone-based coordination in the polyagent construct permits each ghost to adjust its behavior based on many possible alternative behaviors of other entities in the domain.

Similarly, the multiple behaviors contemplated in fictitious play [Lambert et al., 2005] take place against a static model of the rest of the world.

Like the polyagent, ant-colony optimization [Dorigo and Stuetzle, 2004] uses pheromones to integrate the experiences of parallel searchers. The polyagent's advance is the notion of the avatar as a single point of contact for the searchers representing a single domain entity.

Polyagents and Simulation

The need to consider multiple possible futures is widely recognized. Polyagents address, in a more systematic fashion, three simulation techniques that have been used by others: repeated stochastic simulation, recursive simulation, and multitrajectory simulation.

Reference [Tolk, 2005] cites an unpublished study at the University of the German Federal Armed Forces showing that repetitions of a stochastic model offer comparable accuracy to a deterministic model. This approach reflects the stochastic nature of ghost reasoning, but each entity in a run still experiences only one possible trajectory of the other entities, rather than the distribution over possible trajectories that the digital pheromone field presents to polyagents.

Recursive simulation [Gilmer and Sullivan, 2000] occurs when one simulation invokes another in support of a local decision. Previous implementations of this idea invoke recursive simulation only in support of some agent decisions, and the lower-level simulation is another instance of the same simulation model that invokes it, with restricted scope. Polyagent avatars base all of their decisions on lower-level simulations, which run in a stigmergic world of reduced dimensionality to enable computational efficiency.

Polyagents compute multiple futures concurrently in a single run. This technique, called multitrajectory simulation, has also been explored by Gilmer and Sullivan [Gilmer and Sullivan, 1998]. Their approach evaluates possible outcomes at each branch point stochastically, selects a few of the most likely alternatives, and propagates them. This selection is required by the high cost of following multiple paths, but avoiding low-probability paths violates the model's ergodicity and compromises accuracy [Gilmer and Sullivan, 2001]. Polyagents avoid this problem, in three ways. 1) They sample the futures it extends from the distribution of all possible futures, and so does not automatically prune low-probability paths. 2) The highly efficient numerical execution of ghosts lets them explore more paths than can qualitative simulators. 3) Newly developed methods for analyzing the effective state space of a polyagent model let us monitor the model's coverage dynamically and correct it if it becomes skewed.

It is important to situate the polyagent construct with regard multimodels [Fishwick et al., 1994; Yilmaz and Ören, 2005], which Fishwick et al. [Fishwick et al., 1994] define as employing "more" than one model, each derived from a *different* perspective, and utilizing correspondingly *distinct* reasoning and simulation strategies (emphases ours). The several ghosts representing a single avatar are identical, other than (possibly) their weight vectors. When the weight vectors vary, one might view the ghosts as having "different perspective(s)," and thus as forming a multimodel. However, they all use the same "reasoning and simulation strategies," and (unlike multimodels) can be deployed with identical weight vectors, in order to explore alternative futures that can arise from trajectory divergence. Multimodels emphasize differences within the models themselves, while polyagents emphasize differences in behavior resulting from nonlinearities in interactions among the agents. This difference corresponds to the broader distinction between systems of multiple BDI agents, where the intelligence resides in the individual agents, and swarming systems, where the intelligence emerges from interactions among the agents.

The Term "Polyagent"

The term "polyagent" is our neologism for a set of software agents that collectively represent a domain entity and its alternative behaviors. The term is used in three other contexts that should not lead to any confusion.

In medicine, "polyagent therapy" uses multiple treatment agents (notably, multiple drugs combined in chemotherapy).

Closer to our domain, but still distinct, is the use of the term by K. Kijima [Kijima, 2001] to describe a game-theoretic approach to analyzing the social and organizational interactions of multiple decision-makers. For Kijima, the term "poly-agent" makes sense only as a description of a system, and does not describe a single agent. In our approach, it makes sense to talk about a single modeling construct as "a polyagent."

The term "polyagent" is also used in theoretical archaeology to describe (roughly) an entity that participates in multiple processes [Normark, 2006].

4.3 Some Applications of Polyagents

In this section, we describe some specific applications of polyagents and delegate MASs.

4.3.1 Factory Scheduling and Control

The first application of polyagents was to real-time job-shop scheduling [Brueckner, 2000]. We prototyped a self-organizing multi-agent system with three species of agents: processing resources, work-pieces, and policy agents. Avatars of processing resources with different capabilities and capacities and avatars of work-pieces with dynamically changing processing needs (due to re-work) jointly optimize the flow of material through a complex, high-volume manufacturing transport system. In this application, only the avatars of the work-pieces actually deploy ghosts. The policy agents and avatars of the processing resources (machines) are single agents in the traditional sense.

In a job shop, work-pieces interact with one another by blocking access to the resources that they occupy, and thus delaying one another. Depending on the schedule, different work-pieces may interact, in different orders. Polyagents explore the space of alternative routings and interactions concurrently in a single model.

Work-piece avatars currently loaded into the manufacturing system continuously deploy ghosts that emulate their decision processes in moving through various decision points in the manufacturing process. Each of these decisions is stochastic, based on the relative concentration of attractive pheromones in the neighborhood of the next decision point. These pheromones are actually deposited by the policy agents that try to optimize the balance of the material flow across the transport network, but they are modulated by the ghosts. Thus, an avatar's ghosts modulate the pheromone field to which the avatar responds, establishing an adaptive feedback loop into the future.

The avatars continuously emit ghosts that emulate their current decision process. The ghosts travel into the future without the delay imposed by physical transport and processing of the work-pieces. These ghosts may find the next likely processing step and wait there until it is executed physically, or they may emulate the probabilistic outcome of the step and assume a new processing state for the work-piece they are representing. In either case, while they are active, the ghosts contribute to a pheromone field that reports the currently predicted relative load along the material flow system. When ghosts for alternative work-pieces explore the same resource, they interact with one another through the pheromones that they deposit and sense.

By making stochastic decisions, each ghost explores an alternative possible routing for its avatar. The pheromone field to which it responds has been modulated by all of the ghosts of other work-pieces, and represents multiple alternative routings of those work-pieces. Thus the ghosts for each work-piece explore both alternative futures for that work-piece, and multiple alternative interactions with other work-pieces.

Policy agents that have been informed either by humans or by other agents of the desired relative load of work-pieces of specific states at a particular location in turn deposit attractive or repulsive pheromones. Thus, through a local adaptive process, multiple policy agents supported by the flow of ghost agents adapt the appropriate levels of pheromone deposits to shape the future flow of material as desired.

By the time the avatar makes its next routing choice, which is delayed by the physical constraints of the material flow through the system, the ghosts and the policy agents have adjusted the appropriate pheromones so that the avatar makes the "right" decision. In effect, the policy agents and the ghosts control the avatar as long as they can converge on a low-entropy pheromone concentration that the avatar can sample.

Factory scheduling has also been a fertile application area for delegate MAS's [Holvoet and Valckenaers, 2006a,b; Valckenaers et al., 2006; Valckenaers and Holvoet, 2006]. The issuing agents ("avatars" in the polyagent model) are PROSA (Product-Resource-Order-Staff Agent) [Valckenaers and Van Brussel, 2005] agents . All PROSA agents have counterparts in reality, which facilitates integration and consistency (indeed, reality is fully integrated and consistent). The main PROSA agents in the MAS are:

- Resource agents reflecting the actual factory. They offer a structure in cyber space on which other agents can virtually travel through the factory.
- Order agents reflecting manufacturing tasks.
- Product agents reflecting task/product types.

Both resource agent and order agents issue delegate MAS. A single agent may have several delegate MAS. Each delegate MAS has its own responsibility.

Resource agents use a delegate MAS to make their services known throughout the manufacturing system. Ant agents collect information about the processing capabilities of resources while virtually traveling through the factory. These *feasibility ants* deposit this information (pheromone) at positions in cyber space that correspond to routing opportunities.

Each order agent is an issuing agent for a delegate MAS in which *exploring ants* scout for suitable task execution scenarios. In addition, each order agent is an issuing agent for a second delegate MAS that informs the resource agents of its intentions: *Intention ants* regularly reconfirm bookings for slots at resources (Figure 4.3). Specific manufacturing execution systems employ additional delegate MAS to deliver case-specific services [Valckenaers et al., 2006].

4.3.2 Vehicle Routing

Two pressures require that path planning for robotic vehicles be an ongoing activity. 1) The agent typically has only partial knowledge of its environment, and must adapt its behavior as it learns by observation. 2) The environment is dynamic: even if an agent

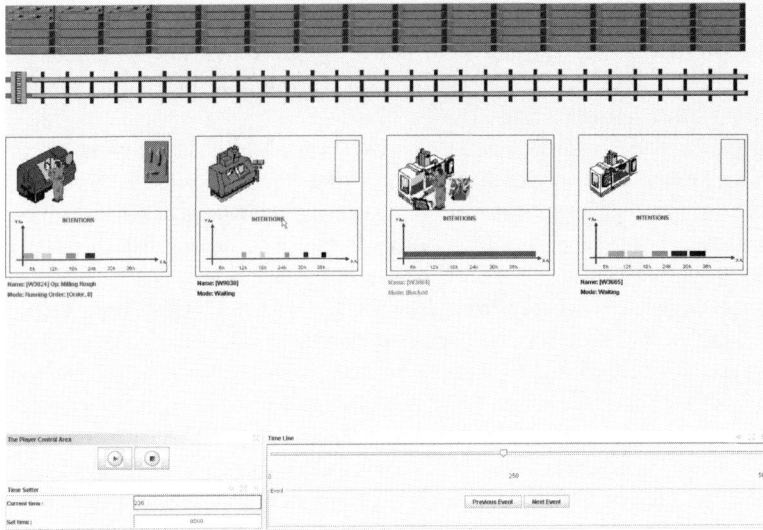

FIGURE 4.3 Intention ants notify resource agents about the intentions of their respective order agent to occupy resources at a specific time slots in the near future. Resource agents use this information to self-schedule.

has complete knowledge at one moment, a plan based on that knowledge becomes less useful as the conditions on which it was based change. These problems are particularly challenging in military applications, where both targets and threats are constantly appearing and disappearing.

In the DARPA JFACC program, we approached this problem by imitating the dynamics that ants use in forming paths between their nests and food sources [Parunak, 1997]. The ants search stochastically, but share their discoveries by depositing and sensing nest and food pheromone. Ants that are searching for food deposit nest pheromone while climbing the food pheromone gradient left by successful foragers. Ants carrying food deposit food pheromone while climbing the nest pheromone gradient. The initial pheromone trails form a random field, but quickly collapse into an optimal path as the ants interact with one another's trails.

The challenge in applying this algorithm to a robotic vehicle is that the algorithm depends on interactions among many ants, while a vehicle is a single entity that only traverses its path once. We use a polyagent to represent the vehicle (in our case, an aircraft) whose route needs to be computed [Parunak et al., 2004a; Sauter et al., 2005]. As the avatar moves through the battlespace, it continuously emits a swarm of ghosts, whose interactions mimic the ant dynamics and continuously (re)form the path in front of the avatar. These ghosts seek targets and then return to the avatar. They respond to several digital pheromones:

- RTarget is emitted by a target.
- GNest is emitted by a ghost that has left the avatar and is seeking a target.
- GTarget is emitted by a ghost that has found a target and is returning to the avatar.
- RThreat is emitted by a threat (e.g., a missile battery).

Ideally, the digital pheromones are maintained in a distributed network of unattended ground sensors dispersed throughout the vehicle's environment, but they can also reside on a central processor, or even on multiple vehicles. In addition, we provide each ghost with

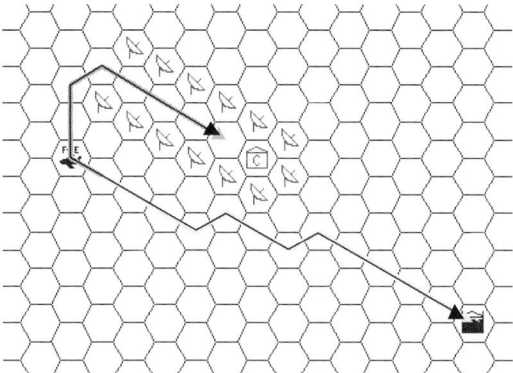

FIGURE 4.4 Gauntlet Routing Problem.

Dist, an estimate of how far away the target is.

In general, ghosts are attracted to RTarget pheromone and repelled from RThreat pheromone. In addition, before they find a target, they are attracted to GTarget pheromone. Once they find a target, they are attracted to GNest pheromone. A ghost's movements are guided by the relative strengths of these quantities in its current cell and each neighboring cell in a hexagonal lattice. It computes a weighted combination of these factors for each adjacent cell and selects stochastically among the cells, with probability proportionate to the computed value.

Each ghost explores one possible route for the vehicle. The avatar performs two functions in overseeing its swarm of ghosts:

1. It *integrates* the information from the several ghosts in their explorations of alternative routes. It observes the GTarget pheromone strength in its immediate vicinity, and guides the robot up the GTarget gradient. GTarget pheromone is deposited only by ghosts that have found the target, and its strength in a given cell reflects the number of ghosts that traversed that cell on their way home from the target. So the aggregate pheromone strength estimates the likelihood that a given cell is on a reasonable path to the target.

2. It *modulates* its ghosts' behaviors by adjusting the weights that the ghosts use to combine the pheromones they sense. Initially, all ghosts used the same hand-tuned weights, and differences in their paths were due only to the stochastic choices they made in selecting successive steps. When the avatar randomly varied the weights around the hand-tuned values, system performance improved by more than 50%, because the ghosts explored a wider range of routes. We then allowed the avatar to evolve the weight vector as the system operates, yielding an improvement nearly an order of magnitude over hand-tuned ghosts [Sauter et al., 2002].

We tested this system's ability to route an aircraft in simulated combat [Parunak et al., 2004a]. In one example, it found a path to a target through a gauntlet of threats (Figure 4.4). A centralized route planner seeking an optimal path by integrating a loss function and climbing the resulting gradient was unable to solve this problem without manually introducing a waypoint at the gauntlet's entrance. The polyagent succeeded because some of the ghosts, moving stochastically, wandered into the gauntlet, found their way to the target, and then returned, laying pheromones that other ghosts could reinforce.

Another experiment flew multiple missions through a changing landscape of threats and targets. The figure of merit was the total surviving strength of the Red and Blue forces. In two scenarios, the aircraft's avatar flew a static route planned on the basis of complete

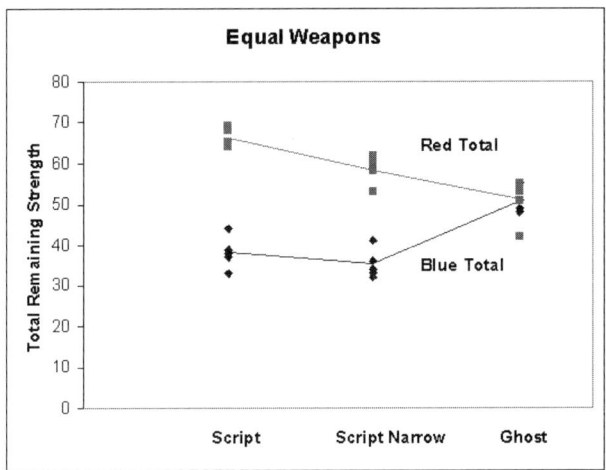

FIGURE 4.5 Real-Time vs. Advance Planning. "Script" is a conservative advance route based on complete knowledge. "Script narrow" is a more aggressive advance route. "Ghost" is the result when the route is planned in real time based on partial knowledge.

knowledge of the location of threats and targets, without ghosts. The routes differed based on how closely the route was allowed to approach threats. A third case used ghosts, but some threats were invisible until they took action during the simulation. Figure 4.5 compares these three cases. The polyagent's ability to deal with partial but up-to-date knowledge both inflicted more damage on the adversary and offered higher survivability than preplanned scripts based on complete information.

Route planning shows how a polyagent's ghosts can explore alternative behaviors concurrently, and integrate that experience to form a single course of action. Since only one polyagent is active at a time, this work does not draw on the ability of polyagents to manage the space of possible interactions among multiple entities.

In a similar vein, delegate MAS's have been applied to the problem of traffic control for conventional vehicles [Huard et al., 2006; Weyns et al., 2007], in an experimental traffic control system that proactively tries to predict and avoid road congestion. Each car in the system is represented by a task agent, while road segments and crossroads are represented by resource agents. Task agents use resource agents to book a certain road in advance trying to avoid road congestion. Three types of delegate MAS are used: (i) resource agents issue feasibility ants to gather information about the underlying environment (which roads lead to which destinations); (ii) task agents issue exploration ants to gather information about the costs of possible routes; (iii) task agents issue intention ants to book the best possible route. A booking must be refreshed regularly to maintain the reservation. The delegate MAS approach has been applied to the Leonard crossroad, a well-known Belgian congestion point between the Brussels Ring and the E411 Motorway (Figure 4.6). Tests for a realistic morning peak scenario show a reduction of 26% of congestion time for an increase of only 4% of extra traveled distance.

4.3.3 Prediction

The DARPA RAID program [Kott, 2004] focuses on the problem of characterizing an adversary in real-time and predicting its future behavior. Our contribution to this effort [Parunak and Brueckner, 2006] uses polyagents to evolve a model of each real-world entity (a group

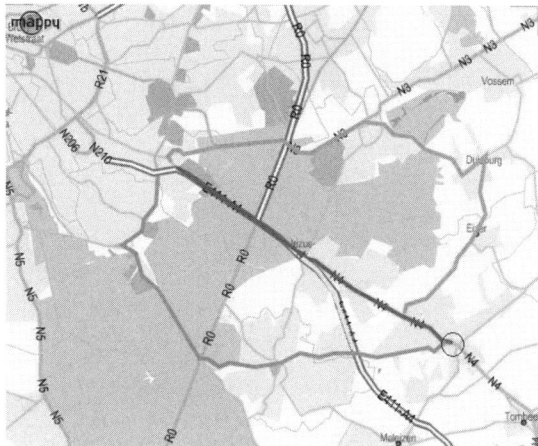

FIGURE 4.6 A car at the circle (at the bottom right) has three possible routes: straight ahead, left, and straight on and then right. In the current situation, the car follows the route straight ahead, which has the minimal cost.

of soldiers known as a fire team) and extrapolate its behavior into the future. Thus we call the system "the BEE" (Behavior Evolution and Extrapolation).

Figure 4.7 is an overview of the BEE process. Ghosts live on a timeline indexed by that begins in the past at the insertion horizon and runs into the future to the prediction horizon. τ is offset with respect to the current time t. The timeline is divided into discrete "pages," each representing a successive value of τ. The avatar inserts the ghosts at the insertion horizon. In our current system, the insertion horizon is at $\tau - t = -30$, meaning that ghosts are inserted into a page representing the state of the world 30 minutes ago. At the insertion horizon, the avatar samples each ghost's rational and emotional parameters (desires and dispositions) from distributions to explore alternative personalities of the entity it represents. The avatar is also responsible for estimating its entity's goals (using a belief network) and instantiating them in the environment as pheromone sources that constrain and guide the ghosts' behavior. In estimating its entity's goals and deliberately modulating the distribution of ghosts, the avatar reasons at a higher cognitive level than do the pheromone-driven ghosts.

Each page between the insertion horizon and $\tau = t$ ("now") records the historical state of the world at its point in the past, represented as a pheromone field generated by the avatars (which at each page know the actual state of the entities they are modeling). As ghosts move from page to page, they interact with this past state, based on their behavioral parameters. These interactions mean that their fitness depends not just on their own actions, but also on the behaviors of the rest of the population, which is also evolving. Because τ advances faster than real time, eventually $\tau = t$ (actual time). At this point, the avatar evaluates each of its ghosts based on its location compared with the actual location of its corresponding real-world entity.

The fittest ghosts have three functions:

1. The avatar reports personality of the fittest ghost for each entity to the rest of the system as the likely personality of the corresponding entity. This information enables us to characterize individual warriors as unusually cowardly or brave.
2. The avatar breeds the fittest ghosts genetically and reintroduces their offspring at the insertion horizon to continue the fitting process.

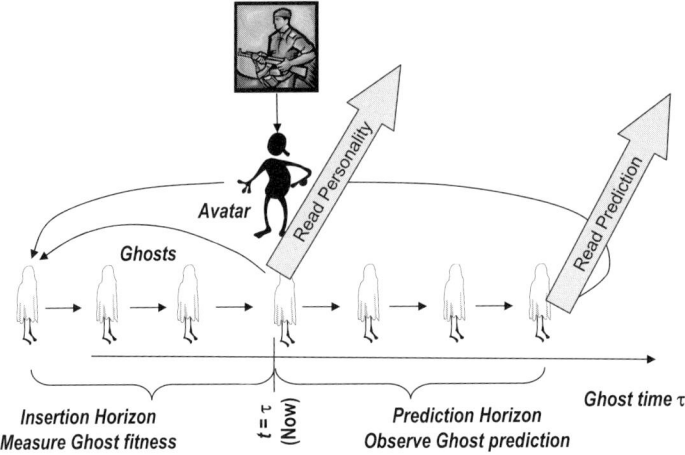

FIGURE 4.7 Behavioral Emulation and Extrapolation. Each avatar generates a stream of ghosts that sample the personality space of its entity. They evolve against the entity's recent observed behavior, and the fittest ghosts run into the future to generate predictions.

3. The fittest ghosts for each entity run past the avatar's present into the future. Each ghost that runs into the future explores a different possible future of the battle, analogous to how some people plan ahead by mentally simulating different ways that a situation might unfold. The avatar analyzes the behaviors of these different possible futures to produce predictions of enemy behavior and recommendations for friendly behavior. In the future, the pheromone field with which the ghosts interact is generated not by the avatars, but by the ghosts themselves. Thus it integrates the various possible futures that the system is considering, and each ghost is interacting with this composite view of what other entities may be doing.

The first and third functions are analogous to the integrating function of the avatars in route planning, while the second is analogous to the modulation function.

This model has proven successful both in characterizing the internal state of entities that we can only observe externally, and in predicting their future behavior. [Parunak and Brueckner, 2006] details the results of experiments based on multiple wargames with human participants. We can detect emotional state of entities as well as a human observer, but faster. Our prediction of the future of the battle is superior both to a human and to statistical and game-theoretic predictors [Parunak, 2007].

4.4 Future Research

At present, polyagents are an effective technique for agent-based modeling and simulation, but their theoretical foundations offer a rich and open field for future study, and they have the potential for even broader application.

4.4.1 Theoretical Opportunities

As with any novel software technology, initial applications of polyagents rely heavily on experimentation. Our work so far has identified a number of directions for theoretical and engineering development to make their application more systematic.

Pheromones as Probabilities. Polyagents enable parallel exploration of multiple possible behaviors of the avatar because ghosts interact with digital pheromones laid down by ghosts of other avatars. These pheromones may be interpreted (up to a normalizing constant) as a probability field over the avatar's possible behavior. This interpretation relies heavily on the boundedness of pheromone fields under constant deposit and evaporation [Brueckner, 2000], which makes global normalization well-defined. In addition, for many ghost decisions (selecting among alternative destinations), all that matters is relative probabilities across the neighborhood of cells being considered, which requires only local knowledge of pheromone concentrations in those cells. Further theoretical exploration of the notion of pheromones-as-probabilities will open the door to integrating polyagents with more traditional probabilistic reasoning systems.

Parameter Analysis. Polyagents are rife with parameters, all of which need to be tuned. Some can be tuned by the avatar, as it modulates the generation of ghosts. Others need to be configured when the system is initialized. Still others may be varied by the ghosts themselves. A disciplined analysis of the kinds of parameters, the mechanisms available for tuning them, and the sensitivity of an application to their values is an ongoing need. A foundation for this agenda is well-defined metrics that can be applied to the behavior of ghosts. We have made some progress in this area in defining formal similarity metrics over bundles of trajectories [Parunak et al., 2007b], and features such as the option set entropy over ghost decisions [Brueckner and Parunak, 2003].

Dynamics. Any look-ahead mechanism in a highly non-linear domain is subject to divergence of trajectories, imposing a prediction horizon beyond which further projection is worthless. We have reported preliminary evidence of such horizons [Parunak et al., 2007a], but much work remains to be done. In particular, there appears to be some similarity between the performance of a polyagent over various time horizons and the performance of the minority game, a simple model of resource allocation, over different history lengths. Understanding this similarity may greatly strengthen our application of both models. More generally, the prediction horizon is an example of an emergent behavior [Shalizi and Crutchfield, 2001]. Such behavior invites the application of new methods that are being developed for its detection and characterization [Chen et al., 2007].

Universality. The success of low-dimensional field-based interactions in predicting the behavior of high-dimensional cognitive systems is on first impression unexpected. We can associate it by analogy with universality in statistical physics, the observation that when systems with very different internal structures are similarly constrained, their behaviors can converge [Parunak et al., 2004b]. While we have empirical evidence for such universality, we are far from understanding in detail when it arises and how far we can exploit it, and its theoretical elaboration would be extremely valuable.

Hybrid Symbolic-Subsymbolic Reasoning. Ghosts are lightweight, stigmergic agents, guided by subsymbolic computations. Their behavior is controlled by their internal personalities and the environment in which they are immersed. Symbolic processes that need to interact with them can modulate these two points of contact. This whole matter of the interface between symbolic processes (which are understandable to humans) and sub-symbolic ones (swarming ghosts) is full of potential for hybrid systems [Parunak et al., 2006a] that combine the benefits of both approaches. In principle, the polyagent is already a hybrid system, since the avatar is not constrained to be stigmergic, and could in principle be a BDI agent with full symbolic reasoning. In this approach, the swarming ghosts serve the purpose of the symbolic agent. Other approaches include mixing ghosts and more complex agents as peers, using a symbolic layer as a user interface to swarming agents, or embedding symbolic structure in the environment that the ghosts manipulate subsymbolically (discussed in the next section).

4.4.2 New Applications

Information Fusion. In many domains, multiple reasoning mechanisms are available, each with its champions who extol its superiority over its rivals. There is growing evidence both anecdotal and formal [Page, 2007] that a combination of diverse competent reasoners can often outperform a single reasoner that is superior to any one of them. Combining mechanisms, such as bagging, boosting, and Bayesian model averaging, are available for classifiers [Dietterich, 2001]. Polyagents offer a mechanism to fuse less-structured data and reasoners. In fact, our work on battle prediction (Section 4.3.3) offers a static example of this, using a geospatial statistical model for threat regions and a Bayesian net reasoner that identifies likely enemy objectives to modulate the pheromone field to which ghosts respond.

Individual reasoners can constrain a polyagent simulation, in one or more of three ways.

1. As in Section 4.3.3, they can deposit digital pheromones at specified locations. A field dedicated to a specific predictor enables the ghosts to respond directly to that predictor.
2. By default, avatars generate ghosts with random personalities and rely on evolution to find personalities with predictive value. A predictor can speed up this process by setting default values for ghost personalities directly. If these default values are correct, the evolution will converge more rapidly to high-fitness ghosts.
3. Predictors can change the structure of the simulation environment, adding or removing nodes and links.

Evolution enables ghosts to prefer the pheromones with greatest predictive value at any moment, thus selecting dynamically among the predictors, and dynamically adjusts the population to give appropriate weight to ghost personalities generated by various predictors. Thus polyagent information fusion automatically prefers the reasoners that can make the greatest contribution to the system's objective at any moment, based on recent performance, and shifts to other reasoners dynamically as circumstances change.

While we have demonstrated polyagent information fusion for a fixed set of reasoners, there is ample room for new development to enable an open environment for fusion of an arbitrary set of reasoners in a given domain.

Swarming over Task Networks. Applications of polyagents to factory control and traffic flow on highways already demonstrates their potential in environments that do not have the regular structure of a lattice or manifold. These networks are flat, and do not reflect the hierarchical structure implicit in many planning tasks. We are currently exploring mechanisms for using polyagents to develop coordinated schedules over hierarchical task networks, in particular those represented in an extension of the TAEMS formalism [Horling et al., 2004; Lesser et al., 2004]. A planner may develop such a network to express a set of high-level tasks that need to be accomplished and the individual actions that can support them, but the problem of coordinating a set of agents to develop a schedule over this plan (an assignment of resources and times to specific actions) is combinatorially complex. By embodying complex symbolic relationships in the environment, we bring the highly efficient subsymbolic processing of stigmergic agents to bear on intrinsically symbolic problems. In effect, we have externalized the symbolic aspect of the problem, moving it from within a single agent to the space within which agents interact.

4.5 Conclusion

Polyagents are a thorough-going approach to using simulation in support of agent decisions. They have been applied successfully to a range of problems, and provide the stimulus for a rich array of ongoing research topics that we and others are pursuing.

Perhaps the most interesting feature of polyagents is their multidisciplinary imitative origin. For example:

- Their underlying software structure is based on multi-agent simulation.
- The use of stigmergic coordination is modeled on pheromone systems in social insects.
- The training process used to adapt the ghosts' weight vectors to the recent observed behavior of the entity being modeled is a genetic algorithm, inspired by biological evolution.
- The motivation of considering multiple trajectories comes from the realization in modern physics of the importance of trajectory divergence resulting from sensitivity to initial conditions.
- The reasonableness of using a low-dimensional model to approximate a high-dimensional reality is inspired by the phenomenon of universality in solid state physics.

The convergence of these various inspirations yields a construct of great flexibility and breadth of potential application.

One important alternative to such an imitative approach is a derivational approach grounded in mathematical formalism. For example, artificial intelligence was for many years closely identified with programming languages rooted in formal logic, such as Lisp (a direct expression of Church's lambda calculus) and Prolog (based on Horn clauses). A number of the alternative approaches to agent decision-making summarized in Section 4.1 reflect this more formal orientation.

At base, this distinction reflects the age-old tension between "neats" and "scruffies" in the AI community. Yet in fact the imitative nature of polyagents opens a door to address this tension. The various disciplines that inspire this construct have all been the object of some degree of formal modeling. As polyagents demonstrate their value to an increasingly broad range of applications, these theories become available to inform design decisions and interpret the results of polyagent models. The foundation of polyagents is indeed in empirical science rather than abstract formalisms, but the mathematics to develop disciplined engineering methods for polyagents can be drawn from the same disciplines that inspired the construct initially, and much of our ongoing research agenda is directed to this end.

References

S. Brueckner. *Return from the Ant: Synthetic Ecosystems for Manufacturing Control.* PhD thesis, Humboldt University Berlin, Department of Computer Science, Berlin, 2000. URL dochost.rz.hu-berlin.de/dissertationen/brueckner-sven-2000-06-21/PDF/Brueckner.pdf.

S. Brueckner and H. Parunak. Information-driven phase changes in multi-agent coordination. In *Proceedings of Autonomous Agents and Multi-Agent Systems (AAMAS 2003)*, pages 950–951. ACM, 2003. URL www.newvectors.net/staff/parunakv/AAMAS03InfoPhaseChange.pdf.

C.-C. Chen, S. Nagl, and C. Clack. Specifying, detecting and analysing emergent behaviours in multi-level agent-based simulations. In *Proceedings of The Summer Computer Simulation Conference (SCSC 2007)*, pages 969–976, 2007. URL www.cs.ucl.ac.uk/staff/C.Clack/research/MultiLevelABM.pdf.

T. Dietterich. Ensemble methods in machine learning. In *Proceedings of International Workshop on Multiple Classifier Systems*, pages 1–15, New York, NY, 2001. Springer.

M. Dorigo and T. Stuetzle. *Ant Colony Optimization.* MIT Press, Cambridge, MA, 2004.

P. Fishwick, N. Narayanan, J. Sticklen, and A. Bonarini. A multi-model approach to reasoning and simulation. *IEEE Transactions on Systems, Man, and Cybernetics*, 24(10):1433–1449, 1994.

K. Forbus, P. Nielsen, and B. Faltings. Qualitative kinematics: A framework. In *Proceedings of International Joint Conference on Artificial Intelligence*, pages 430–435, 1987.

K. Forbus, P. Nielsen, and B. Faltings. Qualitative spatial reasoning: The clock project. *Artificial Intelligence*, 51(1-3):417–471, 1991.

M. Garey and D. Johnson. *Computers and Intractability.* W.H. Freeman, San Francisco, CA, 1979.

J. Gilmer and F. Sullivan. Recursive simulation to aid models of decision making. In *Proceedings of the 32nd Conference on Winter Simulation*, pages 958–963, San Diego, CA, 2000. SCS International.

J. Gilmer and F. Sullivan. Alternative implementations of multitrajectory simulation. In *Proceedings of the 30th Conference on Winter Simulation*, pages 865–872, Los Alamitos, CA, 1998. IEEE Computer Society Press.

J. Gilmer and F. Sullivan. Study of an ergodicity pitfall in multitrajectory simulation. In *Proceedings of the 33nd Conference on Winter Simulation*, pages 797–804, Washington, D.C., 2001. IEEE Computer Society.

P.-P. Grassé. La reconstruction du nid et les coordinations inter-individuelles chez *bellicositermes natalensis et cubitermes sp.*la théorie de la stigmergie: Essai d'interprétation du comportement des termites constructeurs. *Insectes Sociaux*, 6:41–84, 1959.

T. Holvoet and P. Valckenaers. Beliefs, desires and intentions through the environment. In *Proceedings of 5th International Joint Conference on Autonomous Agents and Multiagent Systems (AAMAS2006)*, pages 1052–1054, 2006a.

T. Holvoet and P. Valckenaers. Exploiting the environment for coordinating agent intentions. In *Proceedings of Third International Workshop on Environments for Multi-Agent Systems (E4MAS06)*, 2006b.

J. Hopcroft and J. Ullman. *Introduction to Automata Theory, Languages, and Computation.* Addison-Wesley, Reading, MA, 1979.

B. Horling, V. Lesser, R. Vincent, T. Wagner, A. Raja, S. Zhang, K. Decker, and A. Garvey. The taems white paper, 2004. URL dis.cs.umass.edu/research/taems/white/.

E. Huard, D. Gorissen, and T. Holvoet. Applying delegate multi-agent sytems in a traffic control system. Technical report, CW 467, Katholieke Universiteit Leuven, Leuven, Belgium, 2006. URL www.cs.kuleuven.be/publicaties/rapporten/cw/CW467.abs.html.

C. Jacob. *Illustrating Evolutionary Computation With Mathematica*. Morgan Kaufmann, San Francisco, 2001.

H. Kantz and T. Schreiber. *Nonlinear Time Series Analysis*. Cambridge University Press, Cambridge, UK, 1997.

K. Kijima. Why stratification of networks emerges in innovative society: Intelligent poly-agent systems approach. *Computational and Mathematical Organization Theory*, 7(1 (June)):45–62, 2001.

A. Kott. Real-time adversarial intelligence and decision making, 2004. URL dtsn.darpa.mil/ixo/programdetail.asp?progid=57.

T. Lambert, M. Epelman, and R. Smith. A fictitious play approach to large-scale optimization. *Operations Research*, 53(3 (May-June)), 2005.

P. Laplace. *Essai Philosophique sur les Probabilités*. Krieger, 1820.

V. Lesser, K. Decker, T. Wagner, N. Carver, A. Garvey, B. Horling, D. Neiman, R. Podorozhny, M. Prasad, A. Raja, R. Vincent, P. Xuan, and X. Zhang. Evolution of the GPGP/TAEMS domain-independent coordination framework. *Autonomous Agents and Multi-Agent Systems*, 9(1-2):87–143, 2004.

F. Michel. *Formalisme, méthodologie et outils pour la modélisation et la simulation de systémes multi-agents*. PhD thesis, Université des Sciences et Techniques du Languedoc, Department of Informatique, 2004. URL leri.univ-reims.fr/~fmichel/thesis/index.php.

J. Normark. *The Roads In-Between: Causeways and Polyagentive Networks at Ichmul and Yo'okop, Cochuah Region, Mexico*. PhD thesis, Göteborg University, Department of Archaeology and Ancient Historye, 2006.

S. Page. *The Difference: How the Power of Diversity Creates Better Groups, Firms, Schools, and Societies*. Princeton University Press, Princeton, NJ, 2007.

C. Parent, S. Spaccapietra, and E. Zimanyi. Spatio-temporal conceptual models: Data structures + space + time. In *Proceedings of The 7th ACM Symposium on Advances in GIS*, 1999.

H. Parunak. 'go to the ant': Engineering principles from natural agent systems. *Annals of Operations Research*, 75:69–101, 1997.

H. Parunak. Making swarming happen. In *Proceedings of Swarming and Network-Enabled C4IS*, 2003.

H. Parunak. Real-time agent characterization and prediction. In *Proceedings of International Joint Conference on Autonomous Agents and Multi-Agent Systems (AAMAS07), Industrial Track*, pages 1421–1428. ACM, 2007. URL www.newvectors.net/staff/parunakv/AAMAS07Fitting.pdf.

H. Parunak and S. Brueckner. Extrapolation of the opponent's past behaviors. In A. Kotte and W. McEneany, editors, *Adversarial Reasoning: Computational Approaches to Reading the Opponent's Mind*, pages 49–76. Chapman and Hall/CRC Press, Boca Raton, FL, 2006.

H. Parunak, S. Brueckner, and J. Sauter. Digital pheromones for coordination of unmanned vehicles. In *Proceedings of Workshop on Environments for Multi-Agent Systems (E4MAS 2004)*, pages 246–263. Springer, 2004a.

H. Parunak, S. Brueckner, and R. Savit. Universality in multi-agent systems. In *Proceedings of Third International Joint Conference on Autonomous Agents and Multi-Agent Systems (AAMAS 2004)*, pages 930–937. ACM, 2004b. URL www.newvectors.net/staff/parunakv/AAMAS04Universality.pdf.

H. Parunak, P. Nielsen, S. Brueckner, and R. Alonso. Hybrid multi-agent systems. In *Proceedings of the Fourth International Workshop on Engineering Self-Organizing Systems (ESOA'06)*. Springer, 2006a. URL www.newvectors.net/staff/parunakv/ESOA06Hybrid.pdf.

H. Parunak, R. Savit, and R. Riolo. Agent-based modeling vs. equation-based modeling: A case study and users' guide. In *Proceedings of Multi-agent systems and Agent-based Simulation (MABS'98)*, pages 10–25. Springer, 2006b. URL www.newvectors.net/staff/parunakv/mabs98.pdf.

H. Parunak, T. Belding, and S. Brueckner. Prediction horizons in polyagent models. In *Proceedings of sixth International Joint Conference on Autonomous Agents and Multi-Agent Systems (AAMAS07)*, pages 930–932, 2007a. URL www.newvectors.net/staff/parunakv/AAMAS07AAMAS07PH.pdf.

H. Parunak, S. Brophy, and S. Brueckner. Comparing agent trajectories. In *Proceedings of Agent 2007*, pages 41–50, 2007b. URL agent2007.anl.gov/2007pdf/Paper4-Parunak_CAT.pdf.

H. Parunak, S. Brueckner, D. Weyns, T. Holvoet, and P. Valckenaers. E pluribus unum: Polyagent and delegate mas architectures. In *Proceedings of Eighth International Workshop on Multi-Agent-Based Simulation (MABS07)*. Springer, 2007c. URL www.newvectors.net/staff/parunakv/MABS07EPU.pdf.

J. Sauter, R. Matthews, H. Parunak, and S. Brueckner. Agent-based modeling vs. equation-based modeling: A case study and users' guide. In *Proceedings of Autonomous Agents and Multi-Agent Systems (AAMAS02)*, pages 434–440. ACM, 2002. URL www.newvectors.net/staff/parunakv/AAMAS02Evolution.pdf.

J. Sauter, R. Matthews, H. Parunak, and S. Brueckner. Performance of digital pheromones for swarming vehicle control. In *Proceedings of Fourth International Joint Conference on Autonomous Agents and Multi-Agent Systems*, pages 903–910. ACM, 2005. URL www.newvectors.net/staff/parunakv/AAMAS05SwarmingDemo.pdf.

C. Shalizi and J. Crutchfield. Computational mechanics: Pattern and prediction, structure and simplicity. *Journal of Statistical Physics*, 104:817–879, 2001.

N. Shnerb, Y. Louzoun, E. Bettelheim, and S. Solomon. The importance of being discrete: Life always wins on the surface. *Proc. National Acad. Science USA*, 97(19 (September 12)):10322–10324, 2000. URL www.pnas.org/cgi/reprint/97/19/10322.

A. Tolk. An agent-based decision support system architecture for the military domain. In G. Phillips-Wren and L. Jain, editors, *Intelligent Decision Support Systems in Agent-Mediated Environments*, volume 115 of *Frontiers in Artificial Intelligence and Applications*, pages 187–205. IOS Press, Boca Raton, FL, 2005.

N. Tryfona, R. Price, and C. Jensen. Conceptual models for spatio-temporal applications. In T. Sellis, A. Frank, S. Grumbach, and R. Guting, editors, *Spatiotemporal Databases: The Chorochronos Approach*, volume 2520 of *Lecture Notes in Computer Science*, pages 79–116. Springer-Verlag, Berlin Heidelberg New York, 2003.

P. Valckenaers and T. Holvoet. The environment: An essential abstraction for managing complexity in mas-based manufacturing control. In *Environments for Multi-Agent Systems II*, volume 3830 of *LNAI*, pages 205–211. Springer, Berlin Heidelberg New York, 2006.

P. Valckenaers and H. Van Brussel. Holonic manufacturing execution systems. *CIRP Annals of Manufacturing Technology*, 54(1):427–432, 2005.

P. Valckenaers, K. Hadeli, B. Saint Germain, P. Verstraete, and H. Van Brussel. Emergent short-term forecasting through ant colony engineering in coordination and control systems. *Advanced Engineering Informatics*, 20(3 (July)):261–278, 2006.

D. Weyns, T. Holvoet, and A. Helleboogh. Anticipatory vehicle routing using delegate multiagent systems. In *Proceedings of The 10th International IEEE Conference on Intelligent Transportation Systems*. IEEE, 2007.

T. Wittig. *ARCHON: An Architecture for Multi-agent Systems*. Ellis Horwood, New York, 1992.

S. Wolfram. *A New Kind of Science*. Wolfram Media, Champaigne, IL, 2002.

L. Yilmaz and T. Ören. Discrete-event multimodels and their agent-supported activation and update. In *Proceedings of The Agent-Directed Simulation Symposium of the Spring Simulation Multiconference (SMC'05)*, pages 63–72, 2005. URL www.site.uottawa.ca/~oren/pubs/2005/0503-ADS-MMs.pdf.

5

Combining Simulation and Formal Tools for Developing Self-Organizing MAS

Luca Gardelli
Università di Bologna

Mirko Viroli
Università di Bologna

Andrea Omicini
Università di Bologna

5.1 Introduction... 133
5.2 The A&A Meta-Model for Self-Organizing Systems... 136
   The Role of Environment in Self-Organizing Systems • Overview of the A&A Meta-Model • An Architectural Pattern
5.3 Methodological Issues Raised by Self-Organizing Systems... 139
5.4 A Methodological Approach for Engineering Self-Organizing MAS 140
   Overview • Modeling • Simulation • Verification • Tuning
5.5 Using the PRISM Tool to Support the Method... 144
   Modeling • Simulation • Verification • Tuning
5.6 Case Study: Plain Diffusion........................ 145
   Problem Statement • Modeling Plain Diffusion • Simulating Plain Diffusion • Verifying Plain Diffusion • Tuning Plain Diffusion • Preliminary Scalability Analysis
5.7 Conclusion 158
References 161

5.1 Introduction

Self-organization is increasingly being regarded as an effective approach to tackle the complexity of modern systems. The self-organization approach allows the development of systems exhibiting complex dynamics and adapting to environmental perturbations without requiring a complete knowledge of the surrounding conditions to come. Systems developed according to self-organization principles are composed by simple entities that, locally interacting with others sharing the same environment, collectively produce the target global patterns and dynamics by emergence. Many biological systems can be effectively modeled using a self-organization approach: well-known examples include food foraging in ant colonies, nest building in termites societies, comb pattern in honeybees, brood sorting in ants [Parunak, 1997; Bonabeau et al., 1999; Camazine et al., 2001]. They have inspired the development of many artificial systems, such as decentralized coordination for automated guided vehicles [Sauter et al., 2005; Weyns et al., 2005], congestion avoidance in circuit switched telecommunication networks [Steward and Appleby, 1994], manufacturing scheduling and control for vehicle painting [Cicirello and Smith, 2004], and self-organizing peer-to-peer infrastructures [Babaoglu et al., 2002]. Furthermore, principles of self-organization are currently investigated in several research initiatives that may have industrial relevance in a near future: notable examples include Amorphous Computing [Abelson et al., 2000], Autonomic Computing [Kephart and Chess, 2003] and Bioinformatics.

The first appearance of the term self-organization with its modern meaning seems to be on a 1947 paper by the English psychiatrist William Ross Ashby [Ashby, 1947]. Ashby defined self-organization as the *ability of a system to change its own internal organization, rather being changed from an external force.* In this definition the focus is placed on the *self* aspect, that is the control flow driving the system must be internal. As far as the organization part is concerned, there are different viewpoints in the research community: organization may be interpreted as spatial arrangement of components, system statistical entropy, differentiation of tasks among system components, establishment of information pathways, and the like. It can be observed that all these cases include some sort of relation between components, either topological, structural or functional. Hence, we call a system "self-organizing" if *it is able to re-organize itself by managing the relations between components, either topological, structural or functional, upon environment perturbations solely via the interactions of its components, with no requirement of external forces.* This definition implies that four key features are found in every self-organizing system (SOS):

Autonomy: As previously discussed for the term "self", control must be located within the system and should be shielded from environmental forces: it is worth noting that here we do not mean closed systems, since most of the systems will rely on the environment resources in the shape of information, energy, and matter.

Organization: In the literature self-organizing systems are often described as increasing their own organization: actually, we see no point for this continual increase. Indeed, it is often the case that self-organizing systems stabilize in sub-optimal solutions and their organization varies over time both decreasing and increasing. Hence, we prefer just to say that there should be some re-organization, not necessarily toward a better solution.

Dynamic: Self-organization should always be intended as a process and not as a final state.

Adaptive: Perturbations may happen within the system as well as in the environment: a self-organizing system should be able to compensate for these perturbations. Centralized systems characterized by a single-point-of-failure can hardly be considered self-organizing because of the high sensitivity with respect to perturbations.

Self-organization is closely related to emergence, although being distinct concepts [De Wolf and Holvoet, 2005] Intuitively, the notion of emergence is linked to novelty and an abstraction gap observed between the actual behavior of a system components and a global system property. Although the origin of the term emergence can be traced back to Greeks, the first scientific investigation of emergence is due to philosophers of the British Emergentism in the 1870s, including J. S. Mill, but flourished only in the 1920s with the work of Alexander and Broad [Goldstein, 1999]. Mill (1872) recognized that the notion of emergence was highly relevant to chemistry: Chemistry and Biochemistry provide a wide range of examples involving emergent phenomena, indeed, many chemical compounds exhibit properties that cannot be inferred from the individual components. According to a modern definition of emergence discussed in [Müller, 2004], a phenomenon is emergent if and only if we have

> (i) a system of entities in interaction whose expression of the states and dynamics is made in an ontology or theory D, (ii) the production of a phenomenon, which could be a process, a stable state, or an invariant, which is necessarily global regarding the system of entities, (iii) the interpretation of this global phenomenon either by an observer or by the entities themselves via an inscription mechanism in another ontology or theory D.

Although being a specific class of distributed system, the development of SOS is driven by different principles. For instance, engineers typically design systems as the composition of smaller elements, being either software abstractions or physical devices, where composition rules depend on the reference paradigm (e.g., the object-oriented one), but typically produce predictable results. Conversely, SOS display non-linear dynamics, which can hardly be captured by deterministic models, and, although robust with respect to external perturbations, are quite sensitive to changes on inner working parameters. In particular, engineering SOS poses two big challenges: How do we design the individual entities to produce the target global behavior? And, can we provide guarantees of any sort about the emergence of specific patterns? Even though the existence of these issues is generally acknowledged, few efforts have been devoted to the study of an engineering support either from methodologies and tools—except for a few explorations in the MAS (multi-agent system) community [De Wolf and Holvoet, 2007; Bernon et al., 2004].

While simulation is effectively exploited in complex system analysis, its potentialities in software engineering are typically overlooked [von Mayrhauser, 1993; Tichy, 1998]: although there are examples of simulation in software development, these approaches involve simulation *after* the software has already been designed and developed—that is, simulations are performed afterward for profiling purposes. Conversely, we promote the use of simulation techniques specifically at the early stages of software development. Simulation allows us to preview overall qualitative system dynamics and devise a coarse set of working parameters before actually implementing the system.

Specifically, in this chapter, we focus on methodological aspects concerning the early-design stage of SOS built relying on the agent-oriented paradigm: in particular, we refer to the A&A meta-model, where MAS are composed by agents and *artifacts* [Omicini et al., 2008]. Then, we describe an architectural pattern extracted from a recurrent solution in designing self-organizing systems: this pattern is based on a MAS environment formed by artifacts, modeling non-proactive resources, and environmental agents, acting on artifacts so as to enable self-organizing mechanisms. In this context, we propose an approach for the engineering of self-organizing systems based on simulation at the early design stage: in particular, the approach is articulated in four stages, modeling, simulation, formal verification, and tuning. In this approach, simulations of an abstract system model are used to drive design choices until the required quality properties are obtained, thus providing guarantees that the subsequent design steps would lead to a correct implementation. However, system analysis exclusively based on simulation results does not provide sound guarantees for the engineering of complex systems: to this purpose, we envision the application of formal verification techniques, specifically *model checking*, in order to exactly characterize the system behaviors. Given a formal specification a probabilistic model checker determines whether a specific property is satisfied or not or the actual likelihood value: properties are specified using different flavors of temporal logic, depending on the model type, e.g., probabilistic, stochastic or non-deterministic. Unfortunately, the applicability of model checking techniques is hindered by the explosion of state space: nonetheless, in those cases model checking still a valuable tool for validating simulation results on small problem instances.

As a case to clarify the approach, we analyze a self-organizing solution to the problem of *plain diffusion* [Babaoglu et al., 2006; Gardelli et al., 2007; Canright et al., 2006]: given a networked set of nodes hosting information, we have to homogeneously distribute the information across the nodes. In order for a solution to be self-organizing, control must be decentralized and entities have to interact locally: hence, we consider solutions having at least one agent for each node whose goal consist in sending information only to neighboring nodes and agent knowledge is restricted to the local node. We devise a strategy solving this problem following the methodological approach sketched above.

The main contribution of the chapter consists in the systematic use of simulation and formal tools in the early design phase of the development of self-organizing systems. The main limitation of the approach derives from the impossibility of formally verifying arbitrarily large systems, hence at the moment it is feasible to verify only small instances of systems: however this is still useful to have an overall reliability evaluation of simulation results. The approach is intended for the architectural pattern described, although it may be successfully applied to other scenarios. With respect to the case study, we provide and analyze a novel self-organizing strategy for achieving plain diffusion behavior.

The rest of the chapter is organized as follows: in Section 5.2 we describe the role of environment, our reference meta-model and an architectural pattern encoding a recurrent solution when designing self-organizing MAS. In Section 5.3 we discuss the current panorama of MAS methodologies with respect to self-organization aspects. In Section 5.4 we describe our methodological approach to the engineering of self-organizing MAS, which is an iterative process articulated in modeling, simulation, verification and tuning. As far as formal tools are concerned, we rely on the PRISM Probabilistic Model Checker [PRISM, 2007] for the entire process, namely, modeling, simulation and verification: so, Section 5.5 describes how to use the PRISM software tool to support the whole process. In Section 5.6 we apply the methodology to the case study of plain diffusion, and conclude in Section 5.7.

5.2 The A&A Meta-Model for Self-Organizing Systems

In this section, we describe the A&A meta-model, which is our reference meta-model when designing self-organizing MAS. In particular, we start by clarifying the role of environment in both natural and artificial self-organizing systems. Then, we continue describing the actual meta-model featuring agents and artifacts, the latter wrapping environmental resources. With respect to the meta-model, we describe an architectural pattern featuring environmental agents as artifacts manager: furthermore, we consider environmental agents as the most appropriate locus for encapsulating self-organizing mechanisms. The meta-model combined with the architectural pattern forms the basis for our methodological approach that will be the subject matter of Section 5.4.

5.2.1 The Role of Environment in Self-Organizing Systems

From the analysis of natural self-organizing systems [Parunak, 1997; Bonabeau et al., 1999; Camazine et al., 2001; Solé and Bascompte, 2006] and existing experience in prototyping artificial ones [Weyns et al., 2005; Sauter et al., 2005; Casadei et al., 2007; Gardelli et al., 2007; Mamei and Zambonelli, 2005], it is recognized that environment plays a crucial role in the global SOS dynamics. A typical explanatory example is the case of stigmergy: as pointed out by [Grassé, 1959], among social insects workers are driven by the environment in their activity. Indeed, in animal societies self-organization is typically achieved by the interplay between individuals and the environment, such as the deposition of pheromone by ants or the movement of wooden chips by termites [Camazine et al., 2001]. In particular, these interactions are responsible for the establishment and sustainment of a feedback loop: in the case of ant colonies, positive feedback is provided by ants depositing pheromones, while negative feedback is provided by the environment through evaporation [Camazine et al., 2001].

When moving to artificial systems, and to MAS in particular, there are a few questions that need to be answered. The first one is where to embed self-organizing mechanisms. The above discussion promotes the distribution of concerns between active components and the environment—in the MAS context, between agents and the environment. This partially frees agents from the burden of system complexity, and provides a more natural mapping for those non-goal-oriented behaviors. The second question is how to find the minimum requirements for an environment to support self-organization. From the definition of self-organization provided by [Camazine et al., 2001] we can identify some basic requirements: (i) the environment should support indirect interactions among the components of a system, (ii) the environment should support some notion of locality, and (iii) locality should affect interactions, e.g., promoting local ones. Moreover, specific self-organizing mechanisms may require an *active environment*, i.e. the presence of active processes in the environment making environment evolve to a suitable state: e.g., in pheromone-based systems, the environment may either provide a reactive *evaporation service*, or proactively act upon pheromone-like components to emulate the effect of evaporation.

5.2.2 Overview of the A&A Meta-Model

Software conceived according to the MAS paradigm is modeled as a composition of agents (autonomous entities situated in a computational or physical environment) that interact with each other and with environmental resources to achieve either individual or social goals. Traditionally, the environment consists of a deployment context that provides communication services and access to physical resources: in this context, MAS engineers design agents while the environment is just an output of the analysis stage. Recently, the environment has been recognized as an actual design dimension: then, MAS engineers can hide system complexity behind environmental services, freeing agents from specific responsibilities [Weyns et al., 2007; Viroli et al., 2007]. In this chapter, we adopt the latter notion of environment, i.e. the part of MAS outside agents that engineers should design so as to reach the objectives of the application at hand.

In order to describe the environment, we have to provide suitable abstractions for environmental entities. As pointed out by [Molesini et al., 2009], despite most of the current AOSE (agent-oriented software engineering) methodologies and meta-models provide little or no environment support, it is useful to adopt the A&A meta-model where a MAS is modeled by two fundamental abstractions: *agents* and *artifacts* [Omicini et al., 2008]. Agents are autonomous pro-active entities encapsulating control and driven by their internal goal/task. When developing a MAS, sometimes entities require neither autonomy nor pro-activity to be correctly characterized. This is typical of entities that serve as tools to provide specific functionalities: these entities are the so-called artifacts. Artifacts are passive, reactive entities providing services and functionalities to be exploited by agents through a *usage interface*. It is worth noting that artifacts typically realize those behaviors that cannot or do not require to be characterized as goal-oriented [Omicini et al., 2006; Ricci et al., 2006]. Artifacts mediate agent interactions, support coordination in social activities, and embody the portion of the environment that is designed and controlled to support MAS activities.

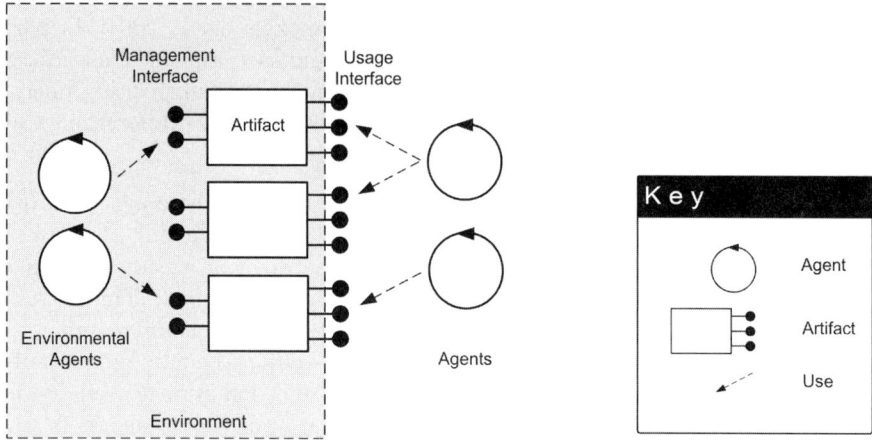

FIGURE 5.1 An architectural pattern for self-organizing MAS featuring environmental agents responsible for sustaining feedback loop by properly managing artifacts.

5.2.3 An Architectural Pattern

The components of modern computational systems often need to interact with an environment populated by *legacy systems*. Hence, the environment can be either completely or partially given: this is subject to investigation during the analysis phase [Omicini, 2001]. In the MAS context, during the design phase, resources are assigned to artifacts, providing a uniform way for agents to exploit resources. Unfortunately, in a scenario involving legacy systems, we may only have partial control on the environment, thus making it difficult to embed self-organizing mechanisms within artifacts. Then, to inject self-* properties in MAS, we need to add a layer on top of existing environmental resources.

To this purpose, we rely on the notion of *environmental agent*: such agents are responsible for managing artifacts to achieve the target self-* property. Hence, environmental agents are seen as distinct from standard agents, also called *user agents*, which exploit artifact services to achieve individual and social goals. This recurrent solution has been encoded in the form of an architectural pattern [Gardelli et al., 2007] with reference to the A&A meta-model. As shown in Figure 5.1, environmental agents act upon artifacts through a management interface: this interface may be public, i.e. accessible to all agents or, most likely, restricted and allowing access to operations typically granted to system administrators. A similar approach to achieve self-organization, involving managers and managed entities, has been adopted also in the Autonomic Computing community [Kephart and Chess, 2003].

Adopting the architecture encoded in this pattern provides several advantages. When working with legacy environmental resources – e.g., provided by an existing infrastructure – relying on additional environmental agents is the only viable solution to add new properties and behaviors, due to a limited control on environmental resources. When developing systems from scratch, the use of this pattern allows different mechanisms to be isolated, thus achieving a finer control on the overall system. Furthermore, we are able to identify and suppress conflicting dynamics that may arise when exploiting different self-organizing mechanisms at the same time [Gardelli et al., 2007]. It is worth noting that environmental agents differentiate from user agents because they play a special role in the system, bound within the environment as perceived by user agents. This is not in contradiction with previous works such as [Omicini et al., 2008], but rather one step beyond: in fact, here user agents perceive the environment exactly in the same way, that is, populated only by artifacts, whereas environmental agents cannot interact with user agents, since they are somehow "encapsulated" within MAS environment.

This pattern can be successfully applied to embed self-organizing mechanisms in MAS environments, especially to environmental services that do not natively support all the self-organization features required. From a methodological viewpoint, when dealing with self-organizing MAS relying on the A&A meta-model and the architectural pattern, the designer focuses its attention to the development of strategies for environmental agents. Indeed, environmental agents are the locus for encapsulating self-organizing mechanisms, since we may not have control over environmental resources. In the remainder of the chapter we provide a brief background on existing MAS methodologies and then describe our systematic approach for engineering self-organizing systems: then, we apply the method to the case study of Plain Diffusion.

5.3 Methodological Issues Raised by Self-Organizing Systems

Currently, the development of agent-oriented systems is supported by several software engineering methodologies. Most methodologies were initially conceived to cover specific issues, and then evolved to encompass the whole software process: for instance, the Gaia methodology [Zambonelli et al., 2003] was mostly concerned with intra-agent problems, while the initial target of the SODA methodology [Omicini, 2001; Molesini et al., 2006] was to tackle the social, inter-agent dimension. On the contrary, other methodologies restricted the domain of applicability to a specific class of MAS, like ADELFE for the Adaptive MAS theory [Bernon et al., 2004].

Concerns like embedding self-organizing mechanisms within an existing MAS, or engineering SOS from scratch, raise peculiar issues that are not typical or so crucial in current AOSE methodologies. Indeed, as pointed out in [Molesini et al., 2009], even core elements such as the environment are currently explicitly supported by only a few of the existing AOSE methodologies. Furthermore, AOSE methodologies, as well as object-oriented ones, tend to focus on design-time aspects rather than run-time ones: in fact, it is common practice to assume that once a system has been designed, its structure will not change and will behave according to the specifications. The Autonomic Computing proposal suggests to consider run-time issues at design-time: then, aspects such as maintenance become a functional requirement of the problem to be solved [Kephart and Chess, 2003], thus increasing the degree of autonomy and adaptiveness of the target system. Along this line, we promote the use of techniques that allow us to preview and analyze global system dynamics at design-time: indeed, when dealing with SOS, more attention should be devoted to observe the emergence of desired properties early in the design stage rather than waiting for the final implementation. In particular, when developing SOS, we have to answer the following question: how can we design the individual environmental agent's behavior in order to ensure the emergence of the desired properties? To tackle this issue, two approaches are typically exploited: (i) devising an *ad-hoc* strategy by decomposition that will solve the specific problem; (ii) observing a system that achieves similar results, and trying to reverse-engineer its strategy. It is generally acknowledged that the former approach is applicable only to a limited set of simple scenarios: due to the non-linearity in entity behaviors, global system dynamics becomes quite difficult to predict. Instead, in the self-organization community, the latter approach is commonly regarded as more fruitful: in nature, it is possible to recognize patterns that are effectively applicable to artificial systems [Babaoglu et al., 2006; De Wolf and Holvoet, 2007; Gardelli et al., 2007; Bonabeau et al., 1999]. Since it is quite unlikely to find a pattern that completely fits a given problem, it is common practice to rely on some modified and adjusted version—in the next section we will elaborate on the implications of these modifications.

Then, once a suitable strategy has been identified and adapted, how can we guarantee that it will behave as expected? Given the specifications of a SOS, how to ensure the emergence of the desired global dynamics is still an open issue. While automatic verification of properties is typically a viable approach with deterministic models, verification becomes more difficult and soon intractable when moving to stochastic models: then, it is useful to resort to a different approach, possibly mixing formal tools and empirical evaluations, so as to support the analysis of the behavior and qualities of a design.

Before describing our approach in the next section, we would like to point out that it is not our goal to develop a brand-new complete methodology for MAS engineering. Instead, we would rather aim at integrating our approach within existing AOSE methodologies, and addressing the peculiar issues raised by self-organizing MAS. For instance, by considering Gaia [Zambonelli et al., 2003], our approach could be seen as a way to direct early design phases: on the one hand, this could help the developer in defining responsibilities for agents and services of the environment by taking inspiration from patterns found in natural systems; on the other hand, it could make it possible to preview the global dynamics of the MAS and tune its behavior before committing to a specific design solution.

5.4 A Methodological Approach for Engineering Self-Organizing MAS

5.4.1 Overview

In this section we describe a systematic approach for the engineering of self-organizing MAS according to the A&A meta-model and our architectural pattern. As previously discussed, since we embed self-organizing mechanisms into environmental agents, our method is mainly focused on the behavior of such agents. Our method should not be considered a full methodology, i.e. encompassing aspects from requirements to maintenance. Conversely, we heavily rely on existing MAS methodologies for many development stages. Indeed, we mainly concentrate in the early design phase, bridging the gap between analysis and the actual design phase. This is probably the most delicate phase when dealing with self-organization and emergence because of the complexity in dynamics.

As an inspiration for our methodology one could easily recognize the statement *"bringing science back into computer science"* [Tichy, 1998]. Indeed, it is heavily based on existing tools commonly used in scientific analysis, especially for complex systems, but that are typically not used in software development. Specifically, our approach is iterative, that is, cycles are performed before actually converging to the final design. During each cycle four steps are performed:

1. *modeling* — proposing a model for the system to engineer taking inspiration from existing natural models and identify concerns for each entity;
2. *simulation* — previewing global system dynamics in different scenarios before continuing with quantitative analysis;
3. *verification* — verifying that the properties of interest holds and identify working conditions;
4. *tuning* — adjusting system behavior and devise a coarse set of parameters for the actual system.

Across the whole process we rely on the use of formal tools and techniques, in order to provide unambiguous specifications and enable automatic processing. In particular, in this work we exploit the PRISM tool [PRISM, 2007; Kwiatkowska et al., 2004], a Probabilistic Symbolic Model Checker developed at University of Birmingham that provides model-checking capabilities along with simulation and model editing integrated within the same software.

Working as a MAS designer, it often happens that, before considering the final system in its actual size, one should first elaborate on a smaller scenario. This is common practice because it is easier to debug systems and observe global dynamics when less entities are involved. In our approach, for verification purposes, we also include model checking despite its well-known problem in tackling large scale systems. Indeed, we envision the use of this technique mostly in the first cycles of development, whereas in the later cycles – when the system is considered in its actual size – we will more rely on simulation techniques. In our opinion, despite its technical limitations, the use of model checking techniques in that context offers several advantages:

- more precise results;
- it is oriented to the property of interest and does not require raw data manipulation: the user queries the model for the satisfaction of a property;
- evaluate the simulation reliability and variance.

Furthermore, technical advancements and research results are providing us with increasing computational power and faster formal verification techniques. Hence, in a near future it will be possible to tackle larger system.

In the remainder of the section we detail each aspect of the approach, and also provide a brief description of the tool capabilities and formalism.

5.4.2 Modeling

In the modeling phase we develop an abstract model of the system, providing a characterization for (i) environmental agents, (ii) artifacts, and eventually (iii) user agents. As far as artifacts are concerned, we can provide an accurate model of their behavior with respect to the usage interface and set of services exposed—though it is often the case that a detailed description of the inner working is not available. Conversely, the repertoire of user agent behavior may be too vast to be accurately modeled. Indeed, in open environments, it is basically impossible to entirely foresee the dynamics of agents to come—self-organization is precisely used to adapt to unpredicted situations. Then, it is necessary to abstract from their peculiarity, resorting to probabilistic or stochastic models of user agent behavior. Models for these agents are developed in terms of usage of resources, i.e. with respect to the observable behavior and by abstracting away from inner processes such as planning and reasoning. The accuracy of the user-agent internal model is not so crucial, since self-organization is built on top of indirect interactions mediated by the environment. Hence, it is sufficient to know how user agents perceive and modify their environment.

Once a suitable model for user agents and artifacts is provided, we move to the core part of modeling, that is, the characterization of environmental agents. A suitable model for environmental agents is typically built on top of the services provided by artifacts, and functionally coupled with user agent behavior in order to establish and sustain a feedback loop: indeed, a feedback loop is a necessary element in every self-organizing system. Consider for example the case of ant colonies, where ants deposit pheromone which diffuses and evaporates in the environment. Then, by perceiving pheromone gradient, ants can coordinate their movement without a priori knowledge of the path to follow.

To find a candidate model for environmental agents, we can take inspiration from known

SOS and look for a model exhibiting or approximating the target dynamics. This step implies the existence of some sort of design-pattern catalog, a required tool for an engineer of SOS. Although this sort of catalog does not exist yet, several efforts by different research groups are moving along this direction. Indeed, several patterns with important applications in artificial systems have already been identified and characterized [Bonabeau et al., 1999; Babaoglu et al., 2006; De Wolf and Holvoet, 2007; Gardelli et al., 2007]. In particular, in the section about the case study, we analyze a problem encoded as the pattern known as *plain diffusion* [Babaoglu et al., 2006; Gardelli et al., 2007]. Hence, even though patterns that perfectly match the target system dynamics can hardly be found, it is still feasible to identify some patterns approximating such a dynamics; however, this typically requires changes of some sort. Given the complex behavior that characterizes SOS, modifications should be carefully evaluated since they require expertise in mechanisms underlying SOS.

Although the model may be provided in several notations, we favor the use of formal languages. In fact, formal languages allow both to devise unambiguous specifications and to perform further automatic analysis, such as simulation and verification. We will discuss more about simulation and verification in the next two sections.

5.4.3 Simulation

In the first cycles we use simulation tools to quickly and easily preview system dynamics before actually performing quantitative analysis. Instead in successive cycles, we eventually rely more on simulation because verification may be unfeasible. However, preliminary verification results act as a sound basis for evaluating reliability of simulation. Before performing simulations, we have to define two key aspects: (i) providing a set of suitable parameters for the model, and (ii) choosing the test instances, i.e. the initial states we consider representative and challenging for the system.

When dealing with self-organizing MAS we mostly rely on stochastic simulation in order to capture both timing and probability aspects. Parameters for this kind of simulation are typically expressed in terms of *rates of action* defined according to suitable statistical distributions. The exponential distribution is typically used because of the *memoryless property*, i.e., to generate new events it is not necessary to know the whole event history but only the current state. Furthermore, the use of exponential distribution allows the mapping to Continuous-Time Markov Chains, which are commonly used in simulation and performance analysis. Then, rates should reflect the conditions in the deployment scenario, otherwise the results of simulation would be meaningless. While parameters for artifacts can be accurately measured, user agents provide a major challenge since we cannot foresee all their possible behaviors. Hence, we have to make assumptions about artifact behaviors both from the qualitative (e.g., rational exploitation of resources) and quantitative (e.g., rate of actions and rate of arrivals/departures) standpoint. Once the parameters of artifacts and agents are defined, we devise an initial set of parameters for environmental agents.

Testing the dynamics of the system in different scenarios and worst case scenarios is very important for a reliable evaluation of the quality and robustness of the model. Typically, we tend to make observations first in very extreme conditions, strictly dependent on the application at hand, since they quickly reveal the presence of faults in the model. Then, we continue the analysis with more real scenarios. In the simulation stage we do not perform quantitative analysis, which is instead a major concern in the verification and tuning stage. This is especially true during the first iteration cycle, since we still have not sufficiently characterized the system model.

5.4.4 Verification

Among the available verification approaches, we are interested in *model checking*, a formal technique for automatically verifying the properties of a target system against its model [Clarke et al., 1999]. The model to be verified is expressed in a formal language, typically in a transition system fashion: the model checker accepts the finite state system specification and translates it into an internal representation, e.g., Binary Decision Diagrams (BDD) [Clarke et al., 1999]. Properties to be verified are expressed using a suitable variant of temporal logic. Then, the properties are automatically evaluated from the model checker for every system execution. Hence, with respect to simulation that tests only a subset of all possible executions, model checking provides more reliable results.

Although model checking was initially targeted to deterministic systems, recent advancements consider also probabilistic as well as stochastic aspects [Kwiatkowska et al., 2007]. A probabilistic model checker uses a temporal logic extended with the suitable operators for expressing probabilities. Then, the output of a probabilistic model checker may be either a boolean answer or the actual probability value for the tested property.

The main drawback of model checking is the *state explosion problem*: since states space typically grow in a combinatorial way, the number of states quickly become intractable as the system grows—and the problem gets even worse with stochastic or probabilistic models. Although modern techniques allows for the reduction of this problem, state explosion is still the main limitation of model checking and abstraction is often required [Delzanno, 2002].

In our approach, model checking is exploited mostly in the first cycles to provide strong guarantees about the emergence of global properties, and in general to precisely characterize the dynamics of the systems. Eventually, if the actual system size is such to prevent verification, in successive cycles we rely mostly on simulation techniques. From the viewpoint of the model checker, emergent properties are not different from ordinary properties. Conversely, from the designer viewpoint emergent properties are quite challenging. A model checker would need properties formulated according to modeled states, however, given the very nature of emergent properties, the states modeled could not be automatically mapped onto the emergent property. Hence, the designer should shift from the global to the individual dynamics and identify the micro properties that are a symptom of the emergent property. This practice will become more clear in Section 5.6.4 when applying the method to the case study.

5.4.5 Tuning

In the *tuning* phase, environmental agent behavior and working parameters are successively adjusted until the desired dynamics are observed. The tuning process is performed exploiting both simulation and verification tools to devise a coarse set of working parameters for the actual system. If not supported by the reference tool, the tuning process can be quite time-consuming. It is worth noting that setting parameters to arbitrary values may lead to unrealistic scenarios. Furthermore, the working rate of environmental agents may affect the actual working rate of artifacts. In a realistic scenario, computational resources are typically limited, hence increasing the working rate of environmental agents may require a decrease in the service rate of artifacts. Without considering this problem, the dynamics of the deployed system may significantly deviate from the expected ones.

At the end of the tuning process, we may realize that the devised set of parameters does not satisfy performance expectations because the values are unrealistic with respect to the execution environment or the system deviates from the desired behavior. In any of these scenarios, we cannot proceed to the actual design phase since the system is not likely to behave properly when deployed. Hence, it is required to perform another iteration for reconsidering the modeling choices or evaluating other approaches. Conversely, when a model meets the target dynamics and the parameters lie within the admissible ranges, we can proceed by providing a more accurate statistical characterization of the system behaviors. This can be performed either by simulation, when the problem instance is too big or does not require strong guarantees, or by model checking which produces more accurate results.

5.5 Using the PRISM Tool to Support the Method

To provide software support to the whole process described in the previous section we evaluated several software tools. In particular, we chose PRISM-Probabilistic Symbolic Model Checker, a software tool developed at University of Birmingham [PRISM, 2007; Kwiatkowska et al., 2004]. PRISM provides facilities supporting most of the stages in our approach, namely, a modeling language, probabilistic/stochastic model checking and simulation. We now analyze the PRISM features with respect to the techniques required at each step.

5.5.1 Modeling

The PRISM modeling language is based on Reactive Modules, and models are specified in a transition systems fashion. The language is able to represent either probabilistic, non-deterministic and stochastic systems using, respectively, Discrete-Time Markov Chains (DTMC), Markov Decision Processes (MDP) and Continuous-Time Markov Chains (CTMC) [Kwiatkowska et al., 2004, 2007]. The components of a system are specified using modules while the state is encoded as a set of finite-values variables. Furthermore, modules are allowed to interact using synchronization in a process algebra style, i.e. labeling commands with actions. Module composition is achieved using the standard parallel composition of Communicating Sequential Processes (CSP) process algebra. As an example consider the specification of a stochastic cyclic counter having base 100

```
ctmc
module counter
    value : [0..99] init 0;
    [] value < 99 -> 1.0 : (value'=value+1);
    [] value = 99 -> 1.0 : (value'=0);
endmodule
```

where `ctmc` declares that the model is stochastic, modeled as a Continuous Time Markov Chain, the module definition is wrapped into the block `module .. endmodule`, `value` is a variable ranging between 0 and 99 initialized to 0, and a transition is expressed according to the syntax `[] guard -> rate : (variable-update)`. For more features and details about the PRISM language syntax please refer to the PRISM documentation [PRISM, 2007].

5.5.2 Simulation

The PRISM built-in simulator provides enough facilities to support our process. The simulator engine makes it possible either to perform step-by-step simulation, or to specify the number of steps to be executed. In particular, the step-by-step mode is very useful for debugging purposes. The simulator traces the values for each variable in the model. It is worth noting that variable range is not very important, conversely to model checking, but can produce unexpected results since ranges may still affect transition guards. PRISM does not provide any plotting capability, however it allows simulation traces to be exported in different formats, so this is not a major concern and can be easily overcome using third-party plotting software. Unfortunately, PRISM simulator does not allow to automatically run experiments, i.e. multiple simulations spanning values for several parameters; hence, the user has to submit several simulation tasks and then collect the results.

5.5.3 Verification

PRISM model checking facilities are very robust: it provides three model checking engines – namely, MTBDD, Sparse and Hybrid – having different memory and computational costs. Properties are expressed according to the temporal logics Probabilistic Computational Tree Logic (PCTL) for DTMC and MDP models, and Continuous Stochastic Logic (CSL) for CTMC models [Kwiatkowska et al., 2004, 2007]. Beyond boolean properties, PRISM allows the computation of actual probability values as well as values for reward-based properties. For instance, referring back to the cyclic counter, the simple property "Which is the steady state probability for the counter to contain the value 10?" is encoded in the CSL formula `S=? [value=10]`, and obviously the result is 0.01. Another example, "Which is the probability for the counter to reach the value 80 within 75 time units?" is encoded in the PCTL formula `P=? [true U<=75 A=80]`, and the result approximated to four decimals is 0.2968.

A very compelling feature of PRISM, because of the CPU-intensive nature of model checking algorithms, is the ability to run model checking experiments. Once the values for parameters range have been defined, the tool automatically performs verification in all the combinations of parameters values.

Another interesting feature is the possibility to evaluate properties with simulation instead of model checking. This allows one to extend model checking results when the instance becomes to large to be formally verified. The user is allowed to set several parameters, e.g., confidence, sample, in order to obtain the desired approximation level.

5.5.4 Tuning

Tuning is performed manually by the user exploiting either simulation and model checking tools. It is mainly a time-consuming trial-and-error approach: given a target set of values, it should be easy to automatize this step exploiting local search algorithms and iterative methods. This would allow the user just to set the desired property and delegate the tool to seek for the right parameters; hence, it worth considering as issues for future works.

5.6 Case Study: Plain Diffusion

In this section we describe a self-organizing strategy for achieving a plain diffusion behavior. The solution is analyzed according to the methodological approach and the tools previously described. Due to space constraints, here we just describe the first iteration of our approach. In order to provide an actual implementation and a more comprehensive characterization several cycles would be required. We decided to consider the case study of plain diffusion mainly for two reasons: on the one hand, our solution to the problem of plain diffusion is very simple but exhibits all the key features of self-organization, hence allowing us to effectively explain our methodological approach; on the other hand, plain diffusion is a key element of many chemical and biological phenomena, e.g., in chemotaxis [Murray, 2002] or in pheromone diffusion in ant colonies [Camazine et al., 2001]. Furthermore, plain diffusion

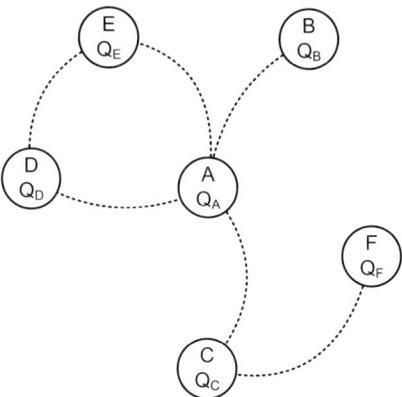

FIGURE 5.2 The reference network topology. This network is interesting because it exhibits features found in real topologies such as cycles, hubs and nodes with limited connectivity.

has been recognized as an important design pattern for self-organizing artificial systems producing gradients and averaging quantities [Babaoglu et al., 2006; Gardelli et al., 2007]. Indeed, despite its simplicity, the diffusion mechanism plays a key role in every digital pheromone-based application, e.g., in the case of Autonomous Guided Vehicles [Sauter et al., 2005; Weyns et al., 2005], and in many distributed systems strategies such as in load-balancing [Canright et al., 2006].

5.6.1 Problem Statement

Consider a networked set of nodes having an arbitrary topology and where each node is labeled with a non-negative quantity. We want to devise a strategy that from an arbitrary initial state eventually evolves into a dynamical state where each node is labelled with the same quantity. In particular, we require the strategy to be self-organizing, i.e. where each node is autonomous and transfers quantities according to local knowledge. In this way our strategy will be independent from the network topology, the distribution of quantities and the overall amount of quantities. A node knows the identities of neighboring nodes and the local quantity, while it has no information about network size and quantities in other nodes.

In order to evaluate our proposal, we have to test it against an actual instance of a network. Specifically, we choose the 6-node topology displayed in Figure 5.2 since it exhibits features commonly found in actual networks. These features include cycles, hubs and nodes with limited connectivity. We believe that the 6-node topology is large enough to clarify the approach, which is our main objective here. Scalability issues are discussed later in Section 5.6.6. For the sake of clarity, from now on when considering system states we use the compact notation $((A, Q_A), ..., (F, Q_F))$. The required dynamics for strategy are the following: each node i having a local quantity Q_i has to eventually reach a state where $Q_{avg} = \frac{\sum_{i=1}^{N} Q_i}{N}$. It is worth noting that because of the limited knowledge available to each node the strategy will never converge since it has no criteria regarding the halting condition. The best result we can achieve is to establish a dynamic equilibrium close to the average value.

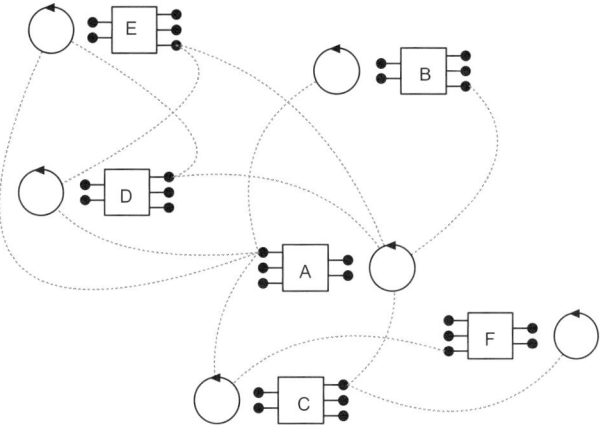

FIGURE 5.3 The picture shows the agents and artifacts diagram equivalent to the previous network topology. It is worth noting that since the agents/artifacts relation is not symmetric, two association are required making the diagram appears more cluttered.

5.6.2 Modeling Plain Diffusion

In this section we provide a solution to the previously described problem with respect to the agents and artifacts meta-model and architectural pattern. The mapping between the network and the A&A equivalent architecture is straightforward, and basically amounts to replacing a node with an artifact and its environmental agent, see Figure 5.3. More precisely,

- each node is represented by an artifact acting as a data repository;
- each artifact is managed by a dedicated environmental agent;
- the connections represent neighborhood information that can be encoded either within environmental agents or artifacts: both approaches are viable thus modeling different constraints.

The constraints and possible actions then become

- an environmental agent can put/remove an item only in the local artifact and remote artifacts in its neighborhood;
- an environmental agent knows the number of items contained within the local artifact;
- an environmental agent knows a limited set of artifacts, which we call neighboring artifacts;
- the agent does know neither the overall number of artifacts nor the the overall number of items in the system.

In our approach the first step consists in finding an existing pattern: here we recognize that the problem can actually be assimilated to Plain Diffusion, which has been recognized as an important design pattern in [Babaoglu et al., 2006; Gardelli et al., 2007]. However, the solution proposed in [Babaoglu et al., 2006] requires the exchange of information between nodes, which is in contrast with our requirements. Hence, we propose a different approach for achieving plain diffusion: to this purpose, we start by considering the two nodes network $((A, 20), (B, 10))$ where B have twice as many items as A. Since they actually do not know the number of items of the other node the only way to reach an equilibrium is via dynamic exchange criteria. Unfortunately if nodes exchange items at the same speed the balance remains unchanged. Hence, as a first proposal we suggest that nodes send items at a speed proportional to the number of items possessed. We notice that this basic strategy does not work for the network $((A, 20), (B, 10), (C, 20))$ where A and C are connected only to B: indeed, we expect a gradient from this situation since B receives items both from A and C. In order to compensate for this gradient, the working rate of each agent should be proportional not only to the number of items but also to the number of neighboring nodes. Hence, the formula for the agent send rate r_i becomes

$$r_i = \frac{Q_i * S_i}{P} \tag{5.1}$$

where Q_i is the local number of items, S_i is the local star, i.e. number of neighboring nodes, and P is a global parameter that scales the overall workload.

From the requirements and the basic strategy we now provide a formal model using the PRISM modeling language: the whole specification is listed in Figure 5.4. Since PRISM language allows the definition of stochastic transition systems [PRISM, 2007], we have to re-interpret the system dynamics in terms of transitions. To the purpose of our model, in this case, we abstract from artifact's details: since in the plain diffusion model we are only interested in an artifact's content, this information can be encoded in a simple variable. Conversely, agents are encoded in modules, that is, a collection of transitions: hence, agents manipulate local and neighboring artifacts by simply modifying the corresponding variable. With respect to the topology defined in Figure 5.2, the definition of the environmental agent A is

```
module agentA
[] tA > 0 & tB < MAX & tC < MAX & tD < MAX ->
rA : (tA'=tA-1) & (tB'=tB+1) +
rA : (tA'=tA-1) & (tC'=tC+1) +
rA : (tA'=tA-1) & (tD'=tD+1) +
rA : (tA'=tA-1) & (tE'=tE+1);
endmodule
```

where `tA` is the local artifact, `tB, tC, tD` are neighboring artifacts, `rA` is the rate of the transition defined by `rA = tA / base_rate`. Each transition models the motion of an item from the local artifact to a neighboring one. The choice between neighbors is probabilistic and in this case all the transitions are equiprobable. It is worth noting that the rate formula does not explicitly take into account the number of neighboring nodes: indeed, this factor is implicitly encoded in the transition rules. Since the model is interpreted as a Markov Chain the overall rate is the sum of all the transition rates. In the previous code sample we have four top-level possible transitions with rate `rA`, hence the overall working rate of `agentA` is `4rA`. The definition of the other agents is very similar to the one of `agentA` but for the number of neighboring artifacts.

5.6.3 Simulating Plain Diffusion

In order to qualitatively evaluate the dynamics of the system, in this section we run some simulations. PRISM allows the execution of simulations directly from the formal specification as long as we provide values for all the parameters. In our model the only parameter is the `base_rate`. This parameter allows the tuning of the system speed according to deployment requirements. Since at the moment we are not interested in performance issues

```
ctmc

const int MAX = 54;
const double base_rate = 100;

formula rA = tA / base_rate;
formula rB = tB / base_rate;
formula rC = tC / base_rate;
formula rD = tD / base_rate;
formula rE = tE / base_rate;
formula rF = tF / base_rate;

global tA : [0..MAX] init 14;
global tB : [0..MAX] init 0;
global tC : [0..MAX] init 8;
global tD : [0..MAX] init 16;
global tE : [0..MAX] init 12;
global tF : [0..MAX] init 4;

module agentA
[] tA > 0 & tB < MAX & tC < MAX & tD < MAX ->
rA : (tA'=tA-1) & (tB'=tB+1) +
rA : (tA'=tA-1) & (tC'=tC+1) +
rA : (tA'=tA-1) & (tD'=tD+1) +
rA : (tA'=tA-1) & (tE'=tE+1);
endmodule

module agentB
[] tB > 0 & tA < MAX ->
rB : (tB'=tB-1) & (tA'=tA+1);
endmodule

module agentC
[] tC > 0 & tA < MAX & tF < MAX->
rC : (tC'=tC-1) & (tA'=tA+1) +
rC : (tC'=tC-1) & (tF'=tF+1);
endmodule

module agentD
[] tD > 0 & tA < MAX & tE < MAX->
rD : (tD'=tD-1) & (tA'=tA+1) +
rD : (tD'=tD-1) & (tE'=tE+1);
endmodule

module agentE
[] tE > 0 & tA < MAX & tD < MAX ->
rE : (tE'=tE-1) & (tA'=tA+1) +
rE : (tE'=tE-1) & (tD'=tD+1);
endmodule

module agentF
[] tF > 0 & tC < MAX ->
rF : (tF'=tF-1) & (tC'=tC+1);
endmodule
```

FIGURE 5.4 The PRISM specification of the plain diffusion strategy for the reference 6-node network topology. It is worth noting that each module represents a node in the network and the respective environmental agent.

we set it to the arbitrary value of 100. We consider now a few scenarios modeling extreme deployment scenarios to evaluate the robustness and adaptiveness of the solution.

The first instance we consider has all the items clustered into a single node, specifically node A, the hub: using the compact notation the system initial state is ((A, 600), (B, 0), (C, 0), (D, 0), (E, 0), (F, 0)). As can be observed from Figure 5.5, all the nodes eventually reach the average value of 100 and then stay close to it. In particular, the node F requires more time to reach the value because it is two hops away from node A, while all the other nodes are just one hop away.

The next instance we consider is the one having all the items clustered into the peripherical node F: specifically, the system initial state is ((A, 0), (B, 0), (C, 0), (D, 0),(E, 0), (F, 600)). As it can be observed from Figure 5.6, all the nodes eventually converge to the average value of 100: in particular, node C converges quickly because it is one hop away from the node F, while all the other nodes are two hops away. It is also worth noting that before node C reaches dynamic equilibrium it goes over the average value. This phenomenon is due to the fact that node A works many times faster than node C which slowly diffuses items to neighboring nodes. With respect to the previous instance, this configuration requires almost twice as much time to establish a dynamic equilibrium. Such a big variance depends on the fact that while on the previous instance the items were clustered in a node with 4 neighbors, in this instance the items were clustered in a peripherical node with only one neighbor, causing a bottleneck.

The next instance we consider is the one having items spread across the nodes, specifically ((A, 50), (B, 150), (C, 200), (D, 0), (E, 50), (F, 150)). As can be observed from Figure 5.7, all the nodes eventually reach the average value of 100 and stay close to it. Since this configuration was more ordered than the previous ones it reaches dynamic equilibrium faster.

The next instance we consider models the situation where a node is dynamically added to the network: specifically the initial state is ((A, 120), (B, 120), (C, 120), (D, 120), (E, 120), (F, 0)) where the nodes from A to E have the same number of items and F is the newly-added node. As can be observed from Figure 5.8, the nodes move from the average value of 120 to the new average value of 100 due to the connection of a new node. It is worth noting that the speed of the adaptation is strongly dependent on the number of connections of the new node, but also weakly dependent on the network topology.

Since in all the simulated scenarios the strategy seems to behave properly we now move to the verification of the desired properties.

5.6.4 Verifying Plain Diffusion

The verification process consists of testing whether the properties of interest hold or not. This process is performed relying on stochastic model checking techniques [Kwiatkowska et al., 2007; Rutten et al., 2004]. As anticipated in Section 5.4.4, model checking techniques suffer from the state explosion problem: considering an instance of the same size as done for the simulations it is just not feasible. Hence, we consider a far smaller instance, specifically, the system instance having 36 items and expecting an average value of 6 items per node. In this section, we always refer to the initial configuration ((A, 4), (B, 0), (C, 10), (D, 12), (E, 6), (F, 4)). Although being a small instance, it is already computationally intensive: specifically, the instance is defined by 749398 states and by 7896096 transitions, and it

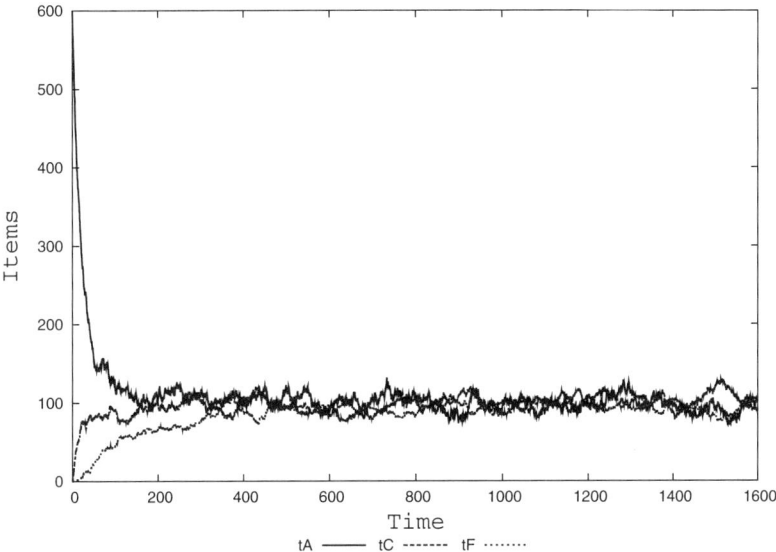

FIGURE 5.5 The evolution of the instance ((A, 600), (B, 0), (C, 0), (D, 0), (E, 0), (F, 0)): as it can be noticed the node F converges more slowly because it is two hops far from node A, while all the other nodes are just one hop away. For the sake of clarity nodes B, D and E have been omitted.

takes about 15 seconds just to compile the model using the PRISM Hybrid Engine*.

We are here interested in verifying a few system properties: the first property is about the quality of the strategy with respect to its goal, i.e., producing an average value. Since the strategy modeled is stochastic, we can provide a statistical characterization of this property: in particular, we can devise the probability** distribution for a node to be in a specific state, that is, being assigned a particular value. In Continuous Stochastic Logic, this property is equivalent to the statement *"Which is the steady-state probability for the variable X to assume the value Y?"* Since we want a probability distribution and not a single value, we run an experiment where Y spans the range [0..36]. Using the PRISM syntax, this property translates to S=? [tA=Y] where S is the steady state operator, tA is the variable containing the actual value and Y is the unbounded constant ranging in the interval 0..36. The chart in Figure 5.9 displays the results of the model checking experiment over the specific node tA: the experiment took about 3 hours using the Hybrid Engine and Jacobi iterative method. Experiments over the other nodes showed an identical probability distribution, providing evidence of the correctness of the strategy and the independence from initial node value and placement in the network. As we expected, the maximum probability peak corresponds to the average value, although being only the 17.59%. The system has a probability of 49.68%

---

*The computer used for all the simulations and verifications has the followings processing capabilities: CPU Intel P4 Hyper Threading 3.0 GHz, RAM 2 GB DDR, System Bus 800 MHz.
**It is worth noting that model checking techniques provide exact probability values rather than estimation as for simulation.

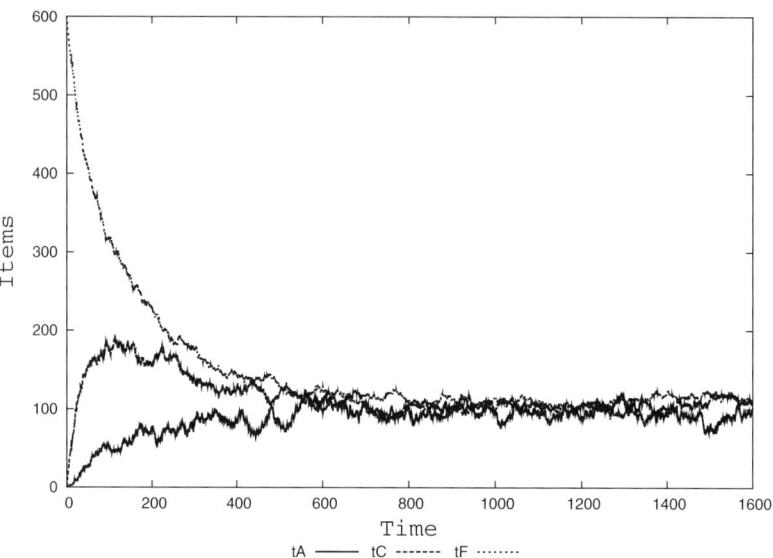

FIGURE 5.6 The evolution of the instance ((A, 0), (B, 0), (C, 0), (D, 0), (E, 0), (F, 600)). As can be noticed, node C converges more quickly because it is one hop away from the node F, while all the other nodes are two hops away. For the sake of clarity nodes B, D and E have been omitted.

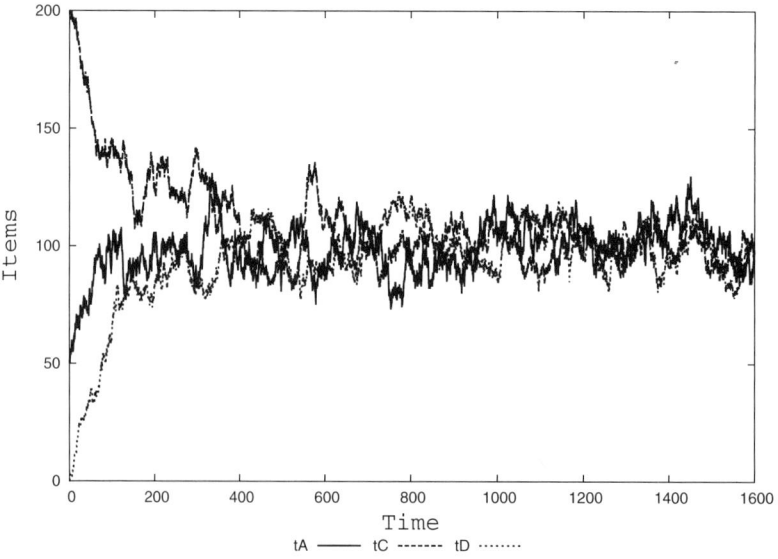

FIGURE 5.7 The evolution of the instance ((A, 50), (B, 150), (C, 200), (D, 0), (E, 50), (F, 150)). As can be noticed all the nodes eventually converge to the average value. For the sake of clarity nodes B, E and F have been omitted.

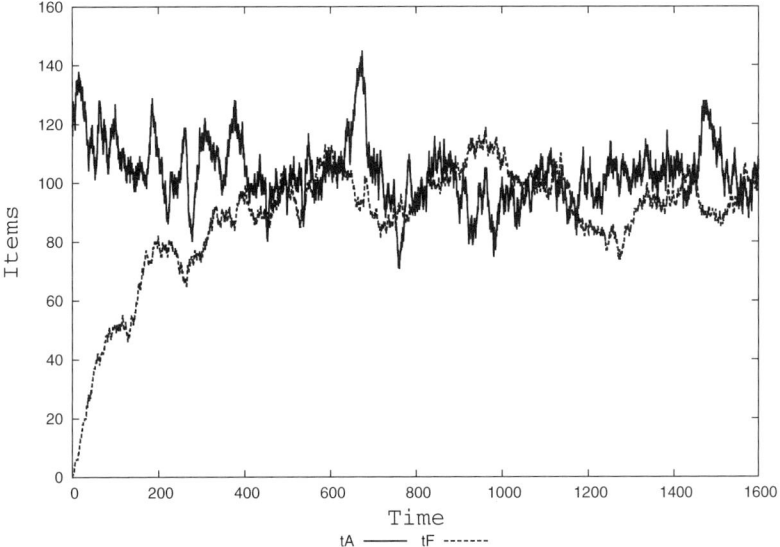

FIGURE 5.8 The evolution of the instance ((A, 120), (B, 120), (C, 120), (D, 120), (E, 120), (F, 0)) modeling the dynamic connection of a new node. As can be noticed, the average value is moved from 120 to 100 and all the nodes eventually reach the new average value. For the sake of clarity nodes B, C, D and E have been omitted.

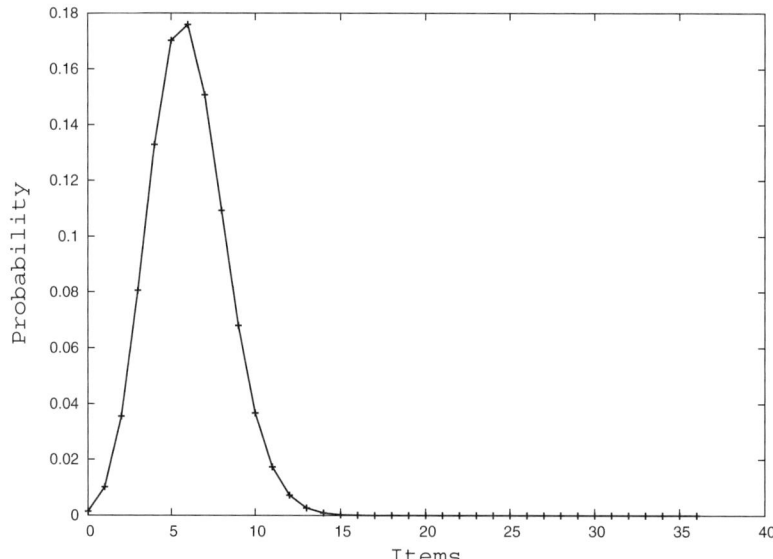

FIGURE 5.9 The chart displays the distribution of the probability for a node to contain a specific number of items; further experiments show that the chart is the same for each node. Notice that the probability peak is located in correspondence with the average value, that is 6. The chart is not symmetrical due to the asymmetry in the range of values.

of being in the range 6 ± 1, while it has a probability of 73.91% of being in the range 6 ± 2. It is worth noting that the chart is not symmetrical with respect to the average value because of the asymmetry of the admissible range.

The next test we consider is about the time for reaching the average value: specifically, considering the previous test instance ((A, 4), (B, 0), (C, 10), (D, 12), (E, 6), (F, 4)) we evaluate the time for tB, having an initial value of 0, to reach the average value of 6. Since CSL does not provide a time operator, we still have to reason in terms of probability. Hence, the query is *"Which is the probability for the node tB to be equal to 6 within Y time steps?"* Using the PRISM syntax, this formula becomes `P=? [true U<=Y tB=6]` where `P` is the probability operator, `true U<=Y` means to be verified in the next `Y` time steps, and `Y` is an unbounded constant. The chart in Figure 5.10 displays the results of model checking for `Y` spanning the range [0..600] at step size equal to 10. Although valid only for node tB, this sort of chart is very useful since it provides a solid basis for tuning and in-depth performance evaluation.

5.6.5 Tuning Plain Diffusion

Since the strategy modeled already exhibits a global dynamics which is compliant to our requirements there is no need for tuning but for performance. In the simulation and verification sections we assumed the arbitrary value of 100 for the `base_rate` parameter. If we want to give guarantees with respect to deployment conditions we have to consider a meaningful parameter value. For example, we can establish a requirement for all the nodes

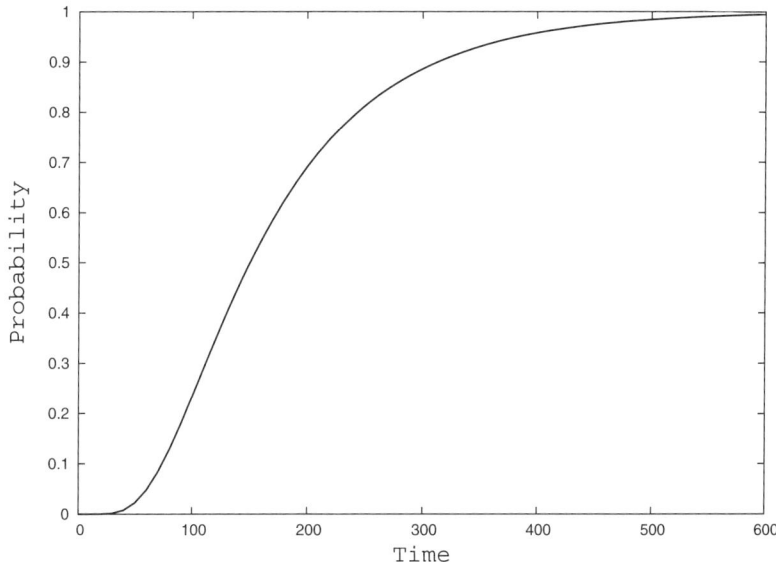

FIGURE 5.10 The chart displays the probability for the node tB to reach the average value from zero: this chart is very useful for the tuning process since it provide a solid basis for performance evaluation.

to have a probability greater or equal to 90% to reach the average value within T time units under specific workloads conditions. Then, by performing several tests we can devise the actual value for the `base_rate` parameter that meets the target requirement.

In order to evaluate the performances, we consider the worst case scenario, which consists in all the items clustered in the peripherical node F. Specifically the initial state is ((A, 0), (B, 0), (C, 0), (D, 0), (E, 0), (F, 36)). As a time constraint, we require the system to reach dynamic equilibrium before 200 time units, while for the probability constraint we set the lower bound to 90%. Hence, by adjusting the value of `base_rate`, we have to test the following property for all nodes: *"Is the probability of reaching dynamic equilibrium condition within 200 time units greater or equals to 90% ?"*. Using the PRISM syntax this property becomes `P>=0.9 [true U<=200 tA=6]` for the node `tA`.

We start by considering the farthest node tB. Since the property must hold for all nodes, testing first the farthest node saves us a lot of computation. As a first exploration, with respect to the property `P=? [true U<=200 tB=6]`, we plot the probability values within the range [10..100]. As can be observed from chart in Figure 5.11, the trend of the probability is non-linear. Nonetheless, we can guess that the desired value for `base_rate` lies in the range [30..40]. Hence, we repeat the experiment zooming in the range [30..40] with unitary step: the results are plotted in Figure 5.12. As we can observe the value for `base_rate` that produces the probability value closest to 90% is 37. Hence, we fix this value for the base rate and tested if the property is also satisfied for the other nodes. Since the node tB is the farthest one, as we expected the property is satisfied also for node tA, tC, tD, tE, but not for tF. Hence, we investigate the property for node tF and results are plotted in Figure 5.13. As can be observed, the property holds for node tF when `base_rate=31` and not 37. This phenomenon can be explained by the fact that we requested a 90% probability for the

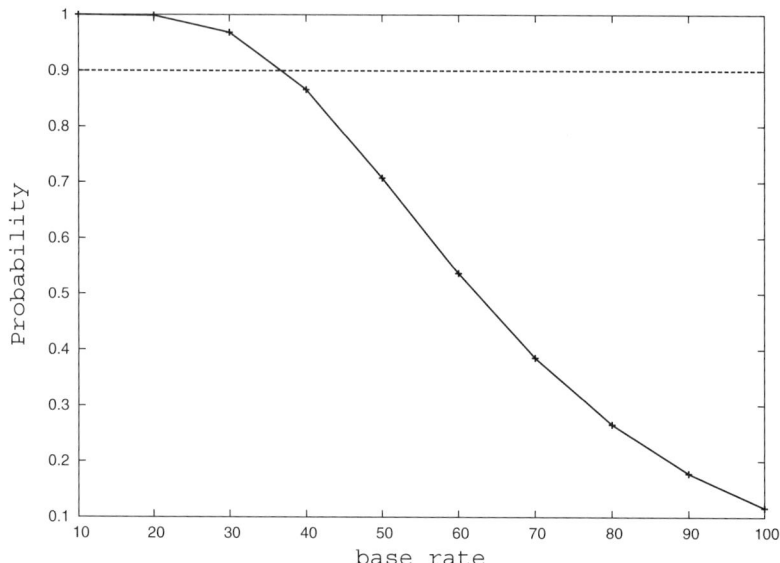

FIGURE 5.11 The chart displays the probability values for the node tB according to the CSL formula $P =?[true U <= 200 tB = 6]$. We can guess that the desired value is within the range [30..40].

nodes to reach the target value; hence, there is a 10% probability for each node not to have reach the target value, which summed up in the source node cause this delay.

Hence, from all the experiments we obtain that the value for the `base_rate` parameter to satisfy the property of $\geq 90\%$ probability of convergence within 200 time units is 31. Our first guess was that the bottleneck should have been the farthest node, conversely it proved out to be the source node. The parameter `base_rate=31` means that initially the system send about 1 item per unit of time per connection, and decreases while converging to about 1 item per 5 units of time per connection.

As far as the number of transfers is concerned, it is not trivial to evaluate it given the very nature of the strategy; indeed, it depends upon neighborhood, local number of items, which changes over time. If there was no dependence on the number of neighbors, assuming a static topology, the global transfer rate could have been evaluated by the simple formula

$$r_g = \frac{Tot.Items}{baserate} \tag{5.2}$$

Conversely, because of the dependence on the number of neighbors, the instantaneous global rate is given by the formula

$$r_g(t) = \sum_{i=1}^{N} \frac{Items_i(t) \times neighbors_i}{baserate} \tag{5.3}$$

where $i = 1..N$ is the node selector. This formula explicitly depends on time, and requires either a simulation or real data to be computed.

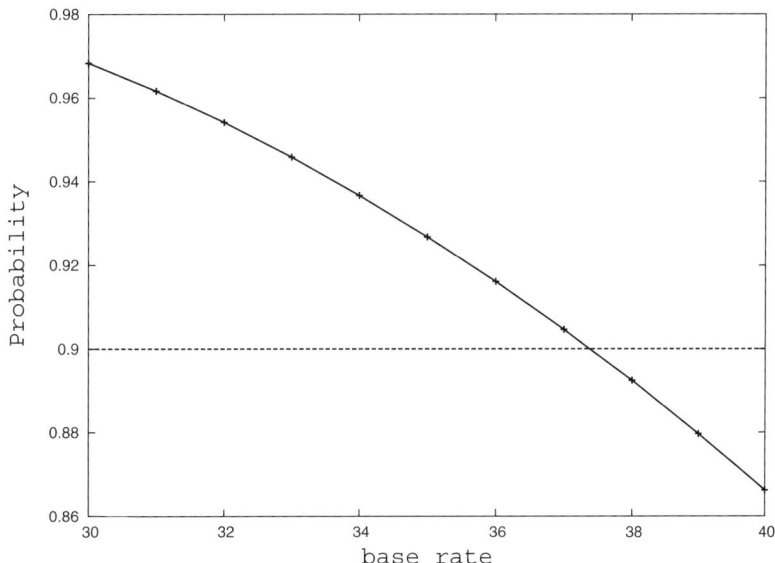

FIGURE 5.12 The chart displays the probability values for the node tB according to the CSL formula $P=?[true U <= 200 tB = 6]$ within the range $[30..40]$ of base rate.

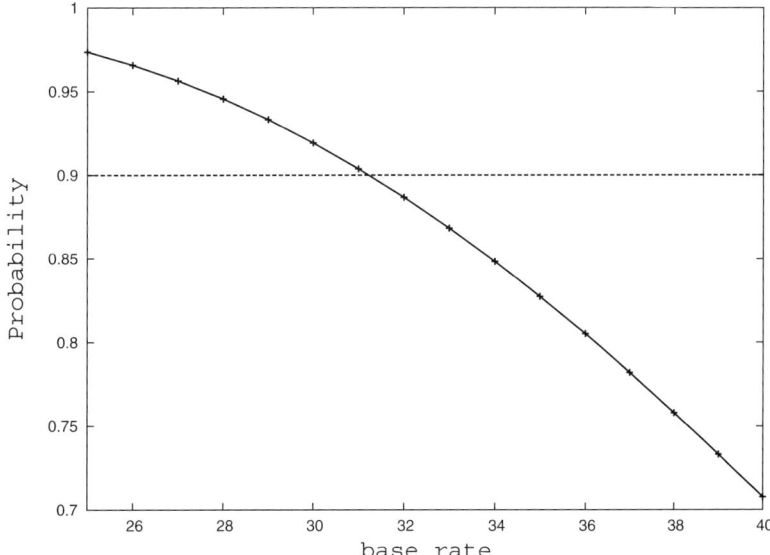

FIGURE 5.13 The chart displays the probability values for the node tF according to the CSL formula $P=?[true U <= 200 tF = 6]$ within the range $[25..40]$ of base rate.

5.6.6 Preliminary Scalability Analysis

In this section we briefly describe how to perform preliminary scalability analysis using model checking techniques. Indeed, it might be the case that we are interested in its performance at the early stages since performance may be a major constraint. Hence, before moving to the problem in its actual size, it could be useful to perform this kind of analysis. In the next iterations, when considering the large system, scalability measures would be necessarily obtained via simulations rather than model checking.

Here, concerning the scalability metric, we consider the time required to converge to the average value with respect to the total number of items within the network. Another interesting scalability measure could be with respect to the network size rather than content. The initial state for each trial consists in all the items placed within node A, i.e. A=X, B=C=D=E=F=0. Since we are using model checking tools to perform such evaluation, we will limit the number of items to 48. Furthermore, to avoid fractional average values we consider only quantities multiple of the number of nodes, specifically the set of values $\{6, 12, 18, 24, 30, 36, 42, 48\}$.

The property we are interested in evaluating is the time required to converge with a probability $\geq 90\%$. Unfortunately PRISM does not allow time queries, hence, we have to run several experiments in order to profile probability of convergence with respect to time. The property we want to verify can be translated into P=? [true U<=T tA=X/6], where T represents the time value and X is the total number of items. Notice that we do not have a formula for describing global convergence, hence, we have to provide a formula for convergence of a single node: we consider the source node since, from previous experiments, we know this is the most restrictive constraint. The charts in Figures 5.14 and 5.15 display the probability for the node tA to reach the average value with respect to time. From these charts it is possible to extrapolate, although with some approximations, the time values for each initial state having a probability $\geq 90\%$ to reach the average value. Data extracted from previous charts is depicted in Figure 5.16. As can be easily noticed the trend is sublinear. It is worth it to point out that these results have been obtained with respect to the specific network topology and initial distribution of items. Hence, the characterization provided here is far from being complete, although we are confident of obtaining similar results with different distributions of items. Conversely, different topologies may heavily affect the scalability of the approach.

5.7 Conclusion

In this chapter we described a systematic approach for driving the early design phase of self-organizing MAS engineering, combining simulation and the use of formal tools. Specifically, it is an iterative approach articulated in four steps—namely, modeling, simulation, verification and tuning. This method wants to promote the adoption of analysis techniques and tools in the self-organizing MAS engineering process. While simulation and formal verification techniques are extensively used for analysis purposes, their advantages are almost overlooked in the software engineering field [Tichy, 1998; von Mayrhauser, 1993; Uhrmacher, 2002]. The limitations on the applicability of the approach are mainly due to the state explosion problem inherent to model checking, which prevents tackling large scale systems. Regardless, the use of model checking techniques is quite useful in the first iterations of the approach on small instances of the problem, allowing also the reliability of the simulations to be evaluated. Furthermore, because of the rapid advancement of computation capabilities and efficiency in verification techniques we believe that such techniques will become more important in the near future.

Combining Simulation and Formal Tools for Developing Self-Organizing MAS 159

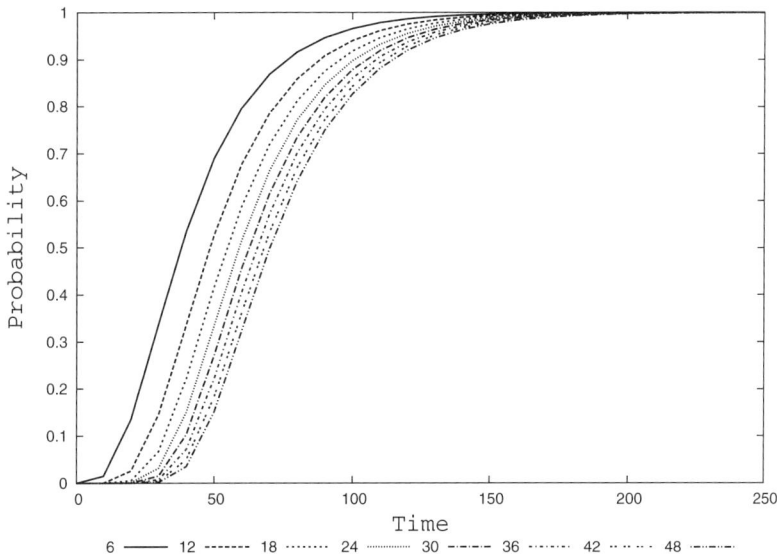

FIGURE 5.14 The chart displays the probability values for the node tA to reach the average value with respect to time and from different initial states.

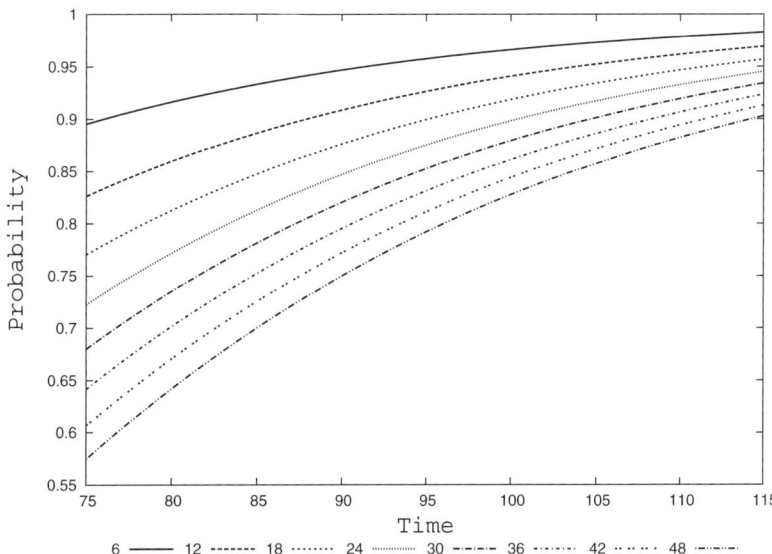

FIGURE 5.15 The chart displays the probability values for the node tA to reach the average value with respect to time and from different initial states: this chart zooms into the range [75..115] with a finer time step.

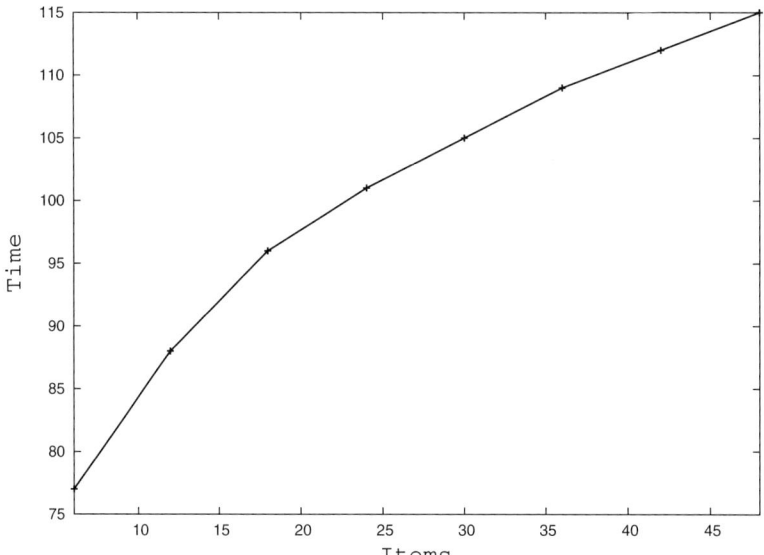

FIGURE 5.16 The chart represents the preliminary scalability analysis performed with model checking techniques: the metric is time to reach dynamic equilibrium with respect to the initial number of items contained in the node tA. From these preliminary results we observe that the trend is sub-linear. Since the results depend on the network topology and distribution of items, further investigation is needed to provide a complete characterization of the strategy.

We have applied the methodology to the case study of plain diffusion. Plain diffusion was chosen because it is a key element of many chemical and biological phenomena, e.g., in chemotaxis [Murray, 2002] or in pheromone diffusion in ant colonies [Camazine et al., 2001], and has been recognized as an important design pattern for self-organizing artificial systems allowing gradients and averaging quantities [Babaoglu et al., 2006; Gardelli et al., 2007], which is the foundation for every digital pheromone-based application [Sauter et al., 2005; Weyns et al., 2005]. To the best of our knowledge, the proposed solution to the plain diffusion problem is novel with respect to the current research context.

The whole process has been supported by the PRISM tool, a software developed at the University of Birmingham. Although mainly targeted at probabilistic model checking, the tool also provides a suitable modeling language and basic simulation capabilities.

Along this research direction, we identify a few future developments:

- Although an earlier version of the approach has been tested on other case studies – e.g., the Collective Sort case [Gardelli et al., 2008] – we could gain more feedback on the suitability of the approach by analyzing further distributed problems before moving to real applications;
- A higher-level language could be developed on top of PRISM in order to ease MAS engineers in the specification process. We believe this is a necessary step to widen the adoption of the approach and the use of formal tools;
- While the state explosion problem of stochastic model checking is the main limitation of the approach, research efforts in the formal methods community in the last few years have been impressive, and new approaches are currently being developed. Applying those improved techniques is also part of the future works;
- Since there exist several well-developed agent-oriented methodologies such as GAIA [Zambonelli et al., 2003] or ADELFE [Bernon et al., 2004], it is worth evaluating the integration of our approach within these methodologies.

References

H. Abelson, D. Allen, D. Coore, C. Hanson, G. Homsy, J. Thomas F. Knight, R. Nagpal, E. Rauch, G. J. Sussman, and R. Weiss. Amorphous computing. *Commununications of the ACM*, 43(5):74–82, May 2000. ISSN 0001-0782. doi: http://doi.acm.org/10.1145/332833.332842.

W. R. Ashby. Principles of self-organizing dynamic systems. *Journal of General Psychology*, 37:125–128, 1947.

O. Babaoglu, H. Meling, and A. Montresor. Anthill: a framework for the development of agent-based peer-to-peer systems. In *22nd International Conference on Distributed Computing Systems (ICDCS'02)*, pages 15–22, Vienna, Austria, July 2002. IEEE Computer Society. ISBN 0-7695-1585-1. doi: 10.1109/ICDCS.2002.1022238. URL http://ieeexplore.ieee.org/xpls/abs_all.jsp?arnumber=1022238.

O. Babaoglu, G. Canright, A. Deutsch, G. A. Di Caro, F. Ducatelle, L. M. Gambardella, N. Ganguly, , M. J. Roberto Montemanni, A. Montresor, and T. Urnes. Design patterns from biology for distributed computing. *ACM Transactions on Autonomous and Adaptive Systems*, 1(1):26–66, Sept. 2006. ISSN 1556-4665. doi: 10.1145/1152934.1152937.

C. Bernon, V. Camps, M.-P. Gleizes, and G. Picard. Designing agents' behaviors and interactions within the framework of ADELFE methodology. In *Engineering So-*

cieties in the Agents World, volume 3071 of *LNCS (LNAI)*, pages 311–327. Springer, 2004. ISBN 978-3-540-22231-6. doi: 10.1007/b98212. 4th International Workshops, ESAW 2003, London, UK, October 29-31, 2003, Revised Selected and Invited Papers.

E. Bonabeau, M. Dorigo, and G. Theraulaz. *Swarm Intelligence: From Natural to Artificial Systems*. Santa Fe Institute Studies in the Sciences of Complexity. Oxford University Press, New York, NY, USA, 1999. ISBN 0195131584.

S. Camazine, J.-L. Deneubourg, N. R. Franks, J. Sneyd, G. Theraulaz, and E. Bonabeau. *Self-Organization in Biological Systems*. Princeton Studies in Complexity. Princeton University Press, Princeton, NJ, USA, 2001. ISBN 0691012113.

G. Canright, A. Deutsch, and T. Urnes. Chemotaxis-inspired load balancing. *Complexus*, 3(1-3):8–23, August 2006. ISSN 1424-8492. doi: 10.1159/000094184.

M. Casadei, L. Gardelli, and M. Viroli. Simulating emergent properties of coordination in Maude: the collective sort case. *Electronic Notes in Theoretical Computer Science*, 175(2):59–80, June 2007. ISSN 1571-0661. doi: doi:10.1016/j.entcs.2007.05.022. 5th International Workshop on the Foundations of Coordination Languages and Software Architectures (FOCLASA 2006).

V. A. Cicirello and S. F. Smith. Wasp-like agents for distributed factory coordination. *Autonomous Agents and Multi-Agent Sytems*, 8(3):237–266, May 2004. ISSN 1573-7454. doi: 10.1023/B:AGNT.0000018807.12771.60. URL http://www.springerlink.com/content/p1126416q70x301w/.

E. M. Clarke, Jr., O. Grumberg, and D. A. Peled. *Model Checking*. The MIT Press, Cambridge, MA, USA, 1999. ISBN 0262032708.

T. De Wolf and T. Holvoet. Emergence versus self-organisation: Different concepts but promising when combined. In S. Brueckner, G. Di Marzo Serugendo, A. Karageorgos, and R. Nagpal, editors, *Engineering Self Organising Systems: Methodologies and Applications*, volume 3464 of *LNCS (LNAI)*, pages 1–15. Springer, May 2005. ISBN 978-3-540-26180-3. doi: 10.1007/11494676_1.

T. De Wolf and T. Holvoet. Design patterns for decentralised coordination in self-organising emergent systems. In *Engineering Self-Organising Systems*, volume 4335 of *LNAI*, pages 28–49. Springer, February 2007. 4th International Workshop on Engineering Self-Organising Applications (ESOA'06), Hakodate, Japan, 9 May 2006.

G. Delzanno. An overview of MSR(C): A CLP-based framework for the symbolic verification of parameterized concurrent systems. *Electronic Notes in Theoretical Computer Science*, 76:65–82, November 2002. ISSN 1571-0661. doi: 10.1016/S1571-0661(04)80786-2. URL http://www.elsevier.com/gej-ng/31/29/23/126/23/show/Products/notes/index.htt. WFLP 2002, 11th International Workshop on Functional and (Constraint) Logic Programming, Selected Papers.

L. Gardelli, M. Viroli, and A. Omicini. Design patterns for self-organising systems. In H.-D. Burkhard, G. Lindemann, R. Verbrugge, and L. Z. Varga, editors, *Multi-Agent Systems and Applications V*, volume 4696 of *LNCS (LNAI)*, pages 123–132. Springer, Heidelberg, 2007. ISBN 978-3-540-75253-0. doi: 10.1007/978-3-540-75254-7_13. 5th International Central and Eastern European Conference on Multi-Agent Systems, CEEMAS 2007, Leipzig, Germany, September 25–27, 2007, Proceedings.

L. Gardelli, M. Viroli, M. Casadei, and A. Omicini. Designing self-organising environments with agents and artefacts: A simulation-driven approach. *International Journal of Agent-Oriented Software Engineering (IJAOSE)*, 2(2):171–195,

2008. ISSN 1746-1375. doi: 10.1504/IJAOSE.2008.017314. URL http://www.inderscience.com/search/index.php?action=record&rec_id=17314. Special Issue on Multi-Agent Systems and Simulation.

J. Goldstein. Emergence as a construct: History and issues. *Emergence*, 1(1):49–72, 1999. doi: 10.1207/s15327000em0101_4.

P.-P. Grassé. La reconstruction du nid et les coordinations interindividuelles chez bellicositermes natalensis et cubitermes sp. la théorie de la stigmergie: Essai d'interprétation du comportement des termites constructeurs. *Insectes Sociaux*, 6(1):41–80, Mar. 1959. ISSN 0020-1812. doi: 10.1007/BF02223791. URL http://www.springerlink.com/content/7855803040371473/.

J. O. Kephart and D. M. Chess. The vision of Autonomic Computing. *Computer*, 36 (1):41–50, January 2003. ISSN 0018-9162. doi: 10.1109/MC.2003.1160055. URL http://www.research.ibm.com/autonomic/manifesto/.

M. Kwiatkowska, G. Norman, and D. Parker. Probabilistic symbolic model checking with PRISM: A hybrid approach. *International Journal on Software Tools for Technology Transfer (STTT)*, 6(2):128–142, September 2004. ISSN 1433-2779. doi: 10.1007/s10009-004-0140-2. URL http://www.springerlink.com/content/9g2xyp0rv0c5bvx0/. Special section on tools and algorithms for the construction and analysis of systems.

M. Kwiatkowska, G. Norman, and D. Parker. Stochastic model checking. In M. Bernardo and J. Hillston, editors, *Formal Methods for the Design of Computer, Communication and Software Systems: Performance Evaluation (SFM'07)*, volume 4486 of *LNCS (Tutorial Volume)*, pages 220–270. Springer, June 2007. ISBN 978-3-540-72482-7. doi: 10.1007/978-3-540-72522-0_6. URL http://www.springerlink.com/content/4717r52705872124/?p=c38dfc068cca4836bceafa278f68bc8f&pi=5. 7th International School on Formal Methods for the Design of Computer, Communication, and Software Systems, SFM 2007, Bertinoro, Italy, May 28-June 2, 2007.

M. Mamei and F. Zambonelli. Programming stigmergic coordination with the TOTA middleware. In *4th International Joint Conference on Autonomous Agents and Multiagent Systems (AAMAS'05)*, pages 415–422, New York, NY, USA, July 2005. ACM Press. ISBN AAMAS'05. doi: http://doi.acm.org/10.1145/1082473.1082537.

A. Molesini, A. Omicini, E. Denti, and A. Ricci. SODA: A roadmap to artefacts. In O. Dikenelli, M.-P. Gleizes, and A. Ricci, editors, *Engineering Societies in the Agents World VI*, volume 3963 of *LNAI*, pages 49–62. Springer, June 2006. ISBN 3-540-34451-9. doi: 10.1007/11759683_4. URL http://www.springerlink.com/link.asp?id=j68184713542525p. 6th International Workshop (ESAW 2005), Kuşadası, Aydın, Turkey, 26–28 Oct. 2005. Revised, Selected & Invited Papers.

A. Molesini, A. Omicini, and M. Viroli. Environment in agent-oriented software engineering methodologies. *Multiagent and Grid Systems*, 5(1), 2009. ISSN 1574-1702. Special Issue on Engineering Environments in Multiagent Systems, In Press.

J.-P. Müller. Emergence of collective behaviour and problem solving. In A. Omicini, P. Petta, and J. Pitt, editors, *Engineering Societies in the Agents World IV*, volume 3071 of *LNCS (LNAI)*, pages 1–21. Springer, June 2004. ISBN 978-3-540-22231-6. doi: 10.1007/b98212. 4th International Workshops, ESAW 2003, London, UK, October 29-31, 2003, Revised Selected and Invited Papers.

J. D. Murray. *Mathematical Biology: An Introduction*, volume 17 of *Interdisciplinary Applied Mathematics*. Springer, 3rd edition, 2002. ISBN 978-0-387-

95223-9. URL http://www.springer.com/east/home/math/biology?SGWID=5-10047-22-2103682-0.

A. Omicini. SODA: Societies and infrastructures in the analysis and design of agent-based systems. In P. Ciancarini and M. J. Wooldridge, editors, *Agent-Oriented Software Engineering*, volume 1957 of *LNCS*, pages 311–326. Springer-Verlag, 2001. ISBN 3-540-41594-7. URL http://www.springerlink.com/content/0pw8rqj5kpbdnflx/. 1st International Workshop (AOSE 2000), Limerick, Ireland, 10 June 2000. Revised Papers.

A. Omicini, A. Ricci, and M. Viroli. *Agens Faber*: Toward a theory of artefacts for MAS. *Electronic Notes in Theoretical Computer Sciences*, 150(3):21–36, 29 May 2006. ISSN 1571-0661. doi: 10.1016/j.entcs.2006.03.003. 1st International Workshop "Coordination and Organization" (CoOrg 2005), COORDINATION 2005, Namur, Belgium, 22 Apr. 2005. Proceedings.

A. Omicini, A. Ricci, and M. Viroli. Artifacts in the A&A meta-model for multi-agent systems. *Autonomous Agents and Multi-Agent Systems*, 17(3):432–456, Dec. 2008. ISSN 1387-2532. Special Issue on Foundations, Advanced Topics and Industrial Perspectives of Multi-Agent Systems.

H. V. D. Parunak. "Go to the ant": Engineering principles from natural multi-agent systems. *Annals of Operations Research*, 75:69–101, Jan. 1997. ISSN 0254-5330. doi: 10.1023/A:1018980001403.

PRISM. PRISM: Probabilistic symbolic model checker, Oct. 2007. URL http://www.prismmodelchecker.org/. Developed at University of Birmingham, UK. Version 3.1.1 available online at http://www.prismmodelchecker.org/.

A. Ricci, M. Viroli, and A. Omicini. Programming MAS with artifacts. In R. P. Bordini, M. Dastani, J. Dix, and A. El Fallah Seghrouchni, editors, *Programming Multi-Agent Systems*, volume 3862 of *LNAI*, pages 206–221. Springer, Mar. 2006. ISBN 3-540-32616-2. doi: 10.1007/11678823_13. URL http://www.springerlink.com/openurl.asp?genre=article&issn=0302-9743&volume=3862&spage=206. 3rd International Workshop (PROMAS 2005), AAMAS 2005, Utrecht, The Netherlands, 26 July 2005. Revised and Invited Papers.

J. J. M. Rutten, M. Kwiatkowska, G. Norman, and D. Parker. *Mathematical Techniques for Analyzing Concurrent and Probabilistic Systems*, volume 23 of *CRM Monograph*. American Mathematical Society, Providence, RI, USA, 2004. ISBN 0821835718. URL http://www.ams.org/bokpages/crmm-23.

J. A. Sauter, R. S. Matthews, H. V. D. Parunak, and S. Brueckner. Performance of digital pheromones for swarming vehicle control. In F. Dignum, V. Dignum, S. Koenig, S. Kraus, M. P. Singh, and M. Wooldridge, editors, *Proceedings of the 4th International Joint Conference on Autonomous Agents and Multiagent Systems (AAMAS 2005)*, pages 903–910, Utrecht, The Netherlands, 25–29 July 2005. ACM. ISBN 1-59593-093-0.

R. V. Solé and J. Bascompte. *Self-Organization in Complex Ecosystems*, volume 42 of *Monographs in Population Biology*. Princeton University Press, Princeton, NJ, USA, 2006. ISBN 0691070407.

S. Steward and S. Appleby. Mobile software agents for control of distributed systems based on principles of social insect behaviour. In *International Conference on Communications Systems (ICCS'94)*, volume 2, pages 549–553, Singapore, Nov. 1994. IEEE. ISBN 0-7803-2046-8. doi: 10.1109/ICCS.1994.474192. URL http://ieeexplore.ieee.org/xpls/abs_all.jsp?arnumber=474192.

W. F. Tichy. Should computer scientists experiment more? *IEEE Computer*, 31

(5):32–40, May 1998. ISSN 0018-9162. doi: 10.1109/2.675631. URL http://ieeexplore.ieee.org/xpl/freeabs_all.jsp?arnumber=675631.

A. M. Uhrmacher. Simulation for agent-oriented software engineering. In W. Lunceford and E. Page, editors, *First International Conference on Grand Challenges*, San Antonio, TX, USA, January 2002. SCS, San Diego. URL http://www.scs.org/getDoc.cfm?id=1328.

M. Viroli, T. Holvoet, A. Ricci, K. Schelfthout, and F. Zambonelli. Infrastructures for the environment of multiagent systems. *Autonomous Agents and Multi-Agent Systems*, 14(1):49–60, Feb. 2007. ISSN 1387-2532. doi: 10.1007/s10458-006-0012-0. Special Issue on Environments for Multi-agent Systems.

A. von Mayrhauser. The role of simulation in software engineering experimentation. In *Experimental Software Engineering Issues: Critical Assessment and Future Directions*, volume 706 of *LNCS*, pages 177–179. Springer, 1993. ISBN 3-540-57092-6. doi: 10.1007/3-540-57092-6_121. URL http://www.springerlink.com/content/75g817r264p18610/?p=b791ac69501a476a95fdbebccc5b3e91&pi=1.

D. Weyns, K. Schelfthout, T. Holvoet, and T. Lefever. Decentralized control of E'GV transportation systems. In M. Pechoucek, D. Steiner, and S. G. Thompson, editors, *Proceedings of the 4rd International Joint Conference on Autonomous Agents and Multiagent Systems (AAMAS 2005) – Special Track for Industrial Applications*, pages 67–74, Utrecht, The Netherlands, 25–29 July 2005. ACM. ISBN 1-59593-093-0.

D. Weyns, A. Omicini, and J. Odell. Environment as a first-class abstraction in multi-agent systems. *Autonomous Agents and Multi-Agent Systems*, 14(1):5–30, Feb. 2007. ISSN 1387-2532. doi: 10.1007/s10458-006-0012-0. URL http://springerlink.metapress.com/content/w571550106301124/. Special Issue on Environments for Multi-agent Systems.

F. Zambonelli, N. R. Jennings, and M. J. Wooldridge. Developing multiagent systems: The Gaia methodology. *ACM Transactions on Software Engineering and Methodology (TOSEM)*, 12(3):317–370, July 2003. ISSN 1049-331X. doi: 10.1145/958961.958963.

6

On the Role of Software Architecture for Simulating Multi-Agent Systems

6.1	Introduction	167
6.2	Background	169
	A System and Its Environment • Characteristics of Multi-Agent Control Systems • Software-in-the-Loop Simulation Mode	
6.3	AGV Transportation System	172
	Physical Setup of an AGV Transportation System • AGV Control System • Requirements of an AGV Simulator	
6.4	Modeling Multi-Agent Control Applications in Dynamic Environments...............................	177
	Overview of Modeling Framework • Simulation Model of the AGV Transportation System	
6.5	Architecture of the Simulation Platform	185
	Requirements • Top-Level Module Decomposition View of the Simulation Platform • Component and Connector View of the Simulated Environment • Component and Connector View of the Simulation Engine • An Aspect-Oriented Approach to Embed Control Software • Component and Connector View of the Execution Tracker	
6.6	Evaluating the AGV Simulator.....................	201
	Flexibility of the AGV Simulator • Measurements of the AGV Simulator • Multi-Agent System Development Supported by the AGV Simulator	
6.7	Related Work	205
	Special-Purpose Simulation Platforms • Embedding the Control Software	
6.8	Conclusions and Future Work	209
	Concrete Directions for Future Research • Closing Reflection	
	References	211

Alexander Helleboogh
Katholieke Universiteit Leuven

Danny Weyns
Katholieke Universiteit Leuven

Tom Holvoet
Katholieke Universiteit Leuven

6.1 Introduction

A control system is a software system connected to an underlying environment. The environment is the part of the external world with which the control system interacts, and in which the effects of the control system will be observed [Jackson, 1997]. The task of a control system is to ensure that particular functionalities are achieved in the environment. A multi-agent control system is a control system of which the software is a multi-agent system, i.e. a system that consists of a number of autonomous software components, called agents, that collaborate to achieve a common goal. Examples of multi-agent control systems include manufacturing control systems [Verstraete et al., 2006; Brueckner, 2000], collective robotic systems [Gu and Hu, 2004; Varshavskaya et al., 2004; Bredenfeld et al., 2006], traffic control systems [Wang, 2005; Roozemond, 1999; Dresner and Stone, 2005] and sensor networks [Sinopoli et al., 2003; DeLima et al., 2006].

Simulation can be used to support the development of multi-agent control systems. Simulation offers a safe and cost-effective way for studying, evaluating and configuring the behavior of a multi-agent control system in a simulated setting [Himmelspach et al., 2003]. In this chapter, we focus on software-in-the-loop simulations for multi-agent control systems in dynamic environments. This family of simulations has the following characteristics: (1) the environment to-be-simulated is dynamic. In a dynamic environment, the operating conditions of a multi-agent control system are continuously changing; (2) the control software of the real multi-agent control system is embedded in the simulation.

Developing software-in-the-loop simulations of multi-agent control systems in dynamic environments is complex. The system-to-be-simulated comprises two parts: a dynamic environment on the one hand and a multi-agent control system embedded in that environment on the other hand. We illustrate two main challenges when building simulations for such systems:

- *Simulating dynamic environments is complex.* In a dynamic environment an agent cannot determine the outcome of its actions a priori [Ferber and Müller, 1996; Helleboogh et al., 2005]. Other activities that are happening in the environment can have a significant impact on the outcome of actions. Consider a robot that was instructed to start driving north. In a dynamic environment, the action of the robot can be affected in different ways. For example, another machine could move into the path of the first robot, blocking it or pushing it aside. Or the robot's path could deviate from the intended path due to jitter in the hardware. Or the robot could run out of energy, causing its movement to stop prematurely. Even a combination of these phenomena could occur. When simulating dynamic environments, it is non-trivial to reproduce the variety of possibly cascading interactions that may occur and the precise way these interactions have an impact on the actions.

- *Integrating the software of a real multi-agent control system in a simulation is complicated.* The devices on which the multi-agent control system is deployed in the real world determine how fast the control software can execute and consequently how much time it takes the software to react to changes in the environment. However, the characteristics of the computer platform on which the simulation is executed, can differ significantly from the devices on which the control software is deployed in the real world. Moreover, a simulation can be executed faster or slower than real time. It is non-trivial to reproduce the real-world timing characteristics of a multi-agent control system in a simulation [Uhrmacher and Kullick, 2000; Anderson, 1997].

Special-purpose modeling constructs and simulation platforms incorporate a large body of expertise on building software-in-the-loop simulations of multi-agent control system in dynamic environments. In simulation platforms this expertise is primarily reified in terms of reusable code libraries and frameworks. In our own research, we have built up knowledge and best practices in several cases, including simulations of the Packet-World [Weyns et al., 2005a], Lego Mindstorms robots [Borgers, 2006], traffic control systems [Weyns et al., 2007] and Automated Guided Vehicles [Helleboogh et al., 2006]. We experienced that we could reuse a lot of expertise across these cases. Nevertheless, a substantial part of this expertise was reused implicitly, as explicit reuse of code libraries and frameworks was rather limited.

In this chapter, we put forward a software architecture for a simulation platform targeted at software-in-the-loop simulation of multi-agent control systems in dynamic environments. This software architecture explicitly documents the knowledge and practice incorporated in a simulation platform in the form of a reusable artifact, clearly distinguished from the code of the simulation platform. More than reusable code libraries and software frameworks, a software architecture captures the essence of a complex software system by identifying key stakeholder concerns and by explicitly specifying how software needs to be structured and behave to address the concerns. The software architecture integrates the essential architectural building blocks for such a simulation platform and explicitly documents the rationale and tradeoffs that underpin its design. Software architecture provides a systematic way to capture and share expertise that was acquired across several cases in a form that has proven its value for software development.

This chapter is structured as follows. After presenting some essential background in Section 6.2, we introduce an industrial case in Section 6.3, i.e. a AGV transportation system that comprises a multi-agent control system for controlling automated guided vehicles (AGVs) that transport loads in a warehouse. This case will be used as an illustration throughout this chapter. In Section 6.4, we summarize our previous work on special-purpose modeling constructs for modeling software-in-the-loop simulations of multi-agent control systems in dynamic environments, and we apply these modeling constructs to describe a simulation model for the AGV transportation system. In Section 6.5, we describe the software architecture for a simulation platform that supports these special-purpose modeling constructs. We document the architecture and explain how important functional and quality requirements are achieved. In Section 6.6, we evaluate a simulation platform that implements this architecture and that was used for conducting simulations of the AGV transportation system. In Section 6.7 we discuss related work on software-in-the-loop simulations of multi-agent control systems in dynamic environments. Finally, we point out directions for future research and we draw conclusions in Section 6.8.

6.2 Background

In this section, we describe the necessary background information. We focus on (1) the relation between a system and its environment, (2) the characteristics of multi-agent control systems, and (3) software-in-the-loop simulation.

6.2.1 A System and Its Environment

Software systems are designed to satisfy particular functional and quality requirements. These requirements are issued by the group of stakeholders involved. For systems that interact with the external world via sensors and actuators, these requirements do not directly concern the software system [Jackson, 1997]. The requirements primarily concern the environment in which the system will be installed [Hayes et al., 2003]. The task of a system is to ensure that particular functionalities are achieved in the environment.

Jackson defines the environment as the part of the external world with which the system interacts, and in which the effects of the system will be observed and evaluated [Jackson, 1997]. The distinction between the environment and the system is partly a distinction between what is given and what is to be constructed.

Figure 6.1 depicts the relation between a system and its environment. The system interacts with the environment by means of shared phenomena that are directly accessible via sensors and actuators. However, influencing phenomena that are private to the environment can only be done in an indirect manner: using sensors and actuators, the system tries to bring about causal chains to observe and affect private phenomena in the environment.

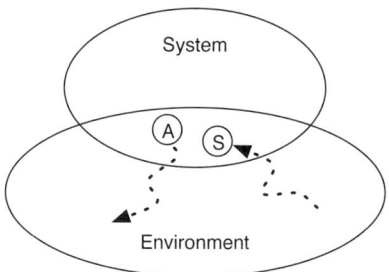

FIGURE 6.1 The system and its environment [Jackson, 1997].

As an example, consider a car's cruise control system. The environment of the system consists of the car, its driver, the atmospheric conditions, the road the car drives on, etc. The cruise control system interacts with its environment through a sensor that can be used to observe the car's speed and an actuator that can be used to adjust the car's throttle. The requirements of the cruise control system are expressed in terms of phenomena in its environment. For example, the cruise control system should ensure that the car drives at a constant speed across the road. The cruise control system can only affect the car's speed indirectly, i.e. by relying on a causal chain between manipulations of the throttle actuator and alterations in the speed of the car.

Today, the environments in which software systems have to operate are typically *dynamic* [Issarny et al., 2007]. A dynamic environment is an environment that changes frequently. In a dynamic environment, the operating conditions of a system are continuously changing. For example, the environment of the cruise control system is dynamic: the road may go uphill or downhill, turbulence or wind may arise that causes additional or reduced drag. These phenomena affect the causal chains by means of which the cruise control system affects the environment. For example, in case the road goes uphill, changing the throttle will affect the car's speed in a different manner compared to the case that the road goes downhill.

As a dynamic environment can significantly affect the software system, it is generally considered good practice to capture properties and assumptions regarding the environment in an explicit model [Jackson, 1997; Hayes et al., 2003]. Such a model of the environment includes assumptions about the frequency and nature of the changes in the environment, the accuracy and latency of sensors and actuators, the assumptions about the causal chain from activation of an actuator to the changes in the actual environment, etc. Such a model describes essential characteristics and assumptions about the environment that must be checked for proper deployment of the system.

6.2.2 Characteristics of Multi-Agent Control Systems

A multi-agent control system is a distributed software application that continuously and autonomously acts in, and reacts to, an underlying environment. Examples of multi-agent control systems include manufacturing control systems [Verstraete et al., 2006; Brueckner, 2000], collective robotic systems [Gu and Hu, 2004; Varshavskaya et al., 2004; Bredenfeld et al., 2006], traffic control systems [Wang, 2005; Roozemond, 1999; Dresner and Stone, 2005] and sensor networks [Sinopoli et al., 2003; DeLima et al., 2006]. Figure 6.2 gives a schematic overview of a multi-agent control system in an environment.

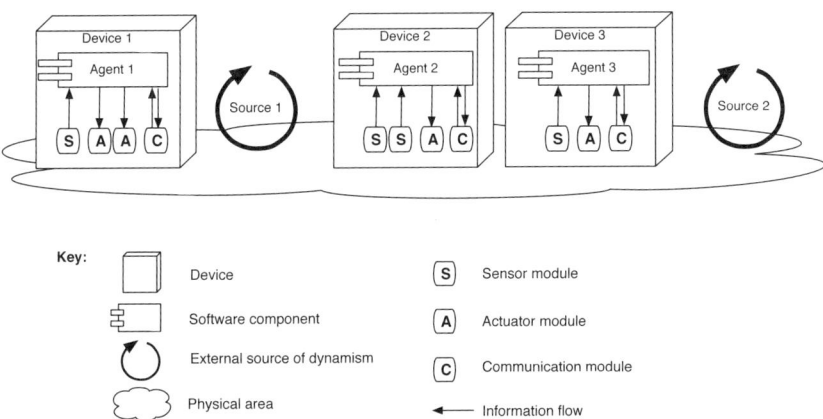

FIGURE 6.2 Schematic view of a multi-agent control system in an environment.

A multi-agent control system consists of several agents. Agents are autonomous software components that are distributed in the environment and that cooperate to solve a particular problem in the environment. In Figure 6.2, three agents are depicted: `Agent 1`, `Agent 2` and `Agent 3`.

The agents of a multi-agent control system are deployed on particular devices in the environment. A device consists of a software and a hardware part. The software part is one of the agents that constitutes the multi-agent control system, whereas the hardware part comprises sensor, actuator and communication modules. An agent can use the sensor, actuator and communication modules of its device to interact with the environment. Figure 6.2 depicts three devices in the environment: `Device 1`, `Device 2` and `Device 3`.

The environment of a multi-agent control system is the part of the external world in which the problem resides and in which the effects of the control system, once installed and set in operation, will be observed [Hayes et al., 2003]. Typically, a multi-agent control system operates in a *dynamic* environment, i.e. an environment where other sources of dynamism are present, e.g., other systems, processes or even humans. These sources of dynamism are external to the multi-agent control system. Figure 6.2 depicts two external sources of dynamism present in the environment: `Source 1` and `Source 2`. The operation of external sources of dynamism can have a significant impact on a multi-agent control system.

Designing and testing a multi-agent control system is complex as it requires an integrated approach that takes into account the environment in which the application is situated [Hu and Zeigler, 2005]. A multi-agent control system should take into account dynamism originating from other systems, processes or humans in the environment and react appropriately to their presence.

6.2.3 Software-in-the-Loop Simulation Mode

Software-in-the-loop simulation mode denotes simulations in which the software of the real control system is embedded in the simulation loop. Software-in-the-loop simulation mode is depicted in Figure 6.3. The simulation contains parts of the real system, i.e. the control software, together with simulated parts, i.e. the device hardware and the environment. The executable code of the real control system is directly embedded in the simulation. In software-in-the-loop simulation mode, the software of the real control system is deployed on simulated devices that reside within a simulated environment with simulated sources of dynamism.

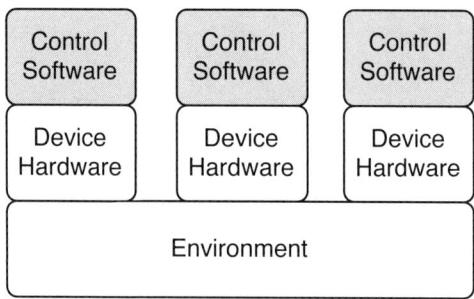

FIGURE 6.3 Schematic view of software-in-the-loop simulation mode. White blocks are simulated parts. Grey blocks are parts of the real system that are integrated in the simulation loop.

Software-in-the-loop simulation is typically used during the late stages of application development, i.e., after the software of the multi-agent control system (or parts thereof) has been implemented. Software-in-the-loop simulation enables experimenting with the agents of a multi-agent control system on simulated devices before deployment on real devices. Software-in-the-loop simulation is extensively used for the development of control systems for robots. For example, software-in-the-loop simulations enable testing the robustness and fault-tolerance of control systems for robots [Bräunl et al., 2006; Finkenzeller et al., 2003] or can facilitate parameter estimation of a control systems for robots [Velez and Agudelo, 2006].

6.3 AGV Transportation System

We introduce a real-world case that will be used as an example throughout this chapter. The case comprises the development of a multi-agent control system that controls automated guided vehicles (AGVs) in warehouse environments. An AGV is an unmanned, computer-controlled transportation vehicle using a battery as its energy source (see Figure 6.4). AGVs have to perform transports. A transport consists of picking up a load at a particular spot in the warehouse and bringing it to its destination. A load ranges from raw materials (e.g., wood, rolls of paper) to completed products (e.g., tires, cheese).

An AGV control system was developed in the EMC^2 (Egemin Modular Controls Concept) project. Egemin N.V. is a Belgian manufacturer of AGVs, and develops control software for

FIGURE 6.4 An AGV in a cheese factory.

automating logistics in warehouses and manufactories using AGVs.

6.3.1 Physical Setup of an AGV Transportation System

Figure 6.5 shows a three-dimensional view on an AGV transportation system. The hardware of an AGV comprises the following. An AGV contains engines to move and turn and a lift to pick and drop loads. An AGV has sensors to observe its position and battery level. Finally, each AGV has a computer platform on which control software can be deployed. The computer platform of an AGV uses wireless communication.

The warehouse is a storage or manufacturing facility that contains various loads positioned at various locations across the warehouse. Loads are typically stored in racks. Racks are used to hold loads and are positioned across the warehouse, usually according a geometrical pattern that combines easy accessibility of the loads, as well as efficient use of the available room for storage purposes. Typically, also one or several battery chargers for the AGVs are positioned at particular locations across the warehouse.

To support AGVs, the warehouse is usually customized. This typically involves a custom configuration of the racks. In addition, a complex layout of magnet strips is built into the warehouse floor to guide the AGVs to move from one spot in the warehouse to another. This *magnet track* allows AGVs to maneuver in an accurate manner according to predefined pathways. Moreover, as magnets are inexpensive and can be installed easily, magnet-guided navigation is relatively cost-effective.

6.3.2 AGV Control System

An AGV control system is a software system that controls a set of AGVs. We discuss the main functionalities of an AGV control system and elaborate on the AGV steering system that can be used by an AGV control system to instruct AGVs.

Functionalities of an AGV Control System

The main functionality of an AGV control system is handling transports, i.e. moving loads from one place to another. Transports are typically generated by order management software, but a transport can also be introduced manually by employees or operators. Ab-

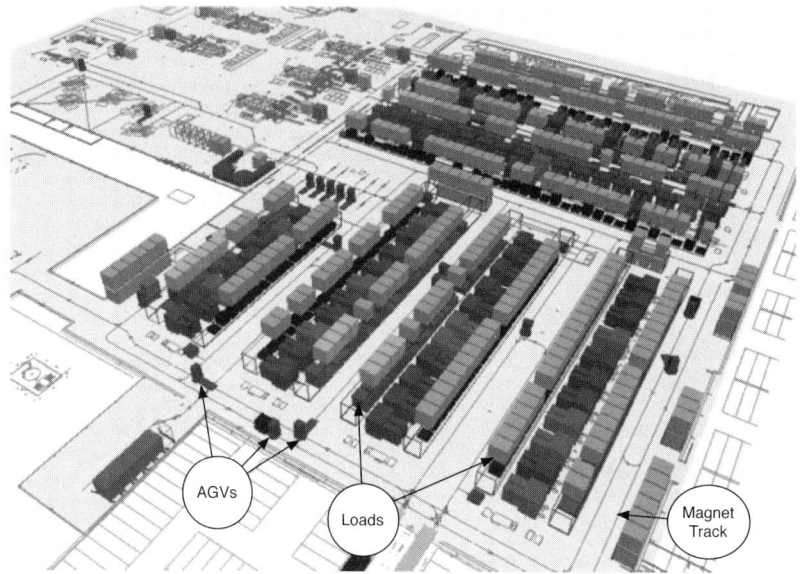

FIGURE 6.5 Three-dimensional view on an AGV transportation system.

stracting from the origin of the transports, systems that generate transports for the AGV control system are called client systems. Client systems input transports to the AGV control system, and expect a confirmation from the AGV control system when the transport is done.

In order to handle transports, the AGV control system has to use the AGVs under its control efficiently. The main functionalities to be performed are the following:

- *Transport assignment:* transports originating from client systems must be assigned to an appropriate AGV. The goal is to assign transports in such a way that overall, transports are handled in an efficient and timely manner.
- *Routing:* in order to carry out transports, AGVs need to move to certain places. For the movement of all AGVs, efficient routes through the warehouse must be determined. Although the road network determined by the magnet track is static, the best route for an AGV is in general dynamic, and depends on the current conditions in the system. For example, the shortest route in distance may take a long time because there is a "traffic jam". So, routing in general may need to be adapted dynamically.
- *Collision avoidance:* while moving around, AGVs may not collide with each other. Collisions do not exclusively occur at intersections of paths; AGVs also need to avoid collisions while passing each other on closely located parallel paths.
- *Deadlock avoidance:* since AGVs cannot divert from their path, they are relatively constrained in their movement. Therefore, deadlocks can occur when a number of AGVs are in a situation where no AGV can move anymore without operator intervention. For example, on a bidirectional path AGVs may be standing head on toward each other. Since AGVs in general cannot drive in reverse, none of the two AGVs can move forward or backward. The AGV control system must ensure that manual intervention for such situations is avoided.

Besides handling transports efficiently, the AGV control system must also ensure the continued operation of the AGVs.

- *Maintenance:* AGVs need regular maintenance, which is typically scheduled in fixed time intervals. Furthermore, AGVs may need to calibrate their positioning system regularly.
- *Battery charging:* when an AGV's battery runs out, it must drive to a charging station. Either the AGV must wait until an operator exchanges the old battery for a full battery, or the battery is charged using contact points in the warehouse floor.
- *Resource saving:* AGVs are expensive and must be used as efficiently as possible. AGVs that are idle must save their resources, and get out of the way of the active AGVs. Therefore, idle AGVs are parked at park nodes.

AGV Steering System

To control an individual AGV, it is equipped with an AGV steering system developed by Egemin, called E'nsor*. E'nsor handles the low-level control of an AGV on the level of reading out sensors and driving actuators. Main functionalities of E'nsor are keeping the AGV on a path, turning, determining the AGV's current position, reading out the battery level, etc.

E'nsor can handle a number of actions on its own. These actions are called jobs. For example, picking up a load is a pick job, dropping it is a drop job and moving over a specific distance is a move job. A transport typically starts with a pick job, followed by a series of move jobs and ends with a drop job. The AGV control system gives jobs to E'nsor, which in turn controls the AGV to handle the jobs autonomously.

To be able to indicate to E'nsor where to pick and move, the layout (i.e. all the possible paths the AGVs can follow in the system) of the warehouse is divided into logical elements: segments and nodes. Segments determine the path an AGV can follow through the warehouse, and can be either straight or curved with lengths of typically three to five meters. A segment can either be unidirectional or bidirectional. In the latter case AGVs can drive over the segment in both directions. Nodes are at the beginning or end of segments. Nodes are the places where an AGV can stand still, or do an action like picking up a load. In normal operation, an AGV can only be at rest when standing on a node. Each segment and node is given a unique identifier. E'nsor is able to steer an AGV on a segment per segment basis. An AGV can stop on every node, possibly to change direction. E'nsor can handle five jobs. None of these jobs route the AGV, so the segment given as argument must be accessible from the node on which the AGV is currently standing.

- Move(segment): this instructs E'nsor to drive the AGV over the given segment.
- Pick(segment): instructs E'nsor to drive the AGV over the given segment and pick up the load at the node at the end of the segment.
- Drop(segment): the same as pick, but drops a load the AGV is carrying.
- Park(segment): instructs E'nsor to drive the AGV over the given segment and park at a park node at the end of the segment.
- Charge(segment): instructs E'nsor to drive the AGV over a given segment to a

---

*E'nsor is an acronym for Egemin Navigation System On Robot.

battery charging node and start charging batteries there.

Furthermore, E'nsor allows the readout of sensor values of the AGV, of which the most important are battery level; position in coordinates on the floor; position in terms of segment and node on the layout; orientation; speed.

6.3.3 Requirements of an AGV Simulator

In the context of the EMC^2 project, we developed an *AGV simulator*. The AGV simulator enables (1) safe experimentation and testing of AGV control systems without risk of damaging the real AGVs, (2) executing experiments faster than real-time, which is essential when investigating long-term scenarios (3) setting up and monitoring experiments in a less costly way, e.g., without the cost of buying AGVs or building particular warehouse layouts.

We elaborate on the requirements of an AGV simulator that was developed in the context of EMC^2. The goal of the AGV simulator is to support evaluating new or altered features of a multi-agent AGV control system by means of software-in-the-loop simulation of AGV agents in a simulated warehouse environment. Software-in-the-loop simulation enables evaluating the actual implementation (or parts thereof) of the AGV agents.

The AGV simulator focuses on evaluating routing, collision avoidance, transport assignment and battery charging.

- *Support for routing.* To evaluate or compare routing behaviors of AGV agents, the AGV simulator should simulate the movements of real AGVs and realistic layouts of the warehouse. This enables monitoring the path followed by an AGV, its travel time, the appearance of traffic jams, etc.
- *Support for collision avoidance.* To evaluate the appropriateness of collision avoidance techniques of AGV agents, the AGV simulator should simulate the movements of AGVs on a warehouse layout and detect situations in which AGVs could collide. Moreover, to test the robustness of collision avoidance techniques, the AGV simulator should simulate unreliable communication between AGVs. This enables a developer to evaluate the adequacy of collision avoidance techniques under a variety of circumstances.
- *Support for transport assignment.* To evaluate transport assignment among AGV agents, the AGV simulator should simulate several transport profiles generated by client systems.
- *Support for battery charging.* To evaluate the charging strategy of AGV agents, the AGV simulator should simulate the energy consumption of an AGV, the charging of its battery at a charging station and the interruption of an AGV's operation in case it runs out of energy.

When building an AGV control system, these functionalities are typically developed iteratively. The AGV simulator should enable testing partial AGV control systems in which some of these functionalities are present and others not yet (fully) operational.

To support evaluating different functionalities of AGV control systems in a variety of settings, modifying core parts of the model of the AGV simulator should be relatively easy and the impact of such modifications should be as local as possible. Important modifications that should be supported include altering the AGV agent software; the layout of the warehouse; the number of AGVs and their characteristics (e.g., characteristics of movements, energy consumption, etc.); the quality of service of communication between AGVs; the accuracy of collision detection; the transport profile of the client systems.

6.4 Modeling Multi-Agent Control Applications in Dynamic Environments

In this section, we (1) summarize our previous work on a modeling framework for software-in-the-loop simulations of multi-agent control systems in dynamic environments, and (2) apply this modeling framework to formulate a simulation model for the AGV transportation system.

6.4.1 Overview of Modeling Framework

The modeling framework offers special-purpose modeling constructs for formulating a simulation model for software-in-the-loop simulations of multi-agent control systems in dynamic environments. The modeling framework captures core characteristics of these simulations in a first-class manner.

The foundation for the constructs of the modeling framework is twofold. On the one hand, the modeling framework results from our own experience of building simulations for multi-agent control systems in dynamic environments. Examples include simulations of the Packet-World [Weyns et al., 2005a], Lego Mindstorms robots [Borgers, 2006] and Automated Guided Vehicles [Helleboogh et al., 2006]. On the other hand, the modeling constructs are underpinned by existing practice on modeling dynamic environments of multi-agent control systems. For a detailed motivation, discussion and a formal description of all modeling constructs as well as of the evolution of the model, we refer to [Helleboogh et al., 2007; Helleboogh, 2007]

The modeling framework comprises two complementary parts: an environment part and a control software part. We give a brief brief overview of the modeling constructs in each part of the modeling framework.

Modeling Dynamic Environments

The environment part of the modeling framework comprises modeling constructs that capture in an explicit manner a number of key characteristics and relations that are pertinent for modeling dynamic environments of multi-agent control systems.

Figure 6.6 gives a graphical overview of the modeling framework for dynamic environments, depicting all modeling constructs and the relations between the constructs. The modeling constructs are organized in four groups:

1. Constructs to represent the *structure of the environment* in the simulation model.
2. Constructs to represent *dynamism in the environment* in the simulation model.
3. Constructs to represent the *manipulation of dynamism in the environment* in the simulation model.
4. Constructs to represent the *sources of dynamism in the environment* in the simulation model.

We give an overview of the modeling constructs in each group.

Structure of the Environment.

A first group of modeling constructs captures the structure of the environment. To capture the constituting parts of the environment in a simulation model, we put forward the modeling constructs `Environmental Entity` and `Environmental Property`. Examples of environmental entities are all sorts of objects in the environment, such as the robots on

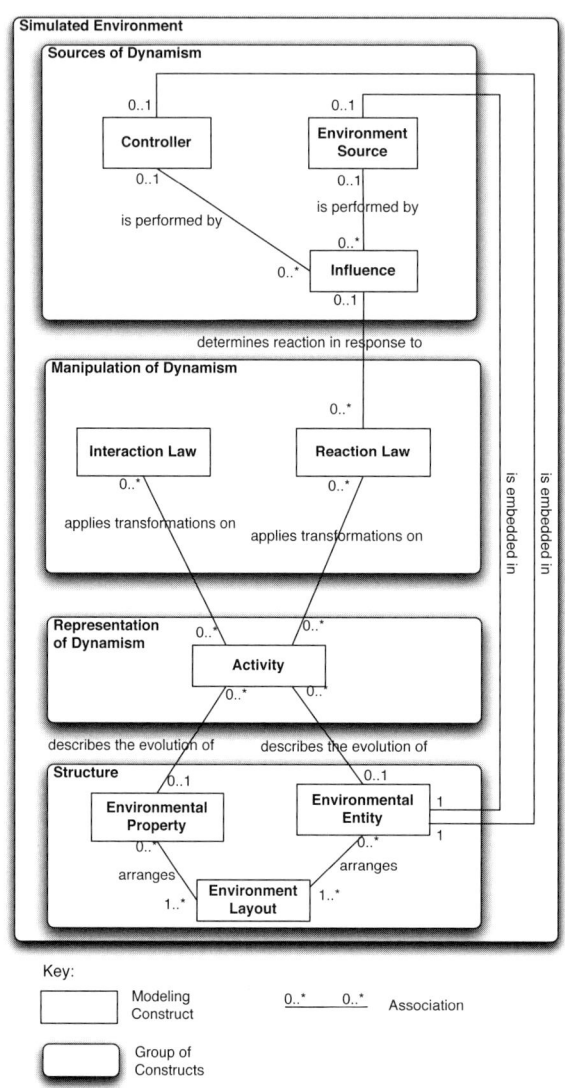

FIGURE 6.6 Overview of the constructs in the environment part of the modeling framework.

which a multi-agent control system is deployed. An example of an environmental property is the temperature in the environment. To represent a physical or logical structure that arranges the different environmental entities and environmental properties with respect to each other, we put forward the modeling construct `Environment Layout`. An example of an environment layout is a two-dimensional geometrical arrangement of the entities.

Dynamism in the Environment.

A second group of modeling constructs captures dynamism in the environment in an explicit manner. To represent dynamism explicitly in the simulation model, we put forward an `Activity` as a modeling construct. The association between `Activity` and `Environmental Entity` and the association between `Activity` and `Environmental Property` expresses that an activity describes a particular evolution of a particular environmental entity or property over a particular time interval. Examples of activities are the movement of a robot or the rolling of a ball.

Manipulation of Dynamism.

A third group of modeling constructs captures the way dynamism in the environment can alter, i.e. the way activities arise, interact and terminate. We put forward the modeling constructs `Reaction Law` and `Interaction Law` to capture the way activities in the environment are manipulated.

A `Reaction Law` is a modeling construct that specifies what happens in the environment in reaction to a particular trigger of a source of dynamism. An example is a reaction law that specifies what happens in the environment in reaction to the trigger of an agent to start the engines of a robot. The reaction law specifies what kind of activity is created, e.g., a movement of that robot characterized by a particular velocity in a particular direction.

An `Interaction Law` is a modeling construct to specify the way dynamism can interact in the environment. For example, an interaction law can specify what happens in case a robot involved in a movement activity hits a wall or another robot.

The associations between `Reaction Law` and `Activity` on the one hand, and between `Interaction Law` and `Activity` on the other hand, express that reaction laws and interaction laws alter the activities present in the environment.

Sources of Dynamism.

A fourth group of modeling constructs captures the sources of dynamism in the environment. We put forward the modeling constructs `Controller` and `Environment Source` to represent the behavior of the various sources of dynamism present in the environment.

The `Controller` is a source of dynamism that is part of the multi-agent control system. An example of a controller is an agent program that controls a particular robot. An `Environment Source` is a source of dynamism that resides in the environment and that is external to the multi-agent control system. An example of an environment source is the behavior of a machine in the environment that is controlled by a human. Controllers and environment sources are embedded in some of the environmental entities. For example, a robot contains a source of dynamism, i.e. its controller, whereas a ball is passive and does not contain a source of dynamism.

Controllers and environment sources can initiate, terminate or alter dynamism in the environment. We put forward an `Influence` as a modeling construct to capture the *attempt* of the controller or of an environment source to affect the environment. An example of an influence is the attempt of an agent to start or stop the movement of a robot. The association between `Environment Source` and `Influence` and between `Controller` and `Influence` represents that dynamism can only be manipulated indirectly, i.e. by means of performing influences in the environment. Reaction laws determine the actual reaction of the environment in response to influences. This is represented by the association between `Reaction Law` and `Influence`.

Modeling the Software of a Multi-Agent Control System

In software-in-the-loop simulations, the software of the real controllers of the multi-agent control system is embedded in a simulated environment. The control software part of the modeling framework comprises modeling constructs that capture key characteristics of the software of the multi-agent control system that is embedded in the simulation.

Figure 6.7 is a detailed view on the group of modeling constructs to represent the sources of dynamism in Figure 6.6, with additional modeling constructs for the controller. We give

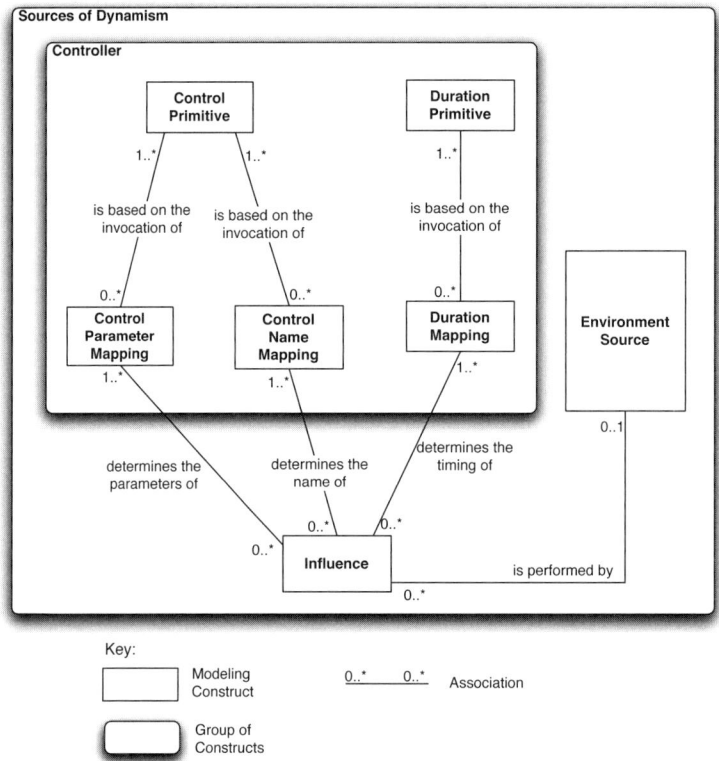

FIGURE 6.7 Overview of the modeling constructs for the control software and their associations.

an overview of the modeling constructs in the controller. The modeling constructs focus on representing the following characteristics of the control software explicitly in the simulation model:

- *Representing the real-world execution time of the software in the simulation model.* The real-world execution time of a controller is the amount of wallclock time that elapses until that controller triggers its next action. The execution time of a controller determines the timing of its actions. In a dynamic environment, the timing of actions is crucial as opportunities come and go.

  To capture the real-world execution time of a controller in the simulation model, we put forward the modeling constructs Duration Primitive and Duration Mapping. A duration primitive represents a code segment that takes an amount of execution time in the real world that is pertinent for the simulation. An example of a duration primitive is a particular java method *foo()* in the software of a particular controller. A duration mapping is a modeling construct that specifies the execution time for invocations of duration primitives of a controller. For example, a duration mapping can specify that invoking the method *foo()* takes 0.338 seconds.

- *Capturing the interaction of the software with the environment.* The software of a controller interacts with its environment. Consequently, the execution of the software of a controller will trigger particular things to happen in the environment. When integrating the software of the controllers with the simulated environment, it is crucial to identify the set of software instructions that are used by the controller to interact with the environment, and to specify the precise consequences in the environment that result from triggering these instructions.

To capture the interaction of the software with the environment in a simulation model, we put forward the modeling constructs `Control Primitive`, `Control Name Mapping` and `Control Parameter Mapping`. A control primitive represents a particular software instruction that can be used by the control software to interact with its environment. An example of a control primitive is a java method *bar()* that triggers the engine of a robot to start running at full power. The modeling constructs `Control Name Mapping` and `Control Parameter Mapping` specify the name and the parameters of the influence that result from invoking a control primitive. A control name mapping and control parameter mapping decouple the signature of control primitives from the specific representation of influences that is used in the simulated environment. For example, a control name mapping specifies that an invocation of *bar()* corresponds to an influence with name *startDriving*, whereas a control parameter mapping specifies that the invocation of *bar()* results in the value 10 to be associated with the parameter of the *startDriving* influence to indicate the speed.

6.4.2 Simulation Model of the AGV Transportation System

We apply the modeling framework to formulate a simulation model for an AGV simulator. The AGV simulator supports the evaluation of new or altered features of a multi-agent control system that controls automated guided vehicles in warehouse environments.

The constructs of the modeling framework are used to capture key characteristics of the AGV system in a first-class manner. This enables a developer to adapt the model of the AGV simulator to the needs of a particular simulation study by activating, deactivating and customizing first-class elements of the simulation model.

We start with an overview of the simulation model of the warehouse environment. Afterward, we focus on the simulation model for integrating the AGV control system. For a detailed discussion of the simulation model of the AGV Transportation System, we refer to [Helleboogh, 2007].

Simulation Model of the Warehouse Environment

Figure 6.8 gives a graphical overview of the environment part of the simulation model of the AGV simulator. This figure shows specific instantiations of the modeling constructs of Figure 6.6. The simulation model is organized in four parts, in analogy with Figure 6.6. We discuss a number of examples for each of the parts of the simulation model for warehouse environments.

Structure of the Warehouse Environment.
The structure of the simulated warehouse environment is modeled in terms of `environmental entities` and an `environmental layout` that arranges the entities with respect to each other. We give examples of environmental entities. *Stations* are locations that connect adjacent segments. Each station can be used for one or several purposes, i.e. as routing location, as storage location for loads, as parking location and/or as battery charging location; A *WiFi access point* enables communication between AGVs and transport bases or among several AGVs. A *transport base* is a computer that can be used to broadcast new transport tasks to the AGVs. A transport generator is embedded in a transport base.

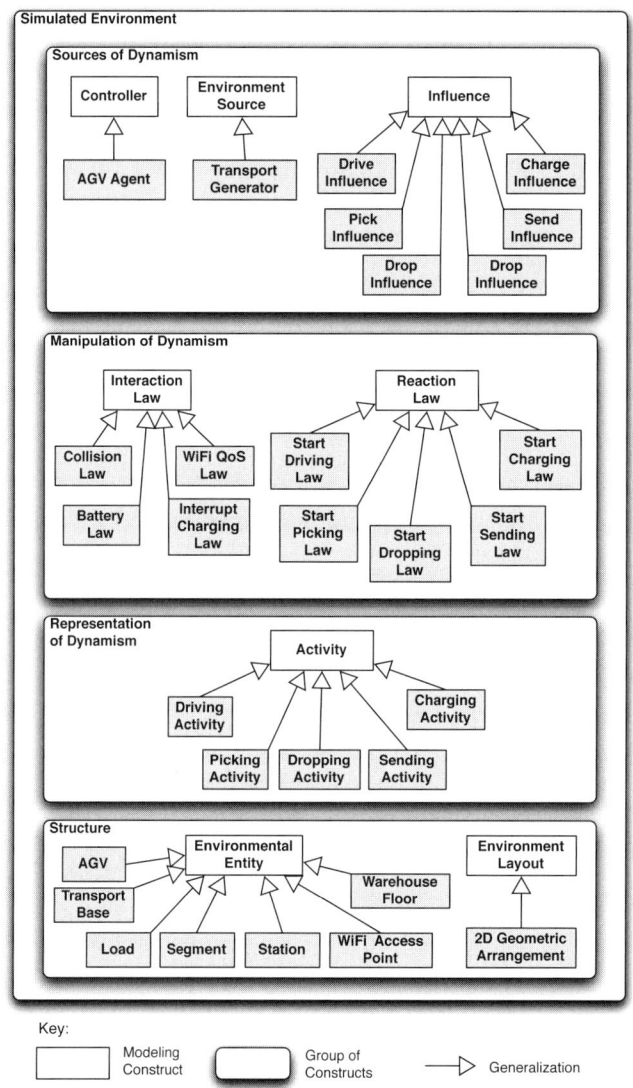

FIGURE 6.8 Overview of the simulation model of the simulated warehouse environment. The grey parts are specific instantiations of the modeling constructs for the AGV simulator.

We arrange the entities in the simulated warehouse environment according to a continuous *2D-geometric layout*. This layout expresses the spatial positioning of all entities with respect to each other.

Dynamism in the Warehouse Environment.

Dynamism in the simulated warehouse environment is modeled in terms of `activities`. For example, *driving activities* represent the driving of an AGV across a segment on the warehouse floor until the station at the other end of that segment is reached. Each driving activity is characterized by the AGV involved in the movement, the segment over which the AGV moves, the direction of the movement over that segment, the time interval during which the movement takes place and the acceleration profile of the AGV during that movement. *Sending activities* represent that a WiFi access point is used to transmit messages from a transport base to AGVs or among AGVs.

Sources of Dynamism in the Warehouse Environment.

In the warehouse environment, several sources of dynamism reside. We make a distinction between `controllers` and `environment sources`. *AGV agents* are the control software that is embedded in an AGV. AGV agents constitute the AGV control system. Each AGV agent is responsible for controlling an AGV and for coordinating with other AGVs for routing, collision avoidance, transport assignment and battery charging. A *transport generator* broadcasts transport tasks to the AGVs. A transport generator generates transports according to a transport profile that specifies the characteristics of the stream of transport tasks that should be handled by the AGVs. Transport generators are external to the AGV control system. Consequently, transport generators are environment sources of dynamism. A transport generator is deployed on a transport base.

Sources of dynamism can manipulate the environment by means of performing `influences`. For example, a *drive influence* represents that attempt of an AGV agent to start driving over a given segment in a given direction. A *send influence* represents the attempt of an AGV agent or a transport generator to send a message.

Manipulation of Dynamism in the Warehouse Environment.

The way dynamism in the warehouse environment can be manipulated is modeled by means of `reaction laws` and `interaction laws`.

An example of a reaction law is a *start driving law*. Start driving law defines the reaction of the environment in response to a *drive influence* or a *park influence*. A real AGV does not always start driving when it is instructed to do so. Therefore, start driving law checks a number of conditions before adding a new *driving activity*. These conditions are that the AGV is not already involved in a driving, picking or dropping activity at the time of the influence; that the segment is adjacent to the station of the AGV; that the AGV is allowed to drive over the given segment in the given direction (as segments can be unidirectional). Start driving law does not define an activity in case one of these conditions does not hold, to reflect that E'nsor discards the instruction in these cases.

An example of an interaction law is a *collision law*. A collision law defines what happens if AGVs in the warehouse environment collide. Based on the driving activities, a collision law determines whether AGVs collide. In case the collision law detects a collision, it transforms the driving activity/activities involved with driving activity/activities that stop at the time the collision occurs. A *battery law* enforces that all activities of an AGV are preempted in case it runs out of energy. Based on the energy consumption of driving, picking and dropping activities, a battery law preempts all activities as soon as an AGV runs out of energy.

Simulation Model for Integrating the AGV Agent Software

Figure 6.9 gives a graphical overview of the control software part of the simulation model of the AGV simulator. This figure shows specific instantiations of the modeling constructs of Figure 6.7. We elaborate on the way the AGV agent software interacts with the environment and the way the execution time of the AGV agent software is captured in the simulation model.

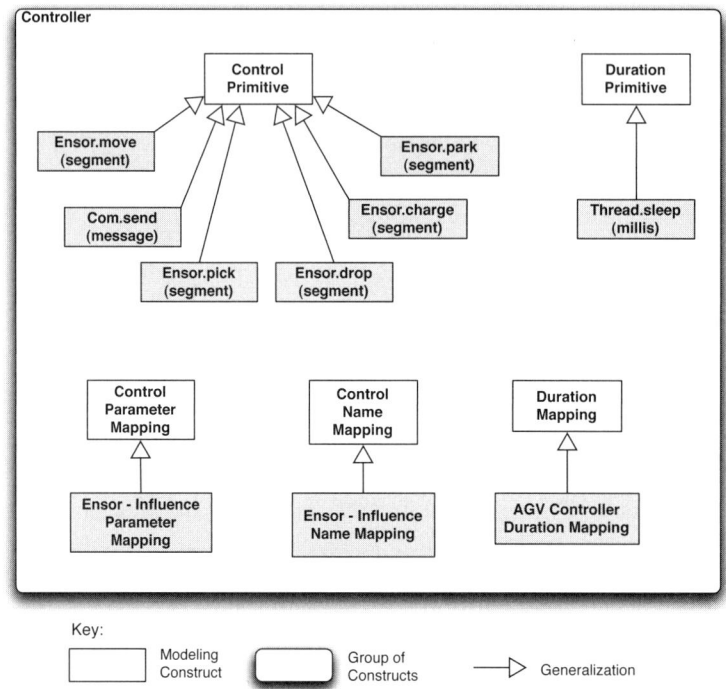

FIGURE 6.9 Overview of the simulation model for integrating the AGV control software in a simulation. The grey parts are specific instantiations of the modeling constructs for the AGV simulator.

Control Interface of AGV Agents.

The interaction of the AGV agent software with the warehouse environment is modeled in terms of `control primitives` and a `control name mapping` and `control parameter mapping`.

We discuss two examples of control primitives. *Ensor.move(segment)* is an E'nsor control primitive that instructs E'nsor to drive the AGV over the given segment. *Com.send(message)* is a control primitive that instructs an AGV's onboard wireless communication module to send a message.

We employ an *Ensor-influence name mapping* to determine the name of the influences that result from the control primitive invocations. The mapping between control primitive invocations and influences is a straightforward one-to-one mapping. For example, invocations of the control primitive *Ensor.charge* will be mapped on *charge influences*.

We employ an *Ensor-influence parameter mapping* to determine the parameters of the influences that result from the control primitive invocations. For example, for all control primitives that take a *segment* as argument, the corresponding influence requires two parameters: the *segment* on the one hand, and one of both end stations of that segment on the other hand (to indicate the direction in which an AGV will drive over that segment). For invocations of the control primitive *Com.send(message)*, the corresponding *send influence* requires the receiver which is encapsulated in the message as an explicit parameter. The Ensor-influence parameter mapping takes care of determining all parameters needed for the influences.

Execution Time of AGV Agents.

The execution time of AGV agent software is modeled in terms of `duration primitives` and a `duration mapping`.

The duration primitives are typically dependent upon the AGV agent software. Therefore, the developer should specify custom duration primitives for the AGV control software that is to be embedded in the simulation. There is one default duration primitive captured in the simulation model: *Thread.sleep(millis)*. *Thread.sleep(millis)* suspends the execution of an AGV agent for the specified number of milliseconds. This control primitive is used extensively in controllers, as the real environment typically evolves several orders of magnitude slower than the control system.

We employ an *AGV agent duration mapping* to specify the duration of invocations of duration primitives. By default, the *AGV agent duration mapping* only associates a duration to invocations of the duration primitive *Thread.sleep(millis)*. That duration corresponds to the amount of time specified by the argument *millis*.

6.5 Architecture of the Simulation Platform

In this section, we describe the software architecture of a simulation platform that supports the modeling constructs of the modeling framework described in Section 6.4.

The software architecture of a system is defined as "the structure or structures of the system, which comprise software elements, the externally visible properties of those elements, and the relationships among them" [Bass et al., 2003]. The software architecture captures the essence of a complex software system by identifying key stakeholder concerns and by explicitly specifying how software needs to be structured and behave to address the concerns. As such, a software architecture is a reusable artifact for the creation of such a simulation platform. We developed a simulation platform that implements this architecture and we applied this simulation platform to support software-in-the-loop simulations for evaluating, comparing and integrating several functionalities of a multi-agent control system for steering AGVs.

We use several architectural views to document the architecture of the simulation platform. A view is a representation of a coherent set of architectural elements and the relations among them [Bass et al., 2003]. Each view presents a particular perspective on the architecture or a part thereof. For each view, we start with a general explanation of the goal of the view and the software elements and relations between elements that are considered in that view. Afterward, we document each view for the simulation platform using a graphical notation and we explain how important quality requirements are realized.

This section is structured as follows. In Section 6.5.1, we put forward the functional and quality requirements of the simulation platform. In the following sections, we elaborate on the different architectural views. We start with a top-level module decomposition view in Section 6.5.2. We describe a component and connector view of the functionality to support dynamic environments in Section 6.5.3. We elaborate on a component and connector to explain the simulation engine that synchronizes all parts of the simulation in Section 6.5.4. In Section 6.5.5, we put forward an aspect-oriented approach to integrate the control software in the simulation. We describe a component and connector view of the functionality that keeps track of the execution time of an agent in Section 6.5.6.

6.5.1 Requirements

The goal of the simulation platform is to provide run-time support for software-in-the-loop simulations of multi-agent control systems in dynamic environments, of which the simulation model is described in terms of the modeling constructs proposed in Section 6.4. We discuss the functional and quality requirements that are the main drivers for the architecture of the simulation platform.

The main functional requirements for the simulation platform are the following:

- *Support the modeling constructs for dynamic environments.* The simulation platform should encapsulate the functionality to support the modeling constructs for dynamic environments described in Section 6.4.1. This functionality includes (1) managing the sources of dynamism and the influences that result from their execution (2) applying the appropriate reaction laws to determine the reaction of the environment to the various influences (3) handling all activities in the environment during a simulation run, and (4) applying the interaction laws to enforce interactions between activities.

- *Support the modeling constructs for the control software.* The simulation platform should encapsulate the functionality to support the modeling constructs for embedding the software of real controllers, described in Section 6.4.1. This functionality includes (1) keeping track of the duration primitives invoked by each of the controllers (2) keeping track of the control primitives invoked of each of the controllers, (3) deriving the nature and the timing of the influences that result from executing the controllers.

- *Support consistent simulation runs.* The simulation platform should encapsulate the functionality to carry out simulation runs that are consistent with the described simulation model. The simulation model specifies the causal relations between all influences, activities, reaction and interaction laws by means of simulation time, e.g., by means of specifying the duration of the various controllers in simulation time, specifying the start and duration of each activity in simulation time, etc. To obtain causal relations in accordance to the specification of the simulation model, the progress of all parts of the simulation, i.e. the progress of the various controllers and environment sources of dynamism and of applying the various reaction and interaction laws, should happen in the order of increasing simulation time. Given the unpredictable delays introduced by the underlying execution platform on which the simulation platform runs, an explicit synchronization between all parts of the simulation is necessary to regulate their relative progress.

We describe the main quality requirements of the simulation platform:

- *Flexibility of embedding the software of the control system.* The simulation platform should provide support for embedding the software of the control system in a flexible way, i.e., with minimal effort from the developer. The fact that some simulation concerns crosscut with the control system's functionality hampers embedding the real controllers in a flexible way. We rely on state-of-the-art software engineering technology to modularize crosscutting simulation concerns in order to insert and remove them in the control system in a plug-and-play manner.

- *Modifiability of the simulation platform.* Modifying core parts of the simulation platform should be relatively easy, and the impact of such modifications should be as local as possible. Core parts of the simulation platform include the simulation

engine and the functionality to support simulated environment.
- *Performance of the simulation platform.* The simulation platform should support *as-fast-as-possible simulation*, to enable executing simulation runs faster than real time. Simulation platforms that support software-in-the-loop simulations are typically limited to *real-time simulation*, i.e. simulation time advances in pace with wallclock time during a simulation run.

6.5.2 Top-Level Module Decomposition View of the Simulation Platform

The goal of a module decomposition view is to show how the simulation platform is decomposed into manageable software implementation units. A module decomposition view is a static view on a system's architecture. The elements depicted in a module decomposition view are *modules*. A module is an implementation unit of software that provides a coherent unit of functionality. The relationship between the modules is *is-part-of* that defines a part/whole relationship between a submodule and the aggregate module. Modules are recursively refined, revealing more details in each decomposition step. The basic criteria for module decomposition is the achievement of quality requirements. For example, parts of a system that are likely to change, are encapsulated in separate modules to support modifiability. Another example is the separation of functionality of a system that has higher performance requirements from other functionality.

The module decomposition view includes a description of the interfaces of each module that documents how a module is used in combination with other modules. The interface description distinguishes between provided and required interfaces. A provided interface specifies what functionality the module offers to other modules. A required interface specifies what functionality the module needs from other modules; it defines constraints of a module in terms of the services a module requires to provide its functionality.

The top-level module decomposition view of the simulation platform is depicted in Figure 6.10. We first discuss the main elements and their properties. Afterward, we describe their interfaces and explain how important qualities are realized.

Elements and Their Properties

The simulation system is decomposed in two main subsystems: Controller and Simulation Platform.

- `Controller` is a software module of the real multi-agent control system that is embedded in the simulation platform in order to test or configure it. A multi-agent control system consists of several controllers, i.e. agents, working in parallel and cooperating to solve a problem in the environment. A controller has well-defined ways to sense the environment and to act upon it. An example of a controller is an agent program to controls a particular robot in a manufacturing plant.
- `Simulation Platform` is the medium in which controllers of a multi-agent control system are embedded in order to test or configure them. The main responsibilities of the simulation platform are:
  - To simulate the real dynamic environment of the multi-agent control system.
  - To manage the execution of all controllers of the multi-agent control system according to the specified duration model.

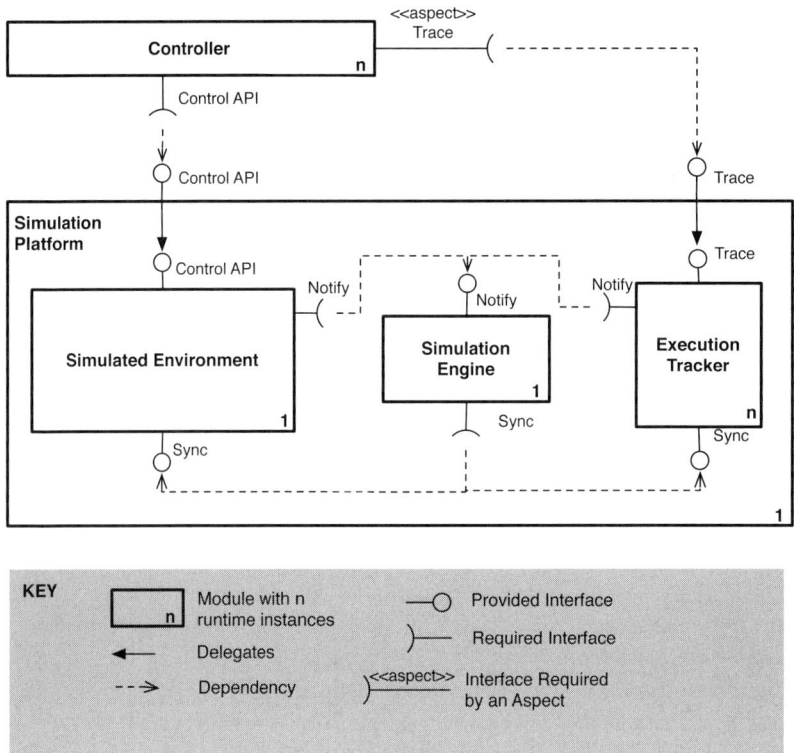

FIGURE 6.10 Top-level module decomposition view of the simulation platform.

- To execute simulation runs as-fast-as-possible, thus enabling simulations faster than real time.

The simulation platform is further decomposed into three different modules: Simulated Environment, Simulation Engine and Execution Tracker.

- Simulated Environment is responsible for managing a simulation model of the real environment of the multi-agent control system. The Simulated Environment encapsulates all functionality to support the modeling constructs described in Section 6.4.1.
- Simulation Engine is responsible for managing the evolution of all parts of the simulation in correspondence to the specifications of the simulation model. The simulation engine encapsulates all functionality to synchronize the progress of the simulated environment with the progress of all execution trackers of the controllers that are embedded in the simulation. This guarantees correct causal relations in correspondence to the specifications of the simulation model.
- Execution Tracker is responsible for tracing the execution of a particular controller of the multi-agent control system. This module encapsulates all functionality to support the modeling constructs for the control software described in Section 6.4.1. Tracing the execution of a controller includes (1) determining the execution time consumed by a particular controller of the multi-agent control system according to the duration mapping, and (2) synchronizing the execution of that controller with the simulation engine, which is necessary to enable as-fast-

as-possible simulations. At runtime, there is an instance of the execution tracker module for each controller.

Interface Descriptions

The simulation platform module provides two interfaces to the controller: `Control API` and `Trace`.

- `Control API` supports the *application* concerns of a controller. `Control API` is the control interface required by the controller to interact with its environment. The `Control API` provided by the simulation platform is identical to the control interface the controller uses to interact with its sensors and actuators in the real environment. By providing the `Control API` interface, the simulation platform cannot be distinguished from the real environment from point of view of a controller.
- `Trace` supports the *simulation* concerns for a controller. `Trace` is the interface provided by the simulation platform to manage the execution of a controller. The `Trace` interface enables (1) monitoring the execution time consumed by the controller and (2) intercepting and synchronizing the execution of the controller with the simulation engine. The `Trace` interface is further explained in Section 6.5.5.

The simulation platform module delegates the `Control API` interface to the simulated environment module, and the `Trace` interface to the execution tracker module.

The simulation engine governs the progress of the simulation by means of the provided `Notify` and required `Sync` interfaces. We elaborate on Notify and Sync in Section 6.5.3.

Architectural Rationale

Each module in the decomposition encapsulates a particular functionality of the simulation platform. By minimizing the overlap of functionality among modules, the architect can focus on one particular part of functionality. Allocating different functionalities of the simulation platform to separate modules results in a clear design. It helps to accommodate change and to update one module without affecting the others, and it supports reusability. We elaborate on the core architectural decisions.

Low Coupling between Controller and Simulation Platform.

As we are concerned with software-in-the-loop simulations, one of the main architectural decisions is a low coupling between the control software on the one hand, i.e. the controllers, and the simulation platform in which it is embedded on the other hand. The `Control API` interface enables all communication, sensing and acting to be directed to the simulation platform transparently. The `Trace` interface connects the controller with a dedicated Execution Tracker in the Simulation Platform. Section 6.5.5 illustrates an aspect-oriented approach to provide existing controllers with support for the `Trace` interface.

Two advantages of a low coupling between Controller and Simulation platform are (1) reuse, i.e. the simulation platform can be reused for testing various controllers, and (2) modifiability, i.e. the controllers can be modified without affecting the simulation platform.

Low Coupling between Simulated Environment and Simulation Engine.

In the simulation platform, we make an explicit distinction between the simulated environment on the one hand, and the simulation engine on the other hand. The simulated environment maintains a model of the real environment. The simulation engine manages the simulation main loop, i.e. advancing simulation time by synchronizing the progress of all parts of the simulation. Simulated Environment and Simulation Engine are coupled by means of well-defined interfaces, i.e. `Notify` and `Sync`. This enables (1) the Simulated Environment to make abstraction of how and with whom synchronization is required, and (2) the Simulation Engine to focus on reliable and efficient synchronization, without knowledge of the internal working of each party that needs synchronization.

Two advantages of the low coupling between Simulated Environment and Simulation Engine are (1) reuse, i.e. it facilitates the integration of a different simulation engine into the simulation platform, and (2) manageability, i.e. the design of the Simulated Environment is facilitated because abstraction can be made of all synchronization issues.

Explicit Support for as-Fast-as Possible Simulations.

In as-fast-as possible simulations, there is no fixed relation between the progress of the simulation engine and wallclock time. This enables simulation runs faster than real time. To support as-fast-as-possible simulation, the execution of each controller must be synchronized explicitly with the simulation engine in the simulation platform. Execution Trackers and the Trace interface encapsulate the functionality to Trace and synchronize the execution of the controllers with the simulation engine in an explicit manner.

6.5.3 Component and Connector View of the Simulated Environment

A component and connector view [Clements et al., 2002; Ivers et al., 2004] shows a system as a set of cooperating units of execution. A component and connector view is a run-time view on a system's architecture. The elements of the component and connector view are run-time elements of computation and data storage, such as repositories and components. Components are run-time instances that perform calculations that typically require data from one or more data repositories. Data repositories store data and mediate the interactions among components. A data repository can provide a trigger mechanism to signal data consumers of the arrival of interesting data. Besides reading and writing data, a data repository may provide additional support, such as support for concurrency and persistency. The relationship between elements within a component and connector view are connectors. A connector is a path for communication that links connecting ports on two or more elements. A port is an interaction point on a run-time element through which data is sent and received according to a specific interface. A port is similar to an interface in that it describes how an element interacts with its environment, but is different in that each port is a distinct interaction point of its element [Ivers et al., 2004].

The component and connector view of the Simulated Environment is depicted in Figure 6.11. This view gives a detailed perspective on the Simulated Environment module of Figure 6.10. The Simulated Environment supports the modeling constructs for dynamic environments described in Section 6.4.1. We first discuss the main elements and their properties. Afterward, we describe how they are connected and explain how important qualities are realized.

Elements and Their Properties

The Simulated Environment contains various components that are connected to five possible repositories: State, Activities, Influences, Reaction Laws and Interaction Laws. We elaborate on each of the five repositories. Afterward, we describe the components they are connected to.

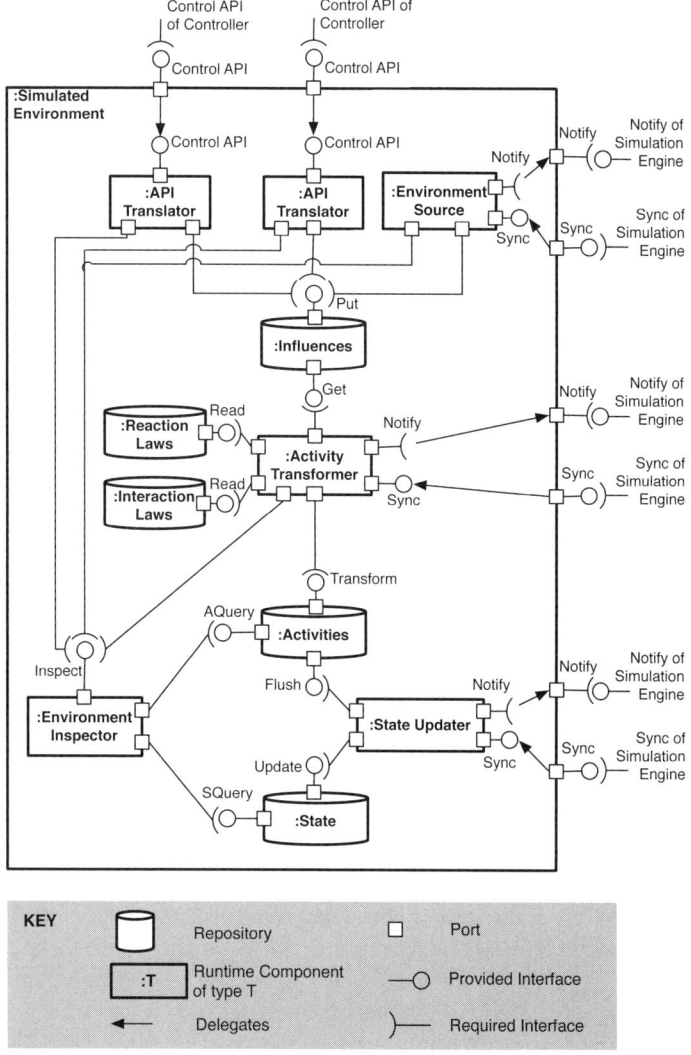

FIGURE 6.11 Component and connector view of the simulated environment.

- **State** repository contains values for all variables to describe the state of the environment. The values of the state describe a snapshot of the environment at a particular instant of simulation time, i.e. the snapshot time. The state of the environment includes the state of all environmental entities and properties of the environment. Examples are the position and battery level of each robot in the environment, the position of various objects in the environment, the temperature of the environment.
- **Activity** repository maintains the activities as first-class elements. Activities describe the evolution of the state of the environment over time. Activities are always expressed relative to the snapshot of the state stored in the States repository. Examples of activities are an activity that describes the driving of a robot and an activity describing the rolling of a ball in a RoboCup Soccer

environment.

- **Influence** repository contains the influences as first-class elements. Influences are attempts to start, stop or alter activities. Influences originate from the controllers of the multi-agent control system on the one hand, and from environment sources external to the multi-agent control system on the other hand.
- **Reaction Law** repository maintains the reaction laws of the environment model as first-class elements. The reaction laws determine the way influences have an impact on the activities in the simulation.
- **Interaction Law** repository maintains the interaction laws of the environment model as first-class elements. The interaction laws determine the way activities may interact in the environment.

The components are runtime instances of corresponding modules within the Simulated Environment:

- **Environment Inspector** acts as the facade that regulates all inspections of both state and dynamics of the environment. This includes functionality to retrieve the state of a part of the environment at any particular point in simulation time, based on the actual content of the **State** repository and **Activity** repository.
- **State Updater** prevents activities from piling up in the Activity repository during a simulation run. The **State Updater** periodically flushes activities from the **Activity** repository and updates the corresponding values in the **State** repository such that they represent the state at a later snapshot time.
- **Activity Transformer** is responsible for applying all laws present in the **Reaction Law** repository and **Interaction Law** repository. The laws are blackbox elements for the **Activity Transformer**, which only orchestrates applying all laws. Applying the laws includes (1) checking whether laws are applicable and (2) manipulating the activities in the **Activity** repository in correspondence to the applicable laws. To check whether laws are applicable, the **Activity Transformer** verifies for each law whether its conditions are satisfied. For interaction laws, this involves contacting the **Environment Inspector**; for reaction laws, this this also involves contacting the **Influence** repository besides the **Environment Inspector**. To apply a reaction law, the **Activity Transformer** removes the respective influences from the **Influence** repository, and performs the activity transformation proposed by the reaction law on the activities in the **Activity** repository. To apply an interaction law, the **Activity Transformer** performs the activity transformation proposed by the interaction law on the activities in the **Activity** repository.
- **API Translator** is responsible for translating a controller's invocations on the **Control API** interface into the concepts of the environment model. More specifically, the **API Translator** maps all triggering of actuators (e.g., (de-)activating motors or sending communication messages) into influences that are stored in the **Influence** repository, according to the control parameter mapping and control name mapping (see Section 6.4.1). The **API Translator** realizes all triggering of sensors (e.g., readout of sensor values or received communication messages) by querying the **Environment Inspector**. In Figure 6.11, two **API translators** are depicted. Each **API translator** is connected to a controller of the multi-agent control system.
- **Environment Source** is responsible for mimicking the behavior of a source of

dynamism in the environment that is external to the multi-agent control system. `Environment Sources` are capable of performing influences and sensing the environment. Examples of `Environment Sources` are other machines or humans that reside in the environment of the multi-agent control system. In Figure 6.11, one instance of an `Environment Source` is depicted.

Interface Descriptions

Figure 6.11 depicts the interconnections between the repositories and the internal components of the simulated environment.

The State repository provides two interfaces:

- `SQuery` is the interface provided by the State repository to read the current values of the variables.
- `Update` is the interface provided by the State repository to enable updating the state to a new snapshot time.

The Activity repository provides three interfaces:

- `AQuery` is the interface for inspecting activities. Inspection is based on matching: the requester specifies a condition that must hold for all activities that are returned.
- `Flush` is the interface to (partially) empty the Activity repository. The requester specifies a point in simulation time. `Flush` returns all activities that finish before the specified time instant. In contrast to the `AQuery` interface, the activities returned by the `Flush` interface are removed from the Activity repository.
- `Transform` is the interface to manipulate the activities in the Activity repository. The requester specifies an activity transformation to be performed on the activities.

The Reaction Law repository and Interaction Law repository provide one interface:

- `Read` is the interface that can be used to access the laws in the corresponding repository.

The Influence repository provides two interfaces:

- `Put` is the interface for storing new influences in the Influence repository.
- `Get` is the interface for returning influences out of the Influence repository. The requester can specify (1) a condition that must be satisfied by each influence that is returned, and (2) whether the returned influences should be removed from the Influence repository.

The Environment Inspector provides the `Inspect` interface that assembles a snapshot of the state of the environment at any particular instant of simulation time. The `Notify` and `Sync` interfaces are described in Section 6.5.4, when the simulation engine is discussed.

Architectural Rationale

Low Coupling Due to Data Repositories.

The use of data repositories decouples the various components within the simulated environment. Low coupling improves modifiability (as the changes in one element do not affect other elements), and reuse (as elements are not dependent on too many other elements).

Decoupled elements do not require detailed knowledge about the internal structures or operations of other elements. Furthermore, decoupled elements are easier to understand due to clear and coherent responsibilities.

For example, the `Influence` repository gathers all influences, regardless of whether these influences originate from controller actions that are translated by `API translators`, or from `Environment Sources` external to the multi-agent control system. As such, the `Influence` repository decouples the `Activity Transformer` from the various sources of influences, i.e. `API Translators` and various `Environment Sources`.

Decoupling Synchronization from the Simulated Environment.

The elements that need synchronization are `Environment Source`, `Activity Transformer` and `State Updater`. All synchronization is delegated to the simulation engine by means of the `Notify` and `Sync` interfaces. Synchronization will be discussed in Section 6.5.4.

Customizable Presentation of the Environment State.

The `Environment Inspector` acts as a facade to hide the internal representation of the environment state in terms of state and activities. As such, the internal representation is decoupled from the way the state is presented toward other components, such as the `API Translators`, `Environment Sources` and the various laws managed by the `Activity Transformer`. The `Inspect` interface enables the use of a custom representation for each component it is connected to.

Customizable State Updating Strategy.

The strategy to update the state is encapsulated in the `State Updater`. This offers the developer the ability to apply a custom updating strategy. For example, in case the execution trace should be logged, the `State Updater` can be deactivated easily, such that all activities are aggregated in the Activity repository and can be inspected afterward.

Reusable Infrastructure.

Finally, we emphasize the reusability of the architecture of the simulated environment. The internal components and repositories of the simulated environment comprise the infrastructure necessary to handle influences, activities, reaction laws and interaction laws. This infrastructure can be reused for all simulation studies whose simulation models are described in terms of these constructs.

6.5.4 Component and Connector View of the Simulation Engine

The Simulation Engine is responsible for advancing simulation time by synchronizing all parts of the simulation such that everything happens in the order of increasing simulation time. This guarantees correct causal relations in correspondence with the specifications of the simulation model.

We focus on the way the Simulation Engine regulates the progress of all parts of the simulation. The component and connector view of the Simulation Engine is depicted in Figure 6.12. We first discuss the main elements and their properties. Afterward, we describe how they are connected and explain how important qualities are realized.

Elements and Their Properties

The `Simulation Engine` is responsible for executing a simulation run by synchronizing the progress of various components. Synchronization is necessary to ensure causality, i.e., to

On the Role of Software Architecture for Simulating Multi-Agent Systems

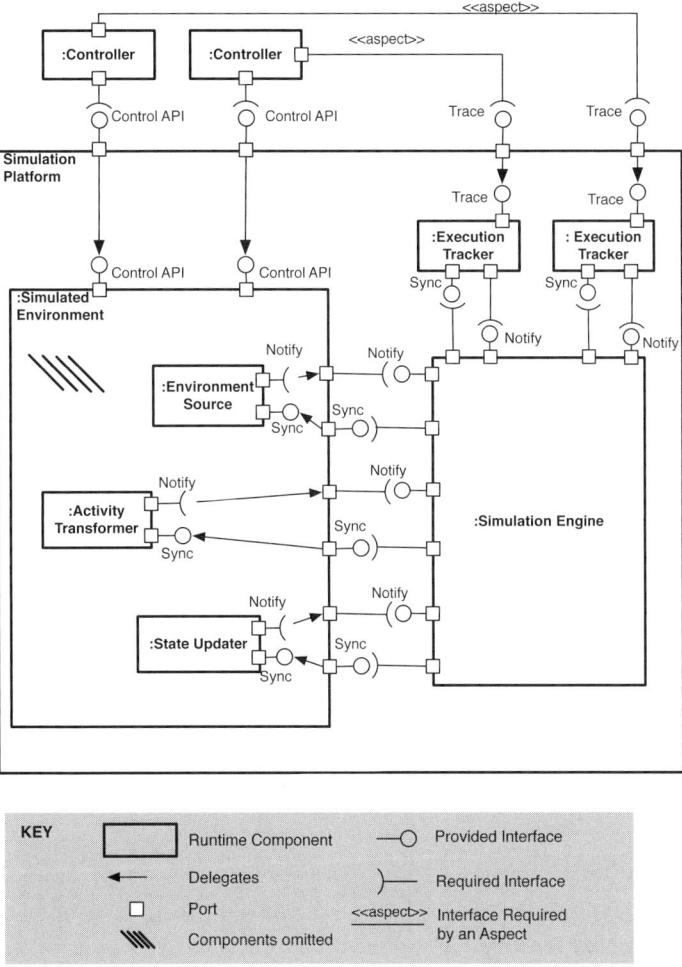

FIGURE 6.12 Component and connector view of the simulation engine.

enforce that everything happens in the order of increasing simulation time. The components that need synchronization are the following: the various sources of dynamism that act in parallel, i.e., the `Controllers` of the multi-agent control system and the `Environment Sources` external to the multi-agent control system, the `Activity Transformer` that applies the reaction and interaction laws and the `State Updater` that updates the state to a new snapshot time.

We discuss each component, explain why it needs synchronization and the way it relies on the `Simulation Engine` for synchronization.

- `Environment Source` (see Section 6.5.3). An `Environment Source` can access the environment in several ways, i.e. by performing an influence or sensing the environment. To ensure correct causal relations between its environment access and other things happening in the simulation, an `Environment Source` synchronizes its execution with the `Simulation Engine`: before performing an influence or sensing the environment, an `Environment Source` notifies the `Simulation Engine` at what moment in simulation time it wants to access the environment and suspends its execution until it is granted permission to proceed by the `Simulation Engine`.

- `Execution Tracker` (see Section 6.5.5). An `Execution Tracker` manages the execution of a `Controller`. An `Execution Tracker` keeps track of the execution time consumed by a `Controller` to deduce at what moment in simulation time a `Controller` accesses the environment. To determine whether a controller accesses the environment, an execution tracker keeps track of all control primitive invocations on the `Control API`. Synchronization is necessary to ensure correct causal relations between the access of a controller to the environment and other things happening in the simulation, Each time a control primitive of the `Control API` is invoked, the `Execution Tracker` notifies the `Simulation Engine` and suspends that `Controller`'s execution until it is granted permission to proceed by the `Simulation Engine`.
- `Activity Transformer` (see Section 6.5.3). An `Activity Transformer` changes the activities in the `Activity` repository by applying the reaction and interaction laws. To ensure correct causal relations between activity transformations and other things happening in the simulation, the `Activity Transformer` notifies the `Simulation Engine` before applying an activity transformation and suspends its execution until it is granted permission to proceed by the `Simulation Engine`.
- `State Updater` (see Section 6.5.3). A `State Updater` updates the `State` repository to a new snapshot time by flushing activities. To guarantee correct causal relations with the rest of the simulation, the `State Updater` notifies the `Simulation Engine` of the new snapshot time it wants to update the `State` repository to and suspends its execution until permission to proceed is granted by the `Simulation Engine`.

Interface Descriptions

Figure 6.12 illustrates how the various components are connected with the `Simulation Engine`. The synchronization of all components happens through a uniform interface:

- `Notify` is the interface provided by the `Simulation Engine` to enable components to publish new events. To notify the Simulation Engine of a new event, a component specifies (1) the simulation time stamp of the event and (2) a callback identifier of the component. The callback identifier is used for granting permission to that component when it is safe to execute that event.
- `Sync` is the interface required by the `Simulation Engine` to grant permission to a component for executing an event.

Architectural Rationale

The Simulation Engine Encapsulates All Synchronization.

An important architectural decision is that the main components within the `Simulation Platform` can make abstraction of all synchronization with other components. The `Simulation Engine` encapsulates the actual synchronization algorithm (in our case a conservative discrete event synchronization algorithm [Chandy and Misra, 1981]) that maintains and manages all synchronization partners. The `Simulation Engine` uses the `Notify` and `Sync` interfaces to synchronize various components. As such, the `Simulation Engine` does not depend upon the internal working and functionality of these components.

6.5.5 An Aspect-Oriented Approach to Embed Control Software

We explain the way the software of the real controllers is embedded in the simulation platform. Recall that a `Controller` is connected to the `Simulation Platform` by means of two interfaces: the `Control API` interface the `Trace` interface, as is depicted in Figures 6.10 and 6.12. The `Trace` interface enables monitoring the execution of a controller by tracking its duration primitive invocations and control primitive invocations (see Section 6.4.1). However, in contrast to the `Control API` interface, the `Trace` interface is not a native interface of a controller. The `Trace` interface is solely necessary for simulation purposes, i.e. to enable synchronizing the execution of a controller with the simulation. Consequently, embedding a controller in the simulation platform would require the developer to modify the design of the controller such that it supports the `Trace` interface. This would be a time-consuming and error-prone job, which we would like to avoid.

We describe an approach to extend a multi-agent control system transparently, i.e., without requiring the developer to perform changes in the design of the controllers. Our approach uses aspect-oriented programming to achieve this. We first introduce aspect-oriented programming in Section 6.5.5. The way aspect-oriented programming is used in the simulation platform is described in Section 6.5.5. We emphasize how important qualities are realized in Section 6.5.5.

Aspect-Oriented Programming

Tracking the execution of a controller is a *crosscutting concern*, i.e. the functionality to do this crosscuts a multi-agent control system's basic functionality. The problem of crosscutting concerns is that they can not be modularized with traditional object oriented techniques. This forces the functionality to monitor the execution of a controller to be scattered throughout the code of the multi-agent control system, resulting in "tangled code" that is excessively difficult to develop and maintain. Aspect-oriented programming [Kiczales et al., 1997, 2001] handles crosscutting concerns by providing *aspects* for expressing these concerns in a modularized way. An aspect is a modular unit of crosscutting implementation. Aspect-oriented programming does not replace existing programming paradigms and languages, but instead, it can be seen as a co-existing, complementary technique that can improve the utility and expressiveness of existing languages. It enhances the ability to express the separation of concerns which is necessary for well-designed, maintainable software systems.

A language extension to Java which supports aspect-oriented programming, is AspectJ. In AspectJ, defining an aspect is based on two main concepts: pointcuts and advice. A *pointcut* is a language construct in AspectJ that selects particular join points, based on well-defined criteria. Each *join point* represents a particular point in the execution flow of a program where the aspect can interfere, e.g., a point in the flow when a particular method is called. As such, pointcuts are a means to express the crosscutting nature of an aspect. *Advice* on the other hand is a language construct in AspectJ that defines additional code that runs at join points specified by an associated pointcut. An aspect encapsulates a particular crosscutting concern and can contain several pointcut and advice definitions. The process of inserting all crosscutting code of an aspect at the appropriate join points within the original program code, is called *aspect weaving*. Aspect weaving is performed at compile-time in AspectJ.

Providing Support for Tracing a Controller's Execution through Aspect Weaving

We describe a way to flexibly embed a multi-agent control system in the simulation platform, i.e., without requiring the developer to alter the design of the controllers. We use aspect-oriented programming technology to plug and unplug into a multi-agent control system all functionality required for simulation purposes.

To embed the controller in a simulation platform, the controller must be extended with the following tracing functionality:

- *Tracing duration primitive invocations.* The simulation platform tracks the execution time consumed by each controller according to the duration mapping, so it must be able to monitor all duration primitives invocations of a controller. To support this kind of monitoring, the controller should be extended with functionality that notifies the simulation platform each time the controller executes a duration primitive.
- *Tracing control primitive invocations.* The simulation platform synchronizes a controller's invocations on the `Control API` with the rest of the simulation (see Section 6.5.4). To support such synchronization, the simulation platform must be capable of intercepting a control primitive invocation and of temporarily suspending a controller's execution.

Figure 6.13 depicts the way the above tracing functionality is inserted in the controller software. This figure shows an example `Controller` that consists of a `Decision Taker` module and a `Plan Library` module.

The left-hand side of the figure depicts the original controller. Note that this controller does not support the `Trace` interface. The left-hand side of the figure also depicts an `Aspect`. The `Aspect` is a separate module that encapsulates all tracing functionality. The `Aspect` is generated from the specification of the duration primitives and control primitives. The *pointcut* definition of the aspect specifies all duration primitive invocations and control primitive invocations as join points. The *advice* of the aspect comprises a call to the `Trace` interface to notify the simulation platform.

The black arrow on the figure illustrates the process of aspect weaving. Aspect weaving happens at compile time, and automatically extends the `Controller` with all tracing functionality necessary to embed it in the simulation platform.

The right hand side of Figure 6.13 depicts the outcome of the weaving process. Within the `Controller`, the `Decision Taker` and `Plan Library` modules are now extended with additional tracing functionality that is the result of weaving the aspect's advice. The added tracing functionality crosscuts the modules of a controller, as depicted by the grey blocks. Note that due to aspect weaving, the controller now supports the `Trace` interface at the appropriate locations without requiring the developer to perform manual modifications to the control software.

Architectural Rationale

Flexibility of Embedding a Multi-Agent Control System.

Aspect weaving supports flexibly embedding the multi-agent control system in a simulation: the developer is no longer bothered to modify a multi-agent control system and manually insert or remove all code necessary for tracing its execution.

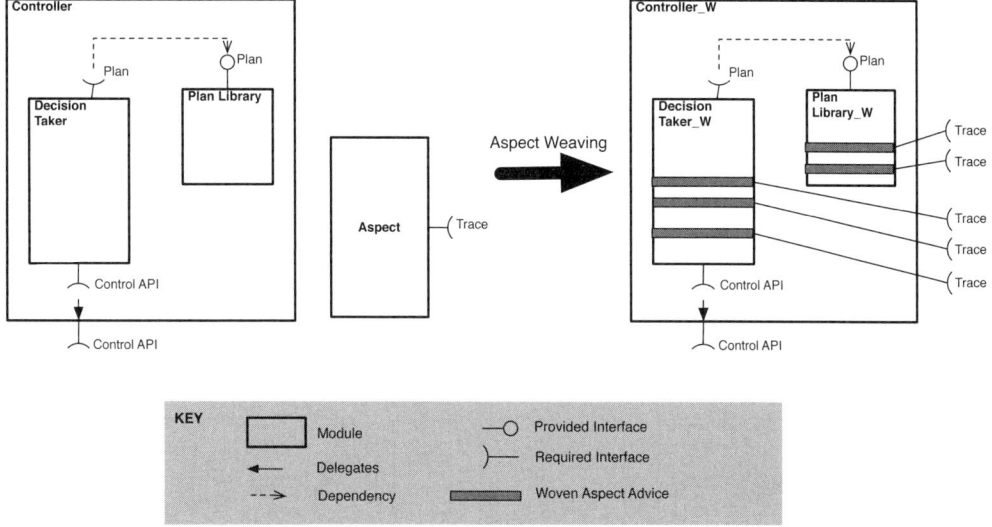

FIGURE 6.13 Controller before (left) and after (right) aspect weaving.

Separating Simulation from Application Concerns.

Aspect technology enables modularizing simulation concerns that crosscut the multi-agent control system's functionality. This leads to a clean separation between application concerns and simulation concerns, as both are encapsulated in separate modules (as depicted on the left-hand side of Figure 6.13).

6.5.6 Component and Connector View of the Execution Tracker

We focus on the `Execution Tracker`. An Execution Tracker is responsible for tracing the execution of a particular controller of the multi-agent control system. Tracing the execution of a controller includes (1) determining the execution time consumed by a particular controller of the multi-agent control system according to the duration mapping, and (2) synchronizing the execution of that controller with the simulation engine, which is necessary to enable as-fast-as-possible simulations.

Figure 6.14 depicts a component and connector view of two controllers embedded in the simulation platform. The focus is on the Execution Trackers and the way they interact with a controller on the one hand, and with the simulation engine on the other hand. We first discuss an Execution Tracker's main elements and their properties. Afterward, we describe how they are connected and explain how important qualities are realized.

Elements and Their Properties

Figure 6.14 depicts two `Execution Trackers`, each connected to a `Controller`. Each Execution Tracker comprises the following components and repositories:

- `Clock Manager` is responsible for managing the simulation clock of a particular `Controller`. The simulation clock indicates how much execution time that particular `Controller` consumed. As a `Controller` executes, the `Clock Manager` is notified of the duration primitives invocations performed by that controller, and advances the simulation clock with the duration that is specified by the duration

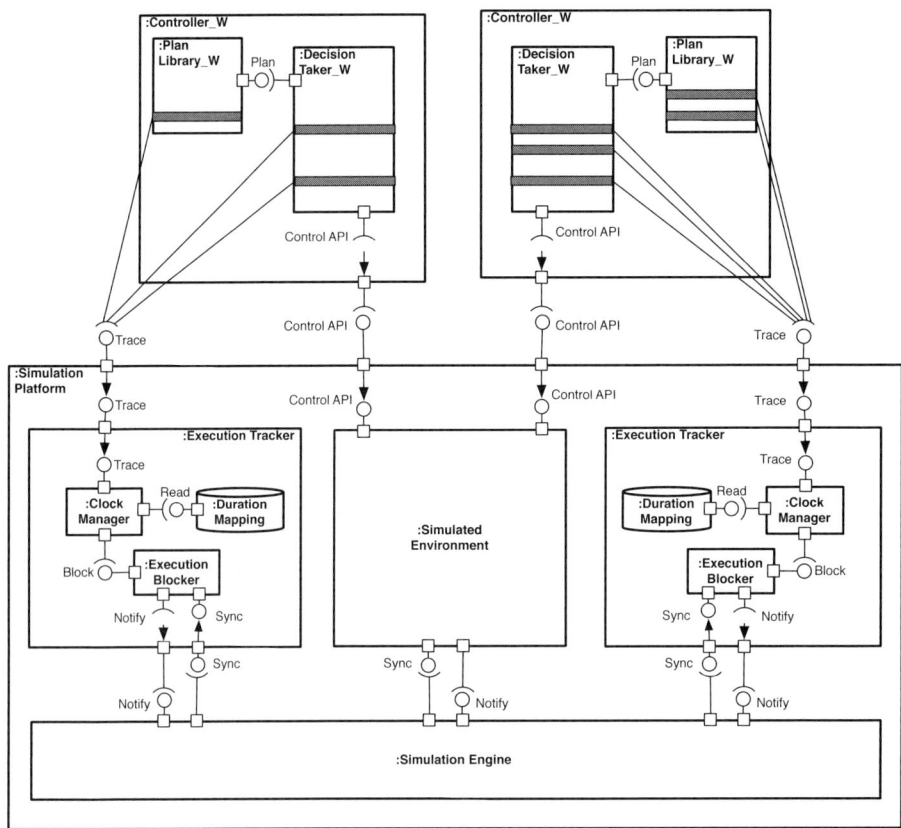

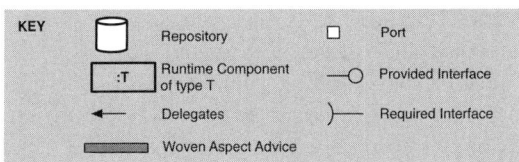

FIGURE 6.14 Component and connector view of controllers and execution trackers.

mapping (see Section 6.4.1). As such, the simulation clock of the `Clock Manager` is kept up-to-date with the execution time of the `Controller`.

- `Duration Mapping` repository is responsible for maintaining the duration mapping of a particular controller. For each duration primitive invocation, the `Duration Mapping` repository specifies a duration in simulation time.
- `Execution Blocker` is responsible for synchronizing the execution of a `Controller` with the `Simulation Engine`. For each control primitive invocation, the `Execution Blocker` can temporarily suspend the execution of a `Controller` until access is granted by the `Simulation Engine`.

Interface Descriptions

Figure 6.14 depicts the interconnections between the various elements of an `Execution Tracker`:

- `Trace` is the interface provided by the `Clock Manager` to keep track of the execution of a `Controller`. By means of aspect weaving (see Section 6.5.5) a `Controller`'s execution is intercepted and redirected to the `Clock Manager` each time a duration primitive invocation or control primitive invocation is performed. The caller of the `Trace` interface specifies the characteristics of the duration primitive invocation or control primitive invocation (see Section 6.4.1).
- `Read` is the interface provided by the `Duration Mapping` repository to enable retrieving the duration in simulation time of various duration primitive invocations. The caller specifies the characteristics of the duration primitive invocation. `Read` returns the associated duration for that duration primitive invocation according to the duration mapping.
- `Block` is an interface provided by the `Execution Blocker` to synchronize the execution of a `Controller` with the `Simulation Engine`. The caller of the `Block` interface specifies a simulation time instant until which the execution of the `Controller` should be suspended. The `Clock Manager` calls the `Block` interface with the current value of its simulation clock in case it traces a control primitive invocation. This guarantees synchronization with the `Simulation Engine` each time a `Controller` accesses the environment.

Architectural Rationale

Separating Monitoring from Synchronization.

The `Clock Manager` encapsulates all functionality to monitor the execution time of a `Controller`. The `Execution Blocker` encapsulates the functionality to synchronize the execution of a `Controller` with the `Simulation Engine`. Because both components have low coupling, a `Clock Manager` can make abstraction of synchronizing the execution of a `Controller`, whereas the `Execution Blocker` can make abstraction of monitoring a `Controller`'s execution time.

Reuseable Infrastructure.

We emphasize the reusability of the architecture of the `Execution Tracker`. The internal components of the `Execution Tracker` are independent of the specified duration mapping. The duration mapping is encapsulated in the `Duration Mapping` repository, where it can be adapted easily.

6.6 Evaluating the AGV Simulator

We developed a simulation platform that implements this architecture and we applied this simulation platform to support software-in-the-loop simulations for evaluating, comparing and integrating several functionalities of a multi-agent control system for steering AGVs. We now focus on evaluating flexibility and performance of the AGV simulator.

In Section 6.6.1, we discuss the flexibility of the AGV simulator. We demonstrate both flexibility and performance of the AGV simulator by means of experiments in Section 6.6.2. Finally, in Section 6.6.3 we discuss research on multi-agent control systems in the EMC^2 project that was supported by the AGV simulator.

6.6.1 Flexibility of the AGV Simulator

To use the AGV simulator, the developer specifies the characteristics of the AGV control software and the simulated warehouse environment.

- To embed the AGV control software in the simulation, the developer specifies the execution time of the control software. This is done by identifying duration primitives and configuring a duration mapping for these primitives. The control primitives and the control mapping are predefined for all E'nsor control primitives.
- The developer can specify the characteristics of the simulated warehouse environment in which the control software is embedded. Besides the physical setup of the warehouse (i.e. the number and positioning of AGVs, nodes, segments, etc.), the developer can also select appropriate environment sources of dynamism (e.g., the transport generator), reaction laws and interaction laws.

Flexibility of the AGV simulator is important to enable experiments with AGV control software of which the functionality is not yet fully operational. We give a number of examples of core parts of the AGV simulator that can be customized to suit the needs of a particular simulation study.

- The *battery law* can be disabled when performing tests with AGV control software of which the battery charging functionality is not yet operational. This prevents AGVs from running out of energy.
- The quality of service of the communication channel can be adjusted by means of the *WiFi QoS law*. Disabling this law ensures reliable transmission of all messages. To simulate degraded quality of service of the communication channel, the law can be configured with the desired behavior, e.g., reduced communication range, message loss, message delay, etc.
- Collision detection can be configured by means of the *collision law*. The collision law can be configured with the accuracy that is required for detecting collisions. By deactivating the collision law, AGVs can drive across the warehouse without affecting each other.
- The activities can be customized to reflect the physical characteristics of the AGVs. For example, *driving activities* encapsulate the specific velocity or acceleration profile of the AGVs.
- The transport profile of the transport generator can be customized to suit the needs of a particular simulation study.

6.6.2 Measurements of the AGV Simulator

We measure the performance of the AGV simulator and demonstrate its flexibility. We focus the collision law, as this law is a dominant factor for the performance of the AGV simulator.

Setup of the Experiments

The goal of the experiments is to illustrate both flexibility and performance of the AGV simulator. We performed experiments with 4 different configurations with respect to the collision law:

1. The collision law *deactivated*. In this particular configuration, the collision law is not used in the simulation. This setting is typically used in simulation studies in

which collision avoidance is out of focus.
2. The collision law configured with an accuracy of *10 centimeters*. As AGVs drive at a maximum speed of 1 meter per second, it takes an AGV 0.1 seconds to move over 10 centimeters. In case two AGVs travel at top speed, their relative position changes at a maximum rate of 2 meters per second. Consequently, to detect collisions with an accuracy of 10 centimeters, the snapshot frequency to detect collisions is 0.05 seconds.
3. The collision law configured with an accuracy of *25 centimeters*. This corresponds to a snapshot frequency of 0.125 seconds.
4. The collision law configured with an accuracy of *100 centimeters*. This corresponds to a snapshot frequency of 0.5 seconds.

The setup of the experiments is the following:

- The warehouse consists of 40 stations connected by 69 segments over an area of 1400 by 900 meters.
- The number of AGVs varies from 2 to 12. These are typical sizes of AGV warehouse transportation systems.
- We use lightweight AGV agents. This enables us to measure the computation time consumed by the AGV simulator itself, with minimal bias from the controllers that are embedded in it. Each AGV agent is programmed to poll the status of its AGV every second. AGVs drive around randomly: as soon an AGV agent notices it has reached the next station, it randomly selects a next segment to drive on. AGVs rely on segment locking for avoiding collisions.

The simulations are executed on the following computer platform[*]: Intel Pentium 4, 2.8GHz, 512MB of memory, Java 1.5.0.

Measurements

Figure 6.15 shows the measured performance of the AGV simulator for each of the four configurations of the collision law discussed above. Each configuration of the collision law was tested in 11 different settings, i.e., from 2 to 12 AGVs. Each point in the graph is the average of 40 measurements, of which the 99% confidence interval is depicted. We discuss a number of observations.

From the measurements it is clear that the AGV simulator is not limited to *real-time simulation*, but supports *as-fast-as-possible simulations*. For example, for detecting collisions of 12 AGVs with an accuracy of 10 centimeters, the *simulation speedup* is about factor 5, i.e. to simulate 100 seconds of (simulation) time in the AGV transportation system, the computer consumes about 20 seconds of wallclock time.

From the measurements it is clear that the collision law dominates the performance of the AGV simulator. The configuration in which the collision law is deactivated scales linearly as the number of AGVs increases, whereas all configurations with the collision law activated scale quadratically with the number of AGVs. This is within the line of expectations, as the complexity of the collision law is $O(n^2)$, with n the number of AGVs.

---

[*]SciMark 2.0 benchmark score: 174.5 Mflops

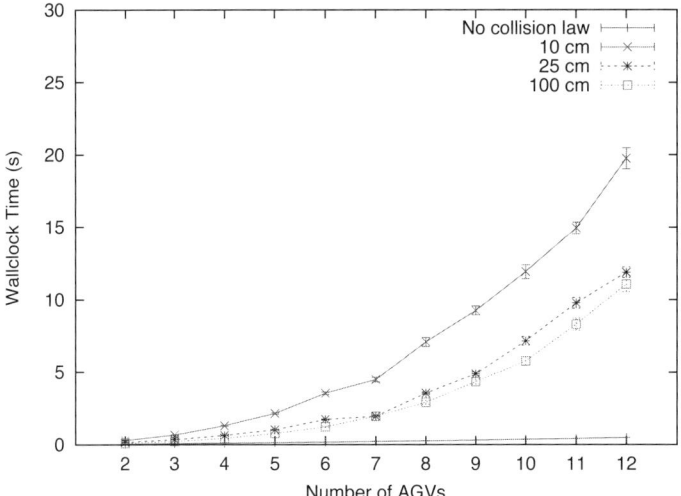

FIGURE 6.15 Performance (in seconds of wallclock time) for simulating 100 seconds of simulation time with the AGV simulator. The four lines correspond to four different configurations of the collision law: the collision law deactivated and the collision law detecting with an accuracy of 10 centimeters, 25 centimeters and 100 centimeters respectively. Each point in the graph is the average of 40 measurements, of which the 99% confidence interval is depicted.

6.6.3 Multi-Agent System Development Supported by the AGV Simulator

The AGV simulator was extensively used during the development of a multi-agent control system in the EMC^2 project. The AGV simulator provides the necessary support for a developer to evaluate different functionalities an AGV control system in isolation, or to compare alternative solutions. We give a number of examples.

- *Virtual environment based routing* [Weyns et al., 2005b]. In this approach, AGV agents use a middleware, called virtual environment, for routing purposes. The virtual environment provides a graph-like map of the paths through the warehouse that the AGV agents use for routing. Signs on the map specify the cost for the AGVs to drive to a given destination. To warn other AGVs that certain paths are blocked or have a long waiting time, AGV agents mark segments with a dynamic cost on the map in the virtual environment. The middleware ensures consistency of the state of the virtual environment on neighboring AGVs. The simulated warehouse environment enables AGV agents to drive over the warehouse layout and it handles the exchange of messages of the middleware.

- *Hull-based collision avoidance* [Weyns et al., 2005d]. AGV controllers avoid collisions by coordinating with other AGVs using the virtual environment. AGV agents mark the path they are going to drive using hulls in their virtual environment. The hull of an AGV is the physical area the AGV occupies. A series of hulls then describes the physical area an AGV occupies along a certain path. If the area is not marked by other hulls (the AGV's own hulls do not intersect with others), the AGV can move along and actually drive over the reserved path. Afterward, the AGV removes the markings in the virtual environment. In case of a conflict, the virtual environments execute a mutual exclusion protocol to determine which

of AGVs involved can move on. The simulated warehouse environment handles the exchange of messages between virtual environments.
- *Field-based transport assignment* [Weyns et al., 2006; Schols, 2005]. In this approach, transport tasks emit fields into the virtual environment that attract idle AGVs. To avoid multiple AGVs driving toward the same transport, AGVs emit repulsive fields. AGVs combine received fields and follow the gradient of the combined fields that guide them toward locations of transports. The AGVs continuously reconsider the situation in the environment and task assignment is delayed until the load is picked, which improves the flexibility of the system. The simulated warehouse environment provides the infrastructure to add new tasks in the system and it handles the exchange of messages to spread fields in the virtual environment.
- *Protocol-based transport assignment* [Weyns et al., 2008]. Besides field-based transport assignment, a dynamic version of the Contract Net protocol [Smith, 1980] was developed to assign transports to AGVs. This protocol, called DynC-NET, allows AGV agents to reconsider the assignment of transports while they drive toward a transport. An extensive series of simulation tests with real world warehouse layouts and order profiles show that both approaches have similar performance characteristics.

The AGV simulator also supports evaluating the integration of different functionalities of an AGV control system. For example, a modular AGV agent [Delbaere and Lamberigts, 2007] was developed that manages combinations of functionalities. A combination consists of a particular approach for routing, a particular approach for collision avoidance, a particular approach for transport assignment and/or a particular approach for battery charging.

6.7 Related Work

We focus our discussion of related work on two facets. First, in Section 6.7.1, we compare our approach with existing simulation platforms that are specifically aimed at software-in-the-loop simulation of multi-agent control systems in dynamic environments. Second, in Section 6.7.2, we zoom in on integrating the control software in a simulation and compare our work with existing approaches. For both these facets, we start with an overview of related approaches, and afterward we compare these approaches with our work.

6.7.1 Special-Purpose Simulation Platforms

Simulation platforms that are specifically aimed at software-in-the-loop simulation of multi-agent control applications in dynamic environments include:

- XRaptor [Bruns et al.] is a simulation platform that supports two- or three-dimensional continuous environments to study the behavior of a large number of agents. XRaptor offers a number of abstractions to support simulations of mobile devices in an environment: an *agent* is either a point, a circular area or a spherical volume that contains a *sensor unit* for observing the world, an *actuator unit* for performing actions and a *control kernel* for action selection. Ordinary differential equations are used for modeling movements.
- SPARK [Obst and Rollmann, 2004] is a simulation platform for physical multi-agent systems in three dimensional environments. *Agent programs* are external

processes for the SPARK simulator. There are *bodies* (i.e. mass and a mass distribution) for the physical simulation. SPARK relies on the Open Dynamics Engine [Smith, 2006], a free library for simulating rigid body dynamics. Additionally, agents possess perceptors and effectors. *Perceptors* provide sensory input to the agent program associated with the representation of the agent in the simulator, and the agent program uses the *effectors* to act in its environment. Other objects in the simulation and the physics of the system can affect the situation of agents; this is reflected in the respective aspects by changing the positions or velocities. SPARK relies on SPADES [Riley and Riley, 2003] to track an agent's execution time between sense and act events.

- Webots [Michel, 2004] is a commercial agent simulation platform that offers support for mobile robots. Like Spark, it uses the Open Dynamics Engine for simulation of physical movements. The focus of Webots is on simulating existing robot platforms. Webots incorporates fine-grained models of low-level sensors and actuators that match their real life counterparts.

- Übersim [Browning and Tryzelaar, 2003] is a multi-robot simulation engine for simulating games of robot soccer for the RoboCup small-size soccer league. Übersim is a simulator specifically designed as a robot development tool. It provides a set of predefined robot models. Like SPARK, Übersim is an Open Source project and uses the Open Dynamics Engine.

- Player/Stage/Gazebo [Gerkey et al., 2003; Koenig and Howard, 2004] is a distributed multi-robot simulator and can simulate a variety of different robots, with a range of conventional sensors, interacting in a complex environment. The Player part of the simulator supports interfaces for the integration of the control software for a variety of robot hardware models. The Stage part of the simulator is a 2D environment with low-fidelity dynamics models that are computationally cheap. The Gazebo part of the simulator is a 3D environment with high-fidelity dynamics based on the Open Dynamics Engine. Stage and Gazebo devices present a standard Player interface.

We make the following observations when comparing these simulation platforms with our approach.

First, similarly to our approach, the aforementioned simulation platforms rely on special-purpose modeling constructs that are targeted at software-in-the-loop simulations of multi-agent control systems in dynamic environments. However, in contrast to our approach, the semantics of the modeling constructs supported by the aforementioned simulation platforms is only described in an informal manner. Consequently, formulating a simulation model that complies with these simulation platforms requires detailed knowledge of the design and implementation of these simulation platforms. In contrast, our approach relies on modeling constructs that are formally specified to unambiguously define their meaning and relations, which is crucial to decouple the simulation model from the simulation platform that is used to execute the model.

Second, the aforementioned simulation platforms support one particular way to simulate dynamism in the environment, either customly developed or by reusing an existing physics library, e.g., the Open Dynamics Engine. Whereas the Open Dynamics Engine simulates dynamism in an accurate way, its high level of detail entails a trade-off in terms of modeling effort and computational efficiency. By putting forward explicit modeling constructs for dynamism (activities, reaction and interaction laws, etc.) we support the modeler to capture dynamism at a level of detail that can be customized to fit the needs of a particular simulation study.

Third, in contrast to the aforementioned simulation platforms, we take a rigorous, view-based approach to documenting the software architecture of our simulation platform. We put forward this software architecture as a reusable artifact, clearly distinguished from the code. The software architecture explicitly documents the knowledge and practice incorporated in the simulation platform. In contrast to code libraries and software frameworks, a software architecture *explicitly* documents (1) stakeholder concerns, (2) how software needs to be structured and behave to address the concerns, and (3) the rationale and tradeoffs that underpin the design of the system. As such, a software architecture captures the essence of the simulation platform and provides a systematic way to capture and share expertise in a form that has proven its value for software development.

6.7.2 Embedding the Control Software

Simulation platforms use various approaches to integrate the software of a (multi-agent) control system in a simulation [Uhrmacher et al., 2003]. We focus on the way simulation platforms support the execution time of a control system. We make a distinction between approaches that incorporate execution time based on direct measurement and approaches that rely on a specification of execution time.

Measurement of Execution Time

A first group of approaches rely on a direct measurement of the execution time during a simulation run. Examples include:

- Player/Stage/Gazebo [Gerkey et al., 2003; Koenig and Howard, 2004] supports software-in-the-loop simulations in which the execution time in taken into account implicitly. The controllers of the distributed control application run on remote hosts and interact with the simulated environment over a network connection. The simulation proceeds in real-time. The execution time is taken into account implicitly: the timing of the actions of the controllers is determined by their arrival time at the host that manages simulated environment. This means that the execution time of a controller is influenced by the performance of the remote host on which the controller is deployed, but also by the latency of the computer network.
- DGensim [Anderson, 2000] supports software-in-the-loop simulations in which wall clock time stamps are used to measure the execution time. Each controller runs on a remote host and interacts with the simulated environment over a network connection. At fixed time intervals, perceptions are given to the controllers, and a controller has a fixed window of time to react to the perception. Before transmitting the actions to the simulated environment, the remote host attaches a time-stamp with the wall clock time of each action. At the host of the simulated environment, all actions within a time window are arranged according to their time stamp in wall clock time. The use of wall clock time stamps reduces the effect of network latencies on the ordering of actions. However, problems arise in case network latencies cause actions do not reach the simulated environment within the time window.
- SPADES [Riley and Riley, 2003] supports software-in-the-loop simulations with a direct measurement of execution time. Each controller of the distributed control application runs on a dedicated host together with a SPADES communication server, which sends the actions of that controller to the simulated environment.

The SPADES communication server supports low-level performance monitoring by means of *perfctr*, a linux kernel driver that offers low-level performance monitoring with per-process CPU-cycle counters. The controller operates in a sense-think-act cycle, and notifies the SPADES communication server of the start and end of each cycle. The simulation time of the actions corresponds to applying an linear scale factor to the performance measurement of the *perfctr* driver.

Measurement is an easy and intuitive way to incorporate the execution time of controllers of a distributed control application in a simulation. Nevertheless, compared to an explicit model of the execution time, measuring the execution time during a simulation has a number of drawbacks [Anderson, 1997].

First, measurements are platform dependent. The computer platform on which a distributed control application is deployed for simulation purposes typically differs from the (heterogeneous) devices on which the controllers are deployed in the real world. Consequently, the measurement of the execution time of a controller is not necessarily a decent estimate of the execution time of that controller in the real world.

Second, measurements are not selective. A measurement takes into account auxiliary code for debugging, logging to file, configuration, interfacing with the user, although this auxiliary code is removed from the distributed control application before it is deployed in the real environment. Auxiliary code can significantly affect the execution time that is measured of a particular controller.

Third, measurements jeopardize repeatable simulation runs. The measurements that are employed are non-deterministic, i.e. small random variations are possible when measuring the execution time. In simulation, non-determinism must always be supported in a controlled manner, i.e. in a simulation all non-determinism should be based on random numbers originating from a random number generator with a known seed. Using the same seed for the random number generator then guarantees the same trace of random numbers during a simulation run, which is a prerequisite to obtain simulation results that can be repeated over and over again. However, measuring the execution time of a controller during a simulation is an example of supporting non-determinism in an uncontrolled manner. As the trace of measurements of the execution time during a simulation run cannot be controlled, it can be extremely difficult or even impossible to reproduce the same simulation result twice.

Specification of Execution Time

A second group of approaches specify the execution time of a distributed control application instead of using measurements. Examples include:

- MESS [Anderson and Cohen, 1996] supports software-in-the-loop simulation of controllers written in the Common Lisp programming language. To model the execution time of a controller, individual language instructs of Common Lisp are associated with a particular duration. MESS relies on TCL (Timed Common Lisp) to derive the execution time of a controller. TCL is an extended version of Common Lisp that advances a clock upon execution of each Common Lisp primitive. The duration for each primitive can be specified by the modeler. Auxiliary code can be annotated such that its duration is not taken into account.

- EyeSim [Bräunl et al., 2006] supports software-in-the-loop simulations of controllers for robotic systems based on the RoBIOS, a list of library functions for motor control, sensor feedback and multi-tasking. To incorporate the execution time of a controller, EyeSim employs a duration for each of the RoBIOS system calls. The duration of all code besides the function calls to the RoBIOS library

is disregarded.
- The Packet-World [Weyns et al., 2005a] employs a very coarse-grained model to specify the execution time of a controller. Each controller has a fixed, constant execution time between consecutive actions, irrespective of the amount of computation it needs to determine its next action. This is a suitable model in case the execution time of a controller in the real world does not vary a lot, or in case only a rough estimate is sufficient.
- In Webots [Michel, 2004], the code that deals with execution time is tangled with the functionality of the control software. The control software must proceed in a step-like fashion. After each step, the controller must return the amount of time consumed during that step of the control loop.

We make the following observation when comparing these approaches with our work. By relying on aspect-oriented programming, our approach has an increased flexibility compared to the aforementioned approaches. A modeler can declaratively define the duration primitives of the control software in a single aspect, clearly separated from the rest of the code. We rely on aspect weaving to automatically enforce the tracing of these duration primitives during the simulation. Aspect weaving avoids that the code to trace the invocation of duration primitives needs to be written and inserted by hand each time the control software is embedded in a simulation and each time the duration primitives are adjusted by the modeler. The tradeoff of our approach is that the granularity of the duration primitives that can be defined is constrained by the expressiveness of the pointcut language, i.e. by he granularity of the join points that can be specified in AspectJ.

6.8 Conclusions and Future Work

In this concluding section, we first put forward concrete suggestions for future research with respect to our own work in Section 6.8.1. Afterward, we reflect on the way our work could stimulate future research on multi-agent simulation in a broader setting in Section 6.8.2.

6.8.1 Concrete Directions for Future Research

We suggest two main areas for future research in the context of our own work: extending the modeling framework and extending the software architecture.

Extending the Modeling Framework.
  The modeling framework for dynamic environments could be extended with additional constructs. We suggest a number of avenues for future research.

- *Supporting perception in dynamic environments.* Currently, the modeling framework does not provide modeling constructs to capture the way agents sense or perceive a dynamic environment. In a dynamic environment, perception is not limited to a static state snapshot of a part of the environment, but closely related with dynamism. For example, sensors can be capable of registering the movement of entities in the environment, rather than their momentary position. Investigating the relation between perception and dynamism is an interesting challenge.
- *Supporting sources of dynamism in the environment.* Currently, the modeling framework does not provide explicit modeling constructs to model the internals of a source of dynamism in the environment. The internal machinery of a source of dynamism in the environment is black box, and its behavior is specific for

a particular simulation study. More elaborate support for sources of dynamism could focus on modeling constructs for various kinds of behaviors, such as reactive [Brooks, 1991; Weyns and Holvoet, 2006], behavior-based [Maes, 1991; Weyns et al., 2005c] or cognitive behaviors [Haddadi and Sundermeyer, 1996; Rao and Georgeff, 1995].

Extending the Software Architecture.

We indicate a direction for extending the software architecture.

- *Distribution of the Simulated Environment.* The simulated environment can become a bottleneck in large-scale simulations involving many agents. Currently, the architecture does not incorporate support for distribution of the simulated environment. The challenges of distributing the simulated environment are not of pure technical nature: as the parts of the simulated environment are explicitly synchronized with the simulation engine, they could technically be distributed across different hosts. The main challenge is determining which distribution scheme is most suitable for a particular simulation study. On the one hand, distribution adds computing power which speeds up a simulation, on the other hand, distribution requires synchronization to happen over a network, which slows down a simulation. Distribution of simulations should be supported in a flexible manner [Ewald et al., 2006], with distribution schemas that can be adapted or self-adapt to a particular simulation study. The distribution scheme of the architecture can be documented using deployment views.

6.8.2 Closing Reflection

As demand for multi-agent control applications increases, more and more simulations are built to support their development. The way such simulations are built becomes common knowledge. In recent research, we observe two trends to consolidate common knowledge on developing such simulations.

With respect to building simulation models, research puts forward special-purpose modeling constructs to reify common knowledge. Special-purpose modeling constructs support the modeler by capturing key characteristics of such systems in a first-class manner.

With respect to building simulation platforms, common knowledge is typically reified in reusable code libraries and software frameworks. We put forward software architecture in addition to such code libraries and software frameworks. A software architecture captures the essence of a simulation platform in an artifact that amplifies reuse beyond traditional code libraries and software frameworks. The software architecture we propose explicitly captures (1) functional and quality requirements (2) how software needs to be structured to address the requirements, and (3) the rationale and tradeoffs that underpin the design of the software. We developed a simulation platform that implements this architecture and we applied this simulation platform to support software-in-the-loop simulations for evaluating, comparing and integrating several functionalities of a multi-agent control system for steering AGVs.

We strongly believe that multi-agent simulation can benefit from a more systematic approach to software architecture. Software architecture supports consolidating and sharing expertise in the domain of multi-agent simulation in a form that has proven its value for software development.

References

J. Anderson. A generic distributed simulation system for intelligen agent design and evaluation. In *10th International Conference on AI, Simulation, and Planning in High Autonomy Systems*, pages 36–44, 2000.

S. D. Anderson. Simulation of multiple time-pressured agents. In *WSC '97: Proceedings of the 29th conference on Winter simulation*, pages 397–404, 1997. ISBN 0-7803-4278-X. doi: http://doi.acm.org/10.1145/268437.268515.

S. D. Anderson and P. R. Cohen. Timed Common Lisp: the duration of deliberation. *SIGART Bull.*, 7(2):11–15, 1996. ISSN 0163-5719. doi: http://doi.acm.org/10.1145/242587.242590.

L. Bass, P. Clements, and R. Kazman. *Software Architecture in Practice, Second Edition*. Addison-Wesley Longman, Amsterdam, April 2003. ISBN 0321154959.

J. Borgers. "Hoe realistisch is een simulatie?": Studie aan de hand van LEGO Mindstorms. Master's thesis, K.U.Leuven, Department of Computer Science, 2006.

T. Bräunl, A. Koestler, and A. Waggershauser. Fault-tolerant robot programming through simulation with realistic sensor models. *International Journal of Advanced Robotic Systems*, 3(2):99–106, 2006.

A. Bredenfeld, A. Jacoff, I. Noda, and Y. Takahashi, editors. *RoboCup 2005: Robot Soccer World Cup IX*, volume 4020 of *Lecture Notes in Computer Science*, 2006. Springer. ISBN 3-540-35437-9.

R. A. Brooks. Intelligence without reason. In J. Myopoulos and R. Reiter, editors, *Proceedings of the 12th International Joint Conference on Artificial Intelligence (IJCAI-91)*, pages 569–595, Sydney, Australia, 1991. Morgan Kaufmann publishers Inc.: San Mateo, CA, USA. ISBN 1-55860-160-0.

B. Browning and E. Tryzelaar. Ubersim: A realistic simulation engine for robot soccer. In *Proceedings of Autonomous Agents and Multi-Agent Systems, AAMAS'03*, Australia, July 2003.

S. A. Brueckner. *Return From The Ant - Synthetic Ecosystems For Manufacturing Control*. PhD thesis, Humboldt University Berlin, Department of Computer Science, 2000.

G. Bruns, P. Mössinger, D. Polani, R. Schmitt, R. Spalt, T. Uthmann, and S. Weber. Xraptor - a simulation environment for continuous virtual multi-agent systems - user manual. URL `citeseer.ist.psu.edu/bruns03xraptor.html`.

K. M. Chandy and J. Misra. Asynchronous distributed simulation via a sequence of parallel computations. *Commun. ACM*, 24(4):198–206, 1981. ISSN 0001-0782. doi: http://doi.acm.org/10.1145/358598.358613.

P. Clements, F. Bachmann, L. Bass, D. Garlan, J. Ivers, R. Little, R. Nord, and J. Stafford. *Documenting Software Architectures: Views and Beyond*. Addison-Wesley Professional, September 2002. ISBN 0201703726.

W. Delbaere and B. Lamberigts. Ontwikkeling van een gedecentraliseerd controle systeem voor autonome voertuigen. Master's thesis, Katholieke Universiteit Leuven, Belgium, 2007.

P. DeLima, G. York, and D. Pack. Localization of ground targets using a flying sensor network. *sutc*, 1:194–199, 2006. doi: http://doi.ieeecomputersociety.org/10.1109/SUTC.2006.84.

K. Dresner and P. Stone. Multiagent traffic management: An improved intersection control mechanism. In F. Dignum, V. Dignum, S. Koenig, S. Kraus, M. P. Singh, and M. Wooldridge, editors, *The Fourth International Joint Conference on Autonomous Agents and Multiagent Systems*, New York, NY, July 2005. ACM Press.

R. Ewald, J. Himmelspach, and A. M. Uhrmacher. A non-fragmenting partitioning algorithm for hierarchical models. In *WSC '06: Proceedings of the 37th conference on Winter simulation*, pages 848–855. Winter Simulation Conference, 2006. ISBN 1-4244-0501-7.

J. Ferber and J. Müller. Influences and reaction: A model of situated multiagent systems. In *Proceedings of the Second International Conference on Multi-agent Systems*, pages 72–79. AAAI Press, 1996.

D. Finkenzeller, M. Baas, S. Thüring, S. Yigit, and A. Schmitt. Visum: a vr system for the interactive and dynamics simulation of mechatronic systems. In *Virtual Concept 2003*, Biarritz, France, November 2003.

B. P. Gerkey, R. T.Vaughan, and A. Howard. The player/stage project: Tools for multi-robot and distributed sensor systems. In *ICAR 2003*, pages 317–323, Coimbra, Portugal, June 2003.

D. Gu and H. Hu. Teaching robots to coordinate their behaviours. In *Proceedings of IEEE International Conference on Robotics and Automation*, New Orleans, LA, May 2004. Riverside Hilton & Towers.

A. Haddadi and K. Sundermeyer. Belief-desire-intention agent architectures. *Foundations of distributed artificial intelligence*, pages 169–185, 1996.

I. J. Hayes, M. Jackson, and C. B. Jones. Determining the specification of a control system from that of its environment. In K. Araki, S. Gnesi, and D. Mandrioli, editors, *FME*, volume 2805 of *Lecture Notes in Computer Science*, pages 154–169. Springer, 2003. ISBN 3-540-40828-2.

A. Helleboogh. *Simulation of distributed control applications in dynamic environments*. Phd, Department of Computer Science, K.U.Leuven, Leuven, Belgium, May 2007. 169 + xxxi pages.

A. Helleboogh, T. Holvoet, and Y. Berbers. Simulating actions in dynamic environments. In *Conceptual Modeling and Simulation Conference, CMS2005, Track on Agent Based Modeling and Simulation in Industry and Environment*, 2005.

A. Helleboogh, T. Holvoet, and Y. Berbers. Testing AGVs in Dynamic Warehouse Environments. In D. Weyns, V. Parunak, and F. Michel, editors, *Environments for Multiagent Systems II*, volume 3830 of *Lecture Notes in Computer Science*, pages 270–290. Springer-Verlag, 2006.

A. Helleboogh, G. Vizzari, A. Uhrmacher, and F. Michel. Modeling dynamic environments in multi-agent simulation. *Autonomous Agents and Multi-Agent Systems: Special issue on environments for multi-agent systems*, 14(1):87–116, February 2007.

J. Himmelspach, M. Röhl, and A. M. Uhrmacher. Simulation for testing software agents - an exploration based on JAMES. In *Proc. of the 2003 Winter Simulation Conference*, New Orleans, USA, December 2003.

X. Hu and B. P. Zeigler. A simulation-based virtual environment to study cooperative robotic systems. *Integrated Computer-Aided Engineering (ICAE)*, 12(4):353 – 367, 2005.

V. Issarny, M. Caporuscio, and N. Georgantas. A Perspective on the Future of Middleware-Based Software Engineering. In *Future of Software Engineering, 29th International Conference on Software Engineering*, Minneapolis, USA, 2007.

J. Ivers, P. Clements, D. Garlan, R. Nord, B. Schmerl, and J. R. O. Silva. Documenting component and connector views with uml 2.0. Technical Report CMU/SEI-2004-TR-008, Software Engineering Institute, 2004.

M. Jackson. The Meaning of Requirements. *Annals of Software Engineering*, 3:5–21, 1997.

G. Kiczales, J. Lamping, A. Menhdhekar, C. Maeda, C. Lopes, J.-M. Loingtier, and J. Irwin. Aspect-oriented programming. In M. Akşit and S. Matsuoka, editors, *Proceedings European Conference on Object-Oriented Programming*, volume 1241, pages 220–242. Springer-Verlag, Berlin, Heidelberg, and New York, 1997. URL citeseer.ist.psu.edu/kiczales97aspectoriented.html.

G. Kiczales, E. Hilsdale, J. Hugunin, M. Kersten, J. Palm, and W. Griswold. Getting started with AspectJ. *Commun. ACM*, 44(10):59–65, 2001. ISSN 0001-0782. doi: http://doi.acm.org/10.1145/383845.383858.

N. Koenig and A. Howard. Design and use paradigms for gazebo, an open-source multi-robot simulator. In *IEEE/RSJ International Conference on Intelligent Robots and Systems*, pages 2149–2154, Sendai, Japan, Sep 2004. URL http://cres.usc.edu/cgi-bin/print_pub_details.pl?pubid=394.

P. Maes. The agent network architecture (ana). *SIGART Bull.*, 2(4):115–120, 1991. ISSN 0163-5719. doi: http://doi.acm.org/10.1145/122344.122367.

O. Michel. Webots: Professional mobile robot simulation. *Journal of Advanced Robotics Systems*, 1(1):39–42, 2004. URL http://www.ars-journal.com/ars/SubscriberArea/Volume1/39-42.pdf.

O. Obst and M. Rollmann. SPARK – A Generic Simulator for Physical Multiagent Simulations. In G. Lindemann, J. Denzinger, I. J. Timm, and R. Unland, editors, *Multiagent System Technologies – Proceedings of the MATES 2004*, volume 3187, pages 243–257. Springer, Sept. 2004.

A. S. Rao and M. P. Georgeff. BDI agents: From theory to practice. In *Proc. of 1st International Conference on Multi-Agent Systems (ICMAS)*, pages 313–319. AAAI Press/MIT Press, 1995.

P. Riley and G. Riley. SPADES — a distributed agent simulation environment with software-in-the-loop execution. In S. Chick, P. J. Sánchez, D. Ferrin, and D. J. Morrice, editors, *Winter Simulation Conference Proceedings*, volume 1, pages 817–825, 2003.

D. Roozemond. Using intelligent agents for urban traffic control control systems. In *Proceedings of the International Conference on Artificial Intelligence in Transportation Systems and Science*, pages 69–79, 1999.

W. Schols. Gradient Field Based Order Assignment in AGV Systems. Master's thesis, Katholieke Universiteit Leuven, Belgium, 2005.

B. Sinopoli, C. Sharp, L. Schenato, S. Schaffert, and S. Sastry. Distributed control applications within sensor networks. In *Proceedings of the IEEE, Special Issue on Sensor Networks and Applications*, August 2003.

R. Smith. *Open Dynamics Engine: User Guide*, 2006. URL http://www.ode.org/ode-latest-userguide.html.

R. G. Smith. The contract net protocol: High-level communication and control in a distributed problem solver. *IEEE Transactions on Computers*, C-29(12):1104–1113, 1980.

A. Uhrmacher and B. Kullick. "Plug and Test" - software agents in virtual environments. In *Proceedings of the 2000 Winter Simulation Conference*, volume 2, pages 1722–1729. Wyndham Palace Resort & Spa, Orlando, Florida, USA, December 2000.

A. M. Uhrmacher, M. Röhl, and J. Himmelspach. Unpaced and paced simulation for testing agents. In SCS, editor, *Proc. of the 15th European Simulation Symposium*, pages 71–80, 2003. ISBN 3-936150-28-1.

P. Varshavskaya, L. P. Kaelbling, and D. Rus. Learning distributed control for modular robots. In *International Conference on Intelligent Robots and Systems*, Sendai, Japan, 2004.

C. M. Velez and A. Agudelo. Control and parameter estimation of a mini-helicopter robot using rapid prototyping tools. *WSEAS Transactions on Systems*, 5(9): 2250–2257, September 2006.

P. Verstraete, B. Germain, P. Valckenaers, and H. Van Brussel. On applying the prosa reference architecture in multiagent manufacturing control applications. In *Multiagent Systems and Software Architecture*, 2006. URL http://people.mech.kuleuven.be/~pverstra/papers/On%20applying%20the%20PROSA%20reference%20architecture%20in%20multi-agent%20manufacturing%20control%20applications_camera_ready.pdf.

F.-Y. Wang. Agent-based control for networked traffic management systems. *IEEE Intelligent Systems*, 20(5):92–96, 2005. ISSN 1541-1672. doi: http://dx.doi.org/10.1109/MIS.2005.80.

D. Weyns and T. Holvoet. From reactive robotics to situated multiagent systems: A historical perspective on the role of environment in multiagent systems. In *Engineering Societies in the Agents World VI, Revised Selected and Invited Papers*, volume 3963 of *Lecture Notes in Computer Science*, pages 63–88. Springer, 2006. Invited paper.

D. Weyns, A. Helleboogh, and T. Holvoet. The Packet-World: A testbed for investigating situated multiagent systems. In *Software Agent-Based Applications, Platforms, and Development Kits*, Whitestein Series in Software Agent Technologies, pages 383–408. Birkhauser Verlag, Basel - Boston - Berlin, Sept. 2005a.

D. Weyns, K. Schelfthout, and T. Holvoet. Exploiting a virtual environment in a real-world application. In D. Weyns, V. Parunak, and F. Michel, editors, *2nd International Workshop on Environments for Multiagent Systems*, pages 1–18, Utrecht, The Netherlands, 2005b.

D. Weyns, K. Schelfthout, T. Holvoet, and O. Glorieux. Towards adaptive role selection for behavior-based agents. In *Adaptive Agents and Multi-Agent Systems III: Adaptation and Multi-Agent Learning*, volume 3394 of *Lecture Notes in Computer Science*, pages 295–314. Springer-Verlag, GmbH, 2005c.

D. Weyns, K. Schelfthout, T. Holvoet, and T. Lefever. Decentralized control of E'GV transportation systems. In *4th Joint Conference on Autonomous Agents and Multiagent Systems, Industry Track*, Utrecht, The Netherlands, 2005d. ACM Press, New York, NY, USA.

D. Weyns, N. Boucké, and T. Holvoet. Gradient Field Based Transport Assignment in AGV Systems. In *5th International Joint Conference on Autonomous Agents and Multi-Agent Systems, AAMAS, Hakodate, Japan*, 2006.

D. Weyns, T. Holvoet, and A. Helleboogh. Anticipatory vehicle routing using delegate multi-agent systems. In *Intelligent Transportation Systems Conference, 2007 (ITSC 2007)*, pages 87–93. IEEE, 2007.

D. Weyns, N. Boucké, and T. Holvoet. A Field-Based Versus a Protocol-Based Approach for Adaptive Task Assignment. *Autonomous Agents and Multi-Agent Systems*, 2008. (in press).

7

Replicator Dynamics in Discrete and Continuous Strategy Spaces

7.1	Introduction	215
7.2	Elementary Concepts from Game Theory	217
	Strategic Games with Discrete Strategy Sets • Strategic Games with Continuous Strategy Spaces	
7.3	Replicator Dynamics	224
	Replicator Dynamics in Discrete Strategy Spaces • Replicator Dynamics in Continuous Strategy Spaces	
7.4	Evolutionary Dynamics in Discrete Strategy Spaces	228
	Analysis of the Evolutionary Dynamics of the Categorization of Games • Relating Evolutionary Dynamics in Discrete Strategy Spaces with Reinforcement Learning	
7.5	Evolutionary Dynamics in Continuous Strategy Spaces	232
	Mutation as Engine for Diffusion in Continuous Strategy Spaces • Non-Isotropic Mutations and Evolutionary Flows in Strategy Spaces • Draining of Pay-Off Streams in Strategy Space	
7.6	Example of the Resulting Dynamics of the Continuous Replicator Equations	237
7.7	Conclusions	240
	References	240

Karl Tuyls
Eindhoven University of Technology

Ronald Westra
Maastricht University

7.1 Introduction

This chapter surveys and investigates the evolutionary dynamics of multiple agents playing strategic games, in the sense of Game Theory (GT), in discrete and continuous strategy spaces. We take the biological perspective towards GT, i.e, known as Evolutionary Game Theory (EGT), to analyze repeated interactions, without perfect information, in multi-agent settings. More precisely, by the application of GT to biology, John Maynard-Smith invented EGT and relaxed the premises behind traditional GT. Classical GT is a normative theory, in the sense that it expects players or agents to be perfectly rational and behave accordingly [von Neumann and Morgenstern, 1944; Weibull, 1996; Redondo, 2003]. In classical GT, interactions between rational agents are modeled as games of two or more players that can choose from a set of strategies and the corresponding preferences. GT is thus the mathematical study of interactive decision making in the sense that the agents involved in the decisions take into account their own choices and those of others. Choices are determined by stable preferences concerning the outcomes of their possible decisions, and strategic interaction whereby agents take into account the relation between their own choices and the decisions of other agents. Players in the classical setting have a perfect knowledge of the environment and the payoff tables, and try to maximize their individual payoff. However, under the biological circumstances considered by Maynard-Smith, it becomes impossible to judge what choices are the most rational. Instead of figuring out, a priori, how to optimize its actions, the question now facing a player becomes how to learn to optimize its behavior and maximize its return, and it does this based on local knowledge and through a process of trial and error. This learning process matches the concept of evolution in biology, and forms the basis of EGT. In contrast to classical GT, then, EGT is a descriptive theory, describing this process of learning, and does not need the assumption of complete knowledge and perfect rationality.

Building models of agents that evolve requires insight into the type and form of these agents' interactions with the environment and other agents in the system. In much work on multi-agent evolution and learning, this modeling is very similar to that used in a standard game theoretical model: players are assumed to have complete knowledge of the environment, are hyper-rational and optimize their individual payoff disregarding what this means for the utility of the entire population. In contrast, the basic properties of multi-agent systems (MAS) seem to correspond well with those of EGT. First of all, a MAS is a distributed dynamic system which makes it hard to model using a static theory like GT. Secondly, a MAS consists of actions and interactions between two or more independent agents, who each try to accomplish a certain, possibly cooperative or conflicting, goal. No agent has the guarantee to be completely informed about the other agents' intentions or goals, nor has it the guarantee to be completely informed about the complete state of the environment. Furthermore, EGT offers us a solid basis to understand dynamic iterative situations in the context of strategic interactions and this fits well with the dynamic nature of a typical MAS. Not only do the fundamental assumptions of EGT and MAS seem to fit each other rather well, there is also a formal relationship between the replicator equations of EGT and reinforcement learning (RL) [Borgers and Sarin, 1997; Tuyls et al., 2003, 2006; Panait et al., 2008].

In this chapter we investigate the key concepts from EGT in multi-agent games, i.e., we look at the stability of strategies in different types of dynamic models, known as the Replicator Dynamics (RD). We do this for games in which the players can choose from a discrete set and continuous set of strategies. Whereas the former has been extensively studied before, the latter has not yet been so thoroughly explored in the context of EGT and Multi-Agent Systems. The RD are formalized as a system of differential equations. Each replicator represents one (pure) strategy available to a player. EGT assumes that players will gradually adjust their strategy over time in response to repeated observations of their own and others' payoffs. The RD control this learning, specifying the frequency with which different pure strategies should be played depending on the mix of strategies played by the remainder of the population of agents playing the game.

In the first part of this chapter we will survey and simulate evolutionary dynamics for the entire class of 2×2 games and unravel the formal relationship between RD and RL using discrete strategy spaces for the players. We will draw the connection with multi-agent Q-learning and illustrate in some experiments where the basins of attraction lie for different parameter settings of the learning process.

In the second part of this chapter we will have a closer look at the RD in iterated games with continuous strategy spaces. In many contexts, rather than a discrete set of pure strategies, there is a continuum of strategies, called the continuous strategy space. An example is, for instance, robot soccer. The decisions faced by the players, such as when and where to kick the ball, or where to move when not in possession of the ball are continuous in nature. A typical approach then is to only allow a small number of discrete actions instead of one, continuous parameter. For instance in the soccer case this would mean that a player could only move in a few predefined directions, i.e., the pure or discrete strategies, whereas opposed to a continuous approach one would allow the player to move in a continuum of directions represented by one direction-parameter. Classical RD describes the evolution of a mixture of the discrete strategies, whereas in the case of a continuous parameter it is necessary to describe a probability distribution over this parameter. Consequently, the RD must be substituted for a spatiotemporal Partial differential Equation (PDE) in strategy space and time. The solution of this PDE provides the evolution of the distribution function over the strategies in time. This ultimately converges to the steady state corresponding to the stationary solution of the PDE. We focus on a continuous time model with an infinite population in a 2-person game-interaction, such that each agent has a continuum of available strategies. This approach allows for modeling the effects of evolution by involving deterministic mutations. Such an approach has been proposed and studied before by Hofbauer and Sigmund [Hofbauer and Sigmund, 1998], and Ruijgrok and Ruijgrok [Ruijgrok and Ruijgrok, 2005], based on the classical RD extended with a simple isotropos diffusion term containing the mutation rate. We will specifically mimic the effect of anisotropos deterministic mutations, and starting from a continuous time with discrete strategy set RD with mutations derive a different formulation for the continuous strategy space. The dynamics of this approach is even more complex as different mutation rates provide entirely different stationary solutions.

The remainder of this chapter is structured as follows. We start by introducing the basic concepts from GT and EGT in Section 7.2 necessary to understand models of evolutionary dynamics. After this we introduce the different RD models under study, in both discrete and continuous time and strategy spaces, in Section 7.3. Next we continue to elaborate on evolutionary dynamics in discrete strategy spaces and draw the formal connection with reinforcement learning in Section 7.4. Section 7.5 discusses evolutionary dynamics in continuous strategy spaces. More precisely, we will introduce and discuss continuous models extended with a mutation term. Section 7.6 shows some simulations with continuous strategy spaces. Finally we conclude.

7.2 Elementary Concepts from Game Theory

In this section we introduce elementary concepts from game theory and evolutionary game theory necessary to understand the remainder of this chapter. The familiar reader can easily skip this section.

This section starts by introducing normal form games with discrete strategy sets. Then we explain the basic concepts, such as Nash equilibrium, Pareto optimality, Evolutionary Stable Strategies, using discrete strategy spaces. After this we extend normal form games to the setting of continuous strategy spaces. For the connection between these concepts we refer the interested reader to [Tuyls and Nowe, 2005; Redondo, 2003; Weibull, 1996].

7.2.1 Strategic Games with Discrete Strategy Sets

Normal Form Games

In this section we define n-player normal form games as a conflict situation involving gains and losses between n players. In such a game n players repeatedly interact with each other by all choosing an action (or strategy) to play. All players choose their strategy at the same

$$A = \begin{pmatrix} a_{11} & a_{12} \\ a_{21} & a_{22} \end{pmatrix} \qquad B = \begin{pmatrix} b_{11} & b_{12} \\ b_{21} & b_{22} \end{pmatrix}$$

	B	
	b_{11}	b_{12}
A a_{11}		a_{12}
	b_{21}	b_{22}
a_{21}		a_{22}

FIGURE 7.1 The left matrix (A) defines the payoff for the row player; the right matrix (B) defines the payoff for the column player; and the strategic form of the game, right, summarizes this information.

time. For reasons of simplicity, we limit the pure strategy set of the players to 2 strategies. A strategy is defined as a probability distribution over all possible actions. In the 2-pure strategies case, we have: $s_1 = (1,0)$ and $s_2 = (0,1)$. A mixed strategy s_m is then defined by $s_m = (x_1, x_2)$ with $x_1, x_2 \neq 0$ and $x_1 + x_2 = 1$.

Defining a game more formally we restrict ourselves to the 2-player 2-action game. A game $G = (S_1, S_2, P_1, P_2)$ is defined by the payoff functions P_1, P_2 and their strategy sets S_1 for the first player and S_2 for the second player. In the 2-player 2-strategies case, the payoff functions $P_1 : S_1 \times S_2 \to \Re$ and $P_2 : S_1 \times S_2 \to \Re$ are defined by the payoff matrices, A for the first player and B for the second player, see Figure 7.1. The payoff tables A, B define the instantaneous rewards. Element a_{ij} is the reward the row-player (player 1) receives for choosing pure strategy s_i from set S_1 when the column-player (player 2) chooses the pure strategy s_j from set S_2. Element b_{ij} is the reward for the column-player for choosing the pure strategy s_j from set S_2 when the row-player chooses pure strategy s_i from set S_1.

If now $a_{ij} + b_{ij} = 0$ for all i and j, we call the game a *zero sum game*. This means that the sum of what is won by one agent (positive) and lost by another (negative) equals zero. This corresponds to a situation of *pure competition*. In case that $a_{ij} + b_{ij} \neq 0$ for all i and j we call the game a *general sum game*. In this situation it might be very beneficial for the different agents to cooperate with one another.

Categorization and Examples

The family of 2×2 games is usually classified in three subclasses, as follows [Redondo, 2003]:

Subclass 1: if $(a_{11} - a_{21})(a_{12} - a_{22}) > 0$ or $(b_{11} - b_{12})(b_{21} - b_{22}) > 0$, at least one of the 2 players has a dominant strategy, therefore there is just 1 strict equilibrium.

Subclass 2: if $(a_{11} - a_{21})(a_{12} - a_{22}) < 0, (b_{11} - b_{12})(b_{21} - b_{22}) < 0$, and $(a_{11} - a_{21})(b_{11} - b_{12}) > 0$, there are 2 pure equilibria and 1 mixed equilibrium.

Subclass 3: if $(a_{11} - a_{21})(a_{12} - a_{22}) < 0, (b_{11} - b_{12})(b_{21} - b_{22}) < 0$, and $(a_{11} - a_{21})(b_{11} - b_{12}) < 0$, there is just 1 mixed equilibrium.

The first subclass includes those type of games where each player has a dominant strategy, as for instance the prisoner's dilemma. However it includes a larger collection of games since only 1 of the players needs to have a dominant strategy. In the second subclass none of the players has a dominated strategy. But both players receive the highest payoff by both playing their first or second strategy. This is expressed in the condition $(a_{11} - a_{21})(b_{11} - b_{12}) > 0$. The third subclass only differs from the second in the fact that the players don't receive their highest payoff by both playing the first or the second strategy. This is expressed by the condition $(a_{11} - a_{21})(b_{11} - b_{12}) < 0$.

Replicator Dynamics in Discrete and Continuous Strategy Spaces

$$
\begin{array}{c|cc}
 & \multicolumn{2}{c}{B} \\
 & D & C \\
\hline
D & 1\quad 1 & 0\quad 5 \\
A & & \\
C & 5\quad 0 & 3\quad 3 \\
\end{array}
$$

FIGURE 7.2 The strategic form of the Prisoner's dilemma between the row player A and the column player B. D is Defect and C is Cooperate.

$$
\begin{array}{c|cc}
 & \multicolumn{2}{c}{B} \\
 & F & O \\
\hline
F & 1\quad 2 & 0\quad 0 \\
A & & \\
O & 0\quad 0 & 2\quad 1 \\
\end{array}
$$

FIGURE 7.3 The strategic form of the Battle of the sexes game. Strategy F is choosing Football and strategy O is choosing the Opera.

We now have a closer look at some examples. As an example of subclass 1 we consider the famous Prisoner's Dilemma (PD) game. In this game 2 prisoners, who committed a crime together, have a choice to either cooperate with the police (to defect) or work together and deny everything (to cooperate). If the first criminal (row player) defects and the second one cooperates, the first one gets off the hook (expressed by a maximum reward of 5) and the second one gets the most severe punishment. If they both defect, they get the second most severe punishment one can get (expressed by a payoff of 1). If both cooperate, they both get a minimum sentence. So each prisoner must make the choice whether he will betray the other prisoner (defect) or will remain silent (cooperate). Of course it is tempting to defect, because this possibly yield the highest reward, i.e., freedom. On the other hand, if the other prisoner also defects they both get a severe sentence. Cooperate is a good alternative, at least if the other prisoner also plays cooperate, in that case they both get a minimal sentence. The rewards are summarized in payoff Figure 7.2. The first matrix has to be read from a row perspective and the second one from a column perspective. For instance if the row player chooses to Defect (D), the payoff has to be read from the first row. It then depends on the strategy of the column player what payoff the row player will receive.

As an example of subclass 2 we consider the battle of the sexes game [Gintis, 2001; Weibull, 1996]. In this game, a married couple loves each other so much they want to do everything together. One night the husband wants to see a football game and the wife wants to go to the opera. This situation is described by the payoff matrices of Figure 7.3. If they both do their activities separately they receive the lowest payoff. This type of games is known as coordination games.

As an illustration of subclass 3 we consider the matching pennies game. In this game two children both hold a penny and independently choose which side of the coin to show (Head or Tails). The first child wins if both coins show the same side, otherwise child 2 wins. This is an example of a zero-sum game as can be seen from payoff Figure 7.4. Whatever is lost

	B	
	H	T
H	-1, 1	1, -1
T	-1, 1	1, -1

(A on left side)

FIGURE 7.4 The strategic form of the Matching Pennies game. Strategy H is choosing Head and strategy T is choosing Tail.

by one player, must be won by the other player.

Nash Equilibrium

In traditional game theory it is assumed that the players are rational, meaning that every player will choose the action that is best for him, given his beliefs about the other players' actions. A basic definition of a Nash equilibrium goes as follows. If there is a set of strategies for a game with the property that no player can increase its payoff by changing his strategy unilaterally (i.e., while the other players keep their strategies unchanged), then that set of strategies and the corresponding payoffs constitute a Nash equilibrium.

Formally, a Nash equilibrium is defined as follows. When two players play the strategy profile $s = (s_i, s_j)$ belonging to the product set $S_1 \times S_2$ then s is a Nash equilibrium if

$$P_1(s_i, s_j) \geq P_1(s_x, s_j) \quad \forall\, x \in \{1, ..., n\}$$
$$P_2(s_i, s_j) \geq P_2(s_i, s_y) \quad \forall\, y \in \{1, ..., m\}.$$

For a definition in terms of best reply or best response functions we refer the reader to [Weibull, 1996].

We illustrate the definition using our examples of the previous section. In the prisoner's dilemma defecting is the dominant strategy for both players and therefore always the best reply toward any strategy of the opponent. So the Nash equilibrium in this game is for both players to defect. For the second subclass we considered the battle of the sexes game [Gintis, 2001; Weibull, 1996]. In this game there are 2 pure strategy Nash equilibriums, i.e. (*football*, *football*) and (*opera*, *opera*). There is also 1 mixed Nash equilibrium, i.e. where the row player (the husband) plays *football* with 2/3 probability and *opera* with 1/3 probability and the column player (the wife) plays *opera* with 2/3 probability and *football* with 1/3 probability. The third class consists of the games with a unique mixed equilibrium. In the case of the matching pennies game, the Nash equilibrium is for both players to play their action with a uniform distribution $(0.5, 0.5)$.

Pareto Optimality

The concept of Pareto optimality is named after the Italian economist Vilfredo Pareto (1848-1923). Intuitively a Pareto optimal solution of a game can be defined as follows: a combination of actions of agents in a game is Pareto optimal if there is no other solution for which all players do at least as well and at least one agent is strictly better off. More formally we have: a strategy combination $s = (s_1, ..., s_n)$ for n agents in a game is Pareto optimal if there does not exist another strategy combination s' for which each player receives at least the same payoff P_i and at least one player j receives a strictly higher payoff than P_j.

A related concept is that of *Pareto dominance*. An outcome of a game is Pareto dominated if some other outcome would make at least one player better off without hurting any other player. That is, some other outcome is weakly preferred by all players and strictly preferred by at least one player. If an outcome is not Pareto dominated by any other, than it is Pareto optimal.

We discuss an example at the end of the next subsection.

Evolutionary Stable Strategies

The core equilibrium concept of evolutionary game theory is that of an *evolutionary stable strategy* (ESS), introduced by Maynard-Smith and Price in 1973 [Maynard-Smith and Price, 1973]. One way to understand this idea is to imagine a population of agents playing some game. After a number of games, the agents breed, producing new agents that play the same strategy, where the number of offspring of an agent depends on the payoff it has obtained. Now, in this setting, consider that initially all agents play the same strategy, but then a small number of agents switch to play a second strategy. If the payoff obtained by this new strategy is smaller than the payoff obtained by the original one, the second strategy will be played by fewer agents each generation, and will eventually disappear. In this case we say that the original strategy is evolutionarily stable against this new appearing strategy. More generally, we say a strategy is ESS if it is robust against evolutionary pressure from any strategy that appears. (In the context of a strategic game, "any strategy" is simply any of the other pure or mixed strategies that are available to an agent).

Formally an ESS is defined as follows. Suppose that a large population of agents is programmed to play the (mixed) strategy s, and suppose that this population is invaded by a small number of agents playing strategy s'. The population share of agents playing this mutant strategy is $\epsilon \in\,]0,1[$. When an individual is playing the game against a random chosen agent, chances that he is playing against a mutant are ϵ and against a non-mutant are $1 - \epsilon$. The payoff for the first player, being a non-mutant is:

$$P(s, (1-\epsilon)s + \epsilon s')$$

and the payoff for being a mutant is:

$$P(s', (1-\epsilon)s + \epsilon s')$$

Now we can state that a strategy s is an ESS if $\forall\ s' \neq s$ there exists some $\delta \in\,]0,1[$ such that $\forall\ \epsilon:\ 0 < \epsilon < \delta$,

$$P(s, (1-\epsilon)s + \epsilon s') > P(s', (1-\epsilon)s + \epsilon s')$$

holds. The condition $\forall\ \epsilon:\ 0 < \epsilon < \delta$ expresses that the share of mutants needs to be sufficiently small.

Note that one can frame the idea of ESS without the need to consider agents breeding. Instead one can consider agents being able to observe the average payoff of agents playing different strategies, and making a decision as to which strategy to adopt based on that payoff. In such a case, the "mutant" strategy is simply a new strategy adopted by some agents — if other agents observe that they do well, they will switch to this strategy also, while if the new strategy does not do comparatively well, the agents that play it will switch back to the original strategy.

As an example, let's now determine whether the Nash equilibrium of the prisoner's dilemma is also an ESS. Suppose $\epsilon \in [0,1]$ is the number of cooperators in the population.

The expected payoff of a cooperator is $3\epsilon + 0(1-\epsilon)$ and that of a defector is $5\epsilon + 1(1-\epsilon)$. Since for all ϵ,
$$5\epsilon + 1(1-\epsilon) > 3\epsilon + 0(1-\epsilon)$$
defect is an ESS. So the number of defectors will always increase and the population will eventually only consist of defectors. In Section 7.4 this dynamical process will be illustrated by the replicator equations.

This equilibrium which is both Nash and ESS, is not a Pareto optimal solution. This can be easily seen if we look at the payoff tables. The combination (*defect, defect*) yields a payoff of $(1, 1)$, which is a smaller payoff for both players than the combination (*cooperate, cooperate*) which yields a payoff of $(3, 3)$. Moreover the combination (*cooperate, cooperate*) is a Pareto optimal solution. However, if we apply the definition of Pareto optimality, then also (*defect, cooperate*) and (*cooperate, defect*) are Pareto optimal. But both these Pareto optimal solutions do not Pareto dominate the Nash equilibrium and therefore are not of interest to us. The combination (*cooperate, cooperate*) is a Pareto optimal solution and Pareto dominates the Nash equilibrium. Finally, we elaborate on the inter-relationship between the different solution concepts.

7.2.2 Strategic Games with Continuous Strategy Spaces

Traditional 2-player game theory assumes that the agents in the game employ one of a limited and discrete set of n so-called 'pure' strategies, which can be enumerated as: $\mathcal{S} = \{s_1, s_2, \ldots, s_n\}$. If the other agent in the game, indicated as player 2, employs the same set $\mathcal{S}$ of pure strategies, in the standard approach a payoff matrix A defines the payoff of agent 2 using strategy s_j against agent 1 using strategy s_i as: $a_{i,j}$.

However, there are numerous realistic situations in which the set of strategies is *continuous* rather than discrete. For example, consider the case where the agents can select a strategy that can be indicated by a continuous variable $x \in [0, 1]$, such that each value of x defines a different, pure strategy that the agent can employ. The set of pure strategies here is a one-dimensional set $X = [0, 1]$. If two agents interact using the same set of pure strategies X, in analogy to the discrete case it is possible to define a payoff-matrix A, such that $A(x, y)$ is the payoff of agent 2 playing pure strategy $y \in X$ to agent 1 playing pure strategy $x \in X$. If a player employs a mix of pure strategies from X, it is similarly possible to define a distribution function ϕ on X, such that $\phi(x)$ indicates the probability that a pure strategy x is employed. This implies that $\int_X \phi(x) dx = 1$. In the same way it is possible that a strategy is indicated by a whole set of D continuous variables rather than one single continuous variable. In this case, each single strategy can be indicated by a vector $\mathbf{x} = (x_1, x_2, ..., x_D)^T \in X \subset R^D$. The associated distribution function over the pure strategies is now $\phi(\mathbf{x})$ with $\int_X \phi(\mathbf{x}) d\mathbf{x} = 1$. Similarly, the payoff matrix from player 2 playing strategy $\mathbf{y}$ to player 1 using strategy $\mathbf{x}$ is $A(\mathbf{x}, \mathbf{y})$.

As an example, consider an interaction in which there are two continuously-valued parameters, indicated as: $\mathbf{x} = (x, y)^T \in [0, 1]^2$. Each point in the (x, y)-plane defines one individual pure strategy. Let us now define the following 'banana'-game: two monkeys have to divide one banana. In order to come to a fair division, the first monkey M_1 will cut the banana in two parts, say of relative lengths $x_1 \in [0, 1]$, and $(1-x_1)$ respectively. Subsequently the second monkey M_2 will choose one of these two parts. Let us assume that monkey M_2 selects the largest part, i.e. $\texttt{max}(x_1, 1-x_1)$ with a probability $y_2 \in [0, 1]$, and therefore the smallest part $\texttt{min}(x_1, 1-x_1)$ with a probability $(1-y_2)$. The direct payoff-matrix A_1 to monkey M_1 with (x_1, y_1) from monkey M_2 with parameters (x_2, y_2) then becomes:

$$A_1(x_1, y_1, x_2, y_2) = y_2.\texttt{min}(x_1, 1-x_1) + (1-y_2).\texttt{max}(x_1, 1-x_1) \qquad (7.1)$$

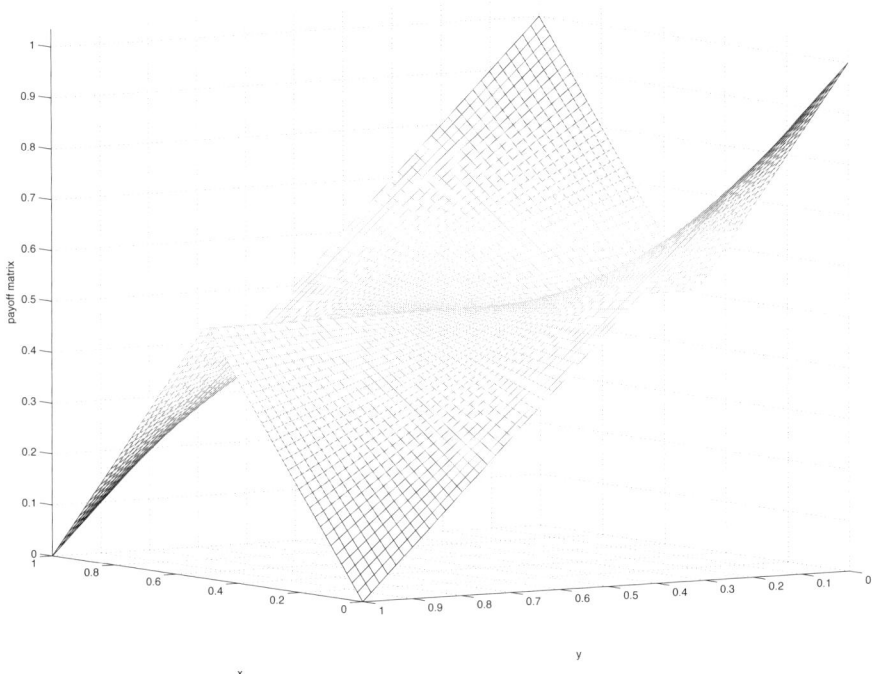

FIGURE 7.5 The banana-payoff matrix $A_2(x_1, y_1, x_2, y_2) = y_2.\mathtt{max}(x_1, 1 - x_1) + (1 - y_2).\mathtt{min}(x_1, 1 - x_1)$ in the (x_1, y_2)-plane.

Similarly, the direct payoff-matrix A_2 from monkey M_1 (the 'divider') to monkey M_2 (the 'selector') is:

$$A_2(x_1, y_1, x_2, y_2) = y_2.\mathtt{max}(x_1, 1 - x_1) + (1 - y_2).\mathtt{min}(x_1, 1 - x_1) \quad (7.2)$$

This is illustrated in Figure 7.5.

These payoffs represent the expected fraction of the banana for each monkey respectively. Another way to measure the advantage of a certain strategy is to look to the *difference* between the fractions that the monkey receive, so for the first monkey (the divider) this difference is:

$$A_{12}(x_1, y_1, x_2, y_2) = A_1(x_1, y_1, x_2, y_2) - A_2(x_1, y_1, x_2, y_2) \quad (7.3)$$

and for the second monkey (the selector):

$$A_{21}(x_1, y_1, x_2, y_2) = A_2(x_1, y_1, x_2, y_2) - A_1(x_1, y_1, x_2, y_2) \quad (7.4)$$

For each type of player this expresses the advantage of its strategy, defined in terms of its position (x, y).

We will use this example later to show the complex behavior of the continuous RD. Note that this game is similar to the famous ultimatum game, in which the divider can propose a certain amount of the banana to the acceptor. If the acceptor refuses the offer, none of them gets a single piece. In our instance of the banana game we eliminated this social dilemma such as present in the ultimatum game. The reason for this is that we do not want to complicate things too much at once, and focus our work on how we can derive and implement continuous replicator dynamics (rather than analyzing the social dilemma itself).

In this chapter we will not discuss ESS in the context of continuous strategy spaces, simply because we want to focus on RD. We refer the interested reader for a discussion of ESS in continuous spaces to [Cressman and Hofbauer, 2005; Cressman et al., 2006].

7.3 Replicator Dynamics

The Replicator Dynamics (RD) [Taylor and Jonker, 1978; Zeeman, 1979, 1981] can be discussed along three dimensions, i.e., time, strategy space and the number of populations. In this section we will derive the RD as well in discrete strategy spaces as continuous strategy spaces and start of by showing how RD can be represented in discrete time with discrete strategy spaces using one population. From this representation we will compute a time limit leading to the continuous time version of the RD in discrete strategy spaces. Next, we will generalize this representation to multiple populations. After this we continue the elaboration on RD by deriving the equations for continuous strategy spaces.

7.3.1 Replicator Dynamics in Discrete Strategy Spaces

Agents playing a strategic game each hold a vector of proportions over possible actions, indicating the proportion of "individuals" or "replicators" in an infinite "population" (the agent) which have adopted a given action. Each timestep, the proportions of actions for each agent are changed based on the rewards in the payoff tables as well as on the current probabilities of other agents' choosing their actions.

An abstraction of an evolutionary process usually combines two basic elements: selection and mutation. Selection favors some population actions over others, while mutation provides variety in the population. The most basic form of RD highlights only the role of selection, i.e., how the most fit actions in a population are selected.

Discrete Time Replicator Dynamics

Suppose there is a single population of actions (or replicators) and consider a discrete time process $t = 1, 2, \ldots$. Let $A = (a_{ij})_{i,j=1}^{n}$ be the reward matrix.

Let us assume that the individuals in the population represent different actions that the agent can perform. The state of the population can be described by a vector $x(t) = (x_1(t), \ldots, x_n(t))$, where $x_i(t)$ denotes the proportion of individual i in the population.

At each time step t, the state $x(t)$ of the population changes according to the fitness values of the different individuals. More precisely, the expected number of offspring for a single individual representing action i equals the ratio between the expected payoff for such an individual, $\sum_{j=1}^{n} a_{ij} x_j(t)$, and the average payoff $\sum_{k=1}^{n} x_k(t) \left(\sum_{j=1}^{n} a_{kj} x_j(t) \right)$ for all individuals in the population. Therefore, the ratio $x_i(t+1)$ of individuals representing action i at the next time step equals:

$$x_i(t+1) = x_i(t) \frac{\sum_{j=1}^{n} a_{ij} x_j(t)}{\sum_{k=1}^{n} x_k(t) \left(\sum_{j=1}^{n} a_{kj} x_j(t) \right)}$$

Dividing both sides by $x_i(t)$ leads to:

$$\frac{x_i(t+1)}{x_i(t)} = \frac{\sum_{j=1}^{n} a_{ij} x_j(t)}{\sum_{k=1}^{n} x_k(t) \left(\sum_{j=1}^{n} a_{kj} x_j(t) \right)}$$

Reordering the fractions leads to the general discrete time replicator dynamics:

$$\frac{\Delta x_i(t)}{x_i(t)} = \frac{\sum_{j=1}^{n} a_{ij} x_i(t) - x(t) A x(t)}{x(t) A x(t)}$$

This system of equations expresses Darwinian selection as follows: the rate of change Δx_i of a particular action i is the difference between its average payoff, $\sum_{j=1}^{n} a_{ij} x_i(t)$, and the average payoffs for all actions, $x(t) A x(t)$.

From a mathematical viewpoint, it is usually more convenient to work in continuous time. Therefore we now derive the continuous time version of the above replicator equations. To do this we make time infinitesimal. More precisely, suppose that the amount of time that passes between two periods is given by δ with $\delta \in [0, 1]$. We also assume that during a time period δ, only a fraction δ of the individuals die and generate offspring. Then the above equation can be rewritten as follows:

$$\frac{\Delta x_i(t)}{x_i(t)} = \frac{x_i(t+\delta) - x_i(t)}{x_i(t)} = \frac{\sum_{j=1}^{n} \delta a_{ij} x_i(t) - x(t) \delta A x(t)}{x(t) \delta A x(t) + (1-\delta)}$$

Now, at the limit when δ approaches 0, the continuous replicator equations are:

$$\frac{dx_i}{dt} = \left[\sum_{j=1}^{n} a_{ij} x_i - x \cdot Ax \right] x_i$$

which can be rewritten as:

$$\frac{dx_i}{dt} = [(Ax)_i - x \cdot Ax] x_i \tag{7.5}$$

In Equation 7.5, x_i represents the density of action i in the population, and A is the payoff matrix that describes the different payoff values that each individual replicator receives when interacting with other replicators in the population. The state x of the population can be described as a probability vector $x = (x_1, x_2, ..., x_n)$ which expresses the different densities of all the different types of replicators in the population. Hence $(Ax)_i$ is the payoff that replicator i receives in a population with state x and $x \cdot Ax$ describes the average payoff in the population. The growth rate $\frac{dx_i/dt}{x_i}$ of the proportion of action i in the population equals the difference between the action's current payoff and the average payoff in the population. [Gintis, 2001; Hofbauer and Sigmund, 1998; Weibull, 1996] detail further information on this issue.

Multi-Population Replicator Dynamics

Usually, we are interested in models of multiple agents that evolve and learn concurrently, and therefore we need to consider multiple populations. For simplicity, the discussion focuses on only two such learning agents. As a result, we need two systems of differential equations, one for each agent. This setup corresponds to an RD for asymmetric games, where the available actions or strategies of the agents belong to two different populations.

This translates into the following coupled replicator equations for the two populations:

$$\dot{p}_i = [(Aq)_i - p \cdot Aq]p_i \qquad (7.6)$$

$$\dot{q}_i = [(Bp)_i - q \cdot Bp]q_i \qquad (7.7)$$

Equations 7.6 and 7.7 indicate that the growth rate of the types in each population is additionally determined by the composition of the other population, in contrast to the single population (learner) case described by Equation 7.5.

Replicator Dynamics with Mutation

In [Hofbauer and Sigmund, 1998] and also in [Ruijgrok and Ruijgrok, 2005] the RD with an additional mutation term were derived. The most straightforward approach for including interaction between adjacent strategies is by including a diffusion term in the RD. This formulation allows for the migration from densely populated areas in strategy space to less populated ones nearby. This procedure is originally proposed by Hofbauer and Sigmund [Hofbauer and Sigmund, 1998], for which Ruijgrok and Ruijgrok [Ruijgrok and Ruijgrok, 2005] have provided a rigorous mathematical foundation and subsequent analysis, which we will follow here. The diffusion process itself can be interpreted as the mutation of strategies to slightly different ones. In this approach, in addition to the discrete strategy RD of Equation 7.5, Ruijgrok and Ruijgrok add a transition probability per unit time q_{ij} for the spontaneous transfer from a strategy i to a strategy j. This extra term describes the mutation between strategies at a rate q_{ij}. In this way they derive the discrete replicator equation with mutation as:

$$\frac{dx_i}{dt} = \left[\sum_{j=1}^{n} a_{ij}x_i - x \cdot Ax\right]x_i + \sum_{j=1}^{N}(q_{ij}x_j - q_{ji}x_i) \qquad (7.8)$$

Now in the course of time, the fraction x_i changes at a rate which is proportional to the difference between the payoff to x_i and the average payoff for the whole population. In addition we assume that for each player of s_j there is a transition probability per unit time to make a spontaneous transfer to strategy s_i at a rate given by q_{ji}.

7.3.2 Replicator Dynamics in Continuous Strategy Spaces

Traditional RD, as seen above, describes the evolution of populations of agents participating in some kind of interaction, where each individual agent can employ a well-defined, finite and discrete set of strategies. Essential in this approach is the assumption that for each possible combination of strategies the gain and loss of each agent can be quantified explicitly as 'pay-off' matrices.

However, multi-agent systems, especially in analogy with our biological world, are not discrete. More precisely, sets of discrete strategies result from basic underlying microscopic processes that have a more continuous character. Ultimately, any living organism is governed by microscopic processes on the molecular level. In this setting, the effects of environmental variables such as light and temperature, the signals of receptor molecules, and the interactions between genes, all types of RNAs, proteins, and other molecules on the behavior of the organism are continuously varying variables. As a closer example of MAS consider for instance robot soccer. The decisions faced by the players, such as when and where to kick the ball, or where to move when not in possession of the ball are continuous in nature. A typical approach then is to only allow a small number of discrete actions instead of one, continuous parameter. For instance in the soccer case this would mean that a player could only move in a few predefined directions, i.e., the pure or discrete strategies, whereas opposed to a continuous approach one would allow the player to move in a continuum of directions represented by one direction-parameter This conclusion has implications for the formulation of a correct fundamental interaction model. We will now study how the RD model can be extended to include a continuum –rather than a discrete set– of strategies.

As a first step we will now define some basic concepts. Suppose that the relation between the agent and its environment can be modeled mathematically such that:

i The agent can be fully described by a set of D continuous parameters $\mathbf{x} = (x_1, x_2, \ldots, x_D)$, with $\mathbf{x} \in \Theta \subset R^D$, and:
ii The interaction between two agents P and P', described by the parameter sets $\mathbf{x}$ and $\mathbf{x}'$ respectively, results in a pay-off matrix $A(\mathbf{x}, \mathbf{x}')$ from agent P' to P.

In this setting we can interpret the parameter set $\mathbf{x}$ of the agent as an instance of a its strategy, or, in other words, there is a continuum of possible strategies each of which is indexed by its set of values in the $\mathbf{x}$-vector. In case of an entire population of such agents, we can define the distribution over all possible strategies as a probability distribution function $\phi(\mathbf{x}, t)$, depending on $\mathbf{x}$ and continuous time t. In the well-studied case of a distribution x over a *discrete* set of D strategies, the continuous time propagation of x is given by the aforementioned RD as:

$$\dot{p}_i = p_i(\sum_j a_{ij} p_j - \mathbf{p}^T A \mathbf{p}) \equiv p_i(V_i - E) \qquad (7.9)$$

Here the 'mixed' strategy p_i represents the fraction of 'players' with 'pure' strategy i, and therefore $\sum_i p_i = 1$. Moreover, A is the $D \times D$ 'payoff' matrix such that a_{ij} indicates the amount of payment the 'row player' using pure strategy i receives from the 'column player' using pure strategy j. It is clear that the evolution of a strategy i is governed entirely by its difference between its average payoff V_i and the overall average payoff E. In our context, rather than a discrete set of pure strategies indexed by i, there is a continuum denoted by $\mathbf{x}$, and consequently the distribution function $\phi(\mathbf{x}, t)$ assumes the role of $\mathbf{p}$, and summation over i becomes integration over $\mathbf{x}$. The procedure of how to translate a formulation on a discrete set of strategies to a continuous strategy space is well-known in the context of Statistical Mechanics. Following this procedure the continuum limit of Equation 7.9 becomes a Partial Differential Equation (PDE):

$$\frac{\partial \phi}{\partial t} = (V(\mathbf{x}, t) - E(t))\phi \qquad (7.10)$$

with:

$$\int_\Theta \phi(\mathbf{x}, t) d\mathbf{x} = 1 \qquad (7.11)$$

In this PDE we used the following definitions:

$$V(\mathbf{x}, t) = \int_\Theta A(\mathbf{x}, \mathbf{x}')\phi(\mathbf{x}', t) d\mathbf{x}' \qquad (7.12)$$

$$E(t) = \int_\Theta V(\mathbf{x}, t)\phi(\mathbf{x}, t) d\mathbf{x} \qquad (7.13)$$

where Θ denotes the set of allowed values of the parameter-vector $\mathbf{x}$, and $A(\mathbf{x}, \mathbf{x}')$ the payoff matrix from strategy $\mathbf{x}'$ to strategy $\mathbf{x}$. Equation 7.10 was obtained and studied by Cressman

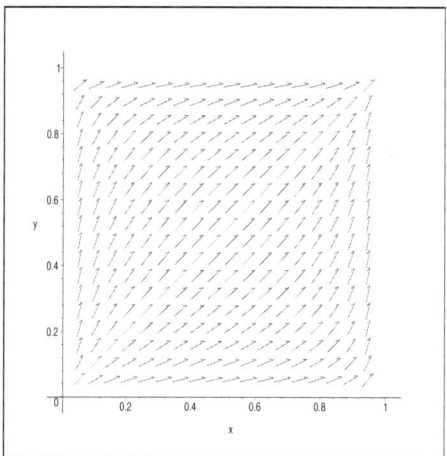

FIGURE 7.6 Direction field plot of the prisoner's dilemma game.

[Cressman., 2005], and Oechsler and Riedel [Oechsler and Riedel, 2001]. This PDE shows that the evolution of the fraction $\phi(\mathbf{x})$ of the population in strategy $\mathbf{x}$ is determined solely by the difference between the local 'potential' $V(\mathbf{x})$ and the global 'total energy' E. This formulation, therefore, does not include any direct interaction between different strategies, as the fraction of the population in each strategy $\mathbf{x}$ only evolves relative to its advantage –or disadvantage– relative to the average payoff E.

7.4 Evolutionary Dynamics in Discrete Strategy Spaces

In this section we first analyze the dynamics of the different types of games introduced in Section 7.2.1. More precisely, we will plot and discuss the direction fields of the RDs applied to the different type of games. This allows us to analyze the different basins of attraction.

Secondly, we will draw the formal connection between the RD and reinforcement learning algorithms in Section 7.4.2. We will illustrate the strength of this connection for multi-agent systems with some well chosen experiments.

7.4.1 Analysis of the Evolutionary Dynamics of the Categorization of Games

Given the RD models of the previous section we can now analyze the evolutionary behavior of these models in the different type of categories of games by plotting their respective direction fields. A direction field is a very elegant and excellent tool to understand and illustrate a system of differential equations. The direction fields presented here consist of a grid of arrows tangential to the solution curves of the system. It is a graphical illustration of the vector field indicating the direction of the movement at every point of the grid in the state space. Filling in the parameters for each game in Equations 7.6 and 7.7, allows to plot this field.

Lets analyze the game dynamics for each category from Section 7.2.1. In Figure 7.6 we plot the RD of the prisoner's dilemma game. The x-axis represents the probability with which the first player will play defect and the y-axis represents the probability with which the second player will play defect. So the Nash equilibrium and the ESS lie at coordinates (1,1). As you can see from the field plot all the movement goes toward this equilibrium and

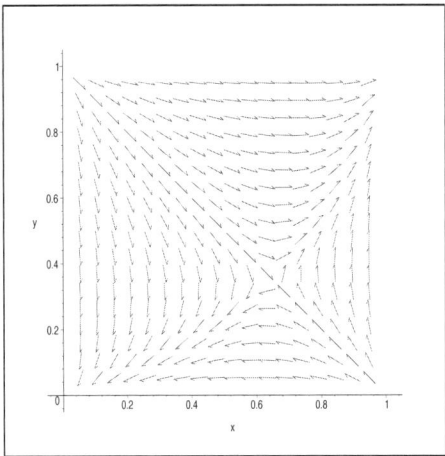

FIGURE 7.7 Direction field plot of the battle of the sexes game.

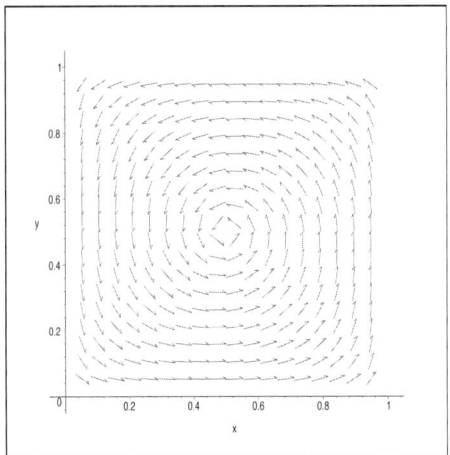

FIGURE 7.8 Direction field plot of the matching pennies game.

it is immediately clear from the plot that the Nash equilibrium is stable and robust against evolutionary pressure.

The field plot of the battle of the sexes game is illustrated in Figure 7.7. Recall that this game has 2 pure Nash equilibria and 1 mixed Nash equilibrium. These equilibria can be seen in the figure at coordinates $(0,0), (1,1), (2/3, 1/3)$. The 2 pure equilibria are ESS as well. This is also easy to verify from the plot, more precisely, any small perturbation away from the equilibrium is led back to the equilibrium by the dynamics. The mixed equilibrium, which is Nash, is not ESS.

Finally, the field plot of the matching pennies game can be found in Figure 7.8. The plot shows the mixed Nash equilibrium at $(0.5, 0.5)$ which is not evolutionarily stable. Typical for this class of games is that the interior trajectories define closed orbits around the equilibrium point as can be seen from the plot.

7.4.2 Relating Evolutionary Dynamics in Discrete Strategy Spaces with Reinforcement Learning

In this section we draw the formal connection between replicator dynamics and reinforcement learning (RL). We limit ourselves to one-state games, or non-associative reinforcement learning tasks. Only recently the connection between multi-state RL problems and EGT has been drawn. This will not be a topic of this chapter, however we refer the interested reader to [Vrancx et al., 2008].

Reinforcement Learning (RL) is an established and profound theoretical framework for learning in stand-alone systems. Yet, extending RL to MAS does not guarantee the same theoretical grounding. As long as the environment an agent experiences is stationary, and the agent can experiment enough, RL guarantees convergence to the optimal strategy. In a MAS however, the reinforcement an agent receives usually depends on the actions taken by the other agents present in the system. Obviously in these environments, the convergence results of RL are lost.

Therefore it is important to fully understand the dynamics of reinforcement learning and the effect of exploration in MAS. For this aim we employ the RD. In this chapter we summarize the dynamics of Q-learning. For an elaboration on a different reinforcement learning algorithm, i.e. Learning Automata, we refer the reader to [Tuyls and Nowe, 2005]. As will be shown, these dynamics open a new perspective in understanding and fine tuning the learning process in games and more general in MAS.

Q-Learning

Common reinforcement learning methods, which can be found in [Sutton and Barto, 1998] are structured around estimating value functions. A value of a state or state-action pair, is the total amount of reward an agent can expect to accumulate over the future, starting from that state. One way to find the optimal policy is to find the optimal value function. If a perfect model of the environment as a Markov decision process is known, the optimal value function can be learned with an algorithm called value iteration. Q-learning is an adaptive value iteration method [Sutton and Barto, 1998], which bootstraps its estimate for the state-action value $Q_{t+1}(s, a)$ at time $t + 1$ upon its estimate for $Q_t(s', a')$ with s' the state where the learner arrives after taking action a in state s:

$$Q_{t+1}(s, a) \leftarrow (1 - \alpha)Q_t(s, a) + \alpha(r + \gamma \, max_{a'} Q_t(s', a')) \qquad (7.14)$$

With α the usual step size parameter, γ a discount factor and r the immediate reinforcement.

The players of the games considered in this chapter can therefore use the algorithm of (7.14) where the state information s is removed. This is called non-associative Q-learning.

In the next subsection we will concisely discuss the Q-learning dynamics derived in [Tuyls et al., 2003; 't Hoen and Tuyls, 2004; Tuyls et al., 2006].

The Dynamics of Q-Learning in Games

In this section we briefly describe the formal relation between Q-learning and the RD.* More precisely we present the dynamical system of Q-learning. These equations are derived by constructing a continuous time limit of the Q-learning model, where Q-values are interpreted as Boltzmann probabilities for the action selection. For reasons of simplicity we consider

---

*The reader who is interested in the complete derivation, we refer to [Tuyls et al., 2003, 2006].

games between 2 players. The equations for the first player are,

$$\frac{dx_i}{dt} = x_i \alpha \tau ((A\mathbf{y})_i - \mathbf{x} \cdot A\mathbf{y}) + x_i \alpha \sum_j x_j ln(\frac{x_j}{x_i}) \qquad (7.15)$$

analogously for the second player, we have,

$$\frac{dy_i}{dt} = y_i \alpha \tau ((B\mathbf{x})_i - \mathbf{y} \cdot B\mathbf{x}) + y_i \alpha \sum_j y_j ln(\frac{y_j}{y_i}) \qquad (7.16)$$

Equations 7.15 and 7.16 express the dynamics of both Q-learners in terms of Boltzmann probabilities. Each agent(or player) has a probability vector over his action set, more precisely $x_1, ..., x_n$ over action set $a_1, ..., a_n$ for the first player and $y_1, ..., y_m$ over $b_1, ..., b_m$ for the second player.

For a complete discussion on this equations we refer to [Tuyls et al., 2003, 2006]. Comparing (7.15) or (7.16) with the RD in (7.5), we see that the first term of (7.15) or (7.16) is exactly the RD and thus takes care of the selection mechanism, see [Weibull, 1996]. The mutation mechanism for Q-learning is therefore left in the second term, and can be rewritten as:

$$x_i \alpha \sum_j x_j ln(x_j) - ln(x_i) \qquad (7.17)$$

In Equation 7.17, we recognize 2 entropy terms, one over the entire probability distribution x, and one over strategy x_i.

Relating entropy and mutation is not new. It is a well-known fact [Schneider, 2000; Stauffer, 1999] that mutation increases entropy. In [Stauffer, 1999], it is stated that the concepts are familiar with thermodynamics in the following sense: the selection mechanism is analogous to *energy* and mutation to *entropy*. So generally speaking, mutations tend to increase entropy. Exploration can be considered as the mutation concept, as both concepts take care of providing variety.

The Q-Learning Experiments

We only describe the experiments of subclass 2. All the experiments can be found in [Tuyls et al., 2003, 2006]. The important aspect is that obtaining convergence to a Nash equilibrium with Q-learning is more cumbersome than with other learning algorithms. In Figure 7.9 the direction field plot of the differential equations of this game is plotted. The fields are plotted for 3 values of τ, more precisely $1, 2, 10$. In the first 2 plots τ isn't big enough to reach for one of the three Nash equilibria. Only in the last one the dynamics attain the Nash equilibria (the 3 attractors in the last plot) for the game at the coordinates $(1,1)$, $(0,0)$ and $(\frac{2}{3}, \frac{1}{3})$. The mixed equilibrium though is very unstable. Any small perturbation away from this equilibrium will typically lead the dynamics to one of the 2 pure equilibria.

In Figure 7.10 we also plotted the Q-learning process for the same game with the same settings as for the system of differential equations. At the chosen points a learning path starts and converges to a particular point. If you compare the plots with the direction field plots for the same value of τ you can see that the sample paths of the learning process approximate the paths of the differential equations. The instability of the mixed equilibrium is the reason why this equilibrium doesn't emerge from the learning process.

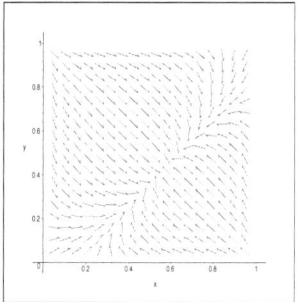 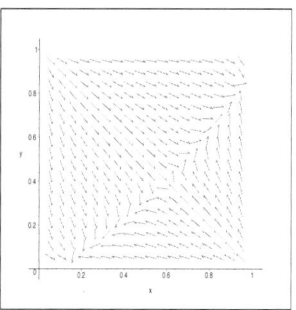 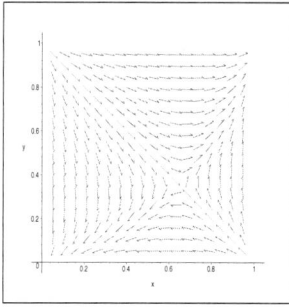

FIGURE 7.9 The direction field plots of the battle of the sexes (subclass 2) game with $\tau = 1, 2, 10$.

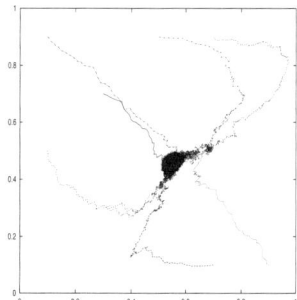 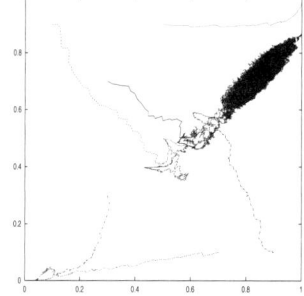 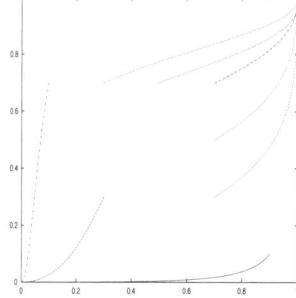

FIGURE 7.10 The Q-learning plots of the battle of the sexes (subclass 2) game with $\tau = 1, 2, 10$.

7.5 Evolutionary Dynamics in Continuous Strategy Spaces

In this section we start by elaborating on how to include mutation in continuous strategy RD. We start of by follow the approach of Hofbauer et al. [Hofbauer and Sigmund, 1998] and Ruijgrok et al. [Ruijgrok and Ruijgrok, 2005] by introducing a diffusion term.

Next, we drop the strong assumption of the diffusion term which assumes that mutations only occur between adjacent strategies. We will introduce an anisotropos continuous replicator model with isotropos and anisotropos mutation matrix.

7.5.1 Mutation as Engine for Diffusion in Continuous Strategy Spaces

In Section 7.3.1 we derived the RD in discrete strategy spaces with an additional mutation term. We will now transfer Equation 7.8 to a continuous strategy space version according to the description of [Ruijgrok and Ruijgrok, 2005]. Recall from Section 7.3.1 that in addition to the discrete strategy RD of Equation 7.9, Ruijgrok and Ruijgrok add a transition probability per unit time q_{ij} for the spontaneous transfer from a strategy i to a strategy j. This extra term describes the mutation between strategies at a rate q_{ij}. In this way the discrete replicator equations with mutation look as follows:

$$\dot{p}_i = p_i(V_i - E) + \sum_{j=1}^{N}(q_{ij}p_j - q_{ji}p_i) \qquad (7.18)$$

Now we will focus on the continuum limit of the mutation term –which we shall denote as $M(\mathbf{x}, t)$ – in this equation:

$$\sum_{j=1}^{N}(q_{ij}p_j - q_{ji}p_i) \to M(\mathbf{x}, t) \equiv \int_{\Theta}[m(\mathbf{x} \mid \mathbf{x}')\phi(\mathbf{x}', t) - m(\mathbf{x}' \mid \mathbf{x})\phi(\mathbf{x}, t)]d\mathbf{x}' \qquad (7.19)$$

With the restriction that only infinitesimally small changes in strategy space are allowed, Ruijgrok and Ruijgrok follow the method of van Kampen [van Kampen, 1992] to derive the Fokker-Planck equation from a master equation. For small $\|\mathbf{x}' - \mathbf{x}\|$, the Taylor-expansion of the mutation term $M(\mathbf{x}, t)$ results to:

$$M(\mathbf{x}, t) = \nabla\{Q_1(\mathbf{x})\phi(\mathbf{x}, t)\} + \frac{1}{2}\nabla^2\{Q_2(\mathbf{x})\phi(\mathbf{x}, t)\} \qquad (7.20)$$

with:

$$Q_1(\mathbf{x}) = \int_\Theta (\mathbf{x}' - \mathbf{x}) m(\mathbf{x}', \mathbf{x}) d\mathbf{x}' \qquad (7.21)$$

and:

$$Q_2(\mathbf{x}) = \int_\Theta (\mathbf{x}' - \mathbf{x})(\mathbf{x}' - \mathbf{x})^T m(\mathbf{x}', \mathbf{x}) d\mathbf{x}' \qquad (7.22)$$

Equation 7.21 represents the average resulting change in strategy due to mutations, and Equation 7.22 indicates the covariance matrix between the mutations. By assuming that there is no net change we obtain that $Q_1 = 0$, and by assuming that the covariance is an isotropic matrix–and therefore a constant, say 2μ, the continuum limit of the mutation term becomes a pure diffusion term:

$$M(\mathbf{x}, t) = \mu \nabla^2 \phi(\mathbf{x}, t) \qquad (7.23)$$

Using this approach, Ruijgrok and Ruijgrok arrive at the same continuous replicator equation as Hofbauer an Sigmund:

$$\frac{\partial \phi}{\partial t} = \mu \nabla^2 \phi + (V(\mathbf{x}) - E)\phi \qquad (7.24)$$

Integration of this equation over strategy space Θ yields:

$$\frac{\partial N(t)}{\partial t} = \mu \int_\Theta \nabla^2 \phi d\mathbf{x} + E(1 - N(t)) \qquad (7.25)$$

where:

$$N(t) = \int_\Theta \phi(\mathbf{x}, t) d\mathbf{x} \qquad (7.26)$$

The condition that the distribution ϕ is normalized, so that Equation 7.11 holds, is satisfied when the Neumann boundary condition holds, and $\nabla \phi$ is perpendicular to the boundary of Θ:

$$\mathbf{n}(\mathbf{x})^T \cdot \nabla \phi(\mathbf{x}) \mid_{\partial \Theta} = 0 \qquad (7.27)$$

where $\mathbf{n}(\mathbf{x})$ is a local normal vector to the boundary $\partial \Theta$ to the domain Θ. Basically, this equation allows undirected and unbiased evolution over short distances in strategy space. Ruijgrok and Ruijgrok study this equation for a number of games that have been transformed to a continuous setting.

More realistic effects can be obtained by adding stochastic terms in this equation.

There is, however, no reason to assume that mutations are only restricted to adjacent strategies, or in other words, that strategy space is a good representation of the genotypes. In fact, in most cases strategy space is a straightforward representation of phenotype space.

In the following sections, we will consider a few alternatives for this approach.

7.5.2 Non-Isotropic Mutations and Evolutionary Flows in Strategy Spaces

The continuous replicator equation 7.24 was derived under a number of restrictions:

1. Only infinitesimally small changes in strategy space are allowed;
2. The average change in strategy due to mutations is equal to zero, so $Q_1 = 0$ (equation 7.21);
3. The covariance matrix is a constant: $Q_2 = 2\mu$ (equation 7.22);
4. The Neumann (a.k.a. reflecting) boundary condition holds (equation 7.27).

Lifting the second and third condition leads to an anisotropic diffusion equation:

$$\frac{\partial \phi}{\partial t} = \nabla^T \cdot \mu(\mathbf{x})\nabla \phi + (V(\mathbf{x}) - E)\phi \quad (7.28)$$

where $\mu(\mathbf{x})$, the spatial diffusion coefficient, is a $D \times D$ spatially dependent symmetric matrix, such that its components $\mu_{ij}(\mathbf{x})$ indicate the rate of mutation from a pure strategy relating to parameter x_i to a pure strategy relating to parameter x_j at a position $\mathbf{x}$ in strategy space.

This may introduce a bias in the propagation of ϕ as certain directions in strategy space are 'blocked' or preferred by matrix μ at position $\mathbf{x}$. This condition occurs, for instance, in the evolution of genetic networks, where certain mutations are less likely for physical or chemical reasons.

Integration of Equation 7.28 over strategy space Θ again yields a condition similar to Equation 7.25:

$$\frac{\partial N(t)}{\partial t} = \int_{\Theta} \nabla^T \cdot \mu(\mathbf{x})\nabla \phi d\mathbf{x} + E(1 - N(t)) \quad (7.29)$$

Therefore, $N(t) = 1$ when the reflecting boundary condition holds, so that $\mu \nabla \phi$ is perpendicular to the boundary of Θ:

$$\mathbf{n}(\mathbf{x})^T \cdot \mu \nabla \phi(\mathbf{x})|_{\partial \Theta} = 0 \quad (7.30)$$

and therefore the distribution ϕ is normalized. This implies that there is no net flux into or out of the strategy space Θ, and therefore the integral probability inside Θ is conserved to 1.

Equation 7.28 is a diffusion-type of partial differential equation, reminiscent of the Fitzhugh-Nagumo model for the propagation of electrophysiological waves over cardiac walls. However, in this formulation, certain optimal solutions are not 'found' from a given initial position, as there exists no viable path in the direction field of phase space that is governed by $\mu(\mathbf{x})$. This contrasts with the continuous replicator Equation 7.24, that exhibits isotropous diffusion. This again is known in the Fitzhugh-Nagumo model as a 'block' that blocks the propagation of a electrophysiological wave over the cardiac walls, as in atrium fibrilation, possibly leading to arrhythmias.

7.5.3 Draining of Pay-Off Streams in Strategy Space

There exists an alternative formulation for the discrete RD that models the influence of nearby strategies on the evolution due to mutation from one pure strategy to another. This formulation is provided by Hofbauer and Sigmund in [Hofbauer and Sigmund, 1998]. Here, we provide a modification of this formulation that is consistent with the master equation for the strategy probability distribution, as it also allows for migration *from* the studied state:

$$\dot{p}_i = \sum_{j,k} a_{jk} p_k \{q_{ji} p_j - q_{ij} p_i\} - p_i(\mathbf{p}^T A \mathbf{p}) \qquad (7.31)$$

Here the same variables are employed as in Equation 7.18. Rather than diffusion due to high probability densities, here the propagation of ϕ is driven by the average payoff-stream between the states i and j: $p_i a_{ij} p_j$.

Following the continuum approach described in Section 7.5.3, and allowing for anisotropic diffusion, Equation 7.31 can be reformulated as a continuous RD:

$$\frac{\partial \phi(\mathbf{x},t)}{\partial t} = \nabla^T \cdot \mu(\mathbf{x}) \nabla \{V(\mathbf{x})\phi(\mathbf{x},t)\} + (V(\mathbf{x}) - E)\phi(\mathbf{x},t) \qquad (7.32)$$

Similar to the approach described above for the integral $N(t)$ of equation 7.32 over strategy space Θ yields:

$$\frac{\partial N(t)}{\partial t} = \int_\Theta \nabla^T \cdot \mu(\mathbf{x}) \nabla \{V(\mathbf{x})\phi(\mathbf{x},t)\} d\mathbf{x} + E(1 - N(t)) \qquad (7.33)$$

This implies that the distribution ϕ is conserved to 1, i.e. $N(t) = \int_\Theta \phi d_D \mathbf{x} = 1$, when the reflecting boundary condition holds:

$$\mathbf{n}(\mathbf{x})^T \cdot \mu(\mathbf{x}) \nabla \{V(\mathbf{x})\phi(\mathbf{x},t)\} \mid_{\partial \Theta} = 0 \qquad (7.34)$$

Our new formulation is able to capture certain aspects of an evolution driven by game theoretic interactions that are absent in the original formulation of Hofbauer, Sigmund, Ruijgrok and Ruijgrok in Equation 7.24. This is mainly caused by following two aspects of this formulation.

i. The inclusion of the scalar 'potential' $V(\mathbf{x})$ into the Laplacean term in equation 7.32. Remember that the potential V was defined in equation 7.12 as an integral over strategy space of the product $A(\mathbf{x}, \mathbf{x}')\phi(\mathbf{x}', t)$. This function $V(\mathbf{x})$ therefore denotes locations $\mathbf{x}$ in D-dimensional strategy space Θ where high payoffs can be obtained, given the payoff matrix A and the current distribution ϕ. In equation 7.32, the gradient $\nabla\{V\phi\}$ points toward increasing values of the product $V\phi$. This means that it is not so much pointing to densely populated areas in Θ, but to areas with current potentially high payoffs – the product of the population density ϕ and the expected payoff potential V. Ignoring the anisotropic and non-local effects of $\mu(\mathbf{x})$, this reasoning also applies to the divergence ∇^T that acts on this gradient. So, the combined term $\nabla^T . \mu . \nabla\{V\phi\}$ is high at the borders of the areas where the product $V\phi$ is high. This forces ϕ to grow especially at the borders of (relative) 'hills' in the potential reward landscape, i.e., where $V(\mathbf{x})\phi(\mathbf{x})$ is high, and likewise to decrease at the borders of (relative) 'depressions' in the same landscape.

ii. The addition of the non-isotropic local mutation $D \times D$ matrix $\mu(\mathbf{x})$ rather than a global scalar μ that we proposed in Equation 7.28. As described above, this may distract the flow of the evolution. In terms of the aforementioned Fitzhugh-Nagumo equation, Equation 7.32 can be reformulated as the flow of an electro-physiological wave on a cardiac substrate:

$$\frac{\partial \phi}{\partial t} + \nabla^T \cdot \mathbf{J} = \rho(\mathbf{x}, t) \qquad (7.35)$$

with
$$(a) \rho = (V - E)\phi,$$
$$(b) \mathbf{F} = -\nabla\{V\phi\}, \tag{7.36}$$
$$(c) \mathbf{J} = \mu \mathbf{F}.$$

Equation 7.35 is the conservation law for probability ϕ and probability flow $\mathbf{J}$, with a local source/sink term $\rho(\mathbf{x}, t)$ that according to Equation 7.36a gives the surplus of potential payoff relative to the global average. Equation 7.36b denotes the 'field strength vector' $\mathbf{F}(\mathbf{x})$ as derived from the potential reward $V(\mathbf{x})\phi$. Equation 7.36c describes an analogon of the famous 'Law-of-Ohm': '$\Delta V = I \times R$' between the field strength $\mathbf{F}$ (compare to ΔV), and the resulting density flow $\mathbf{J}$ (compare to I), where $\mu(\mathbf{x})$ acts as the analogon of the local conductivity, i.e., inverse resistance (compare to $1/R$). Here the conductivity is local and anisotropous, meaning that the resulting flow $\mathbf{J}$ is not necessarily directed according to the field strength $\mathbf{F}$.

Together with the conservation of probability in Equation 7.11 these factors make that probability ϕ flow according to the local direction field–induced by the local column-space of μ–from areas where the potential reward is below-average to areas where it is above-average.

Note that from Equations 7.11, 7.12, 7.13, and 7.32, after some calculation, it follows that the rate of change of the total energy E satisfies:

$$\dot{E} = \int_\Theta \phi(\mathbf{y}, t)\{A(\mathbf{x}, \mathbf{y}) + A(\mathbf{y}, \mathbf{x})\} T(\mathbf{x})\phi(\mathbf{x}, t) d\mathbf{x} d\mathbf{y} \tag{7.37}$$

where $A(\mathbf{x}, \mathbf{y})$ is the payoff matrix, and $T(\mathbf{x})\phi(\mathbf{x}, t)$ is a differential operator defined as:

$$T(\mathbf{x})\phi(\mathbf{x}, t) = \nabla^T \cdot \mu(\mathbf{x})\nabla\{V(\mathbf{x})\phi(\mathbf{x}, t)\} \tag{7.38}$$

Equation 7.37 can be written more concisely as an inner-product of the strategies $|\phi\rangle$ and $|\phi'\rangle$ (using the so-called bra-ket notation of Dirac):

$$\dot{E} = \langle \phi \,|\, (A + A^T) T \,|\, \phi' \rangle \tag{7.39}$$

This shows clearly that E is a constant for all games that obey $A = -A^T$. These are the so-called symmetric or zero-sum games. In this notation it follows that: $E = \langle \phi \,|\, A \,|\, \phi' \rangle = \langle \phi \,|\, A^T \,|\, \phi' \rangle = \frac{1}{2}\langle \phi \,|\, A + A^T \,|\, \phi' \rangle = 0$, so that for symmetric games the value of constant E is indeed zero. Moreover, it also appears that the energy is conserved, i.e. $\dot{E} = 0$, when:

$$T(\mathbf{x})\phi(\mathbf{x}, t) = \nabla^T \cdot \mu(\mathbf{x})\nabla\{V(\mathbf{x})\phi(\mathbf{x}, t)\} = 0 \tag{7.40}$$

In that case Equation 7.32 reduces to Equation 7.10, and there is no direct interaction between the strategies.

In the case that μ is isotropic and homogeneous, i.e. $\mu(\mathbf{x}) = \mu I$ for some scalar constant μ and identity matrix I, we find that

$$\frac{\partial \phi}{\partial t} = \mu \nabla^2\{V\phi\} + (V - E)\phi \tag{7.41}$$

This expression is comparable to Equation 7.24, but where now the diffusion does not originate from high-probability areas, but from areas with a potentially high reward.

7.6 Example of the Resulting Dynamics of the Continuous Replicator Equations

The continuous replicator equations in Equation 7.32 are mathematically similar to the non-linear Schrödinger equation and the Fitzhugh-Nagumo model, which are both known for their complex and intricate non-linear dynamics. In order to obtain an impression of the complex behavior exhibited here, a number of numerical simulations is performed, with different scenarios for the payoff matrix and different values for the mutation tensor μ.

We will now follow the example of the 'banana-game' as defined in Section 7.2.2, and consider a population of 'dividers' that evolve according to the continuous replicator dynamics. This population is represented by its distribution function $\phi(x, y)$ that measures the fraction of the population that employs a strategy indexed by parameter x in the role of a 'divider' and a strategy indexed by parameter y in the role of a 'selector'. We will now compare the four scenarios for the Continuous Replicator Dynamics (CRD):

i. model CRD1: The continuous replicator equation without diffusion as in equation 7.10,

ii. model CRD2: the Hofbauer-Sigmund-Ruijgrok & Ruijgrok continuous replicator model defined in equation 7.24,

iii. model CRD3: our anisotropous continuous replicator model as in equation 7.32 with isotropous mutation matrix $\mu(\mathbf{x}) \equiv \begin{pmatrix} 1 & 0 \\ 0 & 1 \end{pmatrix}$.

iv. model CRD4: again equation 7.32 but now with the diagonal reflection mutation matrix $\mu(\mathbf{x}) \equiv \begin{pmatrix} 0 & 1 \\ 1 & 0 \end{pmatrix}$.

The payoff matrix that is employed here is A_{12} in equation 7.3, that gives the advantage for the divider to employ strategy 'divide in parts x and $1 - x$', when playing against itself. From first principles it follows that this payoff matrix measures the advantage for the divider and not the disadvantage for the selector. This means that a fair selection, $x = \frac{1}{2}$ provides the best solution. Let us furthermore assume that the population starts from an evenly distributed over the strategy space $[0, 1]^2$, it is: $\phi(x, y, 0) \equiv 1$ as this ensures that $\int_0^1 \int_0^1 \phi(x, y, 0)\, dx dy = 1$.

The continuous replicator dynamics in all scenarios is observed to converge to an equilibrium state. These equilibrium states ϕ_e are obtained by setting $\partial \phi / \partial t = 0$ in the associated Equations 7.10, 7.24, and 7.32, and solving the obtained spatial partial differential equation in $\mathbf{x}$ (but not in t). The results of these experiments are given in Figures 7.11, 7.12, 7.13, and 7.14.

From these experiments it is obvious that only model CRD3 in Equation 7.32 is able to find the correct solution $x = \frac{1}{2}$. Model CRD1 converges to a situation where $\phi_e(\mathbf{x})$ is a constant (when starting from a uniform distribution), as is obvious from equating $\partial \phi / \partial t = 0$ in Equation 7.10. Model CRD2 finds a solution that is symmetrical around $x = \frac{1}{2}$, but which is not the ideal solution that is found by model CRD3. Furthermore, using a mutation matrix as in scenario iv thwarts the model CRD3 from finding the optimal solution as the flow of mutations is directed in the 'wrong' direction. The latter may occur in biochemical reaction networks where a certain mutation is not possible for physical or chemical reasons.

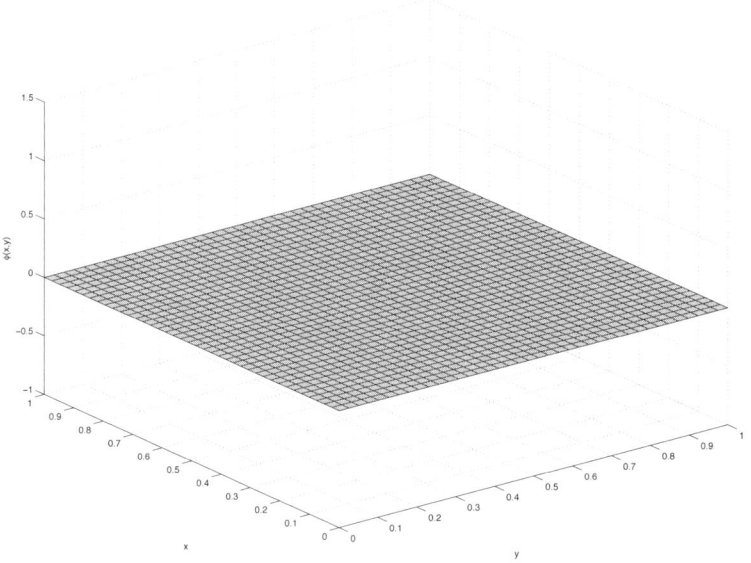

FIGURE 7.11 The equilibrium solution $\phi_e(x, y)$ for model CRD1 for the banana game for the 'divider' with payoff matrix A_{12} in Equation 7.3.

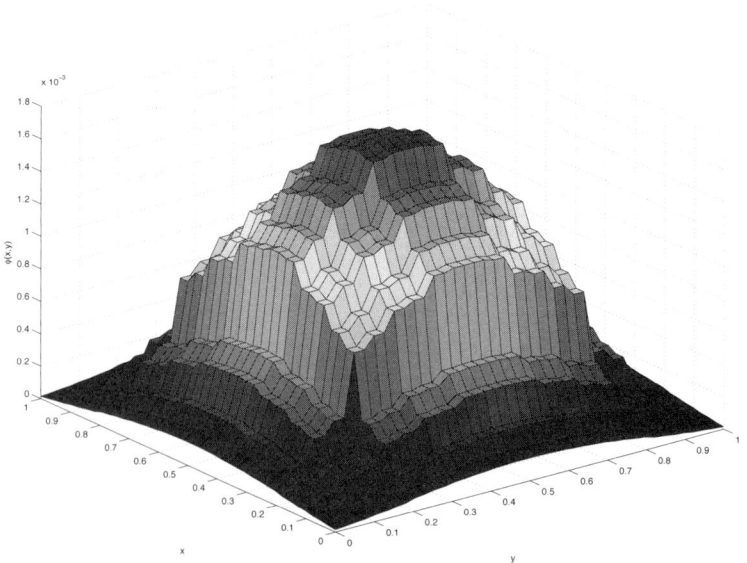

FIGURE 7.12 The equilibrium solution $\phi_e(x, y)$ for model CRD2 for the banana game for the 'divider' with payoff matrix A_{12} in Equation 7.3.

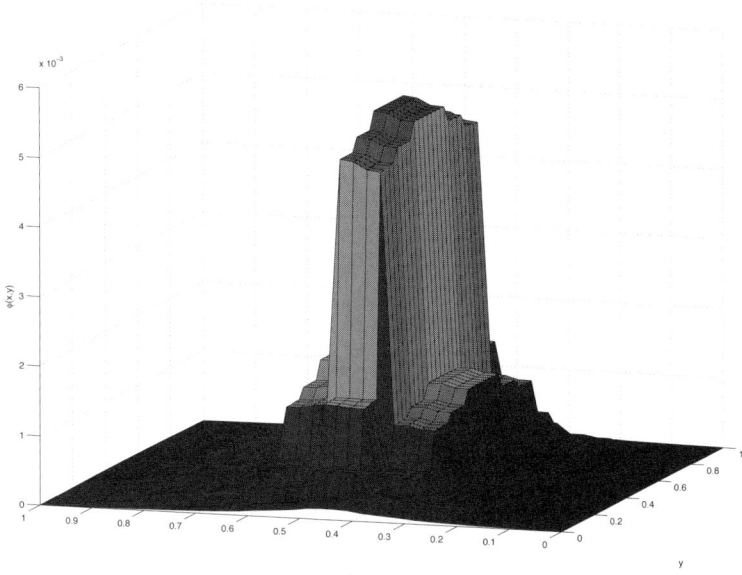

FIGURE 7.13 The equilibrium solution $\phi_e(x,y)$ for model CRD3 for the banana game for the 'divider' with payoff matrix A_{12} in Equation 7.3 with isotropic mutation matrix.

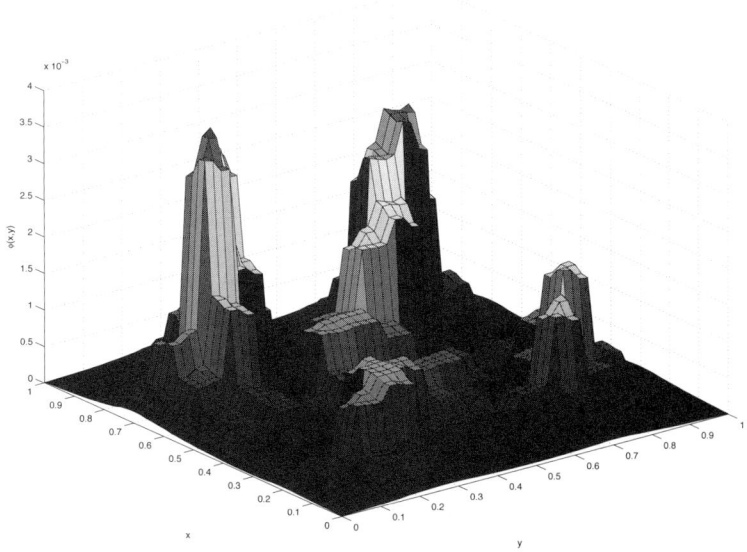

FIGURE 7.14 The equilibrium solution $\phi_e(x,y)$ for model CRD3 for the banana game for the 'divider' with payoff matrix A_{12} in Equation 7.3 with the diagonal reflection matrix.

7.7 Conclusions

In this chapter we provided an in-depth survey of evolutionary dynamics (or replicator dynamics) from Evolutionary Game Theory in games with discrete and continuous strategy spaces. More precisely, we provided an introduction to the basic notions of EGT necessary to understand this chapter. Next we derived the replicator equations along three dimensions, i.e., time, strategy space and multiple populations. Moreover, we also considered, besides selection, a formal description of the mutation mechanism. Then we illustrated the use of the RD in a variety of games with discrete strategy sets and we laid out the important link with Reinforcement Learning. This link is critical for the successful application of learning in multi-agent systems. Finally, we gave a current state of affairs on the application of the replicator equations in games with continuous strategy spaces. Moreover we introduced a new model to describe mutations in this setting, i.e., a anisotropos continuous replicator model with isotropos and anisotropos mutation matrix. Our initial experiments show that our model is more accurate with respect to finding an optimal solution in the banana game, as opposed to the other models. In future work, we plan to do more investigation of this new model in a wide variety of games like for instance the continuous strategy prisoner's dilemma.

In current research we also explore the possibility of applying these models on real world data of 'no limit' poker games. What we see is that by applying our RD models to this real world data, we get a view on the behavior and success of the different strategies employed by human players. This is interesting for several reasons: one, it is possible to analyze how likely it is that a certain strategy will be successful by examining the sizes of the basins of attraction; two, you can compare the dynamics of different learning models in real world examples, just by analyzing the phase plots; and three, it gives a means to examine the evolutionary stability of the present Nash equilibria.

The games presented in this study are essentially stateless (or have one state). Currently, we are also trying to incorporate switching dynamics in our study, which allows to analyze multi-state problems or games. In theory, this would allow to represent and analyze simplified versions of for instance robot soccer in the same manner as done for the stateless multi-agent games.

References

T. Borgers and R. Sarin. Learning through reinforcement and replicator dynamics. *Journal of Economic Theory*, 77:115–153, 1997.

R. Cressman. Dynamic stability of the replicator equation with continuous strategy space. *Math. Soc. Sci*, 50:127–147, 2005.

R. Cressman and J. Hofbauer. Measure dynamics on a one-dimensional continuous trait space: theoretical foundations for adaptive dynamics. *Theoretical Population Biology*, 67:47–59, 2005.

R. Cressman, J. Hofbauer, and F. Riedel. Stability of the replicator equation for a single species with a multi-dimensional continuous trait space. *Theoretical Biology*, 239: 273–288, 2006.

H. Gintis. *Game Theory Evolving: A Problem-Centered Introduction to Modeling Strategic Interaction*. Princeton University Press, Princeton, NJ, 2001.

J. Hofbauer and K. Sigmund. *Evolutionary Games and Population Dynamics*. Cambridge University Press, Cambridge, UK, 1998.

J. Maynard-Smith and J. Price. The logic of animal conflict. *Nature*, 146:15–18, 1973.

J. Oechsler and F. Riedel. Evolutionary dynamics on infinite strategy spaces. *Economic Theory*, 17(1):141–162, 2001.

L. Panait, K. Tuyls, and S. Luke. Theoretical advantages of lenient learners: An evolutionary game theoretic perspective. *Journal of Machine Learning Research (JMLR)*, 9:423–457, 2008.

F. V. Redondo. *Economics and the Theory of Games*. Cambridge University Press, Cambridge, UK, 2003.

M. Ruijgrok and T. W. Ruijgrok. Replicator dynamics with mutations for games with a continuous strategy space. In *eprint arXiv:nlin/0505032*, 2005.

T. Schneider. Evolution of biological information. *journal of NAR*, 28:2794–2799, 2000.

D. Stauffer. Life, love and death: Models of biological reproduction and aging. *Institute for Theoretical physics, Köln, Euroland*, 1999.

R. Sutton and A. Barto. *Reinforcement Learning: An Introduction*. Cambridge, MA: MIT Press, 1998.

P. 't Hoen and K. Tuyls. Engineering Multi-Agent Reinforcement Learning using Evolutionary Dynamics. In *Proceedings of the 15th European Conference on Machine Learning*, volume LNAI 3201, pages 168–179, 2004.

P. Taylor and L. Jonker. Evolutionary stable strategies and game dynamics. *Math. Biosci.*, 40:145–156, 1978.

K. Tuyls and A. Nowe. Evolutionary game theory and multi-agent reinforcement learning. *The Knowledge Engineering Review*, 20:63–90, 2005.

K. Tuyls, K. Verbeeck, and T. Lenaerts. A Selection-Mutation model for Q-learning in Multi-Agent Systems. In *The second International Joint Conference on Autonomous Agents and Multi-Agent Systems. ACM Press, Melbourne, Australia*, pages 693–700, 2003.

K. Tuyls, P. 't Hoen, and B. Vanschoenwinkel. An evolutionary dynamical analysis of multi-agent learning in iterated games. *The Journal of Autonomous Agents and Multi-Agent Systems*, 12:115–153, 2006.

N. van Kampen. *Stochastic Processes in Physics and Chemistry*. Elsevier Science Publishers, Amsterdam, 1992.

J. von Neumann and O. Morgenstern. *Theory of Games and Economic Behavior*. Princeton University Press, Princeton, NJ, 1944.

P. Vrancx, K. Tuyls, R. Westra, and A. Nowe. Switching dynamics of multi-agent learning. In *Proceedings of the seventh joint conference on autonomous agents and multi-agent systems, Estoril, Portugal*, pages 307–314, 2008.

J. W. Weibull. *Evolutionary Game Theory*. MIT Press, Cambridge, MA, 1996.

E. Zeeman. Dynamics of the evolution of animal conflicts. *preprint*, 1979.

E. Zeeman. Dynamics of the evolution of animal conflicts. *Journal of Theoretical Biology*, 89:249–270, 1981.

8
Stigmergic Cues and Their Uses in Coordination: An Evolutionary Approach

Luca Tummolini
Institute of Cognitive Sciences and Technologies - CNR

Marco Mirolli
Institute of Cognitive Sciences and Technologies - CNR

Cristiano Castelfranchi
Institute of Cognitive Sciences and Technologies - CNR

8.1	Introduction...................................	243
	Overview	
8.2	Stigmergy: Widening the Notion but Not Too Much..	245
	A Tale of a Wrong Story • Is It Communication? • Stigmergic Cues as Practical Behavioral Traces	
8.3	Two Uses of Stigmergy in Coordination..........	248
	Coordination and Cues of Interference • Communication and Signals • Behavioral Communication and Implicit Signals • Stigmergic Self-Adjustment and Stigmergic Communication	
8.4	Stigmergy in Cooperation and Competition.......	251
	Cooperative and Competitive Interference • Collaborative and Conflictual Stigmergic Coordination	
8.5	Why Pheromonal Communication Is Not Stigmergic....................................	252
	Pheromones Are Not Stigmergic Cues • The Role of Stigmergy in the Evolution of Pheromonal Communication	
8.6	Understanding Stigmergy through Evolution.....	253
	The Basic Model • Evolution of Practical Behavior • Evolution of Stigmergic Self-Adjustment and Indirect Coordination • Evolution of Stigmergic Communication	
8.7	Future Work.....................................	260
8.8	Conclusion......................................	261
8.9	Acknowledgments.................................	262
References...		262

8.1 Introduction

Since its origins, the field of Multi-Agent Systems (MAS) has been entrenched with the notion of *direct communication* [Hewitt, 1977; Smith, 1980; Bond and Gasser, 1988; Durfee, 1988].

In particular, the shift from blackboard systems [Engelmore and Morgan, 1988], as proto-MAS, to proper MAS was characterized by abandoning a form of indirect coordination in favor of a more direct strategy. The 'agents' in a blackboard system used to leave their partial solutions to a problem in a data structure shared with other 'agents' and each reacted to the product of the work of other agents by producing new work (i.e. new partial solutions). At the time, a single thread of control in writing in the blackboard created a bottleneck in problem solving. This constraint was overcome, at the beginning, by adopting direct communication between multiple blackboard systems (via message passing) and then, eventually, by dropping the indirect approach altogether. Though a new seminal approach was born, the baby, or at least a useful toy of his, was thrown out with the bath water.

From there on, in fact, the crucial coordination mechanism for Multi-Agent systems has been modeled on the most important communication tool available to humans: verbal language. With the help of a shared conventional language, humans in fact coordinate their activities thanks to verbal exchanges in the forms of speech acts [Searle, 1969]. Inspired by their crucial role in human societies, speech acts have been studied and approximated also for the coordination of artificial agents in distributed systems [Cohen and Levesque, 1990; Sadek, 1992]. Direct communication soon began to be considered as necessary to achieve an effective MAS: in order to build autonomous agents capable of coordinating with each other, such agents must be provided with an Agent Communication Language to exchange messages (e.g., FIPA-ACL *) on the background of shared ontologies.

However, direct communication by means of an agent language is just one possible mechanism to design an effective MAS that, together with several benefits, bears also many computational costs. To overcome these limitations many practitioners are nowadays turning their attention to approaches based on self-organizing techniques. In fact, taking inspiration from the biological sciences, agents have been molded on social insects and their capacity to interact "indirectly" through environmental modifications (a phenomenon known as *stigmergy*, see [Grassé, 1959]). The importance of this mechanism has been investigated both in biological [Wenzel, 1991; Karsai and Theraulaz, 1995; Camazine et al., 2001] and artificial organisms [Parunak, 1997; Bonabeau et al., 1998; Parunak et al., 2005; Mamei and Zambonelli, 2007]. Though self-organization and stigmergy are orthogonal phenomena [Theraulaz and Bonabeau, 1999], the narrow and unclear definition of what stigmergy is and its too close association with self-organizing approaches has also impeded the detachment of this coordination mechanism from the original domains of investigation (i.e. collaborative collective behaviors), and so has limited the impact of this approach for a wider range of problems that can be relevant for the MAS community.

Somehow ironically, indirect coordination is quite reminiscent of the blackboard system approach that was stigmergic in its very essence. By bringing the best of these two worlds together, the rehabilitation of indirect forms of coordination might be a critical step toward the next generation of agent technologies.

8.1.1 Overview

The general aim of this chapter is to clarify the notion of stigmergy and its uses for coordination by means of conceptual and operational models obtained with a set of evolutionary agent-based simulations.

In a series of recent papers we have argued that stigmergy is best understood as a peculiar form of *communication* [Castelfranchi, 2006b; Tummolini et al., 2005; Tummolini

---

*See the Communicative Act Library Specification of the Foundation for Intelligent Physical Agents (FIPA): http:// www.fipa.org

and Castelfranchi, 2007]. In particular, stigmergic behavior has been analyzed as a kind of *indirect communicative behavior*, i.e. asynchronous communication enabled by a shared environment used as a repository for signals. Here we partially correct our previous view by distinguishing the notion of *stigmergic cue* from its uses in coordination: *stigmergic self-adjustment* and *stigmergic communication*.

In the first part of the chapter, we review current definitions adopted by leading scholars in different fields. Though the core properties of stigmergic behavior have been acknowledged, a clear-cut conception is still missing. Occasionally the phenomenon is classified as a kind of coordination, other times as communication. To better clarify what stigmergy is as well as when it is communication and when it is not, we introduce the general theory of behavioral implicit communication and we clarify what is peculiar of the stigmergic case.

In the second part of the chapter, we explore the evolution of stigmergic behavior adopting a simulative approach. In particular, we have simulated a population of artificial agents living in a virtual environment containing safe and poisonous items ('fruits'): eating safe fruits increases the fitness of an individual, while eating poisonous ones decreases it. The behavior of the agents is governed by artificial neural networks whose free parameters (i.e. the weights of the networks' connections) are encoded in the genome of the agents and evolve through a genetic algorithm. Agents interact with their environment and between each other through the traces that their behaviors leave in such an environment.

Biological plausibility aside, the simulations are designed to provide an operational model of stigmergic cues together with a principled way to understand their possible uses. By making explicit the transition from (1) a multi-agent system in which agents individually look for their resources to one in which (2) each agent indirectly coordinates with what the other agents do (stigmergic self-adjustment) and, finally, (3) to a situation in which each agent acts also to send a message about what kind of resources are available in a risky environment (stigmergic communication), the simulations offer a precise analysis of the difference between traces that are just signs with a behavioral content and traces that are signals with a behavioral message.

8.2 Stigmergy: Widening the Notion but Not Too Much

8.2.1 A Tale of a Wrong Story

The term *stigmergy* has been coined by the French entomologist Pierre-Paul Grassé to explain the behavior of termites during nest-construction [Grassé, 1959]. The term comes from the combination of two Greek words: "stigma", meaning a puncture or sting (by extension, whatever provokes a reaction), and "ergon", meaning work. Stigmergy, then, refers to the role of work in stimulating a reaction of other agents or, with Grassé: "the stimulation of the workers by the very performances they have achieved" [Grassé, 1959]. Stigmergy is considered by Grassé as *a peculiar kind of stimulation*: the consequence of previously accomplished work. It is such consequence that drives and orients additional construction: "the significant stimuli imperatively direct workers' responses" [Grassé, 1967].

The concept was introduced as the key factor to explain how complex structures such as termites' nests [Grassé, 1959] (or a complex problem-solving like rebuilding a tunnel-gallery after a sudden damage [Grassé, 1967]) can be regulated without, clearly, any complete plan being known by the termites. The complexity of a nest is astonishing and requires the collaborative coordination of a huge number of workers. Exploiting the stimulating factor of accomplished work was considered as the key to success.

Though many were able to replicate the original findings of Grassé, soon the importance of stigmergy was also somehow challenged (see [Downing and Jeanne, 1988] for a review of these critiques). Several researchers remained skeptical about the fact the stigmergy could be sufficient for the coordination of complex building tasks, especially when the construction behavior was not strictly sequential (i.e., composed of an ordered succession of phases, one causally connected with the other in a sequential fashion). Other cues were also suggested to be necessary in order to shift between different activities or to stop the whole endeavor once completed.

Curiously enough, in the very same year in which Grassé published his considerations on termites' construction behavior, another great entomologist, Edward O. Wilson, had discovered that ants, by leaving pheromones in their environment, create odor trails that attract other ants [Wilson, 1959].

It is, however, due to the work of Bruinsma [Bruinsma, 1979] that the role of pheromones also in coordinating termites' nest construction was finally unraveled. Bruinsma discovered that both pheromone trails and the impregnation of soil pellets with pheromone are critical factors in attracting other termites and orienting their activities. The existence of an initial deposit of soil pellets *impregnated with pheromone* stimulates workers to accumulate more material. This reaction originates a positive feedback mechanism because the accumulation of material reinforces the attractivity of deposits through the pheromones emitted by those materials.

The emergence of pillars, walls and royal chambers in termite nests can be accounted for once the *self-organizing properties* of this kind of interactions are understood. In particular it has been suggested that termites' behavior follows just a single pattern (i.e. picking up and depositing a soil pellet, if pheromone is present) [Bonabeau et al., 1997] and this building behavior is modulated by several environmental conditions (e.g., when an air stream drives molecules of pheromones in a given direction the pattern is influenced), not last the fact that the environment is changed by work of other termites. It is self-organization that explains the emergence of a structure at the global level from interactions among its low level components, without any explicit coding of this process in the individual agents [Garnier et al., 2007].

The basic intuition of Grassé was in the end right: no sophisticated cognitive processing is performed by the termites in order to construct their nest. However it is the self-organizing dynamics enabled by the properties of pheromone (i.e. its additivity and its decadence rate) that regulate and modulate termites' behaviors.

Unfortunately, it seems that stigmergy, defined by Grassé himself as a particular kind of stimulation, does not play a distinct role in this process: *it is not the nest under construction that functions as significant stimulus for other termites*. Environmental modifications due to the construction activity do play a crucial role in modulating termites' behavioral patterns but not *as* stimuli for the termites.

If this fact is too embarrassing, one can, as Theraulaz and Bonabeau has done [Theraulaz and Bonabeau, 1999], *change* the original conception of Grassé to include also the influence of pheromone when left in the environment as a kind of stigmergy.

Indeed, as it has been explicitly recognized by Holland and Melhuish [Holland and Melhuish, 1999], the modern practice is to extend the definition of stigmergy by replacing the sense of "work" crucial for Grassé (stigmergy as the product of work functioning as a significant stimulus) to that of "any environmental change produced by the animal".

Though for us this apparently innocent extension is to be resisted (see Sections 8.2.3 and 8.5.2 below), it has been a fortunate one. So much that, today, the fact that there exist two main varieties of stigmergy is commonly accepted: the first being characterized by leaving special markers in the environment (i.e. pheromones), the second being the exploitation of the results of the work so far achieved by the agents [Theraulaz and Bonabeau, 1999; Dorigo et al., 2000; Parunak, 2006].

8.2.2 Is It Communication?

Though Grassé never mentioned it, stigmergy is also usually considered as a form of communication.

Wilson has been the first to explicitly link stigmergy to a communicative process and coined the name *sematectonic communication* (from the Greek "sema" for sign or token and "tekton" for craftsman, builder) [Wilson, 1975]. Reviewing different communication strategies in social animals, Wilson naturally conceived stigmergy as a particular kind of communication exploiting the "structures built by animals".

More recently, many researchers have been explicit in considering stigmergy as a peculiar form of indirect communication, and more precisely: "communication by altering the state of the environment in a way that will affect the behaviors of others for whom the environment is a stimulus" [Kennedy et al., 2001] (see also [Dorigo et al., 2000; Mataric, 1995; Tummolini and Castelfranchi, 2007]).

Though clearly, there are cases that can be defined as *stigmergic communication*, stigmergy by itself is not necessarily so (see Section 8.3.4 and the simulation results described in 8.6.4). The main reason why stigmergy is so commonly seen as a part of a communicative process lies in the fact that stigmergy has been always approached in the limited context of collaborative activities (i.e. nest construction, collective sorting, collaborative foraging, division of labor, clustering etc.), and collaborative tasks such these immediately create the conditions for communicative behaviors in which an agent aims (functionally or intentionally) to influence another one.

Though stigmergic communication is useful as a coordination mechanism in cooperative behavior, there is room for a different use which we name *stigmergic self-adjustment*. In Section 8.3.4 by providing a principled distinction between self-adjustment and communication as two ways to coordinate one's behavior, we will fill that room.

8.2.3 Stigmergic Cues as Practical Behavioral Traces

Apart from his classification of stigmergy as a kind of communication, Wilson also noticed that the role of stigmergy in eliciting additional work was clearly a special case and offered the example of male crabs that build structures to attract females for reproductive purposes. He felt that Grassé's characterization was too restricted, and wanted stigmergy "to denote the evocation of any form of behavior or physiological change by the evidences of work performed by other animals, including the special case of the guidance of additional work" [Wilson, 1975].

Though this is certainly sensible, the use of work in a strict sense is dispensable too. Why do we want to restrict the phenomenon to the construction of something? Why excluding a simple footprint?

On the other hand, Holland and Melhuish [Holland and Melhuish, 1999] suggestion ("any environmental change produced by the animal") is too broad, because it would overgeneralize the original intuition of the French entomologist. Stigmergy would simply refer to any kind of indirect influence, and any road sign would qualify as a case of stigmergy.

How can we reasonably generalize the notion of stigmergy?

Here is our proposal. The basic intuition of Grassé was that the traces of "work" left in the environment might become significant stimuli by themselves for an agent. A *trace* is the effect of any behavior persisting in the environment. But what behaviors are the relevant ones?

At a first approximation, we can say that relevant behaviors are those whose goal or function is a practical one: i.e. any behavior that is not adapted just to influence another agent counts as a relevant practical one. Hence, from this perspective, *stigmergy occurs whenever another agent's behavioral trace (a persistent effect of a practical behavior whatsoever) is used as guide for one's own future behavior.*

In other words, there is a case of stigmergy whenever a behavioral trace of an agent is also a *cue* for another organism. Cue is a notion used by biologists precisely to refer to any (animate or inanimate) feature in the environment, that an agent is adapted (thanks to evolution or learning) to use as a guide for future actions [Hasson, 1994]. Though many stimuli are in principle detectable by the sensory organs of an animal, only a subset of them are actually cues (i.e. are guides for action).

Very close to the spirit of Grassé's original intuition, in our approach, *stigmergy refers to a subset of cues and, in particular, to the practical behavioral 'traces' we leave around whenever we act in a physical world.*

The heuristic value of this precise definition, of course, is to be judged against its pragmatic consequences, that is, by the way it helps to identify and understand phenomena that would be, otherwise, obscured. The next two sections are thus devoted to explore such consequences.

8.3 Two Uses of Stigmergy in Coordination

The most dramatic fact of living in the same world is that the actions of an agent can interfere with the success of those undertaken by others [Castelfranchi, 1998]. Sociality is first of all due to this very basic condition. In fact, beyond one's own individual qualities, there is a social component behind the evolutionary and practical success of an organism. And this social component is mainly due to *interference*, and the opportunities and obstacles it creates. How do agents deal with interference? Is communication necessary to solve interference problems?

In order to set the stage for understanding two distinct uses of stigmergic cues, in this section we clarify the relationship between coordination and communication.

8.3.1 Coordination and Cues of Interference

At the simplest level *coordination* is a form of individual social action [Castelfranchi, 1997]. Coordination is coping with positive and negative interference by means of adaptation (thanks to evolution, learning, reasoning or design). In particular, to coordinate in a social context, an individual agent takes into account the interference created by another agent in one's own action, or better, *coordination is adaptation to another agent's behavior in order to increase one's success**.

While in negative interference the behavior is adapted to an obstacle, in positive interference the behavior is adapted to an opportunity. In cases of reciprocal coordination, the agents' behaviors are adapted to each others.

---

*In this chapter, we are interested only in "social" coordination between the behaviors of different agents. Non-social kinds of coordination such as the coordination of distinct behaviors of a single agent are therefore excluded.

More precisely, the agents need to detect some cue in the environment which predicts such interfering conditions. As stated above, a cue is whatever one is adapted to use as a guide for future actions [Hasson, 1994]. On this basis coordination can be more clearly defined as *adaptation by reading cues of interference*.

There are, however, two fundamental ways to adapt by means of these cues of interference: (1) the agent changes his own behavior to *avoid* the obstacle or to *exploit* the opportunity (self- adjustment) or (2) the agent changes his behavior in order to influence the behavior of the interfering agent either by *impeding* or *inducing* such behavior (hetero-adjustment or influence). See Table 8.1.

TABLE 8.1 Two strategies for solving interference problems

	Coordination by self-adjustment	Coordination by hetero-adjustement or influence
Negative Interference	Avoidance	Impediment
Positive Interference	Exploitation	Inducement

Finally, though coordination is based on such cues of interference, this does not mean that coordination can only be *direct*: via the detection of those cues that are the interfering behaviors themselves. *Indirect coordination* is possible *whenever acting on the basis of some cue other than the behavior of some agent, one is also adapting one's own behavior to those of others*. In traffic, for example, each of us individually act on the visual cues provided by the traffic lights, while at the same time, we also coordinate indirectly with all the others whose behavior we do not directly perceive.

8.3.2 Communication and Signals

We have seen that an agent can solve an interference problem in a more active way than by mere self-adjustment. The agent, in fact, can actively eliminate an obstacle or create an opportunity. In other words, one can *influence* the others to one's own benefit. Coercion is an obvious way to do so. Another is communication.

The most general approach to communication can be found in biology, where it is usually approached as the transfer of a *signal* from an agent to another one. When communication is defined in this way, of course, all the explanatory burden is on a clear notion of signal, which two leading scholars, Maynard-Smith and Harper, define as: "any act or structure which alters the behavior of other organisms, which evolved because of that effect, and which is effective because the receiver's response has also evolved" [Maynard-Smith and Harper, 2003].

This way of understanding communication as signaling has the immediate advantage of making explicit that, at its core, communication is a form of influence, hence a peculiar kind of coordination. Communication, however, is distinct from coercion because the "the receiver's response has also evolved". Being the product of evolution, learning, reasoning or design, receiver's response is in her own interests, while in coercive coordination this is not the case. If a stag pushes another one backwards, this behavior is not a signal. If the stag roars and the other animal retreats it is a signal because the retreating is evolved (or learned, decided or designed).

Moreover, the fact that the action or structure is also evolved *because* of its influencing effect makes clear that the action or structure is not only a cue. While a cue is defined only relatively to the agent using it, a signal is defined relative to its influencing effect on the behavior of others which benefit, first of all, the agent who produces or manifests it.

Given this definition of communication, it is also simple to distinguish its *direct* and *indirect* varieties. In direct communication, the signaler's behavior or structure itself is the signal which is interpreted by another agent. Differently, indirect communication exploits signals that persist in the environment and that are effective in their influencing function notwithstanding the absence of the signaler itself.

8.3.3 Behavioral Communication and Implicit Signals

The above definition of communication, however, also suggests that signals are essentially products of cooperation in which both parties benefit from their exchange. Though a reliable communicative system is one that is evolutionary stable, an unstable but deceitful system, in which only the signaler benefits from the exchange, is still a communicative system after-all.

What is sufficient for a signal, for us, is just that it is defined relatively to the signaler's benefit and that it is at least a cue for some other agent who has evolved to respond to it as a cue but not necessarily as a signal: *a signal is any behavior or structure that alters the behavior of other organisms, which evolved because of that effect, and which is effective because that behavior or structure is a cue for the receiver*. The evolutionary advantages of the signaler and the receiver might diverge.

This is the kind of signal that we call *implicit*, and that characterizes a peculiar kind of communication: *behavioral implicit communication* [Castelfranchi, 2000, 2006a].

Usual practical behaviors in fact are meaningful and can become cues for other agents to be used as guides for future actions. Interference is the reason why agents living in a shared world will become attuned to such cues, and will become behavior readers to improve their coordination abilities. However, once agents are behavior readers, those, whose behavior is read, can start exploiting this fact to their own advantage. Though their behaviors maintain their practical aim (i.e., in traffic, a U-turn behavior), by being read and understood by other agents observing it, the practical behavior can be done also for influencing other agents by the very production of this behavioral cue (i.e. an agent can wait to see if other drivers are noticing the U-turn behavior before doing it because he intends to use this behavior also as a signal with the content: "I'm doing a U-turn" and so provoking their slowing down). As this example clarifies, beyond being a sub-case of coordination, a specific use of communication can also be a coordination mechanism. Since coordination is in general any adaptation by means of cues of interference, communication can also be an instrument for coordination by self-adjustment. In fact, when communication is focused on the production of cues of interference (e.g., by signaling an action one is going to perform), then communication is clearly one of the most powerful coordination mechanisms that we, and our agents, can take advantage of.

8.3.4 Stigmergic Self-Adjustment and Stigmergic Communication

It is now possible to specify two distinct uses of stigmergy as analyzed in Section 8.2.3: stigmergic self-adjustment and stigmergic communication.

Stigmergic self-adjustment amounts to the process of coordination merely by means of stigmergic cues. That is, *adaptation by the detection of practical behavioral traces*, i.e. the effects of practical behaviors registered in the environment. Because the cue is the effect of a behavior and not the behavior itself, stigmergic self-adjustment is a kind of indirect coordination.

Differently, *stigmergic communication* is *influence by an evolved (or learned, reasoned, designed) production of these stigmergic cues*. Given the notion of behavioral implicit communication, it is also clear that stigmergic communication is *implicit* in the sense that the stigmergic signals are effective because the effects of practical actions are cues for other agents (i.e. traces left and accessible in a shared environment that function as implicit signals). In stigmergic communication, the messages that are transmitted are *behavioral* ones, i.e. the implicit signals primarily refer to the agents' behaviors or what a behavior can be a diagnostic or prognostic sign of (for an analysis of the kinds of meanings that can be transmitted with stigmergic communication see [Tummolini and Castelfranchi, 2007]). Finally, because the signal is an effect of a behavior and not the behavior itself, stigmergic communication is a kind of indirect communication.

As for the more general case, stigmergic communication is also a powerful coordination mechanism for at least two reasons: firstly, because it is indirect, and, secondly, because it can exploit practical behavioral traces as cue of interference. While the former property is almost universally acknowledged, the latter has been too quickly forgotten. One of the main aims of this chapter is precisely to rectify such mistake.

8.4 Stigmergy in Cooperation and Competition

Stigmergy has been discovered in the particular context of collaborative nest construction. However, the focus on the role of stigmergy in collaboration has turned to be a bias originated from this initial domain of investigation. Limiting the attention to collaborative scenarios have impeded to notice, for example, that there are natural uses of stigmergic cues also in competitive and conflict situations.

8.4.1 Cooperative and Competitive Interference

In particular, when the goals or the fitness of the agents are such that each benefits from others' success there is room for *cooperation*. More precisely, when interference is positive this entails that the behavior of an agent facilitates the success of the others such that acting for a common goal (i.e. cooperate) can be mutually advantageous. On the contrary, very often it is the case that an increase in fitness or the fulfillment of one's goals compromises the success of others (negative interference). In this case, the agents acting in ways that are reciprocally destructive are, even if unaware, in *competition*.

Of these very basic kinds of interference, there also more specific instances. Cooperation, for example, can be obtained thanks to collaboration, which means that not only the agents functionally or intentionally favor each other, but also that they adopt the goal of each other. *Goal adoption* is the process of internalizing the goal of another agent as one's own and acting *for* another agent [Conte and Castelfranchi, 1995]. Hence, *collaboration* is a specific way to achieve cooperation: it is cooperation via mutual goal adoption.

Differently, competition can originate conflict which is, again, a specific way in which agents might compete. In *conflict* in fact, agents are motivated by the fact that the others should not fulfill their goals. As in the case of collaboration, also in conflict the agents take into consideration the goals of the others but in order to frustrate them.

In-between pure cases of cooperation and competition, there are of course many situations that are partly cooperative and partly competitive but a simpler picture is enough for the aims of this chapter.

8.4.2 Collaborative and Conflictual Stigmergic Coordination

Notwithstanding its relevance for cooperation and collaboration, it is evident that there is plenty of competitive and conflict situations in which stigmergy is used.

Collaborative activities, where the agents are ready to help each other to reach a common goal, are contexts in which stigmergy is obviously natural. Take for example the cooperative task in which two agents collaborate in answering customers' orders in an Italian bar. It is common, during breakfast for example, that many people ask at the same time for very different things: a coffee, a cappuccino and a sandwich for instance. How can the different orders be correctly processed? Who has ordered what? Which is the guy who ordered the cappuccino? A very diffuse strategy to solve this problem is to employ stigmergic collaborative communication. While, for instance, a waiter executes the orders and starts preparing the coffee, the cappuccino and picking the sandwich, the other adaptively changes the environment. Given that the shapes of the dishes where the three items are delivered to the customers are distinctively different, one waiter immediately reacts to the orders by associating each customer with her appropriate dish. In this way, when the items are ready, the other waiter can easily associate each customer with the correct order. While no verbal communication is usually employed, the collaborative task is smoothly solved relying only on indirect communication through stigmergic cues.

Differently, consider the role of fences. Clearly such constructions are there to function as physical obstacles for strangers. Fences are built to avoid that strangers enter in a protected area, hence they are conflictual structures. Though they provide physical interference if somebody intends to trespass them, more than often they are not unavoidable obstacles. How can a fence be effective? The answer turns to be simple: by being also a stigmergic signal. In fact, as a practical behavioral trace left there by somebody, every fence has also a behavioral content related to the conflictual action of erecting it. The combination of its practical function and its being a stigmergic cue, for those that understand it, is often enough to achieve its intended purpose.

8.5 Why Pheromonal Communication Is Not Stigmergic

At the end of Section 8.2.1, we have suggested that the common practice of acknowledging the exploitation of special markers left in the environment is a misleading extension of the notion of stigmergy. Pheromones are probably the most diffuse example of stigmergy between the practitioners in fields like Artificial Life, Swarm Intelligence, Collective Robotics and, more recently, Multi-Agent Systems. Being so widely accepted as the prototypical case, our insistence on their being not stigmergic will arise much skepticism.

In this section, we intend to clarify this issue and to explore the relevance of stigmergy in the evolution of pheromonal communication.

8.5.1 Pheromones Are Not Stigmergic Cues

The use of pheromones is not a case of stigmergy because these chemical molecules are not practical behavioral cues. A pheromone in fact is not only a cue. It is a chemical molecule whose sole function is to excite another agent (again from the Greek, "pherein", to carry or transfer and "hormon", to excite or stimulate), that has evolved for this influencing effect, and which is effective because the response has also evolved [Karlson and Luscher, 1959; Wyatt, 2003]. Hence a pheromone is, by definition, an *explicit signal*.

What pheromones and stigmergic cues have in common is that they are traces, i.e., products of behavior that persist in the environment and this property can have useful consequences, not least by enabling self-organization or self-assembly. But widening the notion to include all kinds of indirect communication is, on the other hand, losing sight of what is peculiar of the stigmergic case.

Consider a familiar example. When driving in traffic, it is one thing to stop because there is a red light in front of you, it is another one to stop because the car in front of you has stopped. While the red light is an explicit signal left in the environment by somebody, the car in front of you is just a cue of a practical behavioral kind, a stigmergic kind.

Beside restoring Grassé's first conceptualization (see Section 8.2.3), our more restrictive notion avoids collapsing stigmergic communication into any kind of indirect communication.

8.5.2 The Role of Stigmergy in the Evolution of Pheromonal Communication

Understanding stigmergy from this perspective is also a contribution for a better explanation of explicit communicative systems themselves.

How in fact can an explicit communicative system evolve if not from an implicit one? As it has been suggested in the biological study of animal communication, a signaling system can evolve from a process of "ritualization" of a behavior that did not evolve for communication at the beginning [Tinbergen, 1952] but was already used as a cue by other individuals to gain information. And the same is true for pheromonal communication.

As stated in Section 8.5, pheromones are not stigmergic cues but chemical ones emitted only for communicative purposes. Pheromones are explicit signals which, most probably, *derived from stigmergic cues.* In [Sorensen and Stacey, 1999], it is, for instance, offered a model of the evolution of pheromonal communication between goldfishes for reproductive purposes. In this model it is speculated that, at the beginning, hormonal products were released by females to water just as a side effect of ovulation. Given the relevance of these hormonal traces as predictors of the biological state of a female, the male soon became attuned to these traces as reproductive cues. Finally, by exploiting this fact, evolution has driven toward a specialization in the production of these hormones just for influencing purposes, and turning the hormones into real pheromones. Hormonal cues are in our approach stigmergic because they are traces of a process other than communication (i.e., ovulation). When specialized in pheromones, they become explicit signals.

Though the evolutionary transition from implicit to explicit signals is a matter for further research, keeping stigmergic communication clearly apart from other kinds of indirect communication is of extreme importance if we want to understand explicit communication system themselves (i.e., their evolution).

8.6 Understanding Stigmergy through Evolution

In order to clarify even further the notions so far discussed, in this section we describe a set of agent-based simulations intended as operational models of stigmergic cues and their uses in coordination. In particular, we simulate a population of artificial agents living in a virtual environment containing safe and poisonous items ('fruits'): eating safe fruits increases the energy (fitness) of an individual, while eating poisonous ones decreases it. If nothing in the physical properties of these items can be used for discriminating the good resources from the bad ones, how can the agents be successful in such a risky environment?

Knowledge of this properties is the prototypical kind of knowledge that is acquired by direct experience with the external environment. However when the risks associated with experience are high, an evolutionary pressure exists to learn indirectly exploiting the experience made by others. Omnivores, for example [Visalberghi and Addessi, 2000], are characterized by food neophobia, i.e., a form of hesitancy to eat novel food. Neophobia is considered as an efficient behavioral strategy to avoid the risk of ingesting poisonous substances. It is also known, however, that social context affects the acceptance of novel food by moderating such instinctive hesitancy.

In this Section, we explore a possible strategy to solve this problem by showing that, given specific circumstances, the individual capacity to categorize some natural kinds as 'safe' or 'poisonous' might rely on stigmergic cues left by other agents that, moreover, act in order to facilitate such individual categorization.

In our simulations, the behavior of the agents is governed by artificial neural networks whose free parameters (i.e., the weights of the networks' connections) are encoded in the genome of the agents and evolve through a genetic algorithm. Agents interact with their environment and between each other through the traces that their behaviors leave in such an environment. By way of such an agent-based model, we explore the evolution of three kinds of behaviors:

1. the basic practical behavior;
2. the use of practical behavioral traces as cues of positive interference (stigmergic self-adjustment);
3. the use of such stimergic cues as implicit signals (stigmergic communication).

Though the simulations are offered only as operational definitions of our domain of investigation (i.e., stigmergy) and not as a plausible biological explanation of this form of social facilitation, they may also suggest that a communicative process might be discovered, whereas today biologists seem to overlook this possibility.

8.6.1 The Basic Model

We simulate a population of 100 agents living in an environment composed of 5 'islands' (Figure 8.1). Each island contains 2 'trees' with 20 'fruits' each. One of the trees produces edible fruits, while the fruits produced by the other tree are poisonous. Eating an edible fruit increases an agent's fitness (i.e., its probability of reproduction) of 1 unit, while eating a poisonous fruit decreases an agent's fitness of 2 units. Poisonous and edible fruits (and trees) are perceptually identical, so agents cannot rely on direct perceptual cues to decide from which tree to feed. Each agent visits all the 5 islands where it can stay for at most 4 time steps. In each time step an agent can decide whether to eat a fruit from one of the two trees or to leave the island.

Agent's behavior is governed by an artificial neural network with 4 groups of input units sending connections to a single group of 5 hidden units which in turn connects to a group of three action units (Figure 8.2). The three possible actions are: (a) eat a fruit from the first tree, (b) eat a fruit from the second tree, or (c) leave the island.

Actions are decided according to a winner-takes-all mechanism: in each time-step the agent performs the action corresponding to the action unit with the maximal activation. When the agent decides to eat a fruit from a given tree, a random fruit disappears from that tree and the fitness of the agent is updated according to the quality of the eaten fruit (+1 for edible fruits, −2 for poisonous ones). The first two groups of input units are composed by 20 units each, with each unit representing the presence/absence of a particular fruit in one of the two trees of an island.

Activation is binary: if the fruit is present, the corresponding unit is activated (1), otherwise it is not (0). The third group of input units is composed by 2 units representing the

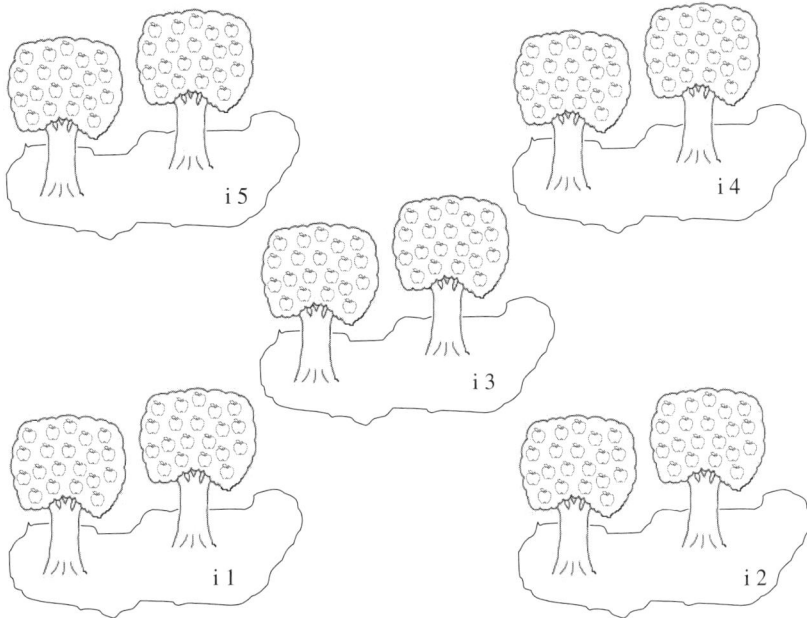

FIGURE 8.1 The environment. The five islands (i 1-5) with two trees each (one with edible and one with poisonous fruits. At the beginning of a trial each tree has 20 fruits). See the main text for details.

feedback from the agent's 'body' regarding the quality of the last eaten fruit: during the first time step in an island, no fruit has been eaten hence both units are silent (0); during the successive time steps, one of the two units is highly activated (10) depending on whether the last eaten fruit was edible or poisonous. The last group of 3 input units constitutes a memory of the last performed action: each unit corresponds to one action and it is highly activated (10) if that action has been the last to be performed and silent otherwise (0). (During the first time step in an island, no action has been performed; hence, all the three units are silent.)

We simulate a population of 100 individuals evolving through a genetic algorithm. Each individual lives for 5 trials, where a trial consists in visiting all the five island (eaten fruits re-grow between one trial and the other). After all individuals have lived their lives, their fitness is calculated according to the following formula:

$$f(x) = \frac{\sum_{t=1}^{T} \sum_{i=1}^{I} en_e - pn_p}{TIC} \qquad (8.1)$$

where n_e and n_p are, respectively, the number of edible and poisonous fruits eaten by x, e and p are two constants (set to 1 and 2, respectively), and T, I, and C are the number of trials (5), islands (5), and maximum cycles spent in an island (4), respectively.

The genome of individuals contains all the connection weights of the neural network, with each free parameter being encoded as an 8-bits string, whose value is then uniformly projected in the range $[-5.0, +5.0]$. The 20 best individuals of each generations are selected for reproduction and generate 5 offsprings each, which inherit their parent's genome with a 0.04 probability for each bit of being replaced with a new randomly selected value. The evolutionary process lasts 250 generations.

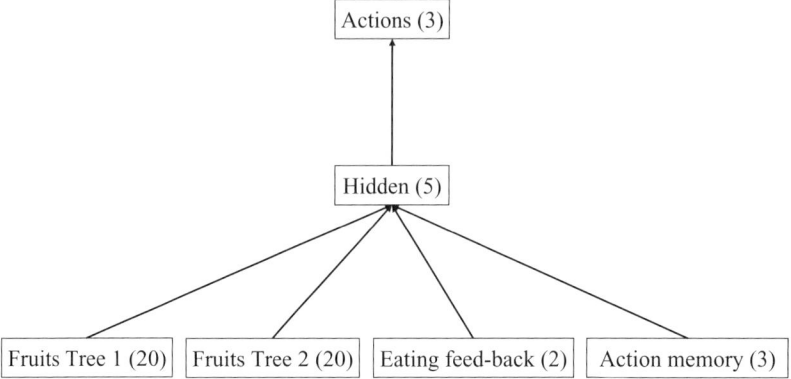

FIGURE 8.2 The neural network. Each block corresponds to a group of neurons. Numbers in parentheses correspond to the numbers of units of the group. Arrows represent all-to-all connections between groups.

8.6.2 Evolution of Practical Behavior

In this kind of simulation, which we will call "base-line," individuals never interact with each other, so they cannot evolve any form of social behavior. Since at their arrival on an island the two trees are perceptually identical and contain the same number of fruits, agents cannot but choose randomly from which tree to feed.

On the other hand, once they have eaten a fruit from a tree, agents can decide whether to keep on eating from the same tree or to change tree depending on the bodily feedback about the quality of the eaten fruit: if the fruit was poisonous agents have to change tree, while if the fruit was edible agents have to keep on eating from the same tree.

Since there is no learning during individuals' life, this kind of behavioral strategy must evolve phylogenetically. This is in fact what happens in our simulations. Figure 8.3 shows the average results of 10 replications (with different random initial conditions) of this base-line simulation. Figure 8.3 (left) shows average fitness and the fitness of the best individual during the 250 generations, while Figure 8.3 (right) shows average number of edible and poisonous fruits eaten by the best individual of the last generation during each staying on an island.

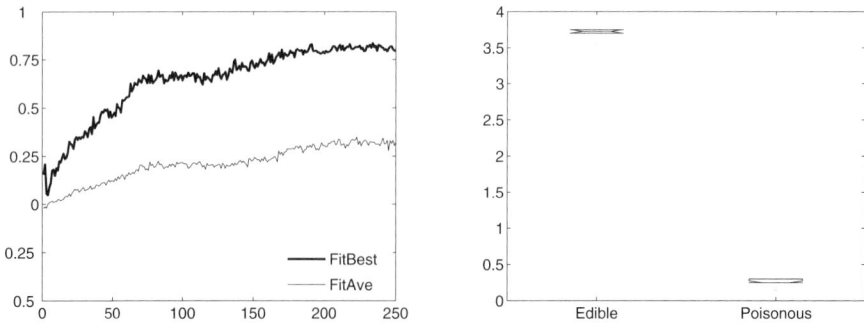

FIGURE 8.3 Results of the base-line simulation. Left: Average and best fitness along the 250 generations. Right: Average number of edible and poisonous fruits eaten by the best individual of the last generation during each staying on an island. Average results of 10 replications of the simulation.

Agents possessing the optimal behavioral strategy discussed above will eat on average 3.5 edible fruits and 0.5 poisonous ones during each visit on an island (0.5 poisonous fruits on average when they have to randomly guess, and 0.5 + 3 edible, since during the second, third and fourth choices the information about the previous action together with the bodily feedback can be used for eating the fruits from the right tree).

Hence, the average expected maximal fitness would be $(3.5 + 0.5 \cdot (-2))/4 = 0.625$.

The described optimal behavioral strategy did manage to evolve, indeed: at the end of the 250 generations the best evolved individual has eaten on average about 3.7 edible fruits and about 0.3 poisonous ones (Figure 8.3, right), thus reaching a fitness of about 0.8 (Figure 8.3, left).

The reason why the best individual reaches a fitness which is higher than the maximal expected one is just chance: in each generation the best individual will happen to be the one which, beyond possessing an optimal behavioral strategy, is also the luckiest, having discovered the edible tree at the first choice more than the expected 50% of the times.

8.6.3 Evolution of Stigmergic Self-Adjustment and Indirect Coordination

Living in a solipsistic world, agents of the base-line simulation can evolve only individual practical abilities. In order to test whether our agents might evolve even social abilities, we run a second set of simulations in which individuals are allowed to live in a social world.

This 'social' simulation runs exactly like the base-line one but for the following modification. Each individual of the population shares its environment with 4 of its own clones (i.e., individuals possessing the same genome and hence the same neural controller). We make agents interact (through their environmental modifications) only with their own clones because we are interested in the emergence of (stigmergic) communication, but we are not interested in the problems of altruism which are typically posed by the evolution of communicative behaviors [Mirolli and Parisi, 2005, 2008]. Allowing interactions only between clones assures us that no problem of altruism can arise, since in this way any possible altruistic behavior will favor only individuals possessing the same altruistic genes which code for the behavior, thus guaranteeing that the behavior can pass through generations (this is the strongest possible form of kin selection, [Hamilton, 1964]).

At the beginning of each trial, each clone is put in a different island where it stays for a maximum of 4 cycles (less if it decides to leave the island earlier). After that, all the clones change their islands so that during each round each island is visited by a different clone and, after 5 rounds, each clone has visited each island. Like in the base-line simulation, the same process is applied for 5 trials.

What makes this simulation interesting for our purposes is that, while in-between different trials all the fruits of all the trees re-grow, within each trial they do not. The consequence is that *in this simulation the environment in which the agents live registers the traces of the behavior of other agents.*

In particular, in all but the very first round within a trial, the trees of an island will lack the fruits eaten by the agents which have already visited that island. Since, as shown by the results of the previous simulation, individuals do not eat randomly, but rather they tend to eat much more edible fruits than poisonous ones, *practical behavioral traces are meaningful*: the tree lacking more fruits is the edible one.

The results of this simulation (Figure 8.4) show that agents in the social condition are in fact able to exploit the information provided by the traces left in the environment by other individuals. In fact, in this condition the best evolved individual reaches a fitness of about 0.9 (Figure 8.4, left), which is significantly higher than the fitness reached by the best individual of the base-line, individual condition.

This fitness increase is explained by the fact that the best individual of the social condition eats an average of about 3.9 edible fruits and only about 0.1 poisonous fruits for each staying on an island (Figure 8.4, right). This is in turn due to the fact that, in the social condition, agents have to choose randomly which tree to feed to only during the first round of each trials. During the other four rounds they can eat from the tree containing less fruits thus avoiding more poisonous fruits and eating more edible fruits than the agents of the individual condition.

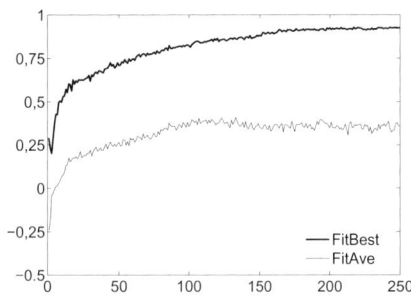 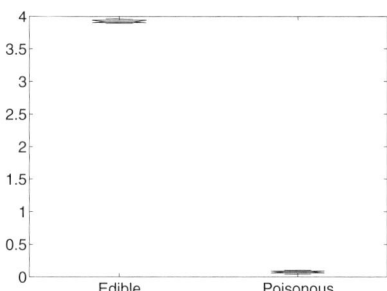

FIGURE 8.4 Results of the 'social' simulation. Left: Average and best fitness along the 250 generations. Right: Average number of edible and poisonous fruits eaten by the best individual during each staying on an island. Average results of 10 replications of the simulation.

This ability of exploiting the traces left in the environment by the actions of other agents is a paradigmatic example of what we call *stigmergic self-adjustment*. In fact, the traces left from previous eating behaviors become cues, stigmergic cues to which agents' behaviors are now sensitive. By acting on these cues, the agents indirectly coordinate their behaviors toward the edible items, and avoiding the poisonous ones.

8.6.4 Evolution of Stigmergic Communication

In the social simulation just described the traces in the environment left by the practical behaviors of eating fruits become stigmergic cues because they carry valuable information which is exploited by the agents who perceive the traces so to increase their own individual fitness.

On the other hand, they are still not (implicit) *signals* in that these cues are produced *exclusively* for the practical purposes of isolated individuals, and not *also* for social ones (i.e., in order to influence the behavior of others), as required by our definition of implicit signals.

The practical behavior which produces the cues, i.e., the behavior of eating from the tree with edible fruits whenever you can, is not affected *in any way* by the fact that there are other agents which might use the traces of that behavior as cues.

In fact, agents of the social condition differ from the agents of the individual condition just for the way they react to environmental cues. When there are no environmental cues, i.e., in the first cycle of the first round of a trial, the behavior of the 'social' agents is exactly the same as the behavior of the isolated agents: i.e., choose a random tree. And agents of the two conditions behave exactly in the same way also when the information about the quality of the fruits is directly available to an agent from its own body. For instance, after a fruit has been eaten, the behavioral rule of both kinds of agents is 'change tree if the fruit was poisonous while keep on feeding from the same tree if the fruit was edible'.

To distinguish proper stigmergic *communication* from mere stigmergic *self-adjustment*, it is necessary that not only the perceiver of the trace modifies its behavior in order to exploit the information provided by the trace, but also *the trace producing behavior must, in some way, be modified* (for influencing). This does not necessarily mean that the behavior itself will change in its morphology. It is also possible that the behavior remains the same, while it is the *conditions under which exactly the same behavior is produced* which may vary.

In order to show how this is be possible we run a third set of simulations in which we slightly modify the set-up of the social simulation in the following way.

In this new simulation, the increase of fitness provided by eating an edible fruit is no more constant, but is proportional to the *hungriness* (h) of the eating individual. Hungriness is 1 each time the individual arrives in an island and decreases of 0.5 each time the individual eats an edible fruit. Furthermore, each eating action has a cost (c) of 0.1 on individual fitness. Hence, the fitness function used in this condition, which we will call the 'hungriness' simulation, is the following:

$$f(x) = \frac{\sum_{t=1}^{T} \sum_{i=1}^{I} hen_e - pn_p - cn_a}{TIC} \tag{8.2}$$

where n_a is the number of eating actions performed by the individual.

The results of this simulation are shown in Figure 8.5. The results in terms of fitness (Figure 8.5, left) are not surprising: both average and best fitness significantly decrease with respect to the other two conditions because in this case edible fruits tend to provide less energy (depending on hungriness) and because there is always a cost to be paid when eating a fruit.

What is interesting here are the results regarding the average number of fruits eaten by the best individual during each staying on an island (Figure 8.5, right). Given the conditions of this simulation, an individual should never eat more than 2 edible fruits for each island: in fact, after it has eaten 2 edible fruits an individual hungriness has decreased to 0, and hence eating a third fruit would not increase fitness at all. On the contrary, it would in fact *decrease* fitness since the individual would pay the cost associated to the eating action. But the best individual of the last generation of this 'hungriness' simulation eats an average of 2.5 edible fruits for each island.

This pattern of eating behavior clearly cannot be explained by referring only to individual advantages. The only function that eating more than 2 edible fruits can have is a social function: that is, *leaving more informative traces to forthcoming agents*. If during the first round on an island an individual would eat only two edible fruits before leaving the island, then the next agent would have to discriminate between trees differing only of one or two fruits (depending on whether the previous agent had chosen the edible tree first or had eaten also a poisonous fruit). This would be of course theoretically possible, but it is evidently too difficult for the simple neural networks possessed by our agents. In fact, consider that, beyond this discrimination ability, the neural networks also have to encode all the other behavioral rules described above, and that fruits are eaten randomly from a tree. Hence, the discrimination ability must be general to all the positions of fruits in the trees.

The number of poisonous fruits eaten by the best individual of the hungriness simulation is significantly less than that of the best individual of the individual simulation (i.e., about 0.5), and quite the same as that of the best individual of the social simulation (i.e., about 0.1).

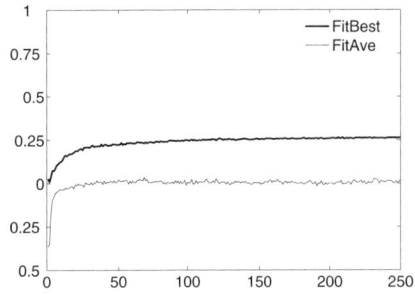 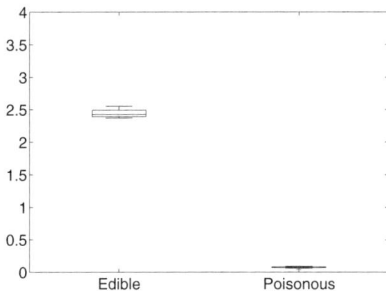

FIGURE 8.5 Results of the 'hungriness' simulation. Left: Average and best fitness along the 250 generations. Right: Average number of edible and poisonous fruits eaten by the best individual during each staying on an island. Average results of 10 replications of the simulation.

This means that the individuals of this condition do indeed exploit the information provided by the traces left by other agents as efficiently as the individuals in the social condition. But in this case the traces are proper *signals* and not mere *cues* because the behavior responsible for the trace production has *not only* the practical function of increasing individual fitness *but also* the social function of producing more informative traces for the individuals to come*.

Hence, this is a paradigmatic case of *indirect behavioral implicit* (i.e., *stigmergic*) *communication*. It is indirect because it exploits the traces left in the environment; it is behavioral because these traces are the results of practical actions; it is implicit because the behavior of producing traces has not lost its practical individual function (agents still need to eat, and, in general, do eat for the individual selective advantage provided by eating); it is communication because the same practical behavior (in this case the conditions of its production) has been slightly modified for influencing purposes, thus acquiring also the communicative function of leaving more readable traces.

8.7 Future Work

Stigmergic approaches are nowadays gaining consensus in several distinct fields of computer science. Due to its relevance for situatedness, stigmergy has been straightforwardly adopted in robotics, and especially by bio-inspired approaches such as Swarm Intelligence or Behavior-based Robotics. As a general mechanism for indirect interaction, it is now raising interest also in Multi-Agent Systems where open environments and huge numbers of heterogeneous agents require new coordination mechanisms. To be fully integrated as first-level abstraction, stigmergy, however, must be more clearly understood than is today. We contend in fact that a principled approach to stigmergy will be essential for providing a new engineering framework for MAS in which both direct and indirect communicative behaviors will be possible at the same time.

---

*The communicative behavior of eating more fruits than expected from a purely individualistic point of view so to leave more informative traces for the agents to come is clearly an altruistic behavior. The evolution of such a behavior is explained by the fact that interacting individuals are clones, meaning that the producer and the receiver of the altruistic behavior always share the same (altruistic) genes. Hence, the altruistic behavior can arise just because the cost for the agent eating one more edible fruit is inferior to the advantage for the forthcoming agent of finding more informative traces.

In this chapter, we have contributed to a more advanced understanding of this very basic phenomenon by providing a precise definition of what stigmergy is, by distinguishing stigmergic self-adjustment from stigmergic communication, and by presenting operational definitions of these phenomena by means of a set of evolutionary agent-based simulations.

However, indirect and direct communicative systems still need further research in order to understand their specific roles and possibly their mutual advantages and disadvantages. Effective design of computational systems employing these mechanisms needs more research on how these mechanisms might co-exist together by supplementing each the limits of the other one.

Similarly, explicit and implicit kinds of communication will also be essential at the same time because each of them fits different needs as natural societies, both of human and animal agents, clearly show. In fact, natural language heavily relies upon more implicit communicative capacities to expand the kind of meaning that can be communicated (i.e. behavioral implicit communication is needed to formalize the Gricean speaker's meaning). The interplay between implicit and explicit communicative behaviors is however still an open challenge that is left out for future research.

Even more than this, one of the major future challenges for the MAS community lies, in our opinion, in designing self-organizing multi-agent systems capable of evolving by themselves their explicit communicative protocols on the basis of more primitive communicative capacities like the ones we have explored in this chapter.

While in this chapter we have mainly adopted an evolutionary perspective with very simple reactive agents as prototypical models, stigmergy and its uses are available also for "cognitive" agents of the Belief-Desire-Intention kind. The relative benefits of enabling stigmergy between deliberative agents is still however an open issue.

Finally, we are convinced that in human-human, human-agent, human-robot, agent-agent, and robot-robot interaction the possibility of communicating through an action and its effects and products could be extremely relevant. In fact, behavioral communication for coordination (in particular stigmergy) has some nice properties and advantages that deserve to be stressed. It is naturally and intrinsically 'situated' in space and time, and thus it transmits information in a perceptual, non-mediated way without any special, arbitrary codification of this kind of information which will necessary be somewhat 'abstract'. This information has also the very nice feature of not being discrete, digitalized, a characteristic that sometimes might be critical.*

8.8 Conclusion

Good old commercial airplanes were flown by the captain and the first officer by means of physically connected control wheels. When the captain was in charge of conducting the plane, the first officer was aware of his steering decisions just by sensing the movements through his own wheel. Differently, modern cockpits took advantage of the advancements in electronics, and have been designed with much more nice and small control wheels very similar to contemporary joysticks. Invisible one from the other, both the captain and first officer now have independent wheels that control the airplane once the corresponding officer is in charge. An unfortunate consequence of this technological innovation has been, however, that the quality of the situation awareness in the cockpit has been compromised

---

*We thank Stefano Nolfi with whom we discussed this point.

[Norman, 1993]. In fact, the officers in these new cockpits need to rely much more on verbal communication to understand what is happening, while situation awareness was guaranteed before by means of a very simple case of stigmergic communication.

This story taken from aviation bears few similarities with what we have discussed in the introduction. The shift from blackboard systems to multi-agent systems has been analogously characterized by the abandonment of a form of stigmergic communication between the agents to adopt the computational variety of verbal communication, that is Agent Communication Languages.

Adopting the perspective defended in this chapter, Multi-Agent Systems are, in fact, characterized by the use of *direct explicit communication* between the agents, that, as it is nowadays widely contested, are difficult to employ in open systems with an unpredictable number of heterogeneous agents. Shared and fixed ontologies and common communication protocols cannot always cope with this complex dynamics.

In this chapter, we have argued for the relevance of a different coordination mechanism: *indirect implicit communication via practical behavioral traces* (i.e., stigmergic communication). To do so, we have provided a clear definition of stigmergic cues as practical behavioral traces. We have also distinguished two varieties of stigmergic uses in self-adjustment and communication, and clarified the difference between explicit and implicit communicative systems.

Though direct and explicit communication and indirect and implicit communication seem, at a first glance, incompatible and alternative strategies, the problems in aviation teach us an important morals: they are not! Having understood the importance of stigmergic communication has helped aviation designers to exploit the relative advantages of both communicative systems without relying only on the properties of one of them. Today's cockpits are designed to take advantage both of implicit and explicit communication between the crew and exploit useful redundancies for safety reasons.

Similarly, the next generation of multi-agent systems should take advantage of all these mechanisms in order to combine them in ways that today are still not foreseeable.

8.9 Acknowledgments

This research has been supported by the project SOCCOP (The Social and Mental Dynamics of Cooperation), funded by the European Science Foundation under the TECT scheme.

References

E. Bonabeau, F. Henaux, S. Guérin, D. Snyers, P. Kuntz, and G. Theraulaz. Routing in telecommunications networks with ant-like agents. In *IATA '98: Proceedings of the second international workshop on Intelligent agents for telecommunication applications*, pages 60–71, London, UK, 1998. Springer-Verlag.

A. Bond and L. Gasser, editors. *Readings in Distributed Artificial Intelligence*. Morgan Kaufman, San Mateo, CA, 1988.

E. Bonabeau, G. Theraulaz, J. Deneubourg, N. Franks, O. Rafelsberger, J. Joly, and S. Blaco. A model for the emergence of pillars, walls and royal chambers in termite nests. *Philosophical Transactions of the Royal Society of London*, 353:1561–1576, 1997.

O. H. Bruinsma. *An analysis of building behaviour of the termite Macrotermes subhyalinus (Rambur)*. PhD thesis, Landbouwhogeschool, Wageningen, 1979.

S. Camazine, J. Deneubourg, N. Franks, J. Sneyd, G. Theraulaz, and E. Bonabeau. *Self-Organization in Biological Systems*. Princeton University Press, Princeton, NJ, 2001.

C. Castelfranchi. When doing is saying. The theory of behavioral implicit communication. Technical report, ISTC-CNR, 2000. URL http://www.istc.cnr.it/doc/62a_20050131162326t_WhenDoingIsSayingADVANCED.rtf.

C. Castelfranchi. Silent agents: From observation to tacit communication. In J. Sichman, H. Coelho, and S. Rezende, editors, *Advances in Artificial Intelligence - IBERAMIA-SBIA 2006*, volume 4140 of *Lecture Notes in Computer Science*, pages 98–107. Springer, 2006a.

C. Castelfranchi. Principles of individual social action. In T. R. and H. G., editors, *Contemporary Action Theory*. Kluwer, Dordrecht, 1997.

C. Castelfranchi. Modelling social action for AI agents. *Artificial Intelligence*, 103: 157–182, 1998.

C. Castelfranchi. From conversation to interaction via behavioral communication. In S. Bagnara and G. Crampton Smith, editors, *Theories and Practice in Interaction Design*, pages 157–179. Erlbaum, New Jersey (USA), 2006b.

P. Cohen and H. Levesque. Intention is choice with commitment. *Artificial Intelligence*, 42:213–261, 1990.

R. Conte and C. Castelfranchi. *Cognitive and Social Action*. UCL Press, London, 1995.

M. Dorigo, E. Bonabeau, and G. Theraulaz. Ant algorithms and stigmergy. *Future Generation Computer Systems*, 16:851–871, 2000.

H. Downing and R. Jeanne. Nest construction by the paper wasp, polistes: a test of stigmergy theory. *Animal Behavior*, 36:1729–1739, 1988.

E. H. Durfee. *Coordination of Distributed Problem Solvers*. Kluwer Academic, Boston, MA, 1988.

R. Engelmore and T. Morgan, editors. *Blackboard Systems*. Addison-Wesley, Reading, MA, 1988.

S. Garnier, J. Gautrais, and G. Theraulaz. The biological principles of swarm intelligence. *Swarm Intelligence*, 1(1):3–31, 2007.

P. Grassé. La reconstruction du nid et les coordinations inter- individuelles chez bellicositermes natalensis et cubitermes. La théorie de la stigmergie: Essai d'interprétation du comportement des termites constructeurs. *Insectes Sociaux*, 6:41–81, 1959.

P. Grassé. Nouvelles experiériences sur le termite de Muller (macrotermes mulleri) et considérations sur la théorie de la stigmergie. *Insectes Sociaux*, 14(1):73–102, 1967.

W. D. Hamilton. Genetic evolution of social behavior. *Journal of Theoretical Biology*, 7(1):1–52, 1964.

O. Hasson. Cheating signals. *Journal of Theoretical Biology*, 167:223–238, 1994.

C. Hewitt. Viewing control structures as patterns of message passing. *Artificial Intelligence*, 8(3):323–364, 1977.

O. Holland and C. Melhuish. Stigmergy, self-organization, and sorting in collective robotics. *Artificial Life*, 5:173–202, 1999.

P. Karlson and M. Luscher. Pheromones: a new term for a class of biologically active substances. *Nature*, 183:55–56, 1959.

I. Karsai and G. Theraulaz. Nest building in a social wasp: postures and constraints. *Sociobiology*, 26:83–114, 1995.

J. Kennedy, R. C. Eberhart, and Y. Shi. *Swarm Intelligence*. Morgan Kaufmann Publishers, San Francisco, CA, 2001.

M. Mamei and F. Zambonelli. Pervasive pheromone-based interaction with rfid tags.

ACM Transactions on Autonomous and Adaptive Systems, 2(2):4, 2007.

M. Mataric. Issues and approaches in the design of collective autonomous agents. *Robotics and Autonomous Systems*, 16:321–331, 1995.

J. Maynard-Smith and D. Harper. *Animal Signals*. Oxford University Press, Oxford, 2003.

M. Mirolli and D. Parisi. How can we explain the emergence of a language which benefits the hearer but not the speaker? *Connection Science*, 17(3-4):325–341, 2005.

M. Mirolli and D. Parisi. How producer biases can favour the evolution of communication: An analysis of evolutionary dynamics. *Adaptive Behavior*, 16(1):27–52, 2008.

D. Norman. *Things that make us smart*. Perseus Publishing, Cambridge, MA, 1993.

V. Parunak. A survey of environments and mechanisms for human-human stigmergy. In D. Weyns, V. Parunak, and F. Michel, editors, *Environments for Multiagent Systems II*, number 3830 in Lecture Notes in Computer Science, pages 163–186. Springer-Verlag, Berlin / Heidelberg, 2006.

V. Parunak. Go to the ant: Engineering principles from natural agent systems. *Annals of Operations Research*, 75:69–101, 1997.

V. Parunak, S. Brueckner, and J. Sauter. Digital pheromones for coordination of unmanned vehicles. In D. Weyns, V. Parunak, and F. Michel, editors, *Environments for Multiagent Systems I*, number 3374 in Lecture Notes in Computer Science, pages 179–183. Springer-Verlag, Berlin / Heidelberg, 2005.

M. Sadek. A study in the logic of intention. In *Proceedings of Knowledge Representation and Reasoning*, pages 462–473, 1992.

J. R. Searle. *Speech Acts: An Essay in Philosophy of Language*. Cambridge University Press, Cambridge, 1969.

R. Smith. The contract net protocol: High-level communication and control in a distributed problem solver. *IEEE Transactions on Computers C*, 29(12):1104–1113, 1980.

P. Sorensen and N. Stacey. Evolution and specialization of fish hormonal pheromones. In J. et al., editor, *Advances in Chemical Signals in Vertebrates*. Kluwer Academic, New York, 1999.

G. Theraulaz and E. Bonabeau. A brief history of stigmergy. *Artificial Life*, 5(2): 97–116, 1999.

N. Tinbergen. Derived activities: their causation, biological significance, origin and emancipation during evolution. *Quarterly Review of Biology*, 27:1–32, 1952.

L. Tummolini and C. Castelfranchi. Trace signals: The meanings of stigmergy. In D. Weyns, V. Parunak, and F. Michel, editors, *Environments for Multi-Agent Systems III*, number 4389 in Lecture Notes in Artificial Intelligence, pages 141–156. Springer-Verlag, Berlin / Heidelberg, 2007.

L. Tummolini, C. Castelfranchi, A. Ricci, M. Viroli, and A. Omicini. Exhibitionists and voyeurs do it better: A shared environment approach for flexible coordination with tacit messages. In D. Weyns, V. Parunak, and F. Michel, editors, *Environments for Multiagent Systems I*, number 3374 in Lecture Notes in Computer Science, pages 215–231. Springer-Verlag, Berlin / Heidelberg, 2005.

E. Visalberghi and E. Addessi. Seeing group members eating familiar food enhances the acceptance of novel foods in capuchin monkeys. *Animal Behaviour*, 60:69–76, 2000.

J. Wenzel. Evolution of nest architecture. In K. Ross and R. Matthews, editors, *The Social Biology of Wasps*, pages 480–519. Cornell University Press, 1991.

E. Wilson. Source and possible nature of the odor trail of fire ants. *Science*, 129(3349):

643, 1959.

E. Wilson. *Sociobiology: The New Synthesis.* Harvard University Press, Cambridge, MA, 1975.

D. Wyatt. *Pheromones and Animal Behaviour: Communication by Smell and Taste.* Cambridge University Press, Cambridge, 2003.

III

MAS for Simulation

9 **Challenges of Country Modeling with Databases, Newsfeeds, and Expert Surveys** 271
   Barry G. Silverman, Gnana K. Bharathy, and G. Jiyun Kim
   Introduction • Cognitive Agent Modeling • Social Agents, Factions, and the FactionSim Testbed • Overview of Some Existing Country Databases • Overview of Automated Data Extraction Technology • Overview of Subject Matter Expert Studies/Surveys • Overview of Integrative Knowledge Engineering Process • Concluding Remarks

10 **Crowd Behavior Modeling: From Cellular Automata to Multi-Agent Systems** 301
   Stefania Bandini, Sara Manzoni, and Giuseppe Vizzari
   Introduction • Pedestrian Dynamics Context: An Overview • Guidelines for Crowds Modeling with Situated Cellular Agents Approach • A Pedestrian Modeling Scenario • From a SCA Model to Its Implementation • Conclusions • Discussion and Future Research Directions

11 **Agents for Traffic Simulation** 325
   Arne Kesting, Martin Treiber, and Dirk Helbing
   Introduction • Agents for Traffic Simulation • Models for the Driving Task • Modeling Discrete Decisions • Microscopic Traffic Simulation Software • From Individual to Collective Properties • Conclusions and Future Work

12 **An Agent-Based Generic Framework for Symbiotic Simulation Systems** 357
   Heiko Aydt, Stephen John Turner, Wentong Cai, and Malcolm Yoke Hean Low
   Introduction • Concepts of Symbiotic Simulation • Different Classes of Symbiotic Simulation Systems • Workflows and Activities in Symbiotic Simulation Systems • Agent-Based Framework • Applications • Conclusions • Future Work

13 **Agent-Based Modeling of Stem Cells** 389
   Mark d'Inverno, Paul Howells, Sara Montagna, Ingo Roeder, and Rob Saunders
   Introduction • The Biological Domain – HSC Biology • Stem Cell Modeling • Drawbacks of Existing Models and Why Agents • Overview of Our Agent Modeling Framework • Agentifying Existing Approaches • From Agent Model to Simulation • Discussion

Multi-agent systems are used as a metaphor in modeling and simulation. At the level of modeling it supports a view of the system under study as a community of interacting autonomous entities. At the level of simulation it helps the design of simulation systems as distributed concurrent entities that interact with their physical environment. The former constitutes a consequent progression of micro-oriented modeling, whereas the latter builds on distributed and online simulation approaches.

The agent metaphor has particularly propelled research on human behavior modeling. Technical questions such as how to facilitate the modeling and simulation of multiple deliberative agents, have received a lot of attention (see also Part IV). However, often the more urgent problem is to find the knowledge and data upon which developing and validating these models can be based. This problem is addressed in the chapter "Challenges of Country Modeling Databases, Newsfeeds, and Expert Surveys" by Barry G. Silverman, Gnana K. Bharathy, and G. Jiyun Kim. Three different sources of knowledge are exploited, i.e., web corpus of empirical materials, specific domain databases, and questionnaires, to construct realistic profiles for the actors and issues at play in ethno-political conflict regions. Whereas in these scenarios detailed models of individual actors play a central role, other applications are content with comparatively coarse behavior models which describe many individuals. To those belong crowd and traffic simulation.

Crowd simulation is concerned with analyzing the behavior of many individuals in too little space. This behavior is determined by individual and collective behavior patterns, a mix of competition for shared space, and collaboration due to, not necessarily explicit but shared, social norms. The purpose of simulation studies ranges from a better understanding of certain emergent phenomena to the concrete management of crowds in certain settings. In the later case, we can further distinguish between offline analysis of the effects of different infrastructures on crowd behavior and online support to handle specific strategies in crises situations. Whereas traditionally cellular automata have been exploited for crowd simulation, situated agent-based approaches offer new possibilities as they allow one to treat the environment and the entities that inhabit it as first class citizens as Stefania Bandini, Sara Manzoni, and Giuseppe Vizzari describe in the chapter "Crowd Behavior Modeling: From Cellular Automata to Multi-Agent Systems".

Vehicular traffic is another classical example of a multi-agent system: Autonomous agents (i.e., the drivers) operate in a shared environment which is given by the road infrastructure. Typically, agent-based models are built on a microscopic modeling approach describing the motion of each individual vehicle. Scientists described the physical propagation of traffic flows by means of dynamic models already in the 1950s. Since then the microscopic approach has grown gradually in popularity because of higher computational power. This development has been accelerated in the last decade by agent oriented modeling and simulation. In the chapter "Agents for Traffic Simulation" Arne Kesting, Martin Treiber and Dirk Helbing provide an overview of the state-of-the-art in traffic modeling and simulation. Detailed driver models and the implications for an effective and efficient simulation of these micro models are presented. Entirely new simulation approaches are required when it comes to using vehicle to vehicle communication and predicting traffic based on incomplete knowledge quasi online in vehicles.

Symbiotic simulation is one particular form of online simulation, which stresses the joint benefit of a close interaction between simulation and physical environment for both. This benefit is due to the data that allow a simulation to adapt itself to more precisely predict dynamics, the physical environment in turn will benefit from these more reliable predictions. As this type of simulation is driven by data, there is also a close relation to dynamic data driven application systems. In the chapter "An Agent-Based Generic Framework for Symbiotic Simulation" Heiko Aydt, Stephen John Turner, Wentong Cai, and Malcolm Yoke

Hean Low present a range of symbiotic simulation approaches. At the core of designing these different systems we find the agent metaphor, as the systems comprise multiple, concurrently active units, whose composition and interaction is highly adaptive and which are in frequent interaction with the environment.

Whereas in the above chapters, modeling and simulation of human behavior is at the core of interest, the chapter on "Agent-Based Modeling of Stem Cells" by Mark d'Inverno, Paul Howells, Sara Montagna, and Rob Saunders is dedicated to a new emerging area for modeling and simulation methods: Systems Biology. Systems Biology brings together researchers from diverse areas such as Biology, Medicine, Physics, and Computer Science. Thereby, in-silico experimentation is aimed at complementing wet-lab experimentation. Whereas continuous, deterministic macro models still prevail, discrete micro models are gaining ground as well. In this context also agent-oriented approaches receive increasingly attention, as they allow interpreting biological systems as a community of interacting entities that can be individually traced. Particularly, in combination with suitable visualization techniques, agent-based approaches allow a comprehensive and realistic modeling of biological systems and can reveal new insights into the behavior of complex cellular systems.

9
Challenges of Country Modeling with Databases, Newsfeeds, and Expert Surveys

9.1	Introduction	271
9.2	Cognitive Agent Modeling	274
	Major PMF Models within Each PMFserv Subsystem	
9.3	Social Agents, Factions, and the FactionSim Testbed	278
9.4	Overview of Some Existing Country Databases	282
9.5	Overview of Automated Data Extraction Technology	284
9.6	Overview of Subject Matter Expert Studies/Surveys	289
9.7	Overview of Integrative Knowledge Engineering Process	290
9.8	Concluding Remarks	296
	Acknowledgment	297
	References	298

Barry G. Silverman
University of Pennsylvania

Gnana K. Bharathy
University of Pennsylvania

G. Jiyun Kim
University of Pennsylvania

9.1 Introduction

According to expert practitioners and researchers in the field of human behavior modeling ([Silverman et al., 2002; Pew and Mavor, 1998; Ritter et al., 2003]), a common central challenge now confronting designers of HBM (human-behavior-modeling) applications is to increase the realism of the synthetic agents' behavior and coping abilities. It is well accepted in the HBM (human-behavior-modeling) community that cognitively detailed, "thick" models are required to provide realism. These models require that synthetic agents be endowed with cognition and personality, physiology, and emotive components. (We will hereafter refer to these rich models as "cognitively detailed models" or "thick agents.") To make these models work, one must find ways to integrate scientific know-how from many disciplines, and to integrate concepts and insights from hitherto fragmented and partial models from the social sciences, particularly from psychology, cultural studies, and political science. One consequence of this kind of integration of multiple and heterogeneous concepts and models is that we frequently end up with a large feature space of parameters that then need to be filled in with data.

In recent years, modeling methodologies have been developed that help to construct models, integrate heterogeneous models, elicit knowledge from diverse sources, and also test, verify, and validate models (see [Bharathy, 2006] for example). However, these methodologies have required extensive use of manual labor to develop each model. The development of automatic techniques would significantly improve the efficiency of this process. In this chapter, we will explore how a modeler can navigate, sift, and harvest the vast ocean of data that today can be accessed with a keystroke in certain cases. We will use country data as our theme. By "country data" we mean all possible sources of information that can be used to instantiate agent-based models of complex social systems. Country data are especially important when we integrate several models in order to build a realistic complex social system. This kind of integration of models tends to produce a large feature space of parameters that then needs to be filled in with data. There is no dearth of country data. However, the challenge lies in finding the right data for the right slots.

The state-of-the-art for extracting relevant data is summarized in Figure 9.1, which shows three parallel extraction pathways including (1) webscraping of newsfeeds, (2) extraction of data from country databases, and (3) self-explanatory expert survey forms. These pathways are the focus of this chapter. We examine what they consist of, how they may be utilized, and what issues and challenges arise as we exploit them to rapidly generate agent-based models. This is an exciting time to be working in this field; important breakthroughs seem possible. At the same time, there are also many unsolved issues, and we review these in this chapter as well. In the end, we believe that there is no single best route to obtaining our information of interest, namely the information we need to determine our model parameters. Instead, it seems wisest to fully utilize all three routes and to try to elicit the best possible information through a careful triangulation. Accordingly, Figure 9.1 thus shows how we conceptualize this triangulation.

To understand these issues, one must take an in-depth look at an exemplary agent-based model and consider its data needs in relation to what the three previously mentioned automated extraction pathways can readily produce. As we explore our example, we will ask: In what ways can the modeling and simulation community best marshal the volumes of country data now being made available in databases assembled by social scientists, area studies specialists, and various governmental agencies, and international organizations (databases that track not only the socio-demographics and politico-economic data, but also significant events and the needs/values/preferences/norms of populations of interest)? Currently, these databases- consisting, variously, of expert and mass surveys and opinion polls, conflict and event databases, socio-cultural and politico-economic indicators, human terrain systems [Kipp et al., 2006], automated scraping of newsfeeds and websites, and more-are not all updated frequently enough to capture the most up-to-date information and may not be user-friendly enough to develop a unified database under a common format. Nonetheless, they are collected and maintained by regional and subject-area experts with in-depth local knowledge and wisdom using cutting-edge survey methods and other reliable data collection methodologies. Moreover, the growing interest in the development of various automated data extraction and consolidation techniques (e.g., General Inquirer [Stone et al., 1966], Kansas Event Data System (KEDS) [Schrodt and Gerner, 1996], Opinion Analysis System (OASYS) [Cesarano et al., 2006], Profiler+ [Young, 2001], ReadMe [Hopkins et al., 2007], STORY [Fayzullin et al., 2007], to mention just a few) highlights both the possibility and the promise of making these databases more user-friendly and the actual process of data collection more efficient, especially with regard to capturing real-time news feeds from various (web-based) sources around the world.

These automated techniques currently complement and can substitute to a certain extent most standard, labor-intensive data collection efforts. Our chapter provides a cursory

9

Challenges of Country Modeling with Databases, Newsfeeds, and Expert Surveys

Barry G. Silverman
University of Pennsylvania

Gnana K. Bharathy
University of Pennsylvania

G. Jiyun Kim
University of Pennsylvania

9.1	Introduction	271
9.2	Cognitive Agent Modeling	274
	Major PMF Models within Each PMFserv Subsystem	
9.3	Social Agents, Factions, and the FactionSim Testbed	278
9.4	Overview of Some Existing Country Databases...	282
9.5	Overview of Automated Data Extraction Technology	284
9.6	Overview of Subject Matter Expert Studies/Surveys	289
9.7	Overview of Integrative Knowledge Engineering Process..	290
9.8	Concluding Remarks................................	296
	Acknowledgment	297
	References	298

9.1 Introduction

According to expert practitioners and researchers in the field of human behavior modeling ([Silverman et al., 2002; Pew and Mavor, 1998; Ritter et al., 2003]), a common central challenge now confronting designers of HBM (human-behavior-modeling) applications is to increase the realism of the synthetic agents' behavior and coping abilities. It is well accepted in the HBM (human-behavior-modeling) community that cognitively detailed, "thick" models are required to provide realism. These models require that synthetic agents be endowed with cognition and personality, physiology, and emotive components. (We will hereafter refer to these rich models as "cognitively detailed models" or "thick agents.") To make these models work, one must find ways to integrate scientific know-how from many disciplines, and to integrate concepts and insights from hitherto fragmented and partial models from the social sciences, particularly from psychology, cultural studies, and political science. One consequence of this kind of integration of multiple and heterogeneous concepts and models is that we frequently end up with a large feature space of parameters that then need to be filled in with data.

In recent years, modeling methodologies have been developed that help to construct models, integrate heterogeneous models, elicit knowledge from diverse sources, and also test, verify, and validate models (see [Bharathy, 2006] for example). However, these methodologies have required extensive use of manual labor to develop each model. The development of automatic techniques would significantly improve the efficiency of this process. In this chapter, we will explore how a modeler can navigate, sift, and harvest the vast ocean of data that today can be accessed with a keystroke in certain cases. We will use country data as our theme. By "country data" we mean all possible sources of information that can be used to instantiate agent-based models of complex social systems. Country data are especially important when we integrate several models in order to build a realistic complex social system. This kind of integration of models tends to produce a large feature space of parameters that then needs to be filled in with data. There is no dearth of country data. However, the challenge lies in finding the right data for the right slots.

The state-of-the-art for extracting relevant data is summarized in Figure 9.1, which shows three parallel extraction pathways including (1) webscraping of newsfeeds, (2) extraction of data from country databases, and (3) self-explanatory expert survey forms. These pathways are the focus of this chapter. We examine what they consist of, how they may be utilized, and what issues and challenges arise as we exploit them to rapidly generate agent-based models. This is an exciting time to be working in this field; important breakthroughs seem possible. At the same time, there are also many unsolved issues, and we review these in this chapter as well. In the end, we believe that there is no single best route to obtaining our information of interest, namely the information we need to determine our model parameters. Instead, it seems wisest to fully utilize all three routes and to try to elicit the best possible information through a careful triangulation. Accordingly, Figure 9.1 thus shows how we conceptualize this triangulation.

To understand these issues, one must take an in-depth look at an exemplary agent-based model and consider its data needs in relation to what the three previously mentioned automated extraction pathways can readily produce. As we explore our example, we will ask: In what ways can the modeling and simulation community best marshal the volumes of country data now being made available in databases assembled by social scientists, area studies specialists, and various governmental agencies, and international organizations (databases that track not only the socio-demographics and politico-economic data, but also significant events and the needs/values/preferences/norms of populations of interest)? Currently, these databases- consisting, variously, of expert and mass surveys and opinion polls, conflict and event databases, socio-cultural and politico-economic indicators, human terrain systems [Kipp et al., 2006], automated scraping of newsfeeds and websites, and more-are not all updated frequently enough to capture the most up-to-date information and may not be user-friendly enough to develop a unified database under a common format. Nonetheless, they are collected and maintained by regional and subject-area experts with in-depth local knowledge and wisdom using cutting-edge survey methods and other reliable data collection methodologies. Moreover, the growing interest in the development of various automated data extraction and consolidation techniques (e.g., General Inquirer [Stone et al., 1966], Kansas Event Data System (KEDS) [Schrodt and Gerner, 1996], Opinion Analysis System (OASYS) [Cesarano et al., 2006], Profiler+ [Young, 2001], ReadMe [Hopkins et al., 2007], STORY [Fayzullin et al., 2007], to mention just a few) highlights both the possibility and the promise of making these databases more user-friendly and the actual process of data collection more efficient, especially with regard to capturing real-time news feeds from various (web-based) sources around the world.

These automated techniques currently complement and can substitute to a certain extent most standard, labor-intensive data collection efforts. Our chapter provides a cursory

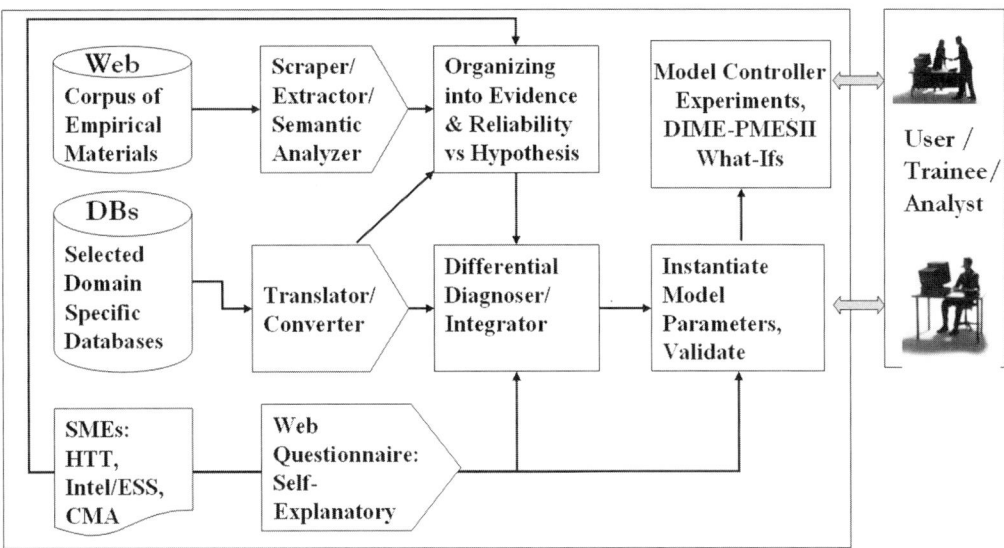

FIGURE 9.1 Overview of New Approaches to Data Extraction and Rapid Generation of Agent-Based Models. (Note: SMEs– Subject Matter Experts; DIME– Diplomatic, Informational, Military, and Economic actions; PMESII– Political, Military, Social, Economic, Informational, and Infrastructure effects; HTT– Human Terrain Tool; Intel/ESS– Intelligence collections/Every Soldier a Sensor– a military program; CMA– Civil Military Affairs. Also, all arrows in all figures of this chapter indicate process direction and/or flow of information.)

overview of these databases and techniques and suggests the next steps for the use of this resource. These datasets and techniques are key assets, as we argue, for those interested in the synthesis of two major agent-based modeling paradigms – the cognitive and the social. Consequently, the modeling and simulation community loses a significant opportunity if it fails to tap into this valuable resource. We pursue this argument by means of a case study integrating a cognitive agent environment (PMFserv) and a social agent environment (FactionSim), which we then apply to various countries, regions, and topics of interest (Iraq, Southeast Asia, the Crusades) to assess their validity and realism. Using the information from these databases to populate such models with realistic agents improves their realism and facilitates their refinement. Some information will also be set aside for later empirical testing of these models and their observable implications with a view to achieving external validity

As we explore this new frontier of (auto-generated) agent-based modeling using country databases and newsfeeds, we ask: What can the field of modeling and simulation add to the conventional studies of countries and regions typically performed by social scientists and area specialists? Country databases have been assembled in order to add depth to the study of countries and regions. In order for the data requirements of modeling and simulation to be met by any one of these sources, a dialog must be initiated to determine what sorts of data that the models actually need versus what is now collected. It is also worth studying whether the modeling and simulation community can seamlessly exchange data with various social science communities. Building on our experience at viewing countries as complex social systems, we aim to outline what agent-based simulation might offer. That is, if we use the data from country databases to help model the "parts" and their micro-decision processes, can we observe macro-behaviors emerging that will aid the work of country analysts? We recognize that, if our aim is to model and simulate a social system from the bottom up, then we need to approach this system with agent technology that covers both the social processes that influence people and also the cognitive processes individuals use as they reason and as they experience emotion. That is, we are interested in discovering what socio-cognitive agents can offer to the study of specific countries or social systems, and we wish particularly to model how diplomatic, intelligence, military, and economic (DIME) actions might affect the political, military, economic, social, informational, and infrastructure (PMESII) systems of a given country of interest.

Finally, as Sun [Sun, 2006] points out in his useful survey of the respective fields of social agents and cognitive agents, there are very few environments that straddle both topics and, consequently, provide socio-cognitive architectures. In this chapter, we illustrate one such architecture to provide insight into its operations, its uses, and the validity of its outputs. More importantly, we argue that this particular socio-cognitive architecture can serve as an ultimate test-bed for evaluating numerous paper-based theories regarding the operations (political, economic, and more) of our countries of interest. We further suggest that all paper-based theories should be tested and implemented in relation to this architecture. While this framework is relatively mature Commercial Off The Shelf (COTS) software, we close with a discussion of future research needs focused on making new software tools better able to support varied analyses of the PMESII (Political, Military, Economic, Social, Informational, and Infrastructure) systems of the country of interest.

This chapter consists of eight additional sections following the introduction and corresponding to many of the blocks of Figure 9.1: Section 9.2 – Cognitive Agent Modeling and Major PMF Models; Section 9.3 – Social Agents, Factions, and the FactionSim Testbed; Section 9.4 – Overview of Some Existing Country Databases; Section 9.5 – Overview of Automated Data Extraction Technology; Section 9.6 – Overview of Subject Matter Expert Studies/Surveys; Section 9.7 – Overview of the Integrative Knowledge Engineering Process (evidence tables, differential diagnosis); and Section 9.8 – Concluding Remarks. In its broadest reach, this chapter introduces and explores a new direction for the modeling and simulation community aimed at capitalizing on a potentially rich symbiotic relationship with the social science/area studies community.

9.2 Cognitive Agent Modeling

We will illustrate the data issue using PMFserv, a COTS (Commercial Off The Shelf) human behavior emulator that drives agents in simulated gameworlds. This software was developed over the past eight years at the University of Pennsylvania as an architecture to synthesize many best available models and best practice theories of human behavior modeling. PMFserv agents are unscripted, but use their micro-decision making, as described below, to react to actions as they unfold and to plan out responses.

A performance moderator function (PMF) is a micro-model covering how human performance (e.g., perception, memory, or decision-making) might vary as a function of a single factor (e.g., sleep, temperature, boredom, grievance, and so on). PMFserv synthesizes dozens of best available PMFs within a unifying mind-body framework and thereby offers a family of models where micro-decisions lead to the emergence of macro-behaviors within an individual. None of these PMFs are "home-grown"; instead they are culled from the literature of the behavioral sciences. Users can turn on or off different PMFs to focus on particular aspects of interest. These PMFs are synthesized according to the inter-relationships between the parts and with each subsystem treated as a system in itself.

9.2.1 Major PMF Models within Each PMFserv Subsystem

The unifying architecture in Figure 9.2 shows how different subsystems are connected. For each agent, PMFserv operates what is sometimes known as an observe, orient, decide, and act (OODA) loop. PMFserv runs the agents perception (observe) and then orients the entire physiology and personality/value system to determine levels of fatigues and hunger, injuries and related stressors, grievances, tension buildup, impact of rumors and speech acts, emotions, and various mobilizations and social relationship changes since the last tick of the simulator clock. Once all these modules and their parameters are oriented to the current stimuli/inputs, the upper right module (decision-making/cognition) runs a best response algorithm to try to determine what to do next. The algorithm it runs is determined by its stress and emotional levels. In optimal times, it is in vigilant mode and runs an expected subjective utility algorithm that reinvokes all the other modules to assess what impact each potential next step might have on its internal parameters. When very bored, it tends to lose focus (perception degrades) and it runs a decision algorithm known as unconflicted adherence mode. When highly stressed, it will reach panic mode, its perception basically shuts down and it can only do one of two things: cower in place or drop everything and flee. In order to instantiate or parameterize these modules and models, PMFserv requires that the developer profile individuals in terms of each of the module's parameters (physiology, stress thresholds, value system, social relationships, etc.).

As an illustration of one of the modules in Figure 9.2 and of some of the best-of-breed theories that PMFserv runs, let us consider "cognitive appraisal" (Personality, Culture, Emotion module) – the bottom left module in Figure 9.2. This is where an agent (or person) compares the perceived state of the real world to its value system and appraises which of its values are satisfied or violated. This in turn activates emotional arousals. For the emotion model, we have implemented one as described in [Silverman et al., 2006b]. Implementing a person's value system requires every agent to have its goals, standards, and preference (GSP) trees filled out. Most significant from the perspective of data production are GSP trees. These are multi-attribute value structures where each tree node is weighted with Bayesian importance weights. A Preference Tree represents an agent's long-term desires for world situations and relations (for instance, no weapons of mass destruction, an end to global warming, etc.) that may or may not be achieved within the scope of a scenario. Among our agents, this set of "desires" translates into a weighted hierarchy of territories and constituencies (e.g., no tokens of leader X in resource Y of territory Z).

The Standards Tree defines the methods an agent is willing to employ to attain his/her preferences. The Standard Tree nodes merge several best available personality and culture profiling instruments such as, among others, Hermann traits governing personal and cultural norms [Hermann, 1999], standards from the GLOBE study [House et al., 2004], top-level guidelines related to Economic and Military Doctrine, and sensitivity to life (humanitarianism). Personal, cultural, and social conventions render inappropriate the purely Machiavellian action choices ("One shouldn't destroy a weak ally simply because they are currently useless"). It is within these sets of guidelines that many of the pitfalls associated with shortsighted Artificial Intelligence (AI) can be sidestepped. Standards (and preferences) allow for the expression of strategic mindsets.

Finally, the Goal Tree covers short-term needs and motivations that drive progress toward preferences. In the Machiavellian [Machiavelli, 1965, 1988] and Hermann-profiled [Hermann, 2005] world of leaders, the Goal Tree reduces to the duality of growing/developing versus protecting the resources in one's constituency. Expressing goals in terms of power and vulnerability provides a high-fidelity means of evaluating the short-term consequences of actions. For non-leader agents (or followers), the Goal Tree also includes traits covering

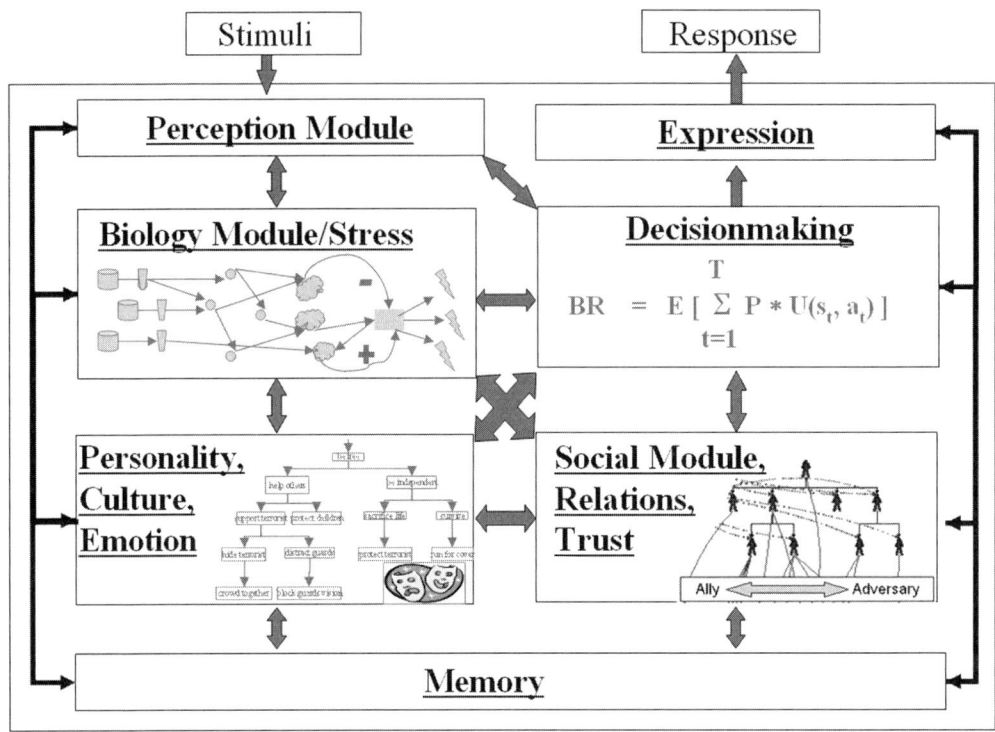

FIGURE 9.2 PMFserv, an example of Cognitive Architecture.

basic Maslovian type needs.

Figure 9.3 not only graphically lists some of the example* performance moderator functions (PMFs) in the collection, but also shows how these different functions are synthesized to create the whole (PMFserv). In this sense, Figure 9.3 is simply a more detailed representation of Figure 9.2. The details of these PMFserv models are beyond the scope of this chapter. Interested readers should consult [Silverman et al., 2006b, 2007] for details.

PMFserv has been deployed in a number of applications, gameworlds, and scenarios. A few of these are listed below in Table 9.1.** To facilitate the rapid composition of new casts of characters we have created an Integrated Development Environment (IDE) in which one

---

*It is worth noting that because our research goal is to study best available performance moderator functions (PMFs), we avoid committing to particular performance moderator functions. Instead, every performance moderator function explored in this research must be readily replaceable. The performance moderator functions that we synthesized are workable defaults that we expect our users will research and improve on as time goes on. From the data and modeling perspective, the consequence of not committing to any single approach or theory is that we have to come up with ways to readily study and then assimilate alternative models that show some benefit for understanding our phenomena of interest. This means that any computer implementation we embrace must support plugin/plugout/override capabilities, and that specific performance moderator functions as illustrated in Figure 9.3 should be testable and validatable against field data such as the data they were originally derived from.

**Many of these previous applications have movie clips, tech reports, and validity assessment studies available at http://www.seas.upenn.edu/~barryg/hbmr. Several historical correspondence tests indicate that PMFserv mimics decisions of the real actors/population with a correlation of approximately 80% (see [Silverman et al., 2006a, 2008]). Some of these applications are discussed in greater detail in the subsequent section.

Challenges of Country Modeling with Databases, Newsfeeds, and Expert Surveys 277

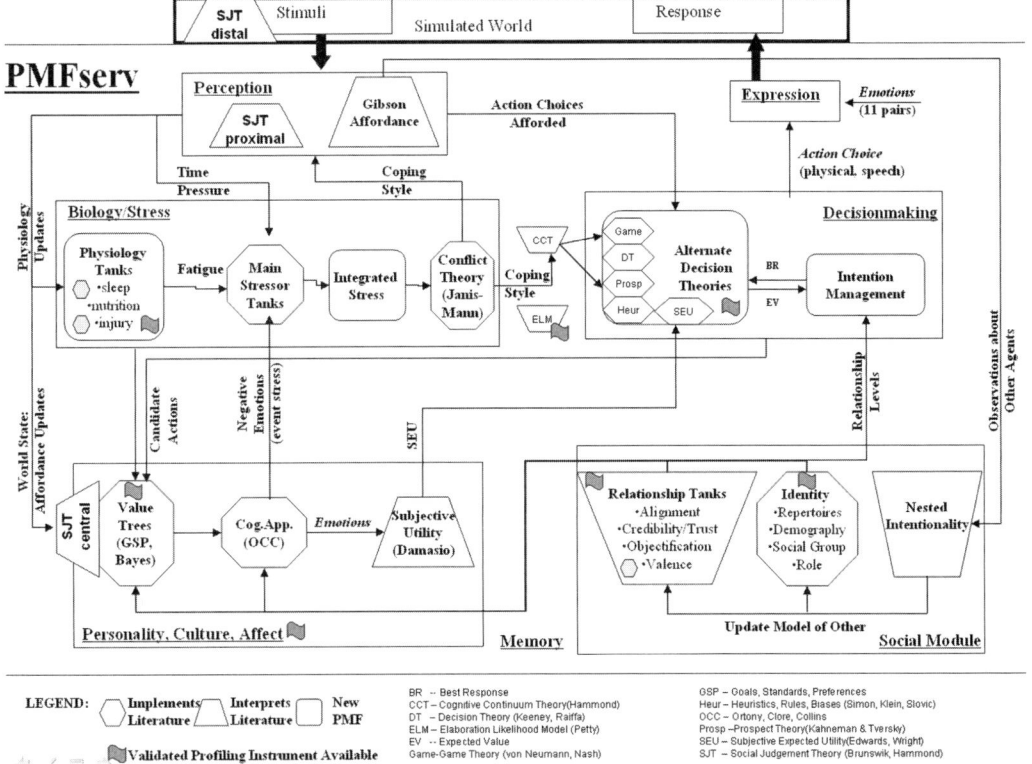

FIGURE 9.3 Summary of Implemented Theories in PMFserv (Note: NfC – Need for Cognition).

knowledge engineers archetypical individuals (leaders, followers, suicide bombers, financiers, etc.) and assembles them into casts of characters useful for creating or editing scenarios.

TABLE 9.1 PMFserv and Its Applications

Domestic Applications	International Applications

• Consumer modeling
    – buyer behavior
    – ad campaign
• Petworld
    – pet behavior
• Gang members
    – hooligans
• Crowd Scenes
    – milling
    – protesting
    – rioting
    – looting
	• Intifadah Recreation (leaders, followers) – Roadmap sim
	• Somalia Crowds – Black Hawk Down (males, females, trained militia, clan leaders
	• Thailand recreation (Buddhists vs. Muslims – radicalization)
	• Iraq DIME-PMESII sim – three ethnic groups, parliament (leaders and 15,000 followers)
	• Urban Resolve 2015 – Sim-Red (multiple insurgent cell)
	• Many world leaders profiled

9.3 Social Agents, Factions, and the FactionSim Testbed

The previous section overviewed the modules of a cognitive agent and some of the components that give it a social orientation. In this section we turn to additional modules that turn the cognitive agent into a socio-cognitive one. Specifically, we introduce FactionSim, an environment that captures a globally recurring socio-cultural "game" that focuses upon inter-group competition for control of resources (Security/Economics/Political Tanks). This environment implements PMFserv within a game theory/PMESII (Political, Military, Economy, Social, Informational, and Infrastructure) Campaign framework. Many of the applications listed above have this game embedded in them. Each group of agents manages the following set of models:

Security Model (Skirmish, Urban Lanchester)
    Power-Vulneraility Computations [Johns, 2006]
    Skirmish Model (force size, training, etc.)
    Urban Lanchester Model (probability of kill)
Economy Model (Harrod-Domar model [Harrod, 1960])
    Black Market
    Undeclared Market [Lewis, 1954; Schneider and Enste, 2000]
    Formal Capital Economy
Political Model (loyalty, membership, mobilization, etc.) [Hirschman, 1970]

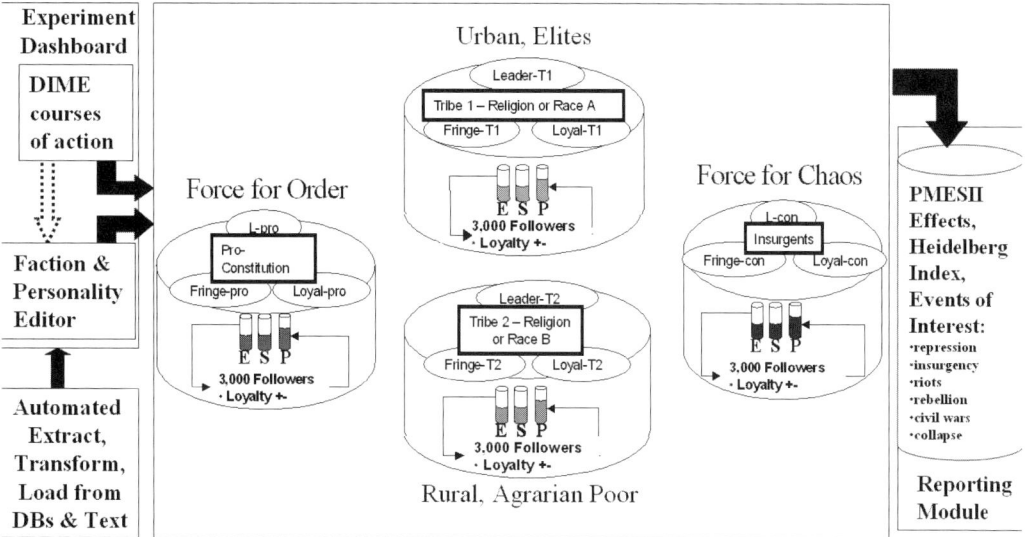

FIGURE 9.4 Models and Components that must be synthesized for a FactionSim Testbed (Note: DIME–Diplomatic, Informational, Military, and Economic; DBs– Databases; Heidelberg Index [Heidelberg Conflict Barometer]; PMESII–Political, Military, Economic, Social, Informational, and Infrastructure).

Institution Sustainment Dynamics

Follower Social Network – Cellular Automata [Axelrod, 1998; Epstein, 2002; Lustick et al., 2004]

Small World Theory/Info Propagation [Milgram, 1967]

This environment facilitates the codification of alternative theories of factional interaction and the evaluation of policy alternatives. FactionSim is a tool that allows conflict scenarios to be established in which the factional leader and follower agents all run autonomously and are free to employ their micro-decision making as the situation requires. A single human player interacts with the environment and attempts to employ a set of DIME (Diplomatic, Informational, Military, and Economic) actions to influence outcomes and PMESII (Political, Military, Economy, Social, Informational, and Infrastructure) effects.

Factions are modeled as in the center of Figure 9.4 where each typically has a leader, two sub-faction leaders (loyal and fringe), a set of starting resources (Economy, E, Security, S, and Political support, P), and a representative set of over 1,000 follower agents. A leader is assumed to manage his faction's E- and S- tanks so as to appeal to his followers and to each of the other tribes or factions he wants in his alliance. Each of the leaders of those factions, however, will similarly manage their own E and S assets in trying to keep their sub-factions and memberships happy. Followers determine the level of the P-tank by voting their membership level. A high P-tank means that there are more members to recruit for security missions and/or to train and deploy in economic ventures. As a result, leaders often find it difficult to move to alignments and positions that diverge very far from the motivations of their memberships.

FactionSim allows one to edit the profiles of all the factions of interest to match a given scenario including:

Faction = { Properties {name, identity repertoire, demographics, salience-entry, salience-exit, other} }

Alignments {alignment-matrix, relationship valence and strength, dynamic alliances}

Roles {leader, sub-leader, loyal-follower, fringe-follower, population-member}

Resources (R) = Set of all resources, r: {econ-tank, security-tank, political support-tank}

$r_{r,f}$ = {Resource level for resource r owned by faction f, $r_{r,f}$ ranges from 1 to 100}

$\Delta_{r(a,b)}$ = {Change in r on group a by group b} = Δ_r

T = Time horizon for storing previous tank values

Dev-Level = {Maturity of a resource where 1=corrupt/dysfunctional, 3=neutral, 5=capable/effective}

Actions (A) = { Leader-actions (target) = {Speak (seek-blessing, seek-merge, mediate, brag, threaten), Act (attack-security, attack-economy, invest-own-faction, invest-ally-faction, defend-economy, defend-security)} }

Follower-actions (target) = {Go on Attacks for, Support (econ), Vote for, Join Faction, Agree with, Remain-Neutral, Disagree with, Vote against, Join Opposition Faction, Oppose with Non-Violence (Voice), Rebel-against/Fight for Opposition, Exit Faction}

Despite efforts at simplicity, stochastic simulation models for domains of this sort rapidly become complex. The strategy space for each leader facing only two other leaders grows impossibly large to explore. As a result, FactionSim's Experiment Dashboard (left side of Figure 9.4) permits inputs ranging from one course of action to a set of parameter experiments the player is curious about. On the bottom left is the profile editor governing the personalities of the leaders and sub-leaders, and of the key parameters that define the starting conditions of each of the factions and sub-factions. Certain actions by the player that are thought to alter the starting attitudes or behavior of the factions can flow between these two components, e.g., a discussion beforehand that might alter the attitudes of certain key leaders (Note: this action is often attempted in settings with real Subject Matter Experts, or SMEs, and diplomats playing our various games).

All data from PMFserv and the socio-cultural game is captured into log files. At present we are developing an after-action report summary module, as well as analytical capabilities for design of experiments, for repeated Monte Carlo trials, and for outcome pattern recognition and strategy assessment.

Now, with this framework in mind, let us look at different types of actors required to construct the kind of social system models we have built. Frequently, we create two different types of individual actors:

- individually named personae, such as leaders, who could be profiled, and
- archetypical members* of the society or of a particular group whose model parameters are dependent on societal level estimates.

In addition, we also have groups (collections of agents with leaders and followers) that display some emergent properties of their own that are more than the sum of their parts.

---

*For each archetype, what's interesting is not strictly the mean behavior pattern, but what emerges from the collective. To understand that, one expects to instantiate many instances of each archetype where each agent instance is a perturbation of the parameters of the set of PMFs whose mean values codify the archetypical class of agent they are drawn from. This means that any computerization of PMFs should support stochastic experimentation of behavior possibilities. It also means that individual differences, even within instances of an archetype, will be explicitly accounted for.

We also model institutions and resources including institutional infrastructures and support plug-in of more detailed models of these dimensions. Typical institutions include the economy (markets, jobs, banking), educational system, the health system, the judicial system, the police and security forces, the utilities/infrastructure (e.g., energy sector, the transportation system, and communication systems), as well as various institutions of civil society.

Types of parameters for typical social system models in PMFserv entities are given below:

Agents (Decision Making Individual Actors):
- Value System/ GSP Tree: Hierarchically organized values such as short term goals, long term preferences and likes, and standards of behavior including sacred values and cultural norms
- Ethno-Linguistic-Religious-Economic/Professional Identities
- Level of Education
- Level of Health
- Level of Wealth
- Savings Rate
- Contribution Rate
- Extent of Authority over the Group

Groups:
- Philosophy
- Leadership
- Relationship to other groups
- Barriers to exit and entry
- Group Level Resources such as Political, Economic and Security Strengths
- Institutional infrastructures owned by the group
- Access to institutional benefits for the group members (Level Available to Group)
- Fiscal, Monetary and Consumption Philosophy
- Disparity

Institutions:
- Capital Investment
- Damage / Decay
- Level of Corruption (indicates usage vs. misuse)

A toolset such as FactionSim (and PMFserv) will only be useful to the extent that it can offer valid recreations of the actual leaders, followers, and populations of interest. In terms of the validity of the current socio-cognitive agent synthesis, this research has tried hard to examine its robustness and cross-sample fitness. FactionSim agents passed validity assessment tests in both of two conflict scenarios attempted to date, as described fully in [Silverman et al., 2008]. In the first scenario, a group of 21 named Iraqi leader agents in 5 factions (FactionSim agents) passed a Turing Test after extensive subject matter expert evaluation by US military personnel, and in the second a separatist movement recreation involving a SE Asian leader (Buddhist) and his Muslim followers (also, FactionSim agents) passed separate correspondence tests (with correlations of over 79% to real world counterparts). In the version of the Turing test we employed, a group of domain experts attempted to distinguish between the behaviors generated using the models of the agents from those in fact generated by the corresponding actual actors. Consequently, this validation procedure may count as both a rigorous face validation test as well as a Turing test. Validity is a difficult goal to achieve, and one can always devise new tests. A strong test, however, is the out-of-sample test that these agents also passed. Thus the SE Asian leader and his followers were trained on different data than they were tested against. Further, a complete model of leader behavior was originally derived from earlier studies of the ancient Crusades [Silverman et al., 2005] and this model was applied to and evolved into the SE Asian and Iraqi domains. The only features updated were the values of the weights for the value trees and various other group relations and membership parameters – all derived from open sources. So the structure of the leader model also survived scrutiny and passed two out-of-sample tests relative to the Crusades dataset. While these may not be definitive tests, they are sufficient for our purposes at this point as we establish that our descriptive agents are useful components for computational what-if experiments, for training worlds, and to drive agents in third party simulators.

In the subsequent three sections, elaborating on what was presented schematically in Figure 9.1, we will overview the three main sources of empirical information we rely on when building complex social systems using our socio-cognitive agent-based model. The three main sources are: (1) country databases in Section 9.4, (2) empirical materials from the world wide web in Section 9.5, and (3) subject matter experts in Section 9.6. In Section 9.5, we will focus on surveying the kinds of automated data extraction technologies that are available today to obtain empirical materials from the web.

9.4 Overview of Some Existing Country Databases

Existing country databases, broadly speaking, fall into one of two categories.* The first consists of event databases that record significant events of interest in numerous countries around the world. These event databases are valuable resources in terms of providing information about parties and factions, their relative alignment, and the resources on which they can draw in various internal conflict and crisis situations that include civil wars, coup d'états, crackdowns, democratic and non-democratic extrications and internal power transitions, mass killings, terrorist activities, and revolutions. The most up-to-date and extensive event database (in terms of the scope and the extent of database coverage) arguably is the Uppsala Conflict Database (UCD) [Uppsala Conflict Data Program], an expanded version of its predecessor, the Correlates of War (COW) [Sarkees, 2000] database. Both UCD (Uppsala Conflict Database) and COW (Correlates of War) contain both inter- and also intra-state conflict information. Given our present goal of studying complex social systems at the country-level, the discussion will focus on the intra-state event databases. COW (Correlates of War)'s intra-state war data has been known as "the granddaddy of all intra-state conflict datasets" and identifies intra-state wars and their participants between 1816 and

---

*Of course, this is only one of the many possible ways to categorize existing country databases. For example, see [Cioffi-Revilla and O'Brien, 2007] for another possible taxonomy of existing databases and, more generally, a good overview of the use of computational analysis in defense and foreign policy research.

1997. The UCD (Uppsala Conflict Database) database significantly improves its coverage in comparison with COW (Correlates of War) by lowering the threshold for conflict identification (from COW (Correlates of War)'s 100 annual conflict deaths to 25) and by more frequently updating its database to cover current developments around the world. Any researcher wanting to gather good snapshots of the histories of significant events of interest from around the world should be able to do so with the combined use of COW (Correlates of War) and UCD (Uppsala Conflict Database). At a minimum, these two databases provide some necessary information on relevant faction identification, relative alignment, and some relative resource estimates. Additional information from other excellent event databases such as the Political Instability Task Force (PIT) [Esty et al., 1998], Minorities at Risk (MAR) [Minorities at Risk Project], Atrocities Event Data (AED) [Political Instability Task Force], Intra-state conflict and Interventions (ICI) [Regan, 2000], and Global Terrorism databases (GTD) [Lafree and Dugan, 2007] can certainly supplement and improve the quality of information for events of interest at the intra-state level. In addition to these significant events of interest, some databases such as Protest and Coercion (PCD) [University of Kansas] and Ethnic Conflict and Civil Life (ECC) [Varshney and Wilkinson, 2006] databases even record events of smaller scale (or involving less violence) such as acts of civil disobediences, demonstrations, rallies, riots, sit-ins, work-stoppages, and strikes. Despite the respectable quality and availability of the aforementioned event databases, these event databases by themselves do not provide sufficient information to populate models such as our joint PMFserv/FactionSim mainly because the unit of analysis used by these databases is events rather than factions. The utility of these event databases is thus limited in terms of providing us with quantified snapshots of our events of interest with readily available faction identification, alignment, and strength information, and we need additional information from a second type of database that focuses on country specific opinion polls and mass attitude surveys.

Populating our joint PMFserv/FactionSim framework with realistic agents requires comprehensive and reliable socio-cognitive information about the people of a particular country at the level of our factions of interest. Gathering sufficient information for one PMFserv subsystem of our Value Systems Module, namely, GSP trees, requires a high level of detailed information about people's goals, standards, and preferences, and obtaining this information can be a daunting task. To our relief, we have access to an extensive collection of survey results complied by survey researchers around the world. The three main publicly available databases in this field are the World Values Survey (WVS) [European Values Study Group and World Values Survey Association], the Global Barometer Surveys (GBS) [Global Barometer Surveys Program], and the Comparative Study of Electoral Systems (CSE) [Comparative Study of Electoral Systems Secretariat]. Both WVS (World Values Survey) and GBS (Global Barometer Surveys) are surveys that are administered in more than 50 countries around the world to measure public opinion on cultural, social, political, and economic issues. The key difference between the two surveys lies in the fact that WVS (World Values Survey) uses a standardized survey questionnaire, while GBS (Global Barometer Surveys) is administered more frequently. Similar to WVS (World Values Survey) and GBS (Global Barometer Surveys), CSE (Comparative Study of Electoral Systems) also tracks public opinion, except that it is held only in countries where there are periodic and reasonably fair elections and focuses on micro-level information on vote choice, candidate and party evaluations, and other relevant information regarding voters' attitudes and values, in addition to standardized socio-demographic measures and aggregate level information on electoral returns and turnouts. It seems, then, that we should be in a position to extract the information we need from these three surveys. Yet, there are two difficulties we face in using the results from these survey instruments for our purpose. The first difficulty lies in the fact that it is hard to find a one-to-one correspondence between a survey questionnaire item and a parameter of, say, our GSP tree. This is an obvious and unavoidable difficulty given that survey researchers did not design their surveys with our GSP tree parameters in mind. This difficulty, however, is not insurmountable; with some effort, we can select survey questionnaire items that can serve as proxy measures for our parameters of interest. The second difficulty lies in the fact that the unit of analysis for these public opinion surveys is countries while, for many modelers of complex social systems-such as countries-use a unit of analysis that is smaller than a whole country.* For our joint socio-cognitive PMFserv/ FactionSim framework, the appropriate unit of analysis is at the faction level. Again, this difficulty that results from the discordance in the unit of analysis can be overcome simply by cross tabulating and sorting these survey databases according to properties that categorize survey respondents into specific groups that match our interests. The surveys are sufficiently detailed to allow us, for example, to infer information about whether an average supporter of a particular political party has a more or less materialistic vision of life than another average supporter of another political party or a different faction. We may even be able to infer a more "elite-level" information (leader-level information within out PMFserv/FactionSim framework) by cross-tabulating our proxy survey items for a particular parameter of interest with the socio-economic information about the respondents, given that leaders are more likely to have higher educational attainment and income and/or are more likely to spend more time in a particular organizational grouping.

In sum, the existing country databases – both the event and the survey ones – are great assets for those of us in the modeling and simulation community who are committed to using realistic agent types to populate our simulated world. However, as noted previously, using existing databases at this stage of their development requires efforts by the modeling and simulation community to study the structure of the available databases and to take into account their strengths, weaknesses, terminology, and idiosyncrasies. In this regard, we need to be creative in finding proxy measures that can reasonably approximate our parameters of interest, and we would need to be imaginative in restructuring our databases in ways that are conducive to extracting the information we want at the level of analysis we want. Finally, it is important to note that the preceding overview of some existing country databases is not exhaustive. There are obviously more event and survey databases than the ones mentioned, not to mention other specialized economic, social, demographic, human capabilities, and political violence databases. We guide interested readers to the Penn Conflict Database Catalogue [Kim and Bharathy, 2007] for a more comprehensive overview of the existing databases that may be of relevance to the modeling and simulation community.

9.5 Overview of Automated Data Extraction Technology

As discussed in the preceding overview of existing country databases, our use of these databases is not as efficient and convenient as we might like it to be; nonetheless, these databases are invaluable to our project of building realistic agents and validating our models. Still, it is important to note that we may not be able to gather all the empirical information we need from these databases alone, and we will at times be forced to collect

---

*One can make an argument that the unit of analysis can also be individual survey respondents in particular countries. In this case, many modelers of complex social systems – such as countries – would be using a unit of analysis that is larger than an individual of a particular country.

TABLE 9.2 A Summary of Some Existing Country Databases

Atrocities Event Data (AED)	Event	A database collected by Kansas Event Data System to provide a systematic sample of atrocities occurring around the world.
Correlates of War (COW)	Event	This database is known as "the granddaddy of all conflict datasets" and most other conflict event datasets either extend or improve this pioneering database.
Comparative Study of Electoral Systems (CSE)	Survey	This database provides individual level information on vote choice, candidate and party evaluations, current and retrospective economic evaluations, etc. as well as aggregate level information on electoral returns and turnouts.
Ethnic Conflict and Civil Life (ECC)	Event	This database provides comprehensive information on all Hindu-Muslim riots reported in the major Indian newspapers in India and Pakistan. The database is expanding to cover other countries in Southeast Asia.
Global Barometer Surveys (GBS)	Survey	This database is one of the two most comprehensive survey databases of public opinion of people around the world. Administered by regions and decentralized.
Global Terrorism Database (GTD)	Event	This database contains information on both domestic and international terrorist events around the world since 1970. For each event, information is available on the date and location of the incident, the weapons used and nature of the target, the number of casualties, and the identity of the perpetrator.
Intrastate Conflict Interventions (ICI)	Event	Unlike some other event databases, this conflict database contains information about external/foreign interveners who play a role in intrastate conflicts in addition to the standard information (time, location, involved actors, etc.) in an events database.
Minorities at Risk (MAR)	Event	This database provides comprehensive information on the status and plight of politically-salient communal/minority groups in countries around the world.
Protest and Coercion Data (PCD)	Event	This database provides comprehensive information on protest and coercion in 32 countries around the world. Information concerning date, day, action type, location, protest groups, targets, and the strength of the protesters are provided. This database is collected using the Kansas Event Data System.
Political Instability Task Force (PIT)	Event	This database contains standard event information on political instability events in most countries around the world. The political instability events include ethnic wars, revolutionary wars, genocides and politicides, and adverse regime changes. Adverse regime changes do not include transitions to more open and democratic forms of governance.
Uppsala Conflict Database (UCD)	Event	This database probably is one of the most comprehensive and widely used event databases by social scientists. It contains information on the location of the conflict, the source of incompatibility, opposition organizations, date and duration, conflict intensity level, the nature of conflict settlement, the quantity and quality of arms, and the duration of peace settlement.
World Values Survey (WVS)	Survey	This database is the other most comprehensive survey of databases of public opinion around the world. Unlike GBS, it has a standardized survey questionnaire but is administered less frequently.

some information by ourselves. Instead of going through the typically labor-intensive data collection process with an army of undergraduate and graduate research assistants, we have a new set of tools and technologies that can streamline our data collection efforts, making them less arduous and more efficient. The most promising approach seems to be the use of various automated semantic analysis tools such as Automap [Carley et al., 2006], General Inquirer, Cultural Simulation Model (CSM) by IndaSea [Park and Fables, 2007], Profiler+, Kansas Event Data System (KEDS), ReadMe, The Resource Description Framework Extractor (T-Rex) [Subrahmanian, 2007], OASYS, and STORY to extract our information of interest from various newsfeeds and web scrapings.

The majority of these automated content analysis tools work according to a similar underlying logic. They contain a list of terms of interest together with their synonyms. They then count the frequency with which these terms and their synonyms appear in an actual text to provide us with usable data. For example, if we want to collect information from various newsfeed and web scripts about a faction leader's propensity to use violence, we would build a list consisting of both generic and also highly specific words pertaining to the use of violence by this leader (words ranging from "killing" to "tire necklace," for example) and let these programs count the frequency of such words in various texts. This procedure requires some degree of simplification of the phenomenon under study. However, no matter how sophisticated, all automated data extraction tools follow this essential underlying logic. Many have specialized search algorithms that allow the program to look for more fine-grained information of interest.

There are more than two dozen available automated content analysis tools. We briefly survey a few of them that we have used or are planning to use for our data collection efforts. AutoMap is an extraction tool that specializes in collecting information about key actors, their relationships, and their relation to an event or a set of events of interest. Automap also provides the attributes of actors including roles (leaders / followers), psychological factors, and resources. On the basis of such information extracted using AutoMap, we can then extract further information concerning groups and the entire structure of social networks of individuals and groups with the use of additional tools such as Organizational Risk Analyzer [Carley and Reminga, 2004]. T-Rex (The Resource Description Framework Extractor) uses cultural, economic, political, social, and religious variables provided by social scientists in conjunction with other data sources such as surveys and event databases and automatically extracts relevant data from news outlets, blogs, newsgroups, and wikis. STORY crawls the web at 50,000 pages/day and extracts facts and schemas. OASYS (Opinion Analysis System) is a specialized content analysis tool that is designed to extract information in real time from over 100 news sites in 8 languages and 12 countries regarding actors' opinions about any given topic, together with a measure of the intensity of these opinions. General Inquirer and ReadMe are more generic and less specialized examples of automated content analysis software that takes a set of text documents as input and processes these texts into various categories chosen by the user. The seminal tool of this field, KEDS (Kansas Event Data System), is specialized for generating event data and has the most extensive databases constructed using its system while Profiler+ performs leadership style analysis by looking for specific words that indicate leadership traits. Table 9.3 summarizes these tools by specialty.

The prospect of using these tools is exciting. However, there are at least six challenges of varying degrees of difficulty that confront potential users. As a test case, IndaSea helped us to use their CSM (Cultural Simulation Model) tool to profile President Musharraf of Pakistan. One strand of the results is shown in Table 9.4. Here we are looking at one of the standards of the GSP trees dealing with military doctrine – specifically, the tendency to shun violence.

TABLE 9.3 A Summary of Some Automated Data Extraction Tools

Name	Specialty
AutoMap	Relevant actor identification and relationship extraction
CSM by IndaSea	Generic
General Inquirer	Generic
KEDS	Event data extraction
OASYS	Opinion on any given topic
Profiler+	Leadership style/trait extraction
ReadMe	Generic
STORY	Generic
T-REX	Generic

TABLE 9.4 Example Output of One of the Automated Web-Scraping Tools Targeting a Personality Trait of an Illustrative Leader (Musharraf's tendency to shun violence)

Standards: Shun Violence	
Keywords	Peace (Some search terms for the opposite of shunning violence: military, army, tactics, strategy, operation, operations, tactic)
Positive Evidence	**05-06-2007** You are saying that this problem will not be solved by fighting, but it will not be solved by peace either. This problem will not be solved unless the main sources and centres of this movement – the Pakistani ISI – is not shut down. **06-18-2007** In addition to disturbing the people in Afghanistan, the ISI has another programme which is to defame Islam across the world. It also comments on the Islamic instructions which urge people to make every possible effort to ensure peace and stability and prevent bloodshed.
Negative Evidence	**04-01-2007** According to the Washington Post, officers within Pakistan's intelligence agency, the Inter-Services Intelligence Agency, proposed the following idea to address the vulnerability of its nuclear weapons to an Indian attack: "Let's hide them in Afghanistan-the Indians will never be able to attack them there." **04-02-2007** There have been violent clashes in the tribal area in the past few days between tribal militia groups, which are difficult to distinguish from the Taleban linked to the ISI, and those groups that cooperate with foreign terrorists. In all likelihood, the ISI is behind this development, because an ISI patrol has been ambushed for the very first time, almost certainly in retaliation. **04-10-2007** According to her, in the fall of 1993 her assassination was ordered and the "chosen assassin was a Pakistani with ties to the ISI during the Afghan Jihad. His name was Ramzi Yusef. He had participated in the first attack on the World Trade Centre in New York earlier that same year, on February 26."

(1) Coverage, as already mentioned, is a concern with the databases, but not any more than it is with the newsfeeds. Any given country may or may not have an open, free press, so the viewpoints available and indeed, the veracity of what is published may be called into question. Where there is a free press, one must be sure that all views across the political spectrum are captured and appropriately tagged. These issues may render the newsfeed extraction problematic for certain of the parameter sets of interest. In Table 9.4, Musharraf's case, while there may not be a totally free press within Pakistan, the Pakistani president is a high visibility individual, and there is coverage at least by the Western press, a press with its own worldview.

(2) Another challenge in using automated content analysis tools lies in building the catalog that contains the necessary categories of key words and their combinations, both (or all) of which represent our model parameters. The main snag we face on this front lies in building a truly comprehensive and accurate catalog of keywords for the machine to use in extracting information from the exponentially growing quantity of available machine-readable text. Programming software that looks for such keywords and their combinations and counts their frequency is not difficult to build, and such programs already exist. There are also readily available generic catalogs or "dictionaries" that contain categories of words for us to import into a program and use for content analysis. However, these categories are sometimes too broad and generic for our purpose of extracting very specific information of interest, and thus these available "dictionaries" may be of only limited use. Table 9.5 shows a simple search with a single keyword ("peace").

(3) In addition to having proper keyword synonyms, it also may be that a schema or model of a given parameter has to be constructed to accommodate the interpretation and transformation of proxy variables. For example, in the row designated "positive evidence" in Table 9.4, the second item found is not necessarily proof that Musharraf (via the ISI (Inter-Services Intelligence)) shuns violence so much as it is evidence that he has a program to defame Islam. This suggests that perhaps a schema of typical human actions that do and do not constitute "shunning violence" might be able to weed out such an item by classifying it as "inconclusive." It is possible that this can be learned automatically if we provide our prior hand-coded models. However, that is still an untested assumption.

(4) One needs to test the error rates of all the extraction tools. This implies assembling a test corpus in addition to a training data set where all the ground truth is known. One can then measure precision and recall rates and determine if the extraction tools are doing a credible job or not. While they may work well on other test sets for which they were designed for, one must always recalibrate their performance for the types of searches and extractions of interest to a given model. For example, Table 9.4 shows the retrieval of two bits of positive evidence and three bits that are negative. Are these all the items to be found? If we extend the keyword list and add a model/schema of the behavior, what happens to precision and recall? This is an aspect of our project that merits some research.

(5) Seamless integration is a desirable objective and one would like whatever extraction technology is adopted to be invisible to end users. In many cases there are setup issues and challenges for using the output of these tools. This is mostly a question of effort needed to embed these tools so the end user will not need to deal directly with them.

(6) Even if all the other challenges are eliminated, a remaining issue is how to weigh all the evidence collected assess its reliability, and transform it into actual parameter estimates. For example, how do we get the computer to summarize Table 9.4? Is it just a matter of adding two positives and three negatives? Obviously that would be simplistic and misleading, particularly when some of the positives are mild or inconclusive whereas some of the negatives are extreme items such as assassination attempts that may or may not be tied to Musharraf's ISI (Inter-Services Intelligence). How do we combine such bits of

evidence? Some of the more difficult aspects of a text and knowledge extraction tool have to do with understanding a personality and determining its underlying motivations. This is a hard problem, and human analysts who work at a "country desk" or a "leader desk" tackle it according to a well-developed methodology that we recap in Section 9.7.

9.6 Overview of Subject Matter Expert Studies/Surveys

The most obvious and intuitive method of obtaining information we need for our model parameters is to simply to ask subject matter experts (SMEs) to provide this information in our preferred format for our countries of interest. Let us suppose that we are modifying and populating our joint PMFserv/FactionSim framework to build a virtual country for the purpose of, say, better understanding and simulating potential political instabilities in this particular country of interest. In this scenario, we would be particularly interested in modeling and analyzing how Diplomatic, Intelligence, Military and Economic (DIME) actions might affect the Political, Military, Economic, Social, Informational, and Infrastructure (PMESII) systems of the country, and, given the importance of this kind of project, we would like to use the most up-to-date and accurate information for our country of interest. Knowing the limitations of the two previously discussed means of extracting information–namely, country databases and automated data extraction tools–in the short term at least, we might in fact be better off by gathering information directly from the best available country experts, tapping their expertise by means of a survey questionnaire to them or by conducting open-ended interviews. For our purposes, administering a structured, self-explanatory web survey tailored to elicit exactly the information we need would in most cases be preferable to conducting unstructured, open-ended interviews (partly because these interviews would elicit a wealth of information that would then need to be sorted and coded).

There are three main difficulties associated with using subject matter experts to elicit the information we need. First, administering interviews with experts in either form – expert survey or open – ended interview-requires significant financial and human resources. This method of collecting information costs at least as much as – and in most cases considerably more than – the other previously discussed options. Unless we are fortunate enough to have high quality SMEs available to us on a pro-bono basis, seeking their expertise for a task such as filling the more than fifty parameters for the GSP Tree alone may be prohibitively complicated and expensive. Second, subject matter experts, by definition of being subject matter experts and by virtue of being human and therefore fallible, may sometimes provide us with biased and, from time to time, even blatantly incorrect information: e.g., see [Tetlock, 2005; Heuer, 1999]. To limit this bias, we would probably want to consult more than one subject matter expert on any particular country or topic. More importantly, being a country expert does not mean that one has complete and comprehensive knowledge; a country expert does not know everything there is to know about a country. Third and finally, simply finding subject matter experts for a particular country of interest may by itself pose a significant challenge. Social scientists, historians, and area studies scholars with specific country expertise are not in short supply, but their expertise is not evenly distributed around the globe; given the structural constraints that exist in academia, certain parts of the world and certain countries receive disproportionate attention, while others are relatively neglected (for example, there is a glut of available expertise on China, but much less expertise on countries such as Bangladesh and Fiji). In sum, while at first this most direct route of getting parameters from experts looks easy and straightforward, it is also beset with difficulties.

Authoring a survey (or a set of slider-bar GUI screens) that is self-explanatory and has validated questions about each parameter needed in a socio-cognitive agent model is time consuming, but not intellectually difficult. Such a survey is needed for eliciting knowledge from country or leader desk experts. A different approach that is an extension of this one involves a distributed set of experts, each knowledgeable about a sub-part of the ethno-political region to be modeled. The US military today, for example, currently plans for three sets of multi-person teams to perform this task for an area of operations. These three types of teams were listed earlier in Figure 9.1 as the Human Terrain Team (HTT), Intel and Every Soldier a Sensor (ESS) team, and Civil Military Affairs (CMA) team. The Human Terrain team includes anthropologists and social experts who collect data that is directly relevant to profiling agent personas, their clan structures, attributes, and kinship links. The Intel and Every Soldier a Sensor group tends to collect data pertaining to biometrics, demographics, intent, and information flow patterns in the target region. The Civil Military Affairs team focuses on quantitative estimates of resources, facilities, jobs, economic activity, infrastructure, and the like. For a model like FactionSim-PMFserv, all of this data is important. At present most of it is collected and entered into databases. In the future, one can envision the agent-based models as being the main repositories of such information. This would both improve the data collection focus and provide tools for the analyst and trainee that are sensitized to the DIME-PMESII issues of the area of operation. Getting to that point may be a grand challenge worthy of a DARPA (Defense Advanced Research Projects Agency) style program given the scale-up entailed by such a distributed activity.

9.7 Overview of Integrative Knowledge Engineering Process

Some of the more difficult aspects of a text and knowledge extraction tool have to do with understanding a personality and determining its underlying motivations. This is a hard problem, and human analysts who work at a "country desk" or a "leader desk" approach it according to a well-developed methodology, though even they are subject to errors of omission and commission, biases, or slipups. We have studied that methodology during the years of assembling the Athena's Prism diplomatic role playing game [Silverman et al., 2005] and have adopted our own version of it for the leaders and followers we have profiled. We published an account of that methodology in [Silverman and Bharathy, 2005] and recap it very briefly here since it is the essence of the automated knowledge extraction workbench we are trying to assemble.

Since multiple sources of data are involved, a process is required to integrate and bring all the information together. We employ a process centered around differential diagnosis. This design is also based on the fact that directly usable numerical data are limited and one has to work with qualitative, empirical materials. Therefore, in the course of constructing these models, there is the risk of contamination by cognitive biases and human error.

The burden of this integrative modeling process is to systematically transform empirical evidence, tacit knowledge and expert knowledge from diverse sources into data for modeling; to reduce, if not eliminate, the human errors and cognitive biases (for example, for confirming evidence); to ensure that the uncertainties in the input parameters are addressed; and to verify and validate the model as a whole, and the knowledge base in particular.

For lack of a better term, the process has been conveniently referred to as a Knowledge Engineering (KE) process due to extensive involvement of knowledge engineering techniques and construction of the knowledge models. A diagrammatic representation of the knowledge

Challenges of Country Modeling with Databases, Newsfeeds, and Expert Surveys

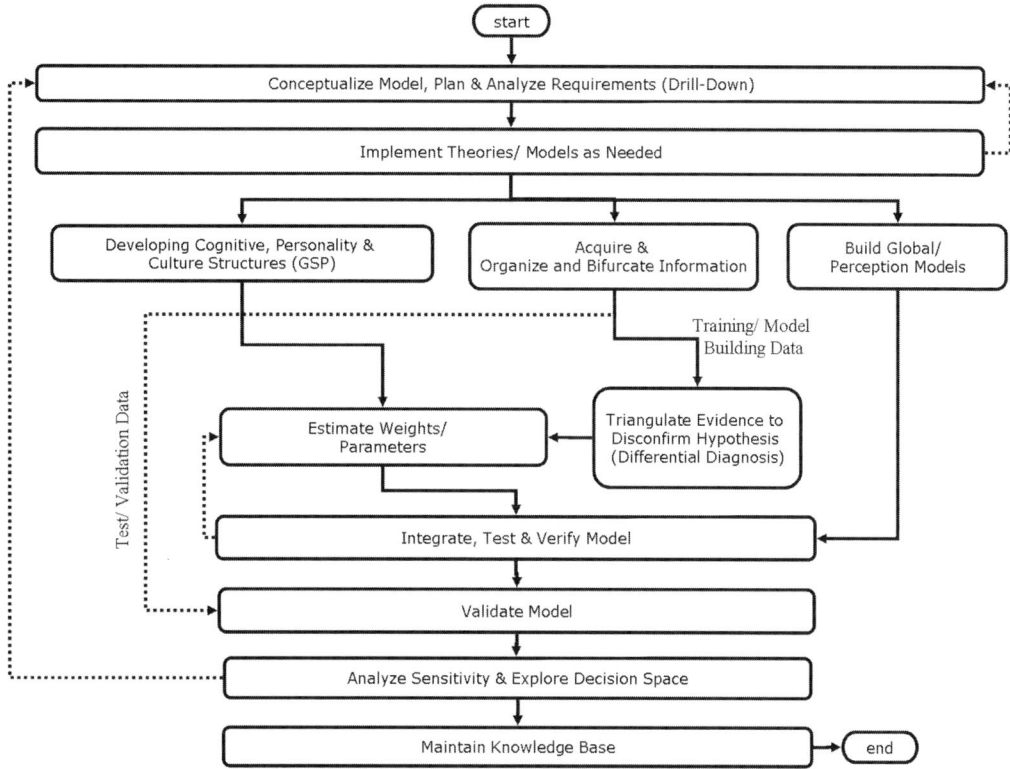

FIGURE 9.5 Knowledge Engineering Process Summary.

engineering process is given in Figure 9.5.* The details of the process are beyond the scope of this chapter, but a summary of the methodology has been given below in Figure 9.5. Let us describe the salient features of the method.

Firstly, the body or corpus of qualitative information from different sources is aggregated and thematically organized in an evidence table. The input from experts and country database output, which directly pertains to the parameters, may be employed to help set the initial parameter values in the model, while anecdotal expert inputs and tangential estimates from the country databases are also incorporated in the evidence table. In order to ensure separation of model building (training and verification) and validation data, the empirical materials concerned are longitudinally divided into two different parts. One part is set aside for validation. The model is constructed and verified using the remaining part.

Since organizing information from otherwise diverse or amalgamated sources is critical to the success of the remaining modeling activities, a modified content analysis process is employed to collate and organize the evidence thematically. The themes of relevance are obtained from the high-level goals, standards, and preferences as well as from people and general potential behaviors of interest in the domain. The body of materials describing behavior in the model is split up into records, with each representing one and only one theme. Then, these are assigned theme codes, relevance, and reliability (subjective estimate of the source or info), and sorted according to the themes. The output is organized information

---

*This is a simplified view. Full details can be found in [Bharathy, 2006].

in tabular form with additional attributes such as reliability, frequency of occurrence, and relevance. Alternative hypotheses are selected at this juncture. The following table shows an excerpt from the evidence table, pertaining to the behavior of Richard the Lionheart in the Crusades.

TABLE 9.5 Sample Evidence Table

Theme	Evidence	Reliability	Relevance
Economy	Amasses wealth in battles	Very High	...
Military	Conquers territory 1, etc.	Very High	...

Having collected the data, one must integrate the data to arrive at the estimate of the parameters. Several tools and techniques have been devised for this purpose. Among them are tools for differential diagnosis (differential diagnoser) and pairwise comparison, which help elicit parameters in the graphical models through a systematic, defensible and transparent process. These tools accompany a mathematical framework, and contain provisions for estimating uncertainties in the expert inputs and empirical evidence.

When constructing models of behavior from evidence (be it from empirical evidence taken from the literature or from expert input), the modelers (or experts) employ the cognitive process of determining the motives of someone's behavior. This is subject to several biases [Kahneman et al., 1982; Gilovich and Griffin, 2002]. Foremost among these is confirmation bias, which may also subsume other biases such as availability bias and attribution bias. Simultaneous evaluation of multiple, competing hypotheses is very difficult to do and is against the cognitive bias of the human mind. Without an instrument designed to counter this fundamental bias, a modeler attempting to build Value Tree Models may easily be misled. The tendency to build models by confirming a plausible but favorite hypothesis will have enormous and grave implications at the next stage, when collating evidence provided by experts and empirical materials and building Value Tree Models. The human mind works though a "satisfying strategy." The process of selecting a favorite hypothesis is highly influenced by ones own conditioning, and the tendency is to see what one is looking for, and to overlook alternatives. Assessing evidence and attributing behavioral traits should ensure that no external cause explains the same behavior, that other competing traits do not explain the same behavior, and that confirmation bias is eliminated by a disconfirming hypothesis [Gorman and Gorman, 1984; Heuer, 1999].

In order to minimize the risk of not considering alternatives and considering non-diagnostic evidence, we have provided a tabular design, to carry out Differential Diagnosis, for generating and weighting alternative hypotheses, explanations or conclusions. Accordingly, the process forces one to look at competing hypotheses, and methodically disconfirm these alternatives rather than simply confirming a first hypothesis using available evidence. Once again, testing the usefulness and effectiveness of this differential diagnosis tool will be an important part of this project. Later, we will look at how this tabular structure can be exploited quantitatively.

The hypotheses in this case are parameters such as the nodes in the Value Trees (goals, standards and preferences of the characters being assessed). On the left, the framework includes the Organizer containing key evidence, thematically coded and attributed with reliability and relevance. However, the tool for Differential Diagnosis is centered on the hypotheses and evidence. Essentially, the hypotheses are pitted against the evidence through this matrix. If reliable evidence rejects a hypothesis, then the likelihood of that hypothesis is diminished significantly. We advocate including reliability and relevance for each piece of evidence. Relevance (from the Organizer) identifies which items are most helpful in judging the relative likelihood of the hypotheses, and helps control the time spent on what appears to be irrelevant evidence. Using the Bayesian framework, we have developed a quantitative technique for the differential diagnosis, which attributes higher weight (an order of magnitude or more) to disconfirming evidence. We also established that otherwise rare events weigh more when they do occur as evidence. Essentially, the hypotheses are pitted against the evidence through this matrix.

The approach we suggest is to take all competing hypotheses (H_j) that explain a set of evidence and then pit them against this evidence (E_i). We find it best to work with a confirmation index that weighs disconfirming evidence about an order of magnitude higher than confirming evidence. Let us call this process of estimation based on disconfirming evidence "differential diagnosis," a term found in medical decision-making. Differential diagnosis is nothing but a triangulation technique. This technique much used in the field of medicine, where observing and discovering evidence leads gradually to a consideration of a short list of illnesses most likely to be behind a particular set of symptoms. While we share with medical diagnosis the same intent of unearthing the most likely hypotheses, there are some minor differences in terms of means and ends. Our purpose is not to identify a specific cause, but instead to attribute behavioral evidence to hypotheses of causes. While medical diagnosis tends to favor testing hypotheses largely serially and in a qualitative fashion, our models involve running several tests simultaneously by considering a set of competing hypotheses and triangulating a large set of evidence, as quantitatively (or quasi-quantitatively) as possible.

While both methods give more weight to disconfirming evidence, we give some (but lesser) consideration to confirming or supporting evidence. While disconfirmation is a much more powerful technique compared to confirmation, the latter provides some weak, yet economical, diagnosis in the absence of disconfirming evidence. We also take into consideration the reliability of data and typicality of events. The main difference might be in our employment of an explicit and simple tool that is amenable to both Bayesian analysis and also simple, score-based decision support.

This results in the following simplified expression for a metric called Confidence Index. Mathematically, Confidence Index (CI_{Avg}) for a given Hypothesis (H_j) may be defined as the weighted average measure of all the confirmations (and disconfirmations) associated with a hypothesis (with the subscript denoting that it is an average index over the given hypothesis):

$$CI_{Avg}(H_j) = \frac{1}{n} \times \sum_{i=1}^{n} K \times C_{ij} \times R_i \times \frac{f_i}{f_{R_i}}$$

where $K = \{w_1$ when $C_{ij} \geq 0$, and w_2 when $C_{ij} < 0\}$.

Essentially, K is used to assign a higher weight (say an order of magnitude) to disconfirming evidence ($w2 \gg w1$). We have used w_1 value of 1 and w_2 value of 20. f_i is the frequency of the evidence, if the evidence given summarizes separate occurrences of behavioral evidence. Similarly, f_{R_i} is the typicality of the evidence (indicator of frequency of seeing that type of evidence in the real world. That is a measure of $P[E_i]$). Reliability (R_i) is subjectively estimated based on the source of the evidence as well as the confidence with which the evidence has been outlined by the source. For the sake of illustration, ignoring f_i and f_{R_i}, the expression for CI_{Avg}, this may also be simplified as:

$$\frac{1}{n} \times \sum_{i=1}^{n} K \times C_{ij} \times R_i$$

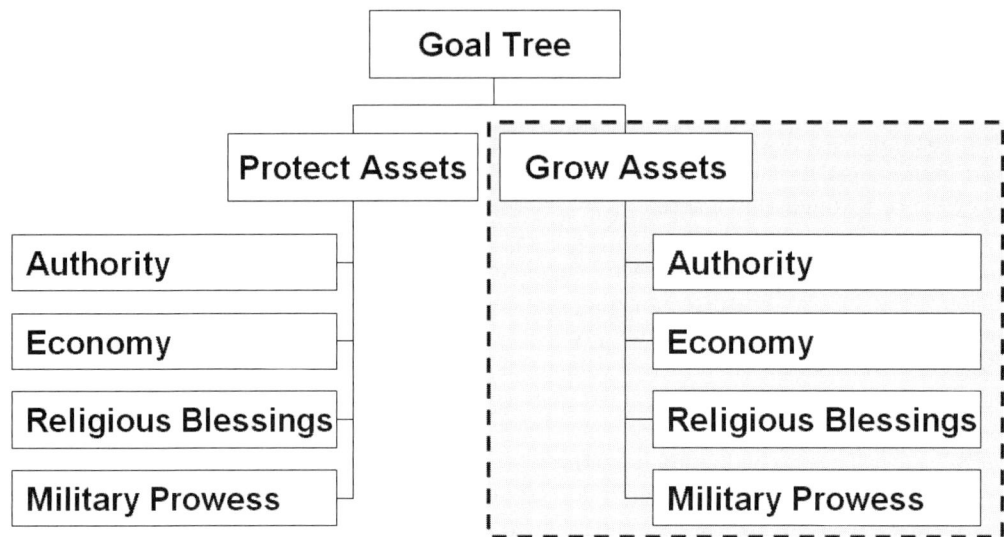

FIGURE 9.6 Segment of Goal Tree.

The competing hypothesis that has the highest positive confidence wins only if the hypotheses are mutually exclusive, if the difference in CI is significant ($CI_{Avg} > 1.0$), and if the variance is small. For hypotheses which are not mutually exclusive, ordinal ranking might be obtained. When mutually exclusive hypotheses cannot be clearly distinguished by their confidence score, multiple competing hypotheses might have to be entertained during the course of the sensitivity analysis. Differential diagnosis allows one to consider all relevant evidence at once, and also gives higher weight to disconfirming evidence as described above. It allows one to find out whether these hypotheses could be ranked in the context of all available evidence. The details pertaining to the derivation and use of differential diagnosis have been taken from [Bharathy, 2006].

Now, let us consider the following cases to illustrate this technique. Differential diagnosis in the Crusade example has been illustrated through the following stylized cases. Note that simplifications have been made to introduce and illustrate the technique.

Example Question Again, consider the character of Richard the Lionheart. There are a few hypotheses (that could form the basis for some selected nodes of the GSP Tree) offered to explain Richard's spending time on a number of wars. Is Richard's inclination to grow any of the following resources (expansion of empire, wealth, religious blessings, or military prowess) more influential than other inclinations in explaining his behavior? Could his inclinations be ranked?

Let us formulate the above questions into the following competing hypotheses: Growth and expansion of the empire (Authority: H1), wealth (Economy: H2), religion (Religion: H3), or whether he loves warfare for its own sake (Military prowess: H4).

H1: Grow Authority, Expand and Rule: Richard is an expansionist and wanted to expand his kingdom and authority.

H2: Protect Authority & Govern: Richard fought to protect his authority from enemies, and the wars were thrust upon him.

H3: Fill the Coffers/ Max Econ Benefits: Richard wants to grow his economic assets through fighting wars.

TABLE 9.6 Stylized Example of Differential Diagnosis

Theme Code	Evidence (Ei) √ = Evidence supports, or can be made to support, the hypothesis × = Evidence rejects the hypothesis	Reliability (Ri)	H1: Expansionism	H2: Protecting the Kingdom	H3: Economy	H4: Religious Duty	H5: Loves wars	H7: Loss of Control/ Due to External Circumstances/ Unexplainable	Frequency of sighting this type of event: P(E(e)) ('Typical?')
M1	Amasses wealth in battles	V. High				√			High
M2	Conquers territory 1, territory 2 etc.	V. High	√	√	√	√	√	×	Low
E2	Sells territory conquered	V High	×	×	√				Medium
A1	Seldom governs the lands he has authority over.	V. High	×	×				×	Low
E1	Spends excessively on battles.	V. High	√		×		√	×	High
E3/A3	Known to have said: "If I could find a buyer, I would mortgage London to raise money for battles"	Medium	×	×	√		√		Medium
A4	Spends most of his 9 year rein outside England on Crusade	V. High	√	×		√	√	√	Low
R1	Has not obeyed other religious laws	Moderate				×		×	High
R2	Fought against his father first	V. High	√			×	√	×	Low
E4	Died fighting over a treasure	V. High			√				Low
	Confidence Index								

H4: Religious Duty: Richard wants to protect and grow his religious blessings.

H5: War for the War's Sake: loves the battlefields and wants to fight wars for the war's sake.

Then, we construct the table (Tables 9.6, 9.7)* and pit all of these hypotheses against the available sets of evidences. As one can see in Tables 9.6, 9.7, a number of rows of evidence disconfirm Richard's religious inclination, while there is little that contradicts Richard's inclination to grow military assets. It should not surprise the reader that Richard seems most inclined to grow his military prowess, followed by his desire for wealth, and his desire to govern, in that order. Therefore, this is a behavior that may not provide much additional information for identifying and sifting through his values. However, his other behaviors begin to contradict some of the existing the hypotheses.

In addition to the above use of differential diagnosis, where we illustrated the process of disconfirming hypothesis with available evidence, the same technique could be employed in different forms. For example, an expert could be encouraged to come up with different plausible scenarios. Once such a set of scenarios has been gathered and recorded, the expert could be asked to carry out differential diagnosis using these scenarios. The expert then attempts to disconfirm the hypothesis using the scenarios he or she has generated. This thought experiment could work as a powerful technique.

Introspection, Revision and Dialog: The degree of disagreement can be used to generate feedback to the experts themselves, and their assumptions can thereby be made transparent. Then the expert can redesign the GSP Tree while con-

---

*For the sake of simplicity, I have used the expression that $CI_{Avg} = 1/n \sum_{i=1}^{n} K \times C_{ij} \times R_i$, where $K = w_1$ when $C_{ij} \geq 0$ and $K = w_2$ with $C_{ij} < 0$. Essentially, K is used to assign a higher weight (say an order of magnitude) to disconfirming evidence ($w_2 \gg w_1$). Other forms of this relationship are being investigated. Additionally, thought experimenting the plausibility of generated scenarios disconfirming hypotheses may be employed as another input in this process.

TABLE 9.7 Stylized Example of Differential Diagnosis

	Stylized Example of Differential Diagnosis		Hypotheses						
Theme Code	Evidence (Ei) +ve Entry = Evidence supports, or can be made to support, the hypothesis -ve Entry = Evidence rejects the hypothesis	Reliability (Ri)	H1: Expansionism	H2: Protecting the Kingdom	H3: Economy	H4: Religious Duty	H5: Loves wars	H7: Loss of Control/Due to External Circumstances/Unexplainable	Frequency of sighting this type of event: P(E|c) (Typical?)
M1	Amasses wealth in battles	0.9			+0.3				High
M2	Conquers territory 1, territory 2 etc.	0.9	+0.9	+0.6	+0.3	+0.8	+0.9	-0.8	Low
E2	Sells territory conquered	0.9	-0.9		+0.9				Medium
A1	Seldom governs the lands he has authority over.	0.9	-1.0	-1.0				-0.5	Low
E1	Spends excessively on battles.	0.9	+0.5		-0.9		+0.8	-0.8	High
E3/A3	Known to have said: "If I could find a buyer, I would mortgage London to raise money for battles"	0.5	-0.9	-1.0	+0.5		+0.9		Medium
A4	Spends most of his nine-year reign outside England on Crusade	0.9	+0.7			+0.7	+0.8	+0.5	Low
R1	Has not obeyed other religious laws	0.5				-0.6		-0.5	High
R2	Fought against his father first	0.9	+0.5			+0.8	-0.8		Low
E4	Died fighting over a treasure	0.9			+0.9	-0.2			High
	Confidence Index (with K=1 when Cij>=0, K=20 when Cij <0)		-4.1	-2.8	-1.4	-0.8	0.3	-5.7	

sciously bracketing one or more assumptions. This kind of exercise can also be useful in group sessions to discuss the differences. In essence, it can create introspection and dialog, which will often focus attention on the root of the actual problem being studied. In a more superficial treatment, a structure could be adopted through a consensus seeking process, or by bootstrapping, or differently weighing expert and lay designs.

Uncertainty Estimation: The estimates can also provide estimates of the uncertainty (or confidence) in the GSP Tree.

We have employed this process manually in the past to create several models of leaders, followers, crowd members, rebels, agitators in conflict situations. We have been able to validate our integrative process under naturalistic conditions by testing, verifying and validating these models. As mentioned earlier, the process does get very laborious when constructing multiple models by hand. Therefore, we are in the process of automating the previously described manual process, incorporating text-mining, semantic analysis as well as Bayesian update.

9.8 Concluding Remarks

Our community would be remiss if it did not try to respond to the ideas of leaders in military and diplomatic circles who are now facing the challenge of promoting deeper thought, creating rehearsal environments, and developing analytic capability about cultural issues and local population needs/wants around the globe. They have funded programs that collect country data and conduct link analysis and social network studies. At the same time, they may lack the experience or expertise to appreciate the tools that the field of human behavior modeling currently has to offer, or is now in the process of developing.

In this chapter, we have argued that the available country datasets are an invaluable resource that will permit us in the human behavior M&S (Modeling and Simulation) field to more realistically profile factions, and their leaders and followers. This in turn will help us to develop tools for those interested in analyzing alternative competing hypotheses for DIME-PMESII (Diplomatic, Informational, Military, and Economic actions – Political, Military, Economic, Social, Informational, and Infrastructure effects) studies. At the same time, there are significant growing pains and challenges involved in trying to put the country data to use. This chapter reviewed those challenges by looking at three pathways for extracting and parameterizing the data – webscraping of newsfeeds, extracting and translating data from country databases, and (semi-) automated surveying (i.e., web questionnaires with data translation and model instantiation capacity) of subject matter experts. In each of these areas there are significant challenges and obstacles to seamless integration, not the least of which is that profiling individuals and groups is difficult even for the smartest humans. By using a triangulation of the three approaches, and a knowledge engineering approach that mimics how country and leader experts currently do the job (alternative competing hypotheses), we believe that one can move ahead as outlined in this chapter.

This chapter examined how to use this approach with the help of a case study involving a socio-cognitive agent architecture (FactionSim-PMFserv). The hope is that the automated extraction will speed the development of gameworlds and scenarios with a tool like this. This push seems doubly pertinent since a parallel development in recent years has been the scientific struggles of those working to unify multi-resolution frameworks that permit modeling "deep" modeling of a small number of cognitively-detailed agents able to interact with and influence large numbers of "light" socio-political agents. This work is necessary if we are to have more realistic "socio-cognitive" agents, ones that are useful for the types of analysis and training/rehearsal M&S worlds envisioned here. This is part of the wider effort to have more realistic agents and detailed worlds that influence their decisions.

The validity of the models and theories inside the agents has not been a focus of this chapter. However, "correctness" is in equal parts about the data used and the generative mechanisms inside the agents. Both of these are finally more important than whether any particular predictions turn out to be accurate. Much of this chapter dealt with how to obtain the best possible data. We should close by also pointing out that if the generative mechanisms are roughly or in principle "correct," then one can trust that experiments with the agents will yield useful insights about various policies and how these policies in turn will influence the agents. That is why one attempts to equip social agents with more and more advanced cognitive capabilities. This work suggests some words of advice and also caution to those attempting simulations with various country databases – start with best available models (with higher internal validity), then conduct adequacy tests, validity assessments, and replication of results across samples. Even after all that, social system simulations will rarely yield precise forecasts and predictions. Rather, their utility lies in exploring the possibility space and in understanding mechanisms and causalities so that one can see how alternative DIME (Diplomatic, Informational, Military, and Economic) actions might lead to the same or unexpected PMESII (Political, Military, Economic, Social, Informational, and Infrastructure) effects.

Acknowledgment

This research was partially supported by AFOSR, DARPA, and the Beck Fund, though no one except the authors is responsible for any statements or errors in this manuscript.

References

R. Axelrod. *The Complexity of Cooperation: Agent-Based Models of Competition and Collaboration.* Princeton University Press, Princeton, NJ, 1998.

G. K. Bharathy. *Agent Based Human Behavior Modeling: A Knowledge Engineering Based Systems Methodology for Integrating Social Science Frameworks for Modeling Agents with Cognition, Personality and Culture.* PhD thesis, University of Pennsylvania, 2006.

K. M. Carley and J. Reminga. ORA: Organization risk analyzer. Technical Report CMU-ISRI-04-106, Carnegie Mellon University, School of Computer Science, Institute for Software Research International, 2004.

K. M. Carley, J. Diesner, and M. D. Reno. AutoMap user's guide. Technical Report CMU-ISRI-06-114, Carnegie Mellon University, School of Computer Science, Institute for Software Research International, 2006.

C. Cesarano, B. Dorr, A. Picariello, D. Reforgiato, A. Sagoff, and V. S. Subrahmanian. OASYS: An opinion analysis system. In *Proceedings of AAAI-2006 Spring Symposium on Computational Approaches to Analyzing Weblogs*, pages 21–26, 2006.

C. Cioffi-Revilla and S. P. O'Brien. Computational analysis in US foreign and defense policy. Paper prepared for the *First International Conference on Computational Cultural Dynamics*, University of Maryland, College Park, MD, August 2007.

Comparative Study of Electoral Systems Secretariat. Comparative study of electoral systems 2001-2006. Computer File, 2004. ICPSR version. Ann Arbor, MI: University of Michigan, Center for Political Studies [producer] 2003. Ann Arbor, MI: Inter-university Consortium for Political and Social Research [distributor].

J. M. Epstein. Modeling civil violence: An agent-based computational approach. *Proceedings of the National Academy of Sciences*, 99(3):7243–7250, 2002.

D. C. Esty, J. A. Goldstone, T. R. Gurr, B. Harff, P. T. Surko, A. N. Unger, and R. Chen. The state failure project: Early warning research for US foreign policy planning. In J. L. Davies and T. R. Gurr, editors, *Preventive Measures: Building Risk Assessment and Crisis Early Warning Systems*, chapter 3. Rowman and Littlefield, Boulder, CO and Totowa, NY, 1998.

European Values Study Group and World Values Survey Association. European and world values surveys four-wave integrated data file, 1981-2004, 2006. v.20060423.

M. Fayzullin, V. S. Subrahmanian, M. Albanese, C. Cesarano, and A. Picariello. Story creation from heterogeneous data sources. *Multimedia Tools and Applications*, 33(3):351–377, 2007.

T. Gilovich and D. W. Griffin. Heuristic biases: Then and now. In T. Gilovich, D. W. Griffin, and D. Kahneman, editors, *Heuristics and Biases: The Psychology of Intuitive Judgment*, pages 1–18. Cambridge Univeristy Press, Cambridge, UK, 2002.

Global Barometer Surveys Program. Global barometer surveys. http://www.globalbarometer.net/index.htm. Accessed November 2007.

M. E. Gorman and M. E. Gorman. A comparison of disconfirmatory, confirmatory and control strategies on wason's 2-4-6 task. *The Quarterly Journal of Experimental Psychology Section A*, 36(4):629–648, 1984.

R. F. Harrod. A second essay in dynamic theory. *Economic Journal*, 70:277–293, 1960.

Heidelberg Conflict Barometer. Heidelberg conflict barometer. http://www.hilk.de/en/konflikbarometer/index.html. Accessed May 2008.

M. G. Hermann. Who becomes a political leader? Leadership succession, generational change, and foreign policy. http://www.allacademic.com/meta/p69844_index.html, 2005. Accessed November 2007.

M. G. Hermann. Assessing leadership style: A trait analysis. http://socialscienceautomation.com/, 1999. Accessed November 2007.

R. J. Heuer. *Psychology of Intelligence Analysis*. Center for the Study of Intelligence, Central Intelligence Agency, Washington, DC, 1999.

A. O. Hirschman. *Exit, Voice, and Loyalty: Responses to Decline in Firms, Organizations and States*. Harvard University Press, Cambridge, MA, 1970.

D. Hopkins, G. King, M. Knowles, and S. Melendez. ReadMe: Software for automated content analysis. http://gking.harvard.edu/readme/docs/readme.pdf, 2007.

R. J. House, P. J. Hanges, M. Javidan, P. W. Dorfman, and W. Gupta, editors. *Culture, Leadership, and Organizations: The GLOBE Study of 62 Societies*. Sage Publications, Thousand Oaks, CA, 2004.

M. Johns. *Deception and Trust in Complex Semi-Competitive Environments*. PhD thesis, University of Pennsylvania, 2006.

D. Kahneman, P. Slovic, and A. Tversky, editors. *Judgment Under Uncertainty: Heuristics and Biases*. Cambridge University Press, Cambridge, UK, 1982.

G. J. Kim and G. K. Bharathy. Penn conflict database catalogue. http://www.seas.upenn.edu/~jiyunkim/pcdc.pdf, 2007. Accessed November 2007.

J. Kipp, L. Grau, K. Prinslow, and D. Smith. The human terrain system: A CORDS for the 21st century. *Military Review*, 86(5):8–15, September-October 2006.

G. Lafree and L. Dugan. Global terrorism database, 1970-1977. Computer File, 2007. College Park, MD: University of Maryland [producer], 2006. Ann Arbor, MI: Inter-university Consortium for Political and Social Research [distributor], 04-04.

W. A. Lewis. Economic development with unlimited supplies of labour. *Manchester School*, 22:139–191, 1954.

I. S. Lustick, D. Miodownik, and R. J. Eidelson. Secessionism in multicultural states: Does sharing power prevent or encourage it? *American Political Science Review*, 98(2):209–229, 2004.

N. Machiavelli. *The Chief Works and Others*. Duke University Press, Durham, 1965.

N. Machiavelli. *The Prince*. Cambridge University Press, Cambridge, UK, 1988.

S. Milgram. The small world problem. *Psychology Today*, pages 60–67, May 1967.

Minorities at Risk Project. Minorities at risk project. http://www.cidcm.umd.edu/mar/, 2003. Accessed November 2007.

J. Park and W. Fables. Perception based synthetic vision system. In *International Conference on Complex Systems*, Boston, MA, October-November 2007.

R. W. Pew and A. S. Mavor, editors. *Modeling Human and Organizational Behavior: Application to Military Simulations*. National Academy Press, Washington, DC, 1998.

Political Instability Task Force. Political instability task force worldwide atrocities dataset. http://web.ku.edu/keds/data.dir/atrocities.html. Accessed November 2007.

P. M. Regan. *Civil Wars and Foreign Powers: Interventions and Intrastate Conflict*. University of Michigan Press, second edition, 2000.

F. E. Ritter, M. N. Avraamides, and I. G. Councill. Validating changes to a cognitive architecture to more accurately model the effects of two example behavior moderators. In *Proceedings of the Eleventh Conference on Behavior Representation in Modeling and Simulation, SISO*, Orlando, 2003.

M. R. Sarkees. The correlates of war data on war: An update to 1997. *Conflict Management and Peace Science*, 18(1):123–144, 2000.

F. Schneider and D. H. Enste. Shadow economies: Size, causes, and consequences. *Journal of Economic Literature*, 38:77–114, 2000.

P. A. Schrodt and D. J. Gerner. The kansas event data system: A beginner's guide illustrated with a study of media fatigue in the palestinian intifada. http://web.ku.edu/keds/home.dir/guide.html, 1996. Accessed 1996.

B. G. Silverman and G. K. Bharathy. Modeling the personality & cognition of leaders. In *Proceedings of the Thirteenth Conference on Behavioral Representations in Modeling and Simulations, SISO*, May 2005.

B. G. Silverman, M. Johns, R. Weaver, K. O'Brien, and R. Silverman. Human behavior models for game-theoretic agents. *Cognitive Science Quarterly*, 2(3/4):273–301, 2002.

B. G. Silverman, R. L. Rees, J. A. Toth, J. G. Cornwell, K. O'Brien, M. Johns, and M. Caplan. Athena's prism – a diplomatic strategy role playing simulation for generating ideas and exploring alternatives. In *Proceedings of the First International Conference on Intelligence Analysis*, MacLean, VA, 2005.

B. G. Silverman, G. K. Bharathy, K. O'Brien, and J. Cornwell. Human behavior models for agents in simulators and games: Part II: Gamebot engineering with PMFserv. *Presence: Teleoperators and Virtual Environments*, 15(2):163–185, 2006a.

B. G. Silverman, M. Johns, J. G. Cornwell, and K. O'Brien. Human behavior models for agents in simulators and games: Part I: Enabling science with PMFserv. *Presence: Teleoperators and Virtual Environments*, 15(2):139–162, 2006b.

B. G. Silverman, G. Bharathy, B. Nye, and R. J. Eidelson. Modeling factions for "effects based operations": Part I – leaders and followers. *Computational & Mathematical Organization Theory*, 13(4):379–406, 2007.

B. G. Silverman, G. K. Bharathy, B. Nye, and T. Smith. Modeling factions for "effects based operations": Part II – behavioral game theory. *Computational & Mathematical Organization Theory*, 14(2):120–155, 2008.

P. J. Stone, D. C. Dunphy, M. S. Smith, and D. M. Ogilvie. *General Inquirer: A Computer Approach to Content Analysis*. MIT Press, Cambridge, MA, 1966.

V. S. Subrahmanian. Cultural modeling in real time. *Science*, 317(5844):1509–1510, 2007.

R. Sun. *Cognition and Multi-Agent Interaction*. Cambridge University Press, 2006.

P. E. Tetlock. *Expert Political Judgment: How Good Is It? How Can We Know?* Princeton University Press, Princeton, NJ, 2005.

University of Kansas. European protest and coercion data. http://web.ku.edu/ronfran/data/index.html. Accessed November 2007.

Uppsala Conflict Data Program. Uppsala conflict database. http://www.pcr.uu.se/database/. Accessed November 2007.

A. Varshney and S. Wilkinson. Varshney-wilkinson dataset on hindo-muslim violence in india, 1950-1995. Computer File, 2006. Cambridge, MA: Harvard University, WCFIA [producer], 2004. Ann Arbor, MI: Inter-university Consortium for Political and Social Research [distributor] 02-17.

M. D. Young. Building WorldView(s) with Profiler+. In M. D. West, editor, *Applications of Computer Content Analysis*, volume 17 of *Progress in Communication Sciences*. Ablex Publishing, 2001.

10

Crowd Behavior Modeling: From Cellular Automata to Multi-Agent Systems

Stefania Bandini
Complex Systems and Artificial Intelligence Research Center

Sara Manzoni
Complex Systems and Artificial Intelligence Research Center

Giuseppe Vizzari
Complex Systems and Artificial Intelligence Research Center

10.1	Introduction...	301
10.2	Pedestrian Dynamics Context: An Overview......	303
	Pedestrians as Particles • Pedestrians as States of CA • Pedestrians as Autonomous Agents	
10.3	Guidelines for Crowds Modeling with Situated Cellular Agents Approach..................................	306
	Spatial Infrastructure and Active Elements of the Environment • Pedestrians	
10.4	A Pedestrian Modeling Scenario...................	310
	The Scenario • The Modeling Assumptions • The Environment • The Passengers • Simulation Results	
10.5	From a SCA Model to Its Implementation........	315
	Supporting and Executing SCA Models	
10.6	Conclusions...	319
10.7	Discussion and Future Research Directions........	321
	Acknowledgements.......................................	321
	References...	321

10.1 Introduction

Crowds of pedestrians are complex entities from different points of view, starting from the difficulty in providing a satisfactory definition of the term "crowd". *"(Too) many people in (too) little space"* [Kruse, 1986] is a pedestrian crowd definition aggregating several disciplinary interpretations, and the range of this definition sites its fuzzy borders in the traditional opposition between humanistic and scientific cultures in these studies. The range of individual and collective behaviors that take place in a crowd, the composite mix of competition for the shared space but also collaboration due to, not necessarily explicit but shared, social norms, the possibility to detect self-organization and emergent phenomena; they are all indicators of the intrinsic complexity of a crowd. Nonetheless, the relevance of human behavior, and especially of the movements of pedestrians in a built environment in normal and extraordinary situations (e.g., evacuation), and its implications for the activities of architects, designers and urban planners are apparent (see, e.g., [Batty, 2001] and [Willis et al., 2004]), especially given recent dramatic episodes such as terrorist attacks, riots and fires, but also due to the growing issues in facing the organization and management of public events (ceremonies, races, carnivals, concerts, parties/social gatherings, and so on) and in designing naturally crowded places (e.g., stations, arenas, airports). Crowd models and simulators are thus increasingly being investigated in the scientific context, sold by firms*, and used by decision makers. In fact, even if research on this topic is still quite lively and far from a complete understanding of the complex phenomena related to crowds of pedestrians in the environment, models and simulators have shown their usefulness in supporting architectural designers and urban planners in their decisions by creating the possibility to envision the behavior/movement of crowds of pedestrians in specific designs/environments, to elaborate what-if scenarios and evaluate their decisions with reference to specific metrics and criteria.

The current state of the art in crowd simulation approaches comprises very different types of models, ranging from a physical approach to the representation of pedestrians (viewed in terms of particles subject to forces generated by points of reference/interest of the environment and by pedestrians themselves; see, e.g., [Helbing et al., 1997]), to a discrete modeling of the environment in terms of a lattice in which pedestrians are viewed as particular states of a cell, in a Cellular Automata (CA) approach (see, e.g., [Schadschneider et al., 2002]), to agent based approaches (see, e.g., [Klügl and Rindsfüser, 2007]), more clearly separating the representations of the environment and the entities that inhabit it, acting and interacting according to their perceptive capabilities and behavioral specification. This significant difference in the ways pedestrians are represented in terms of properties, internal state, motivations, behaviors, also reflects the growing range of the applications of the simulators based on these models. The encouraging results obtained by relevant approaches to modeling and simulating pedestrians in normal and egress situations [Helbing et al., 2001] have in fact led to consider, on the one hand, the possibility of extending the range of applications of these models to other application areas, for instance to simulate the behaviors of pedestrians in shopping centers (see, e.g., [Kitazawa and Batty, 2004; Dijkstra et al., 2007]). On the other hand, the possibility of integrating the current pedestrian models (i) with more complex interpersonal interaction mechanisms (e.g., from those that lead people moving in pairs and small groups, to those guiding the spreading of emotional states [Adamatzky, 2005]), (ii) with models related to additional relevant phenomena (e.g., smoke or fires diffusion**), (iii) with more complex perception mechanisms, and in general with more complex models of pedestrian motivations, goals and agendas [Dijkstra et al., 2005]. It must be noted that these developments call for field data and (whenever possible), experimental studies to make models more realistic; however, it must also be noted that in several of the aforementioned studies the role and relevance of the single pedestrian, of his/her goals, characteristics, relationships with the others in a given simulation scenario, are all factors that make the agent-based approach and Multi-Agent Systems (MAS) [Ferber, 1999] a natural way of analyzing, describing and modeling crowds as systems of pedestrians, with their own features, their environment, the ways they interact with it and with other pedestrians.

The MAS approach to the modeling and simulation of complex systems has been applied in very different contexts, ranging from the study of social systems [Axtell, 2000], to biological systems (see, e.g., [Christley et al., 2007]), and it is considered as one of the most

---

*See, e.g., Legion Ltd. (http://www.legion.com), Crowd Dynamics Ltd. (http://www.crowddynamics.com/), Savannah Simulations AG (http://www.savannah-simulations.ch).
**Like in the EXODUS evacuation tool (http://fseg.gre.ac.uk/exodus/index.html)

successful perspectives of agent–based computing [Luck et al., 2005], even if this approach is still relatively young, compared, for instance, to analytical equation-based modeling. The main aim of this chapter is to introduce a formal and computational framework based on MAS principles and related guidelines for its adoption to represent and simulate crowds of pedestrians. The described approach is rooted on basic principles of discrete modeling approaches based on Cellular Automata, but it clearly separates the active entities (i.e., pedestrians) from the environment they inhabit. This approach has two main long term goals, namely (i) the definition of a MAS model for crowds capable of generating phenomena that are currently described by more analytical and CA-based approaches, but that can also be integrated with more complex models of pedestrian motivations, perceptive capabilities and behaviors, as well as with other models related to phenomena that can take place in the environment, and (ii) to realize an integrated and extensible instrument supporting experts in their studies and analysis on crowd phenomena and pedestrian behaviors in crowded spaces, without requiring particular technical skills, such as proficiency in software engineering.

This chapter describes the current advancement of this research project covering the different aspects of the above mentioned framework. In particular, after an analysis of the related works in particular within pedestrian dynamics research context, Section 10.3 introduces Situated Cellular Agents (SCA) [Bandini et al., 2006b], the MAS-based model adopted for this project. This section also introduces the guidelines of the proposed modeling approach and Section 10.4 describes a case study related to the simulation of complex underground station crowd dynamics. Section 10.5 discusses one of the roles of tools supporting; conclusions and future trends of research in this specific application area will end the chapter.

10.2 Pedestrian Dynamics Context: An Overview

It is not a simple task to provide a compact yet comprehensive overview of the different approaches and models for the representation and simulation of crowd dynamics. In fact, entire scientific interdisciplinary workshops and conferences are focused on this topic (see, e.g., the proceedings of the first edition of the International Conference on Pedestrian and Evacuation Dynamics [Schreckenberg and Sharma, 2001]). However, most approaches can be classified according to the way pedestrians are represented and managed, and in particular:

- pedestrians as particles subject to forces of attraction/repulsion;
- pedestrians as particular states of cells in a CA;
- pedestrians as autonomous agents, situated in an environment.

10.2.1 Pedestrians as Particles

Several models for pedestrian dynamics are based on an analytical approach, representing pedestrians as particles subject to forces, modeling the interaction between pedestrian and the environment (and also among pedestrians themselves, in the case of *active walker* models [Helbing et al., 1997]). Forces of attraction lead the pedestrians/particles towards their destinations (modeling thus their goals), while forces of repulsion are used to represent the tendency to stay at a distance from other points of the environment. Figure 10.1 shows a diagram exemplifying the application of this approach to the representation of an intersection that is being crossed by three pedestrians. In particular, the velocity of the gray

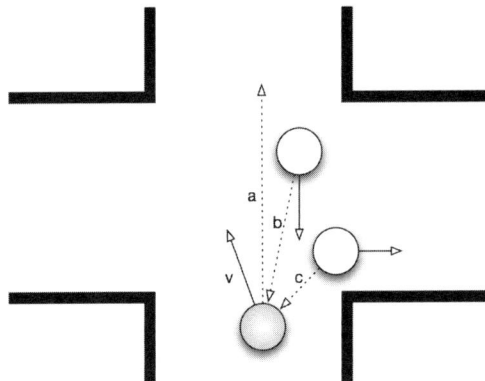

FIGURE 10.1 A diagram exemplifying an analytical model for pedestrian movement: the gray pedestrian, in the intersection, has an overall velocity v that is the result of an aggregation of the contributions related to the effects of attraction by its own reference point (a), and the repulsion by other pedestrians (b and c).

pedestrian is determined as an aggregation of the influences it is subject to, that are the attraction to its reference point (the top exit) and the repulsion from the other pedestrians. This kind of effect was introduced by a relevant and successful example of this modeling approach, the *social force* model [Helbing and Molnár, 1995]; this approach introduces the notion of social force, representing the tendency of pedestrians to stay at a certain distance from one another; other relevant approaches take inspiration from fluid-dynamic [Helbing, 1992] and magnetic forces [Okazaki, 1979] for the representation of mechanisms governing flows of pedestrians.

While this approach is based on a precise methodology and has provided relevant results, it represents pedestrian as mere particles, whose goals, characteristics and interactions must be represented through equations, and thus it is not simple to incorporate heterogeneity and complex pedestrian behaviors in this kind of model.

10.2.2 Pedestrians as States of CA

A different approach to crowd modeling is characterized by the adoption of Cellular Automata (CA) [Wolfram, 1986], with a discrete spatial representation and discrete time-steps, to represent the simulated environment and the entities it comprises. The cellular space thus includes both a representation of the environment and an indication of its state, in terms of occupancy of the sites it is divided into, by static obstacles as well as human beings. Transition rules must be defined in order to specify the evolution of every cell's state; they are based on the concept of neighborhood of a cell, a specific set of cells whose state will be considered in the computation of its transition rule. The transition rule, in this kind of model, generates the illusion of movement, that is mapped to a coordinated change of cells state. To make a simple example, an atomic step of a pedestrian is realized through the change of state of two cells, the first characterized by an *"occupied"* state that becomes *"vacant"*, and an adjacent one that was previously *"vacant"* and that becomes *"occupied"*. Figure 10.2 shows a sample effect of movement generated by the subsequent application of a transition rule in the cellular space. This kind of application of CA-based models is es-

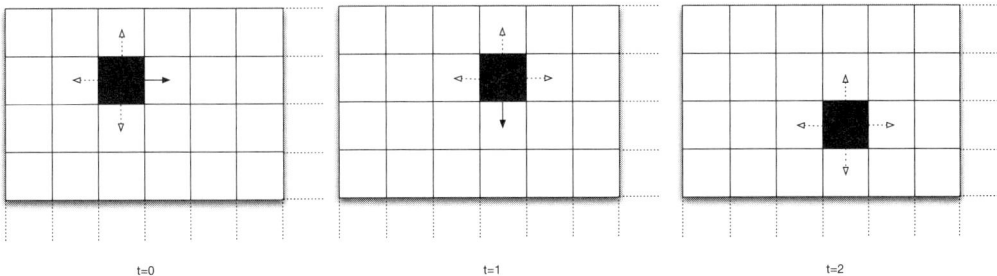

FIGURE 10.2 A diagram showing a sample effect of movement generated through the coordinated change of state of adjacent cells in a CA. The black cell is occupied by a pedestrian that moves to the right in turn 0 and down in turn 1, but these effects are obtained through the contemporary change of state among adjacent cells (previously occupied becoming vacant and vice versa).

sentially based on previous works adopting the same approach for traffic simulation [Nagel and Schreckenberg, 1992].

Local cell interactions are thus the uniform (and only) way to represent the motion of an individual in the space (and the choice of the destination of every movement step). The sequential application of this rule to the whole cell space may bring to emergent effects and collective behaviors. Relevant examples of crowd collective behaviors that were modeled through CAs are the formation of lanes in bidirectional pedestrian flows [Blue and Adler, 2000a], the resolution of conflicts in multidirectional crossing pedestrian flows [Blue and Adler, 2000b]. In this kind of example, different states of the cells represent pedestrians moving towards different exits; this particular state activates a particular branch of the transition rule causing the transition of the related pedestrian to the direction associated to that particular state. Additional branches of the transition rule manage conflicts in the movement of pedestrians, for instance through changes of lanes in case of pedestrians that would occupy the same cell coming from opposite directions.

It must be noted, however, that the potential need to represent goal driven behaviors (i.e., the desire to reach a certain position in space) has often led to extend the basic CA model to include features and mechanisms breaking the strictly locality principle. A relevant example of this kind of development is represented by a CA based approach to pedestrian dynamics in evacuation configurations [Schadschneider et al., 2002]. In this case, the cellular structure of the environment is also characterized by a predefined desirability level, associated to each cell, that, combined with more dynamic effects generated by the passage of other pedestrians, guide the transition of states associated to pedestrians.

10.2.3 Pedestrians as Autonomous Agents

Recent developments in this line of research (e.g., [Henein and White, 2005; Dijkstra et al., 2006]), introduce modifications to the basic CA approach that are so deep that the resulting models are effectively agent-based and Multi Agent Systems (MAS) models exploiting a cellular space representing spatial aspects of agents' environment. A MAS is a system made up of a set of autonomous components which interact, for instance according to collaboration or competition schemes, in order to contribute in realizing an overall behavior that could not be generated by single entities by themselves. As previously introduced, MAS models have been successfully applied to the modeling and simulation of several situations characterized by the presence of autonomous entities whose action and interaction determines the evolution of the system, and they are increasingly being adopted also to model crowds of pedestrians [Batty, 2001; Gloor et al., 2004; Toyama et al., 2006]. All these approaches are characterized by the fact that the agents encapsulate some form of behavior inspired by the above described approaches, that is, forms of attractions/repulsion generated by points of interest or reference in the environment but also by other pedestrians.

Other agent based approaches to the modeling of pedestrians and crowds were developed with the primary goal of providing an effective 3D visualization of the simulated dynamics. The approaches described in [Musse and Thalmann, 2001] and in [Shao and Terzopoulos, 2007] are characterized by a very composite model of pedestrian behavior, including basic reactive behaviors as well as a cognitive control layer; moreover, actions available to agents are not strictly related to their movement, but they also allow forms of direct interaction among pedestrians and interaction with objects situated in the environment. Another relevant approach, described in [Murakami et al., 2003], is less focused on visual effectiveness of the simulation dynamics, and it supports a flexible definition of the simulation scenario also without requiring the intervention of a computer programmer. However, these virtual reality focused approaches to pedestrian and crowd simulation were not tested in paradigmatic case studies, modeled adopting analytical approaches or cellular automata and validated against real data.

10.3 Guidelines for Crowds Modeling with Situated Cellular Agents Approach

The activity of modeling complex simulation scenarios to study crowds and their dynamic phenomena with a MAS-based approach consists in a complex activity, that generally, at least requires a deep knowledge of the domain, proficiency in computational modeling (with specific reference to MAS models), the skills of a programmer, a deep acquaintance with the simulation platform. Rarely such heterogeneous competences can be found in the same person. Even in these rare cases, the difficulty of building a software simulation directly from a domain level theory or model without specific methodologies and implementation tools, with the risk of hiding the complexity of the overall process, several often necessary intermediate choices and assumptions. In the following we describe a methodology, a set of guidelines directed to experts but also to non-experts, developed to tackle these issues and provide users that wish to create models for pedestrian (and crowds) dynamics with a formal and computational reference framework. We describe the process of building a model based on Situated Cellular Agent (SCA) of a system of pedestrians, basically defined by the specification of three elements: the spatial abstraction in which the simulated entities are situated, the relevant elements of this structure which are able to shape and influence crowd behavior, and the behavioral specification of moving entities. Figure 10.3 summarizes the overall process, in which starting from an abstract scenario, a computational model is built with the reference guidelines of a formal model. Software tools, designed and developed according to the same reference framework, can effectively support model implementation and execution to run specific experiments.

The Situated Cellular Agent (SCA) model is a specific class of Multilayered Multi-Agent Situated System (MMASS) [Bandini et al., 2002] and, among MAS-based approaches to complex systems modeling, it provides an explicit structured spatial representation of the agents, environment and space-dependent agent behavioral and interaction mechanisms. A thorough description of the model is out of the scope of this chapter; the syntax and semantics of SCA basic elements will be briefly introduced when required for sake of clarity. A *Situated Cellular Agent* is defined by the triple $\langle Space, F, A \rangle$ where *Space* models the environment where the set A of agents is situated, acts autonomously and interacts at-a-

Crowd Behavior Modeling: From Cellular Automata to Multi-Agent Systems

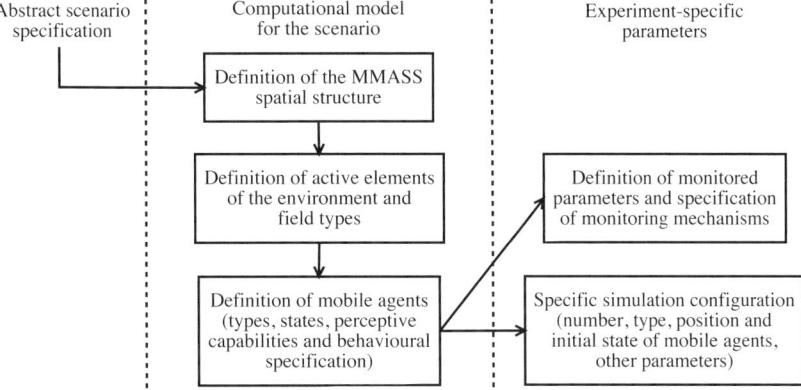

FIGURE 10.3 A diagram showing the main phases of the modeling approach.

distance, through the propagation of the set F of fields, and locally through synchronous reaction operator. Figure 10.4 shows a diagram of the two interaction mechanisms provided by the model. More precisely *Space* consists of a set P of sites arranged in a network (i.e., an undirected graph of sites). The structure of the space can be represented as a neighborhood function, $N : P \longrightarrow 2^P$ so that $N(p) \subseteq P$ is the set of sites adjacent to $p \in P$; the previously introduced *Space* element is thus the pair $\langle P, N \rangle$. Focusing instead on the single basic environmental elements, a site $p \in P$ can contain at most one agent and is defined by the 3–tuple $\langle a_p, F_p, P_p \rangle$ where:

- $a_p \in A \cup \{\bot\}$ is the agent situated in p ($a_p = \bot$ when no agent is situated in p that is, p is empty);
- $F_p \subset F$ is the set of fields active in p ($F_p = \varnothing$ when no field is active in p);
- $P_p \subset P$ is the set of sites adjacent to p (i.e., $N(p)$).

A SCA agent is defined by the 3–tuple $\langle s, p, \tau \rangle$ where τ is the *agent type*, $s \in \Sigma_\tau$ denotes the *agent state* and can assume one of the values specified by its type (see below for Σ_τ definition), and $p \in P$ is the site of the *Space* where the agent is situated. As previously stated, agent *type* is a specification of agent state set, perceptive capabilities and behavior. It is defined by the 3–tuple $\langle \Sigma_\tau, Perception_\tau, Action_\tau \rangle$. Σ_τ defines the set of states that agents of type τ can assume. $Perception_\tau : \Sigma_\tau \to [\mathbf{N} \times W_{f_1}] \ldots [\mathbf{N} \times W_{f_{|F|}}]$ is a function associating to each agent state a vector of pairs representing the *receptiveness coefficient* and *sensitivity thresholds* for that kind of field. $Action_\tau$ represents instead the behavioral specification for agents of type τ. Each SCA agent is thus provided with a set of sensors (i.e.,

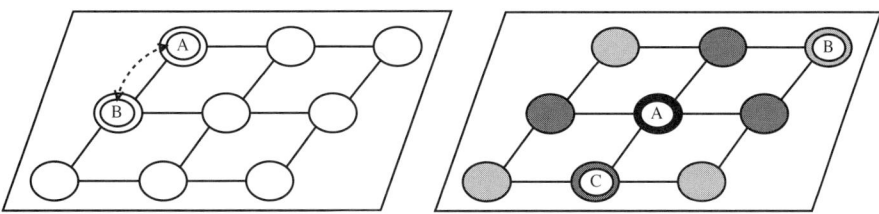

FIGURE 10.4 A diagram showing the two interaction mechanisms provided by Situated Cellular Agent: on the left, two reacting agents in a coordinated way change their states, and, on the right, an agent emitting a field that diffuse within the space structure.

defined in terms of perception function) that allows its interaction with the environment and other agents. At the same time, agents can be the source of fields that diffusing within the structured SCA space can be perceived by other agents endowed with suitable sensors (e.g., noise emitted by a talking agent that can be perceived at-a-distance).

The behavior of each type of SCA agent can be specified using a language that defines the following primitives:

- $emit(s, f, p)$: to *start the diffusion of a field* f from site p, where the agent is situated;
- $react(s, a_{p_1}, a_{p_2}, \ldots, a_{p_n}, s')$: it allows the specification of a *coordinated change of state* among adjacent agents. In order to preserve agents' autonomy, a compatible primitive must be included in the behavioral specification of all the involved agents; moreover when this coordination process takes place, every involved agents may dynamically decide to effectively agree to perform this operation;
- $transport(p, f, q)$: it allows one to *define agent movement* from site p to site q (that must be adjacent and vacant);
- $trigger(s, f, s')$: it specifies that an agent must *change its state* when it senses a particular condition in its local context (i.e., its own site and the adjacent ones); this operation has the same effect of a reaction, but does not require a coordination with other agents.

10.3.1 Spatial Infrastructure and Active Elements of the Environment

SCA agents' actions take place in a discrete and finite space. In order to obtain an appropriate *abstraction of space* suitable for the SCA model, a discrete abstraction of the space in which the pedestrian dynamics has to be studied must be defined as an undirected graph: nodes represent the positions that can be occupied by single pedestrians. SCA space represents thus an abstraction of a walking pavement, but it can be sufficiently detailed to be considered an approximation of the real environment surface, and it allows a realistic representation of the movements and paths that individuals would follow. The scale of discretization can vary, but according to [Schadschneider et al., 2002] a cell dimension of 40 × 40 cm^2 is adequate to represent the typical space occupied by a pedestrian in a dense crowd. Since *active elements of the environment* can be perceived and thus influence, or even determine, the movement of pedestrians, SCA approach suggests representing them as agents endowed with the ability of emitting a sort of presence field that can be perceived by all agents situated on sites reached by its diffusion and endowed with a suitable perceptive ability (i.e., perception function). Typically the latter are objects of the environment which constrain agent movement (e.g., gateways, doors), but also objects that can transmit some kind of conceptual information (e.g., exit signs or indications). To adopt field emission–diffusion–perception mechanism as a basic instrument to model at-a-distance influences between agents, a specific field type must be defined (i.e., diffusion, composition and comparison functions). This mechanism allows one to represent several types of fields; for instance, visual and acoustic perception of a signal may be modeled taking into account the different influence of obstacles in their diffusion (e.g., a sound can pass through a door even if at reduced intensity). A library of signal types can easily be build to support this phase. By selecting a predefined field type the modeler is actually specifying the 4-tuple $\langle W_t, Diffusion_t, Compare_t, Compose_t \rangle$, that defines a field type (i.e., the set of values it can assume, how it diffuses within the spatial structure, how different emissions of the same field type combine and how to perform comparisons, e.g., to evaluate if a given field overcomes a threshold when non-numerical values are allowed).

The behavioral specification of an agent representing an active element of the environment always include the following action:

```
action    : emit(p, f_t)
condition : a = ⟨s, p, τ⟩
effect    : added(f_t, p)
```

where f_t is a field of type t and $a = \langle s, p, \tau \rangle$ specifies that the agent a of type τ is in state s and occupying site p. The effect of this emit action (i.e., $added(f, p)$) is a modification in a set of sites in the space (determined by function $Diffusion_t$) to notify its presence to other agents. In particular, the set P_e of sites that will be affected by this action is $P_e = \{q \in P \mid Diffusion_t(p, f_t, q) \neq 0\}$ (i.e., the set of sites for which the diffusion function is not the null field). Given $q \in P_e$, the set F'_q of fields active in it after the diffusion will be:

- $F_q \cup \left\{ \langle Diffusion_t(p, f_t, q), Compose_t, Compare_t \rangle \right\}$ if F_q (i.e., the set of fields that were active in the site q before the emission) does not include fields of type t;

- $\left(F_q - \{f'_t\} \right) \cup \left\{ \langle Compose_t(Diffusion_t(p, f_t, q), w'_t), Compose_t, Compare_t \rangle \right\}$ where $f'_t = \langle w'_t, Compose_t, Compare_t \rangle$ was the *unique* non null field of type t active in q before the emission.

More complex behaviors for active elements of the environment can also be defined if required, for instance to model active elements whose emission starts or stops only under specific conditions.

10.3.2 Pedestrians

The modeling of *pedestrians* populating the environment whose spatial structure and relevant elements have been specified as above can take into account the possibility of modeling non homogeneous systems, where pedestrians with different behavioral specifications, perceptive abilities and capabilities can interact thanks to homogeneous interaction mechanisms specified by SCA (i.e., field emission-diffusion-perception and local reaction). SCA model in fact supports the definition of heterogeneous agent systems thanks to the notion of agent type.

The specification of agent behaviors can for instance represent different preferences of pedestrians toward one of multiple movement directions but also support more complex behavioral models in which reasoning, planning, scheduling and other abilities has to be properly integrated in the model. The basic agent behavioral specification of pedestrians is based on *transport* action that is, in the specification of how agents select next destination site:

```
action    : transport(p, q)
condition : a = ⟨s, p, τ⟩, A_q = ⊥, q ∈ P_p, best(s, q)
effect    : a = ⟨s, q, τ⟩, A_p = ⊥
```

where $a = \langle s, p, \tau \rangle$ specifies that the agent a (for which the action is specified) is in state s, is occupying site p, and is of type τ; $q \in P_p$ belongs to the set of sites adjacent to p and $best(s, q)$ is verified if, for state s, $\nexists r \in P_p \mid utility(s, r) > utility(s, q) \wedge a_r = \bot$. A possible way to define agent behavior, coherently with traditional notion of agents utility in Artificial Intelligence, can be: $utility(s,r) = \sum_{t \in T} w_t(s) \cdot fval(t,r)$ where $w_t(s)$ denotes the weight associated to fields of type t for agents with a given *attitude* (i.e., the desirability of that kind of signal represented by or as function of agent state) and $fval(t,r)$ denotes the value of field of type t in site r. In case of more sites having the same utility value the agent can make a non deterministic choice among them or adopt other strategies. The effect of the action is to free site p, and correspondingly change the position of agent a to site q.

For each agent type, the modeler can specify fields emitted by the agent and the sensitivity to fields emitted by other agents. A change in agent's attitude can be defined either by a *trigger* or a *react* primitive. Both primitives in fact determine an agent state change. The first option can be used when the change of attitude can be mapped to a specific condition associated to a field, such as the fact that the current intensity of a signal exceeds a given threshold (i.e., the agent is close enough to a given point of interest):

action   : $trigger(s_1, f_g, s_2)$
condition: $a = \langle s_1, p, \tau \rangle, perceive(f_g), compare(f_g, f_t)$
effect   : $a = \langle s_2, p, \tau \rangle$

where f_t represents the above introduced threshold. The second option, the *react* primitive, can be adopted when two agents coordinate themselves to state change:

action   : $react(s_1, b, s_2)$
condition: $a = \langle s_1, p, \tau_1 \rangle, b = \langle s_b, q, \tau_2 \rangle, q \in P_p, agreed(b)$
effect   : $a = \langle s_2, p, \tau_1 \rangle$

where the $agreed(b)$ specifies that agent b has agreed to perform a coordinated change of state, a necessary condition for the reaction to take place. It must be noted that a compatible *react* action must be specified in another agent (that could be of the same type or not). The effect of both actions is to change the state of agent a from s_1 to s_2, but in different conditions and exploiting different primitives. It must be noted that these are just sample action specifications, and that additional conditional elements can be defined to better fit the specific situation.

10.4 A Pedestrian Modeling Scenario

10.4.1 The Scenario

An underground station is an environment where various crowd behaviors take place. Passengers' behaviors are difficult to predict, because crowd dynamics emerge from interactions between passengers, and between single passengers and parts of the environment, such as signals (e.g., current stop indicator), doors, seats and handles. The behavior of passengers changes noticeably in relation to the different priorities that characterize each phase of their trips. That means, for instance, that passengers close to each other may display very different behaviors because of their distinct aims in that moment. Passengers on board may have to get off and thus try to reach for the door, while other ones are instead looking for a seat or standing beside a handle. Moreover when trains stop and doors open very complex crowd dynamics happen, as people that have to get on the train have to allow the exit of passengers that are getting off. Passengers have to match their own priority with the obstacles of the environment, with the intentions of other passengers, and with implicit behavioral rules that govern the social interaction in those kind of transit stations, in a mixture of competition and collaboration, to avoid stall situations. Given the complexity of

Crowd Behavior Modeling: From Cellular Automata to Multi-Agent Systems

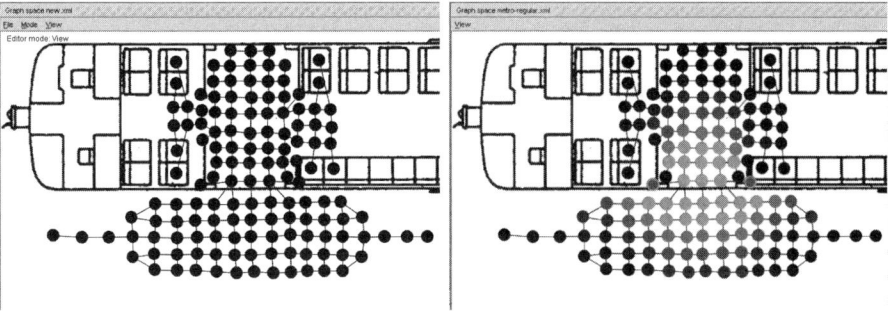

FIGURE 10.5 Discretization of a portion of the environment and extension of fields associated to the doors of the wagon, in the tool supporting the definition of the simulated space.

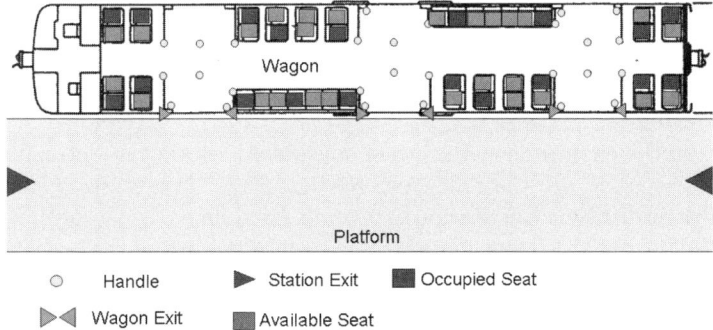

FIGURE 10.6 Immobile active elements of the environment.

the overall scenario, we decided to focus on a specific portion of this environment in which some of the most complex patterns of interaction take place: the part of the platform in the presence of a standing wagon from which some passengers are attempting to get off while other waiting travelers are trying to get on.

10.4.2 The Modeling Assumptions

To build up our simulation we made some behavioral assumptions, now we will make some brief examples of the kind of behaviors we wanted to capture. Passengers that do not have to get off at a train stop tend to remain still, if they do not constitute an obstacle to the passengers that are descending. Passengers will move only to give way to a descending passenger, to reach some seat that has became available, or to reach a better position like places at the side of the doors or close to the handles. On the other hand in very crowded situations it often happens that people that do not have to get off can constitute an obstacle to the descent of other passengers, and they "are forced to" get off and wait for the moment to get on the wagon again. Passengers that have to get off have a tendency to go around still agents to find their route toward the exit, if it is possible. Once the train is almost stopped the waiting passengers on the platform identify the entrance that is closer to them, and try to move toward it. If they perceive some passengers bound to get off, they first let them get off and then get on the wagon.

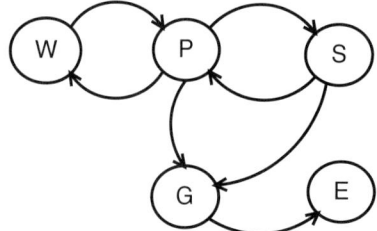

FIGURE 10.7 A diagram showing various attitudes for agent type passenger and the allowed transitions between attitudes.

10.4.3 The Environment

In reference to the modeling approach stated in the previous paragraph, to build an environment suitable for SCA platform, first of all we need to define a discrete structure representing the actual space in which the simulation is set. In our case study we started from an available diagram of an underground wagon. A discrete abstraction of this map was defined, devoting to each node the space generally occupied by one standing person, as shown in Figure 10.5.

The elements of the environment that were considered relevant in determining the crowd dynamics of this scenario are the following: *Station Exits*, *Doors*, *Seats* and *Handles* (see Figure 10.6 for their disposition). Station exits emit fixed fields, constant in intensity and in emission, that will be exploited by agents headed toward the exit of the station, that perceive them as attractive. Agent-doors emit another type of field which can guide passengers that have to get off the wagon, toward the platform, and passengers that are on the platform and are bound to get in the wagon. Seats may instead have two states: occupied and free. In the second state they emit a field that indicates their presence, and that is perceived as attractive by passengers, and they become occupied by reacting with agents that effectively occupy them. Handles also emit a field type very similar to the one emitted by seats, whose attractive effect is however less intense.

10.4.4 The Passengers

The above introduced elements support the definition of agents able to move in this environment evaluating the related signals according to their attitudes. We have identified the following attitudes for agent of type passengers: *waiting* (w), *passenger* (p), *get-off* (g), *seated* (s), *exiting* (e). In relation to its attitude, an agent will be sensitive to some fields, and not to others, and attribute different relevance to the perceived signals. In this way, the changing of attitude will determine a change of priorities. Attitude w is associated to an agent that is waiting to enter in the wagon. In this condition, agents perceive the fields generated by the doors as attractive, but they also perceive as repulsive the fields generated by passengers that are getting off, in other words those in attitude g. In attitude w the agent "ignores" (is not sensitive to) the fields generated by other active elements of the environment, such as exits' attractive fields, chairs attractive field and so on. Once inside the wagon, w agents change their attitude to p (passenger), through a *trigger* action activated by the perception of the maximum intensity of field generated by agent-door type. Agent in attitude p is attracted by fields generated by seats and handles, and repulsed by fields related to passengers that are getting off. In attitude g the agent will instead emit a field warning other agents of its presence, while it is attracted by fields generated by the doors. Once passed through the wagon door a g agent changes its attitude to e (exiting) and its priority will become to find the exits of the station. Figure 10.7 summarizes the various agent attitudes and the allowed transitions among them (that are modeled by means of *trigger* or *react* actions).

Table 10.1 describes instead the sensitiveness of the passenger to various fields in relation to their attitude. The table's cells provide also the indication about if the perceived field is considered attractive or repulsive, as well as the relevance associated to that field type. This table can be used as a guide for the definition of the *utility* function associated to the *transport* action characterizing passenger agents.

It must be noted that all passengers, except those in state g, emit a field of type *Presence* that generally has a repulsive effect, but a much lesser one with respect to the one generated by fields of type *Exit pressure* emitted by agents in get-off state.

10.4.5 Simulation Results

A simulator implementing the previously introduced model was realized exploiting the SCA model: only a subset of the overall introduced model was implemented, and more precisely active objects of the environment and passenger agents in state w, g, e, p. State s was omitted, to focus on the conflicts between agents in state w and g, which represent the most characteristic element of the overall system dynamic.

Figure 10.8 shows a screen-shot of this simulation system, in which waiting agents move to generate room for passenger agents that are going to get off the train. The system is synchronous, meaning that every agent performs one single action per turn; the turn duration is about one-fourth of second of simulated time. The goal of a small experimentation as this one is to qualitatively evaluate the modeling of the scenario and the developed simulator. The execution and analysis of several simulations shows that the overall system dynamics and the behavior of the agents in the environment is consistent with a realistic scenario, and fits with our expectations. To determine this evaluation, we executed over 100 simulations in the same starting configuration, which provides 6 passengers located on a metro train in state g (i.e., willing to get off), and 8 agents that are outside the train in state w (i.e., waiting to get on). A campaign of tests was necessary since in this specific application pedestrians perform a non-deterministic choice whenever they can move to different sites characterized by the same utility value.

In all simulations the agents achieved their goals (i.e., get on the train or get out of the station) in a number of turns between 40 and 60, with an average of about 50 turns. Nonetheless we noticed some undesired transient effects, and precisely oscillations, "forth and back" movements and in few simulations static forms providing "groups" facing themselves for a few turns, until the groups dispersed because of the movement of a peripheral element. These phenomena, which represent minor glitches under the described initial conditions, could lead to problems in case of high pedestrian density in the simulated environment.

Instead of modifying the general model, in order to introduce a sort of agent "facing" (not

TABLE 10.1 The table shows, for every agent state, the relevance of perceived signals.

State	Exits	Doors	Seats	Handles	Presence	Exit press.
W (getting on)	not perc.	attr. (2)	not perc.	not perc.	rep. (3)	rep. (1)
P (on board)	not perc.	not perc.	attr. (1)	attr. (2)	rep. (3)	rep. (2)
G (getting off)	not perc.	attr. (1)	not perc.	not perc.	rep. (2)	not perc.
S (seated)	not perc.	attr. (1)*	not perc.	not perc.	not perc.	not perc.
E (exiting)	attr. (1)	not perc.	not perc.	not perc.	rep. (2)	not perc.

* = The door signal also conveys the current stop indication.

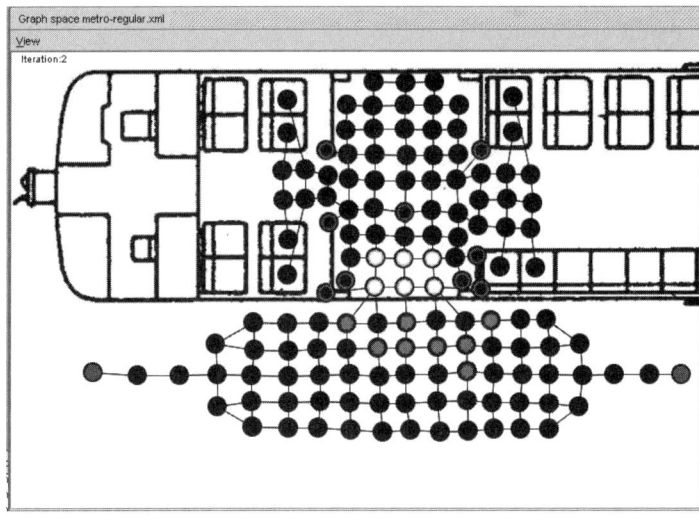

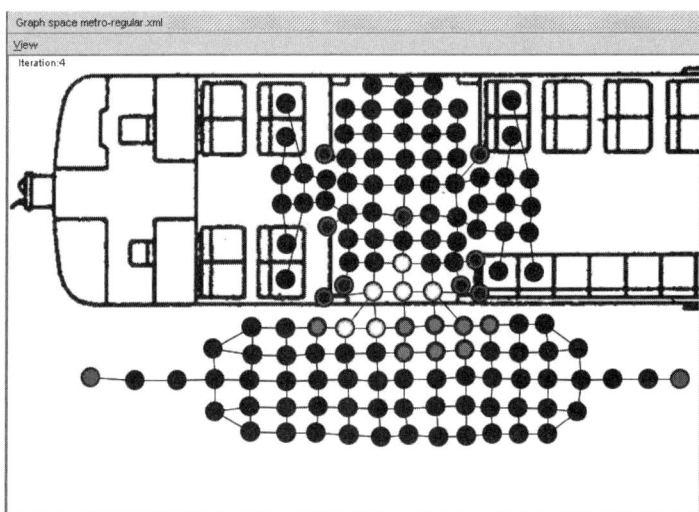

FIGURE 10.8 Two screenshots of the underground station simulation. On the first one 6 light gray agents are inside the train and going to get off, while 8 dark agents are standing outside and are going to get on. On the second, the latter have made some room for the former to get off.

provided by the SCA model), we allowed agents to keep track of their previous position, in order to understand if a certain movement is a step back. The utility of this kind of movement can thus be penalized. Instead, in order to avoid stall situations, the memory of the past position can also be exploited to penalize immobility, lowering the utility of the site currently occupied by the agent whenever it was also its previous position. These correctives were introduced in the behavioral specification of mobile agents, and a new campaign of tests was performed to evaluate the effect of these modifications in the overall system dynamics. By introducing these correctives, the occurrence of oscillating agent movement was drastically reduced, and the penalization of immobility simplified the solution of stall situations among facing groups. In all simulations the agents were able to achieve their goals, but the number of turns required to complete agents movement is between 20 and 40, with an average of about 30 turns. The analysis and identification of other significant

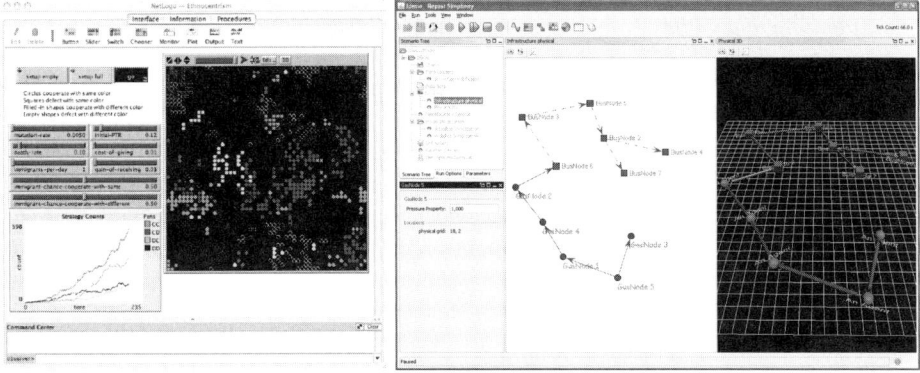

FIGURE 10.9 A screenshot of a Netlogo simulation applet, on the left, and a Repast simulation model, on the right.

parameters to be monitored, in this specific simulation context and in general for crowding situations, are objects of future developments.

10.5 From a SCA Model to Its Implementation

To effectively realize a simulator based on the introduced MAS model, even before considering the possibility of realizing tools supporting the specific aspects of its application to pedestrian dynamics, we considered several existing software frameworks for agent–based simulation systems.

This kind of framework comprises abstractions and mechanisms for the specification of agents and their environments, to support their interaction, but also additional functionalities such as the management of the simulation (e.g., set-up, configuration, turn management), its visualization, monitoring and the acquisition of data about the simulated dynamics. We will not introduce a detailed review of the current state of the art in this sector, but we will introduce some classes of instruments that have been considered for the implementation of SCA-based simulations.

A first category of these tools provides *general purpose frameworks* in which agents mainly represent *passive abstractions*, sort of data structures that are manipulated by an overall simulation process. A successful representative of such tools is *NetLogo*[*], a dialect of the Logo language specifically aimed at modeling phenomena characterized by a decentralized, interconnected nature. NetLogo does not even adopt the term agent to denote individuals, but it rather calls them *turtles*; a typical simulation consists in a cycle choosing and performing an action for every turtle, considering its current situation and state. The choice of a very simple programming language that does not require a background on informatics, the possibility to deploy in a very simple way simulations as Java applets, and the availability of simple yet effective visualization tools, made NetLogo extremely popular.

A second category of these tools is frameworks that are developed with a similar rationale, providing for very similar support tools, but these instruments are based on *general purpose*

---

[*]http://ccl.northwestern.edu/netlogo/

programming languages (generally object oriented). *Repast*[*] [North et al., 2006], represents a successful representative of this category, being a widely employed agent-based simulation platform based on the Java language. The object-oriented nature of the underlying programming language supports the definition of computational elements that make these agents more autonomous, closer to the common definitions of agents, supporting the encapsulation of state (and state manipulation mechanisms), actions and action choice mechanism in the agent's class. The choice of adopting a general purpose programming language, on one hand, makes the adoption of these instruments harder for modelers without a background in informatics but, on the other, it simplifies the integration with external and existing libraries. Repast, in its current version, can be easily connected to instruments for statistical analysis, data visualization, reporting and also Geographic Information Systems.

While the above introduced frameworks offer functionalities that are surely important in simplifying the development of an effective simulator, it must be noted that their neutrality with respect to the specific adopted agent model leads to a necessary preliminary phase of adaptation of the platform to the specific features of the model that is being implemented. If the latter defines specific abstractions and mechanisms for agents, their decision making activities, their environment and the way they interact, then the modeler must in general develop proper computational supports to be able to fruitfully employ the platform. These platforms, in fact, are not endowed with specific supports to the realization of agent deliberation mechanisms or infrastructures for interaction models, either direct or indirect (even if it must be noted that all the above platforms generally provide some form of support to agent environment definition, such as grid-like or graph structures).

These considerations are the main motivations for the efforts that lead to the development of tools belonging to a third category of frameworks, providing a *higher level linguistic support* in an effort aimed at reducing the distance between agent-based models and their implementations. The latest version of Repast, for instance, is characterized by the presence of a high level interface for "point-and-click" definition of agent's behaviors, that is based on a set of primitives for the specification of agent's actions. *SimSesam*[**] [Klügl et al., 2003] defines a set of *primitive functions* as basic elements for describing agents' behaviors, and it also provides visual tools supporting model implementation. At the extreme border of this category, we can mention efforts that are specifically aimed at supporting the development of simulations based on a precise agent model, approach and sometimes even for a specific area of application.

Given the overall aims of this research effort, the availability of a formal model that can represent a proper level for the specification of agents and their behaviors, our choice for the implementation of a simulator based on the SCA model was to realize an ad hoc framework, specifically suited to this model, the introduced abstractions and mechanisms.

10.5.1 Supporting and Executing SCA Models

The framework that we developed for this project is essentially a library developed in C++ providing proper classes to realize notions and mechanisms related to the SCA and MMASS models; this library is based on a previous experience [Bandini et al., 2006a], in which a Java framework for MMASS models was developed. The reason for this change in the adopted programming language was mainly the need of integrating the simulation

---

[*]http://repast.sourceforge.net/
[**]http://www.simsesam.de/

Crowd Behavior Modeling: From Cellular Automata to Multi-Agent Systems 317

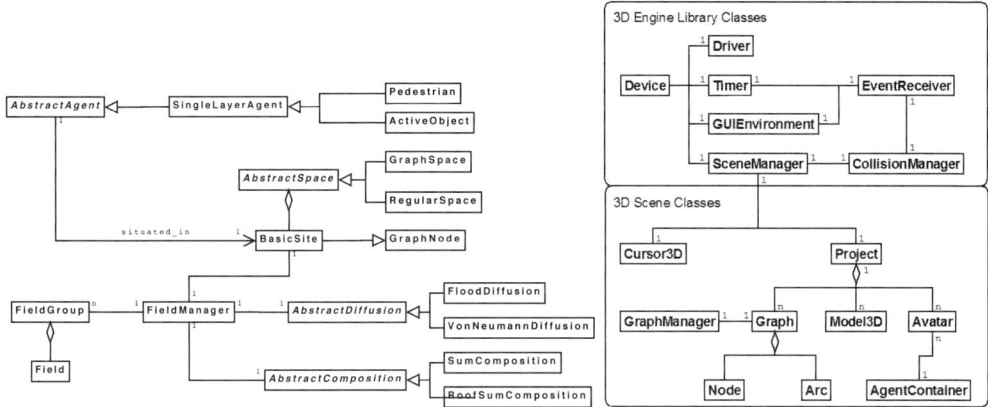

FIGURE 10.10 Simplified class diagrams respectively related to the part of the framework devoted to the realization of MMASS concepts and mechanisms, on the left, and to the management of the visualization of the dynamics generated by the model, on the right.

model with an effective realtime 3D engine, to realize a module for an effective visualization of the dynamics generated by the simulation models. In particular, to realize the second component we adopted Irrlicht*, an open-source 3D engine and usable in C++ language. It is cross-platform and it provides a performance level that we considered suitable for our requirements. It provides a high level API that was adopted for several projects related to 3D and 2D applications like games or scientific visualizations.

In particular, a simplified class diagram of the framework is shown in Figure 10.10 on the left. The lower part of the diagram is devoted to the environment, and it is built around the BasicSite class. The latter is essentially a graph node (i.e., it inherits from the GraphNode class) that is characterized by the association with a FieldManager. The latter provides the services devoted to field management (diffusion, composition and comparison, defined as abstract classes). An abstract space is essentially an aggregation of sites, whose concretizations define proper adjacency geometries (e.g., regular spaces characterized by a Von Neumann adjacency or possibly irregular graphs). An abstract agent is necessarily situated in exactly one site. Concrete agents defined for this specific framework are active objects (that are used to define concrete points of interest/reference to be adopted in a virtual environment) and pedestrians (that are basic agents capable of moving in the environment). Actual pedestrians and mobile agents that a developer wants to include to the virtual environment must be defined as subclasses of Pedestrian, overriding the basic behavioral methods and specifically the action method. Figure 10.11 shows the methods defined by the AbstractAgent class, on the left, and the pseudo-code description of the Pedestrian agent *action* method, on the right. This particular method encapsulates the sub-steps that are carried out by an agent every time it evaluates what action should be performed. In particular, in the underground simulation example, the agent must (i) process the relevant perceptions (i.e., essentially gather the perceivable fields in the site it occupies and on the adjacent ones), (ii) evaluate the possibility to alter its own attitude, (iii) choose the destination site and (iv) move to it. It must be noted that in this application, the pedestrian agents automatically emit a presence field every turn, before their actions are effectively triggered.

---

*http://irrlicht.sourceforge.net/

```
┌─────────────────────────────────────────────┐  ┌──────────────────────────────────────────────┐
│ AbstractAgent basic API                     │  │ Pedestrian Agent Action method               │
│                                             │  │ (pseudo-code)                                │
│ - int tryToMoveTo(BasicSite* a);            │  │                                              │
│ - int moveTo(BasicSite* a);                 │  │ void action(){                               │
│ - BasicSite *getSite();                     │  │     processPercepts();                       │
│ - vector <BasicSite> getAdjs(string fieldtype); │     evaluateChangeAttitude();              │
│ - vector <BasicSite> getMaxAdjs(string fieldtype); │  bestSite=maxUtilityAdjSite()          │
│ - vector <BasicSite> getMinAdjs(string fieldtype); │  moveTo(bestSite);                     │
│ - BasicSite *getRandomMaxAdj(string fieldtype); │ }                                          │
│ - BasicSite *getRandomMinAdj(string fieldtype); │                                            │
│ - BasicSite *getRandomAdj();                │  │                                              │
│ - Field perceiveField(string fieldtype);    │  │ N.B. In the metro simulation, passengers    │
│ - vector <Field> perceiveAll();             │  │ automatically emit a presence field every turn│
│ - void emit(Field* f);                      │  │                                              │
└─────────────────────────────────────────────┘  └──────────────────────────────────────────────┘
```

FIGURE 10.11 Methods defined by the AbstractAgent class, on the left, and the pseudo-code description of the Pedestrian agent *action* method, on the right.

While the previous elements of the framework are devoted to the management of the behaviors of autonomous entities and of the environment in which they are situated, another relevant part of the described framework is devoted to the visualization of these dynamics. More than entering in the details of how the visualization library was employed in this specific context, we will now focus on how the visualization modules were integrated with the previously introduced MMASS framework in order to obtain indications on the scene that must be effectively visualized. Figure 10.10, on the right, shows a simplified class diagram of the main elements of the 3D Engine Library. The diagram also includes the main classes that are effectively in charge of inspecting the state of the MMASS environment and agents, and of providing the relevant information to the SceneManager that will translate it into a scene to be visualized. The Project class acts as a container of the 3D models providing the graphical representation of the virtual environment (Model3D objects), as well as the graph related to the adopted discretization of this physical space (a Graph object visually representing the previously discussed physical layer). It also includes a set of Avatar objects, that are three dimensional representations of Pedestrian objects (introduced in the previous subsection).

The framework must be able to manage in a coordinated way the execution of the model defined for the specific virtual environment and the updating of its visualization. To manage this coordinated execution of different modules and procedures three main operative modes have been defined and are supported by the framework. The first two are characterized by the fact that agents are not provided with a thread of control of their own. A notion of turn is defined and agents are activated to execute one action per turn, in a sequential way or in a conceptually parallel way (as for a Cellular Automaton). In this case, respectively after each agent action or after a whole turn the scene manager can update the visualization. On the other hand, agents might be associated with a thread of control of their own and no particular fairness policy is enforced. The environment, and more precisely the sites of the MMASS space, is in charge of managing possible conflicts on the shared resource. However, in order to support a fluid visualization of the dynamics generated by the execution of the MAS, the Pedestrian object before executing an action must coordinate with the related Avatar: the Pedestrian action will be temporarily blocked until the visualization engine has updated the scene, moving the Avatar according to the previous Pedestrian movement. It must be noted that in all the introduced activation modes the environment is in charge of a regulation function [Bandini and Vizzari, 2007] limiting agents' autonomy for sake of managing the consistency of the overall model or to manage a proper form of visualization.

Crowd Behavior Modeling: From Cellular Automata to Multi-Agent Systems 319

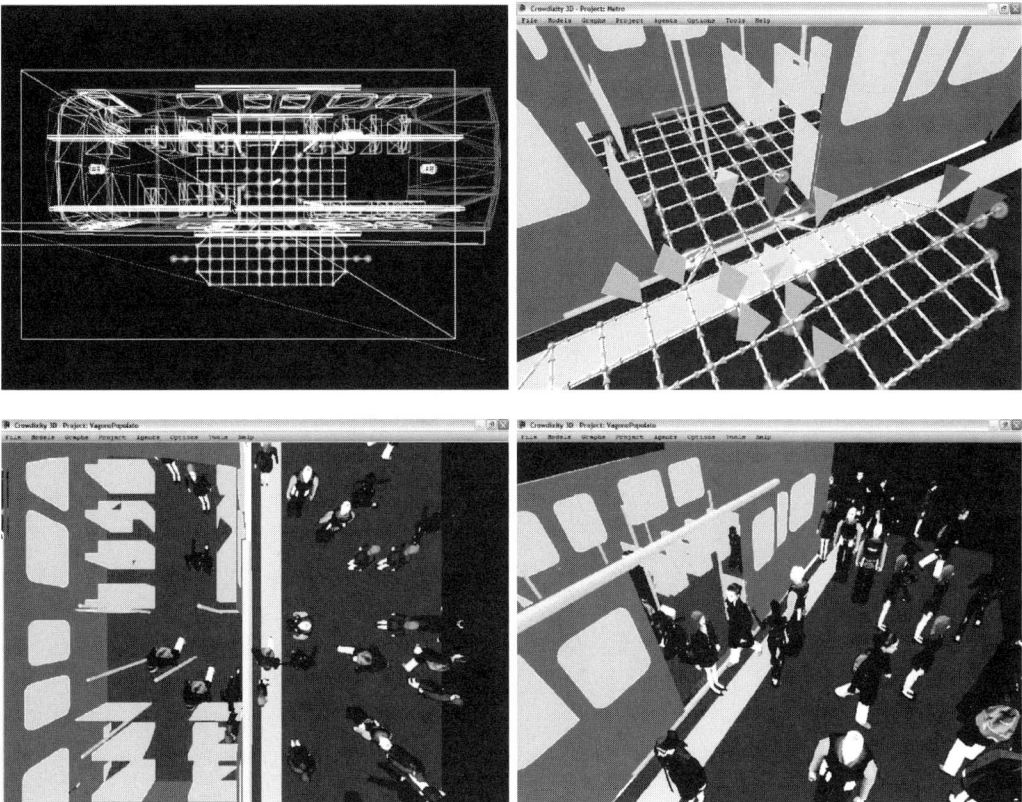

FIGURE 10.12 Four screenshots of the application of the SCA simulation framework to the underground scenario.

The framework was also applied in the underground simulation example; some screenshots of this application are shown in Figure 10.12. The framework comprises tools supporting the definition of the spatial structure of the environment and for positioning and configuring active elements of the environment, as well as agents included in a pre-defined library. In the future, we aim at supporting the dynamic configuration of pedestrian agents' behaviors as well. The visualization can show or hide the discrete structure of the environment, and it can use visually effective or simplified models for pedestrians. Additional examples of applications of this framework are presented in [Vizzari et al., 2008].

10.6 Conclusions

The chapter has introduced and discussed issues and approaches to the modeling and simulation of crowds of pedestrians. A brief discussion of the existing modeling approaches was provided, and a modeling approach for the representation of crowds with SCAs has been described and adopted in the sample simulation scenario. Preliminary results of this case study have also been illustrated and discussed. Finally different approaches to the generation of an effective 3D visualization of simulated dynamics, that can be used for a more effective communication of simulation results, have been described.

The model has been applied to paradigmatic cases and it was able to generate dynamics and phenomena (i.e., formation of lanes in corridors, freezing by heating) that were previ-

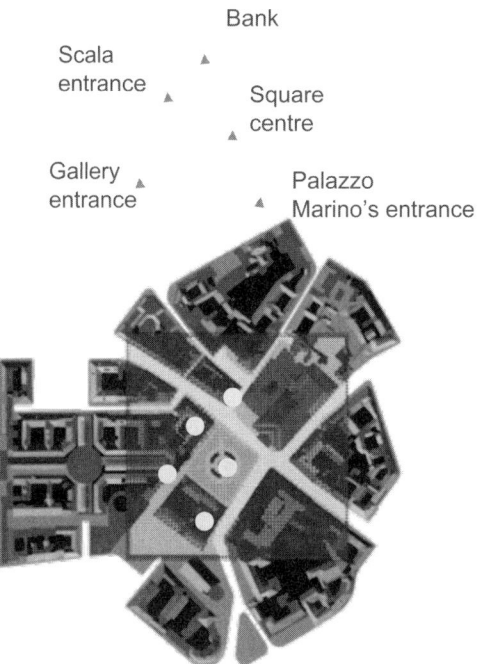

FIGURE 10.13 A diagram illustrating a multilayered environment specification: the bottom layer is a fine grained discretization of Scala Square and the top layer represents its points of interest.

ously by models adopting different approaches [Bandini et al., 2007]. On the other hand, the SCA model supports the definition of heterogeneous agent types, and thus of different types of pedestrians, and it was adopted to design and realize tools supporting the definition and execution of SCA based pedestrian simulations.

In this line of research, we proposed a possible enhancement of the described modeling approach to improve its expressive power, starting from the representation of the environment. In particular, by adopting the MMASS multilayered environmental structure, the different points of interest/reference might be represented on a graph whose links represent proximity or direct reachability relations among the related points, realizing a sort of abstract map of the environment. This layer might be interfaced to the previously introduced more fine-grained representation of the environment (i.e., the physical layer), and it could be the effective source of fields generated by infrastructural elements, that are diffused to the physical layer by means of interfaces. A sample diagram illustrating this approach to the modeling of a physical environment is shown in Figure 10.13: the bottom layer is a fine grained discretization of Scala Square, and the top layer represents its points of interest that are associated with agents emitting a proper distinctive presence field. The abstract map could also be (at least partly) known by an agent, that could thus make decisions on what attitude toward movement should be selected according to its own goals and according to the current context by reasoning on/about the map, instead of following a predefined script. These kinds of considerations do not only emphasize the possible use of a multiple layered representations of the environment, but they also point out the possibility of enhancing the current agents (that are characterized by a reactive architecture) by endowing them with proper forms of deliberation, toward a hybrid agent architecture. A complete definition of these deliberative elements of the situated agents is the object of current and future works.

10.7 Discussion and Future Research Directions

One of the future trends of research related to the topic of pedestrian and crowd modeling and simulation is focused on **data acquisition**: the current landscape on approaches to data acquisition about crowds of pedestrians is quickly evolving, due to the growing availability of innovative localization technologies (GPS, RFID, UWB) and to the improvement of existing techniques based on analysis of video footage. However, proper methodologies for the acquisition of data to characterize crowds of pedestrians in complex situations are still missing.

Moreover, the presence of (i) recent works focusing on the analysis of the (empirical) relation between density and velocity of pedestrians [Seyfried et al., 2005] (also as a very general test for the evaluation of the currently available pedestrian simulation models and platforms) and (ii) contrasting models taking a very different perspective on pedestrians' movement conflicts, some preventing them, by activating asynchronously agents in a random order (i.e., shuffling) [Klüpfel, 2003], other ones considering conflicts a relevant phenomenon to be modeled and considered [Kirchner et al., 2003], suggests the fact that there are still open **methodological issues** to be clarified.

These two lines of research will hopefully converge and provide new results that will support the definition of proper methodological framework for pedestrian modeling and simulation, and more effective **validation techniques** [Klügl, 2008] that would help the field in advancing beyond the already empirically studied paradigmatic cases [Helbing et al., 2001].

Finally, a very lively and promising research trend is focused on the integration of **psycho/sociological considerations** in crowd models. Most models of pedestrian behaviors effectively disregard differences among individuals that make up a crowd. The integration of psycho/sociological considerations on human behavior (such as the inclusion of emotions and their effects influencing pedestrian movement) could provide an improvement to the expressiveness of crowd models that could also extend the range of applications, for instance toward the modeling of consumers' behaviors in shopping spaces.

Acknowledgements

The work presented in this chapter has been partially funded by the University of Milano-Bicocca within the project 'Fondo d'Ateneo per la Ricerca - anno 2007'. We wish to thank Mizar Luca Federici for the fruitful discussions on multi-agent simulation of crowds of pedestrians. We also thank the anonymous referees for their comments and suggestions.

References

   A. Adamatzky. *Dynamics of Crowd-Minds: Patterns of Irrationality in Emotions, Beliefs and Actions*. World Scientific, Singapore, 2005.
   R. Axtell. Why Agents? On the Varied Motivations for Agent Computing in the Social Sciences. *Center on Social and Economic Dynamics Working Paper*, 17, 2000.
   S. Bandini and G. Vizzari. Regulation function of the environment in agent-based simulation. In D. Weyns, H. V. D. Parunak, and F. Michel, editors, *Environments for Multi-Agent Systems III, Third International Workshop, E4MAS 2006, Hakodate, Japan, May 8, 2006, Selected Revised and Invited Papers*, volume 4389 of *Lecture Notes in Computer Science*, pages 157–169. Springer–Verlag, 2007.

S. Bandini, S. Manzoni, and C. Simone. Dealing with space in multi–agent systems: a model for situated MAS. In *Proceedings of the first international joint conference on Autonomous agents and multiagent systems*, pages 1183–1190. ACM Press, New York, 2002. ISBN 1-58113-480-0.

S. Bandini, S. Manzoni, and G. Vizzari. Towards a Platform for Multilayered Multi Agent Situated System Based Simulations: Focusing on Field Diffusion. *Applied Artificial Intelligence*, 20(4–5):327–351, 2006a.

S. Bandini, G. Mauri, and G. Vizzari. Supporting action-at-a-distance in situated cellular agents. *Fundamenta Informaticae*, 69(3):251–271, 2006b.

S. Bandini, M. Federici, S. Manzoni, and G. Vizzari. Pedestrian and Crowd Dynamics Simulation: Testing SCA on Paradigmatic Cases of Emerging Coordination in Negative Interaction Conditions. In V. Malyshkin, editor, *Parallel Computing Technologies, 9th International Conference, PaCT 2007, Pereslavl-Zalessky, Russia, September 3-7, 2007, Proceedings*, volume 4671 of *Lecture Notes in Computer Science*, pages 360–369. Springer–Verlag, 2007.

M. Batty. Agent based pedestrian modeling (editorial). *Environment and Planning B: Planning and Design*, 28:321–326, 2001.

V. Blue and J. Adler. Cellular automata microsimulation of bi-directional pedestrian flows. *Transportation Research Record*, 1678:135–141, 2000a.

V. Blue and J. Adler. Modeling four-directional pedestrian flows. *Transportation Research Record*, 1710:20–27, 2000b.

S. Christley, X. Zhu, S. Newman, and M. Alber. Multiscale agent-based simulation for chondrogenic pattern formation *in vitro*. *Cybernetics and Systems*, 38(7):707–727, 2007.

J. Dijkstra, H. Timmermans, and B. de Vries. Modelling behavioural aspects of agents in simulating pedestrian movement. In *Proceedings of CUPUM 05, Computers in Urban Planning and Urban Management, 30-Jun-2005, London*. CASA Centre for Advanced Spatial Analysis - University College London, 2005.

J. Dijkstra, J. Jessurun, B. de Vries, and H. Timmermans. Agent architecture for simulating pedestrians in the built environment. In *International Workshop on Agents in Traffic and Transportation*, pages 8–15, 2006.

J. Dijkstra, B. de Vries, and H. Timmermans. Empirical estimation of agent shopping patterns for simulating pedestrian movement. In *Proceedings of the 10th Internationa Conference on Computers in Urban Planning and Urban Management*, 2007.

J. Ferber. *Multi–Agent Systems*. Addison–Wesley, Harlow, 1999.

C. Gloor, P. Stucki, and K. Nagel. Hybrid techniques for pedestrian simulations. In P. Sloot, B. Chopard, and A. Hoekstra, editors, *Cellular Automata, 6th International Conference on Cellular Automata for Research and Industry, ACRI 2004*, volume 3305 of *Lecture Notes in Computer Science*, pages 581–590. Springer–Verlag, 2004.

D. Helbing. A fluid–dynamic model for the movement of pedestrians. *Complex Systems*, 6(5):391–415, 1992.

D. Helbing and P. Molnár. Social force model for pedestrian dynamics. *Phys. Rev. E*, 51(5):4282–4286, May 1995. doi: 10.1103/PhysRevE.51.4282.

D. Helbing, F. Schweitzer, J. Keltsch, and P. Molnár. Active walker model for the formation of human and animal trail systems. *Physical Review E*, 56(3):2527–2539, January 1997.

D. Helbing, I. Farkas, P. Molnár, and T. Vicsek. *Pedestrian and Evacuation Dynamics*, chapter Simulation of Pedestrian Crowds in Normal and Evacuation Situations, pages 21–58. Springer-Verlag, Heidelberg, 2001.

C. Henein and T. White. Agent-based modelling of forces in crowds. In P. Davidsson, B. Logan, and K. Takadama, editors, *Multi-Agent and Multi-Agent-Based Simulation, Joint Workshop MABS 2004, New York, NY, USA, July 19, 2004, Revised Selected Papers*, volume 3415 of *Lecture Notes in Computer Science*, pages 173–184. Springer-Verlag, Heidelberg, 2005.

A. Kirchner, K. Nishinari, and A. Schadschneider. Friction effects and clogging in a cellular automaton model for pedestrian dynamics. *Physical Review E*, 67(5), 2003.

K. Kitazawa and M. Batty. *Developments in Design & Decision Support Systems in Architecture and Urban Planning*, chapter Pedestrian Behaviour Modelling – An application to retail movements using a genetic algorithms, pages 111–125. Eindhoven University of Technology, Department of Architecture, Building, and Planning, Eindhoven, The Netherlands, 2004.

F. Klügl. A validation methodology for agent-based simulations. In Wainwright and Haddad [2008], pages 39–43. ISBN 978-1-59593-753-7.

F. Klügl and G. Rindsfüser. Large-scale agent-based pedestrian simulation. In P. Petta, J. Müller, M. Klusch, and M. Georgeff, editors, *Multiagent System Technologies, 5th German Conference, MATES 2007, Leipzig, Germany, September 24-26, 2007, Proceedings*, volume 4687 of *Lecture Notes in Computer Science*, pages 145–156. Springer-Verlag, 2007.

F. Klügl, R. Herrler, and C. Oechslein. From Simulated to Real Environments: How to Use SeSAm for Software Development. In M. Schillo, M. Klusch, J. Müller, and H. Tianfield, editors, *Multiagent System Technologies, First German Conference, MATES 2003, Erfurt, Germany, September 22-25, 2003, Proceedings*, volume 2831 of *Lecture Notes in Computer Science*, pages 13–24. Springer-Verlag, Heidelberg, 2003.

H. Klüpfel. *A Cellular Automaton Model for Crowd Movement and Egress Simulation*. Phd thesis, University Duisburg-Essen, 2003.

L. Kruse. *Changing conceptions of crowd ming and behaviour*, chapter Conceptions of crowds and crowding. Springer, 1986.

M. Luck, P. McBurney, O. Sheory, and S. Willmott, editors. *Agent Technology: Computing as Interaction*. University of Southampton, Southampton, 2005.

Y. Murakami, T. Ishida, T. Kawasoe, and R. Hishiyama. Scenario description for multi-agent simulation. In *The Second International Joint Conference on Autonomous Agents & Multiagent Systems, AAMAS 2003, July 14-18, 2003, Melbourne, Victoria, Australia, Proceedings*, pages 369–376. ACM press, 2003.

S. R. Musse and D. Thalmann. Hierarchical model for real time simulation of virtual human crowds. *IEEE Trans. Vis. Comput. Graph.*, 7(2):152–164, 2001.

K. Nagel and M. Schreckenberg. A cellular automaton model for freeway traffic. *Journal de Physique I France*, 2(2221):222–235, 1992.

M. North, N. Collier, and J. Vos. Experiences creating three implementations of the repast agent modeling toolkit. *ACM Transactions on Modeling and Computer Simulation*, 16(1):1–25, 2006.

S. Okazaki. A study of pedestrian movement in architectural space, part 1: Pedestrian movement by the application of magnetic models. *Transactions of A.I.J.*, (283): 111–119, 1979.

A. Schadschneider, A. Kirchner, and K. Nishinari. CA Approach to Collective Phenomena in Pedestrian Dynamics. In S. Bandini, B. Chopard, and M. Tomassini, editors, *Cellular Automata, 5th International Conference on Cellular Automata for Re-*

search and Industry, ACRI 2002, volume 2493 of *Lecture Notes in Computer Science*, pages 239–248. Springer, Heidelberg, 2002.

M. Schreckenberg and S. D. Sharma, editors. *Pedestrian and Evacuation Dynamics*. Springer–Verlag, Heidelberg, 2001.

A. Seyfried, B. Steffen, W. Klingsch, and M. Boltes. The fundamental diagram of pedestrian movement revisited. *Journal of Statistical Mechanics*, 2005(10):P10002, 2005.

W. Shao and D. Terzopoulos. Autonomous pedestrians. *Graphical Models*, 69(5-6): 246–274, 2007.

M. Toyama, A. Bazzan, and R. da Silva. An agent-based simulation of pedestrian dynamics: from lane formation to auditorium evacuation. In H. Nakashima, M. Wellman, G.Weiss, and P.Stone, editors, *5th International Joint Conference on Autonomous Agents and Multiagent Systems (AAMAS 2006)*, pages 108–110. ACM press, 2006.

G. Vizzari, G. Pizzi, and F. da Silva. A Framework for Execution and 3D Visualization of Situated Cellular Agent Based Crowd Simulations. In Wainwright and Haddad [2008], pages 18–22. ISBN 978-1-59593-753-7.

R. Wainwright and H. Haddad, editors. *Proceedings of the 2008 ACM Symposium on Applied Computing (SAC), Fortaleza, Ceara, Brazil, March 16-20, 2008*, 2008. ACM. ISBN 978-1-59593-753-7.

A. Willis, N. Gjersoe, C. Havard, J. Kerridge, and R. Kukla. Human movement behaviour in urban spaces: Implications for the design and modelling of effective pedestrian environments. *Environment and Planning B*, 31(6):805–828, 2004.

S. Wolfram. *Theory and Applications of Cellular Automata*. World Press, Singapore, 1986.

11
Agents for Traffic Simulation

11.1	Introduction...	325
	Aim and Overview	
11.2	Agents for Traffic Simulation	327
	Macroscopic vs. Microscopic Approaches • Driver-Vehicle Agents	
11.3	Models for the Driving Task	331
	The Intelligent Driver Model • Inter-Driver Variability • Intra-Driver Variability	
11.4	Modeling Discrete Decisions	338
	Modeling Lane Changes • Approaching a Traffic Light	
11.5	Microscopic Traffic Simulation Software	341
	Simulator Design • Numerical Integration • Visualization	
11.6	From Individual to Collective Properties	346
	Emergence of Stop-and-Go Waves • Impact of a Speed Limit • Store-and-Forward Strategy for Inter-Vehicle Communication	
11.7	Conclusions and Future Work	352
	References	352

Arne Kesting
Technische Universitat Dresden

Martin Treiber
Technische Universitat Dresden

Dirk Helbing
ETH Zurich

11.1 Introduction

Efficient transportation systems are essential to the functioning and prosperity of modern, industrialized societies. Mobility is also an integral part of our quality of life, sense of self-fulfillment and personal freedom. Our traffic demands of today are predominantly served by individual motor vehicle travel which is the primary means of transportation. However, the limited road capacity and ensuing traffic congestion have become a severe problem in many countries. Nowadays, we additionally have to balance the human desire for personal mobility with the societal concerns about its environmental impact and energy consumption. On the one hand, traffic demand can only be affected indirectly by means of policy measures. On the other hand, an extension of transport infrastructure is no longer an appropriate or desirable option in densely populated areas. Moreover, construction requires high investments and maintenance is costly in the long run. Therefore, engineers are now seeking solutions to the questions of how the capacity of the road network could be used more efficiently and how operations can be improved by way of intelligent transportation systems (ITS).

In the presence of increasing computing power, realistic microscopic traffic simulations are becoming a more and more important tool for diverse purposes ranging from generating surrounding traffic in a virtual reality driving simulator to large-scale network simulations for a model-based prediction of travel times and traffic conditions [Ministry of Transport,

TABLE 11.1 Subjects in transportation systems sorted by typical time scales involved.

Time Scale	Subject	Models	Aspects
0.1 s	Vehicle Dynamics	Sub-microscopic	Drive-train, brake, ESP
1 s			Reaction time, Time gap
10 s	Traffic Dynamics	Car-following models	Accelerating and braking
1 min		Fluid-dynamic models	Traffic light period
10 min			Period of stop-and-go wave
1 h			Peak hour
1 day		Traffic assignment models	Day-to-day human behavior
1 year	Transport. Planning	Traffic demand model	Building measures
5 years		Statistics	Changes in spatial structure
50 years		Prognosis	Changes in Demography

Energy and Spatial Planning of Nordrhein-Westfalen, 2007]. The primary application for traffic simulations is the evaluation of hypothetical scenarios for their impact on traffic. Computer simulations can be valuable in making these analyses in a cost-effective way. For example, simulations can be used to estimate the impact of future driver assistance systems and wireless communication technologies on traffic dynamics. Another example is the prediction of congestion levels in the future, based on demographic forecasts.

Before going into detail about possible traffic flow models, it is worth mentioning differences between modeling the short-term traffic dynamics on a single road section and the approach used for transportation planning describing behavioral pattern in a network on a larger time scale. Table 11.1 shows typical time scales ranging over nine orders of magnitude including vehicle dynamics, traffic dynamics and transportation planning. While dynamic flow models explicitly describe the *physical propagation of traffic flows* of a given traffic volume in a road network, transportation planning tools deal with the calculation of the traffic demand by considering the *decisions of travelers* to participate in economical, social and cultural activities. The need for transportation arises because these activities are spatially separated. The classical approach in trip-based transportation models is based on a four-step methodology of *trip generation*, *trip distribution*, *mode split* and *traffic assignment* [De Dios Ortúzar and Willumsen, 2001; Schnabel and Lohse, 1997; Daganzo, 1997; Maerivoet and De Moo, 2005; Helbing and Nagel, 2004]. In the fourth step, the origin-destination matrix of trips with a typical minimum disaggregation of one hour (comprising a typical peak-hour analysis) is assigned to routes in the actual (or prospective) transportation network while taking into account the limited capacity of the road infrastructure by means of simplified effective models. Recently, even large-scale multi-agent transportation simulations have been performed in which each traveler is represented individually [Nagel et al., 2000; Raney et al.; Charypar, 2005]. For the purposes of demand-modeling, mobility-simulation and infrastructure re-planning the open-source software MATSIM provides a toolbox to implement large-scale agent-based transport simulations [MATSim, 2008].

11.1.1 Aim and Overview

The chapter will review the state-of-the-art in microscopic traffic flow modeling and the implications for simulation techniques. In Section 11.2, we will introduce the concept of a *driver-vehicle agent* within in the context of common traffic modeling approaches.

In order to perform traffic simulations, we will take a "bottom-up" approach and present concrete models for describing the behavior of an agent. In Section 11.3.1, the Intelligent Driver Model [Treiber et al., 2000] serves as a basic example of a car-following model representing the operational level of driving. As a first contribution, we will give special attention to the heterogeneity in traffic. Different drivers behave differently in the same situation (so called "inter-driver variability") but can also change their behavior over the course of time ("intra-driver variability"). While the first aspect can be addressed by individual parameter sets for the agents (Section 11.3.2), the latter can be modeled by introducing a time-dependency of some parameters (e.g., to model the frustration of drivers after being in a traffic jam for a period, Section 11.3.3).

Realistic simulations of multi-lane freeway traffic and traffic in city networks also require discrete decisions by the agents. For example, lane-changing decisions allow faster cars to pass slower trucks. Another decision is related to the decision process of whether to brake or not to brake when approaching a traffic light turning from green to amber. In Section 11.4.1, we will introduce a general framework for dealing with these discrete decision processes. The presented "meta-model" MOBIL [Kesting et al., 2007a] is an example of how complexity can in fact be reduced by falling back on the agent's model for calculating longitudinal accelerations.

After having modeled the agent's acceleration and lane-changing behavior, we will consider multi-agent simulations. Section 11.5 addresses the design of microscopic traffic simulators. In order to be specific, we will discuss the explicit numerical integration scheme, input and output quantities and visualization possibilities.

In Section 11.6, we will demonstrate the expressive power of the agent-based approach for handling current research questions. Traffic simulations will illustrate the emergence of collective dynamics from local interaction between agents. By way of example, we will show how the desired individual behavior of agents to move forward fast can lead to contrary effects such as the breakdown of traffic and self-organized stop-and-go waves. Another simulation will evaluate the effect of traffic flow homogenization by means of a speed limit (Section 11.6.2). Last but not least, we will discuss an application of inter-vehicle communication for propagating traffic-related information in a decentralized way. Inter-vehicle communication has recently received much attention in the academic and engineering world as it is expected to be a challenging issue for the next generation of vehicle-based Intelligent Transportation Systems (ITS). Finally, in Section 11.7, we will discuss such trends in traffic modeling and simulation.

11.2 Agents for Traffic Simulation

Vehicular traffic is a typical example of a *multi-agent system*: Autonomous agents (the drivers) operate in a shared environment provided by the road infrastructure and react to the neighboring vehicles. Therefore, the activities include both *human interaction* (with the dominant influence originating from the directly leading vehicle) and *man-machine-interactions* (driver interaction with the vehicle and the physical road environment). The *microscopic modeling* or *agent-based* approach describing the motion of each individual vehicle has grown in popularity over the last decade. The following Section 11.2.1 will provide an overview of common mathematical approaches for describing traffic dynamics. In Section 11.2.2, we will introduce the concept of a "driver-vehicle agent" within the context of microscopic traffic modeling.

11.2.1 Macroscopic vs. Microscopic Approaches

The mathematical description of the dynamics of traffic flow has a long history already. The scientific activity had its beginnings in the 1930s with the pioneering studies on the fundamental relations of traffic flow, velocity and density conducted by Greenshields [Greenshields, 1959]. By the 1950s, scientists had started to describing the physical propagation of traffic flows by means of dynamic macroscopic and microscopic models. During the 1990s, the number of scientists engaged in traffic modeling grew rapidly because of the availability

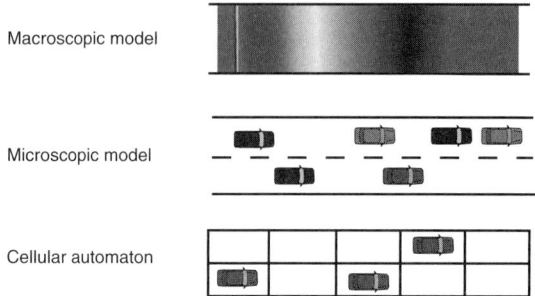

FIGURE 11.1 Illustration of different traffic modeling approaches: A snapshot of a road section at time t_0 is either characterized by *macroscopic traffic quantities* like traffic density $\rho(x, t_0)$, flow $Q(x, t_0)$ or average velocity $V(x, t_0)$, or, *microscopically*, by the positions $x_\alpha(t_0)$ of single driver-vehicle agent α. For cellular automata, the road is divided into cells which can be either occupied by a vehicle or empty.

of better traffic data and higher computational power for numerical analysis.

Traffic models have been successful in reproducing the observed *collective, self-organized traffic dynamics* including phenomena such as breakdowns of traffic flow, the propagation of stop-and-go waves (with a characteristic propagation velocity), the capacity drop, and different spatiotemporal patterns of congested traffic due to instabilities and nonlinear interactions [Helbing, 2001; Kerner, 2004; Kerner and Rehborn, 1996a; Cassidy and Bertini, 1999; Schönhof and Helbing, 2007]. For an overview of experimental studies and the development of miscellaneous traffic models, we refer to the recently published extensive review literature [Helbing, 2001; Chowdhury et al., 2000; Nagatani, 2002; Maerivoet and De Moor, 2005; Hoogendoorn and Bovy, 2001; Leutzbach, 1988].

As mentioned, there are two major approaches to describe the spatiotemporal propagation of traffic flows. *Macroscopic traffic flow models* make use of the picture of traffic flow as a physical flow of a fluid. They describe the traffic dynamics in terms of aggregated macroscopic quantities like the traffic density, traffic flow or the average velocity as a function of space and time corresponding to partial differential equations (cf. Fig. 11.1). The underlying assumption of all macroscopic models is the conservation of vehicles (expressed by the continuity equation) which was initially considered by Lighthill, Whitham and Richards [Lighthill and Whitham, 1955; Richards, 1956]. More advanced, so-called "second-order" models additionally treat the macroscopic velocity as a dynamic variable in order to also consider the finite acceleration capability of vehicles [Kerner and Konhäuser, 1994; Treiber et al., 1999].

By way of contrast, *microscopic traffic models* describe the motion of each individual vehicle. They model the action such as accelerations, decelerations and lane changes of each driver as a response to the surrounding traffic. Microscopic traffic models are especially suited to the study of heterogeneous traffic streams consisting of different and individual types of *driver-vehicle units* or *agents*. The result is individual trajectories of all vehicles and, consequently, any macroscopic information by appropriate aggregation. Specifically, one can distinguish the following major subclasses of microscopic traffic models (cf. Fig. 11.1):

- *Time-continuous models* are formulated as ordinary or delay-differential equations and, consequently, space and time are treated as continuous variables. *Car-following models* are the most prominent examples of this approach [Bando et al., 1995; Treiber et al., 2000; Jiang et al., 2001; Helbing and Tilch, 1998]. In general, these models are deterministic but stochasticity can be added in a natural way [Treiber et al., 2006b]. For example, a modified version of the Wiedemann model [Wiedemann, 1974] is used in the commercial traffic simulation software PTV-VISSIM™.

- *Cellular automata* (CA) use integer variables to describe the dynamic state of the system. The time is discretized and the road is divided into cells which can be either occupied by a vehicle or empty. Besides rules for accelerating and braking, most CA models require additional stochasticity. The first CA for describing traffic was proposed by Nagel and Schreckenberg [K. Nagel and Schreckenberg, 1992]. Although CA lack the accuracy of time-continuous models, they are able to reproduce some traffic phenomena [Lee et al., 2004; Helbing and Schreckenberg, 1999; Knospe et al., 2001]. Due to their simplicity, they can be implemented very efficiently and are suited to simulating large road networks [Ministry of Transport, Energy and Spatial Planning of Nordrhein-Westfalen, 2007].

- *Iterated coupled maps* are between CA and time-continuous models. In this class of model, the update time is considered as an explicit model parameter rather than an auxiliary parameter needed for numerical integration [Kesting and Treiber, 2008b]. Consequently, the time is discretized while the spatial coordinate remains continuous. Popular examples are the Gipps model [Gipps, 1981] and the Newell model [Newell, 1961]. However, these models are typically associated with car-following models as well.

At first glance, it may be surprising that simple (and deterministic) mathematical models aimed at describing the complexity of and variations in the human behavior, individual skills and driving attitudes would lead to reasonable results. However, a traffic flow can (in a good approximation) be considered as a one-dimensional system (with reduced degrees of freedom). Furthermore, traffic models typically assume rational and safe driving behavior as a reaction to the surrounding traffic while at the same time taking into account the fundamental laws of kinematics.

Another aspect concerns the important issue of traffic safety. The traditional models for describing traffic dynamics assume rational drivers that are programmed to avoid collisions.* Therefore, traffic safety simulation belongs to the field of human centered simulation where the perception-reaction system of drivers with all its weak points has to be described. Up to now, a general modeling approach is still lacking.

11.2.2 Driver-Vehicle Agents

Let us now adopt the concept of an agent to implicate the complex human driving behavior into a general modeling framework. We therefore introduce the term *driver-vehicle agent* which refers to the idea that an atomic entity includes internal characteristics of human drivers as well as external properties of a vehicle. Figure 11.2 provides an overview of relevant influences and features affecting human driving. In this context, the relevant time scales are a first characteristic feature: The *short-term operations* are constituted by control tasks such as acceleration and braking, and typically take place in the range of a second. Specific *behavioral attributes* vary between individual drivers and affect the resulting driving characteristics on an intermediate time scale of seconds up to minutes. Finally, a strategic level of driving includes time periods of hours, e.g., the decision to start a trip or to find a route in a network.

---

*Of course, collisions happen in numerical simulations due to instable models and for kinematic reasons. However, these collisions do not have explanatory or predictive power.

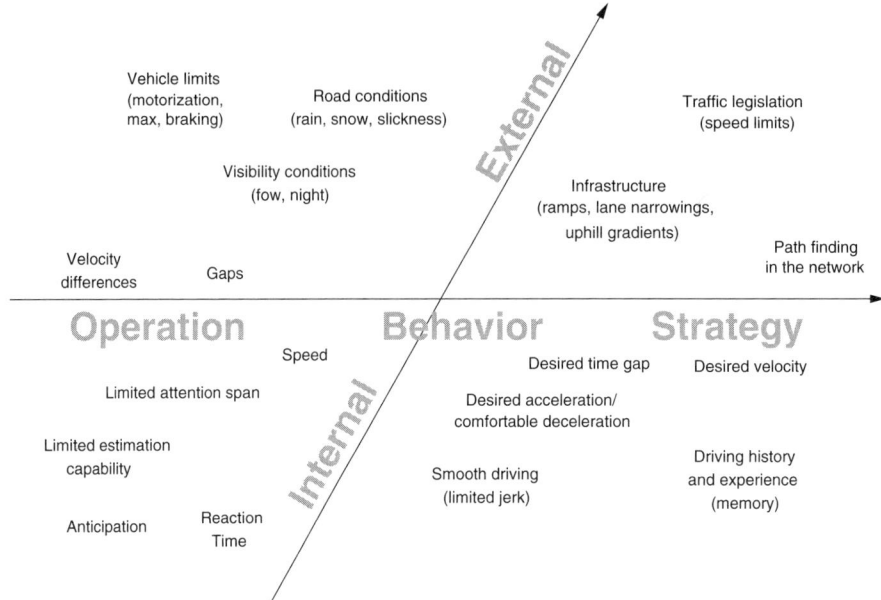

FIGURE 11.2 Characteristics of a driver-vehicle agent. The operation of driving can be classified according the involved time scales ranging from short-term actions in terms of acceleration and braking via intermediate time scales describing behavioral characteristics to long-term strategic decisions. In addition, the agent's behavior is influenced by the physical properties of the vehicle, by interactions with other agents and by the environment.

The driving task can be considered as a cognitive and therefore *internal* process of an agent: The driver's perception is limited to the observable *external* objects in the neighborhood while his or her reaction is delayed due to a non-negligible reaction time as a consequence of the physiological aspects of sensing, perceiving, deciding, and performing an action. On the intermediate time scale, the agent's actions are affected by his or her individual driving behavior which may be characterized in terms of, e.g., preferred time gaps when following a vehicle and smooth driving with a desired acceleration and a comfortable deceleration. Moreover, the individual driving style may be influenced by the experience and history of driving. For example, it is observed that people change their behavior after being stuck in traffic congestion for a period [Brilon and Ponzlet, 1996; Fukui et al., 2003]. Such features can be incorporated by internal state variables corresponding to the agent's "mind" or "memory".

However, short-time driving operations are mainly direct responses to the stimulus of the surrounding traffic. The driver's behavior is externally influenced by environmental input such as limited motorization and braking power of the vehicle, visibility conditions, road characteristics such as horizontal curves, lane narrowings, ramps, gradients and road traffic regulations. In the following sections, we will address a number of these defining characteristics of a driver-vehicle agent.

11.3 Models for the Driving Task

Microscopic traffic models describe the motion in longitudinal direction of each individual vehicle. They model the action of a driver such as accelerations and decelerations as a response to the surrounding traffic by means of an *acceleration strategy* toward a desired speed in the free-flow regime, a *braking strategy* for approaching other vehicles or obstacles, and a *car-driving strategy for maintaining a safe distance* when driving behind another vehicle. Microscopic traffic models typically assume that human drivers react to the stimulus from neighboring vehicles with the dominant influence originating from the directly leading vehicle known as "follow-the-leader" or "car-following" approximation.

By way of example, we will consider the Intelligent Driver Model (IDM) [Treiber et al., 2000] in Section 11.3.1. The IDM belongs to the class of deterministic follow-the-leader models. Like other car-following models, the IDM is formulated as an ordinary differential equation and, consequently, space and time are treated as continuous variables. This model class is characterized by an acceleration function $\dot{v} := \frac{dv}{dt}$ that depends on the actual speed $v(t)$, the gap $s(t)$ and the velocity difference $\Delta v(t)$ to the leading vehicle (see Fig. 11.3). Note that the dot is the usual shorthand notation for the time derivative of a function. The acceleration is therefore defined as the time derivative of the velocity $\dot{v} := dv/dt$.

In Section 11.3.2, we will model inter-driver variability by defining different classes of drivers which is an inherent feature of microscopic agent approaches. A model for intra-driver variability (changing behavior over the course of time) will then be discussed in Section 11.3.3.

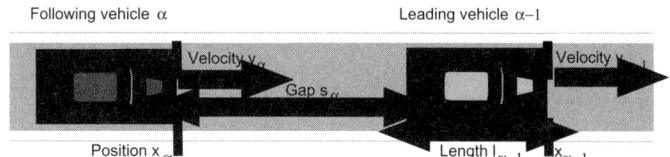

FIGURE 11.3 Illustration of the input quantities of a car-following model: The bumper-to-bumper distance s for a vehicle α with respect to the vehicle $(\alpha - 1)$ in front is given by $s_\alpha = x_{\alpha-1} - x_\alpha - l_{\alpha-1}$, where l_α is the vehicle length and x the position on the considered road stretch. The approaching rate (relative speed) is defined by $\Delta v_\alpha := v_a - v_{\alpha-1}$. Notice that the vehicle indices α are ordered such that $(\alpha - 1)$ denotes the preceding vehicle.

11.3.1 The Intelligent Driver Model

The IDM acceleration is a continuous function incorporating different driving modes for all velocities in freeway traffic as well as city traffic. Besides the distance to the leading vehicle s and the actual speed v, the IDM also takes into account velocity differences Δv, which play an essential stabilizing role in real traffic, especially when approaching traffic jams and avoiding rear-end collisions (see Fig. 11.3). The IDM acceleration function is given by

$$\frac{dv_\alpha}{dt} = f(s_\alpha, v_\alpha, \Delta v_\alpha) = a \left[1 - \left(\frac{v_\alpha}{v_0}\right)^\delta - \left(\frac{s^*(v_\alpha, \Delta v_\alpha)}{s_\alpha}\right)^2 \right]. \tag{11.1}$$

This expression combines the acceleration strategy $\dot{v}_{\text{free}}(v) = a[1 - (v/v_0)^\delta]$ toward a *desired speed* v_0 on a free road with the parameter a for the *maximum acceleration* with a braking strategy $\dot{v}_{\text{brake}}(s, v, \Delta v) = -a(s^*/s)^2$ serving as repulsive interaction when vehicle α comes too close to the vehicle ahead. If the distance to the leading vehicle, s_α, is large, the interaction term $\dot{v}_{\text{brake}}$ is negligible and the IDM equation is reduced to the free-road acceleration $\dot{v}_{\text{free}}(v)$, which is a decreasing function of the velocity with the maximum value $\dot{v}(0) = a$ and the minimum value $\dot{v}(v_0) = 0$ at the desired speed v_0. For denser traffic, the deceleration term becomes relevant. It depends on the ratio between the effective "desired minimum gap"

$$s^*(v, \Delta v) = s_0 + vT + \frac{v\Delta v}{2\sqrt{ab}}, \qquad (11.2)$$

and the actual gap s_α. The *minimum distance* s_0 in congested traffic is significant for low velocities only. The main contribution in stationary traffic is the term vT which corresponds to following the leading vehicle with a constant *desired time gap* T. The last term is only active in non-stationary traffic corresponding to situations in which $\Delta v \neq 0$ and implements an "intelligent" driving behavior including a braking strategy that, in nearly all situations, limits braking decelerations to the *comfortable deceleration* b. Note, however, that the IDM brakes stronger than b if the gap becomes too small. This braking strategy makes the IDM collision-free [Treiber et al., 2000]. All IDM parameters v_0, T, s_0, a and b are defined by positive values. These parameters have a reasonable interpretation, are known to be relevant, are empirically measurable and have realistic values [Kesting and Treiber, 2008a]. We will discuss parameter values in detail in Section 11.3.2 and will use their clear meaning to characterize different driving styles, that is, inter-driver variability.

For a simulation scenario with a speed limit (which we will study in Section 11.6.2), we consider a refinement of the IDM for the case when the actual speed is higher than the desired speed, $v > v_0$. For example, an excess of $v = 2v_0$ would lead to an unrealistic braking of $-15a$ for $\delta = 4$. This situation may occur when simulating, e.g., a speed limit on a road segment that reduces the desired speed locally. Therefore, we replace the free acceleration for the case $v > v_0$ by

$$\dot{v}_{\text{free}}(v) = -b\left[1 - \left(\frac{v_0}{v}\right)^\delta\right]. \qquad (11.3)$$

That is, the IDM vehicle brakes with the comfortable deceleration b in the limit $v \gg v_0$. Further extensions of the IDM can be found in Refs. [Treiber et al., 2006a; Kesting and Treiber, 2008b; Treiber et al., 2006b].

The *dynamic properties* of the IDM are controlled by the maximum acceleration a, the acceleration exponent δ and the parameter for the comfortable braking deceleration b. Let us now consider the following scenario: If the distance s is large (corresponding to the situation of a nearly empty road), the interaction $\dot{v}_{\text{brake}}$ is negligible and the IDM equation (11.1) is reduced to the free-road acceleration $\dot{v}_{\text{free}}(v)$. The driver accelerates to his or her desired speed v_0 with the maximum acceleration $\dot{v}(0) = a$. The acceleration exponent δ specifies how the acceleration decreases when approaching the desired speed. The limiting case $\delta \to \infty$ corresponds to approaching v_0 with a constant acceleration a while $\delta = 1$ corresponds to an exponential relaxation to the desired speed with the relaxation time $\tau = v_0/a$. In the latter case, the free-traffic acceleration is equivalent to that of the Optimal Velocity Model [Bando et al., 1995]. However, the most realistic behavior is expected between the two limiting cases of exponential acceleration (for $\delta = 1$) and constant acceleration (for $\delta \to \infty$). Therefore, we set the acceleration exponent constant to $\delta = 4$.

In Fig. 11.4, acceleration periods from a standstill to the desired speed $v_0 = 120\,\text{km/h}$ are simulated for two different settings of the maximum acceleration (the other model parameters are listed in the caption): For $a = 1.4\,\text{m/s}^2$, the acceleration phase takes approximately

40 s while an increased maximum acceleration of $a = 3\,\text{m/s}^2$ reduces the acceleration period to $\sim 15\,\text{s}$. Notice that the acceleration parameter a of $1.4\,\text{m/s}^2$ ($3\,\text{m/s}^2$) corresponds to a free-road acceleration from $v = 0$ to $v = 100\,\text{km/h}$ within 23 s (10.5 s).

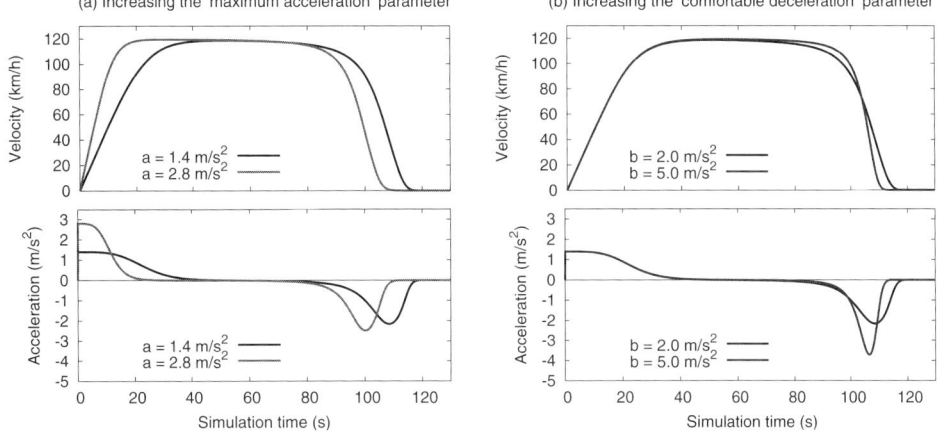

FIGURE 11.4 Simulation of a single driver-vehicle agent modeled by the IDM: The diagrams show the acceleration to the desired speed $v_0 = 120\,\text{km/h}$ followed by braking as a reaction to a standing obstacle located 3000 m ahead for several combinations of the IDM acceleration parameters a [in diagram (a)] and b [in (b)]. The remaining parameters are $a = 1.4\,\text{m/s}^2$, $b = 2.0\,\text{m/s}^2$, $T = 1.5\,\text{s}$, $s_0 = 2\,\text{m}$.

The *equilibrium properties* of the IDM are influenced by the parameters for the desired time gap T the desired speed v_0 and the minimum distance between vehicles at a standstill s_0. Equilibrium traffic is defined by vanishing speed differences and accelerations of the driver-vehicle agents α:

$$\Delta v_\alpha = 0, \tag{11.4}$$

$$\frac{dv_\alpha}{dt} = 0, \tag{11.5}$$

$$\text{and } \frac{dv_{\alpha-1}}{dt} = 0. \tag{11.6}$$

Under these stationary traffic conditions, drivers tend to keep a velocity-dependent *equilibrium gap* $s_e(v_\alpha)$ to the leading vehicle. In the following, we consider a homogeneous ensemble of identical driver-vehicle agents corresponding to identical parameter settings. Then, the IDM acceleration equation (11.1) with the constant setting $\delta = 4$ simplifies to

$$s_e(v) = \frac{s_0 + vT}{\sqrt{1 - \left(\frac{v}{v_0}\right)^4}}. \tag{11.7}$$

The equilibrium distance depends only on the minimum jam distance s_0, the safety time gap T and the desired speed v_0. The diagrams (a) and (b) in Fig. 11.5 show the equilibrium distance as a function of the velocity, $s_e(v)$, for different v_0 and T parameter settings while keeping the minimum distance constant at $s_0 = 2\,\text{m}$. In particular, the equilibrium gap of homogeneous *congested* traffic (with $v \ll v_0$) is essentially equal to the desired gap, $s_e(v) \approx s^*(v, 0) = s_0 + vT$. It is therefore composed of the minimum bumper-to-bumper distance s_0 kept in stationary traffic at $v = 0$ and an additional velocity-dependent contribution vT corresponding to a constant safety time gap T as shown in the diagrams by straight lines. For $v \to 0$, the equilibrium distance approaches the minimum distance s_0. If the velocity is close to the desired speed, $v \approx v_0$, the equilibrium distance s_e is clearly larger than the distance vT according to the safety time gap parameter. For $v \to v_0$, the equilibrium distance diverges due to the vanishing denominator in Eq. (11.7). That is, the free speed is reached *exactly* only on a free road.

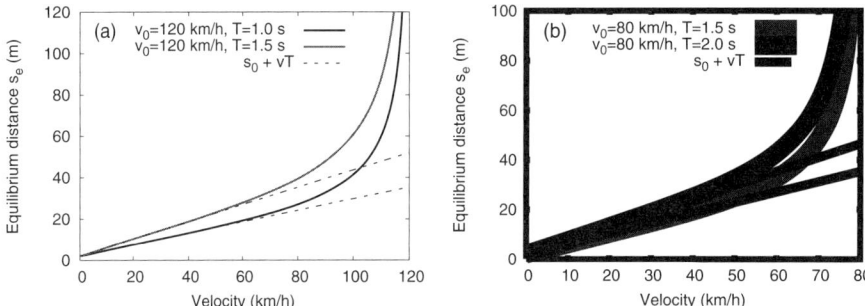

FIGURE 11.5 Equilibrium distance $s_e(v)$ according to Eq. (11.7) as functions of the speed for different settings of the desired speed v_0 and the safety time gap T. The deviations from the dotted lines are discussed in the main text. The other parameters are those listed in the caption of Fig. 11.4.

In the literature, the equilibrium state of homogeneous and stationary traffic is often formulated in macroscopic quantities such as traffic flow Q, (local) average velocity V and traffic density ρ. The translation from the microscopic net distance s into the density is given by the *micro-macro relation*

$$s = \frac{1}{\rho} - l, \qquad (11.8)$$

where l is the vehicle length. In equilibrium traffic, ρ is therefore given by s_e, the mean velocity is simply $V = v_e$ and the traffic flow follows from the hydrodynamic relation

$$Q = \rho V. \qquad (11.9)$$

So, the equilibrium velocity v_e is needed as a function of the distance s_e. An analytical expression for the inverse of Eq. (11.7), that is the *equilibrium velocity* as a function of the gap, $v_e(s)$, is only available for the acceleration exponents $\delta = 1, 2$ or $\delta \to \infty$ [Treiber et al., 2000]. For $\delta = 4$, we only have a parametric representation $\rho(v)$ with $v \in [0, v_0]$ resulting from Eqs. (11.8) and (11.7). Figures 11.6(a) and (b) show the equilibrium velocity-density relation $V_e(\rho)$ for the same parameter settings as in Fig. 11.5. The assumed vehicle length $l = 5\,\text{m}$ together with the minimum jam distance $s_0 = 2\,\text{m}$ results in a maximum density $\rho_{\max} = 1/(s_0 + l) \approx 143\,\text{vehicles/km}$. Using the relation (11.9), we obtain the so-called *fundamental diagram* between the traffic flow and the vehicle density, $Q(\rho) = V\rho(v)$ which is displayed in Fig. 11.6(c) and (d). Notice that Q is typically given in units of vehicles per hour and the density ρ in units of vehicles per km.

According to Eqs. (11.7) and (11.8), the *fundamental relations* of homogeneous traffic depend on the desired speed v_0 (low density), the safety time gap T (high density) and the jam distance s_0 (jammed traffic). In the low-density limit $\rho \ll 1/(v_0 T)$, the equilibrium flow can be approximated by $Q \approx v_0 \rho$. In the high density regime, one has a linear decrease

Agents for Traffic Simulation

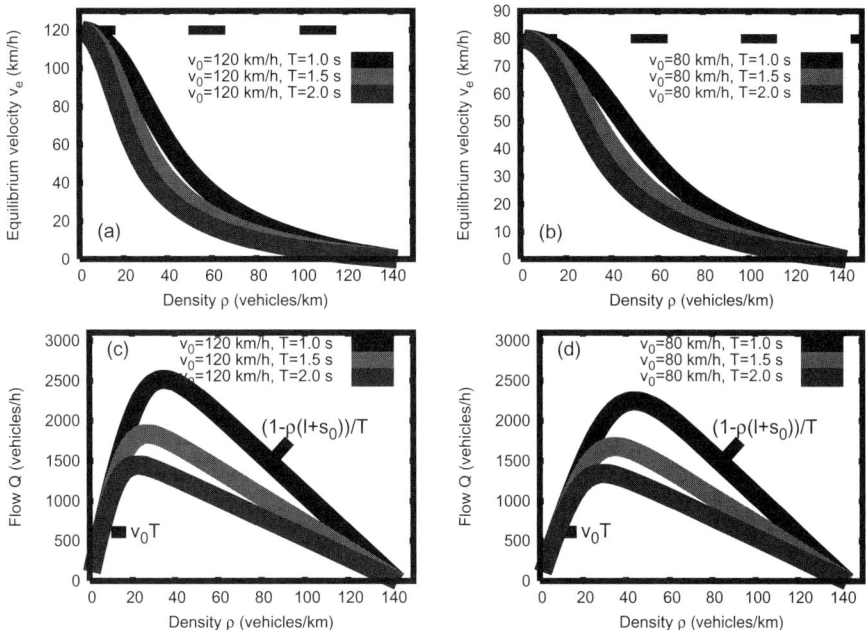

FIGURE 11.6 Equilibrium velocity-density relations of the IDM (top) and corresponding flow-density relations, so-called *fundamental diagrams* (bottom). The equilibrium properties depend on the minimum distance s_0 (here set to 2 m), the desired speed v_0 (here displayed for 120 and 80 km/h) and the time gap T (here 1.0, 1.5 and 2.0 s). The safety time gap is the most important parameter determining the maximum flow (stationary freeway capacity).

of the flow,

$$Q(\rho) \approx \frac{1 - \rho(l + s_0)}{T}, \quad (11.10)$$

which can be used to determine the effective length $l + s_0$ and T. Notice that the vehicle length is not a model parameter but only a scaling quantity that determines the (static) maximum density $\rho_{\max}$ together with the IDM parameter s_0.

11.3.2 Inter-Driver Variability

An important aspect of vehicular traffic is the heterogeneity of agents, a term which includes characteristics of the drivers as well as features of the vehicle. The multi-agent simulation approach is appropriate for representing this heterogeneity on a microscopic level. In order to address *inter-driver variability* (different drivers behave differently in identical traffic situations) and vehicle properties (such as length, width, weight and motorization) we propose to group driver-vehicle agents into classes defining their specific driving styles and vehicle properties. For this purpose, it is advantageous that the parameters of the Intelligent Driver Model do have an intuitive meaning and are directly related to driving behavior. In the following, we discuss the parameter settings for three classes of passenger car drivers representing "normal", "timid" and "aggressive" driving styles. In addition, we model a typical truck driver. The corresponding parameter values are listed in Table 11.2.

- The *desired speed* v_0 is the maximum speed a driver-vehicle agent aims to reach under unobstructed driving conditions. A natural value and upper limit for this

TABLE 11.2 Model parameters of the Intelligent Driver Model for three classes of passenger car drivers and a typical truck driver.

IDM Parameter	Normal	Timid	Aggressive	Truck
Desired speed v_0 in km/h	120	100	140	85
Desired time gap T in s	1.5	1.8	1.0	2.0
Jam distance s_0 in m	2.0	4.0	1.0	4.0
Maximum acceleration a in m/s^2	1.4	1.0	2.0	0.7
Desired deceleration b in m/s^2	2.0	1.0	3.0	2.0

parameter would be the typical (highest) speed on the considered road element. The normal driver chooses for instance 120 km/h on a freeway while a timid driver prefers a lower value and a more aggressive driver likes to go faster. The desired speed could be limited by legislation. In city traffic, the speed is typically limited to 50 km/h (cf. the simulation scenario in Section 11.4.2). In this case, a timid driver likes to drive a bit below this limit while an aggressive driver can easily be modeled by an individual "disobedience factor". Notice that strict speed limits apply to trucks on the whole road network in most countries.

- The *desired time gap* T refers to the preferred distance vT while driving at speed v, cf. Eq. (11.2), and mainly determines the maximum capacity (cf. Figure 11.6). A typical value in dense traffic is about 1.4 s while German road authorities recommend 1.8 s. A common observation on European freeways is that very small time gaps are kept [Treiber et al., 2006b; Knospe et al., 2002].

- The parameter s_0 describes the *minimum bumper-to-bumper distance* at a standstill, cf. Eq. (11.2). Typical gaps in a queue of vehicles standing at traffic lights are in the range between 1 m and 5 m. While a normal driver typically keeps a minimum gap of 2 m, a cautious driver prefers larger gaps and an aggressive driver likes tailgating. It is natural to assume that truck drivers prefer slightly larger gap than the normal car driver due to larger vehicle dimensions. Notice that the vehicle length is not a model parameter. However, it determines the maximum density together with the minimum distance s_0 according to Eq. (11.8). Typical vehicle lengths are for instance 5 m for cars and 12 m for trucks.

- The *desired acceleration* a describes the acceleration behavior of the driver. Notice that the acceleration depends on the actual vehicle speed as shown, for example, in Figure 11.4. Since the acceleration *behavior* is based on a physical movement, the value of a has to respect the limits of motorization. Consequently, a truck has to be modeled by a lower desired acceleration a than a passenger car. An aggressive driver prefers to accelerate fast (e.g., 3 m/s^2) while a timid driver prefers a lower value (e.g., 1 m/s^2). The acceleration exponent $\delta = 4$ is kept constant for all driver classes, cf. Eq. (11.1).

- The *comfortable braking deceleration* b determines the approaching process toward slower leaders or stationary objects such as traffic lights (see Section 11.4.2). As the IDM tries to limit the braking deceleration to b, a low value ($b = 1$ m/s^2) represents a driver who breaks accurately in an anticipative way corresponding to a smooth driving style. By way of contrast, a higher value ($b = 3$ m/s^2) describes an aggressive driver who prefers to approach the leader with a large velocity difference.

Taking these average parameters for each driver class as a starting point, it is straightforward to distribute individual agent parameters randomly within given limits, e.g., according to a uniform distribution with a variation of 20%.

11.3.3 Intra-Driver Variability

Besides reacting to the immediate traffic environment, human drivers adapt their driving style on longer time scales to the traffic situation. Thus, the actual driving style depends on the traffic conditions of the last few minutes which we call *memory effect* [Treiber and Helbing, 2003]. For example, it is observed that most drivers increase their preferred temporal headway after being stuck in congested traffic for some time [Brilon and Ponzlet, 1996; Fukui et al., 2003]. Furthermore, when larger gaps appear or when reaching the downstream front of the congested zone, human drivers accelerate less and possibly decrease their desired speed as compared to a free-traffic situation.

In contrast to inter-driver variability considered in Section 11.3.2, the memory effect is an example of *intra-driver variability* meaning that a driver behaves differently in similar traffic situations depending on his or her individual driving history and experience. Again, the multi-agent approach can easily cope with this extension of driving behavior as soon as one has a specific model to implement. By way of example, we present a model that introduces a time-dependency of some parameters of the Intelligent Driver Model to describe the *frustration* of drivers being in a traffic jam for a period [Treiber and Helbing, 2003].

We assume that adaptations of the driving style are controlled by a single internal dynamical variable $\lambda(t)$ that represents the "subjective level of service" ranging from 0 (in a standstill) to 1 (on a free road). The subjective level of service $\lambda(t)$ relaxes to the instantaneous level of service $\lambda_0(v)$ depending on the agent's speed $v(t)$ with a relaxation time τ according to

$$\frac{d\lambda}{dt} = \frac{\lambda_0(v) - \lambda}{\tau}. \tag{11.11}$$

This means that for each driver, the subjective level of service is given by the exponential moving average of the instantaneous level of service experienced in the past:

$$\lambda(t) = \int_0^t \lambda_0(v(t'))\, e^{-(t-t')/\tau} dt'. \tag{11.12}$$

We have assumed the instantaneous level of service $\lambda_0(v)$ to be a function of the actual velocity $v(t)$. Obviously, $\lambda_0(v)$ should be a monotonically increasing function with $\lambda_0(0) = 0$ and $\lambda_0(v_0) = 1$ when driving with the desired speed v_0. The most simple "level-of-service function" satisfying these conditions is the linear relation

$$\lambda_0(v) = \frac{v}{v_0}. \tag{11.13}$$

Notice that this equation reflects the level of service or efficiency of movement from the agent's point of view, with $\lambda_0 = 1$ meaning zero hindrance and $\lambda_0 = 0$ meaning maximum hindrance. If one models inter-driver variability (Section 11.3.2) where different drivers have different desired velocities, there is no objective level of service, but rather only an individual and an average one.

Having defined how the traffic environment influences the degree of adaptation $\lambda(t)$ of each agent, we now specify how this internal variable influences driving behavior. A behavioral variable that is both measurable and strongly influences the traffic dynamics is the desired time gap T of the IDM. It is observed that, in congested traffic, the whole distribution of time gaps is shifted to the right when compared with the data of free traffic [Treiber and Helbing, 2003; Treiber et al., 2006b]. We model this increase by varying the corresponding IDM parameter in the range between T_0 (free traffic) and $T_{\text{jam}} = \beta_T T_0$ (traffic jam) according to

$$T(\lambda) = \lambda\, T_0 + (1 - \lambda)\, T_{\text{jam}} = T_0\, [\beta_T + \lambda(1 - \beta_T)]. \tag{11.14}$$

Herein, the adaptation factor β_T is a model parameter. A value for the frustration effect is $\beta_T = T_{\text{jam}}/T_0 = 1.8$ which is consistent with empirical observations. A typical relaxation time for the driver's adaptation is $\tau = 5\,\text{min}$. Notice that other parameters of the driving style are probably influenced as well, such as the acceleration a, the comfortable deceleration b or the desired velocity v_0. This could be implemented by analogous equations for these parameters. Furthermore, other adaption processes as well as the presented frustration effect are also relevant [Treiber et al., 2006b].

11.4 Modeling Discrete Decisions

On the road network, drivers encounter many situations where a decision between two or more alternatives is required. This relates not only to lane-changing decisions but also to considerations as to whether or not it is safe to enter the priority road at an unsignalized junction, to cross such a junction or to start an overtaking maneuver on a rural road. Another question concerns whether or not to stop at an amber-phase traffic light. All of the above problems belong to the class of *discrete-choice problems* that, since the pioneering work of McFadden [Hausman and McFadden, 1984], has been extensively investigated in an economic context as well as in the context of transportation planning. In spite of the relevance to everyday driving situations, there are fewer investigations attempting to incorporate the aforementioned discrete-choice tasks into microscopic models of traffic flow, and most of them are restricted to modeling lane changes [Gipps, 1986]. Only very recently have acceleration and discrete-choice tasks been treated more systematically [Toledo et al., 2007; Kesting et al., 2007a].

The modeling of lane changes is typically considered as a multi-step process. On a *strategic* level, the driver knows about his or her route on the network which influences the lane choice, e.g., with regard to lane blockages, on-ramps, off-ramps or other mandatory merges [Toledo et al., 2005]. In the *tactical* stage, an intended lane change is prepared and initiated by advance accelerations or decelerations of the driver, and possibly by cooperation of drivers in the target lane [Hidas, 2005]. Finally, in the *operational* stage, one determines if an immediate lane change is both safe and desired [Gipps, 1986]. While mandatory changes are performed for strategic reasons, the driver's motivation for discretionary lane changes is a perceived improvement of the driving conditions in the target lane compared with the current situation.

In the following, we will present a recently formulated general framework for modeling traffic-related discrete-choice situations in terms of the acceleration function of a longitudinal model [Kesting et al., 2007a]. For the purpose of illustration, we will apply the concept to model mandatory and discretionary lane changes (Section 11.4.1). Furthermore, we will consider the decision process whether or not to brake when approaching a traffic light turning from green to amber (Section 11.4.2).

11.4.1 Modeling Lane Changes

Complementary to the longitudinal movement, lane-changing is a required ingredient for simulations of multi-lane traffic. The realistic description of multi-agent systems is only possible within a multi-lane modeling framework allowing faster driver-vehicle agents to improve their driving conditions by passing slower vehicles.

When considering a lane change, a driver typically makes a trade-off between the expected own advantage and the disadvantage imposed on other drivers. For a driver considering a lane change, the subjective utility of a change increases with the gap to the new leader in the target lane. However, if the speed of this leader is lower, it may be favorable to stay in the present lane despite the smaller gap. A criterion for the utility including *both* situations is the difference between the accelerations after and before the lane change. This is the core idea of the lane-changing algorithm MOBIL [Kesting et al., 2007a] that is based on the expected (dis)advantage in the new lane in terms of the difference in the acceleration which is calculated with an underlying microscopic longitudinal traffic model, e.g., the Intelligent Driver Model (Section 11.3.1).

For the lane-changing decision, we first consider a *safety constraint*. In order to avoid accidents by the follower in the prospective target lane, the safety criterion

$$\dot{v}_{\text{follow}} \geq -b_{\text{safe}} \qquad (11.15)$$

guarantees that the deceleration of the successor $\dot{v}_{\text{follow}}$ in the target lane does not exceed a safe limit $b_{\text{safe}} \simeq 4\,\text{m/s}^2$ after the lane change. In other words, the safety criterion essentially restricts the deceleration of the lag vehicle on the target lane to values below b_{safe}. Although formulated as a simple inequality, this condition contains all the information provided by the longitudinal model via the acceleration $\dot{v}_{\text{follow}}$. In particular, if the longitudinal model has a built-in sensitivity with respect to *velocity differences* (such as the IDM) this dependence is transferred to the lane-changing decisions. In this way, larger gaps between the following vehicle in the target lane and the own position are required to satisfy the safety constraint if the speed of the following vehicle is higher than one's own speed. In contrast, lower values for the gap are allowed if the back vehicle is slower. Moreover, by formulating the criterion in terms of safe braking decelerations of the longitudinal model, crashes due to lane changes are *automatically* excluded as long as the longitudinal model itself guarantees crash-free dynamics.

For discretionary lane changes, an additional *incentive criterion* favors lane changes whenever the acceleration in one of the target lanes is higher. The incentive criterion for a lane change is also formulated in terms of accelerations. A lane change is executed if the sum of the own acceleration and those of the affected neighboring vehicle-driver agent is higher in the prospective situation than in the current local traffic state (and if the safety criterion (11.15) is satisfied of course). The innovation of the MOBIL framework [Kesting et al., 2007a] is that the immediately affected neighbors are considered by the "politeness factor" p. For an egoistic driver corresponding to $p = 0$, this incentive criterion simplifies to $\dot{v}_{\text{new}} > \dot{v}_{\text{old}}$. However, for $p = 1$, lane changes are only carried out if this increases the combined accelerations of the lane-changing driver and all affected neighbors. This strategy can be paraphrased by the acronym *"Minimizing Overall Braking Induced by Lane Changes"* (MOBIL). We observed realistic lane-changing behavior for politeness parameters in the range $0.2 < p < 1$ [Kesting et al., 2007a]. Additional restrictions can easily be included. For example, the "keep-right" directive of most European countries is implemented by adding a bias to the incentive criterion. A "keep-lane" behavior is modeled by an additional constant threshold when considering a lane change.

11.4.2 Approaching a Traffic Light

When approaching a traffic light that switches from green to amber, a decision has to be made whether to stop just at the traffic light or to pass the amber-phase light with unchanged speed. For an empirical study on the stopping/running decision at the onset of an amber phase we refer to Ref. [Rakha et al., 2007]. If the first option is selected, the traffic light will be modeled by a standing "virtual" vehicle at the position of the light. Otherwise, the traffic light will be ignored. The criterion is satisfied for the "stop at the light" option if the own braking deceleration at the time of the decision does not exceed the safe deceleration b_{safe}. The situation is illustrated in Figure 11.7. Denoting the distance to the traffic light by s_c and the velocity at decision time by v_c and assuming a longitudinal model of the form (11.1), the safety criterion (11.15) can be written as

$$\dot{v}(s_c, v_c, v_c) \geq -b_{\text{safe}}. \quad (11.16)$$

Notice that the approaching rate and the velocity are equal ($\Delta v_c = v_c$) in this case. The incentive criterion is governed by the bias toward the stopping decision because legislation requires that one stop at an amber-phase traffic light if it is safe to do so. As a consequence, the incentive criterion is always fulfilled, and Eq. (11.16) is the only decision criterion in this situation.

Similarly to the lane-changing rules, the "stopping criterion" (11.16) will inherit all the sophistication of the underlying car-following model. In particular, when using realistic longitudinal models, one obtains a realistic stopping criterion with only one additional parameter b_{safe}. Conversely, unrealistic microscopic models such as the Optimal Velocity Model [Bando et al., 1995] or the Nagel-Schreckenberg cellular automaton [K. Nagel and Schreckenberg, 1992] will lead to unrealistic stopping-decisions. In the case of the Optimal Velocity Model, it is not even guaranteed that drivers deciding to stop will be able to stop at the lights.

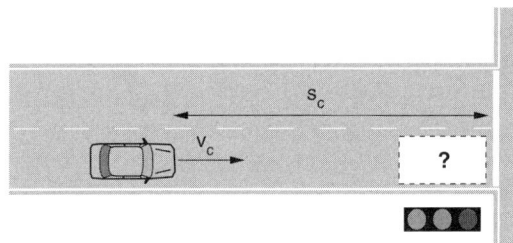

FIGURE 11.7 Approaching a traffic switching from green to amber. The two options of the decision situation are to stop in front of the light or to pass the amber-phase traffic light with unchanged speed.

For the purpose of illustration, we apply the concept to the following situation in city traffic: A car is driving at speed $v_c = 50 \, \text{km/h}$ toward an amber traffic light located at a distance $s_c = 50 \, \text{m}$. Applying the IDM parameters of a "normal" driver listed in Table 11.2 in combination with an adapted desired speed of $v_0 = 50 \, \text{km/h}$, the acceleration function (11.1) results in an initial braking of $\dot{v}(0) \approx 3.6 \, \text{m/s}^2$ at $t = 0 \, \text{s}$. For a safe deceleration equal to the desired deceleration of the IDM, that is $b_{\text{safe}} = b = 2.0 \, \text{m/s}^2$, the MOBIL decision says "drive on". If, however, a safe braking deceleration of $b_{\text{safe}} = 4 \, \text{m/s}^2$ is assumed, the driver agent would decide to brake resulting in the approaching maneuver shown in Figure 11.8. The initial braking stronger than $-2 \, \text{m/s}^2$ makes the situation manageable for the agent. After $2 \, \text{s}$, the situation is "under control" and the vehicle brakes approximately with the comfortable deceleration $b = 2 \, \text{m/s}^2$. In order to reach a standstill in a smooth way, the deceleration is reduced to limit the *jerk* which defines the change in the acceleration. In addition, Figure 11.8 shows the behavior of the second vehicle following the leader. The acceleration time series shows the important feature of the IDM in limiting braking decelerations to the comfortable limit b as long as safety is warranted. From these results it is obvious that the setting $b_{\text{safe}} = b$ is a natural assumption to model the decision process realistically. Notice, however, that a human reaction time of about 1 s [Green, 2000] has to be taken into account as well.

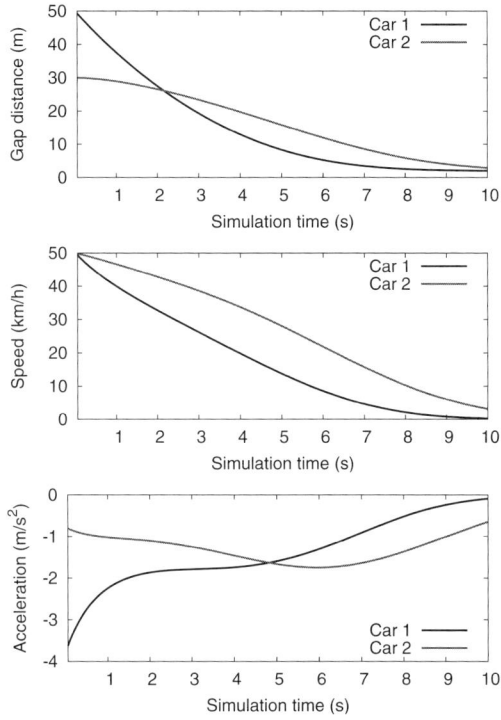

FIGURE 11.8 Maneuver of approaching a traffic light initially 50 m with a speed of 50 km/h according to the Intelligent Driver Model. The braking deceleration is limited to the comfortable braking deceleration (IDM parameter b) whenever possible. The stronger braking of the first car is needed to keep the situation under control. The parameters for the simulation are listed in Table 11.2.

11.5 Microscopic Traffic Simulation Software

So far, we have discussed models describing the longitudinal movement and discrete decisions of individual driver-vehicle agents. Let us now address the issue of a simulation framework that integrates these components into a microscopic multi-lane traffic simulator. Typical relations among functions in a microscopic traffic simulator are shown in Figure 11.9. On the level of input data, simulation settings can be provided by input files, e.g., encoded in XML, by command line or via a graphical user interface (GUI). The main simulation loop is organized by a *Simulation Controller* which keeps track of the program operations and user actions. This central control unit calls the update methods of the road-section objects. We will elaborate on these components in Section 11.5.1. Since the calculation of the vehicle accelerations is the very core of a traffic simulation, we will pinpoint the issue in Section 11.5.2. Simulation results can be written to data files and, in addition, visualized by 2D and 3D computer graphics on the screen (see Section 11.5.3). Furthermore, we will extend the simulator in order to simulate inter-vehicle communication (see Section 11.6.3 below).

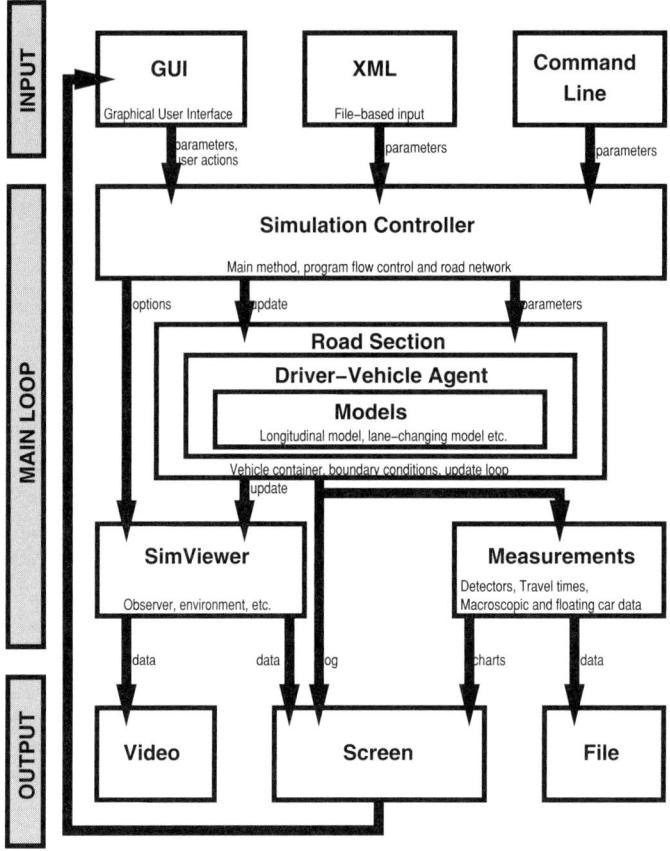

FIGURE 11.9 Illustration of possible relations among functions in a traffic simulation framework. The input data defining a simulation setting can be provided by data files, the command line or a graphical user interface (GUI). The main simulation loop is organized by a "Simulation Controller" which controls the update of the road network, the graphical representation ("SimViewer") and the output functions corresponding to measurements of several microscopic and macroscopic quantities.

There are a number of interactive simulators available publicly. The website [Treiber, 2007] deploys the Intelligent Driver Model [Treiber et al., 2000] introduced in Section 11.3.1 for cars and trucks together with the lane-changing algorithm MOBIL [Kesting et al., 2007a]. This demonstrator simulates typical bottleneck scenarios such as on-ramps, lane-closings, uphill grades and or traffic lights. Another open source simulator for whole traffic networks is SUMO [, Simulation of Urban MObility]. The software uses the Krauss model [Krauss et al., 1997]. Recently, FREESIM has been made available to the public [Miller, 2007]. Furthermore, commercial traffic simulation software tools (for instance VISSIM™, AIMSUN™ or PARAMICS™) offer a variety of additional modules such as emission or pedestrian models and interfaces, e.g., for controlling simulation runs by remote and for implementing additional features. These commercial products incorporate sophisticated virtual environment 3D engines. Note, however, that the underlying models are generally not well documented.

11.5.1 Simulator Design

Next to the functional view shown in Figure 11.9, a hierarchical view can be used to represent the dependencies and inherited properties which makes use of the object-oriented programming paradigm by representing and abstracting functional units as classes. The best example is the representation of a driver-vehicle agent as an abstract class with several possible designs for human drivers, vehicles equipped with adaptive cruise control [Kesting et al., 2007b, 2008] or even driverless ones as recently demonstrated in reality [DARPA].However, each agent has a number of defining properties such as length, width, weight, form and color. Furthermore, each agent requires a model for the lengthwise movement which is in turn an abstract class with the presented IDM as a specific implementation. Further components are required in order to model other aspects of driver behavior such as lane changes, memory, etc. Since each agent is represented by an individual object, it is straightforward to assign individual parameter values to account for driver diversity (Section 11.3.2).

The road network can be represented by connected road sections such as main roads, on-ramps and off-ramps. A road section is defined by its properties like length, number of lanes, etc. In addition, an element may contain attributes representing the concrete infrastructure relevant to the driver-vehicle agents such as lane closures, lane narrowings, speed limits, uphill gradients and/or traffic lights. Notice that the set of attributes which is relevant for the behavior and decision-making has to be available to the agent.

The most detailed view on the innermost update loop of a road section is given in terms of the following pseudo code:

```
updateRoadSection(){
  updateNeigborhood(); // organizing set of vehicles in multiple lanes
  updateInfrastruture(); // active road attributes (e.g., traffic lights)
  updateAgentsRoadConditions(); // attributes affect agents
  calculateAccelerations(); //evaluate longitudinal models of agents
  laneChanges();    // decision making and performing lane changes
  updatePositionsAndSpeeds(); // integration within discrete update
  updateBoundaries(); // inflow and outflow
  updateOutput(); // log observable quantities and update detectors
}
```

11.5.2 Numerical Integration

The explicit integration in the `updatePositionsAndSpeeds` function of all driver-vehicle agents α is the very core of a traffic simulator. In general, the longitudinal movement of the vehicles is described by car-following models which take into account the direct leader and result in expressions for the acceleration function of the form

$$\frac{\mathrm{d}v_\alpha}{\mathrm{d}t} = f\left(s_\alpha, v_\alpha, \Delta v_\alpha\right), \tag{11.17}$$

that is the acceleration depends only on the own speed v_α, the gap s_α, and the velocity difference (approaching rate) $\Delta v_\alpha = v_\alpha - v_{\alpha-1}$ to the leading vehicle $(\alpha - 1)$. Note that we discussed the Intelligent Driver Model (IDM) as an example for a car-following model in Section 11.3.1. Together with the gap $s_\alpha(t) = x_{\alpha-1}(t) - x_\alpha(t) - l_{\alpha-1}$ and the general equation of motion,

$$\frac{\mathrm{d}x_\alpha}{\mathrm{d}t} = v_\alpha, \tag{11.18}$$

Eq. (11.17) represents a (locally) coupled system of *ordinary differential equations* (ODEs) for the positions x_α and velocities v_α of all vehicles α.

As the considered acceleration functions f are in general nonlinear, we have to solve the set of ODEs by means of numerical integration. In the context of car-following models, it is natural to use an explicit scheme assuming constant *accelerations* within each update time interval Δt. This leads to the explicit numerical update rules

$$v_\alpha(t + \Delta t) = v_\alpha(t) + \dot{v}_\alpha(t)\Delta t,$$
$$x_\alpha(t + \Delta t) = x_\alpha(t) + v_\alpha(t)\Delta t + \frac{1}{2}\dot{v}_\alpha(t)(\Delta t)^2, \tag{11.19}$$

where $\dot{v}_\alpha(t)$ is an abbreviation for the acceleration function $f(s_\alpha(t), v_\alpha(t), \Delta v_\alpha(t))$. For $\Delta t \to 0\,\text{s}$, this scheme locally converges to the exact solution of (11.17) with consistency order 1 for the velocities ("Euler update", cf. [Press et al., 1992]) and consistency order 2 for the positions ("modified Euler update") with respect to the L^2-norm.* Because of the intuitive meaning of this update procedure in the context of traffic, the update rule (11.19) or similar rules are sometimes considered to be part of the model itself rather than as a numerical approximation [Kesting and Treiber, 2008b]. A typical update time interval Δt for the IDM is between 0.1 s and 0.2 s. Nevertheless, the IDM is approximately numerically stable up to an update interval of $\Delta t \approx T/2$, that is half of the desired time gap parameter T.

11.5.3 Visualization

Besides the implementation of the simulation controller with the focus on quantitative models, the visualization of vehicle movements is also an important aspect of simulation software. In the case of vehicular traffic it is straightforward to envision the vehicle trajectories over the course of time whether in 2D or 3D. The latter representation is of course more demanding. Figure 11.10 illustrated an example of a "bird's eye view" of a two-lane freeway with an on-ramp, while Figure 11.11 illustrates a "cockpit perspective" of a driving vehicle on the road. Note that the 3D engine was programmed from scratch as an exercise by the authors. However, higher level tools and open source 3D engines for OpenGL are available. For more details on this subject we refer to the chapter "Crowd Behavior Modeling: From Cellular Automata to Multi-Agent Systems" by Bandini, Manzoni and Vizzari, this volume.

The animated visualization demonstrates both the individual interactions and the resulting collective dynamics. In particular, the graphical visualization turns out to be an important tool when developing and testing lane-changing models and other decisions based on complex interactions with neighboring vehicles for their plausibility. In fact, the driving experiences of programmers offer the best measure of realism and also provide stimulus for further model improvements.

Last but not least, scientists and experts have to keep in mind that computer animations have become an important tool for a fast and intuitive knowledge transfer of traffic phenomena to students, decision-makers and the public. In particular, visualization in *real time*

---

*A time-continuous traffic model is mathematically consistent if a unique local solution exists and if a numerical update scheme exists whose solution locally converges to this solution when the update time interval goes to zero. It has the consistency order q if $\|\epsilon\| = O(\Delta t^q)$ for $\Delta t \to 0\,\text{s}$ where ϵ denotes the deviation of the numerical solution for x_α or v_α with respect to the exact solution, and $\|\cdot\|$ is some functional norm such as the L^2-Norm.

Agents for Traffic Simulation

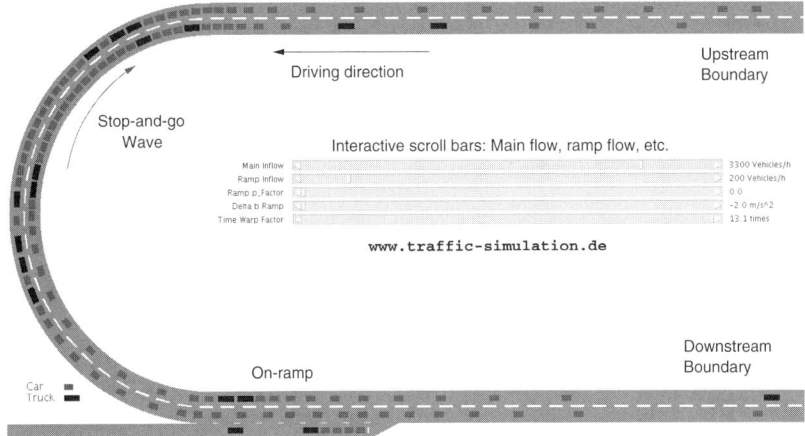

FIGURE 11.10 Example of 2D visualization of a two-lane freeway with an on-ramp from the website [Treiber, 2007]. The screenshot shows the breakdown of traffic flow at the on-ramp serving as a bottleneck. A stop-and-go wave (cluster of vehicles) is propagating against the driving direction. The source code of the simulator is publicly available as an open source.

allows for direct user interaction influencing the simulation run, e.g., by changing the simulation conditions (in terms of inflows and driver population) and parameter settings of the underlying models. In this way, the complexity of simulation techniques (which are based on assumptions, mathematical models, many parameters and implementational details) can become more accessible.

FIGURE 11.11 Example of a 3D animation from the driver perspective. Notice that the "Coffeemeter" visualizes the acceleration and the jerk (the changes of all acceleration with time) which are difficult to visualize by other means. It therefore serves as a measure of the driving comfort.

Finally, we remark that animation is a playground for the programmers. For instance, the full cup of coffee in Figure 11.11 represents not just a comforting habit of the agent but is also a vivid way to illustrate the simulated longitudinal acceleration as well as transverse accelerations due to lane-changing. Moreover, the coffee level is a measure of the riding comfort because it is also sensitive to the derivative of the acceleration which is perceived as a jerk by the driver. The hydrodynamic equations of the coffee surface in the cup with a diameter $2r$ are astonishingly realistically approximated by a harmonious pendulum with two degrees of freedom ϕ_x and ϕ_y denoting the angles of the surface normal:

$$\ddot{\phi}_x + \frac{2\pi}{\tau}\dot{\phi}_x + \omega_0^2 \phi_x + \frac{\ddot{x}}{r} = 0,$$
$$\ddot{\phi}_y + \frac{2\pi}{\tau}\dot{\phi}_y + \omega_0^2 \phi_y + \frac{\ddot{y}}{r} = 0. \qquad (11.20)$$

The second time derivates $\ddot{x}$ and $\ddot{y}$ denote the vehicle accelerations in longitudinal and transversal direction. The angular frequency is $\omega_0 = \sqrt{g/r}$ where g is the gravitational constant. During a coffee break, the damping time τ of Java coffee was empirically determined as $\tau = 12\,\mathrm{s}$ by the authors.

11.6 From Individual to Collective Properties

After having constructed the driver-vehicle agents, let us now adopt them in a multi-agent simulation in which they interact with each other. The process of simulating agents in parallel is one of emergence from the microscopic level of pairwise interactions to the higher, macroscopic level in order to reproduce and predict real phenomena.

In this Section, we will present three simulation applications. In Section 11.6.1, we will demonstrate the *emergence of a collective pattern from individual interactions* between driver-vehicle agents by simulating the breakdown of traffic flow and the development of a stop-and-go wave. The simulation will show the expressive power of the Intelligent Driver Model in reproducing the characteristic backwards propagation speed which is a well-known constant of traffic world wide. In Section 11.6.2, we will apply the traffic simulation framework to analyze the impact of a speed limit as an example of a traffic control task. By way of this example, we will demonstrate the predictive power of microscopic traffic flow simulations.

Last but not least, we apply the simulation framework to study a coupled system consisting of communicating driver-vehicle agents using short-range wireless networking technology in Section 11.6.3. Since the multi-agent approach is a flexible general-purpose tool, one can additionally equip an agent with short-range communication devices that can self-organize with other devices in range into ad-hoc networks. Such inter-vehicle communication has recently gained much attention in the academic and engineering world. It is expected to provide great enhancement to the safety and efficiency of modern individual transportation systems. By means of simulation, we will demonstrate the dissemination of information about the local traffic situation over long distances even for small equipment rates in the vehicle fleet.

11.6.1 Emergence of Stop-and-Go Waves

Let us first study the emergence of a collective traffic phenomenon in a simple ring road scenario as depicted in Figure 11.12(a). Note that this scenario can be used interactively on the website [Treiber, 2007]. Such a closed system is defined by an initial value problem. The control parameter is the homogeneous traffic density which essentially determines the long-term behavior of the system. In the simulation, the initial traffic density is too high to be able to retain free flow conditions.

Agents for Traffic Simulation

In the course of time, a vehicle eventually changes lanes resulting in a smaller gap for the following vehicle, which, in turn, has to brake in order to re-establish a safe distance to the new leader. After the initial braking, the next follower again needs some time to respond to this new situation by decelerating. The perturbation therefore increases while propagating in upstream direction, that is against the driving direction of the vehicle flow (see Figure 11.12(a)). This response mechanism acts like a "vicious circle": Each following driver has to reduce his or her speed a bit more to regain the necessary safety distance. Eventually, vehicles further upstream in the platoon brake to a standstill. Moreover, the time to re-accelerate to the restored speed of the leading vehicle takes even more time due to limited acceleration capabilities. Finally, we observe the emergence of a *stop-and-go wave*.

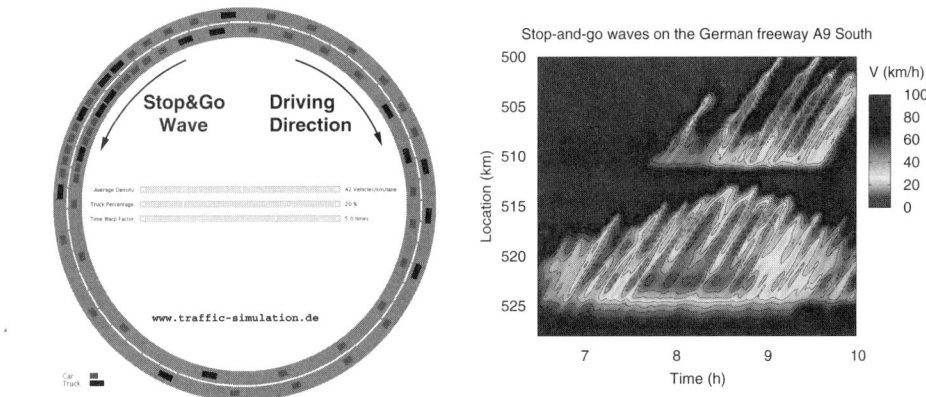

FIGURE 11.12 (a) Emergence of a stop-and-go wave in a simulation and (b) stop-and-go traffic on the German freeway A9 South in the north Munich region. Notice that the spatiotemporal traffic dynamics in diagram (b) have been reconstructed from loop detector data using an interpolation method [Treiber and Helbing, 2002]. The characteristic propagation speed of stop-and-go waves of about 15 km/h against the driving direction is a self-organized "constant" of traffic flow which is reproduced by the Intelligent Driver Model used in the simulation (a).

Stop-and-go waves are also observed in real traffic as shown in Figure 11.12(b) for the German freeway A9 South in the north Munich region. Single stop-and-go waves propagate over more than 10 km leading to their description as "phantom traffic jams". Their propagation speed is arguably constant. From the time-space diagram in Figure 11.12(b), the propagation speed of the downstream front of the stop-and-go wave can be determined as approximately 15.5 km/h. In each country, typical values for this "traffic constant" are in the range 15 ± 5 km/h, depending on the accepted safe time clearance and average vehicle length [Kerner and Rehborn, 1996b]. Consequently, realistic traffic models should reproduce this self-organized property of traffic flow.

11.6.2 Impact of a Speed Limit

Microscopic traffic models are especially suited to the study of heterogeneous traffic streams consisting of different and individual types of driver-vehicle agents.

In the following scenario, we will study the effect of a speed limit for a section of the

German freeway A8 East containing an uphill section around $x = 40\,\text{km}$ [Treiber and Helbing, 2001]. We considered the situation during the evening rush hour on November 2, 1998. In the evening rush hour at about 17 h, traffic broke down at the uphill section. In the simulation shown in Figure 11.13(a), we used lane-averaged one-minute data of velocity and flow measured by loop detectors as upstream boundary conditions reproducing the empirical traffic breakdown. In contrast to the ring road scenario in Section 11.6.1, the inflow at the upstream boundary is the natural control parameter for the open system.

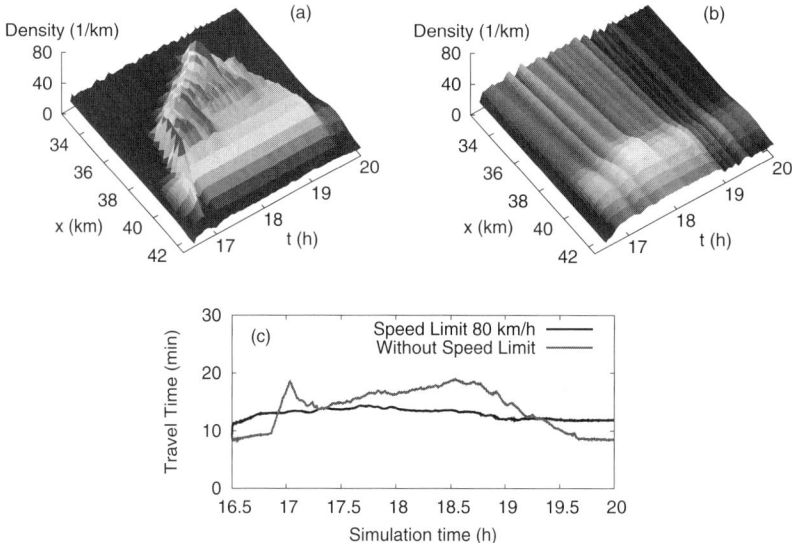

FIGURE 11.13 Realistic simulation of an empirical traffic breakdown caused by an uphill gradient located around $x = 40\,\text{km}$ by using velocity and flow data from loop detectors as upstream boundary conditions (a) without speed limit and (b) with a speed limit of $80\,\text{km/h}$. Diagram (c) shows the travel times corresponding to the scenarios (a) and (b). Notice that the microscopic modeling approach allows for an estimation of the current travel time by simply summing up the travel times derived from the speeds of each driver-vehicle agent simultaneously. In accordance with Treiber and Helbing [Treiber and Helbing, 2001].

We have assumed two vehicle classes: 50% of the drivers had a desired speed of $v_0 = 120\,\text{km/h}$, while the other half had $v_0 = 160\,\text{km/h}$ outside of the uphill region. A speed limit reduces the desired velocities to $80\,\text{km/h}$. Within the uphill region, both driver-vehicle classes are forced to drive at a maximum of $60\,\text{km/h}$ (for example, due to overtaking trucks that are not considered explicitly here).

Figure 11.13 shows spatiotemporal plots of the locally averaged traffic density for scenarios with and without the speed limit. The simulations show the following:

- During the rush hour ($17\,\text{h} \leq t \leq 19\,\text{h}$), the overall effect of the speed limit is positive. The increased travel times in regions without congestion are overcompensated by the saved time due to the avoided breakdown.
- For lighter traffic ($t < 17\,\text{h}$ or $t > 19:30\,\text{h}$), however, the effect of the speed

limit is clearly negative. Note that this problem can be circumvented by traffic-dependent, variable speed limits.

Although the speed limit reduces the velocity, it can improve the quality of traffic flow while in uphill regions it obviously results in a deterioration. To understand this counter-intuitive result, we point out that the desired speed v_0 corresponds to the lowest value of (i) the maximum velocity allowed by the motorization, (ii) the imposed speed limits (possibly with a "disobedience factor"), and (iii) the velocity actually "desired" by the driver. Therefore, speed limits act selectively on the *faster* vehicles, while uphill gradients reduce the speed especially of the *slower* vehicles. As a consequence, speed limits reduce velocity differences, thereby stabilizing traffic, while uphill gradients increase them. For traffic consisting of *identical* driver-vehicle combinations (one driver-vehicle class), these differences are neglected and both speed limits and uphill gradients have in fact the same (negative) effect. Since global speed limits always raise the travel time in off-peak hours when free traffic is unconditionally stable (cf. Figure 11.13(c)), traffic-dependent speed limits are an optimal solution. Note that the impact of a speed limit on the homogenization of traffic flow can be studied interactively on the website [Treiber, 2007] for a lane-closing scenario instead of an uphill bottleneck.

11.6.3 Store-and-Forward Strategy for Inter-Vehicle Communication

Recently, there has been growing interest in wireless communication between vehicles and potential applications. In particular, inter-vehicle communication (IVC) is widely regarded as a promising concept for the dissemination of information on the local traffic situation and short-term travel time estimates for advanced traveler information systems [Jin and Recker, 2006; Yang and Recker, 2005; Schönhof et al., 2006; Schönhof et al., 2007; Wischhof et al., 2005]. In contrast to conventional communication channels which operate with a centralized broadcasting concept via radio or mobile phone services, IVC is designed as a *local service* based on the Dedicated Short Range Communication standard enabling data transmission at a frequency of 5.8 GHz. These devices broadcast messages which are received by all other equipped vehicles within a *limited broadcasting range*. As IVC message dissemination is not controlled by a central station, no further communication infrastructure is needed. For example, wireless local-area networks (IEEE 802.11 a/b/g) have already shown their suitability for IVC with typical broadcasting ranges of 200-500 m [Singh et al., 2002; Ott and Kutscher, 2004].

In the context of freeway traffic, information on the local traffic situation has to be propagated in an upstream direction. In general, there are two transport strategies: Either a message "hops" from an equipped car to a subsequent equipped car within the same driving direction ("longitudinal hopping") or the message is transmitted to an IVC-equipped vehicle in the other driving direction which transports the message upstream and delivers it back by broadcasting it to cars in the original driving direction ("transversal hopping", "cross-transference" or *store-and-forward*). The latter strategy is illustrated in Figure 11.14. Although the longitudinal hopping process allows for a quasi-instantaneous information propagation, the connectivity due to the limited broadcasting range is too weak in the presence of low equipment rates [Schönhof et al., 2006]. A concept using IVC for traffic-state detection must therefore tackle the problem that both the required transport distances into upstream direction and the distances between two equipped vehicles are typically larger than the broadcasting range. The transversal hopping mechanism overcomes this problem by using vehicles in the opposite driving direction as relay stations. Despite the time delay in receiving messages, the messages propagate faster than typical shock waves (which are limited to a speed of -15 km/h, cf. Section 11.6.1).

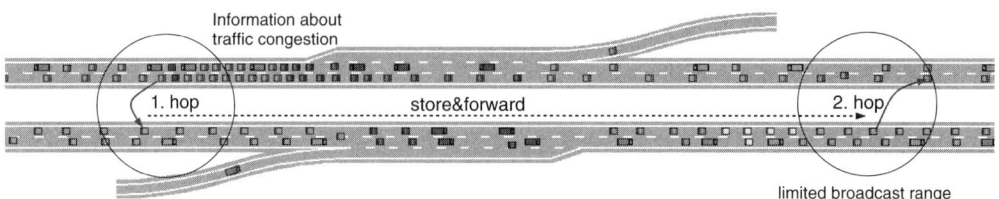

FIGURE 11.14 Illustration of the store-and-forward strategy using the opposite driving direction for propagating messages via short-range inter-vehicle communication in upstream direction. First, a message is generated on the occasion of a local change in speed. The broadcasted message will be picked up by an equipped vehicle in the opposite driving direction (first hop). After a certain traveling distance, the vehicle starts broadcasting the message which can be received by vehicles in the original driving direction (second hop).

The microscopic simulation approach is well suited to coupling traffic and information flows: The movement of vehicles represents a dynamic network of nodes which determines the spread of information on the network. For the purpose of demonstration, let us now simulate the chain of message propagation by means of IVC in an integrated simulation:

1. The generation of traffic-related messages by individual vehicles,
2. the transmission of up-to-date information in upstream direction using store-and-forward strategy via the opposite driving direction and
3. the receipt of the messages for predicting the future traffic situation further downstream.

The object-oriented design of the traffic simulation software (cf. Section 11.5) can be extended in a straightforward way: First, the simulation of the store-and-forward strategy requires two independent freeways in opposite directions. Second, each equipped driver-vehicle agent autonomously detects jam fronts (by means of velocity gradients) and generates traffic-related messages based on locally available time series data. To this end, the design of a vehicle has been extended by a `detection` unit which generates traffic-relevant messages and a `communication` unit for broadcasting and receiving messages. Finally, the exchange of messages has been realized by a `message pool` which organizes the bookkeeping of message broadcast and reception between equipped cars within a limited broadcasting range (corresponding to the outdated ether concept). As the routing in this system is obviously given by the two traffic streams in opposite directions, no further rules are necessary for modeling the message exchange process.

We consider a scenario with an assumed fraction of only 3% communicating vehicles. The resulting trajectories of equipped vehicles in both driving directions together with the generation of messages and their reception by a considered vehicle are illustrated in Figure 11.15. In this scenario, a temporary road blockage has triggered a stop-and-go wave reflected by horizontal trajectory curves in one driving direction while the traffic flow in the opposite driving direction was free. When cars encountered the propagating stop-and-go wave, they started to broadcast messages about the detected position and time of the upstream jam front and the following downstream jam front. The event-driven messages were received and carried forward by vehicles in the other driving direction via the store-and-forward mechanism.

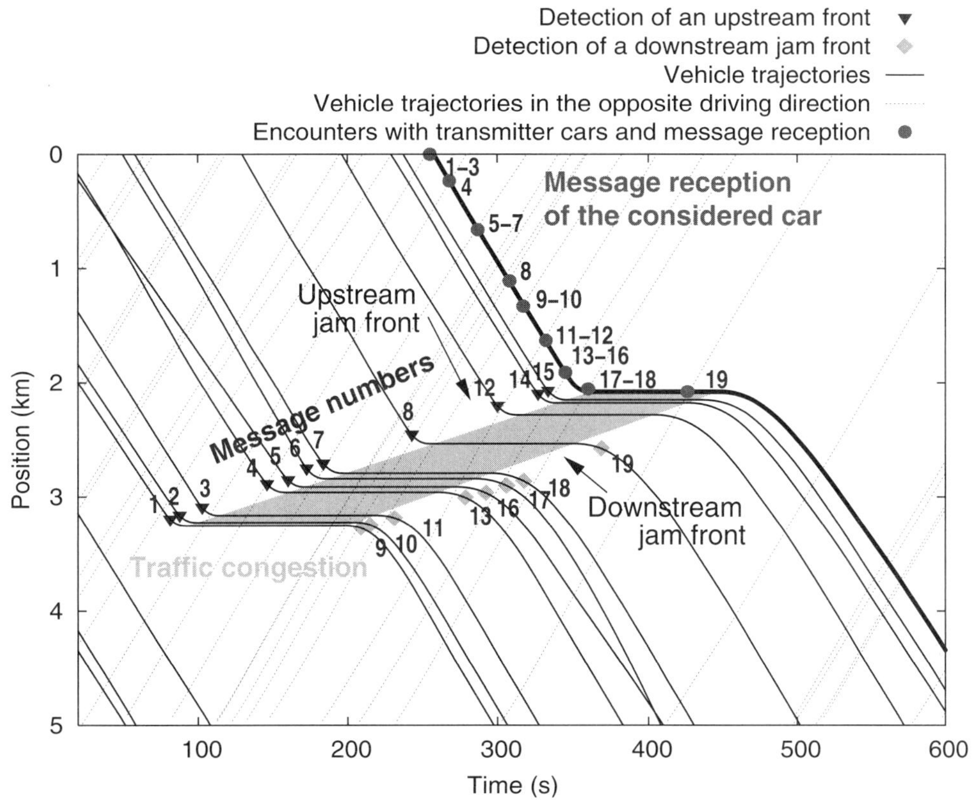

FIGURE 11.15 Space-time diagram of the simulated traffic scenario. The trajectories of the IVC-equipped vehicles (3%) are displayed by solid or dotted lines depending on the driving direction. The vehicles in the opposite driving direction serve as transmitter cars for the store-and-forward strategy. For the purpose of illustration, we have set the maximum broadcasting range to 10 m. When cars pass the upstream or downstream jam front of the moving jam, they broadcast messages (marked by numbers) containing the detected position and time. They are later received by the considered vehicle further upstream (thick solid line). Note that the crossing trajectories of equipped vehicles (e.g., in the upper-left corner of the diagram) reflect passing maneuvers due to different desired velocities.

As shown in Figure 11.15, the considered vehicle already received the first message about the upcoming traffic congestion 2 km before reaching the traffic jam. Further received messages from other equipped vehicles could be used to confirm and update the upcoming traffic situation further downstream. Thus, based on a suitable prediction algorithm, each equipped vehicle could autonomously forecast the moving jam fronts by extrapolating the spatiotemporal information of the messages. In the considered simulation scenario, the upstream jam fronts were already accurately predicted with errors of ±50 m 1 km ahead of the jam, while the errors for the predicted downstream jam amounted to ±100 m. Obviously, the quality of the jam-front anticipation improves with the number and the timeliness of the incoming messages. More details about the used prediction algorithm can be found in Ref. [Schönhof et al., 2007].

11.7 Conclusions and Future Work

Agent-based traffic simulations provide a flexible and customizable framework for tackling a variety of current research topics. Simulation of control systems as a part of traffic operations is an important topic in transport telematics. Due to the interrelation of the control systems with traffic, both the control systems and the driver reactions must be described in a combined simulation framework. Examples are variable message signs and speed limits, on-ramp metering, lane-changing legislation and dynamic route guidance systems. An interesting research challenge is adaptive self-organized traffic control in urban road networks [Lämmer et al., 2008].

Furthermore, traffic simulations are used to assess the impacts of upcoming driver assistance systems such as *adaptive cruise control* systems on traffic dynamics. The microscopic modeling approach is most appropriate because it allows for a natural representation of heterogeneous driver-vehicle agents and for a detailed specification of the considered models, parameters and vehicle proportions [VanderWerf et al., 2001, 2002; Kesting et al., 2007b; Hoogendoorn and Minderhoud, 2002; Minderhoud, 1999; Davis, 2004; Tampère and van Arem, 2001]. The challenging question is whether it is possible to design vehicle-based control strategies aimed at improving the capacity and stability of traffic flow [Kesting et al., 2008].

With rapid advances in wireless communication technologies, the transmission of information within the transportation network is a challenging issue for the next generation of Intelligent Transportation Systems (ITS). Agent-based systems form the basis for a simulation of hybrid systems coupling vehicle and information flow. The decentralized propagation of information about the upcoming traffic situation has been discussed as an application for inter-vehicle communication. Many other applications are conceivable based on the integration of vehicles and infrastructures implying vehicle-to-infrastructure communication technologies. However, realistic and predictive simulations are essential for developing and testing applications of upcoming communication technologies and applications.

References

> M. Bando, K. Hasebe, A. Nakayama, A. Shibata, and Y. Sugiyama. Dynamical model of traffic congestion and numerical simulation. *Physical Review E*, 51(2):1035–1042, Feb 1995. doi: 10.1103/PhysRevE.51.1035.
>
> W. Brilon and M. Ponzlet. Application of traffic flow models. In D. Wolf, M. Schreckenberg, and A. Bachem, editors, *Traffic and Granular Flow*, pages 23–40. World Scientific, Singapore, 1996.
>
> M. Cassidy and R. Bertini. Some traffic features at freeway bottlenecks. *Transportation Research Part B: Methodological*, 33:25–42, 1999.
>
> D. Charypar. Generating complete all-day activity plans with genetic algorithms. *Transportation*, 32(4):369–397, 2005. doi: 10.1007/s11116-004-8287-y.
>
> D. Chowdhury, L. Santen, and A. Schadschneider. Statistical physics of vehicular traffic and some related systems. *Physics Reports*, 329:199, 2000.
>
> C. Daganzo. *Fundamentals of Transportation and Traffic Operations*. Pergamon-Elseviers, Oxford, U.K., 1997.
>
> D. A. R. P. A. (DARPA). Darpa grand challenge prize competition for driverless cars. http://www.darpa.mil/grandchallenge. Accessed December 2007, 2007.
>
> L. Davis. Effect of adaptive cruise control systems on traffic flow. *Physical Review E*, 69(6):066110, 2004. doi: 10.1103/PhysRevE.69.066110.

J. De Dios Ortúzar and L. Willumsen. *Modelling Transport*. Wiley, 2001.

M. Fukui, Y. Sugiyama, M. Schreckenberg, and D. Wolf, editors. *Traffic and Granular Flow '01*. Springer, Berlin, 2003.

P. Gipps. A behavioural car-following model for computer simulation. *Transportation Research Part B: Methodological*, 15:105–111, 1981.

P. Gipps. A model for the structure of lane-changing decisions. *Transportation Research Part B: Methodological*, 20:403–414, 1986.

M. Green. "How long does it take to stop?" Methodological analysis of driver perception-brake times. *Transportation Human Factors*, 2:195–216, 2000.

B. Greenshields. A study of traffic capacity. In *Proceedings of the Highway Research Board*, volume 14, pages 228–477, 1959.

J. Hausman and D. McFadden. Specification tests for the multinomial logit model. *Econometrica*, 52(5):1290–1240, 1984.

D. Helbing. Traffic and related self-driven many-particle systems. *Reviews of Modern Physics*, 73:1067–1141, 2001.

D. Helbing and K. Nagel. The physics of traffic and regional developments. *Contemporary Physics*, 45(5):405–426, 2004.

D. Helbing and M. Schreckenberg. Cellular automata simulating experimental properties of traffic flow. *Physical Review E*, 59(3):R2505–R2508, 1999. doi: 10.1103/PhysRevE.59.R2505.

D. Helbing and B. Tilch. Generalized force model of traffic dynamics. *Physical Review E*, 58(1):133–138, 1998. doi: 10.1103/PhysRevE.58.133.

P. Hidas. Modelling vehicle interactions in microscopic traffic simulation of merging and weaving. *Transportation Research Part C: Emerging Technologies*, 13:37–62, 2005.

S. Hoogendoorn and P. Bovy. State-of-the-art of vehicular traffic flow modelling. *Proceedings of the Institution of Mechanical Engineers, Part I: Journal of Systems and Control Engineering*, 215(4):283–303, 2001.

S. Hoogendoorn and M. Minderhoud. Motorway flow quality impacts of advanced driver assistance systems. *Transportation Research Record*, 1800:69–77, 2002.

R. Jiang, Q. Wu, and Z. Zhu. Full velocity difference model for a car-following theory. *Physical Review E*, 64(1):017101, 2001. doi: 10.1103/PhysRevE.64.017101.

W. Jin and W. Recker. Instantaneous information propagation in a traffic stream through inter-vehicle communication. *Transportation Research Part B*, 40(3): 230–250, 2006.

K. K. Nagel and M. Schreckenberg. A cellular automaton model for freeway traffic. *J. Phys. I France*, 2:2221–2229, 1992.

B. Kerner. *The Physics of Traffic*. Springer, Heidelberg, 2004.

B. Kerner and P. Konhäuser. Structure and parameters of clusters in traffic flow. *Physical Review E*, 50(1):54–83, 1994. doi: 10.1103/PhysRevE.50.54.

B. Kerner and H. Rehborn. Experimental features and characteristics of traffic jams. *Physical Review E*, 53(2):R1297–R1300, 1996a. doi: 10.1103/PhysRevE.53.R1297.

B. Kerner and H. Rehborn. Experimental properties of complexity in traffic flow. *Physical Review E*, 53(5):R4275–R4278, 1996b. doi: 10.1103/PhysRevE.53.R4275.

A. Kesting and M. Treiber. Calibrating car-following models using trajectory data: Methodological study. *Transportation Research Record*, 2008a.

A. Kesting and M. Treiber. How reaction time, update time and adaptation time influence the stability of traffic flow. *Computer-Aided Civil and Infrastructure Engineering*, 23:125–137, 2008b.

A. Kesting, M. Treiber, and D. Helbing. General lane-changing model mobil for car-following model. *Transportation Research Record*, 1999:86–94, 2007a.

A. Kesting, M. Treiber, M. Schönhof, and D. Helbing. Extending adaptive cruise control to adaptive driving strategies. *Transportation Research Record*, 2000:16–24, 2007b.

A. Kesting, M. Treiber, M. Schönhof, and D. Helbing. Adaptive cruise control design for active congestion avoidance. *Transportation Research Part C: Emerging Technologies*, 16:668–683, 2008. doi: 10.1016/j.trc.2007.12.004.

W. Knospe, L. Santen, A. Schadschneider, and M. Schreckenberg. Human behavior as origin of traffic phases. *Physical Review E*, 65(1):015101, 2001. doi: 10.1103/PhysRevE.65.015101.

W. Knospe, L. Santen, A. Schadschneider, and M. Schreckenberg. Single-vehicle data of highway traffic: Microscopic description of traffic phases. *Physical Review E*, 65(5):056133, 2002. doi: 10.1103/PhysRevE.65.056133.

S. Krauss, P. Wagner, and C. Gawron. Metastable states in a microscopic model of traffic flow. *Physical Review E*, 55(5):5597–5602, 1997. doi: 10.1103/PhysRevE.55.5597.

S. Lämmer, R. Donner, and D. Helbing. Anticipative control of switched queueing systems. *The European Physical Journal B - Condensed Matter and Complex Systems*, 63:341–347, 2008.

H. Lee, R. Barlovic, M. Schreckenberg, and D. Kim. Mechanical restriction versus human overreaction triggering congested traffic states. *Physical Review Letter*, 92(23):238702, 2004. doi: 10.1103/PhysRevLett.92.238702.

W. Leutzbach. *Introduction to the Theory of Traffic Flow*. Springer, Berlin, 1988.

M. Lighthill and G. Whitham. On kinematic waves: II. A theory of traffic on long crowded roads. *Proc. Roy. Soc. of London A*, 229:317–345, 1955.

S. Maerivoet and B. De Moor. Transportation planning and traffic flow models. preprint, physics/0507127, 2005.

S. Maerivoet and B. De Moor. Cellular automata models of road traffic. *Physics Reports*, 419:1–64, 2005. doi: 10.1016/j.physrep.2005.08.005.

MATSim. Large-scale agent-based transport simulations toolbox. http://www.matsim.org. Accessed March 2008, 2008.

J. Miller. Freesim. http://www.freewaysimulator.com/. Accessed December 2007, 2007.

M. Minderhoud. *Supported Driving: Impacts on Motorway Traffic Flow*. Delft University Press, Delft, 1999.

Ministry of Transport, Energy and Spatial Planning of Nordrhein-Westfalen. Traffic state prediction for the freeway network. http://autobahn.nrw.de. Accessed May 2007, 2007.

T. Nagatani. The physics of traffic jams. *Reports of Progress in Physics*, 65:1331–1386, 2002.

K. Nagel, J. Esser, and M. Rickert. Large-scale traffic simulations for transportation planning. *Annual Reviews of Computational Physics VII*, 7:151–202, 2000.

G. Newell. Nonlinear effects in the dynamics of car following. *Operations Research*, 9:209, 1961.

J. Ott and D. Kutscher. Drive-thru internet: IEEE 802.11b for "automobile" users. In *Proceedings of Infocom 2004, Twenty-third Annual Joint Conference of the IEEE Computer and Communications Societies*, pages 362–373. 2004.

S. Press, W.H. and Teukolsky, W. Vetterling, and B. Flannery. *Numerical Recipes in C: The Art of Scientific Computing*. Cambridge University Press, Cambridge, 2nd edition, 1992.

H. Rakha, I. El-Shawarby, and J. Setti. Characterizing driver behavior on signalized intersection approaches at the onset of a yellow-phase trigger. *IEEE Transactions on Intelligent Transportation Systems*, 8(4):630–640, 2007.

B. Raney, N. Cetin, A. Völlmy, M. Vrtic, K. Axhausen, and K. Nagel. An agent-based microsimulation model of swiss travel: First results. *Networks and Spatial Economics.*, 3(1):23–41.

P. Richards. Shock waves on the highway. *Operations Research.*, 4:42–51, 1956.

W. Schnabel and D. Lohse. *Grundlagen der Straßenverkehrstechnik und der Verkehrsplanung, Band 2: Verkehrsplanung.* Verlag für Bauwesen, 1997.

M. Schönhof and D. Helbing. Empirical features of congested traffic states and their implications for traffic modeling. *Transportation Science*, 41:1–32, 2007.

M. Schönhof, A. Kesting, M. Treiber, and D. Helbing. Coupled vehicle and information flows: Message transport on a dynamic vehicle network. *Physica A.*, 363:73–81, 2006.

M. Schönhof, M. Treiber, A. Kesting, and D. Helbing. Autonomous detection and anticipation of jam fronts from messages propagated by intervehicle communication. Transportation Research Record, 1999:3–12, 2007.

S. (Simulation of Urban MObility). Open source traffic simulator. URL http://sumo.sourceforge.net.

J. Singh, N. Bambos, B. Srinivasan, and D. Clawin. Wireless LAN performance under varied stress conditions in vehicular traffic scenarios. *IEEE Vehicular Technology Conference.*, pages 743–747, 2002.

C. Tampère and B. van Arem. Traffic flow theory and its applications in automated vehicle control: A review. *IEEE Intelligent Transportation Systems Proceedings*, pages 391–397, 2001.

T. Toledo, C. Choudhury, and M. Ben-Akiva. Lane-changing model with explicit target lane choice. *Transportation Research Record: Journal of the Transportation Research Board*, 1934:157–165, 2005.

T. Toledo, H. Koutsopoulos, and M. Ben-Akiva. Integrated driving behavior modeling. *Transportation Research Part C: Emerging Technologies*, 15(2):96–112, 2007.

M. Treiber. The website provides an interactive simulation of the Intelligent Driver Model in several scenarios including open-source access. http://www.traffic-simulation.de. Accessed May 2007, 2007.

M. Treiber and D. Helbing. Memory effects in microscopic traffic models and wide scattering in flow-density data. *Physical Review E*, 68:046119, 2003.

M. Treiber and D. Helbing. Microsimulations of freeway traffic including control measures. *Automatisierungstechnik*, 49:478–484, 2001.

M. Treiber and D. Helbing. Reconstructing the spatio-temporal traffic dynamics from stationary detector data. *Cooperative Transportation Dynamics*, 1:3.1–3.24, 2002.

M. Treiber, A. Hennecke, and D. Helbing. Derivation, properties, and simulation of a gas-kinetic-based, non-local traffic model. *Physical Review E*, 59:239–253, 1999.

M. Treiber, A. Hennecke, and D. Helbing. Congested traffic states in empirical observations and microscopic simulations. *Physical Review E*, 62:1805–1824, 2000.

M. Treiber, A. Kesting, and D. Helbing. Delays, inaccuracies and anticipation in microscopic traffic models. *Physica A*, 360:71–88, 2006a.

M. Treiber, A. Kesting, and D. Helbing. Understanding widely scattered traffic flows, the capacity drop, and platoons as effects of variance-driven time gaps. *Physical Review E (Statistical, Nonlinear, and Soft Matter Physics)*, 74(1):016123, 2006b. doi: 10.1103/PhysRevE.74.016123.

J. VanderWerf, S. Shladover, N. Kourjanskaia, M. Miller, and H. Krishnan. Modeling effects of driver control assistance systems on traffic.,. *Transportation Research Record*, 1748:167–174, 2001.

J. VanderWerf, S. Shladover, M. Miller, and N. Kourjanskaia. Effects of adaptive cruise control systems on highway traffic flow capacity. *Transportation Research Record*, 1800:78–84, 2002.

R. Wiedemann. Simulation des straßenverkehrsflusses. In *Heft 8 der Schriftenreihe des IfV*. Institut für Verkehrswesen, Universität Karlsruhe, 1974.

L. Wischhof, A. Ebner, and H. Rohling. Information dissemination in self-organizing intervehicle networks. *IEEE Transactions on Intelligent Transportation Systems*, 6:90–101, 2005.

X. Yang and W. Recker. Simulation studies of information propagation in a self-organized distributed traffic information system. *Transportation Research Part C: Emerging Technologies*, 13(5-6):370–390, 2005.

12

An Agent-Based Generic Framework for Symbiotic Simulation Systems

12.1	Introduction..	357
12.2	Concepts of Symbiotic Simulation	359
12.3	Different Classes of Symbiotic Simulation Systems...	361
	Symbiotic Simulation Decision Support Systems (SSDSS) • Symbiotic Simulation Control Systems (SSCS) • Symbiotic Simulation Forecasting Systems (SSFS) • Symbiotic Simulation Model Validation Systems (SSMVS) • Symbiotic Simulation Anomaly Detection Systems (SSADS) • Hybrid Symbiotic Simulation Systems • Symbiotic Simulation Systems at a Glance	
12.4	Workflows and Activities in Symbiotic Simulation Systems...	365
	Workflows • Activities	
12.5	Agent-Based Framework	372
	Architecture Requirements • Discussion of Existing Architectures • Web Services Approach vs. Agent-Based Approach • Capability-Centric Solution for Framework Architecture • Layers and Associated Capabilities	
12.6	Applications...	378
	Proof of Concept Showcase • Conceptual Showcases	
12.7	Conclusions	383
12.8	Future Work	384
References		384

Heiko Aydt
Nanyang Technological University, Singapore

Stephen John Turner
Nanyang Technological University, Singapore

Wentong Cai
Nanyang Technological University, Singapore

Malcolm Yoke Hean Low
Nanyang Technological University, Singapore

12.1 Introduction

Symbiosis has its origins in biology where it refers to organisms of different species which exist in a long-lasting and close association with each other. The partners in a symbiosis, also referred to as symbionts, are highly dependent upon each other in regard to their outcome. Depending on the definition of symbiosis, this relationship is either mutually beneficial to both (i.e, mutualism) or at least beneficial to one of the symbionts involved [Wilkinson, 2001; Douglas, 1994]. Similarly, *symbiotic simulation* describes a paradigm in which a simulation system and a physical system are closely associated with each other. In such a relationship, the simulation system benefits from sensor data while the physical system may benefit from conclusions drawn, based on the simulation results. Symbiotic simulation has been an active field of research since the term was coined at the Dagstuhl Seminar on Grand Challenges for Modeling and Simulation in 2002 [Fujimoto et al., 2002]. Although the original definition of symbiotic simulation is more precisely referring to mutualism we will use the wider definition of symbiotic simulation [Aydt et al., 2008a] which is based on the original meaning of symbiosis [Douglas, 1994]. The extended definition emphasizes the close association between a simulation system and a physical system which is to the benefit of at least one of them. In addition, control feedback to the physical system is optional. This distinguishes the extended definition from the original one, in which a control feedback to the physical system is considered mandatory.

Symbiotic simulation is used in several application domains. This includes the application of symbiotic simulation in semiconductor manufacturing [Low et al., 2005; Aydt et al., 2008b], path planning for unmanned aerial vehicles (UAVs) [Lozano et al., 2006; Kamrani and Ayani, 2007], and social sciences [Kennedy et al., 2007b; Kennedy and Theodoropoulos, 2006]. Related work includes research on dynamic data driven applications systems (DDDAS) [National Science Foundation], which is a paradigm closely related to symbiotic simulation. DDDAS has become popular in recent years because of its abilities to dynamically adapt to new sensor data and to steer the measurement process. While DDDAS is more generally concerned with dynamic data driven applications, symbiotic simulation is more specifically concerned with the simulation of a physical system. This simulation is also dynamic data driven and uses sensor data to improve the quality of the simulation. The similarities and differences between symbiotic simulation and DDDAS have been explained in [Aydt et al., 2008a] and are only summarized here. DDDAS emphasizes the ability of the application to actively steer the measurement process. While this is an essential feature of DDDAS, it is not necessarily required in symbiotic simulation systems. In addition, the control feedback in symbiotic simulation systems aims to affect the physical system. This is not necessarily the case in a DDDAS where control feedback is mainly used to steer the measurement process. Although the focus of both paradigms is different, they do overlap in some respects. The most notable common feature is the dynamic data driven component.

Another paradigm which is closely related to symbiotic simulation is on-line simulation. Some of the concepts used in symbiotic simulation have also been used in on-line simulation applications. For example, on-line simulation has been used in the context of an on-line planning system [Davis, 1998]. In this system, various control policies are generated and evaluated by means of simulation. Multiple scenarios are also used in an on-line simulation system for military networks [Perumalla et al., 2002]. Similar concepts are used in symbiotic simulation (e.g., analysis of what-if scenarios). On-line simulation is described in the context of UAV path planning as a simulation that runs in real-time and in parallel with the physical system and does not necessarily create feedback to the physical system [Kamrani, 2007]. Although this definition is similar to real-time simulation [Fujimoto, 2000] it should not be confused with it. In real-time simulation, advances in simulation time are paced by wallclock time. This is not the case in on-line simulation. Usually on-line simulation refers to a simulation which is initialized and driven by real-time sensor data. Furthermore, we consider an on-line simulation not capable of generating any control feedback to the physical system on its own. Despite the various works on on-line simulation, there is no proper definition. The term "online simulation" can also be found in the context of gaming and education where it is used in an entirely different context. In addition, on-line simulation does not emphasize the close relationship between the simulation system and the physical system as does symbiotic simulation.

An overview of symbiotic simulation and its related paradigms is illustrated in Table 12.1, reflecting the above definitions.

Several applications, which can be found in the literature on DDDAS, use symbiotic sim-

TABLE 12.1 Comparison between symbiotic simulation, on-line simulation, and DDDAS.

Paradigm	Steering of the Measurement Proc.	Control Feedback	Data-Driven App./Sim.	What-if Analysis
Symbiotic Simulation	Optional	Optional	Simulation only	Yes
On-line Simulation	Optional	No	Simulation only	Optional
DDDAS	Mandatory	Optional	Both	Optional

ulation. The increasing popularity, and the lack of a symbiotic simulation standard, has led to a variety of application specific solutions related to symbiotic simulation. A standard would be an advantage as it would make it easier to develop symbiotic simulation applications by providing standard solutions to common problems in this kind of system. With this chapter we intend to make a first step toward a standard by proposing an architecture for an agent-based generic framework for symbiotic simulation systems. The need for such a generic framework was first expressed and briefly explained in [Huang et al., 2006] but not discussed in detail. Here, we introduce our agent-based generic framework for symbiotic simulation systems and discuss its architecture and generic applicability.

This chapter is structured as follows. In Section 12.2 we explain essential concepts of symbiotic simulation. In Section 12.3 we give an overview of the different classes of symbiotic simulation systems. These classes are further analysed in Section 12.4 which is concerned with the various workflows and common activities in symbiotic simulation systems. In Section 12.5 we state key requirements for a generic framework architecture and propose an agent-based framework after discussing existing architectures. We apply the proposed framework to several applications in Section 12.6 and present our conclusions in Section 12.7. We discuss future work in Section 12.8.

12.2 Concepts of Symbiotic Simulation

A symbiotic simulation system is based on simulations of the physical system. An appropriate *model* of the physical system, further denoted by m, has to be used to perform simulations. Model instances are created by a *model function*, further denoted by M. This function can have an arbitrary number of parameters $p_0, p_1, \ldots, p_n$ of which each is associated with a set of possible values. The set of model parameters is further denoted by P_M. Depending on the specific model function and the values used for each of its parameters, different model instances can be created.

For example, let $m = M_{ms}(p_1, p_2, \ldots, p_8)$ be a model function which creates model instances for a manufacturing system, consisting of eight machines to process product units. Each machine can be configured independently. This is reflected by corresponding parameters of the model function. Depending on the values used for each parameter, a specific model instance of the manufacturing system is created. The two model instances $m_i = M_{ms}(0,0,0,0,0,0,0,0)$ and $m_j = M_{ms}(1,0,0,0,0,0,0,0)$, for instance, represent the manufacturing system with two different machine settings.

By using different parameter values, different model instances can be created, of which each represents the same physical system with slightly different settings. However, only one of them can be considered as the *reference model*, further denoted by m_R. The reference model represents the physical system with the highest accuracy among all known model instances. The physical system may change over time and eventually requires the reference model to change as well. A reference model is valid until a better model is found which represents the physical system with higher accuracy.

Alone a model is not sufficient to perform a simulation of the physical system. It is also necessary to initialize the simulation with a specific *state* of the physical system, further denoted by s. The current state of the physical system has to be determined by using real-time sensor data. Similarly, previous states can be determined by using historical sensor data. For example, the state of a manufacturing system may include information about the number of product units which are currently being processed in the system.

A simulation of the physical system can be initialized by using an appropriate model and an initial state. However, this is not enough to drive the simulation. The behavior of the physical system is influenced by its environment. This *external influence*, further denoted by e, has to be considered when simulating the physical system. For example, the behavior of a manufacturing system depends on incoming product orders from customers. Depending on the demand for a certain product, the behavior of the manufacturing system is different. Therefore, it is necessary to provide data about external influence when simulating the manufacturing system. If the current state of the physical system is used to initialize the simulation, predicted data about external influence has to be used because the simulation aims to simulate the future behavior of the physical system. In the case where a previous state is used, it is also possible to use historical data for external influence.

An important concept of symbiotic simulations is that of *what-if scenarios*. They are used to simulate the behavior of the physical system depending on specific assumptions. For example, different what-if scenarios can be used to simulate the performance of a manufacturing system by assuming different combinations of machine settings and expected customer demand for certain products. A simulation of the physical system depends on the model, the initial state, and external influence. Different scenarios are created by using different values for these three elements. Therefore, each scenario is a triple, consisting of a specific model m_i, an initial state s_i, and external influence e_i.

$$S = (m_i, s_i, e_i) \qquad (12.1)$$

The information provided by a scenario is used to initialize and drive a simulation. In order to compare different scenarios with each other or with the physical system, more information is needed. For this purpose, *key indicators* are introduced. They reflect specific characteristics of the physical system which are considered important in the context of the application. By using key indicators, it is possible to compare simulation runs with each other or with the physical system. For example, the throughput of a manufacturing system can be considered as a key indicator in order to evaluate the performance. Data about key indicators are obtained from the physical system by using corresponding sensors. Similarly, data about key indicators have to be provided by the simulation.

Based on simulation results, the physical system can be influenced. Decision making in symbiotic simulation systems relies on the simulation of different what-if scenarios. Each what-if scenario represents an alternative decision, i.e., an alternative way of influencing the system. For simulation purposes it is necessary that the simulation model is able to reflect possible ways of influencing the physical system. For example, if a symbiotic simulation system is used to decide upon machine settings of a manufacturing system, the simulation model needs to be able to reflect different machine settings. Different model instances, of which each reflects alternative machine settings, can then be used to create what-if scenarios. Parameters which can be changed in the physical system are therefore important for the simulation system which aims to exercise influence. These parameters are further referred to as *decision parameters*. The set of decision parameters, further denoted by P_D, contains all parameters of the physical system which can be influenced. Decision parameters are also considered by the model function with corresponding model parameters:

$$P_D \subseteq P_M \tag{12.2}$$

Decision making cannot be based on decision parameters only. It is also important to consider constraints and dependencies between them. For example, when using a particular setting for a machine, possible settings for another machine are restricted. For this purpose, a *decision model* is used which considers decision parameters and constraints and dependencies between them. In addition, the decision model also reflects the ability of the simulation system to implement a specific decision. For example, this may depend on currently available actuators.

12.3 Different Classes of Symbiotic Simulation Systems

Analogous to organic symbionts in biology, there are two symbionts in the context of symbiotic simulation: the *simulation system* and the *physical system*. The simulation system is driven by measured data and is capable of influencing the physical system. The whole system, consisting of the simulation system, sensors, actuators, and other components which are necessary to facilitate a symbiotic relationship, represents a *symbiotic simulation system*.

Three forms of symbiosis are distinguished in biology [Douglas, 1994]. Mutualism refers to a symbiotic relationship in which both symbionts benefit $(+/+)$. In other forms of symbiosis, only one symbiont benefits while the other one is either suffering $(+/-)$ or not affected at all $(+/0)$. In biology, these relationships are referred to as parasitism and commensalism, respectively. Depending on the application, the term predation may be more appropriate than parasitism. For example, in a military application, the simulation system benefits from real-time sensor data while the hostile physical system is suffering from decisions made by the simulation system. In this case, the symbiotic relationship between the simulation system and the physical system is predation rather than parasitism. Commensalism in symbiotic simulation systems is realized by taking away the control feedback from the simulation system to the physical system. In this case, the physical system will not be affected while the simulation system still benefits from real-time sensor data.

DDDAS is an active field of research and is used in the context of a variety of disciplines, from environmental forecasting to decision support for threat management, traffic control, and UAV path planning. Although many applications are based on symbiotic simulation, most papers do not explicitly mention it and focus rather on the particular domain-specific problem. Analysis of related work has shown that, despite the variety of applications, symbiotic simulation is used for only few purposes: decision support/control, forecasting, model validation, and anomaly detection. We define five different classes of symbiotic simulation systems [Aydt et al., 2008a]: *symbiotic simulation decision support systems (SSDSS)*, *symbiotic simulation control systems (SSCS)*, *symbiotic simulation forecasting systems (SSFS)*, *symbiotic simulation model validation systems (SSMVS)*, and *symbiotic simulation anomaly detection systems (SSADS)*. Furthermore, we distinguish between closed-loop and open-loop symbiotic simulation systems. While a control feedback is created in closed-loop symbiotic simulation systems in order to affect the physical system, this is not the case in an open-loop system. An SSDSS/SSCS is therefore considered to be a closed-loop symbiotic simulation system while all other classes are considered to be open-loop symbiotic simulation systems. The remainder of this section is dedicated to explaining the different classes of symbiotic simulation systems.

12.3.1 Symbiotic Simulation Decision Support Systems (SSDSS)

The first class of symbiotic simulation systems is concerned with *decision support*. Simulation in decision support systems is used to predict and evaluate the impact of possible decisions on the physical system. For this purpose, a number of different what-if scenarios are simulated, each representing an alternative decision. The results of these simulations are analyzed and used to support an external decision making process. The influence of the simulation system in an SSDSS is only *indirect* because actual decision making and implementation is performed by an external decision maker.

An SSDSS predicts possible future states of a physical system for a number of what-if scenarios. Simulation results are analyzed and interpreted in order to draw conclusions which are used to support an external decision making process. The simulation system in an SSDSS benefits from real-time sensor data which can be injected into running simulations in order to improve the accuracy. It does indirectly influence the physical system which can be either to the advantage or disadvantage for the physical system. The symbiosis between the simulation system and a physical system in an SSDSS is therefore either mutualism or parasitism/predation.

Examples for symbiotic simulation used for decision making include answering what-if questions asked by fire fighters for situation assessment [Michopoulos et al., 2004], simulation of alternative threat management scenarios upon an incident in a water distribution network [Mahinthakumar et al., 2006], and dynamic path planning of a UAV using what-if scenarios in [Kamrani and Ayani, 2007; Lozano et al., 2006]. In these examples, various scenarios are simulated and analyzed to make appropriate decisions. There are also examples of decision support applications, using symbiotic simulation without mentioning analysis of various what-if scenarios. These examples include traffic control and management [Fujimoto et al., 2006] and electric power transmission systems [McCalley et al., 2007, 2006].

12.3.2 Symbiotic Simulation Control Systems (SSCS)

The second class of symbiotic simulation systems is concerned with *control*. As an extension to a decision support system, the results of symbiotic simulation can be used to control the physical system by means of actuators. Unlike a decision support system, the control system does not depend on an external decision making process to influence the physical system. Therefore, the influence of the SSCS is *direct*.

Like an SSDSS, an SSCS evaluates a number of what-if scenarios by means of simulation. Also, analysis results are interpreted in order to draw conclusions which can be used to influence the system. In contrast to an SSDSS where influence is indirect only, the simulation system in an SSCS can directly influence the physical system. This is the major difference between an SSCS and an SSDSS. The symbiosis between the simulation system and the physical system in an SSCS is either mutualism or parasitism/predation.

Examples for symbiotic simulation control include the use of controlling agents to make necessary modifications to a semiconductor manufacturing system [Low et al., 2005], on-line planning and control in manufacturing [Davis, 1998], implementation of simulated escalation scenarios by activating actuation mechanisms [Michopoulos et al., 2003], control of hydraulic gates in the context of floodwater diversion [Xue et al., 2007], and optimization of a business process workflow in the context of high-tech manufacturing and service networks [Low et al., 2007].

12.3.3 Symbiotic Simulation Forecasting Systems (SSFS)

The third class of symbiotic simulation systems is concerned with *forecasting*. Many applications use symbiotic simulation to predict future states of a system. Almost all applications that use symbiotic simulation for forecasting purposes are involved in some form of decision making. However, there are also some applications which use symbiotic simulation solely for forecasting purposes. Examples for forecasting include short-term wildland fire prediction [Mandel et al., 2007; Douglas et al., 2006; Mandel et al., 2005, 2004] and image guided neurosurgery [Majumdar et al., 2005]. In both cases, data from the physical system is injected into the simulation to improve accuracy.

An SSFS predicts possible future states of a physical system for a number of what-if scenarios. The output of the what-if simulations can be used for visualization purposes, for instance. It is also possible to use the output of the forecast for further analysis in an external process. An SSFS is similar to an SSDSS/SSCS as it predicts possible future states of the physical system. However, it does not analyze the simulation results and does not draw any conclusions in order to create a control feedback to the physical system. The future behavior of the physical system depends on the influence from its environment. External influence is therefore an important factor in forecasting. Various what-if scenarios can be created by making different assumptions for external influence. An SSFS benefits from real-time sensor data which can be injected into running simulations in order to improve the accuracy. It does not influence the physical system, i.e., the physical system is not affected by the symbiotic simulation system. The symbiosis between the simulation system and the physical system in an SSFS is therefore commensalism.

12.3.4 Symbiotic Simulation Model Validation Systems (SSMVS)

The fourth class of symbiotic simulation systems is concerned with *model validation*. Model validation is an important aspect of symbiotic simulation because many applications rely on an accurate reference model of the physical system. Simulations can be used to validate the underlying model against the physical system. Different models are simulated in order to identify the one that describes the current state of the physical system best. Therefore, an SSMVS can be used to find the reference model. Once identified, the reference model can be used for subsequent forecasting or other purposes. Model validation is performed by comparing the simulated behavior and the actual behavior of the physical system. Unlike other symbiotic simulation systems, an SSMVS can make use of historical data to initialize and drive the simulation. In addition, it is possible to perform model validation *on the fly*, by initializing a paced simulation with the current state of the physical system and using real-time sensor data to drive the simulation. The simulated behavior of the physical system is compared with recorded or measured data of the actual behavior. A model which can explain the actual behavior of the physical system accurately enough can further serve as the reference model.

An SSMVS compares the results of various simulations, each using a different possible model, with the physical system in order to determine a model that describes the physical system with sufficient accuracy. An SSMVS benefits from historical and real-time sensor data to initialize and drive simulations. It does not influence the physical system and the symbiosis between the simulation system and the physical system in an SSMVS is therefore commensalism.

It is possible to distinguish between model validation used for *calibration* and *classification* purposes. In the first case, a model is calibrated in order to reflect the behavior of the physical system with higher accuracy. In the second case, different hypotheses can be evaluated in order to find one that explains the behavior of the physical system. Examples for calibration include determination of unknown boundary conditions in fluid-thermal systems [Knight et al., 2007] and prediction of material properties of geological structures [Akcelik et al., 2004]. An example for classification can be found in [Madey et al., 2007]. In this example, various hypotheses about an emergency event are tested.

12.3.5 Symbiotic Simulation Anomaly Detection Systems (SSADS)

The fifth class of symbiotic simulation systems is concerned with *anomaly detection*. Symbiotic simulation can be used to detect anomalies either in the physical system or in the model of the physical system. This is done by comparing the simulated behavior with the actual behavior. A paced simulation is initialized with the current model and the current state of the physical system. Simulation updates are essential for a paced simulation to evolve closely to the physical system. Real-time sensor data is therefore used to drive the simulation. Measured and simulated values of key indicators are compared with each other and discrepancies are interpreted as anomalies. A detected anomaly can either be due to an inaccurate model, or be due to an unexpected or abnormal behavior of the physical system. For example, a sudden failure in the physical system leads to a discrepancy between simulated and actual behavior of the physical system. In this case, the anomaly is due to the physical system. An SSADS does not influence the physical system and the symbiosis between the simulation system and the physical system in an SSADS is therefore commensalism.

An example of symbiotic simulation used to detect anomalies in a physical system is described in [Cortial et al., 2007]. In this example, the model of a healthy structure is used to simulate the wing of a fighter jet. Simulated values and measured values are compared and a detected anomaly is interpreted as structural damage of the wing. An example of symbiotic simulation used to detect anomalies in the context of social sciences is described in [Kennedy et al., 2007b]. In this example, a housing scenario is modeled and used to simulate the moving behavior of households. If simulations turn out to be inconsistent with observations, the anomaly is interpreted as inaccuracy of the model.

12.3.6 Hybrid Symbiotic Simulation Systems

A symbiotic simulation system can consist of a number of subsystems, each representing a symbiotic simulation system by itself. Such a system is referred to as a *hybrid symbiotic simulation system*. Because it is a composition of various subsystems, a hybrid symbiotic simulation system is not considered as an independent class. For example, a system may consist of a forecasting subsystem and a model validation subsystem. The model validation subsystem is responsible for calibrating the model which is used by the forecasting subsystem. It is also possible to have a hybrid symbiotic simulation system with mixed symbiotic simulation and application specific subsystems.

An example of a hybrid forecasting and model validation system can be found in [Madey et al., 2006], where various hypotheses about an emergency are tested and validated. This is done by simulating various anomalies and comparing their behavior to the actual behavior of the physical system (SSMVS). Once validated, a simulation can be used to predict the future development of the emergency (SSFS). An example for a hybrid anomaly detection and forecasting system is described in [Cortial et al., 2007], where symbiotic simulation is used to detect structural damage (SSADS) and to forecast its possible evolution (SSFS). Another example of a hybrid symbiotic simulation system, in the context of semiconductor manufacturing, can be found in [Aydt et al., 2008a].

TABLE 12.2 Overview of the various symbiotic simulation systems.

Class	Purpose	Control Feedback	Meaning of What-if Scenarios	Type of Symbiosis
SSDSS	Support of an external decision maker	Indirect	Decision alternatives	Mutualism/Parasitism
SSCS	Control of a physical system	Direct	Control alternatives	Mutualism/Parasitism
SSFS	Forecasting of a physical system	No	Different assumptions for external influence	Commensalism
SSMVS	Validation of a simulation model	No	Alternative models or different parameters	Commensalism
SSADS	Detection of anomalies either in the physical system or in the simulation model	No	Reference model only	Commensalism

12.3.7 Symbiotic Simulation Systems at a Glance

An overview of the various classes of symbiotic simulation systems described in Sections 12.3.1–12.3.5 is illustrated in Table 12.2. This table lists the five classes of symbiotic simulation systems, their purpose, the type of control feedback (if any), the meaning of the various what-if scenarios, and the type of symbiotic relationship between the simulation system and the physical system.

12.4 Workflows and Activities in Symbiotic Simulation Systems

Each symbiotic simulation system has its own specific workflow. Here, by workflow we mean the sequence in which various activities are performed in a symbiotic simulation system. Initially it may seem as if the various workflows differ significantly from each other. However, there are a relatively small number of common activities which are used by all workflows. In this section, we describe the workflows of the different classes of symbiotic simulation systems, as well as the activities involved.

12.4.1 Workflows

Regardless of its type, a symbiotic simulation system is always observing the physical system by means of corresponding sensors. Therefore the observation activity is performed throughout the time a symbiotic simulation application is executed. In addition, application specific trigger conditions are continuously evaluated. Such a trigger condition can be a sudden drop in performance of a manufacturing system, for instance. In addition, depending on the application, either one or more triggering conditions can be used. If all necessary conditions are fulfilled, the analysis process is triggered. This process depends on the symbiotic simulation system class used, of which each has its own class-specific workflow. For example, in a decision support or control system for a manufacturing system, various alternative machine configurations have to be evaluated in order to solve the performance problem which has triggered the analysis process. Figure 12.1 illustrates the general workflow of a symbiotic simulation system.

All class-specific workflows (except the one for an SSADS) explained in the following

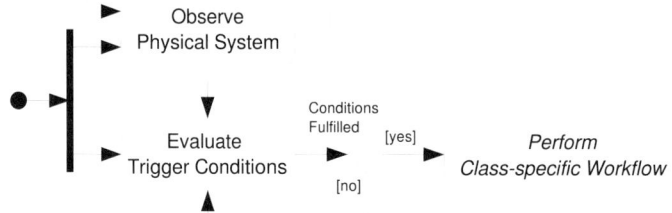

FIGURE 12.1 Activities involved in the general workflow of a symbiotic simulation system.

sections make use of a common workflow for executing a symbiotic simulation. This common workflow is illustrated in Figure 12.2.

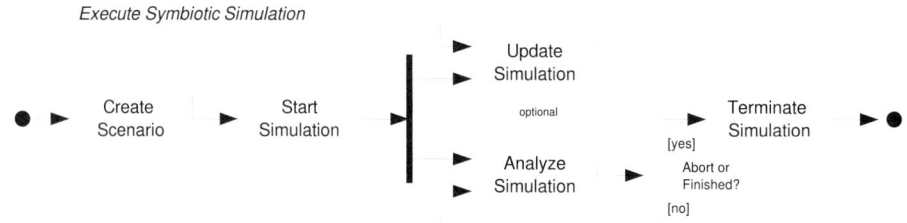

FIGURE 12.2 Activities involved in the common workflow of executing a symbiotic simulation.

SSDSS Workflow

An SSDSS is used to analyze different possible ways of influencing the physical system in order to see whether one leads to the desired behavior. This is done by predicting and evaluating the potential outcome of different what-if scenarios. Several forecasts are performed concurrently, each using a hypothetical what-if scenario which reflects a different way of influencing the physical system. The performance of each what-if scenario can be analyzed during runtime by evaluating the simulated key indicators. It is possible to decide upon premature termination of what-if simulations which perform poorly. For example, the performance of a manufacturing system depends on its configuration. Various configurations are reflected by different what-if scenarios. During the simulation of a what-if scenario, the simulated performance can be analyzed. A simulation which performs poorly can be terminated and replaced by another one, using a different what-if scenario. Once enough scenarios are simulated, the results are analyzed and compared with each other. If one of the simulated scenarios has produced the desired result, it is suggested as a possible solution to the external decision maker. The class-specific workflow of an SSDSS is illustrated in Figure 12.3.

SSCS Workflow

An SSCS is used for the same purpose as an SSDSS. Various possible what-if scenarios are simulated and evaluated in order to determine one which leads to a desired result. The only difference is that solutions are directly implemented by using corresponding actuators, i.e.,

FIGURE 12.3 Activities involved in the class-specific workflow of an SSDSS.

a control system is the extension of a decision support system. Consider the semiconductor manufacturing example from the previous section. Various what-if scenarios, representing alternative configurations are simulated and analyzed. A winning scenario is determined and the winning configuration is implemented by using actuators. In the context of semiconductor manufacturing, an actuator can be a tool controller, for instance. The class-specific workflow of an SSCS is illustrated in Figure 12.4.

FIGURE 12.4 Activities involved in the class-specific workflow of an SSCS.

SSFS Workflow

An SSFS is used to predict the future state of a system based on given scenarios. Unlike other symbiotic simulation systems, the simulated scenarios are not analyzed in order to draw any conclusions. For example, decision support and control systems use the simulation of what-if scenarios to make decisions regarding the physical system. Model validation systems and anomaly detection systems compare the simulation results with the physical system to find a model which reflects the physical system accurately enough, or to detect anomalies, respectively. This is not the case in an SSFS.

Simulation runs are started using the given scenarios and executed concurrently. Depending on the application, simulation updates might not be possible or necessary. The update activity is therefore optional. Even if no simulation update is used, the simulation system is still using sensor data to determine the state of the physical system, which is then used to initialize a simulation run. Therefore, the simulation system still benefits from the symbiotic relationship with the physical system. During its execution, a simulation can be analyzed. Similar to SSDSS/SSCS, it is possible to terminate the simulation run based on an on-line analysis. For example, it might be possible to confirm early predictions of the simulation run by comparing them with observations from the physical system. If this confirmation fails, i.e., if the prediction is not accurate enough, there is no reason to continue the simulation. In this case the simulation can be terminated prematurely. The class-specific workflow of an SSFS is illustrated in Figure 12.5.

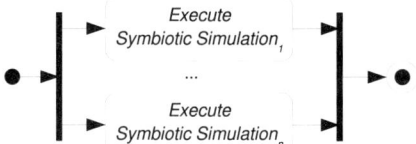

FIGURE 12.5 Activities involved in the class-specific workflow of an SSFS.

SSMVS Workflow

An SSMVS is used to find a model that explains the behavior of the physical system with sufficient fidelity, i.e., it is used to determine the reference model. Various scenarios are simulated and evaluated, each reflecting a different model. In the case of calibration, models are slightly modified variants of the current reference model. In the case of classification, models represent various hypotheses about the physical system and can significantly vary from each other. Similar to decision support and control systems, these scenarios are simulated and analyzed concurrently. If a model is found that describes the behavior of the physical system accurately enough, the current reference model is replaced. How accurate a model has to be in order to serve as reference model depends on the application context. The class-specific workflow of an SSMVS is illustrated in Figure 12.6.

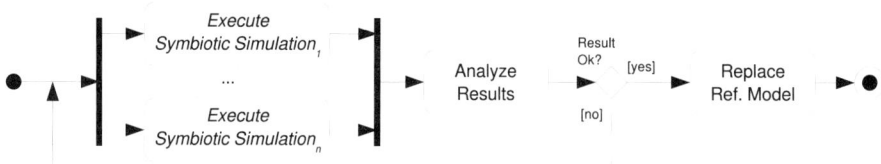

FIGURE 12.6 Activities involved in the class-specific workflow of an SSMVS.

SSADS Workflow

An SSADS is used to identify discrepancies between the simulated behavior and the actual behavior of the physical system. Unlike other symbiotic simulation systems, a paced simulation is used which is driven by measured data about external influence. This sensor data is injected into the simulation as soon it is available. Frequent and highly accurate data injections with low latency allow the simulation to evolve closely with the physical system. The simulation update is essential for the paced simulation to evolve and is therefore not optional, as it is the case in other symbiotic simulation systems. During runtime the simulation is continuously analyzed and compared with the actual behavior of the physical system. Significant discrepancies are interpreted as anomalies. The class-specific workflow of an SSADS, illustrated in Figure 12.7, is almost identical to the common workflow for executing a symbiotic simulation (see Figure 12.2) used by the other classes of symbiotic simulation systems. The only difference is the obligatory update activity. In addition, a paced simulation is started rather than a normal simulation which is executed as fast as possible.

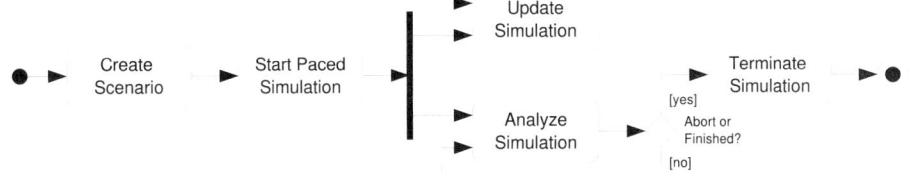

FIGURE 12.7 Activities involved in the class-specific workflow of an SSADS.

12.4.2 Activities

Although the objectives of the various types of symbiotic simulation systems are different, their workflows are composed of the same activities. This results in a relatively small set of essential activities needed by symbiotic simulation systems. In this section, we describe these activities.

Observe Physical System

Observing the physical system is essential for a symbiotic simulation system. All systems collect data about the state of the physical system and external influence. This information is required to initialize and drive a simulation. Except for forecasting systems, all other classes of symbiotic simulation systems also collect information about key indicators. Based on them, different simulation runs are compared with each other or with the physical system. Since an SSFS does not interpret simulation runs, there is no need for key indicators. The observation process in a symbiotic simulation system is illustrated in Figure 12.8.

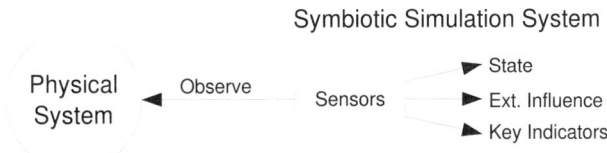

FIGURE 12.8 Observation of a physical system by a symbiotic simulation system.

Evaluate Trigger Conditions

Typically symbiotic simulation systems are triggered by specific conditions depending on the application. For example, the purpose of an SSDSS/SSCS in a manufacturing system is to maintain predefined performance targets. In this context, the symbiotic simulation system is triggered when the performance becomes critical. Similarly, corresponding trigger conditions are used for the other types of symbiotic simulation systems depending on the application.

In a hybrid symbiotic simulation system, consisting of several symbiotic simulation subsystems, the output of one subsystem can be used as trigger for another subsystem. For example, an SSADS is used to detect anomalies between the actual behavior of the physical system and the simulated behavior. Once an anomaly is detected, which is due to an inaccurate model of the physical system, an SSMVS is triggered which is responsible for calibrating the simulation model. Similarly, it is possible to use an SSADS to trigger an

SSDSS/SSCS in order to make decisions regarding the physical system.

It is possible to further distinguish between external and internal triggers. An external trigger resides outside the symbiotic simulation system. For example, a symbiotic simulation system can be triggered manually by a human operator. A symbiotic simulation subsystem which triggers other symbiotic simulation subsystems can also be considered as an external trigger. Internal triggers are used within a symbiotic simulation system. They evaluate data about key indicators and trigger symbiotic simulation accordingly. For example, an SSDSS/SSCS evaluates the performance indicators of a manufacturing system. If the performance drops below a specified threshold, the what-if analysis process is triggered. External and internal triggering of a symbiotic simulation system is illustrated in Figure 12.9.

FIGURE 12.9 External and internal triggering in symbiotic simulation systems.

Create Scenario

An SSFS is used to simulate possible future behavior of the physical system. Each scenario in a forecasting system uses the current reference model and the current state of the physical system to initialize the simulation. Different scenarios are created by using different values for external influence. For example, in the context of a manufacturing system, different values for external influence can be used to create scenarios which reflect cases of high or low demand for a particular product.

An SSDSS/SSCS is used to make decisions regarding how to influence a physical system in order to achieve a desired behavior of the physical system. Various what-if scenarios are simulated to see how the behavior of the physical system develops under certain assumptions. Different what-if scenarios are created by using the reference model with different parameters and different values for external influence. Scenario creation in the context of SSDSS/SSCS is only concerned with the decision parameters of the model. For example, the future performance of a manufacturing system depends on the various machines and their settings as well as on the estimated order volume. Machine settings that perform well under normal workload conditions might perform poorly under high workload conditions. The effect of different machine settings under different load assumptions can be analyzed by using corresponding what-if scenarios.

An SSMVS is used to determine a model which reflects the physical system more accurately than the current reference model. Each scenario in a model validation system reflects a different model. In contrast to SSDSS/SSCS which modifies only decision parameters, scenario creation in an SSMVS is not limited to decision parameters. For example, the throughput of a manufacturing system depends on several factors of which one is the efficiency of workers. Initially, workers are not experienced and thus less efficient. After a while they gain experience and perform their tasks more efficiently. This positively affects the performance of the manufacturing system. To reflect this gain of experience in the simulation, the model has to provide a corresponding parameter. This parameter is then subject to recalibration by an SSMVS.

An SSADS is used to detect anomalies in the behavior of the physical system and the simulation model. In contrast to other classes of symbiotic simulation systems, only one scenario is used in the context of SSADS. This scenario uses the current reference model and the current state of the system. In the case of an SSADS, a paced simulation is performed which is driven by real-time measurements of the physical system and its environment. Therefore, the scenario which is used for anomaly detection does not provide any external influence.

Update Simulation

Simulation updates can be performed in order to increase the accuracy of the simulation. For this purpose, real-time sensor data is injected into the running simulation. Whether a simulation update is possible depends on the capabilities of the particular simulation. A simulation is typically performed as fast as possible and might have advanced a good deal in terms of simulation time when new sensor data becomes available. This kind of simulation is used in forecasting, decision support, and control systems. In this case it has to be decided whether new data samples should be considered or not. Strategies to decide whether simulation update should be performed, should consider the time which is necessary for the simulation to adapt to new sensor data. Depending on the underlying problem and the simulator, an update might require a complete rollback to the time of the update. In other cases, the simulation might be more flexible and is only rolled back in the case where the measured values differs significantly from the anticipated values which are currently used in the simulation. Whether or not to perform simulation updates depends on the ratio of the anticipated gain in simulation accuracy and the time it takes for the simulation to adapt. For example, if strict time constraints have to be considered, a less accurate simulation result obtained in time might be more preferable than a highly accurate but delayed simulation result.

Simulations used in anomaly detection systems are paced simulations. They evolve at the same speed as the real system. External data is used to drive the simulations. Therefore, without continuously updating the simulation by injecting corresponding data, the simulation would not evolve in accordance with the physical system. Hence, a simulation update is essential for paced simulations.

Unlike decision support and control systems, which aim to predict the future state of the physical system, a model validation system is used to predict the current state of the system based on a previous state and driven by historical data about external influence. An update of the simulation run with real-time data is therefore not necessary.

Analyze Simulation

Simulations are analyzed while being executed, i.e., *in vivo*. This analysis process provides information about the simulation to the symbiotic simulation system and its environment. For example the information can be used for notification or visualization purposes. Information about the simulation run can also be used to decide whether a simulation should be terminated prematurely. This could be the case for simulations which perform poorly and are unlikely to produce any useful results. Premature termination of such a simulation would save valuable computation time.

Analyze Results

Results are analyzed once all simulations have finished execution, i.e., *post mortem*. This analysis process aims to draw conclusions based on the simulation results. In decision support and control systems, the analysis is concerned with identifying a scenario which leads to the desired behavior of the physical system. In model validation systems, the analysis is concerned with identifying a scenario which describes the current behavior of the physical system with sufficient accuracy. The model used in this scenario can then be used to replace the current reference model.

Other Activities

Other common activities include starting and terminating simulations. In addition, other class-specific activities are performed by SSDSS, SSCS, and SSMVS. Decision support systems suggest solutions to external decision makers, control systems implement solutions by means of actuators, and model validation systems replace the reference model.

12.5 Agent-Based Framework

A number of common activities were identified in the previous section. These activities have to be performed by functional components of the framework. In this section we propose an agent-based framework which provides all the necessary components to realize the various classes of symbiotic simulation systems. Before proposing our own architecture, we state requirements regarding the architecture, discuss existing architectures, and compare an agent-based approach with a Web service approach.

12.5.1 Architecture Requirements

Requirements regarding an architecture usually depend highly on the application context. For example, an embedded application can be expected to have requirements regarding its real-time behavior and its memory consumption. On the contrary, a large enterprise transaction application typically has enough memory resources for an architecture which may include several large components. An architecture which is suitable for such an enterprise application is most likely not suitable for an embedded system.

When designing a generic framework architecture the application context is unknown. It is therefore not possible to state requirements that are too specific. However, we believe that there are three requirements which are essential for a generic framework architecture for symbiotic simulation systems.

1. A generic framework must be applicable in different disciplines. This includes the use of the framework to realize any of the five different classes of symbiotic simulation systems described in Section 12.3. Therefore, the first requirement for a generic framework architecture is *applicability*.
2. Although the described classes of symbiotic simulation systems cover most applications of symbiotic simulation at the time of writing, this does not rule out the possibility of additional classes and functionality being identified and defined in future. Therefore, the second requirement for a generic framework architecture is *extensibility*.
3. Symbiotic simulation systems can be applied in various disciplines. They rely on potentially computationally intensive simulations. In addition, it might be necessary to simulate a large number of scenarios concurrently. All this might imply that the application of symbiotic simulation is limited to the domain of grid and cluster computing. However, as embedded systems are becoming more and more powerful, we would not be surprised to see the first embedded symbiotic simulation systems soon. We expect that in future, symbiotic simulation appli-

cations will range from large and inherently distributed systems to embedded systems. Therefore, the third requirement for a generic framework architecture is *scalability*.

12.5.2 Discussion of Existing Architectures

Various symbiotic simulation applications are already described in the literature. In the discussion here, we include only those which are relevant for our purposes.

Low et al. [Low et al., 2005] describe a symbiotic simulation system for semiconductor backend assembly and test operation. Their system is implemented using the JADE agent toolkit [Bellifemine et al., 1999]. In addition, the physical system is emulated using a simulation model. The agent-based system consists of a number of agents which are responsible for system management, monitoring, control, simulation, and optimization. Anomalies are detected by the system management agent and, as consequence, what-if analysis is carried out by the optimization agent. Depending on the outcome of this analysis, the physical system is modified by a corresponding control agent. The symbiotic simulation system is primarily used for controlling the physical system. A reference simulation is also mentioned which runs alongside the physical system and is constantly compared with observations from the physical system. Model maintenance is triggered upon detection of discrepancies between simulated and observed data.

Kennedy and Theodoropoulos [Kennedy and Theodoropoulos, 2006] describe an architecture for autonomous DDDAS agents. This kind of agent uses symbiotic simulation to predict states of its environment. Based on this prediction, the sensors of the agent can be redirected if necessary. An assistant DDDAS agent is used to compare simulation predictions and real world data to determine whether they are consistent [Kennedy et al., 2007b]. If not, the simulation model can be revised accordingly. Essentially, a DDDAS agent consists of sensors, a control component, and a model component. DDDAS agents support sensor steering and can be used to realize symbiotic simulation anomaly detection systems. Although it is possible to have multiple agents, each using a different model, the simulation of several scenarios is not directly supported.

Madey et al. [Madey et al., 2007, 2006] describe WIPER, an emergency response system. Their system uses wireless call data to detect any abnormal call patterns or population movements. Upon detection of an anomaly, a simulation and prediction system is triggered. Agent-based simulations are used to test various hypotheses about an anomaly. A decision support system can use validated simulations to predict the evolution of the anomaly. The architecture of the system consists of five components: data source, historical data storage, anomaly detection and alert system (DAS), simulation and prediction system (SPS), and decision support (DSS) with a web interface [Madey et al., 2007]. Once triggered by the DAS, the SPS creates an ensemble of agent-based simulations. These simulation runs are updated by real-time data injections. Although the simulation itself is agent-based, the various components of the WIPER system are realized by Web services. Although WIPER consists of several components, including a component for anomaly detection, symbiotic simulation is only used for prediction and model validation.

Lozano et al. [Lozano et al., 2006] describe a symbiotic simulation based decision support system for UAV path planning. In this system, the path of a UAV is continuously optimized by considering threats, sensor constraints, and changes of the environment. In this application, the best path is determined among a set of alternative UAV paths. Symbiotic simulation is used as part of the path planning algorithm described in [Kamrani and Ayani, 2007]. In this system, each alternative UAV path is evaluated using simulation. Although a detailed description of the architecture is not given, several functional components can be identified from the brief description which can be found in [Lozano et al., 2006]. The symbiotic simulation system consists of several components for sensing and data fusion, scenario creation, what-if simulation, evaluation and analysis, and decision making. Some of these components are very similar to the ones which will be discussed in Section 12.5.5.

All existing architectures are used for a specific application which is concerned with either a single class, or only few classes, of symbiotic simulation systems. To use these application specific architectures for other applications, including other classes of symbiotic simulation systems, it would therefore be necessary to significantly extend their functionality. This is not surprising as none of these architectures was meant to be generic. The applicability of existing architectures is therefore very limited.

Each architecture requires a different degree of effort to extend the current functionality. For example, the architecture described by Low et al. [Low et al., 2005] can possibly be extended by adding new agents to the system. The WIPER architecture can be extended in a similar way. In this case, new functionality has to be realized as a Web service rather than by an agent. However, extending the functionality of the other architectures seems to be more difficult. For example, the DDDAS agent by Kennedy and Theodoropoulos [Kennedy and Theodoropoulos, 2006] contains all required functionality for symbiotic simulation anomaly detection. Extending this agent to support other classes as well is probably not a good solution because it would lead to an agent which is overloaded with functionality. An alternative would be to introduce various types of DDDAS agents, of which each is responsible for a different class of symbiotic simulation system.

All systems were developed with a particular application in mind. Scalability was therefore not necessarily a design criterion. The agent-based approach of Low et al. [Low et al., 2005] and the Web service based approach by Madey et al. [Madey et al., 2007, 2006] should scale better than the other architectures because the various agents and Web services can be distributed on several machines, depending on the application. However, the Web service approach is not suitable for small-scale systems as we will further explain in the next section.

12.5.3 Web Services Approach vs. Agent-Based Approach

Agent-based systems and Web service based systems have already been used successfully in various symbiotic simulation applications. For example, an agent-based symbiotic simulation system was used in [Low et al., 2005] to realize a control system for semiconductor manufacturing. DDDAS agents, described in [Kennedy et al., 2007b], represent another agent-based symbiotic simulation system. The WIPER Emergency Management System [Madey et al., 2007] represents an example of a Web service–based approach.

Web services are a popular implementation of the Service-Oriented Architecture (SOA). SOA is an interesting concept that might be suitable for symbiotic simulation systems. The concept of SOA involves service providers, service brokers, and service requestors. Services can range from relatively small functions to entire business processes. They have clearly defined interfaces, are loosely coupled, and communicate by means of XML messages. This guarantees interoperability of services regardless of their underlying implementation.

When using a Web service approach to realize a framework for symbiotic simulation systems, each of the various activities required for symbiotic simulation can be realized as a Web service. A particular symbiotic simulation system can then be composed of various Web services, for example using the Business Process Execution Language (BPEL). If new functionality is required, the framework can easily be extended by adding new Web services.

Web services are built on existing standards such as HTTP and XML. These standards already dominate the Internet and Web services are becoming increasingly important in the

Internet [Gottschalk et al., 2002]. However, they have certain disadvantages which make them unsuitable for a generic framework for symbiotic simulation systems. We give the following two examples to support our argument.

Web services are aimed at large, enterprise systems which have access to the Internet. A particular service, offered by a service provider, can be accessed by a client over a network. Web services require an infrastructure consisting of a service broker and a service provider. This typically includes several servers which makes them unsuitable for mobile and small-scale symbiotic simulation solutions. In contrast, agent-based systems do not need a large server infrastructure and they are also suitable for small-scale systems.

Symbiotic simulation is also used in mobile applications. For example, a symbiotic simulation system is used for UAV path planning in [Lozano et al., 2006; Kamrani and Ayani, 2007]. Typically mobile applications have no, or only limited, access to the Internet. Web services provided by a service provider that is only reachable over the Internet cannot be used. In order to facilitate self-contained symbiotic simulation systems, all required functional components have to be available locally. Even if an Internet connection is available, using a Web service approach might be undesirable from a performance point of view. This depends on the location of the Web services involved and the degree of communication between them and other components of the symbiotic simulation system. Unlike Web services, agents provide local functionality. Therefore, agents can be used to realize mobile and autonomous symbiotic simulation systems.

The need for an agent-based solution is based on the distributed nature of many applications that use symbiotic simulation, the need for mobile solutions, and the broad field of applications ranging from possibly embedded systems to large enterprise systems. Agent-based systems can be distributed and mobile if this is required by the application. In addition, agent-based systems can be used in embedded systems as well as in large enterprise systems. Nevertheless, using an agent-based approach does not rule out the use of a symbiotic simulation system as part of an enterprise application which is realized with Web service technology. If necessary, an agent-based symbiotic simulation system can be packaged and provided as a Web service which can then be used by other services within the enterprise application.

12.5.4 Capability-Centric Solution for Framework Architecture

Each application has its own requirements regarding the symbiotic simulation system. Depending on this and the class of symbiotic simulation system used, some functionality might not be required. Developers need to be able to tailor the symbiotic simulation system according to their needs by using only functionality that is required by the application.

This can be achieved by modularization of functionality. Each of the different activities in the framework has to be realized as a functional component. In this way, the framework can be extended by adding new functional components. When using an agent-based approach to realize the framework for symbiotic simulation systems, all the required functionality has to be provided by agents. In the context of BDI agent systems, capabilities are used to achieve modularization of functionality. Busetta et al. introduced the concept of capabilities and described a capability as "[...] a cluster of plans, beliefs, events, and scoping rules over them" [Busetta et al., 1999]. A similar concept of capabilities was described in Sabater et al. [Sabater et al., 1999] using the term 'module'. The original concept of capabilities was later extended by Braubach et. al, who explained that "[...] the capability concept addresses most of the five fundamental criteria of modularization from [Meyer, 1997]: decomposability, composability, understandability, continuity and protection". The extended capability concept is realized in Jadex[Pokahr et al., 2005], a BDI agent system. We use the concept of capabilities to realize an extensible architecture for an agent-based framework for symbiotic simulation systems.

Depending on the application, the symbiotic simulation system has to provide a certain functionality. During runtime, this functionality is provided by the various agents of the system. However, the functionality of the agents depends on their capabilities. For example, the functionality needed to simulate a what-if scenario can be realized by a capability. Any agent that is equipped with this capability will be able to perform what-if simulations. It is also possible to have several capabilities of the same type. For example, an application consists of several sensors, each represented by a corresponding sensor capability. Agents can be equipped with an arbitrary number of capabilities. It is therefore possible to have a single agent that is equipped with all capabilities required to realize an entire symbiotic simulation system. In this case the symbiotic simulation system is represented by a single agent. On the other hand, the same symbiotic simulation system can be realized by using more than one agent. In this case every agent is equipped with some of the capabilities required by the system. In the extreme case, every agent is equipped with only one capability, i.e., there are as many agents as there are capabilities in the system.

Applications which are inherently distributed need to distribute the functionality of the symbiotic simulation system on several agents. For example, in a manufacturing application, there might be several machines that have to be monitored. It might be necessary to have several agents to monitor all machines. In another application, the what-if simulations have to be carried out on a computing cluster which is located at a different department. This would require an agent to be located on the cluster in order to carry out the simulations. In contrast to inherently distributed applications, in embedded applications it might not be necessary to have several agents. Therefore, the number of agents that are used depends highly on the application. By using capabilities, it is possible to consider application specific requirements regarding the number of agents.

12.5.5 Layers and Associated Capabilities

The high-level architecture of the framework consists of three layers: a perception layer, a process layer, and an actuation layer. Each layer comes with a number of associated capabilities. The layers and their associated capabilities are illustrated in Figure 12.10. Examples on how these capabilities can be composed to a particular symbiotic simulation system can be found in Section 12.6.

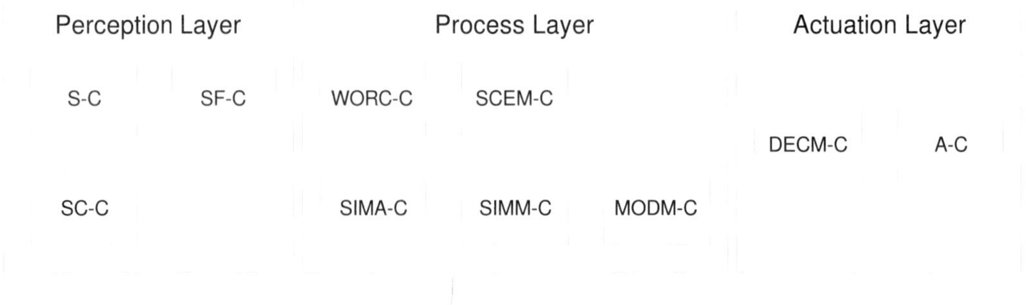

Perception Layer		Process Layer			Actuation Layer	
S-C	SF-C	WORC-C	SCEM-C			
					DECM-C	A-C
SC-C		SIMA-C	SIMM-C	MODM-C		

FIGURE 12.10 Overview of different layers with their associated capabilities.

A symbiotic simulation system is often only a single component within a larger application. It is therefore necessary to allow other application components to interact with the symbiotic simulation system. This can be done by using application specific *adapter capabilities*. They represent the link between the symbiotic simulation system and the rest of the application. Adapter capabilities are not associated with any particular layer of the symbiotic simulation system.

Perception Layer

A symbiotic simulation system needs to be able to observe the system. For this purpose, sensors are used which provide data to the symbiotic simulation system. *Sensor* capabilities (S-C) represent the link between the symbiotic simulation system and the actual sensors. There can be an arbitrary number of sensor capabilities, each providing data about a different aspect of the physical system.

Applications might consist of a large number of sensors, each providing incomplete information about a certain aspect of the physical system. In this case, it makes sense to aggregate all available information about the same aspect and provide complete (or more complete) information about this aspect to other capabilities. The *Sensor Fusion* capability (SF-C) is used for this purpose.

In many applications, sensors are not static but can be steered dynamically. Depending on the application, this includes relocation of sensors and deploying/removing sensors. The *Sensor Control* capability (SC-C) allows other capabilities to control the sensor network of the physical system.

Process Layer

The workflow performed by a symbiotic simulation system is triggered by certain conditions. The *WORkflow Controller* capability (WORC-C) is responsible for evaluating different trigger conditions and invoking corresponding activities. It represents the heart of a symbiotic simulation system, as it coordinates the interplay between other capabilities. Each subsystem in a hybrid symbiotic simulation system has its own WORC-C.

Simulation of various scenarios is an essential concept in symbiotic simulation. The *SCEnario Management* capability (SCEM-C) is responsible for creating meaningful scenarios depending on the application and type of symbiotic simulation system.

Depending on the application, a specific simulator is used. The *SIMulation Management* capability (SIMM-C) represents the link between the framework and the application-specific simulator and allows other capabilities to invoke and terminate simulations. In addition, it is also responsible for updating a simulation run with sensor data, if required.

Simulation runs are analyzed for several purposes. The *SIMulation Analysis* (SIMA-C) capability is responsible for analysing simulation runs a) while they are executed (in-vivo), and b) after they have finished execution (post-mortem). The first analysis activity is used to decide upon termination or replacement of a simulation run. Further, statistic data is collected which can be used later on in the decision making process. The second analysis activity is used to compare various simulation runs and draw conclusions from them.

Several models of the physical system have to be managed in the context of SSMVS and SSADS. The *MODel Management* capability (MODM-C) maintains models of the physical system and provides access to them. By using this capability, other capabilities can request new models or modify existing models. The MODM-C is only required in the context of an SSADS and SSMVS where several models have to be managed. An SSDSS/SSCS only considers the decision model that is provided by the DECM-C.

Actuation Layer

In an SSDSS/SSCS, the scenario creation depends on the decision model. The *DECision Management* capability is responsible for updating the decision model and providing it to other capabilities of the symbiotic simulation system. In an SSCS, possible ways of influencing the physical system depend on available actuators. Therefore, the decision model has to be updated whenever an actuator becomes available or unavailable. In addition, the DECM-C is responsible for propagating decisions to the corresponding actuator capabilities. An SSDSS is not equipped with actuators. Therefore, decisions have to be propagated to an external decision maker rather than to actuators.

Actuators are used to enable a symbiotic simulation system to actively influence the physical system. The *Actuator* capability (A-C) represents the link between the framework and the actual actuator used in the application. An actuator capability is responsible for implementing a decision by exercising corresponding influence on the physical system. In addition it provides information regarding the availability of the actuator to other capabilities. This is important because the DECM-C needs to update the decision model based on available actuators.

12.6 Applications

We apply the proposed framework to an application in the context of semiconductor manufacturing. This example will serve as a proof on concept showcase. In addition, we describe four conceptual showcases, based on the applications described in Section 12.5.2. Each showcase will be concerned with a particular class of symbiotic simulation system.

12.6.1 Proof of Concept Showcase

We have developed a proof of concept application in the context of semiconductor manufacturing where quick decision making is required in order to improve the operational performance of a semiconductor wafer fabrication plant (fab) [Aydt et al., 2008b]. In this showcase, a set of wet bench tools is operated by an SSCS. A wet bench tool is used to clean wafer lots after certain fabrication steps. Each wet bench consists of a number of baths with different chemical liquids. A wafer lot is processed in these liquids strictly according to a certain recipe. A recipe specifies the exact sequence and the exact timing for a wafer lot to be processed in the various baths. Violation of these constraints leads to significantly reduced wafer quality.

Some recipes introduce more particles into the liquids than others. We therefore distinguish between 'clean' and 'dirty' recipes [Gan et al., 2006]. In order not to affect the wafer quality by mixing wafers associated with different types of recipes, wet benches are dedicated to either 'clean' or 'dirty' recipes. A wet bench needs to be reconfigured in order to process wafers that are associated with a different type of recipe. No particular activity is required when changing the configuration from 'clean' to 'dirty'. However, when changing from 'dirty' to 'clean' it is necessary to change the chemical liquids first.

There are typically several wet benches in a fab, equipped with different chemicals in order to be able to process a variety of recipes. The performance of such a wet bench tool set (WBTS) depends on the configuration of the different wet benches and the product mix, i.e., the distribution of recipes in the arriving wafer lots. If the product mix is changing, then it might be necessary to reconfigure several wet benches, i.e., to change their dedication. An SSCS is used to make decisions and change the configuration of the WBTS if necessary. The architecture of the WBTS-SSCS is illustrated in Figure 12.11.

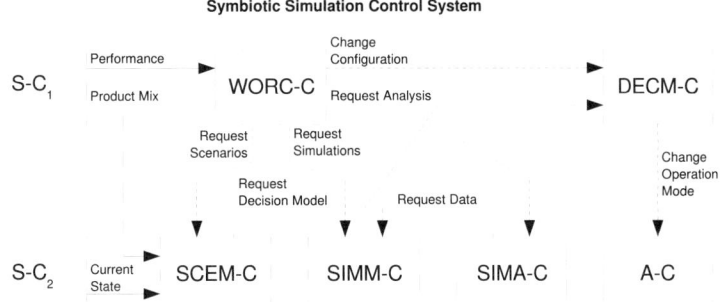

FIGURE 12.11 Architecture for equipment control in semiconductor manufacturing using the proposed framework.

Two sensor capabilities (S-C$_1$ and S-C$_2$) provide information about the performance of the physical system, the product mix, and the current state of the physical system. The performance is observed by the workflow control capability (WORC-C). If a critical condition is observed, the what-if analysis process is triggered. A number of what-if scenarios with alternative wet bench configurations is created by the scenario management capability (SCEM-C). A decision model, provided by the decision management capability (DECM-C), is used for this purpose. The scenarios are simulated and their results analyzed by the simulation management capability (SIMM-C) and simulation analysis capability (SIMA-C), respectively. Based on the simulation results, the best configuration among the various alternatives is determined. The configuration of the WBTS can be changed by instructing the DECM-C to change the operation mode of particular wet benches by using the corresponding actuator, represented by the actuator capability (A-C).

12.6.2 Conceptual Showcases

We intend to provide a generic framework for symbiotic simulation systems. In addition to the proof of concept showcase, discussed in the previous section, we discuss how our framework could be applied in the context of various other applications.

SSDSS Example

The symbiotic simulation system described in [Kamrani and Ayani, 2007] is used for adaptive path planning of UAVs. Information about the location of the UAV and the target is provided by corresponding sensors. Alternative what-if scenarios are created based on a set of alternative paths and simulated concurrently. Each scenario is initialized with the current location of the UAV and a priori estimations about the target. The simulation results are compared in order to determine the best path. Detailed algorithms for path planning and what-if simulations are given in [Kamrani and Ayani, 2007]. As part of the path planning algorithm, the best path is determined using the results from the various what-if simulations. Although the authors mention that it is possible to apply the results directly to the physical system which would indicate a control system, this is not explained. Instead, the results of the what-if simulations are used as input for a decision maker. Therefore, we consider the symbiotic simulation system for UAV path planning to be a decision support system rather than a control system. An illustration of the architecture for this application can be found in the technical report on symbiotic simulation based decision support in [Lozano et al., 2006].

Applying the proposed framework results in a symbiotic simulation system, consisting of two subsystems. The first subsystem represents an SSDSS which is responsible for proposing decisions to the external decision maker which is represented by the second subsystem. The architecture is illustrated in Figure 12.12.

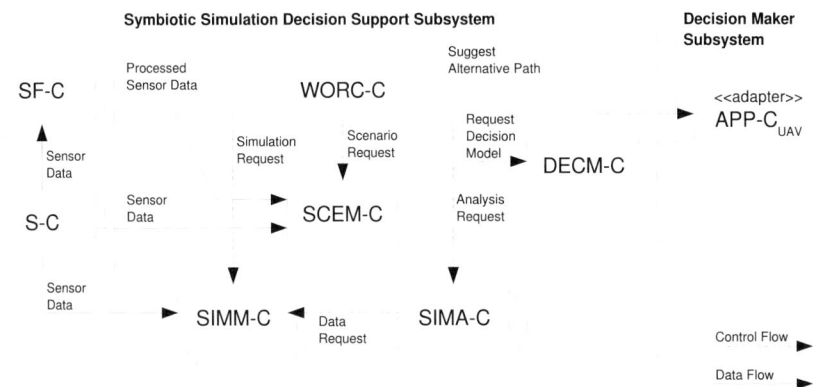

FIGURE 12.12 Architecture for symbiotic simulation decision support in UAV path planning using the proposed framework.

Sensor capabilities (S-C) and sensor fusion capabilities (SF-C) provide data which is important in the context of the application (e.g., geography of the road network, estimated location of the target). The workflow control capability (WORC-C) recurrently triggers a what-if analysis process. A set of alternative UAV paths is created by the scenario management capability SCEM-C, based on the decision model which is provided by the decision model management capability DECM-C. In this particular application, decision making is concerned with the future directions of the UAV, i.e., with the path of the UAV. The decision model describes how possible alternative paths of the UAV can be created. The what-if scenarios also incorporate the current location of the UAV and the estimated location of the target. Each scenario is simulated and analyzed by a simulation management capability (SIMM-C) and a simulation analysis capability capability (SIMA-C), respectively. The best alternative path, which can be determined once all simulation results are available, is proposed to the external decision maker. An adapter capability (APP-C_{UAV}) is used to link the application specific decision maker into the symbiotic simulation system. It is up to this decision maker to consider the alternative path, suggested by the SSDSS.

SSCS Example

The symbiotic simulation system described in [Low et al., 2005] is used to optimize outsourcing decisions by considering the operating cost for a semiconductor back-end factory. To avoid late deliveries and consequent penalties, wafer lots are marked for out-sourcing under certain conditions. Marking is switched on/off if the number of lots in a queue exceeds or falls below a specified upper or lower limit, respectively. An optimal combination of upper and lower queue sizes yield minimum operating cost. In this application, symbiotic simulation is used to find an optimal value for these queue sizes. Sensors are used to retrieve the current state of the factory. This information is used to initialize the what-if simulations. In addition, each what-if scenario uses a different set of upper and lower limits. The results of the what-if analysis can be directly used in the physical system to make decision regarding out-sourcing. In the original architecture, a control agent is responsible for making the necessary modifications. Therefore, applying the proposed framework results in using an SSCS which is responsible for directly implementing decisions regarding the upper and lower queue sizes. The architecture is illustrated in Figure 12.13

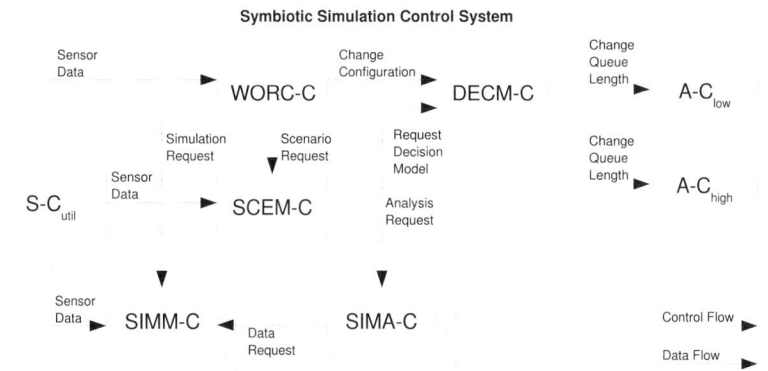

FIGURE 12.13 Architecture for symbiotic simulation of semiconductor backend operation using the proposed framework.

A sensor capability (S-C_{util}) provides data on machine utilization. If the utilization exceeds 80%, the workflow control capability WORC-C triggers a what-if analysis. During the analysis process, a number of what-if scenarios is created by the scenario management capability SCEM-C based on the decision model provided by the decision management capability DECM-C. The simulation management capability SIMM-C is responsible for simulating the various what-if scenarios. These simulations are analyzed by the simulation analysis capability SIMA-C and results are used by the WORC-C to decide upon modifying the settings of the physical system. The new configuration, proposed by the WORC-C, is checked by the DECM-C before being implemented. The actuator capabilities (A-C_{low} and A-C_{high}) are responsible for changing the lower queue size of a machine and the upper queue size, respectively.

SSADS Example

The symbiotic simulation system described in [Kennedy et al., 2007b; Kennedy and Theodoropoulos, 2006] is part of the AIMSS architecture and is used to detect anomalies between simulated and actual behavior in social science applications. The purpose of the simulation is to make predictions in a housing scenario. The model used allows the space to be divided into different regions (e.g., expensive/inexpensive, high/low crime rate). Households, which are represented by agents, will move to different regions depending on affordability, crime level, and other factors. A DDDAS agent is used to interpret and compare simulations with the real world. The agent may suggest model revision if the simulation does not resemble the real world accurately enough. A case study which uses this kind of agent is described in [Kennedy et al., 2007a]. The agent architecture consists of sensors which can be redirected, a description of the world, and a control component which compares measured and simulated data. In addition, the agent may maintain its own model of the world.

Applying the proposed framework results in using an SSADS which is responsible for detecting anomalies between the simulated and actual behavior of the real world. The architecture is illustrated in Figure 12.14.

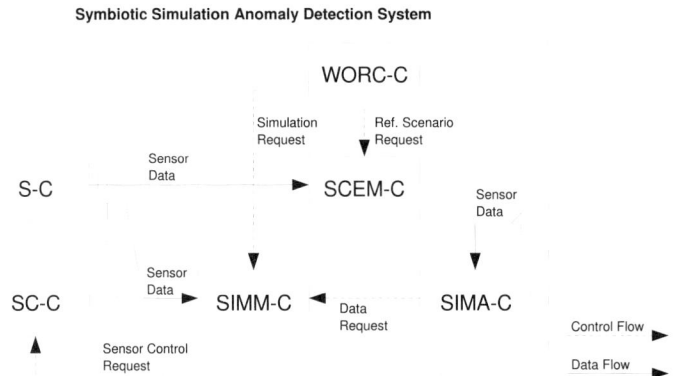

FIGURE 12.14 Architecture for symbiotic simulation in AIMSS assistant agents using the proposed framework.

Sensor capabilities (S-C) are used in the same way as the sensor components in the original architecture. The control component is realized by the workflow control capability WORC-C, the scenario creation management capability SCEM-C, the simulation management capability SIMM-C, and the simulation analysis capability SIMA-C. At the start, the reference scenario, consisting of the reference model and the current state of the physical system, is determined by the SCEM-C and used by the SIMM-C to perform a paced simulation. Measured values and simulated values are continuously compared by the SIMA-C. Detected anomalies may cause a revision of the model, initiated by the WORC-C. In the description of the original architecture, it is indicated that sensors are redirected depending on the predicted state of the simulation. Therefore, sensor redirection is initiated by the SIMA-C and carried out by the sensor control capability SC-C.

SSMVS and SSFS Example

The symbiotic simulation system described in [Madey et al., 2007] is used to classify anomalies by testing various hypotheses. Communication patterns of millions of cell phone users are monitored to recognise anomalies. Sensor data used by WIPER includes activity data (e.g., call initiator and recipient) and spatial location data provided by mobile phone companies, as well as the 30 second CDR (Call Detail Record) tags which identify the closest cell tower to the phone [Madey et al., 2006]. Anomalies are classified by testing various hypotheses about the nature of the anomaly. The one that describes the anomaly best is used to predict its further evolution.

Applying the proposed framework results in using a hybrid symbiotic simulation system, consisting of four subsystems. These subsystems are used for anomaly detection, model validation, forecasting, and decision support. However, symbiotic simulation is only used for model validation and forecasting in this application. Therefore, an SSMVS and SSFS are used in addition to two application specific subsystems which are responsible for anomaly detection and decision support. The architecture is illustrated in Figure 12.15. In this architecture, each symbiotic simulation subsystem has its own SCEM-C and SIMM-C. Depending on the particular implementation of the application it might be possible to share the same capabilities among the two symbiotic simulation subsystems. However, only the model management capability and the sensor capabilities are shared by both symbiotic simulation subsystems here.

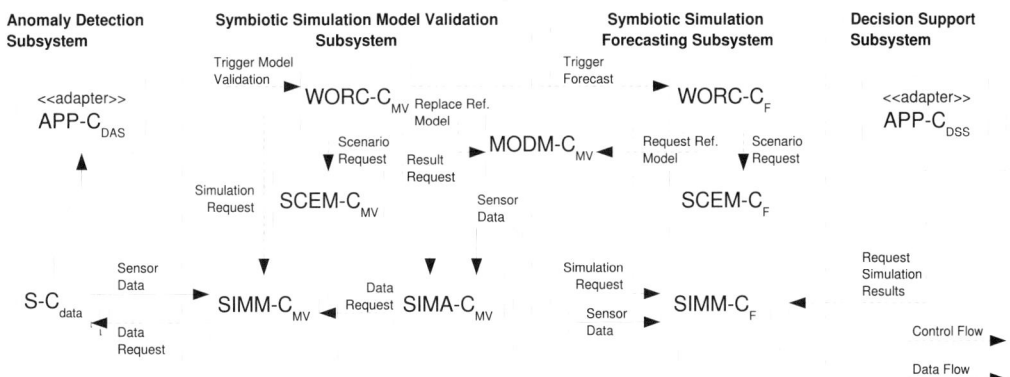

FIGURE 12.15 WIPER architecture using the proposed framework.

Sensor capabilities (S-C$_{data}$) provide data about cell phone activity. Historical data is also provided by these capabilities. The application-specific anomaly detection and decision support subsystem are represented by the adapter capabilities APP-C$_{DAS}$ and APP-C$_{DSS}$, respectively. These adapter capabilities are used to link application specific components with the symbiotic simulation system.

Once an anomaly is detected, the workflow control capability WORC-C$_{MV}$ of the model validation subsystem is triggered. A number of scenarios are created by the scenario management capability SCEM-C$_{MV}$. Each scenario represents a hypothesis about the anomaly and is simulated by the corresponding simulation management capability SIMM-C$_{MV}$. Real-time data, provided by the sensor network, are used to dynamically update the various simulation runs. The simulation analysis capability SIMA-C$_{MV}$ dynamically validates simulation runs against sensor data and decides upon termination of a particular simulation. The reference model, which is managed by the model management capability MODM-C$_{MV}$, is updated once an explanation is found that describes the current behavior of the system.

This reference model can be used by the forecasting subsystem to predict how the anomaly is likely to evolve in future. The subsystem itself consists of a workflow controller capability (WORC-C$_F$), a scenario creation management capability (SCEM-C$_F$), and corresponding simulation management capability (SIMM-C$_F$). The scenarios used for forecasting purposes are all based on the reference model. The actual decision making, based on these forecasts, is performed by the application specific decision support subsystem. For this purpose, forecasts of the anomaly are provided to the decision support subsystem on request.

12.7 Conclusions

In this chapter we have analyzed different classes of symbiotic simulation systems. Although the various classes have a distinct functionality, they are composed of only a few common activities. These activities are performed by a number of distinct components, which are part of our framework for symbiotic simulation systems. In addition, this framework provides generic solutions for each of these components. We have proposed an architecture for an agent-based framework which uses the concept of capabilities to realize the various functional components of the framework. The framework architecture was designed with requirements regarding applicability, extensibility, and scalability in mind. We show the applicability of the framework in the context of various application domains by using practical examples. Extensibility and scalability of the framework is achieved by using capabilities, which makes it possible to add new functional components and facilitates tailoring of symbiotic simulation systems for a specific application.

A reference implementation of the framework has been developed using Jadex as the agent toolkit. The reference framework has been used to realize an SSCS application in the context of semiconductor manufacturing to control a WBTS. Most of the different capabilities discussed in Section 12.5.5 have been implemented and used for this application. The only exceptions are the sensor fusion and control capabilities (SF-C and SC-C) and the model management capability (MODM-C) which is not required in the context of an SSCS.

Symbiotic simulation can be used in a large variety of applications. We envision a standard for symbiotic simulation that provides recommended implementations and methods which can easily be used by other researchers and developers for developing their own symbiotic simulation systems. By providing a framework for these different classes of systems, we believe we have made a first step toward a standard on symbiotic simulation.

12.8 Future Work

An architecture for a generic framework has been introduced in this chapter. The framework provides a number of functional components which are frequently used in a symbiotic simulation system. In future work, we will investigate generic solutions and algorithms for the various functional components of the framework. In this context, efficient scenario creation is an important research issue as it is not feasible to explore all possible scenarios unless the application is trivial. Therefore, solutions are needed which are efficient and generic enough to be used in the framework. In addition, we will investigate how our framework can be used with commercial off-the-shelf simulation packages. Many applications use already existing simulation packages which cannot easily be replaced. An important issue is therefore the definition of an interface which is required in order to use a simulation package in a symbiotic simulation system. This should make it easier for developers to plan strategies on how to integrate existing simulation packages into a symbiotic simulation system.

References

V. Akcelik, J. Bielak, G. Biros, I. Epanomeritakis, O. Ghattas, L. Kallivokas, and E. Kim. A framework for online inversion-based 3D site characterization. In *Proceedings of the International Conference on Computational Science*, pages 717–724, 2004.

H. Aydt, S. Turner, W. Cai, and M. Low. Symbiotic simulation systems: An extended definition motivated by symbiosis in biology. In *Proceedings of the 22nd Workshop on Principles of Advanced and Distributed Simulation*, pages 109–116, 2008a.

H. Aydt, S. Turner, W. Cai, M. Low, P. Lendermann, and B. Gan. Symbiotic simulation control in semiconductor manufacturing. In *Proceedings of the International Conference on Computational Science*, 2008b.

F. Bellifemine, A. Poggi, and G. Rimassa. JADE – a FIPA-compliant agent framework.

In *Proceedings of the Practical Applications of Intelligent Agents and Multi-agent Systems*, pages 97–108, 1999.

P. Busetta, N. Howden, R. Rönnquist, and A. Hodgson. Structuring BDI agents in functional clusters. In *Proceedings of the 6th International Workshop on Agent Theories Architectures and Languages*, pages 277–289, 1999.

J. Cortial, C. Farhat, L. Guibas, and M. Rajashekhar. Compressed sensing and time-parallel reduced-order modeling for structural health monitoring using a DDDAS. In *Proceedings of the International Conference on Computational Science*, pages 1171–1179, 2007.

W. Davis. On-line simulation: Need and evolving research requirements. In J. Banks, editor, *Handbook of Simulation*, pages 465–516. Wiley-Interscience, New York, 1998.

A. Douglas. *Symbiotic Interactions*. Oxford University Press, Oxford, 1994.

C. Douglas, J. Beezley, J. Coen, D. Li, W. Li, A. Mandel, J. Mandel, G. Qin, and A. Vodacek. Demonstrating the validity of a wildfire DDDAS. In *Proceedings of the International Conference on Computational Science*, pages 522–529, 2006.

R. Fujimoto. *Parallel and Distributed Simulation Systems*. Wiley Series on Parallel and Distributed Computing. John Wiley & Sons, Inc., New York, NY, USA, 2000.

R. Fujimoto, D. Lunceford, E. Page, and A. U. (editors). Grand challenges for modeling and simulation: Dagstuhl report. Technical Report 350, Schloss Dagstuhl. Seminar No 02351, August 2002.

R. Fujimoto, R. Guensler, M. Hunter, H. Kim, J. Lee, J. Leonard, M. Palekar, K. Schwan, and B. Seshasayee. Dynamic data driven application simulation of surface transportation systems. In *Proceedings of the International Conference on Computational Science*, pages 425–432, 2006.

B. Gan, P. Lendermann, K. Quek, B. van der Heijden, C. Chin, and C. Koh. Simulation analysis on the impact of furnace batch size increase in a deposition loop. In *Proceedings of the Winter Simulation Conference*, pages 1821–1828, 2006.

K. Gottschalk, S. Graham, H. Kreger, and J. Snell. Introduction to web services architecture. In *IBM Systems Journal*, volume 41, pages 170–177, 2002.

S. Huang, W. Cai, S. Turner, W. Hsu, S. Zhou, M. Low, R. Fujimoto, and R. Ayani. A generic symbiotic simulation framework. In *Proceedings of the 20th Workshop on Principles of Advanced and Distributed Simulation*, pages 131–131, 2006.

F. Kamrani. Using on-line simulation in UAV path planning. Licentiate Thesis in Electronics and Computer Systems, KTH, Stockholm, Sweden, 2007.

F. Kamrani and R. Ayani. Using on-line simulation for adaptive path planning of UAVs. In *Proceedings of the 11th IEEE International Symposium on Distributed Simulation and Real-time Applications*, pages 167–174, Chania, Greece, October 2007.

C. Kennedy and G. Theodoropoulos. Intelligent management of data driven simulations to support model building in the social sciences. In *Proceedings of the International Conference on Computational Science*, pages 562–569, 2006.

C. Kennedy, G. Theodoropoulos, E. Ferrari, P. Lee, and C. Skelcher. Towards an automated approach to dynamic interpretation of simulations. In *Proceedings of the First Asia International Conference on Modelling & Simulation*, 2007a.

C. Kennedy, G. Theodoropoulos, V. Sorge, E. Ferrari, P. Lee, and C. Skelcher. AIMSS: An architecture for data driven simulations in the social sciences. In *Proceedings of the International Conference on Computational Science*, pages 1098–1105, 2007b.

D. Knight, Q. Ma, T. Rossman, and Y. Jaluria. Evaluation of fluid-thermal systems by dynamic data driven application systems - part ii. In *Proceedings of the International Conference on Computational Science*, pages 1189–1196, 2007.

M. Low, K. Lye, P. Lendermann, S. Turner, R. Chim, and S. Leo. An agent-based approach for managing symbiotic simulation of semiconductor assembly and test operation. In *Proceedings of the 4th International Joint Conference on Autonomous Agents and Multiagent Systems*, pages 85–92, New York, NY, USA, 2005. ACM Press. ISBN 1-59593-093-0. doi: http://doi.acm.org/10.1145/1082473.1082809.

M. Low, S. Turner, D. Ling, H. Peng, L. Chai, P. Lendermann, and S. Buckley. Symbiotic simulation for business process re-engineering in high-tech manufacturing and service networks. In *Proceedings of the Winter Simulation Conference*, 2007.

M. Lozano, F. Kamrani, and F. Moradi. Symbiotic simulation (S2) based decision support. Methodology report FOI-R-1935-SE, FOI - Swedish Defence Research Agency, February 2006.

G. Madey, G. Szabó, and A. Barabási. WIPER: The integrated wireless phone based emergency response system. In *Proceedings of the International Conference on Computational Science*, pages 417–424, 2006.

G. Madey, A. Barabási, N. Chawla, M. Gonzalez, D. Hachen, B. Lantz, A. Pawling, T. Schoenharl, G. Szabó, P. Wang, and P. Yan. Enhanced situational awareness: Application of DDDAS concepts to emergency and disaster management. In *Proceedings of the International Conference on Computational Science*, pages 1090–1097, 2007.

K. Mahinthakumar, G. von Laszewski, S. Ranjithan, D. Brill, J. Uber, K. Harrison, S. Sreepathi, and E. Zechman. An adaptive cyberinfrastructure for threat management in urban water distribution systems. In *Proceedings of the International Conference on Computational Science*, pages 401–408, 2006.

A. Majumdar, A. Birnbaum, D. Choi, A. T., S. Warfield, K. Baldridge, and P. Krysl. A dynamic data driven grid system for intra-operative image guided neurosurgery. In *Proceedings of the International Conference on Computational Science*, pages 672–679, 2005.

J. Mandel, M. Chen, L. Franca, C. Johns, A. Puhalskii, J. Coen, C. Douglas, R. Kremens, A. Vodacek, and W. Zhao. A note on dynamic data driven wildfire modeling. In *Proceedings of the International Conference on Computational Science*, pages 725–731, 2004.

J. Mandel, L. Bennethum, M. Chen, J. Coen, C. Douglas, L. Franca, C. Johns, M. Kim, A. Knyazev, R. Kremens, V. Kulkarni, G. Qin, A. Vodacek, J. Wu, W. Zhao, and A. Zornes. Towards a dynamic data driven application system for wildfire simulation. In *Proceedings of the International Conference on Computational Science*, pages 632–639, 2005.

J. Mandel, J. Beezley, L. Bennethum, S. Chakraborty, J. Coen, C. Douglas, J. Hatcher, M. Kim, and A. Vodacek. A dynamic data driven wildland fire model. In *Proceedings of the International Conference on Computational Science*, pages 1042–1049, 2007.

J. McCalley, V. Honavar, S. Ryan, W. Meeker, R. Roberts, D. Qiao, and Y. Li. Auto-steered information-decision processes for electric system asset management. In *Proceedings of the International Conference on Computational Science*, pages 440–447, 2006.

J. McCalley, V. Honavar, S. Ryan, W. Meeker, D. Qiao, R. Roberts, Y. Li, J. Pathak, M. Ye, and Y. Hong. Integrated decision algorithms for auto-steered electric transmission system asset management. In *Proceedings of the International Conference on Computational Science*, pages 1066–1073, 2007.

B. Meyer. *Object-oriented software construction (2nd ed.)*. Prentice-Hall, Inc., Upper Saddle River, NJ, USA, 1997. ISBN 0-13-629155-4.

J. Michopoulos, P. Tsompanopoulou, E. Houstis, J. Rice, C. Farhat, M. Lesoinne, and F. Lechenault. DDEMA: A data driven environment for multiphysics applications. In *Proceedings of the International Conference on Computational Science*, pages 309–318, 2003.

J. Michopoulos, P. Tsompanopoulou, E. Houstis, and A. Joshi. Agent-based simulation of data-driven fire propagation dynamics. In *Proceedings of the International Conference on Computational Science*, pages 732–739, 2004.

National Science Foundation. DDDAS: Dynamic data driven applications systems. Program Solicitation 05-570, http://www.nsf.gov/pubs/2005/nsf05570/nsf05570.htm. Accessed June 2008.

K. Perumalla, R. Fujimoto, T. McLean, and G. Riley. Experiences applying parallel and interoperable network simulation techniques in on-line simulations of military networks. In *Proceedings of the 16th Workshop on Parallel and Distributed Simulation*, pages 88–95, 2002.

A. Pokahr, L. Braubach, and W. Lamersdorf. Jadex: A BDI reasoning engine. In R. Bordini, M. Dastani, J. Dix, and A. Seghrouchni, editors, *Multi-Agent Programming*, pages 149–174. Springer Science+Business Media Inc., USA, 2005.

J. Sabater, C. Sierra, S. Parsons, and N. Jennings. Using multi-context systems to engineer executable agents. In *Proceedings of the 6th International Workshop on Agent Theories Architectures and Languages*, pages 260–276, 1999.

D. Wilkinson. At cross purposes. *Nature*, 412(6846):485, 2001.

L. Xue, Z. Hao, D. Li, and X. Liu. Dongting lake floodwater diversion and storage modeling and control architecture based on the next generation network. In *Proceedings of the International Conference on Computational Science*, pages 841–848, 2007.

13

Agent-Based Modeling of Stem Cells

13.1 Introduction	389
Overview	
13.2 The Biological Domain – HSC Biology	390
The Hematopoietic System • The Hematopoiesis Control Mechanisms	
13.3 Stem Cell Modeling	392
Experimental Limitations • What a Model Can Be Useful For	
13.4 Drawbacks of Existing Models and Why Agents .	394
13.5 Overview of Our Agent Modeling Framework	395
Framework Components • Interfaces between the Framework Components • Behavior of the Framework Components • Extending Our Agent Modeling Framework	
13.6 Agentifying Existing Approaches	400
A Cellular Automata Approach to Modeling Stem Cells • Discussion about the Cellular Automata Approach • Re-formulation Using an Agent-Based Approach • Operation • Roeder-Loeffler Model of Self-Organization	
13.7 From Agent Model to Simulation	407
MASON • Implementation of CELL in MASON • Implementation of Roeder-Loeffler Model in MASON	
13.8 Discussion	412
Concluding Remarks	
Acknowledgments	415
References	415

Mark d'Inverno
Goldsmiths, University of London

Paul Howells
University of Westminster

Sara Montagna
Università di Bologna

Ingo Roeder
University of Leipzig

Rob Saunders
University of Sydney

13.1 Introduction

The multi-agent systems approach has become recognized as a useful approach for modeling and simulating biological complex systems. In this chapter we provide an example of such an approach, which concerns the modeling and simulation of the Hematopoietic Stem Cell (HSC) system in adults. We are specifically interested in how local cell interactions give rise to well understood properties of systems of stem cells, such as the ability to maintain their own population and to maintain a population of fully differentiated functional cells. There is a need to establish key cell mechanisms that can produce self-regulating behavior of HSC systems using different theoretical techniques. It is our belief that modeling the behavior of HSCs in the adult human body as an agent-based system is the most appropriate way of understanding these mechanisms and the consequent process of self-organization.

In recent years there has been a growing debate about how stem cells behave in the human body; whether the fate of stem cells is pre-determined or stochastic, and whether the fate of cells relies on their internal state, or on extra-cellular micro-environmental factors. However, current experimental limitations mean that stem cells cannot be tracked in the adult human body. There is no way of "observing" micro-level behavior. Models and simulations have a crucial role therefore in explaining the relationship of micro-behavior to macro-behavior and it now seems that the importance of computational modeling and simulation for understanding stem cells is beginning to be realized in many wet-labs. There have been several attempts to build formal models of these theories, so that predictions can be made about how and why stem cells behave, both individually or collectively. In this chapter we propose an agent based model which describes at the same time the intracellular behavior of the cell (i.e., intra-cellular networks) and the cellular level where all the systemic interactions are developed. This enables us to build a multi-level model.

13.1.1 Overview

In Section 13.2 we provide an overview of the background biology for this investigation. We then discuss the current experimental limitations and how these motivate the significance for using formal, computational models. We then describe the main approaches of formalizing conceptual models of stem cells, which can be classified as single cell vs population models, stochastic vs deterministic models. By discussing these limitations we are able to clearly promote the advantages of the agent approach. Not only in understanding how cell-cell interactions give rise to overall system behavior but also because they provide an appropriate level of abstraction for collaborating with biologists.

Then we describe out agent modeling framework and how we have used it to take existing formal modeling approaches and "agentify" them to address certain weaknesses in the model. Finally, we will provide details of the simulation and discuss our results to date and plans for future work.

This chapter is intended for an interdisciplinary audience and so we include introductory material about the background biology as well as about why the agent approach is most suitable in this domain.

13.2 The Biological Domain – HSC Biology

Due to the natural life-cycle of cells, a tissue has to be self-renewing: in order to guarantee a continuous replacement of cells that die, naturally or after injury, a cellular population must contain cells that are able to proliferate, generating a mixture of progeny. The population of daughter cells then includes cells that remain undifferentiated, i.e., self-renew the identity of their parent, and cells that differentiate, i.e., change their properties. Cells with the potential for both self-renewal and differentiation are called *stem cells*.

Even though no commonly agreed phenotypic or molecular definition of a stem cell exists, there is a general consent on their functional capabilities. A stem cell is an undifferentiated cell that is able to (i) proliferate, (ii) self-renew, i.e., able to go through numerous cycles of cell division while maintaining the undifferentiated state, (iii) differentiate, i.e., able to produce a progeny of distinct cell types, (iv) recover the tissue after injury or disease. Those capabilities and their importance for a functional definition of tissue stem cells has extensively been discussed by [Potten and Loeffler, 1990] and [Loeffler and Roeder, 2002].

It is still not clear whether stem cell types for different tissues are strictly committed to these tissues or whether the can flexibly adopt features of other tissues under certain circumstances. However, there is increasing experimental evidence for the flexibility and the reversibility of stem cell function and phenotype [Blau and Blakely, 1999; Quesenberry et al., 2001].

Furthermore, it is evident (see citation given above) that the emergent behavior of the stem cell system depends on multiple factors, such as their own actual cellular state, their interaction with other cells or environmental cues.

13.2.1 The Hematopoietic System

Hematopoietis is the entire process of production and maintenance of all types of blood cells which exhibit very different functions, such as transport of oxygen, production of antibodies to fight infection and blood clotting. The life-span of blood cells, however, is limited so that they must be continuously produced throughout the life of the animal. The hematopoietic stem cells (HSCs), which can be found mostly in the bone marrow, are responsible for the constant replacement of blood cells lost to normal turnover processes as well as to illness or trauma.

HSCs are able to generate every lineage found in the hematopoietic system through a successive series of intermediate progenitors. An exhaustive representation of the HSCs lineage tree is published in the *Kyoto Encyclopedia of Genes and Genomes*, see [Kanehisa and Goto, 2000]. Although the process of lineage specification is continuous, it is possible to identify some main phases, the first of which includes common lymphoid progenitors (CLPs), which can generate only B, T, NK cells, and common myeloid progenitors (CMPs) which can generate only red cells, platelets, granulocytes, and monocytes. In the following phases there are more mature progenitors that are further restricted in the number and type of lineages that they can generate. The last phase is the terminally differentiated cell that cannot divide and that undergoes apoptosis, i.e., programmed cell death, after a period of time ranging from hours to years.

In this way, during homoeostasis, a proportion of stem cells is expected to balance the fundamental processes of self-renewal, differentiation, apoptosis and quiescence to maintain (i) a constant flow of short-lived progenitors that can generate enough cells to replace those that are constantly lost during normal turnover and (ii) a constant number of primitive cells to sustain hematopoiesis. Under homoeostatic conditions this results in a number of tissue stem cells, as well as blood cells, that fluctate around a relatively constant average value. During times of physiologic stress, the entire hematopoietic system is then able to provide the mature cells required to fight infections or to replenish cells lost during a hemorrhage. When the stress is resolved the number of cells returns again to normal levels.

A variety of homoeostatic mechanisms allow blood cell production to maintain homoeostasis or to respond quickly to stress such as bleeding or infection; the next section provides a brief account of some of these mechanisms.

13.2.2 The Hematopoiesis Control Mechanisms

The process of hematopoiesis involves a complex interplay between the intrinsic genetic processes of HSCs, progenitors and mature blood cells, that can be either deterministic or stochastic, and their microenvironment dynamics, responsible for cell to cell interactions. As a consequence of these internal and external processes, these cells remain quiescent, differentiate, self-renew, or undergo apoptosis.

Cell intrinsic regulatory processes are cell autonomous and are determined by the cell's developmental state, which translates into specific levels of genes and protein expression. With these processes we refer to intracellular pathways and gene regulatory networks. Depending on the number of molecules involved they appear to be deterministic or stochastic. The intracellular events are in fact determined by the interactions between molecules: as the number of reacting molecules increases the probability of their interaction increases until a threshold over which the process can be considered deterministic.

The role of the microenvironmental dynamics is also crucial. The specific microenvironment of stem cells has been historically called "stem-cell niche" [Schofield, 1978]. A niche is composed of a dynamically changing chemical environment, containing a range of different molecules. Cells maintain this environment by secreting and absorbing molecules, in order to send and receive signals. The information passed in these signals influences the autonomous behavior of the cells, e.g., *growth factors* stimulate cells to divide and differentiate to produce more terminally differentiated cells.

All of the genetic and environmental mechanisms that govern blood cells production operate by affecting the intracellular level dynamics, resulting in different cell actions and behaviors. Under normal conditions, the majority of HSCs and many progenitors are quiescent. In the event of a physiologic stress quiescent progenitors and HSCs are stimulated by a variety of growth factors to proliferate and differentiate into mature white cells, red cells, and platelets. When the bleeding or infection ceases and the demand for blood cells returns to normal, the antiapoptotic and proliferative stimuli decreases and the kinetics of hematopoiesis return to baseline levels [Wichmann and Loeffler, 1985; Smith, 2003; Attar and Scadden, 2004; Wilson and Trumpp, 2006].

13.3 Stem Cell Modeling

To understand the behavior of hematopoietic systems, it is not enough to study the single cell behavior, but we need a complete view of the system, with the overall mechanisms that govern the internal dynamics of the cells and their communications through the microenvironment.

13.3.1 Experimental Limitations

We believe that recent experimental evidence makes it clear that it is increasingly necessary to use formal, computational models to investigate the nature of stem cell systems rather than stem cells in isolation. This is for several key reasons. First, adult stem cells cannot be easily isolated, indeed it may be that it is only by looking at their behavior in a system, not isolated, can we tell what kind of cell we were originally looking at. Second, any attempt to determine the properties of stem cells requires a functional test, which itself feeds back on the cells and changes their properties [Potten and Loeffler, 1990; Loeffler and Roeder, 2002]. Third, even if we were able to track the behavior of a cell in the body, it would only tell us about one of the possible behaviors of the original cell; it tells us nothing about the potentially considerably larger array of behaviors that may have been possible if the environment and the chance elements had been different. Fourth, by removing a cell from its original and natural habitat the new environmental conditions will influence future behavior and lead to misleading results [Theise, 2005]. Fifth, it is the totality of the stem cells as a system in the human body that is important. A key quality of the system is its ability to maintain exactly the right production of cells in all manner of different situations.

In response, therefore, we have developed a formal model that reflects many of the key experimental and recent theoretical developments in stem cell research. Using techniques from multi-agent systems, we are currently building a complex adaptive system to simulate stem cell systems in order to provide a testbed from which to be able to investigate key properties of them in general and to formulate new experiments to identify the underlying physiological mechanisms of tissue maintenance and repair.

We next outline some of the techniques that have been employed up till now.

13.3.2 What a Model Can Be Useful For

The mathematical modeling, conceptualization and simulation of stem cell behavior is beginning to receive a substantial amount of interest from an increasing number of researchers. As has been pointed out by others, predictive models of stem cell systems could provide important new understandings of the self-regulating mechanisms that result in the global properties of stem cells.

There has been a growing debate about how stem cells behave in the human body; whether the fate of stem cells is pre-determined [Nicola and Johnson, 1982] or stochastic [Ogawa, 1999; Thornley et al., 2003], and whether the fate of cells relies on their internal state [Novak and Stewart, 1991], or on extra-cellular micro-environmental factors [Trentin, 1970]. There have been several attempts to build formal models of these theories, so that predictions can be made about how and why stem cells behave as they do, either individually or collectively. Reviews of these formal approaches can be found in recent publications [Sowmya and Zandstra, 2003; Loeffler and Roeder, 2004; Roeder, 2006] and we do not propose to review these models in this chapter.

It is worth noting, however, that the first model we know of was published in 1964 ([Till et al., 1964]) and that there has been surprisingly little work in this field until the last couple of years. There has been a noticeable climate change in this respect, and there is now a growing awareness of the need to use mathematical modeling and computer simulation to understand the processes and behaviors of stem cells in the body.

We summarize what we see are the key reasons for the systematic development of models and simulations to consider hypotheses about the nature and behavior of stem cells.

1. The size and complexity of stem cell systems mean that without simulation, it is practically impossible to consider the whole system. Simulations provide an important tool for understanding the global behavior of complex systems.
2. Clearly any model, and resulting simulation, of stem cells will necessarily incur massive simplifications and abstractions about the machinations of the human body. It is our belief, however, that theoretical simplifications are often key to understanding fundamental properties of natural systems.
3. It is the *potential* of cells to behave in lots of different ways which makes them more or less stem like. It may be that stem cell is a notion rather than a concrete entity and refers to the wide-ranging set of potential behaviors that a cell might have that are influenced by internal, environmental, and stochastic processes. Simulations provide a way of determining which behaviors are essential to stem cells and which are incidental in systems that have been studied in the laboratory.
4. When you consider experimental evidence you have seen only one behavior. This behavior may have been one of many, and it is the potential for cells to behave in certain ways that might be key to defining them. Modeling and simulation is a much more effective device for understanding "behavioral potential" than looking at completed chains of events in the lab.
5. Simulation is cheap.

13.4 Drawbacks of Existing Models and Why Agents

As we have pointed out in the previous sections, we believe that studying the behavior of hematopoietic system needs the comprehension of both intracellular processes and interacting events among cells that are mediated by the cells dynamic microenvironment. In order to understand the evolution of the cellular population, we should look at the system at least, at two levels: the intracellular level and the extracellular one. We therefore require a *multi-level model*, which is the one that best fits with the biological problem we are proposing [Uhrmacher et al., 2005].

Most of the existing approaches to modeling and simulating biological systems can capture details at only one level. The most common approach concerns differential equations, in different forms: ordinary, partially or stochastic differential equations (ODE, PDE, SDE). A huge amount of literature, especially in the systems biology discipline, describes ODE models of metabolic pathways, signalling pathways and gene regulatory networks [Conrad and Tyson, 2006]. The results obtained at this level with this approach gave good insights on biological systems behavior. When the system shows non-deterministic dynamics, then the introduction of stochastic approaches such as chemical master equations (CME), stochastic simulation algorithms (SSA), tau leaping, chemical Langevin equations and so on [Gillespie, 2008] allow the study of stochastic events. Alternatively a lot of other different approaches have been applied: random boolean networks [Kauffman, 1993], Bayesian networks and more recently, computational models such as different process algebras (an interesting example is the work that has been done with stochastic *pi*-calculus [Regev et al., 2001] and its extensions [Dematte et al., 2008]), or petri-nets [Talcott, 2008; Heiner et al., 2008]. An excellent review is given by [De Jong, 2002].

All of these approaches have their peculiarities, advantages and drawbacks, which make them useful under certain circumstances, and for specific biological systems. But they share a property: they hardly can be used for a multi-level model.

In this context we can motivate the agent approach by noting the following,

1. An agent-based approach provides more flexibility than other more limited approaches and so delivers greater potential for modeling more sophisticated, globally emergent, behavior both on the individual cell and on the cell population level.
2. We can explicitly represent an environment.
3. An agent-based approach provides more biological plausibility than existing approaches such as cellular automata and other mathematical approaches. One of the main reasons that biological plausibility is important is to attract biologists to use and work with any models and simulations that are created.
4. We can build multi-layer models that capture micro- and macro-level features of the biological system.
5. Stem cells are a prime example of a self-organizing system where individual cells react to their local physical, chemical and biological environment. The system should therefore be most suitably modeled as a system of interacting reactive agents, where the reaction at the micro level gives rise to the emergent behavior at the system level.
6. Even though we are simulating cells and environment, the Brooksian idea of an agent being something which is both situated and embodied [Brooks, 1991] is a fundamental driving force of our use of agents as the appropriate modeling paradigm. Cells modeled as agents have a physical, chemical and biological pres-

ence and are situated in a physical, chemical and biological environment in which they react. The way in which they react will then influence the way other cells react in the future and so on.

7. Agent-based models and simulation provide us with an easy facility to perform "what if" experiments. By changing the behavior of individual cells or environmental conditions we can perform experiments to see how system behaviors might be affected.
8. Agent systems lend themselves to elegant visualizations which have meaning to and impact on, the stem cell biologist.

In addition, by situating our simulation work in a wider formal framework, the SMART agent modeling framework, based on years of previous investigation [d'Inverno and Luck, 2004; d'Inverno et al., 2004] we can compare and evaluate different models. We believe that this is necessary for this new field to develop in a systematic manner. Moreover, the formal framework allows us to "agentify" existing models, making it very clear what the relationship between the existing version and the agent version is. We shall see this later in the chapter. Finally, by building a formal model of stem cell behavior in an agent framework and using a specification language from software engineering, there are techniques to ensure that the simulation correctly implements the model.

The application of multi-agent based simulation for the analysis of biological systems is quite recent, but already some work has been done in this area, for example, [Merelli et al., 2007; Montagna et al., 2007, 2008].

13.5 Overview of Our Agent Modeling Framework

We begin the overview of our stem cell agent modeling framework by first describing and explaining the purpose of the the main components of the framework: *cell agents*, the *environment* and the *simulation engine*. Next we describe the interfaces between the various components and finally we describe the behavior of the components.

A diagrammatic overview of the framework is given in Figure 13.1. The role of each of these framework components is as follows:

Cell Agents: these are used to model the individual cells of the system being modeled.
Environment: represents the biological, chemical and physical environment that the cells exist in.
Simulation Engine: the purpose of this is to "drive" a simulation of a stem cell system. It does this by updating the environment in response to requests and signals from the individual cell agents.

An important point to note is that the introduction of an environment that includes chemical and topological information allows us to reduce the need to rely on stochastic processes, i.e., probability functions. Moreover, it allows us to actually model the physical movement of cells which clearly provides us with a more sophisticated and detailed model of self-organization in stem cell systems. We will see examples of the advantage of modeling physical aspects in agent-based models later in this chapter as we agentify existing formal models.

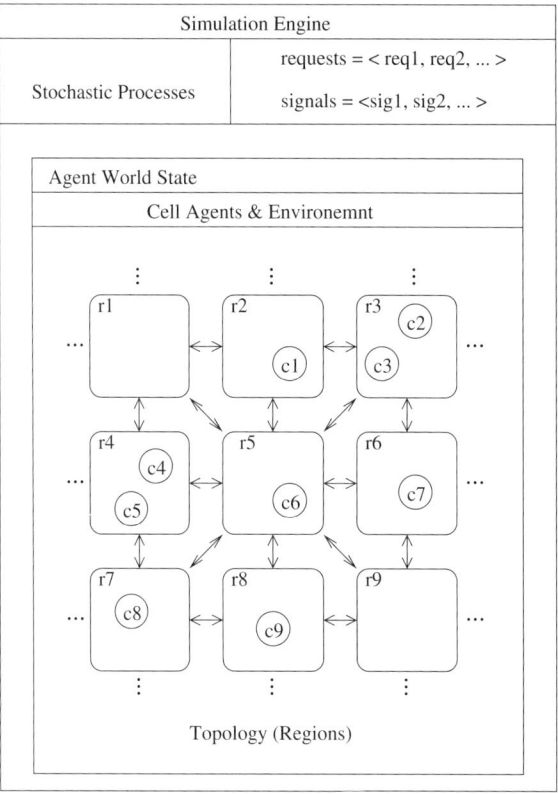

FIGURE 13.1 The Agent Framework for Stem Cell modeling. Where the r_i are regions and c_i are cell agents.

13.5.1 Framework Components

In this section we describe the *state information* that is contained in each of the three framework components: *cell agents*, the *environment* and the *simulation engine*.

These components can be viewed as forming subsystems: the *Agent World State* that comprises the cell agents and the environment; and the *Stochastic World State* which adds the stochastic processes to *Agent World State*.

The *Agent World State* is defined as a combination of the cell agents, the environment and the locations of the cell agents in the environment:

- The agent world state is composed of two components the cell agents and the environment.
- A cell is modeled as an agent and consequently a cell agent's state consists of biological and operational information.
- In addition, a cell agent has a number of physical properties, for example, location, orientation and possibly velocity and momentum as a result of being in the environment.
- The environment is composed of the following sub-components chemical and physical laws and a topology of regions.
- The individual regions of a topology have attributes that can be *perceived* by the cell agents.

The request mechanism, described later in this chapter, allows the modeling of cells to focus on the logic of their operation and leaves questions of physical simulation, e.g., collision detection, to the simulation engine.

Cell Agent

A cell agent is used to model a biological cell, but it is also necessary that it include sufficient attributes to be able to function as an autonomous agent. So in our framework a cell agent's state is composed of four main components:

- The internal state of the cell that represents its biological state.
- The cell agent's percepts, that is the information that a cell can perceive from its environment.
- The cell agent's plan, these are the actions that the cell has decided to attempt.
- A cell may also have a globally defined stochastic component that needs to be added purely for modeling purposes.

The second and third components are concerned with the operational aspects of a cell agent, that is the *perceive*, *plan* and *act* agent behavioral cycle. At this level of model description the precise nature of the plans developed by the agents is left open. Planning may be implemented using approaches like those typically used in Belief-Desire-Intention (BDI) agents or, as in the case of the work described later, using a reactive planner. The last component is not consistent with a true agent approach because it introduces a global level of control but is a necessary component when agentifying existing systems, e.g., the Roeder/Loeffler model [Roeder and Loeffler, 2002].

Environment

We define the notion of an environment using the general notion of a *topology*, which is defined as a set of adjacent *regions*. Cells are located within a region, and each region may contain from zero up to a predetermined maximum number of cells. Further each region has associated with it chemical concentrations, physical forces and so on. So for example, in a particular model a region may contain at most one cell, thus it may be either empty or contain a stem, progenitor, determined or stromal cell.

So the environment is composed of the following components:

- A *topology*, defined in terms of a set of *regions* and an *adjacency* relationship between them. These properties are fixed for a particular model.
- An individual region's *static attributes*, e.g., how many cells it can contain. These properties are fixed for a region in a particular model.
- An individual region's *dynamic attributes*, e.g., chemical concentrations and the cells it contains. These properties can vary for a region in a particular model.
- A mapping which represents the *location* of the cells within the environment, i.e., the region it is currently in.
- Chemical and physical laws, e.g., chemical diffusion rates. These properties are fixed across all models.

Simulation Engine

The simulation engine is able to see the state of all components in the system being modeled, in particular the cell agents and the environment, thus it is able to extract any system state information it requires from these components. In addition, the simulation engine also has the following internal state information:

- The signals and requests it has received from the cell agents.
- The state of any stochastic processes that it is using to provide stochastic inputs to the cell agents.

The simulation engine used in the following implementations works in a discrete step-wise manner, but the high level of description used at the modeling phase permits for other types of simulation engines, including discrete event simulations with asynchronous execution. For more information on different types of simulation engines see the chapter by Theodoropoulos et al. in this book [Theodoropoulos et al., 2008].

13.5.2 Interfaces between the Framework Components

We now describe the *interfaces* between the cell agents, environment and simulation engine.

Cell Agent and Environment

Each cell agent *perceives* information from its environment, that is it inputs the "state" of its current environment. Its current environment is defined as its current location which will consist of at least its current region and possibly the adjacent regions. This information is input in the form of *percepts*. In general these will be chemical, physical and biological information. For example, a cell agent can perceive its current location (region) as well as the chemical concentration in that location.

Cell Agent and Simulation Engine

The cell agents *issue requests* to the simulation engine to perform a "global" event and *send signals* to notify the "environment" that it has performed a unilateral action, for example secrete chemicals.

Examples of cell agent *requests* would be:

- to move to another location; or
- a parent cell to divide and be replaced by two daughter cells; or
- a cell to die and be removed from the environment.

In practice this is achieved by a cell agent sending its request to the simulation engine, which then appends the request to its list of pending requests. Some of a cell agent's planned (internal) actions do not require the simulation engine to do anything, and hence no request is generated.

Depending on the model under consideration a cell agent would be required to input stochastically generated information from the simulation engine.

Simulation Engine and Environment

The simulation engine simply retrieves state information from the environment and updates the state of the environment as a result of processing the cell agents' requests and signals. The simulation of physical processes, such as chemical diffusion, are handled by the simulation engine, rather than being embedded in the environment, allowing freedom of implementation.

13.5.3 Behavior of the Framework Components

In this section we discuss the behavior of the cell agents and the simulation engine. The environment is a passive repository of information and hence does not have an active behavior as such.

Cell Agent Behavior

In agent-based computing the operation of an agent is given by the *perceive*, *plan* and *act* cycle. In particular, what an agent perceives depends on its internal state and that of the local environment; what it plans to do depends on its perceptions and internal state; and what action it actually does depends on what it plans to do and the state of the environment.

The above general agent behavior cycle is refined for a cell agent as follows:

- The *perceive* action is a passive engagement with the environment, that simply involves inputting *percepts*. In general we aim to model chemical, physical and biological perceptual abilities. For example, a cell agent can perceive the chemical concentration in its current location (region).
- A plan to perform one of its possible actions is formulated after the cell agent has perceived its environment, and then which action it decides to add to its plan is determined by these percepts and its internal state. The types of actions that a cell agent can choose to do are remain in its current location, move to a new location, divide, secrete chemicals or ultimately die. In the framework a cell agent's plan consists of several types (see below) of actions: *internal actions* and their corresponding *requests* and *signals* to the simulation engine. The requests and signals are determined by the planned internal action.
- The agent *acts* by performing the first action in its plan, the remaining plan is then just what is left after this action has been removed. The generalized cell agent's *action* behavior can further be divided into the following sub-actions of *send requests*, *send signals* and *do action*. Where these represent the cell agent sending a request and signal respectively to the simulation engine, and the performance of the cell's planned internal action respectively.

Note that in some types of system the cell agent's behavior cycle would include the inputting of stochastically generated information calculated globally by the simulation engine, in a true agent model this would not be necessary. This is not part of the cell agent's perceptions but part of the stochastic modeling technique for certain models. When the cell agent's behavior cycle is implemented we assume that it would be *atomic*, in the sense that after an agent has perceived its environment the information that it has perceived, i.e., percepts, remain true throughout the remainder of its behavior cycle.

In our framework we have identified three types of "actions" that can form a cell agent's plan:

Internal actions: these are actions that a cell agent can perform independently of its environment, these actions only effect its internal state and do not require the participation of its environment.
Signals: these represent a cell agent *signaling* to its environment. These signals do not require the environment's participation in the sense of a joint action, but

only that the environment records the effect of the signal having occurred*. For example, when a cell decides to secrete a chemical, this would be modeled by a signal.

External requests: these are requests that a cell agent makes to its environment. They are associated with the cell's (internal) actions and are used when a cell's chosen action requires the participation of the environment for the complete effect of the action to occur. If the environment is unwilling or unable to collaborate with the action, i.e., grant the request, then it can not be performed, in these circumstances the agent would continue to *wait* for the request to be granted. This allows us to model an action that effects the state of the cell and environment, for example, cell division or death.

Simulation Engine Behavior

In the current model the main purpose of the simulation engine is to "run" the simulation, it does this by processing the action requests and signals generated by the cell agents. For example, each request is processed as appropriate by either:

- moving a cell to a new location (region); or
- deleting a dead cell from the system; or
- deleting the parent cell from the system and creating and adding two new daughter cells; or
- if the request can not be granted taking no action.

In general the simulation engine may have to deal with any conflicts that might arise due to incompatible requests.

The simulation engine is also responsible for:

- Updating the state of the environment, for example, a regions chemical concentrations when a cell agent has secreted a chemical.
- It is also responsible for generating stochastic information using stochastic processes, e.g., probability functions, when this is required by the system under consideration.

13.5.4 Extending Our Agent Modeling Framework

We are not fixed on this model. It is important that as we incorporate different models, perspectives and experimental findings into our approach that the model can develop. One of the ways we validate and grow this model is to apply this agent-based modeling approach to "agentify" other formal models. We discuss two case studies in the next section.

13.6 Agentifying Existing Approaches

To demonstrate the validity and applicability of our agent approach we consider two recent models in more detail.

---

*This is similar to Roscoe's [Roscoe, 1997] view of the process termination event *tick*, in the process algebra CSP.

13.6.1 A Cellular Automata Approach to Modeling Stem Cells

In recent work, Agur et al. [Agur et al., 2002] built a cellular automata (CA) model to show how the number of stem cells in the bone marrow could be maintained and how they could produce a continuous output of determined cells. The bone marrow is considered to be a stem cell *niche* where most biologists believe that the human body's supply of hematopoietic stem cells are situated and maintained.

This work is important because it is one of the few examples where a mathematical model has been used to investigate properties of stem cells that might be required to enable the maintenance of the system's homeostasis. The model demonstrates a possible mechanism that allows a niche to maintain a reasonably fixed number of stem cells, produce supply of mature (determined) cells, and to be capable of returning to this state even after very large perturbations that might occur through injury or disease. The behavior of a cell is determined (equally differentiated) by both internal (intrinsic) factors, e.g., a local counter, and external (extrinsic) factors, e.g., the prevalence of stem cells nearby, as stated by the authors as follows.

1. Cell behavior is determined by the number of its stem cell neighbors. This assumption is aimed at simply describing the fact that cytokines, secreted by cells into the micro-environment are capable of activating quiescent stem cells into proliferation and determination.
2. Each cell has internal counters that determine stem cell proliferation and stem cell transition into determination as well as the transit time of a differentiated cell before migrating to the peripheral blood.

In order to demonstrate this model we will provide a small part of an existing specification of this system. The specification is written in the Z specification language [Spivey, 1992]. The full specification can be found elsewhere [d'Inverno and Saunders, 2006] but we provide a taste since the formal background to our agent-based approach is a key part of this work. In the cellular automata model, the stem cell niche is modeled as a connected, locally finite, undirected graph.

$$graph : Node \leftrightarrow Node$$
$$neighbors : Node \rightarrow (\mathbb{P}\, Node)$$

Any *Node* is either empty, or it is occupied by either a stem cell or a determined cell.

$$TypeAg ::= EmptyAg \mid StemAg \mid DeterminedAg$$

The state of any node is given by the node location, the state, and an internal clock.

$$\begin{array}{l} \underline{NodeStateAg} \\ node : Node \\ type : TypeAg \\ counter : \mathbb{N} \end{array}$$

The set of all such nodes is then given below, and defines the system state. We also define a function that returns the neighboring node states for any given node state.

$$\begin{array}{l} \underline{SystemStateAg} \\ nodes : \mathbb{P}\, NodeStateAg \\ neighborsAg : NodeStateAg \rightarrow (\mathbb{P}\, NodeStateAg) \end{array}$$

There are three constant values, we will call them *LeaveNicheAg*, *CyclingPhaseAg*, and *NeighborEmptyAg* in our specification, that are used to reflect experimental observation. *LeaveNicheAg* represents the time taken for a determined cell to leave the niche. *CyclingPhaseAg* represents the cycling phase of a stem cell; a certain number of ticks of the counter are needed before the cell is ready to consider dividing. Finally, *NeighborEmptyAg* represents the amount of time it takes for an empty space that is continuously neighbored by a stem cell, to be populated by a descendent from the neighboring stem cell.

$$\mid \textit{LeaveNicheAg}, \textit{CyclingPhaseAg}, \textit{NeighborEmptyAg} : \mathbb{N}$$

We now specify how the system changes over time. Whenever there is a change of state in the system, we identify the node that we are considering as *node*. All locations are updated simultaneously.

The rules of this model, which determine what happens at a node based on internal and external factors are described and specified below.

1. Determined cell nodes

    (a) If the internal counter of a node representing a determined cell has reached *LeaveNicheAg* then the cell leaves the niche; the internal counter of the node is reset to 0, and the new state at the node becomes empty.

    In the schema below, the variable *node* represents the node determining its next state, and the variable *newnode* is that new state. We show one operation only, the rest are very similar.

    ```
    ┌─ DeterminedLeaveNicheAg ──────────────────
    │ ΔSystemStateAg
    ├────────────────────────────
    │ node.type = DeterminedAg
    │ node.counter = LeaveNicheAg
    │ newnode.type = EmptyAg
    │ newnode.counter = 0
    └────────────────────────────
    ```

    (b) If the internal counter has not yet reached *LeaveNicheAg* then the internal counter is incremented.

2. Stem cells nodes

    (a) If the internal counter of a node representing a stem cell has reached the constant *CyclingPhaseAg*, and all of the nodes neighbors are stem cells, then the state of the node becomes a determined cell and the internal counter is reset to 0.

    (b) If the internal counter of a node representing a stem cell is equal to *CyclingPhaseAg* but not all the node's neighbors are stem cells then do nothing; leave the internal counter unchanged.

3. Empty nodes

    (a) If the internal counter at an empty node has reached *NeighborEmptyAg* and there is a stem cell neighbor then introduce, i.e., give birth to, a stem cell in that location. The internal counter of the node is reset to 0.

    (b) If the counter at an empty grid has not reached *NeighborEmptyAg* and there is exists a stem cell neighbor then increment the counter by 1.

(c) If there are no stem cell neighbors at all then reset the internal counter to 0.

A simulation of this model, together with its source code, can be found at:

http://doc.gold.ac.uk/~mas02md/cell/simulations/agur/index.html

13.6.2 Discussion about the Cellular Automata Approach

The specification of the cellular automata model reveals the following issues:

1. Empty niche spaces must do computational work, i.e., to maintain counters
2. Stem cell division is not explicitly represented, stem cells are brought into being by empty space
3. Individual stem cells are not explicitly tracked before they give rise to new stem cells
4. The state of the stem cell after division is not defined

The specification clearly reveals that niche spaces, i.e., empty nodes, must have counters for this model to work. In a sense, empty space is having to do some computational work. Clearly this lacks biological feasibility and is against what the authors state about modeling cells, rather than empty locations, having counters.

Stem cell division is not explicitly represented, instead stem cells are brought into being by empty space. More subtly, these stem cells appear when empty nodes have been surrounded by at least one stem cell for a period of time. The location of the neighboring stem cell, however, can vary at each step.

The model details the fact that if a stem cell is next to an empty space long enough then it will divide so that it's descendent occupies this space, however, the rule does not state that the neighboring stem cell must be the same stem cell for every tick of the counter. It states something much weaker; that there must be a neighboring cell, possibly different each time, for each tick of the counter, from 1 to *NeighborEmptyAg*. Biologically, it would seem more intuitive that the same stem cell should be next to an empty niche space for this length of time in order for "division" to occur into the space but the model lacks a "directional component".

The state of a stem cell after division is not defined. Let us for a moment assume that the neighboring stem cell (S) is fixed for all counts from 1 to *NeighborEmptyAg* from some specific location (N). Nothing is said about what happens to S after a new stem cell appears in N. For example, should the counter of S be reset after division? Neither does it give any preconditions on S. For example, does S's local counter need to have reached an appropriate point in its cycling phase for this to happen?

So the basic problem is that this model relies on allowing both unfilled niche locations as well as stem and determined cells to have counters. Moreover, it does not investigate or model the nature of a stem cell before and after division. We now attempt to re-interpret these rules using an agent-based approach that still retains the overall qualities of the model.

13.6.3 Re-formulation Using an Agent-Based Approach

One of the biggest differences between the original cellular automata model and our re-formulation is the change in the role of graph nodes. In the cellular automata model each node represents either a cell or an empty space. In our re-formulation, each node represents

FIGURE 13.2 A comparison of the original Agur cellular automata model and our reformulation as a grid-based agent model. In the original model the nodes maintain the state of the cells, whereas in our re-formulation the nodes contain agents and it is the agents that maintain the state of the cells.

a space that may or may not contain an agent that represents a cell. This difference in the two models is illustrated in Figure 13.2.

With the agent approach we also provide each cell with a unique identifier. We model all cells as having one internal counter as before. In addition, each cell maintains a counter associated with each of its neighboring nodes. The counters associated with neighboring nodes record how long the neighboring location has been empty. Moreover, cells can sense the type of cell at each of its neighbors, although this perception ability is only used by stem cells. If an agent represents a stem cell then it can potentially divide into any location where the counter has reached $NeighborEmptyAg$.

A cell agent is specified formally as:

$[AgentId]$

$\underline{\quad AgentCellAg \quad}$
$id : AgentId$
$type : TypeAg$
$counter : \mathbb{N}$
$nscounter : Node \nrightarrow \mathbb{N}$
$nstype : Node \nrightarrow TypeAg$

$type = StemAg \lor type = DeterminedAg$
$\text{dom } nscounter = \text{dom } nstype$
$\forall n : Node \mid nstype\ n \neq EmptyAg \bullet nscounter\ n = 0$

A stem cell agent is defined as follows. Other types are defined similarly.

$\underline{\quad AgentStemCellAg \quad}$
$AgentCellAg$

$type = StemAg$

The system state consists of the niche where some nodes are filled with cells. The first predicate simply states that the empty nodes are those nodes which do not contain a cell. The second predicate states that the neighbors are defined by the graph to which the cells are attached.

Agent-Based Modeling of Stem Cells 405

$\begin{array}{|l} \hline \text{\textit{AgentSystemStateAg}} \\ \textit{cells} : \textit{Node} \rightarrow \textit{AgentCellAg} \\ \textit{emptynodes} : \mathbb{P}\, \textit{Node} \\ \hline \textit{emptynodes} = \textit{Node} \setminus (\text{dom}\, \textit{cells}) \\ \forall\, n : \textit{Node};\ c : \textit{AgentCellAg} \mid (n, c) \in \textit{cells} \wedge c.\textit{type} = \textit{StemAg} \bullet \\ \quad \text{dom}\, c.\textit{nscounter} = \text{ran}(\{n\} \lhd \textit{graph}) \\ \hline \end{array}$

13.6.4 Operation

Space does not permit us giving a full treatment, but we outline the basic operations here.

1. Cells set/update counters.
2. Mature stem cells that are surrounded by empty neighbors and have neighbor counters that have reached *NeighborEmptyAg* will make a request to the environment to divide into two daughter stem cells.
3. The environment resolves any conflicts where several cells wish to divide into the same node (region) and informs those mature stem cells that can divide and those that are not able to.
4. Mature stem cells that are able to divide do so. Mature stem cells that are surrounded by stem cells become new determined cells. Mature determined cells which are ready to leave the niche do so.

Our agent-based approach to modeling forces us to consider what happens when two stem cells attempt to divide into the same location. In our model, we specify that when the internal counter reaches *CyclingPhaseAg*, it signals to the environment the niche spaces that it is prepared to divide into.

Notice, that this approach is also agent-based in nature. Namely, the agent attempts to do something but the environment is a dynamic and uncertain one. From the perspective of a single cell with its limited sensory abilities the world is no longer deterministic like it was in the cellular automata model, and not all attempts at action will be successful.

The agent-based model not only considers the nature of acting in a dynamic environment but also addresses issues such as the basic physical limitations of the stem cell niche in general. Once again, it's difficult to see how such issues can be considered, at least explicitly, with the cellular automata approach.

A stem cell agent that is ready to divide, signals to the environment those neighbors that have been empty for long enough, and so are able to receive the new cell. Of course the output may be empty.

The environment receives requests from cells to divide, and non-deterministically assigns those cells that can divide and those that have insufficient space around them. There are several safety properties that we can specify here:

1. all agents get a reply (first predicate)
2. no agent can be told to divide and not divide (second predicate)
3. no node ever has more than one agent dividing into it (third predicate)
4. cells only get to divide into a node they have requested (fourth predicate)
5. there is no remaining empty node that has been requested by any of the agents not-granted division (fifth predicate).

Cells that divide get told where they should divide into. We have two alternatives with the assignment of identifiers to the daughter cells: we can either give both daughters new identifiers or allow one of the daughter cells to keep the Id of its parent. For the purposes of tracking the inheritance between parent and daughter cells, we prefer the first of these alternatives and give each daughter cell a unique Id.

We have run many simulations of both the original CA model and of our agent recapitulation to check that the behaviors of our agent model has the same properties of the CA model. As we explained above, the agent model has allowed us to address the issues of biological implausibility.

It is interesting to note that allowing cells to split into all available spaces, i.e., up to four daughters, gives us the closest possible agent-based simulation match to the original CA models, however, any biologically plausibility we may have introduced would be negated by this. Going back to the original CA model we see that this four-way division is supported, further reducing the plausibility of the model. By limiting cell division to result in a maximum of at most two daughter cells we still maintain the integrity of the original cellular automata version.

All these simulations are available on the web site.

13.6.5 Roeder-Loeffler Model of Self-Organization

A recent example of an approach that uses a more sophisticated model that more successfully addresses issues of biological implausibility, is that of Markus Loeffer and Ingo Roeder at the University of Leipzig, who model hematopoietic stem cells using various, but limited, parameters including representing both the growth environment within the marrow, one particular stem cell niche, and the cycling status of the cell [Roeder and Loeffler, 2002; Roeder et al., 2005; Roeder and Loeffler, 2004; Roeder et al., 2006]. In this model, the ability of cells to both escape and re-enter the niche and to move between high and low niche affinities, referred to as within-tissue plasticity, is stochastically determined.

The validity of this model is demonstrated by the fact that it produces results in global behavior of the system that match experimental laboratory observations. The point is that the larger patterns of system organization emerge from these few simple rules governing variations in niche-affinity and coordinated changes in cell cycle.

In recent work it has been shown to model cases of chronic myloid leukemia [Roeder et al., 2006]. Although the Roeder-Loeffler model is is rather sophisticated (it is formal, there is a simulation, it addresses key issues of self-organization and much of the modeling has an agent-like quality to it), there are a number of issues regarding this model that we have addressed by extending it using our agent framework. Most significantly, the use of a global probability function to control the movement of cells between environments, and in the agent-view this is problematic; this probability is calculated from global information relating to the numbers of various cells in the system. Although it useful to assume access to this global information when developing the model of stem cell behavior, no mechanism is known for how stem cells could have access to this information in real biological systems.

We will say more about this model when we discuss the implementation, but to summarize we have extended the Roeder-Loeffler model to produce an agent-based model that increases the biological intuition and plausibility of the model, and allows us to investigate emergence due to the subtle changes in micro-environmental effects for each cell. Modeling cells as agents responding autonomously to their local environment is much more fine grained than the previous model using equations to model cell transitions and allows for a much greater degree of sophistication in the possibilities of understanding how self-organization actually takes place in the adult human body. The main point is that an agent does not rely on getting information about the system state, in keeping with the reactive multi-agent systems approach, and we believe that this gives a more biologically plausible handle on how things might be working at the micro-environmental level.

In general the agent models we produce suffer from none of the drawbacks of other formal systems approaches. However, models of agents by themselves have no real practical value as we do not have any method of determining how the individual behavior of agents affects the overall system behavior. Formality is useful for building consistent clearly defined models but they give us very little clue as to how the overall system of interacting agents will actually evolve. Therefore we need to move from modeling to simulation, and this is considered next.

13.7 From Agent Model to Simulation

To investigate the behavior of our models we have developed computational models to simulation populations of stem cells as a Multi-Agent Based Simulation (MABS). MABS is an approach to simulation that builds on Agent-Based modeling (ABM) and Multi-Agent Systems (MAS) to develop simulations of real-world phenomena as systems of multiple interacting agents. Implementing MAS and MABS models is challenging* and several simulation frameworks have been developed to facilitate this process. In this section we will describe our implementation of the agent-based model as a MABS using MASON [Luke et al., 2004], a MAS/MABS development environment.

Examples of agent modeling tools that span different approaches to modeling multi-agent systems include:

FIPA-compliant tools: The Foundation for Intelligent Physical Agents (FIPA) is the standards organization for agents and multi-agent systems. It has established a standard for inter-agent communication that is the core of the FIPA agent system model. The FIPA approach to modeling agents has been widely adopted by multi-agent system tool developers. FIPA-compliant tools include JADE [Bellifemine et al., 1999], Aglets [Lange and Mitsuru, 1998].

StarLogo: StarLogo was developed as an educational tool by Mitchell Resnick to introduce children to issues of emergence through computational modeling of *turtles* and *patches*. Where patches model the environment and turtles represent agents that inhabit patches. Since its introduction, StarLogo has inspired other environments for multi-agent simulations, most notably NetLogo [Wilensky, 1999].

Sugarscape: In 1996, Epstein and Axtell introduced Sugarscape, an initial attempt to develop a "bottom up" social science [Epstein and Axtell, 1996]. Sugarscape simulates the behavior of artificial people (agents) located on a landscape of a generalized resource (sugar). Fundamental collective behaviors including group formation, cultural transmission, combat, and trade are seen to "emerge" from the interaction of individual agents following simple local rules.

Swarm: In the Swarm multi-agent simulation platform the basic unit of simulation is the *swarm*, a collection of agents executing a schedule of actions. Swarm supports

---

*In the early stages of this project we considered using Erlang [Armstrong et al., 1996], a language based on Haskell, developed by Ericsson, that is well suited to modeling our agent framework but lacks a MABS environment. Sadly, developing such a framework was outside the scope of this project, although we are still very interested in using this powerful language in the future.

hierarchical modeling, such that agents can be composed of swarms of other agents. As a computational modeling approach, Swarm has inspired several other popular simulation platforms, including Repast and MASON.

Several existing multi-agent simulation tools were considered for this part of the project, including Swarm [Minar et al., 1996], NetLogo [Wilensky, 1999], Ascape [Inchiosa and Parker, 2002] and Repast [Tatara et al., 2006]. We chose a relative newcomer to the simulation frameworks, MASON, as we found that its carefully thought out architecture provided the best framework for our research.

13.7.1 MASON

MASON (Multi-Agent Simulator Of Neighborhoods/Networks) is a multi-agent simulation development environment and simulation tool [Luke et al., 2004]. MASON has been developed around a discrete-event multi-agent simulation library written in Java. This flexible foundation allows for a range of simulation approaches, including discrete step and discrete event simulations; so far, we have used it to develop discrete step simulations.

MASON was designed to be the foundation for large custom-purpose Java simulations. MASON contains both the simulation library and an optional suite of visualization tools in 2D and 3D. Features of MASON that were particularly desirable for our development, that affected our decision to use it, include:

Portability: The developers of MASON have taken great care to ensure that cross-platform portability, going as far as implementing their own version of the Merseinne-Twister algorithm for generating random numbers so that stochastic elements of simulations will run identically across machines and platforms.

Speed and size: The MASON library has been developed to be fast and relatively small, allowing many cells to be simulated at once using standard desktop and notebook computers. This has allowed us to simulate relatively large stem cells systems without the need for complex distributed computing solutions.

Separation of model and visualization: Models are completely independent from visualization, which can be added, removed, or changed at any time. This has allowed us to easily develop simulations that can be run either with a visualization for demonstration purposes, or without a visualization for experimentation.

Checkpointing: Models may be checkpointed and recovered, or dynamically migrated across platforms. This allows stem cell models to be tested by first running them on a development machine, and then migrating them to a dedicated server for long-term processing.

Embedding: Models are self-contained and can run inside other Java frameworks and applications. This allows us to embed simulations in other Java applications and applets. In addition, the MASON library sets no limits on the use of other libraries and frameworks, e.g., allowing the graphing libraries to be used to provide numerical feedback as a visualization option.

Visualization: The MASON library includes support for several standard types of 2D and 3D visualizations, using the core Java libraries for 2D and 3D graphics. These have provided a foundation for developing specific visualizations of stem cells and their environments in our models.

Export of rich media: The MASON development environment and the MASON library support the export of images (PNG) and movies (Quicktime) for documenting particular simulation runs. In addition, MASON provides support for the generation of other output data streams, e.g., XML.

Advantages of Using MASON

Combining the powerful features offered by MASON has provided us with an excellent environment for development, presentation and experimentation. During development the MASON development environment allows for rapid testing and evaluation of the simulations. During presentations, the speed and size of the MASON library allow users to demonstrate "live simulations" on standard notebook hardware. During experimentation we run large-scale experiments using simulations without visualizations to automate "batch processing" of stem cell models across clusters of machines.

13.7.2 Implementation of CELL in MASON

The MASON library provides a comprehensive foundation for developing our agent foundation, the basis of which is a thin layer around the `Steppable` interface class. This is described by the Unified Modeling Language (UML) [Rumbaugh et al., 1999] *class diagram* in Figure 13.3.

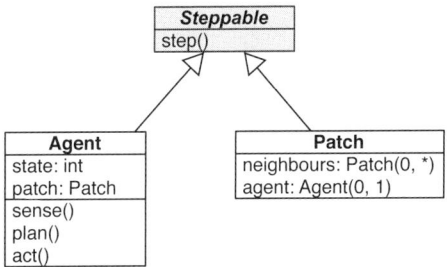

FIGURE 13.3 Our framework is rooted with implementations of `Agent` and `Patch` that implement the MASON interface `Steppable`.

The `Agent` class implements the `Steppable` interface to implement the base class for all cell classes in models. The `Agent` class decomposes the `step()` method into a sequence of three methods, `sense()`, `plan()` and `act()`, controlled by the `state` of the agent. Every instance of `Agent` can be assigned to a single instance of `Patch`. The `Patch` class implements the `Steppable` interface to provide the base class for decomposing space in simulations for the simulation of processes over the space, e.g., chemical diffusion. Each instance of `Patch` contains a set of `neighbors` that define the topology of the space.

The `Patch` class corresponds to a *region* in our framework, although the exact details of a region for a model are left to the individual implementations. In accordance with our formal framework, `Patch` is used as a base for building the environment, but apart from easing certain implementation issues its inheritance from `Steppable`, e.g., allowing the use of existing portrayals for visualization purposes, it acts as a passive element in the implemented systems, simply recording the topology of the environment, as well as dynamic features updated by the simulation engine.

The CELL framework provides a utility class for simulating the diffusion of chemicals, the `DoubleGrid2DDiffuser` class acts on two-dimensional arrays of floating point numbers, i.e., doubles, to diffuse the values across the array. This class implements the Steppable interface as illustrated in Figure 13.4 so that it can be inserted into schedules, just like agents. In the simulations developed so far, the DoubleGrid2DDiffuser class is used to simulate the diffusion of chemicals released by stem cells and stromal cells.

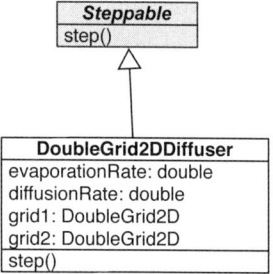

FIGURE 13.4 UML diagram of the Double2DDiffuser class.

In MASON, agents and environments cannot directly render themselves to a visualization, instead simulations are visualized using *portrayals*. The CELL framework defines two convenience portrayal classes for use by simulations, see Figure 13.5. CellPortrayal2D renders cells within a visualization as a simple oval. ChemicalGridPortrayal2D renders chemical concentrations, held in a DoubleGrid2D, using a gradient to indicate the concentration of patches. Similarly to the implementation of Agent, these classes represent thin wrappers around existing MASON classes; OvalPortrayal2D and FastObjectGridPortrayal2D.

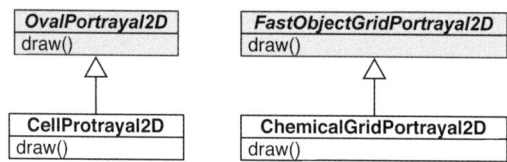

FIGURE 13.5 UML diagram of the Portrayal2D subclasses.

13.7.3 Implementation of Roeder-Loeffler Model in MASON

The small set of classes described above forms the framework, built on MASON, that we implement specific models upon. This section details the classes developed to implement the Roeder-Loeffler model of stem cell organization. This implementation adds specific details of the model of cells and their environment as described.

Figure 13.6 illustrates the implementation of the basic Cell class. The Cell class is a subclass of the Agent class and overrides the sense(), plan(), and act() methods to implement the standard agent cycle within the specific environment described for the Roeder-Loeffler model. Each Cell class maintains an internal state within an instance of the CellStateInternal class.

Cells communicate with the environment, simulation engine, visualization, etc. through the use of CellEvents that CellListeners can subscribe to be notified about. Cells will

Agent-Based Modeling of Stem Cells

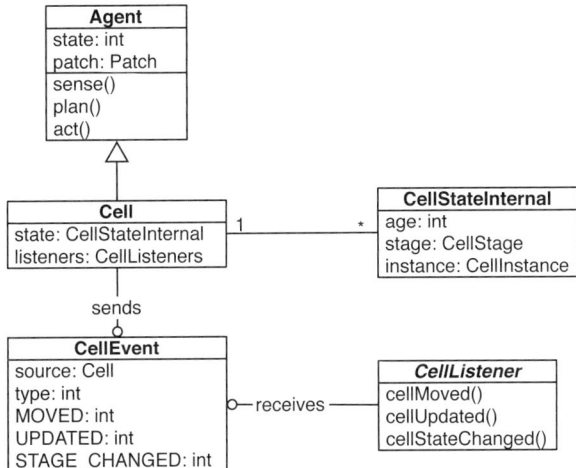

FIGURE 13.6 UML diagram of the `Cell` class.

generate events when they move (`CellEvent.MOVED`), have their internal state updated (`CellEvent.UPDATED`), or when the stage that they are in changes (`CellEvent.STAGE_CHANGED`).

Figure 13.7 illustrates the classes that implement the *cell typing* system in the Roeder-Loeffler model. The `CellType` class is the core of this system and allows different types of cells, e.g., stem cells and stromal cells, to be defined with different *cell cycles*, using the `CellCycle` class, for each *stage* in the cell's life, represented using `CellStage` objects.

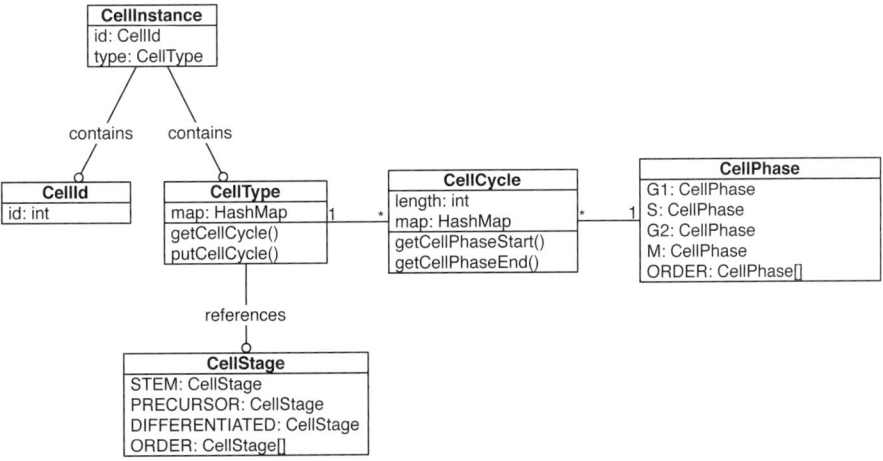

FIGURE 13.7 UML diagram of the `CellType` and related classes.

Possible `CellStages` in this implementation Roeder-Loeffler model are STEM, PRECURSOR and DIFFERENTIATED. Hematopoetic cells in the simulation can potentially pass through all of these stages. Stromal cells begin in the DIFFERENTIATED stage and remain in it throughout simulation runs as defined in the model.

Cell cycles are defined as a sequence of *phases* that cells pass through, the different phases are represented by the `CellPhase` class. The phases that our cells pass through are: G1, S, G2 and M. These are defined in this order in the `CellPhase` class. Different types of cells are defined using the `CellType` class as spending different lengths of time in these phases.

Figure 13.8 illustrates the classes that form the top-level in the agent-based Roeder-Leoffler simulation, i.e., `WorldState` that maintains all of the information about the environment and `AgentRoeder` that implements the top-level simulation engine and stand-alone application. The `AgentRoederWithUI` class wraps the `AgentRoeder` class with a user-interface for displaying visualizations produced using the portrayal classes defined in the CELL framework. Using this pattern of developing a stand-alone application for running on the command-line without a visualization, and a wrapper to add the necessary GUI elements when desired, provides a powerful method for developing the simulations.

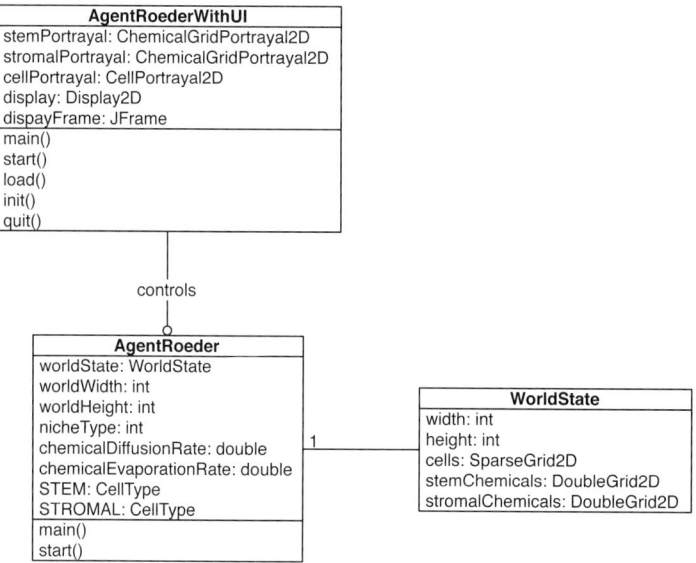

FIGURE 13.8 UML diagram of the `WorldState` container class, the `AgentRoeder` simulation class and the `AgentRoederWithUI` visualization class.

Using the sophisticated library of components provided by MASON has allowed us to build an agent framework comprised of some utility classes, and implement specific models with relatively little code.

13.8 Discussion

We have spoken much in this chapter about why agents are "good" for modeling natural systems but perhaps the best way to see the advantages is in the visualizations of the simulations themselves. Again we encourage the reader to look at the visualizations for themselves on the project web site*, but we present some of the images here.

---

*The CELL Project web site is located at http://doc.gold.ac.uk/~mas02md/cell/

Agent-Based Modeling of Stem Cells

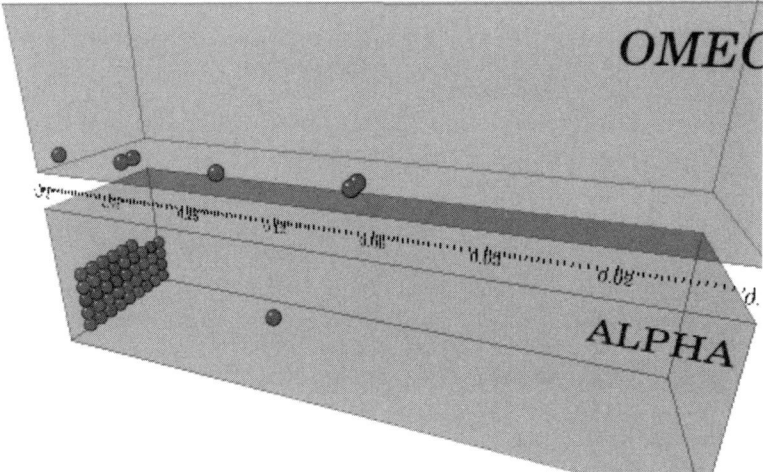

FIGURE 13.9 The visualization of the original model. We see cells switching between two compartments according to the calculation of a probability function.

We have extended the Roeder model to incorporate a model of space so that we can consider cell movement in more detail. The difference between the two approaches can best be seen by looking at the visualizations online.

In Figure 13.9 we see a visualization of the Roeder-Loeffler model, based on an existing visualization, where cells are in one of two compartments. The compartments represent broad environments inside (α) and outside (Ω) the stem cell niche. Movement within the compartments represents changes in niche affinity, as indicated by the scale, not physical movement within the environments. Consequently, the visualization should be read more like a 3D graph than a representation of 3D space. When the visualization moves in time, we see cells switching between the two compartments. The particular simulation being shown is a competition between two populations of stem cells, red and blue, competing for control of the niche. The biological intuition behind what is actually happening is very difficult to unpick.

In our agent view, shown in Figure 13.10, we get an understanding of the actual *mechanics* of what is happening to individual systems and how the system as a whole is interacting. Movements of agents in the simulated environment are represented as movements in the visualization. The secretion of chemicals from stem cells and stromal (niche) cells is what keeps this system continually generating new blood cells and maintaining the population of stem cells. Chemical concentrations are represented in the visualization by the coloring of regions according to the chemicals they contain. From this visualization, it is clear that we can experiment with properties such as the physical shape of the stem cell niche by changing the placement of stromal cells, something that isn't directly possible with the original model. The real value of visualizations like this is that biologists are attracted to them.

One of the predictions that our agent-based visualization makes is that stem cell activity pulses around the niche. Although not verified to date it is a testable prediction made about the nature of stem cell activity. We are currently seeking interested biologists to investigate whether this prediction can be verified.

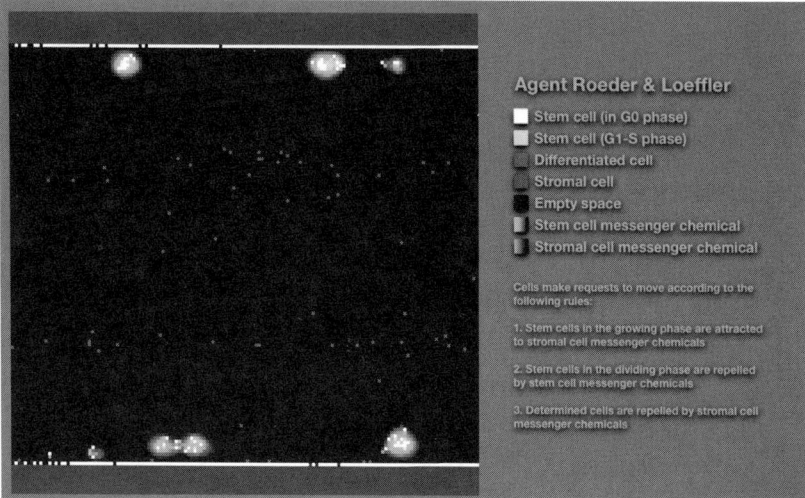

FIGURE 13.10 The visualization of our agent model. In our view a much more biologically intuitive visualization.

Our next goal is to build an agent-based model of chronic myeloid leukemia which will aim at providing an opportunity to evaluate the *mechanisms* which are taking place when this cancer takes hold in the development of blood cells.

13.8.1 Concluding Remarks

Recent medical evidence suggests that the way to understand how stem cells organize themselves in the body is as a self-organizing system, whose global behavior is an emergent quality of the massive number of interactions of cells with each other and of the environment of which they are a part. We claim, therefore, that the multi-agent system approach to modeling is the most suitable one for exploring means to simulate the behavior of stem cells and from resulting simulations [Theise and d'Inverno, 2004; d'Inverno et al., 2006], suggest how tiny changes in individual stem cell behavior might lead to disease at the global, and hence observable from an experimental perspective, system level. We have outlined the benefits of this approach by comparing it to a cellular automata approach in detail.

Modeling and simulating stem cells promises much. Our intention is to build a *common conceptual framework* from which integrated research can ensue. We believe that the *agent* approach to modeling, coupled with the natural feel of the visualizations makes it a valuable currency in this effort.

Indeed, the fact that the work came from an interdisciplinary project (rather than, say, a team of mathematicians working in isolation) means that this work is ideally placed to be the foundation from which we can produce this common conceptual framework so important to harnessing the energies of research from different fields.

One last comment. Perhaps the greatest evidence for the impact of our work has been in the language used by one of our collaborators on a recent paper in *Nature* [Theise, 2005].

> Cells fulfill all the criteria necessary to be considered agents within a complex system: they exist in great numbers; their interactions involve homeostatic, negative feedback loops; and they respond to local environmental cues with limited stochasticity ('quenched disorder'). Like any group of interacting individuals fulfilling these criteria, they self-organize without external planning. What emerges

is the structure and function of our tissues, organs and bodies.

Until working on this project the stem cell researcher Theise new nothing of agents. Now they have become the currency in which he conceptualizes them.

Acknowledgments

Andrea Omicini, Neil Theise, Jane Prophet, Peter Ride, EPSRC, AHRC and the Wellcome Trust.

References

Z. Agur, Y. Daniel, and Y. Ginosar. The universal properties of stem cells as pinpointed by a simple discrete model. *Journal of Mathematical Biology*, 44(1):79–86, 2002.

J. Armstrong, R. Virding, C. Wikström, and M. Williams. *Concurrent Programming in ERLANG*. Prentice Hall, New Jersey, 2nd edition, 1996.

E. Attar and D. Scadden. Regulation of hematopoietic stem cell growth. *Leukemia*, 18 (11):1760–1768, November 2004. URL http://dx.doi.org/10.1038/sj.leu.2403515.

F. Bellifemine, A. Poggi, and G. Rimassa. JADE - a FIPA-compliant agent framework. In *Proceedings of PAAM'99*, pages 97–108, April 1999.

H. M. Blau and B. T. Blakely. Plasticity of cell fate: insights from heterokaryons. *Semin. Cell Dev. Biol.*, 10(3):267–272, 1999.

R. Brooks. Intelligence without representation. *Artificial Intelligence*, 47(1-3):139–159, 1991.

E. Conrad and J. Tyson. Modeling Molecular Interaction Networks with Nonlinear Ordinary Differential Equations. In Z. Szallasi, J. Stelling, and V. Periwal, editors, *System Modeling in Cell Biology - From Concepts to Nuts and Bolts*, chapter 6, pages 97–123. MIT Press, Cambridge, MA, 2006.

H. De Jong. Modeling and simulation of genetic regulatory systems: a literature review. *Journal of Computational Biology*, 9(1):67–103, 2002.

L. Dematte, C. Priami, and A. Romanel. The blenxlanguage: A tutorial. In M. Bernardo, P. Degano, and G. Zavattaro, editors, *Formal Methods for Computational Systems Biology*, volume 5016 of *Lecture Notes in Computer Science*, pages 313–365. Springer, 2008. ISBN 978-3-540-68892-1. URL http://dblp.uni-trier.de/db/conf/sfm/sfm2008.html#DemattePR08.

M. d'Inverno and M. Luck. *Understanding Agent Systems (Second Edition)*. Springer; Berlin, 2004.

M. d'Inverno and R. Saunders. Agent-based modelling of stem cell organisation in a niche. In S. A. Brueckner, G. D. Marzo, A. Karageorgos, and R. Nagpal, editors, *Engineering Self-Organising Systems : Methodologies and Applications*. Lecture Notes in Artificial Inteligence, Springer, Berlin, 2006.

M. d'Inverno, M. Luck, M. Georgeff, D. Kinny, and M. Wooldridge. The dmars architechure: A specification of the distributed multi-agent reasoning system. *Autonomous Agents and Multi-Agent Systems*, pages 5–53, 2004.

M. d'Inverno, N. Theise, and J. Prophet. Mathematical modelling of stem cells: a complexity primer for the stem cell biologist. In C. Potten, J. Watson, R. Clarke, and A. Renehan, editors, *Tissue Stem Cells: Biology and Applications*. Marcel Dekker, New York, 2006.

J. Epstein and R. Axtell. *Growing Artificial Societies: Social Science from the Bottom Up*. Brookings Institute Press, Washington, DC, 1996.

D. T. Gillespie. Simulation methods in systems biology. In M. Bernardo, P. Degano, and G. Zavattaro, editors, *Formal Methods for Computational Systems Biology*, volume 5016 of *Lecture Notes in Computer Science*, pages 125–167. Springer, 2008. ISBN 978-3-540-68892-1. URL http://dblp.uni-trier.de/db/conf/sfm/sfm2008.html#Gillespie08.

M. Heiner, D. Gilbert, and R. Donaldson. Petri nets for systems and synthetic biology. In M. Bernardo, P. Degano, and G. Zavattaro, editors, *Formal Methods for Computational Systems Biology*, volume 5016 of *Lecture Notes in Computer Science*, pages 215–264. Springer, Berlin, 2008. ISBN 978-3-540-68892-1. URL http://dblp.uni-trier.de/db/conf/sfm/sfm2008.html#HeinerGD08.

M. Inchiosa and M. Parker. Overcoming design and development challenges in agent-based modeling using ASCAPE. *Proceedings of the National Academy of Sciences*, 99(90003):7304–7308, 2002. doi: 10.1073/pnas.082081199. URL http://www.pnas.org/cgi/content/abstract/99/suppl_3/7304.

M. Kanehisa and S. Goto. KEGG: Kyoto encyclopedia of genes and genomes. *Nucleic Acids Res.*, 28:27–30, 2000.

S. Kauffman. *The origins of order*. Oxford University Press, New York, 1993.

D. Lange and O. Mitsuru. *Programming and Deploying Java Mobile Agents Aglets*. Addison-Wesley Longman Publishing Co., Inc., Boston, MA, USA, 1998. ISBN 0201325829.

M. Loeffler and I. Roeder. Tissue stem cells: Definition, plasticity, heterogeneity, self-organization and models - a conceptual approach. *Cells Tissues Organs*, 171(1): 8–26, 2002.

M. Loeffler and I. Roeder. Conceptual models to understand tissue stem cell organization. *Curr. Opin. Hematol.*, 11(2):81–87, 2004.

S. Luke, C. Cioffi-Revilla, L. Panait, and K. Sullivan. MASON: A new multi-agent simulation toolkit. In *Proceedings of the 2004 SwarmFest Workshop*, 2004.

E. Merelli, G. Armano, N. Cannata, F. Corradini, M. d'Inverno, A. Doms, P. Lord, A. Martin, L. Milanesi, S. Möller, M. Schroeder, and M. Luck. Agents in bioinformatics, computational and systems biology. *Briefings in Bioinformatics*, 8(1): 45–59, 2007. URL http://dx.doi.org/10.1093/bib/bbl014.

N. Minar, R. Burkhart, C. Langton, and M. Askenazi. The swarm simulation system: a toolkit for building multi-agent simulations. Working Paper 96-06-042, Santa Fe Institute, Santa Fe, 1996.

S. Montagna, A. Omicini, A. Ricci, and M. d'Inverno. Modelling hematopoietic stem cell behaviour: An approach based on multi-agent systems. In F. Allgöwer and M. Reuss, editors, *2nd Conference "Foundations of Systems Biology in Engineering" (FOSBE 2007)*, pages 243–248, Stuttgart, Germany, 9–12 Sept. 2007. Fraunhofer IBR Verlag. ISBN 978-3-8167-7436-5.

S. Montagna, A. Ricci, and A. Omicini. A&A for modelling and engineering simulations in Systems Biology. *International Journal of Agent-Oriented Software Engineering*, 2(2):222–245, 2008. ISSN 1746-1375. doi: 10.1504/IJAOSE.2008.017316. URL http://www.inderscience.com/search/index.php?action=record&rec_id=17316. Special Issue on Multi-Agent Systems and Simulation.

N. Nicola and G. Johnson. The production of committed hemopoietic colony-forming cells from multipotential precursor cells in vitro. *Blood*, 60(4):1019–1029, October 1982.

J. Novak and C. Stewart. Stochastic versus deterministic in haemopoiesis: what is what?

British Journal of Haematology, 78(2):149–154, 1991.

M. Ogawa. Stochastic model revisited. *International Journal Hematology*, 69(1):2–5, 1999.

C. S. Potten and M. Loeffler. Stem cells: attributes, cycles, spirals, pitfalls and uncertainties. lessons for and from the crypt. *Development*, 110(4):1001–1020, 1990.

P. Quesenberry, H. Habibian, M. Dooner, C. McAuliffe, J. F. Lambert, G. Colvin, C. Miller, A. Frimberger, and P. Becker. Physical and physiological plasticity of hematopoietic stem cells. *Blood Cell. Mol. Dis.*, 27(5):934–937, 2001.

A. Regev, W. Silverman, and E. Shapiro. Representation and simulation of biochemical processes using the pi-calculus process algebra. In *Pacific Symposium on Biocomputing*, pages 459–470, 2001.

I. Roeder. Quantitative stem cell biology: computational studies in the hematopoietic system. *Curr. Opin. Hematol.*, 13(4):222–228, 2006.

I. Roeder and M. Loeffler. A novel dynamic model of hematopoietic stem cell organization based on the concept of within-tissue plasticity. *Exp. Hematol.*, 30(8): 853–861, 2002.

I. Roeder and M. Loeffler. Simulation of hematopoietic stem cell organisation using a single cell based model approach. In A. Deutsch, J. Howeard, M. Falke, and W. Zimmermann, editors, *Mathematics and Biosciences in Interaction*, pages 287–293. Birkhaeuser Verlag, Berlin, 2004.

I. Roeder, L. M. Kamminga, K. Braesel, B. Dontje, G. d. Haan, and M. Loeffler. Competitive clonal hematopoiesis in mouse chimeras explained by a stochastic model of stem cell organization. *Blood*, 105(2):609–616, 2005.

I. Roeder, M. Horn, I. Glauche, A. Hochhaus, M. C. Mueller, and M. Loeffler. Dynamic modeling of imatinib-treated chronic myeloid leukemia: functional insights and clinical implications. *Nat. Med.*, 12(10):1181–4, 2006.

A. W. Roscoe. *The Theory and Practice of Concurrency*. Prentice Hall International, New Jersey, 1997. ISBN 0136744095.

J. Rumbaugh, I. Jacobson, and G. Booch. *The Unified Modeling Language Reference Manual*. Addison-Wesley Professional, 2nd edition, 1999. ISBN 0201309980.

R. Schofield. The relationship between the spleen colony-forming cell and the haemopoietic stem cell. *Blood Cells*, 4(1-2):7–25, 1978.

C. Smith. Hematopoietic stem cells and hematopoiesis. *Cancer Control*, 10(1):9–16, Jenuary/February 2003.

V. Sowmya and P. W. Zandstra. Toward predictive models of stem cell fate. *Cyotechnology Review*, 41(2):75–92, 2003.

J. M. Spivey. *The Z Notation: A Reference Manual*. Prentice Hall International Series in Computer Science, 2nd edition, 1992. ISBN 0139785299.

C. L. Talcott. Pathway logic. In M. Bernardo, P. Degano, and G. Zavattaro, editors, *Formal Methods for Computational Systems Biology*, volume 5016 of *Lecture Notes in Computer Science*, pages 21–53. Springer, Berlin, 2008. ISBN 978-3-540-68892-1. URL http://dblp.uni-trier.de/db/conf/sfm/sfm2008.html#Talcott08.

E. Tatara, M. North, T. Howe, N. Collier, and J. Vos. An introduction to repast modeling by using a simple predator-prey example. In *Proceedings of the Agent 2006 Conference on Social Agents: Results and Prospects*. Argonne National Laboratory, Argonne, IL USA, 2006.

N. Theise. Now you see it, now you don't. *Nature*, 7046(435):1165, June 2005.

N. D. Theise and M. d'Inverno. Understanding cell lineages as complex adaptive systems. *Blood, Cells, Molecules and Diseases*, 32:17–20, 2004.

G. Theodoropoulos, R. Minson, M. Lees, and R. Ewald. Simulation Engines for Multi-Agent Systems. In A. Uhrmacher and D. Weyns, editors, *Agents, Simulation and Applications*. Taylor and Francis, Boca Raton, FL, 2008.

I. Thornley, R. Sutherland, R. Wynn, R. Nayar, L. Sung, G. Corpus, T. Kiss, J. Lipton, J. Doyle, F. Saunders, S. Kamel-Reid, M. Freedman, and H. Messner. Early hematopoietic reconstitution after clinical stem cell transplantation: evidence for stochastic stem cell behavior and limited acceleration in telomere loss. *Blood*, 99(7):2387–2396, 2003.

J. Till, E. Mcculloch, and L. Siminovitch. A stochastic model of stem cell proliferation, based on the growth of spleen colony-forming cells. *Proc Natl Acad Sci USA*, 51(1):29–36, 1964.

J. Trentin. Influence of hematopoietic organ stroma (hematopoietic inductive microenvironment) on stem cell differentiation. In A. Gordon, editor, *Regulation of Hematopoiesis*, volume 1, pages 161–168. Appleton-Century-Crofts, New York, 1970.

A. M. Uhrmacher, D. Degenring, and B. Zeigler. Discrete event multi-level models for systems biology. In C. Priami, editor, *Transactions on Computational Systems Biology I*, volume 3380 of *Lecture Notes in Computer Science*, pages 66–89. Springer, Berlin, 2005. ISBN 3-540-25422-6.

H. E. Wichmann and M. Loeffler. *Mathematical modeling of cell proliferation: Stem cell regulation in hemopoiesis.*, volume 1,2. CRC Press, Boca Raton, 1985.

U. Wilensky. *NetLogo*. http://ccl.northwestern.edu/netlogo/, 1999.

A. Wilson and A. Trumpp. Bone-marrow haematopoietic stem-cell niches. *Nature Reviews Immunology*, 6(2):93–106, February 2006.

IV

Tools

14 RoboCup Rescue: Challenges and Lessons Learned 423
Tomoichi Takahashi
Introduction • Needs in Disaster and Rescue Management • Architecture of RoboCup Rescue Simulation System and Usage as MAS Platform • Lessons Learned from RoboCup Rescue Competitions • Examples and Discussions on Experiments Using Real Maps • Analysis of ABSS Based on Probability Model • Discussion and Summary

15 Agent-Based Simulation Using BDI Programming in Jason 451
Rafael H. Bordini and Jomi F. Hübner
Introduction • Programming Languages for Multi-Agent Systems • Programming Multi-Agent Systems Using *Jason* • *Jason* Features for Simulation • Example • Ongoing Projects • Conclusion

16 SeSAm: Visual Programming and Participatory Simulation for Agent-Based Models 477
Franziska Klügl
Introduction • Simulation Study and User Roles • Core SeSAm • Users, Tasks, and SeSAm • Experiences • General Discussion and Future Work

17 JAMES II - Experiences and Interpretations 509
Jan Himmelspach and Mathias Röhl
Introduction • JAMES II • Multi-Agent Modeling and Simulation in JAMES II • Application • The Role of Time • Experiences and Interpretations • Outlook

Agents are used as a metaphor for describing a system of interest. Software agents are evaluated in virtual dynamic environments. This can be done by testing in small environments via so-called test-beds for agents or by testing in large ones, which implies a modeling of the environment the software agents shall dwell in. All require simulation systems. Often these environments are built from scratch. However, the re-use of simulation environments promises less effort, improved repeatability and comparability of experiments. Not surprisingly, quite a few simulation systems for multi-agent systems have been built. However, due to the diversity of agents and objectives of simulation studies, a widely accepted modeling and simulation approach has remained elusive. The current landscape of agent modeling and simulation tools reflects the different needs and background of the systems' developers equally.

Concrete test-beds, that provide dynamic, standardized scenarios in which different agent strategies e.g., of deliberation and commitment, mark the beginning of simulation tools for multi-agent systems in the 80s. With the RoboCupSoccer in 1997 a problem was found that allowed to examine a wide range of technologies in robotics and artificial intelligence. To demonstrate that the strategies and evaluations are also viable in other socially relevant scenarios (a concern that has accompanied work on test-beds from the very beginning) the RoboCupRescue project was initiated in 2000. The areas of promoted research in disaster-rescue comprise multi-agent teamwork coordination, robotics, information infra-structure, evaluation benchmarks for rescue strategies and simulator development. Thus, the objective of simulation research in this project is twofold, i.e. to provide a suitable evaluation platform for those other areas, and to identify requirements and to develop solutions for simulators to be applied in rescue settings. Tomoichi Takahashi provides insight into this research with his chapter "Robocup: Challenges and Lessons Learned".

The aim to support arbitrary scenarios led to the development of general modeling and simulation systems for multi-agent systems. Whereas reactive agents can easily be described in simulation systems that support an object-oriented, modular model design, the effective and efficient handling of multiple deliberative agents provide some challenges for modeling and simulation environments. Rafael Bordini and Jomi Hübner address those in their chapter "Agent-Based Simulation Using BDI Programming in *Jason*" by focusing on the description of deliberative agents in the agent language AgentSpeak. Thereby, they distinguish between agents with their declarative representations of mental attitudes and display of rational, goal-directed behavior and the reactive environment the agents are situated in. The latter is responsible for advancing the simulation.

Developing agent and environmental models requires programming in *Jason*. In contrast, other agent-based simulation systems offer easy to use interfaces to facilitate agent-based simulation for domain experts not familiar with programming - a motivation that also drove the development of expert systems shells in the 1980s. In the chapter "SeSAm (Shell for Simulated Multi-Agent Systems): Visual Programming and Participatory Simulation for Agent-Based Models" Franziska Klügl describes experiences with this type of simulation tool and the efforts required toward achieving this goal. In this context, the role of adequate visual means is examined in several interdisciplinary applications. Whereas many agent simulation systems focus on the encoding of the agents and its environment only, SeSAm and its user interface address the entire modeling and simulation life cycle: from model development, experiment design, to the analysis of simulation results.

The entire modeling and simulation life cycle is supported in the tool JAMES II, which first developed as a tool for multi-agent modeling and simulation turned due to the required flexibility into a "Java multi-purpose environment for simulation". To support different modeling formalisms, simulation engines, experimental settings, random number generators etc. a plug-in architecture has been realized which allows an easy extension and adaptation of the system to new applications. In contrast to other multi-agent simulation systems, JAMES II supports the description of agents in modeling formalisms with a clear operational semantics being realized by different sequential and parallel simulation engines. The modeling formalisms are traditional discrete event modeling formalisms which have been extended to support dynamic patterns of composition and interaction. In the chapter "JAMES II - Experiences and Interpretations" Jan Himmelspach and Mathias Röhl give an overview about the architecture of the system, and exemplarily describe the extension of a modeling formalism, its use for describing agents, and its simulation in JAMES II.

14

RoboCup Rescue: Challenges and Lessons Learned

14.1	Introduction..	423
14.2	Needs in Disaster and Rescue Management.......	425
14.3	Architecture of RoboCup Rescue Simulation System and Usage as MAS Platform...............	427
	RoboCup Rescue Project • Architecture of the Simulation System • Progress of the Simulation • World Model and Representation • Protocol	
14.4	Lessons Learned from RoboCup Rescue Competitions..	434
	Lessons from Agent Competitions • Researches Related to Real Applications	
14.5	Examples and Discussions on Experiments Using Real Maps..	438
	Validity of Agent-Based Simulations • Discussion on Experiments	
14.6	Analysis of ABSS Based on Probability Model...	441
	Agent Behavior Formulation and Presentation • Analysis Results of RoboCup Rescue Competition Logs	
14.7	Discussion and Summary.............................	447
	References	449

Tomoichi Takahashi
Meijo University

14.1 Introduction

Disaster management is one of the most serious social issues, involving very large numbers of heterogeneous agents in a hostile environment. The RoboCup Rescue project was motivated by the Great Hanshin-Awaji earthquake, which hit Kobe City on January 17, 1995. The lessons learned from this earthquake indicated that information systems should be built to support the collection of necessary information and prompt planning for disaster mitigation. We proposed the RoboCup Rescue project in 2000. This rescue project is intended to promote research and development in the disaster rescue domain at various levels: multi-agent teamwork coordination, physical robotic agents for search and rescue, information infrastructure, personal digital assistants, standard simulator and decision support systems, evaluation benchmarks for rescue strategies and robotic systems. The RoboCup Rescue Simulation (RCRS) was designed to simulate the rescue measures taken during the Hanshin-Awaji earthquake disaster. Section 14.2 describes the setting of the RCRS and the need for disaster management in the local government system.

When earthquakes occur in urban areas, various types of causalities and accidents occur. These are related to one another. Collapsed buildings injure civilians and block roads with debris. The rescuers must rush the victims to hospitals, help civilians evacuate to safe areas, and prevent fires, if any, from spreading. The accumulated debris hinders the rescue operations. Fires destroy houses, and the resulting smoke impedes the activities of the fire fighters. In order to simulate such situations, it is necessary to comprehensively integrate the results of disaster simulations and human behaviors in urban areas. The RCRS uses an architecture that simulates disaster events by repeated individual simulation of the component disasters and integration of the results obtained in each simulation step. Thus, this computation process can be distributed among multiple computers linked by a network when the simulation size exceeds the limit that can be supported on a single computer. This architecture allows users to customize the simulations for any city or town by inputting the appropriate GIS data, plugging in disaster simulators with their regional peculiarities, and implementing the prevention plans as agent codes. Section 14.3 describes the system architecture of the RCRS, methodology for implementation, and multi-agent system (MAS) technologies.

The RCRS has been used as a platform for the RoboCup competition since 2001, in which many teams from various countries have participated. The teams have improved their rescuing ability by use of realtime/anytime planning, collaboration among heterogeneous agents, mixed-initiative planning, learning, and so on, and have competed on their performances. They have not only developed their own rescue agent teams but have also improved the RCRS. For example, some teams repeated disaster simulations for the learning of rescue agents, they have developed components such as fire simulators and traffic simulators in a more practical and stable manner than before. Other teams interested in collaboration with the rescue teams have proposed a communication model that allows the agents to select channels for communications among themselves. This function has been implemented in the RCRS. Section 14.4 presents the results and the observations from the RoboCup Rescue agent competitions.

Several attempts have been made to put the RCRS to practical use. A system that employs the RCRS to train rescue teams, a demonstration system to support disaster prevention planning in their countries, and a system that combines data from real world and disaster simulations that involve numerous agents have been presented. Section 14.5 introduces these examples and discusses the experiments using real maps. The problems discussed here are not limited to the disaster rescue domains, but are also related to other simulations of social domains that are realized through the emergence of macro level system properties from the interaction among the agents at the micro level.

It is necessary to demonstrate the validity of the simulation results to users when an agent-based social simulation (ABSS) is applied to domains where it is difficult to obtain data from real cases or conduct experiments. In scientific and engineering fields, the following process has been repeatedly used to increase the fidelity of simulations [Feynman, 1967].
Guess → Compute consequence → Compare experiment results with simulation results In contrast to the physical phenomena that can be explained using the laws of natural science, social phenomena are not generally objectively measured, but are subjectively interpreted by humans [Moss and Edmonds, 2005]. This makes it difficult to systematically analyze the social phenomena and to evaluate the performances of the agents. In most cases, it is difficult to obtain data based on real cases or conduct experiments to verify the ABSS results and analysis results.

Earthquake disaster simulators are composed of disaster simulators and human-behavior simulations. While the components of disaster simulators, such as fire and building collapses, have been programmed on the basis of models developed in civil engineering fields, agent technology is used to simulate human behaviors with these disaster simulations. This makes it difficult to compare the results with the real data and repeat the process of guess, computation, and comparison to increasing the fidelity because we cannot conduct physical experiments on disasters on a real scale or involve humans in the experiments; furthermore the results of agent-based approach simulation have emergent properties. This also raises the question of whether it is realistic to expect agent-based systems to be useful in disaster management. Section 14.6 discusses the assessment of the fidelity of ABSS for use in real life situations as one of the new challenges. The discussion on future research topics and a summary are presented in Section 14.7.

14.2 Needs in Disaster and Rescue Management

We constantly face disasters in various forms, and even though they cannot be accurately predicted, appropriate evacuation instructions could save human lives. Typhoons (hurricanes or cyclones) are anticipated disasters; they are forecasted by meteorological bureaus, which issue warnings to sensitive areas. The exact location and time of earthquake occurrences are, however, difficult to anticipate, although sensor network systems to detect earthquakes are being planned or applied practically. The Hanshin-Awaji earthquake hit Kobe City on January 17, 1995. More than 6,000 people were killed, and at least 300,000 were injured; further, basic infrastructure worth more than 100 billion US dollars was damaged. The experiences gained during the Hanshin-Awaji earthquake revealed that the following functions are necessary for disaster information systems [Disaster Reduction and Human Renovation Institution, 2008].

1. Collection, accumulation, relay, selection, summarization, and distribution of necessary information.
2. Prompt support for action planning for disaster mitigation and search and rescue operations.
3. Reliability and robustness of the system.
4. Continuity from normal times to emergencies.

Recently, some projects and systems have been proposed to ensure prompt planning for disaster mitigation, risk management, and support of IT infrastructures [de Walle and Turoff, 2007][Mehrotra et al., 2008]. Sahana is an example of a system that was originally built by a volunteer group after the Sumatra-Andaman earthquake in 2004 [Currion et al., 2007]. Takeuchi et al. proposed an integrated earthquake disaster simulation system based on the concept Risk-adaptive Regional Management Information System [Takeuchi et al., 2003]. The risk-adaptive system is a system that is designed to be used on a regular basis by local government officers and to be linked to information servers during emergencies [Turoff, 2002]. The Combined Systems project commenced after September 11, 2000, has worked toward decision making in situations involving considerable number of people and organizations working together [Lab, 2006]. The project has developed actor-agent networks that interconnect actors (human beings) and software agents. The network enables people to act and respond more rapidly during disasters.

Before disasters occur, local governments employ a simulator to plan disaster response strategies that suggest how civilians can be evacuated to safe places, where refuges can be built, and how communal facilities such as roads, life lines, and hospitals can be operated. During disasters, information systems that provide relevant and reliable information to the local government and the public are critical to ensure smooth rescue operations. A decision support system (DSS) supports the operational or strategic decision of the rescue centers,

TABLE 14.1 Disaster, components of simulation, and their uses

disasters		components of simulation		uses of simulation	
		simulators	data	metrics of damages	time of application
natural	earthquake	fire	GIS	human lives	before disaster
	tsunami	*smoke			*analysis
	typhoon	collapses	facilities	damages	after disaster
	flood	*building, mudslide		* public property	*planning
man-made	terror	human activity	life lines	* private property	
		*traffic			
		*evacuation			

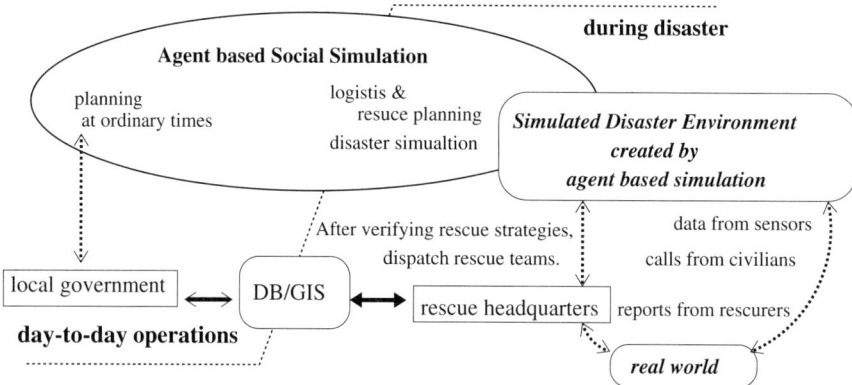

FIGURE 14.1 An image of a disaster management system used by local governments. The system and data used for day-to-day operation will be utilized at rescue headquarters.

and the simulations of disasters play an important role in the DSS. Whether disasters are natural or man-made, they involve various factors. In the case of a fire disaster, the factors are the simulation of fire, firefighting, the influence of smoke, and the interaction among them such as smoke emanating from the fire hampers firefighting efforts. Disaster simulation systems are integration systems. Table 14.1 shows the types of components to be simulated, the data required for these simulations, and the metrics that are calculated from the simulation results.

Disasters are classified based on how they damage human life, namely, how many lives are lost, how many houses were burned down, etc. These are used as metrics of classifications. In order to practically apply the simulations in the DSS at the rescue center, the metrics of disasters calculated from the agent-based social simulations are used for planning response strategies before, during, and after a disaster. Figure 14.1 shows a representation of the disaster-rescue simulation that will be used at the emergency centers run by local governments in the future. The left part shows the day-to-day operations of the local governments. Agent-based social simulations are used for their planning at ordinary times such as traffic flow estimations. When disasters occur, the local government will play a role of rescue headquarters and the system and data will be used for disaster management systems. Disaster situations will be simulated using agent-based simulations and data such as sensor data, phone calls, and TV/media obtained from the real world. The rescue headquarters will simulate various disaster scenarios, analyze their effects, and plan their rescue operations, including the deployment of rescue agents and evacuation of civilians to safe places.

14.3 Architecture of RoboCup Rescue Simulation System and Usage as MAS Platform

We proposed a RoboCup Rescue Simulation (RCRS) based on the experiments of the Hanshin-Awaji earthquake. The RCRS is a comprehensive simulation system that integrates various disaster simulation results and agent actions [Kitano et al., 1999]. The architecture of RCRS is introduced in this section.

14.3.1 RoboCup Rescue Project

The RCRS was designed to simulate the Hanshin-Awaji earthquake and also other disasters by plugging or replacing components. The necessary conditions for the rescue project were set by investigating the disasters in the Nagata Ward, one of the areas severely affected by the earthquake. The specifications during the development of the simulation tool are as follows.

Rescue scenarios: During earthquakes, related disasters occur, for example, fires break out, and rescue agents must rush to help victims or prevent the spread of the disaster. They obtain the necessary information by inspecting the sites or by responding to distress calls from other agents. Civilians are evacuated to safe places or can act as volunteers to help other agents. Local rescue agents are the first to respond to disasters, and help in extinguishing fires and moving the injured to hospitals with the help of other volunteers.

Target sizes: Nagata Ward has an area of 11.47 km^2 and a population of 130,466 people (53,284 families). A total of 7 rescue teams are present at the Nagata fire office. While representing disaster situations or rescue activities, it is required that the target size be displayed at a resolution that can be recognized on the display. The resolution of the display is set to about the sizes of cars. The Geographic Information System (GIS) data are provided at a resolution of 5 m.

Simulation period: The functions of rescue activities change with time. These activities are classified into five stages: chaos stage, initial operation stage, recovery stage, reconstruction stage, and normal stage. In the chaos stage, there is no external aid available, and the survival rate of the injured rapidly decreases after a few days. The main purpose at this stage is to save the victims using locally available facilities; considering these aspects, the simulation period was set to the chaos stage.

The RCRS comprehensively simulates agents who move autonomously in the simulated disaster world, and interactions among disasters and the behaviors of the agents. The simulations are modeled by the following equations.

$$e(t) = f(x(t), u(t), t) \tag{14.1}$$
$$x(t + \Delta t) = g(x(t), e(t)). \tag{14.2}$$

Here, t is the time; Δt is the step size to progress the simulation discretely; and $x(t)$, is a status variable that represents the disaster situation at time t. The variables represent the properties of the entities in the real world, such as the state of a house (whether it is burning or not), the conditions of roads (whether they are buried or not), and the state of a civilian (where he/she is and whether he/she is trapped in the debris or has been carried to a hospital) are included. $u(t)$ represents the external effects. Some of these effects include the wind direction, aftershock, and the amount of water sprayed from fire engines. The function f calculates the effects due to the disasters, and g represents the change of $x(t)$ from t to $t + \Delta t$. The details are explained as follows:

Disaster simulations: Disaster simulations comprise several events. For instance, building collapses affect the spread of fire. The spreads change the actions of the fire fighters to extinguish the fire, and the traffic conditions. The traffic jams, in turn, hamper the timely arrival of the fire fighters.

$$\begin{aligned} e_1(t) &= f_1(x(t), u(t), t) \\ e_2(t) &= f_2(x(t), u(t), t) \\ &\vdots \\ e_n(t) &= f_n(x(t), u(t), t). \end{aligned} \tag{14.3}$$

Here, $f_1, f_2, \cdots, f_n$ represent disaster simulations such as fire simulations or traffic simulations. $e_1, e_2, \cdots, e_n$ are the associated respective effects of these simulations. The mutual relationship among the disasters should be reflected in each simulation. The disaster situations are synchronously updated, and the integrated results are used as the initial conditions of the subsequent calculation.

$$x(t + \Delta t) = g(x(t), e_1(t), e_2(t), \cdots, e_n(t)). \tag{14.4}$$

Agents' behavior including rescue activities simulations: Agents autonomously decide their actions according to the situation; their actions could involve visiting refugees, obtaining assistance, extinguishing fires, rescuing victims, etc. The decision-making process is described in the form of a program as follows:

$$e_a(t) = f_a(x_a(t), s(x(t)), u_a(t), t). \tag{14.5}$$

Here, x_a is a subset of x and also represents properties peculiar to humans, such as their stamina and a record of their activities. The function s filters the input sensory data of circumstances around the agent, for example, the ranges of vision and hearing alter according to the disaster situations. $u_a(t)$ is the input to the agent from the outside world, which includes information through communication with other agents. $e_a(t)$ represents the intended actions of the agents during the simulation step*. These actions change the states of the world, $x(t + \Delta t)$.

$$x(t + \Delta t) = g(x(t), e_1(t), \cdots, e_n(t), e_{a1}(t), \cdots, e_{aN}(t)). \tag{14.6}$$

14.3.2 Architecture of the Simulation System

The RoboCup Rescue simulation is built with a number of modules [Takahashi et al., 2000]. Figure 14.2 shows the architecture of the RCRS, which is composed of the kernel, simulators, viewer, GIS, and agents. $f_1, \ldots f_n$ are the simulators, and f_a corresponds to the agents. g represents kernel. The kernel is the main component of the RCRS, and it combines all the information and updates the status of the world to be simulated. The components are plugged into the kernel via a protocol based upon TCP/UDP.

---

*In the present implementation, the agents can submit commands to the kernel at any time in a given simulation step. The kernel integrates commands from all the agents at a specified time in the simulation step, and the last command from the agents becomes effective.

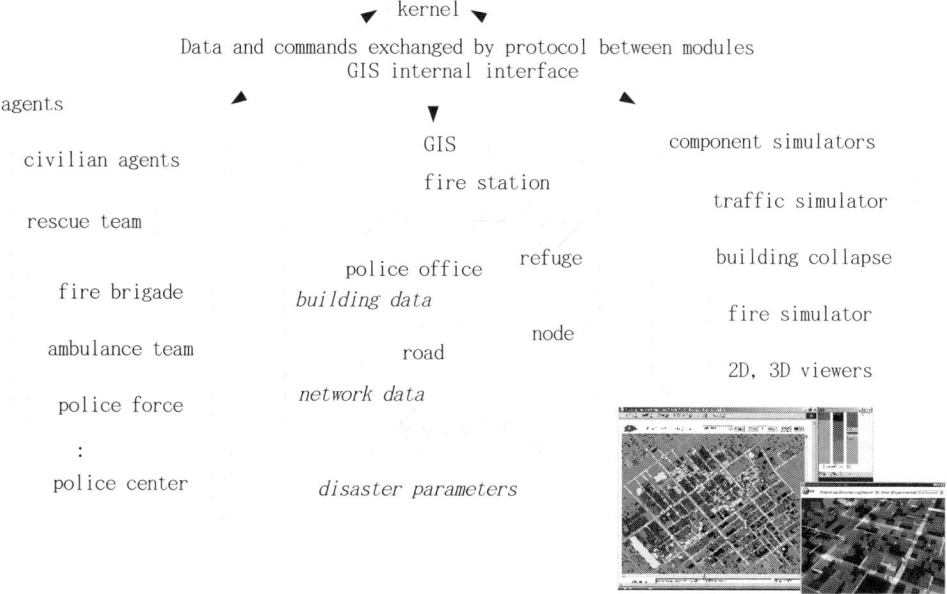

FIGURE 14.2 Architecture overview of RCRS. Components are plugged into kernel via specified protocol.

Agent: Civilians and fire fighters are examples of agents. These individuals are virtual entities in the simulated world. Their decision processes or action-selection rules are programmed in the corresponding agents, which are the RCRS client programs, and their intentions are expressed as commands to the kernel.

Here, an agent is a unit such as a family or a fire brigade in order to avoid making the simulation system too large, although it is desirable that a single person (or a single robot) is represented by an individual agent.

Component simulators: These correspond to disasters such as building collapses and fires. The disaster simulators compute how disasters affect the world. Besides these disaster simulations, there are two default simulations related to humans. One simulator monitors the states of civilian agents: their stamina and health. The stamina of an agent decreases when he/she performs actions. When the agents are trapped in collapsed houses or injured by fire, their physical and mental situation deteriorates. Furthermore, when they take refuge, they slowly recover; thus, their health improves. The other simulator is a traffic simulator, which changes the positions of the agents according to their intentions, and resolves conflicts that arise when numerous agents attempt to do similar actions, for example, to visit the same position.

GIS: This module provides the initial configuration of the world defining where roads and buildings are, and their properties such as width of street, floor area, number of stories, fire-resistant building or not, and so on that are needed for disaster simulations. The roads are represented in the form of network and the data buildings are presented as polygons. The numbers of agents and fires, and their positions are given as parameters of the simulation.

Viewers: Viewers visualize the disaster situations and agent activities on a 2D or 3D viewer and display the images according to predefined matrices obtained from

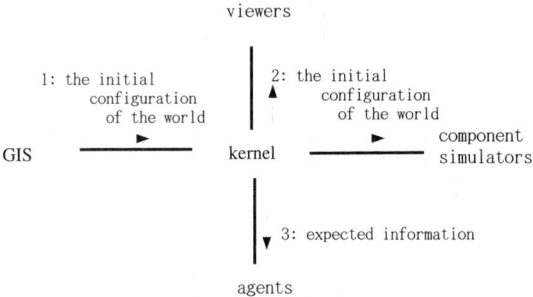

FIGURE 14.3 Initial protocol communication.

the simulation results.

Kernel: The kernel works as a server program to the clients and controls the simulation process and facilitates information sharing among the modules. No two agents can communicate directly with each other. In other words, agents can communicate with one another only via the kernel module, which accepts commands that are admissible in the communication protocol. The kernel module checks the actions. For example, though a fire-fighting agent decides to extinguish a fire, the kernel prohibits the agent from using more water than that present in the tank.

The elements of $x(t)$ cover all the entities; therefore, its size increases with the simulated area. By involving numerous agents, it becomes possible to simulate disaster scenarios on a real scale and to test rescue operations. This requires powerful computation and wide-range communication resources to simulate disaster situations and rescue actions on a real scale. The architecture of the RCRS makes possible to distribute these components among the computers linked by the network. Although the inter-module communication protocol was designed to facilitate multiple agent modules sharing a socket, there is a possibility of a bottle neck or congestion when the number of agents is large or the target area is wide.

Koto and Takeuchi have proposed another distributed computation model by dividing the simulation space [Koto and Takeuchi, 2003].

14.3.3 Progress of the Simulation

At the beginning of the simulation, the GIS transmits the initial configuration of the simulated world to the kernel (Figure 14.3). The kernel then forwards this information to the component simulators and to each agent module. The simulation proceeds by repeating the following steps (Figure 14.4):

1. At the beginning of every cycle, the kernel transmits sensory information to each agent module. This sensory information consists of data that the agent can sense in the simulated world at that time and may contain a certain degree of error. During the simulations, the sensory information transmitted by the kernel to each agent contains only the information in the vicinity of the individual, which can be visually sensed by them*.

---

*The radius parameter is set in a configuration file.

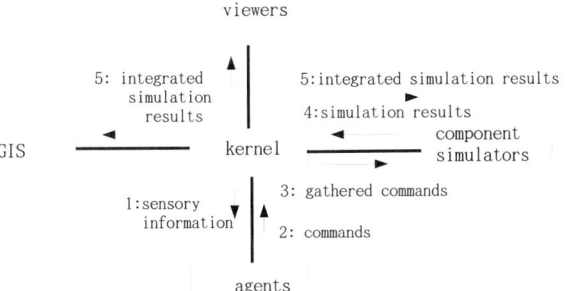

FIGURE 14.4 Protocol communication during simulations.

2. Each agent module decides actions to be performed by the individual, and accordingly, sends commands to the kernel.
3. The kernel gathers all the commands from the agent modules and broadcasts them to the component simulators. The packets are not guaranteed to arrive in timestamp order; therefore, it is possible that the commands are transmitted by an agent module whose corresponding individual is already dead. These packets that arrive behind time are discarded. The kernel, within a fixed period, only accepts commands and forwards them to the component simulators.
4. Based on their internal status and the commands received from the kernel, the component simulators individually compute the changes in the world. These results are then transmitted back to the kernel.
5. The kernel integrates the results received from the component simulators and broadcasts them to the GIS and component simulators. Although some simulators may take a long time to be simulated, the kernel will only integrate those results received within a certain time step. The kernel then increases the simulation clock and notifies the viewers about the update.
6. The viewers receive the updated information about the world, and display the information visually according to various evaluation criteria.

14.3.4 World Model and Representation

Disaster simulators have been developed in fields such as civil engineering and combustion engineering. The models and data structure of the simulations are different, and there is no unified format to represent target areas. For example, some fire simulators may split the world into planar grids, while others may possess a 3D model of a town and its houses.

A simple planar graph structure model with Euclidean metrics is used for inter-module communication. Each module must communicate with others based upon this simplified world model, whether or not it uses a more sophisticated or finer-grained model. Figure 14.5 shows the hierarchy of classes that form the simulation world. The motionless objects are the elements of the graphical model. A road is represented as an edge object that has properties such as width, roadways or sidewalks, and the IDs of adjacent node objects. A node object contains its coordinates and the IDs of the edges that are connected to it. These node objects represent crossroads, the outstanding points of curves, or the points that link roads and building objects.

An ambulance team enters into a house and rescues persons injured in a building collapse. The entrance property of the building indicates a node in the graph where the road

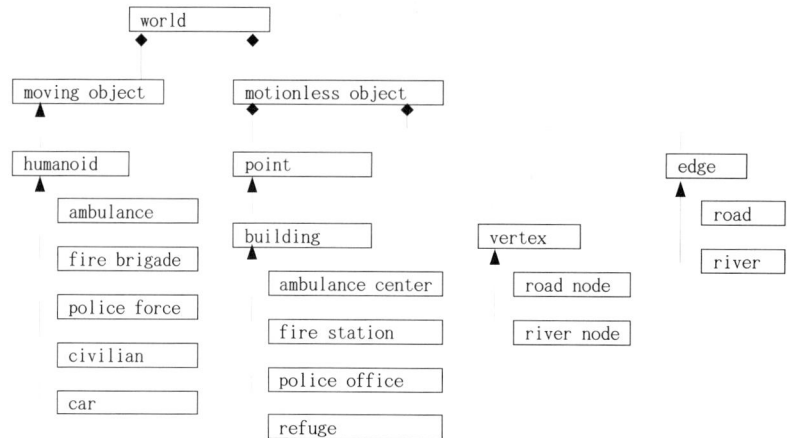

FIGURE 14.5 Class hierarchy of objects in the RCRS disaster area.

is connected to the entrance of the building. There are four special buildings: ambulance center, fire station, police offices, and refuges. The ambulance center, fire station, and police offices can communicate with the fire brigades, ambulance agents, and police agents, respectively. They collect information from the corresponding rescue agents and direct them for collaboration.

During the simulation, the modules send only the necessary or a limited part of the properties of the objects in order to reduce the amount of communication. For example, the simulation results transmitted to the kernel from the component simulators contain only those properties that will be changed during the subsequent step.

14.3.5 Protocol

The modules of the RCRS communicate with one another using a protocol based on TCP/UDP. The protocol is generic, is independent of a particular simulation algorithm, and specifies communications between the kernel and the modules, agents, simulators, GIS, and viewers. Table 14.2 shows the protocols between the agents and the kernel. The protocols between the component simulators are SK_CONNECT, SK_UPDATE, KS_CONNECT_OK etc.* The following pseudo code exemplifies a fire brigade agent. It shows how the agent communicates to the kernel. In `control` method, the agent gets sensing information at every step and plans their action and sends corresponding commands to the kernel by calling `act` method.

```
class FireBrigadeAgent {
    FireBrigadeAgent(int agentType, InetAddress kernelAddress,
                    int kernelPort) {
      // establish communication link to kernel.
        socket = new ProtocolSocket(kernelAddress, kernelPort);
        socket.akConnect(TEMPORARY_ID, VERSION, agentType);
        Object data = socket.receive();
        if (data instanceof KaConnectError)
```

---

*The first two letters, AK and SK, are acronyms of Agent/Simulation to Kernel, while KA and KS are acronyms of Kernel to Agent/Simulation.

TABLE 14.2 Header and the use of commands between kernel and agents

Header	A	K	Use
connection:			
AK_CONNECT	→		request for connection to the kernel
AK_ACKNOWLEDGE	→		acknowledge KA_CONNECT_OK
KA_CONNECT_OK		←	inform of the success of the connection
KA_CONNECT_ERROR		←	inform of the failure of the connection
sensing and communication:			
KA_SENSE		←	send vision information
KA_HEAR		←	send auditory information
AK_TELL	→		submit the intention to convey a message via assigned channel
AK_CHANNEL	→		allocate communication channels for TELL command
rescue operation:			
AK_MOVE	→		submit the intention to move to another position
AK_RESCUE	→		submit the intention to rescue a humanoid
AK_LOAD	→		submit the intention to load a humanoid
AK_UNLOAD	→		submit the intention to unload a humanoid
AK_EXTINGUISH	→		submit the intention to extinguish a fire
AK_CLEAR	→		submit the intention to clear a blockade
AK_REST	→		submit the intention to do nothing
AK_SAY	→		submit the intention to convey a message (obsolete)
AK_TELL	→		submit the intention to convey a message via assigned channel
AK_CHANNEL	→		allocate communication channels for TELL command

A: agent, K: kernel. AK implies that agents issue commands to the kernel.

```
        quit();
      KaConnectOk ok = (KaConnectOk)data;
    // get the initial world model from the kernel.
      initialize(worldModel, ok);
        socket.akAcknowledge(ok.selfId);
    // initialization finish.
        control();
  }
  void control() {
      while (true) {
          // get sensory information per simulation step.
          Object data = socket.receive();
          // visual data or voice data?
          if (data instanceof KaSense) {
              update(worldModel, (KaSense)data);
              } else
              hear((KaHear)data);
          act();
      }
  }
  void act() {
    // decide their intention, and act, in this code;
    // search the nearest fire and extinguish it.
      int[] routePlan;
      // this function returns the array that contains a path to the fire.
      searchNearFire(routePlan);
      // this function creates AK_MOVE commands according to routePlan,
      // and extinguishes the fire when it arrives at the destination.
      MoveExtinguish(self.id, routePlan);
  }
}
```

14.4 Lessons Learned from RoboCup Rescue Competitions

RoboCup Rescue Competition consists of two competitions, agent competition and infrastructure competition. The agent competition is the developing intelligent agents that are provided with the capabilities of the main actors in a disaster response scenario. The infrastructure competition is to develop simulators that form the infrastructure of the simulation system and emulate realistic phenomena predominant in disasters. The teams that participated in the RCRS have improved their rescuing ability. The performances of rescue agents have been measured by metrics that reflect the policies of rescue centers. This section discusses the metrics used in the RCRS competitions.

14.4.1 Lessons from Agent Competitions

The purpose of the RCRS project is to provide emergency decision support by integrating disaster information, prediction, planning, and human interface. Rescuers have several responsibilities during disasters, and varied metrics are used in the reports on disasters published by governments or insurance companies. Metrics such as loss of human life, the number of damaged buildings, and the funds required for restoration activities indicate the extent of the disasters. The metrics are intricately linked to one another, and it is difficult to identify the dominant metric.

During the agent competitions, metrics are required to rank the performances of the rescue teams; these metrics have been set based on the principle that human life is the most valuable. The following definitions of $V1$ and $V2$ were used in 2001 and after 2002, respectively.*

$$V1 = Pint - P + 1 - (H/Hint \times B/Bint) \qquad (14.7)$$
$$V2 = (P + H/Hint) \times \sqrt{B/Bint} \qquad (14.8)$$

P is the number of living civilian agents; H, the HP values** of all the agents; and B, the area of houses not affected by the fire. $Pint$, $Hint$, and $Bint$ are the initial values. All are alive and all houses are not burned. As the disaster spreads, their scores decrease. The lesser the $V1$ value (greater the $V2$ value), the better the rescue operation.

The rescue teams comprise firefighters, ambulances, and police agents. They perform rescue operations in order to obtain better results in terms of the $V2$ values. The lessons learned from the 2004 and 2005 competitions are discussed below.

a. **Search and rescue operations in unknown areas.** Figure 14.6 shows the relative scores of semifinal games of RoboCup 2004. Eight teams (labeled A to H) performed rescue operations under four different disaster situations: Kobe, Virtual City, Foligno, and Random map[†]. The vertical scales are the normalized $V2$ values with the high score for each situation. Table 14.3 shows the number

---

*Because, the efficiency in behavior of H and the firefighting effort of B are underestimated in V1. The rules and competition settings are revised every year.
** Health point (HP) is stamina of an agent. A fixed amount is given as an initial value and the value is decreased by actions or injuries of the agents. The value is recovered by resting at refuges. $H/Hint$ represents how effectively it moves.
[†]Kobe was the most damaged area during the Great Hanshin-Awaji earthquake. Virtual City is a manmade squared area. Foligno is Italian city, where the historical tower, which was erected in the 15th century, was destroyed in an earthquake in 1997. Random map is a computer-generated map.

RoboCup Rescue: Challenges and Lessons Learned

of agents and parameters in the situations. In the case of the Kobe Map, 33 rescue agents (7 ambulance teams, 12 fire brigades, 11 police forces, 1 ambulance center, 1 fire station, and 1 police office) and 72 civilian agents performed rescue operations in the Kobe Map, where two refuges were located.

The rescue performances in the disaster situations were different among different teams. For example, team A performed better in the Virtual City situation than in the Kobe situation, while the performance of team F showed a reverse trend. We expect that it is common for a rescue team to handle one disaster situation in a better manner than the other; however, the result indicates that a performance in one disaster situation does not guarantee similar performances in other situations.

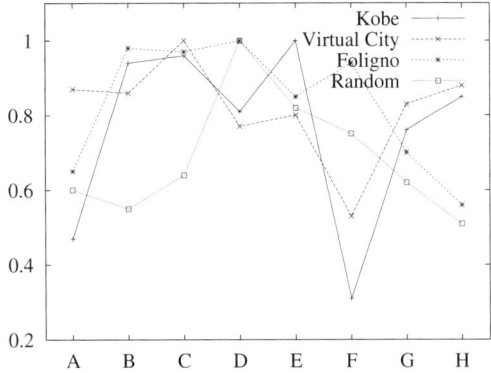

FIGURE 14.6 Normalized scores of 8 teams at RoboCup 2004 games.

TABLE 14.3 Number of agents and parameters of disaster situations in 2004 Competition

Map	Number of agents				Disaster situation	
	Ambulance	Fire	Police	Civilian	No. of Refuge	No. of fire breakout
Kobe	7	12	11	72	2	6
Virutal City	8	12	11	77	1	6
Foligno	7	11	14	81	3	6
Random	7	11	10	87	1	4

b. **Robustness to changing environments.** In 2004, the Rescue Simulation League organized a challenge session to test the robustness of the rescue operations [RoboCup 2004]. The teams perform rescue operations under different sensing conditions in the same disaster situation. On a map that has 1065 roads, 1015 nodes, and 953 buildings, there were 13 fire brigade agents, 12 police agents, 6 ambulance agents, and 89 civilian agents (Figure 14.7).

The first run (labeled Base in Table 14.4) was a simulation result using the normal sensing conditions of the rescue agents. In the second run, the visual ability of the agents was set at half the normal value, and in the third run, the hearing ability of agents was set to half its original value. In the fourth run, both the visual and hearing abilities were set to half their original values (labeled

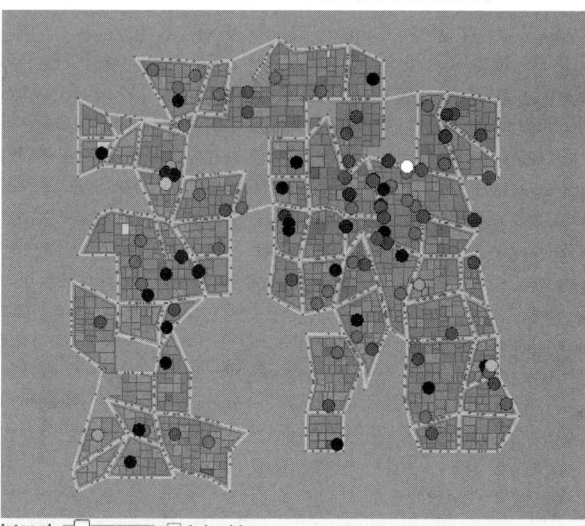

FIGURE 14.7 Snapshot of simulation during 2004 Challenge Game (Map: randomly generated one. The dark squares are houses that are burning or burnt down. The light ones are houses that are not burning or those where the fire has been extinguished. Gray, light gray, white, and black circles indicate fire brigades, civilians, ambulances and dead agents, respectively.).

TABLE 14.4 Changes in rescue performance, $V2$, for sensing abilities

sensing condition	Team X	Team Y	Team Z
Base	78.92	97.69	88.24
Half vision	78.92	35.41	83.30
Half hearing	79.92	83.49	51.45
Both	78.91	90.87	45.76

Both). Table 14.4 shows the scores of the top three teams.

The teams that showed less variation in their scores under different sensing conditions were defined to be robust in the competition. The performance of team X remains consistent, while those of teams Y and Z varied with the sensing conditions. The variations in these two teams were different. According to the definition, team X was the most robust. The results show that a more suitable definition for robust rescue operations is required, and indicates that the agent performance is dependent on the information conveyed for their decisions.

c. **Agent ability and map dependency.** In the 2005 competition, a new map, KobeBig, was adapted. Figure 14.8 shows the burning rate, $B/Bmaxs$, of one session * in the RoboCup 2005 competition. The vertical scale, $B/Bmax$, assumes a value from 0 to 1, and the higher score indicates that less houses are burnt, ie., the fire agents performed better. In this session, three maps were used, the Virtual City map with 530 nodes, 621 roads, and 1,266 buildings, and two

---

*Teams could modify their codes between sessions during the competitions. However, the same code was used during one session.

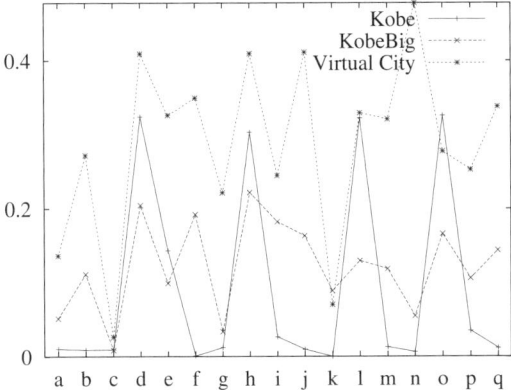

FIGURE 14.8 The changes of burning rate ($B/Bmax$) of teams in 2005 (day 3).

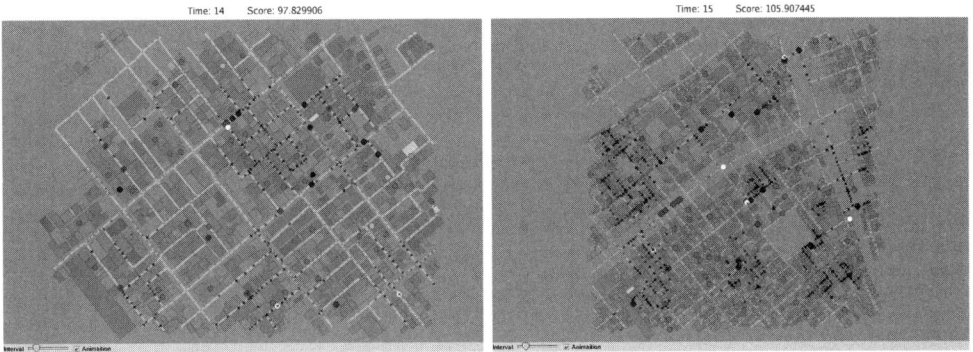

FIGURE 14.9 Snapshots of simulations at Kobe(left: the bottom left part of KobeBig) and KobeBig maps.

Kobe maps of different sizes: Kobe (nodes = 765, roads = 820, buildings = 730) and KobeBig (nodes = 2,143, roads = 2,277, building = 2,156). As shown in Figure 14.9, the bottom left part of KobeBig includes Kobe and its size is about four times that of Kobe. The size of Kobe is of the same order as that of Virtual City, and the former has a network structure similar to that of KobeBig.

The correlation of the burning rate between the performances of 17 teams (labeled a to q) for Kobe and KobeBig was 0.52, while the correlation between the performances of Kobe and Virtual city was 0.34. This result indicates that the ability of agents is maintained with regard to the road network rather than to the network size and the rescue strategy can be applied to areas with the same network rather than the same size.

14.4.2 Researches Related to Real Applications

When fires break out at various places simultaneously, a fire department deploys fire engines to minimize the possible damages. This mission is handled as a task allocation problem. Ohta showed that a fire-extinguishing task that features a collaboration among n homogeneous agents shows more than n times results [Ohta et al., 2001]. His simulation also showed that it was better to deploy fire brigades intensively to one ignition point than to deploy them widely over many ignition points. This corresponds to one of the fire department's methodologies for firefighting. It indicates that disaster and rescue simulations will possibly be used to improve the disaster prevention plans of the local governments.

Schurr et al. applied the RCRS to train human rescue officers who, in RCRS, direct rescue agents as they would direct a real rescuer [Schurr et al., 2005]. The performance of the rescue operations is used to estimate how well the officers direct them to save various towns. In this case, the officers are assumed to be familiar with their towns; in other words, they know the exact location of hospitals and gasoline stations, and have planned the evacuations of the civilians to safe places well in advance.

14.5 Examples and Discussions on Experiments Using Real Maps

It is important to show some validity in the agent-based social simulation results and to have a strong basis for the logical explanation of the results when they are applied to practical usages. This section presents the results of simulations using real maps and compares with simulation results with another method.

14.5.1 Validity of Agent-Based Simulations

In scientific and engineering fields, the following process has been repeatedly applied to increase the fidelity of simulations [Feynman, 1967].
Guess → *Compute consequence* → *Compare experiment results with simulation results*
Once the results of the simulations match the experimental data, they are used as experiment tools that substitute physical experiments.

The components of disaster simulators, such as fire and building collapses, have been programmed on the basis of models developed in civil engineering fields. Agent technology, which is used to simulate human behaviors with these disaster simulations, makes it difficult to compare the results with the real data and follow the above principle. This is because we cannot conduct physical experiments on disasters of a real scale and involve humans as factors. We compared the simulation results of the RCRS* with the data in these local government reports that were estimated by conventional methods.

The Japanese Government has been preparing to face strong earthquakes that are anticipated in the near future and support the local governments to study their respective regions [Non-Life Insurance Rating Organization of Japan, 1998]. The Fire Department of Nagoya, where our university is located, published a report on the possible damage from anticipated earthquakes with 7 to 8 of Japan Meteorological Agency seismic intensity [Nagoya Fire Bureau, 1999][Japan Meteorological Agency, 2008].

TABLE 14.5 GIS properties, damage estimations of Nagoya and simulation results

ward	GIS network		Nagoya City's estimation				RCRS simulation results			
	node	edge	build.	ig.p	burn	burn_F	build.	ig.p	burn	burn_F
Chikusa	5,581	3,711	32,156	0	0.0%	0.0%	9,924	7	2.05%	1.74%
Higashi	2,420	1,690	14,761	1	0.1%	0.1%	4,283	3	2.09%	1.38%
Kita	6,069	3,870	39,302	22	3.9%	3.4%	9,541	7	1.51%	0.99%
Nishi	6,430	4,122	44,773	58	5.8%	4.9%	10,468	7	1.97%	1.74%
Nakamura	6,044	3,766	41,769	45	5.1%	4.5%	8,994	6	2.16%	1.15%
Naka	2,026	2,093	18,726	5	0.9%	0.5%	5,396	4	2.03%	1.12%
Showa	3,795	2,456	28,464	0	0.0%	0.0%	6,325	4	0.78%	0.58%
Mizuho	4,053	2,563	30,092	2	0.5%	0.1%	6,656	4	0.65%	0.47%
Atsuta	2,609	1,760	17,580	3	1.3%	1.0%	4,309	3	4.32%	1.42%
Nakagawa	9,449	6,154	58,612	31	2.6%	1.7%	17,327	13	0.93%	0.87%
Minato	7,127	4,892	38,694	0	0.0%	0.0%	15,269	18	1.32%	1.24%
Minami	5,718	3,710	43,318	1	0.0%	0.0%	10,157	7	2.11%	1.71%
Moriyama	6,651	4,413	39,821	0	0.0%	0.0%	13,077	13	1.80%	1.22%
Midori	8,945	5,996	53,445	0	0.0%	0.0%	18,301	15	1.11%	1.06%
Meito	5,612	3,724	27,554	0	0.0%	0.0%	10,740	8	2.27%	1.66%
Tenpaku	5,986	3,951	29,584	0	0.0%	0.0%	11,259	9	2.03%	1.79%
correlation between Nagoya data and simulations							0.85	-0.03		0.58
correlation between burn and burn_F						1				0.65

Experiment 1: Consistency with other methods

Table 14.5 shows the data from the report and simulation results**. The left-hand column shows the properties of maps that are larger than those used in the RoboCup Rescue competitions. The second column shows the data obtained from the Nagoya report. They provide estimates of the damages incurred by a macro model developed based on past fire incidents.

build.: the number of houses in each ward,

ig.p: the number of expected ignition points, assuming that no fire breaks out in half of the number of wards,

burn: rate of houses burned without firefighting,

burn_F: rate of houses burned after firefighting.

The **burn** and **burn_F** values are the ratios of the estimated number of burnt houses to the total number of houses. The same macro fire model was used to calculate **burn** and **burn_F**, and they simulated the effect of fire fighting by decreasing the number of ignition points. The correlation between them in the Nagoya City report is equal to 1, which confirms this fact.

The right-hand column shows the results of RCRS simulations. Only the fire fighters were taken into consideration in the Nagoya report; therefore, fire brigade agents were involved in this experiment. The number of fire fighters was 40 across all the wards. A program produces data of each house so that the number of generated houses was proportional to the real numbers of the report.*** The number of ignition points was made proportional to the areas, and the ignition positions were uniformly distributed over the areas.

---

*The RCRS (Ver.46) was used for the simulation [RoboCup Rescue Simulation Project, 2008].

** Maps have been converted from the digital maps of Japan that are available as free data on a 1:25.000 scale [Geographical Survey Institute, 2008]. The RCRS uses some random seeds for the simulations. Even the random seeds and the initial conditions are set to be the same, the simulation results fluctuate due to data exchanges over the network. The values in Tables 14.5 and 14.6 are average values over ten simulations.

***Data related to building are essential for fire-spread simulation, building collapse simulation, and so on. Buildings are privately owned properties, and are not included in the public data.

The data of Nagoya City report are also the results of other simulation methods, so their relative values are checked. The **burn_F** values were lower than the **burn** values for every ward; this shows that introducing fire agents leads to lesser damage. The **burn_F** values in the Nagoya report and the simulation show a positive correlation. While the **burn** values do not show such a correlation, this difference may be caused by the difference of fire simulator models. The correlation between the **burn** and the **burn_F** is positive (0.65). From the above, the simulation results of RCRS indicate a similar trend with the data of the report.

Experiment 2: Effect of shape of disaster-struck areas

Disasters occur beyond the confines of the administrative districts of local governments. It is natural to simulate disasters in a continuous area. Figure 14.10 shows an outline of such cases. The figure to the left shows two wards, Nishi and Nakamura (shown in Table 14.5). The central map is a combination of the two adjoining maps. The figure to the right shows the bottom central part, where the houses are clustered. This area is the densest, and is identified to be the most disaster-prone area by the government. Hence, prevention and rescue actions should be also performed in these areas during earthquakes.

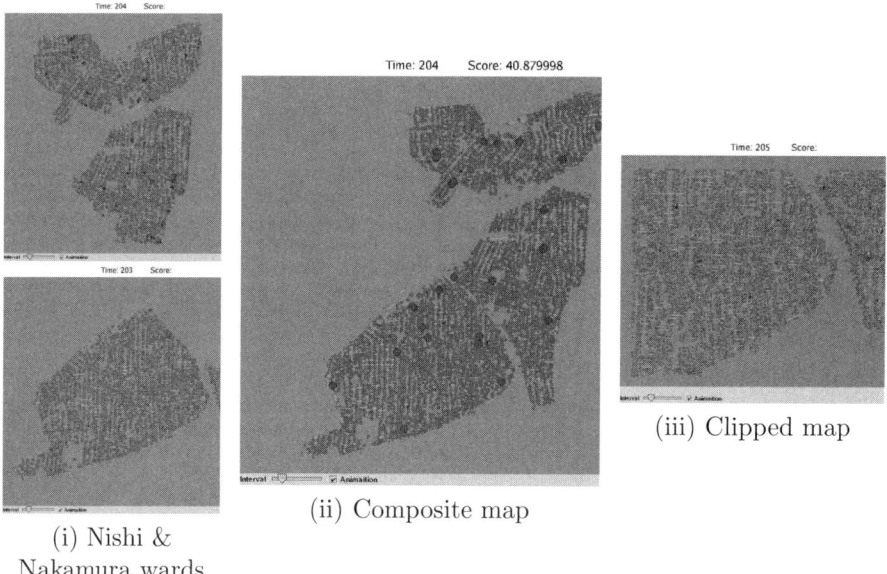

(i) Nishi & Nakamura wards   (ii) Composite map   (iii) Clipped map

FIGURE 14.10 Administrative boundaries of the two wards: The composite map of Nishi and Nakamura wards, and the disaster-prone area (the clipped map) are used for disaster simulation.

Table 14.6 shows the simulation results of the disaster-prone area. The numbers of ignition points were set for two cases, during the daytime (numbers in experiment 1) and at night, as shown in the report. The results show good correlations with the Nagoya report, and the simulation results for the composite map show a similar trend.

TABLE 14.6 Ignition points set equal to the estimated earthquake

ward	no. ignitions		no fire brigade	fire brigade
Nichi	30	(night)	8.53%	8.08%
	58	(day)	13.40%	12.96%
Nakamura	22	(night)	8.90%	8.45%
	45	(day)	15.64%	15.23%
correlation with Nagoya estimation data			0.89	0.92
Clipped area (the disaster-prone area)			7.20%	7.14%

14.5.2 Discussion on Experiments

The results of the experiments qualitatively match the data of the report. The officials were questioned about the possibility of using the RCRS as their tool when we explained the results to them. Their comments were as follows:

- There are no precedents or no theoretical backgrounds.
- The simulation size is significantly different from the real one; for example, the number of agents in the experiments is smaller than the real-world situation.

These comments indicate that the validity of the RCRS is required to persuade potential users to use it as their tool. This requirement is not limited to only the RCRS but is also applicable to the agent-based social simulation (ABSS), which is difficult to be supported by actual examples or theories.

Local governments intend to know why one rescue plan is better than another and what type of agent actions produce good simulation results. In the real world, interviewing the survivors and the rescue members provide the necessary information on rescue strategies to save their town from disasters. By using only metrics such as P, B, H, and $V2$, it is difficult to determine the reason for one team performing better than the other and the desired agent actions for obtaining good results. To link agent behaviors with the metrics, it is necessary to interpret the behaviors of the specific agents from the simulation results and to express them between microscopic and macroscopic levels.

14.6 Analysis of ABSS Based on Probability Model

In this section, we propose a method to analyze the simulation outputs without any metrics related with the application domain and show the interpretation of rescue agents behaviors with the analyzed results.

14.6.1 Agent Behavior Formulation and Presentation

The ABSS outputs complex collective results of the agent behaviors and agent interaction with other components. The MAS is formalized simply as follows [Weiss, 2000].

$$action : S^* \to A, \quad env : S \times A \to \mathcal{P}(S),$$

where $A = \{a_1, a_2, \ldots, a_l\}$ is a set of actions that agents perform; $S = \{s_1, s_2, \ldots, s_m\}$, a set of environment states; and $\mathcal{P}(S)$, the power set of S. An agent at s_i will plan his/her possible actions and execute one a_i according to the prior knowledge and information received. The action changes the situations to the next state. The interaction between an agent i and the environment leads to the next state. This is represented as a history h_i:

$$h_i : s_0 \xrightarrow{a_0} s_1 \xrightarrow{a_1} s_2 \xrightarrow{a_2} s_3 \cdots.$$

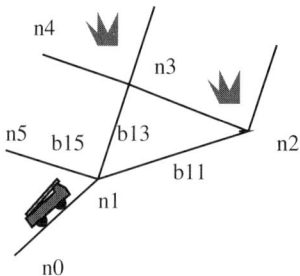

FIGURE 14.11 Illustration of rescue agent behavior model at n1.

Here, $H = \{h_1, \ldots, h_n\}$, where n is the number of agents involved, and the environments are the observable outputs.

Our idea is to extract the features of the agent behavior from H and the properties of the environments based on the probability models. The behaviors of agents with the same aim will lead to the selection of similar actions. However, all the agents do not necessarily act in the same manner because their internal states such as the history of their actions and prior information are different. $\mathcal{P}(S)$, indicates that stochastic processes exist. The behaviors of the agents at state i are presented as a set $\{p_{ij}\}$, where p_{ij} is the probability of the agents performing an action that causes the state to be at state j subsequently.

Figure 14.11 shows a situation in which a rescue agent arrives from n_0 to extinguish fires. The agent knows that there are two fires near n_2 and n_3. At crossing n_1, the agent decides on the fire to be extinguished first and decides to move forward, turn right, or do something else. The behaviors of the rescue agent are observed as $P_1 = \{p_{10}, p_{11}, p_{12}, p_{13}, p_{14}, p_{15}\}$, where p_{1j} is the probability of the agents being at n_j in the subsequent step. From these probabilities, the agent behaviors are analyzed, and the results are used to help users recognize the differences precisely.

A stochastic matrix $\mathbb{P} = \{p_{ij}\}$, where p_{ij} is the probability of the agents at state i, moving to state j after one time step, describes the behavior of the rescue agent by assuming the following:

Assumption 1 *Agents select their actions to attain their goal efficiently and promptly.*

Assumption 2 *States that agents visit more often are more important to them than other states.*

Assumption 3 *Differences between actions with good and bad performances are represented as the differences in their probabilities.*

The stochastic matrix $\mathbb{P}$ has the conditions. $p_{ij} \geq 0$ and $\sum_i p_{ij} = 1$ for all j. The transition probability can be approximated by calculating the ensemble average of the times the agents change states from i to j from the ABSS output. The following conditions pose problems when applying a stochastic matrix directly to interpret the agent behaviors:

- Let m be the size of states, the size of $\mathbb{P}$ is $m \times m$. Knowing m itself indicates that the states of the agents are known before analyzing their behaviors.
- $\sum_i p_{ij} = 1$ for all j implies that transitions to any state can occur. All the states are equally treated, while the values of the states may differ depending on the applied tasks.

RoboCup Rescue: Challenges and Lessons Learned

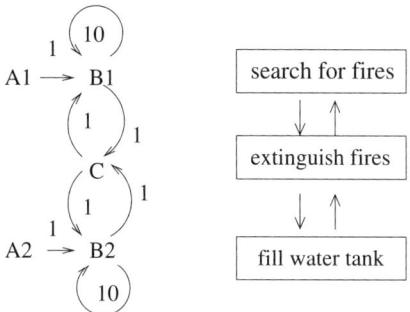

FIGURE 14.12 Action pattern of fire brigade agent and its state transition.

We use a frequency matrix, $\mathbb{F}$, where p_{ij} are normalized as $\sum_{ij} f_{ij} = 1$ instead of $\mathbb{P}$ *. The following properties are indicated by considering $\mathbb{F}$ as a directed graph with f_{ij} s associated with the correspondent edges.

Property 1 $\mathbb{F}$ *indicates agent behavior. The rank of $\mathbb{F}$ is proportional to the range of agent motions.*

Property 2 *Large elements of dominant eigenvectors correspond to places that are important in a social simulation (in the rescue domain, to protect a town from disasters).*

Property 3 *When agents move in separate areas, the components of the eigenvectors are also divided.*

Figure 14.12 shows a pattern of the actions of the rescue agent, in which two fires break out at two places, B1 and B2, and the two agents are at A1 and A2 initially. According to assumption 1, the rescue agents promptly locate victims at the disaster sites and rescue them efficiently. The fire brigade agents move toward the fires and attempt to extinguish them until their water supplies are all gone. When observing the simulation results, the places visited by the agents are assumed to represent their states, namely, when an agent is at Bs, the agent is assumed to be in the state of **extinguish fire**.

Let the tank of the fire engine become empty after ten consecutive extinguishing actions, and let the fire engine move to a water station, C, with a cost of one time step. This corresponds to ten **extinguish fire** states and one **fill water tank** state. In the $\mathbb{F}$ notation, the behaviors of the fire brigade agent are represented as follows.

	A_1	B_1	A_2	B_2	C
A_1		$1/m$			
B_1		$10n/m$			n/m
A_2				$1/m$	
B_2				$10n/m$	n/m
C		n/m		n/m	

$$\xrightarrow{n\to\infty} \begin{pmatrix} 0 & 0 & 0 & 0 & 0 \\ 0 & 10/24 & 0 & 0 & 1/24 \\ 0 & 0 & 0 & 0 & 0 \\ 0 & 0 & 0 & 10/24 & 1/24 \\ 0 & 1/24 & 0 & 1/24 & 0 \end{pmatrix}$$

The representation on the left shows $\{f_{ij}\}$ after repeating n times of fire-extinguishing and tank filling operations. The total number steps is $m = 24n + 2$. By repeating the action patterns, the factors corresponding to the key states become dominant, while the

---

*Let N_i be a number representing how many agents are at state s_i, while N is $\sum_i N_i$. Then $f_{ij} = \frac{N_i}{N} p_{ij}$.

others become less significant. The matrix on the right goes from n goes to infinity, and its rank is 3, which corresponds to the number of dominant places: B1, B2, and C. The three eigenvalues of $\mathbb{F}$ are $0.425, 0.417$, and -0.008, and the eigenvectors $(0, 0.7, 0, 0.7, 0.14)^t$ and $(0, 0.7, 0, -0.7, 0)^t$ correspond to the first and second dominant eigenvalues, respectively. The major components of the eigenvectors correspond to the locations of the fires frequented by the agents, i.e., B1 and B2. This also indicates that the most significant state is that of extinguish fire.

14.6.2 Analysis Results of RoboCup Rescue Competition Logs

The above method was applied to the log files of the RCRS containing the records of the agent positions. When assuming that the places visited by the agents represent their states, $\mathbb{F}$ is a matrix with $NP \times NP$ where NP is the sum of the number of roads, nodes, and buildings. For M simulation steps, there are at the most $M - 1$ non-zero elements in $\mathbb{F}$ for one agent. In the case of $M \ll NP$, $\mathbb{F}$ becomes a sparse matrix. A matrix, $\mathbb{X}$, whose elements are the nonzero values of $\mathbb{F}$ is used in the following discussion. In addition, the analysis results of the behavior of the fire brigade agents are presented below. The first result shows that the analysis method provides more information than the metric V2 introduced in Section 14.4.1. The second result illustrates that the method can be used to show time sequence changes.

2004 Challenge Session

The 2004 challenge session cited in Section 14.4.1 was conducted on a map with 1065 roads, 1015 nodes, 953 buildings, and $\mathbb{F}$ with a size of 3033. Table 14.7 shows the results obtained by analyzing the behaviors of the fire brigades; interesting possibilities that could not be envisioned by V2 are observed in Table 14.4.

- The size/ranks column shows the size of $\mathbb{X}$ and its rank, which indicates the range of activities. It is apparent that there are changes in their behaviors with regard to their sensing abilities, even in the case of team X, whose V2 scores are similar across the four cases. When the visual sensing ability is reduced to half its original value, the size and rank of $\mathbb{X}$ are greater than those when the ability is normal; therefore, there is a tendency for area broadening. In addition, team X moves to a lesser extent when the hearing ability is reduced to half. On the other hand, team Y moves over a broader area in the three conditions with degradation in sensing than under the normal conditions; team Z shows a behavior opposite to that of Y.
- The sizes of the eigenvalues correspond to the importance of the places that the agents visited. When there are two major places, the values of the corresponding eigenvalues are of the same order as that in the case of Figure 14.12. The ratio of the 1st and 2nd eigenvalues is close to 1.0. In case the agents visit one place, the 1st eigenvalue becomes dominant and the ratio decreases to 0.0. Columns 2/1 and 3/1 show the ratios of the dominant eigenvalue to the 2nd and the 3rd values, respectively.

    For example, the agents of team Y are assumed to move mostly according to the dominant eigenvalue under the normal sensing conditions. Under the conditions when the hearing and visual sensing abilities are reduced to half of the original, the agent behaviors are assumed to be divided into three patterns since the magnitudes of the dominant, 2nd, and 3rd eigenvalues are equal. It is interpreted

TABLE 14.7 Size of $\mathbb{X}$ and ratio 2nd & 3rd eigenvalues to 1st eigenvalue

Sensing Condition	Team X			Team Y			Team Z		
	Size/Rank	2/1	3/1	Size/Rank	2/1	3/1	Size/Rank	2/1	3/1
Base	899/820	0.38	0.25	342/254	0.27	0.27	433/391	0.49	0.21
Half vision	920/867	0.31	0.19	650/563	0.85	0.73	372/297	0.58	0.43
Half hearing	787/649	0.80	0.69	626/552	0.85	0.39	372/316	0.69	0.12
Both	685/592	0.85	0.78	575/484	0.96	0.93	332/284	0.79	0.78

2/1, 3/1 stands for 2nd eigenvalue / 1st eigenvalue, 3rd eigenvalue / 1st eigenvalue.

that the agents move inconsistently when the sensing conditions are degraded.

Time Sequence Analysis of Disaster Simulations

Figure 14.13 shows the snapshots of the simulations on the Kobe map. In this simulation, 13 fire fighter agents, 7 ambulance agents, 11 police agents and 85 civilian agents were involved. By visual inspection, the following observations were obtained.

(a) Initially, three fires break out at locations, B1, B2, and B3.
(b) After 50 steps, the fires at B1 and B2 are extinguished, while the fire at B3 spreads. (In Figure 14.13, the black portions indicate the areas that are burned.)
(c) Agents converge to extinguish the fire at B3.
(d,e) After 150 steps, the spread of the fire is prevented by the actions of firefighters.

Table 14.8 shows the time sequence changes in the burning rate and $\mathbb{X}$ of the fire brigade agents; the size, rank, and coefficients of the dominant eigenvectors. From the table, we can make the following observations:

- The rank of $\mathbb{X}$ increases with the simulation steps. This change indicates that the range of the fire brigade agents widens.
- The ratio of the 1st eigenvalue to the 2nd and 3rd values becomes more dominant over time. The ratio of the 1st to 2nd eigenvalue is 0.38 at 50 steps, and it decreases to 0.03 at 300 steps. This indicates that the agents initially moved separately, and they showed a uniform behavior as the simulation proceeded.
- The following changes in the components of the dominant vectors correspond to the changes in the snapshots.
    - refuge_0 is the key building in all the steps,
    - between 100 to 250 steps, a place (road_1) near the fire (B1) is the second dominant,
    - beyond 250 steps, a place (b_3) near another fire (B3) appears as the key place and becomes the 2nd one.

These interpretations correspond to that obtained from visual inspection; this shows that $\mathbb{X}$ can indicate better the process of disaster simulations than the burning rate.

446 *Multi-Agent Systems: Simulation and Applications*

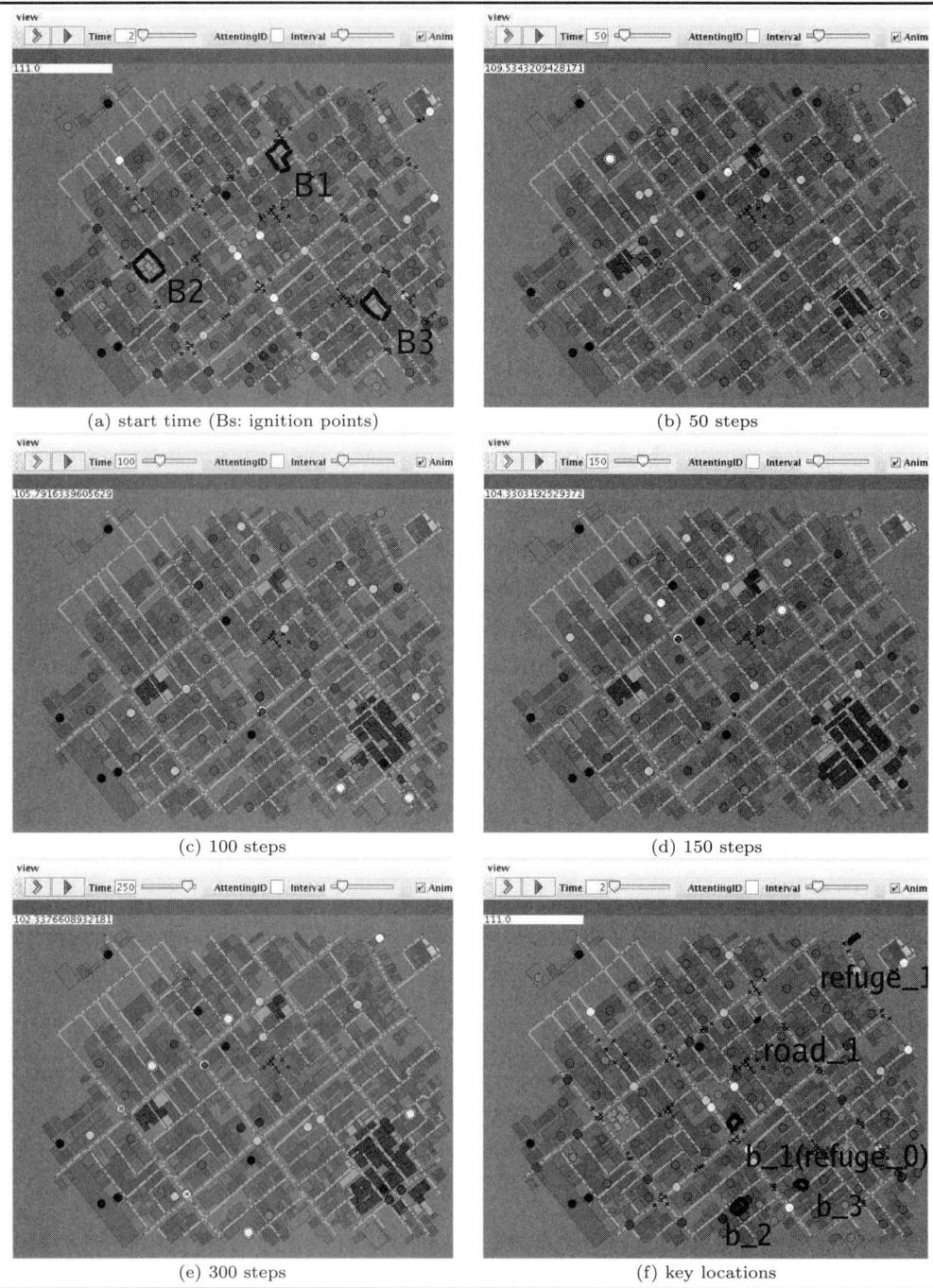

FIGURE 14.13 Time sequences of disaster simulations. The burned areas, the black portions, spread as the simulation proceeds.

TABLE 14.8 Time sequence of burning rate, size of matrix and eigenvectors

step	burning rate	properties of X			key locations corresponding to component of dominant e.vectors		
		size/rank	ratio of E.V.s		1st	2nd	3rd
			e.v(2/1)	e.v(3/1)			
50	2.6%	155/135	0.38	0.38	b_1	b_2	road_1
100	4.0%	246/217	0.11	0.10	b_1	road_1	b_2
150	4.9%	300/271	0.06	0.06	b_1	road_1	b_2
200	5.0%	355/325	0.05	0.04	b_1	road_1	b_2
250	5.0%	422/388	0.04	0.04	b_1	road_1	b_3
300	5.1%	484/442	0.03	0.03	b_1	b_3	road_1

b_1, b_2, b_3, and road_1 are marks of (f) in Figure 14.13

TABLE 14.9 Features of two MAS: Rescue simulation and Soccer simulation

Applied field	Rescue actions	Soccer games
number of agents	100 or more	11 per team
agents in the team	heterogeneous (fire brigade, ambulance) hierarchical (center office, platoon)	homogeneous (FW,GK)
communication	voice, telephone (wire/wireless),	voice
simulators	disasters, traffic	motion (player, ball)
logistics	major issue	-
information access	very bad	reasonably good
representation	symbolic & nonsymbolic	nonsymbolic
control	distributed/semi-central	distributed
filed structure	dynamic (GIS)	static (goal, line)
agent motions	only move along roads	can move to any point in xy plane

14.7 Discussion and Summary

The MAS provides schemes to solve complex problems that are difficult to model [Jennings and Bussmann, 2003]. In this chapter, we discussed the following features that are required to utilize the ABSS practically.

- An architecture to simulate social activities including human behavior.
- A method to assess the simulation results and explain them in a lucid style.

Architecture of the ABSS

Table 14.9 shows the features of two simulation systems: Rescue and Soccer simulation systems. Both have been used in RoboCup Competitions, and have been the subject of interest for many researchers. Since the RCRS system was developed based on the success of the RoboCup Soccer simulation system, their architectures are similar. The architecture of the RCRS has been extended to simulate disasters and human behaviors in the GIS data, as explained in Section 14.3. The kernel controls the simulation process and facilitates information sharing among the modules, and it is expected to manage tens thousands of modules and their communications in real time in the future. This will be one of the most challenging problems in the agent based social simulation system. The RCRS does not support the following features, which are the future requisites:

1. Real-time simulation: When the system is used practically, the data from disaster sites will be input to the system. Simulations using the data will help local governments instruct civilians to evacuate to safe places; therefore, it is required that the data be obtained from the real world and disasters be simulated promptly.
2. Distribution of data and computation for a large city: There are situations where most of the data are derived from one area, while the other areas have few data to be calculated. More computation resources are used for simulations of wider areas with more agents. The distribution over the area and integration would

then become key technologies.

3. Interface for real world: Data from the real world cannot be obtained regularly, and the time of availability of the data triggers the event simulation. Asynchronous timing is required to support such simulations. High Level Architecture, an IEEE standard, supports the asynchronous timing and Fiedrich developed an HLA based disaster simulations [Fiedrich, 2006]. As Koto et al. pointed that HLA has not specified the interface of distributed Runtime Infrastructure at present [Koto and Takeuchi, 2003].

4. Communication among agents: To simulate the ABSS with more than 10,000 agents, not only the communication traffic among them but also the ontology used in the communication becomes important. It is easy to express the intention of the agents by a task-level language rather than by command-level operations.

Assessment of Simulation Results

To practically use a rescue simulation that has no model or experimental data, the assessment of the output becomes a new challenge. There are two types of assessments, one type is a domain-specific assessment and the other is a method to task-independently evaluate agent behavior.

1. We first discussed the need for agent-based disasters and rescue simulation systems based on their usage by local governments.
2. Next, we showed that the simulation results are comparable to those of other traditional methods. The experiments conduct using real GIS data are in good correlation with other methods. This indicates the possibility of the use of ABSS for practical application.
3. Finally, we proposed a method to represent agent behaviors by a probability model and interpret them by using the eigenvalues of the matrix. The task-independent method when applied to the RCRS results shows good behavior interpretations. This supports the ABSS approach for practical applications in the future. We hope that the discussions and our proposed method will help future researchers to study the development of disaster simulations in greater detail.

The task-independent evaluation method provides useful methods to show the validity of the results when the MAS is applied to social tasks. A stochastic matrix has been used in various domains. Recently, it has been applied to fields where the data are not theoretically well formalized but are heuristically presented. For example, the Page Ranking method used in Google in its simple form is identical to the eigenvector centrality of the matrix constructed from the hyperlink data on the WWW [Page et al., 1998]. Our proposed method, which presents the histories of the agent behaviors by a variant of a stochastic matrix, has certain aspects that are common with the Page Ranking method.

Ever since we proposed the RCRS in 2000, many disasters have occurred*. We hope that our simulation technologies will mitigate the damage caused by disasters, and that the development of human behavior models will help analyze the simulation results [Silverman

---

*Hurricane Katrina, the Indian Ocean tsunami, and 2008 Sichuan earthquake; The mechanism of these disasters were different from the Hanshin-Awaji earthquake, however, simulation systems to decrease the damages are desired.

et al., 2008]. We also hope that this work will help researchers investigate future topics on related disaster simulations.

References

P. Currion, C. de Silva, and B. V. de Walle. Open source software for disaster management. *Communications of the ACM*, 50(3):61–65, 2007.

B. V. de Walle and M. Turoff (2007). Emergency response information systems: Emerging trends and technologies. *Communications of the ACM*, 50(3):28–65, 2007.

Disaster Reduction and Human Renovation Institution. Lessons from the great hanshin-awaji earthquake., 2008. http://www.dri.ne.jp/kensyu/pdf/jica_en.pdf, accessed June 2008.

R. P. Feynman. *The Character of Physical Law*. The MIT Press, Cambridge, MA, 1967.

F. Fiedrich. An HLA-based multiagent system for optimized resource allocation after strong earthquakes. In *WSC '06: Proceedings of the 38th conference on Winter simulation*, pages 486–492. Winter Simulation Conference, 2006. ISBN 1-4244-0501-7.

Geographical Survey Institute. Japanese portal site for retrieving geographic information. http://zgate.gsi.go.jp/ch/jmp20/jmp20_eng.html, accessed June 2008, 2008.

Japan Meteorological Agency. Explanation table of jma seismic intensity scale. http://www.jma.go.jp/jma/kishou/know/shindo/explane.html, accessed June 2008, 2008.

N. R. Jennings and S. Bussmann. Agent-based control systems. *IEEE Control Systems Magazine*, 23 (3):61–74, 2003.

H. Kitano, S. Tadokoro, I. Noda, H. Matsubara, T. Takahashi, A. Shinjou, and S. Shimada. Robocup rescue: Search and rescue in large-scale disasters as a domain for autonomous agents research. In *IEEE International Conference on System, Man, and Cybernetics*, volume 6, pages 739–743, 1999.

T. Koto and I. Takeuchi. A distributed disaster simulation system that integrates sub-simulators. In *First International Workshop on Synthetic Simulation and Robotics to Mitigate Earthquake Disaster*, 2003. Online proceedings, http://www.dis.uniroma1.it/~rescue/events/padova03/papers/index.html. Accessed June 2008.

D. Lab. *Combined Systems, Combining More for Crisis Management*. ISBN 90-811158-1-2, 2006.

S. Mehrotra, T. Znatri, and W. T. (edited). Crisis management. *IEEE Internet Computing*, 12(1):14–54, 2008.

S. Moss and B. Edmonds. Towards good social science. *Journal of Artificial Societies and Social Simulation*, 8(4), 2005. ISSN 1460-7425.

Nagoya Fire Bureau. *Nagoya City: Damage Estimate from Earthquake*. Nagoya City, 1999. in Japanese.

Non-Life Insurance Rating Organization of Japan. Report on seismic assumed damage, 1998. http://www.nliro.or.jp/disclosure/q_tyosahoukoku/28.html, accessed June 2008, in Japanese.

M. Ohta, T. Takahashi, and H. Kitano. Robocup-rescue simulation: in case of fire fighting planning. In *RoboCup 2000: Robot Soccer World Cup IV*, pages 351–356. Lecture Notes in Computer Science, Springer, Heidelberg, 2001.

L. Page, S. Brin, R. Motwani, and T. Winograd. The pagerank citation ranking: Bringing order to the web, 1998. http://www.db.stanford.edu/ backrub/pageranksub.ps. Accessed June 2008.

RoboCup 2004. Robocup 2004 rescue simulation league hp. http://robot.cmpe.boun.edu.tr/rescue2004/, accessed June 2008.

RoboCup Rescue Simulation Project. http://sourceforge.net/projects/roborescue, accessed June 2008., 2008.

N. Schurr, J. Marecki, N. Kasinadhuni, M. Tambe, J. Lewis, and P. Scerri. The defacto system for human omnipresence to coordinate agent teams: The future of disaster response. In *AAMAS 2005*, pages 1229–1230, 2005.

B. G. Silverman, G. K. Bharathy, and G. J. Kim. Challenges of country modeling with databases, newsfeeds, and expert surveys. This volume, 2008.

T. Takahashi, I. Takeuchi, F. Matsuno, and S. Tadokoro. Rescue simulation project and comprehensive disaster simulator architecture. In *IEEE/RSJ International Conference in Intelligent Robots and Systems(IROS2000)*, pages 1894–1899, 2000.

I. Takeuchi, S. Kakumoto, and Y. Goto. Towards an integrated earthquake disaster simulation system. In *First International Workshop on Synthetic Simulation and Robotics to Mitigate Earthquake Disaster*, 2003.

M. Turoff. Past and future emergency response information systems. *Communications of the ACM*, 45(4):29–32, 2002.

G. Weiss, editor. *Multiagent Systems - A Modern Approach to Distributed Artificial Intelligence*. The MIT Press, Cambridge, MA, 2000.

15

Agent-Based Simulation Using BDI Programming in *Jason*

15.1	Introduction...	451
15.2	Programming Languages for Multi-Agent Systems...	452
15.3	Programming Multi-Agent Systems Using *Jason* . Language • Interpreter	453
15.4	*Jason* Features for Simulation .... Environments • Execution Modes • Internal Actions • Customized Architectures	459
15.5	Example .... Environment • Agents • Results	464
15.6	Ongoing Projects...................................	471
15.7	Conclusion	473
	References	473

Rafael H. Bordini
University of Durham, UK

Jomi F. Hübner
ENS Mines Saint-Etienne, France

15.1 Introduction

Programming languages for multi-agent system have received enormous research attention in the last few years. Such languages are now very expressive yet with practical interpreters and reasonably user-friendly platforms for developing multi-agent systems based on such languages. One advantage of these languages for developing multi-agent systems is that the language provides constructs for several concepts and abstractions used in designing or specifying sophisticated multi-agent systems. So not only are these languages suitable for implementing designs created with agent-oriented software engineering methodologies, but they are also ideal for developing simulations with cognitive agents. In particular, the use of such languages gives a direct declarative representation for an agent's mental states, for example the beliefs it currently holds about its environment or other agents sharing the environment (e.g., in a particular simulation) or the goals it is currently trying to achieve, as well as how the agent has decided to act upon the environment so as to bring about those goals. As a consequence, human observers of a simulation developed with an agent programming language can check not only how social phenomena emerge but also what are the agents' mental attitudes that emerged at the same time or led to the observed social phenomena. This is of fundamental importance for future trends in social simulation aiming at investigating the micro-macro link problem; that is, how social interaction and organizations affect mental states and vice-versa.

Most available platforms for agent-based simulation offer easy to use interfaces so that scientists can develop agent-based simulation even without much knowledge of programming. However, the consequence is that agents developed with those tools are either reactive or have very simple behavior. The main contribution of our approach (described in this chapter) is that it supports the development of agent-based simulations where agents are potentially much more sophisticated then typically used in current simulations. At the same time, the use of efficient interpreters for agent programming languages means they can run in reasonable time, unlike some AI-based approaches where there is, for example, reasoning from first principles. Although this approach has not been tried extensively yet for social simulation, it offers a completely new approach for developing agent-based simulations where agents have declarative representations of mental attitudes and display rational, goal-directed behavior.

In this chapter, we will concentrate in introducing particularly the use of the ***Jason*** agent platform for (social) simulation. ***Jason*** includes an interpreter for an extended version of the AgentSpeak programming language, which is based on the BDI agent architecture. We will first summarize the main characteristics of BDI agents and how they are implemented in ***Jason*** and then discuss in detail some of the aspects which are particularly important for simulation, for example the execution mode where agents can run synchronously (whereas agents typically run asynchronously, reacting to perceived events by executing plans that include the achievement of long-term goals), the support for developing environments where agents are situated, ***Jason***'s "mind inspector", and the integration with code written in traditional programming languages. We will also describe various ongoing projects which aim to add various functionalities to the ***Jason*** platform; these in turn will have a significant impact in the types of simulations that can be developed with ***Jason***. The chapter also has a running example showing how the various aspects of the platforms could be used in (social) simulations.

15.2 Programming Languages for Multi-Agent Systems

In recent years, there has been an extremely rapid increase in the amount of research being done on agent-oriented programming languages, and multi-agent systems techniques that can be used in the context of an agent programming language. The number and range of different programming language, tools, and platforms for multi-agent systems that have appeared in the literature [Bordini et al., 2006] is quite impressive, in particular logic-based languages [Fisher et al., 2007]. In [Bordini et al., 2005b], some of the languages that have working interpreters of practical use were presented in reasonable detail. Other languages have been discussed in other available surveys, e.g. [Mascardi et al., 2004; Dastani and Gomez-Sanz, 2006]. Even though we here do not discuss existing agent languages and platforms, it is worth giving references to some examples of well known agent-oriented programming and platforms: 3APL [Dastani et al., 2005] (and its recent 2APL variation), MetateM [Fisher, 2004], ConGolog [de Giacomo et al., 2000], CLAIM [El Fallah Seghrouchni and Suna, 2005], IMPACT [Dix and Zhang, 2005], Jadex [Pokahr et al., 2005], JADE [Bellifemine et al., 2005], SPARK [Morley and Myers, 2004], MINERVA [Leite et al., 2002], SOCS [Alberti et al., 2005; Toni, 2006], Go! [Clark and McCabe, 2004], STEAM [Tambe, 1997], STAPLE [Kumar et al., 2002], JACK [Winikoff, 2005].

Quite a few of these languages are based on the BDI agent architecture or at least on the essential notion of *goal*. Explicit representations of long-term goals that the agent is trying to achieve are a fundamental requirement for the implementation of a software entity that displays the kind of behavior that we expect of an "autonomous agent" [Wooldridge, 2002]. It allows agents to take further action when a plan executed to achieve a particular goal fails to do so (e.g., because the environment where the agent is situated is unpredictable);

it also allows agents to reconsider the goals they committed themselves to achieve if new opportunities are perceived in the changing environment. For this reason, it seems that, at least conceptually, agent programming languages subsume the kind of reactive agent behavior: we can write *plans* telling an agent how to behave in reaction to something directly perceived in the environment, but we can also keep track of long-term goals the agent is trying to achieve, and we can then add mechanisms to help the agent make decisions on how to balance both types of behavior. On the other hand, reactive agent approaches rely significantly on aspects of the environments as well as agent behavior; we mention in Section 15.4.1 that existing approaches for environment modeling can be combined with ***Jason***, and indeed with other agent languages too (see, e.g., [Ricci et al., 2008]).

However, it is important to bear in mind that these are just programming languages, they do not provide any "intelligence" for free. What they do provide are abstractions which help *humans* cope with the development of sophisticated (distributed) systems. Most multi-agent programming platforms will work on top of agent-based middleware (e.g., JADE [Bellifemine et al., 2005]) that make certain issues of distributed computing fairly transparent for developers. Another essential characteristic is that they provide direct mechanisms for agents to *act* within an environment, and inter-agent communication uses much higher-level abstractions than in classical distributed systems: agents will typically use knowledge-level communication languages, based on speech-act theory [Austin, 1975] (possibly changing their mental states as a consequence). The other essential features of agent platforms and languages are typically related to agents' mental attitudes. An important aspect in this approach to distributed systems is that agents' mental attitudes are *private*: no other software component can directly access the current mental state of an agent.

15.3 Programming Multi-Agent Systems Using *Jason*

As one would expect from the discussion above, the most important constructs in agent programming languages were created precisely for representing mental attitudes. We will now concentrate on presenting one particular language, the variant of AgentSpeak [Rao, 1996] as interpreted by ***Jason*** [Bordini et al., 2007b], to give a flavor of such language constructs. We will then also discuss important aspects of the interpreter for this language.

15.3.1 Language

The first thing to note about AgentSpeak is that it builds upon logic programming, so the basic representation units are "predicates", very much as in languages such as Prolog (including the convention that identifiers starting with an uppercase character denote logical variables). In ***Jason***, the main (agent) language constructs are:

beliefs, which represent information the agent has about the environment, other agents, or about itself — the term "belief" is used to emphasize the fact that agents might have incorrect and/or incomplete information about the environment and other agents;

goals, which represent state of affairs the agent wishes to bring about (essentially things the agent could potentially come to believe if the goal is achieved); and

plans, which are courses of action that the agent can use to achieve goals or to react to perceived changes in the environment.

An example of a belief is `food(2,3)`, which could mean that the agent believes there is food at coordinates ⟨2,3⟩. Another example is `~own(food,2,3)` to mean that the agent explicitly believes it is *not* the case that it owns the food at those coordinates. Another interesting aspects of beliefs, in **Jason**, is that they can have *annotations* which can be used for example to maintain meta-level information about individual beliefs. In particular, **Jason** automatically annotates the source of all information received by an agent: percept is used to denote information received from sensors (i.e., through perceiving the environment), self is used to denote beliefs created by the agent itself as "mental notes" (e.g., of things it has done in the past), or an agent name is used in the representation of information received from other agents. So, for example, food(2,3)[source(percept),source(ag2),time(10)] would mean that the belief, acquired at time 10, that there is food at those coordinates was both perceived by the agent itself as well as communicated to this agent by agent ag2. The representation of a goal (in particular an *achievement goal*) is the same of a belief expect that it is prefixed by the symbol '!'. For example, !eaten(2,3) could be used to mean that the agent has the goal of achieving a state of affairs where the food at those coordinates is believed to have been eaten.

The agent behavior is determined by the set of plans that the agent has in its *plan library*. The agent program specifies the initial state of the belief base, possibly initial goals of the agent, and also the plans that will be available in the agent's plan library when the agent starts running. Interestingly, the agent behavior can change over time if new plans are acquired (e.g., by communication with other agents). An AgentSpeak plan has the following general structure:

```
triggering_event : context <- body.
```

where the *triggering event* is used to specify the types of events the plan is meant to handle (these are typically changes in mental attitudes, specifically beliefs and goals); the *context* is used the specify the circumstances under which the plan is thought to be suitable for handling that event — the context needs to be a logical consequence of the current state of the agent's belief base for a plan to be considered when the agent is choosing a plan for it to commit to execute in order to handle the event (such plans are called *applicable* plans); and the body is a sequence of actions to be executed or new goals for the agent to achieve (in its basic form, other things are allowed in the **Jason** variant of AgentSpeak). Consider the following example:

```
01 // initial beliefs
02 trust([john,mary]).
03
04 permitted(eat,loc(X,Y))
05    :- .my_name(Me) & own(Me,loc(X,Y))
06     | permission(eat,loc(X,Y))[source(Ag)] & own(Ag,loc(X,Y)) &
07       trust(TA) & .member(Ag,TA).
08
09
10 // initial plans
11
12 +food(X,Y) : hungry & permitted(eat,loc(X,Y))
13    <- !at(X,Y);
14       eat;
15       +eaten(X,Y);
15       !ack(X,Y).
16
```

```
17  +!at(X,Y) : at(X,Y).
18
19  +!at(X,Y)
20     <- move_toward(X,Y);
21        ?at(X,Y).
22
23  -!at(X,Y) : not at(X,Y)
24     <- !at(X,Y).
25
26  +!ack(X,Y) : .my_name(Me) & own(Me,loc(X,Y)).
27
28  +!ack(X,Y) : own(Ag,loc(X,Y))
29     <- .send(Ag,tell,eaten(X,Y)).
```

The initial belief in line 2 is used because when the agent starts running it is meant to know in advance that it should trust agents john and mary. Note that trust is a predicate symbol like any other, the programmer chose this representation, presumably because the agent will need to consider whether a particular agent is trustworthy or not. The belief in line 4 is a Prolog-like rule (just note that the syntax is the same as used in the context part of AgentSpeak plans, which is slightly different from Prolog); it says that the agent can conclude it is permitted to do action "eat" at a particular location of the environment if either itself owns that location or the agent who owns that location has given permission for that, and the agent who gave that permission is trustworthy. This rule allows us to have a more compact context condition for the plan in line 12. The rule uses two "pre-defined internal actions": .my_name and .member. Internal actions are a mechanism to allow legacy code (e.g., written in Java) to be referenced from the high-level agent reasoning as defined by the AgentSpeak code. Unlike *actions*, internal actions do *not* change the state of the environment; they are run to completion within the agent reasoning cycle. Users can provide libraries of internal actions (Section 15.4.3 explains how); **Jason** provides a number of pre-defined internal actions to help with various programming tasks — pre-defined internal actions available with **Jason** are those that start with the '.' character.

The plan in line 12 says that whenever the agent gets to believe (i.e., acquires a *new* belief) that there is food at some coordinates, provided the agent happens to believe it is hungry and also believes it is permitted to eat the food that is available at those coordinates (according to the rule in its belief base which says when the action eat is permitted), the course of action in the plan body is one potential means the agent could commit to in order to handle that event (the event of perceiving new food in the environment). If this particular plan is chosen (it is then called an *intended means*) the agent will, eventually (unless the intention is later reconsidered), have a new goal of being at those coordinates (where food was perceived), then when that goal has been achieved, do the (environment changing*) action of eating the food, then adding a note to self to remember that food was eaten at that position, then having the goal of acknowledging to the owner of the location (when that's the case) that food was eaten there.

The plans in lines 17, 19, and 23 together allow the agent to persist in the goal of being at some particular coordinates until it is achieved. The plan in line 17 says that the agent

---

*Environment actions represent the repertoire of agent capabilities, i.e., things the agent is assumed to be capable of doing to change the environment, as the actions a robot is built to perform, for example.

has nothing else to do when the goal has already been achieved (i.e., the agent already believes to be at those coordinates). The plan in line 19 executes the action to move the agent location within the environment (`move_toward`), then uses a *test goal* (as opposed to achievement goal, a test goal is used to retrieve information from the belief base as part of a course of action) to make sure the goal has been achieved; test goals are prefixed with the '?' symbol. If the agent does not believe to be where it wishes to be, that formula of the plan body fails, so the whole plan also fails. In ***Jason***, when a plan for a goal +!g fails, an event of type -!g is generated. So the plan in line 23 is a "contingency plan", i.e., a plan that programmers can write for when a plan to achieve that goal failed. In this case, the same goal is simply generated again, which in BDI parlance would be the same as saying that this agent is "blindly committed" to achieving this goal. Note also how the agent has various plans with the same triggering event; this is very typical in agent programming: programmers will give various alternative courses of action for the same purpose, as each of those might only work under certain circumstances (and typically an agent can have a number of plans all of which are, as far as the agent know-how is concerned, possible courses of action to take to handle a particular event at a particular moment in time, so only one such plan needs to be chosen for execution).

It is also interesting to note that this plan pattern, where a test goal at the end of the body is used for the same achievement goal in the triggering event, is used to program *declarative goals* in ***Jason*** [Hübner et al., 2006]. The notion of a declarative goal in agent-oriented programming is a goal which is only considered achieved when the predicate in the goal construct is *believed* by the agent (this helps programmers in developing rational behavior, in making agents persist in achieving a goal even if, for example, a plan to achieve the goal finishes executing but fails to achieve the goal, according to the agent's beliefs). A different use of the goal construct is that in the plans in lines 26 and 28. There, a goal is just being used to execute an appropriate plan, much like a procedure name, to help in modularising the code. The plan in line 26 simply says that there is no need to acknowledge eating if the location was owned by the agent itself. The plan in line 28 uses the `.send` internal action to send a message to the agent who gave permission for the food to be eaten. The message uses the `tell` performative to *inform* (i.e., aiming at changing the beliefs of the receiver) the owner that food at that location has been eaten (the message content is `eaten(X,Y)`, and note that variables X and Y are bound at that point and that the source annotation placed by ***Jason*** can be used by the receiver to know who has eaten that food).

15.3.2 Interpreter

The examples above have shown the main language constructs. We now need to introduce also a number of important data structures maintained by the language interpreter (Figure 15.1 briefly illustrates the relations among them). These data structures are essential to allow programmers to implement BDI agents*:

---

*There is often much confusion between idealized BDI agents — where, e.g., means-ends reasoning is done by reasoning from first principles — and practical BDI implementations following on ideas introduced by PRS [Georgeff and Lansky, 1987] — where agents are "reactive planning systems". Reactive planning systems are "reactive systems" in the sense of distributed computing (i.e., systems that are built to run continuously rather than compute a function and terminate) which repeatedly perceive their environment and react to perceived changes by executing plans available in a plan library, determining the courses of actions they take in order to achieve their goals. In fact, this determines an agent *reasoning cycle*, typically starting with perception of the environment, further committing to plans as intended means to achieve goals or react to environment changes, choosing one intention (i.e., a commitment to a course of

Agent-Based Simulation Using BDI Programming in *Jason*

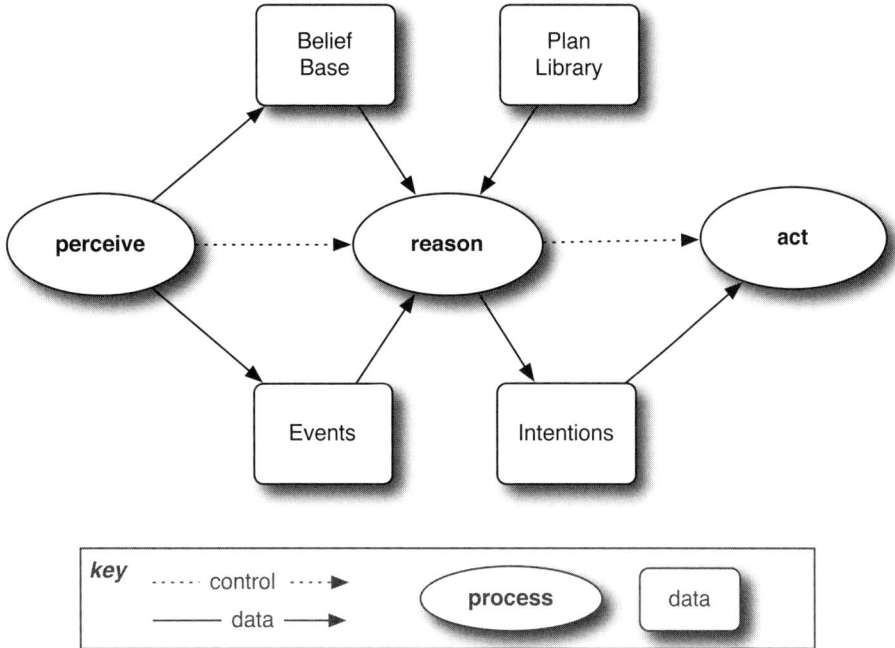

FIGURE 15.1 Main components of the *Jason* interpreter.

belief base: the belief base stores all beliefs currently held by the agent; both closed and open world can be used (the '~' operator is used for strong negation). At every reasoning cycle (unless configured otherwise) *Jason* obtains all percepts available to the agent from the environment model and executes a "belief update function" (both *belief update* and *belief revision* can be customized by the programmer, and in fact the belief base itself can be customized, e.g. to use a database for some of the beliefs, which may be useful in large-scale applications).

set of events: all changes in beliefs (due to belief update or communication) as well as changes in goals (e.g., because a plan being executed created a new goal for the agent to achieve, or because other agents tried to delegate new goals through communication) create new *events* which might trigger the execution of plans whose triggering event matches (unifies with) those events, provided they are applicable at the time the event is chosen for handling at a particular reasoning cycle. The choice of which event to handle in a particular reasoning cycle is made by a method that can be overridden by the programmer (otherwise a FIFO policy is used), so as to allow the use of application-specific information (e.g., on what events are known to have priority for a particular agent).

plan library: this is where the agent know-how is stored. The plan library is initialized with the plans that programmers write in the AgentSpeak code for that agent. While simple agents will have the plan library unchanged throughout

---

action, e.g., to achieve a certain goal) to be executed further and then acting based on such choice. The agent repeatedly executes such reasoning cycles, possibly doing nothing but wait for new changes to be perceived in the environment which could then trigger the agent to act further.

their execution, it is possible to change the agent behavior for example by plan exchange using speech-act based inter-agent communication (this could also be used to communicate with an "agent" wrapper for a planner, thus creating new plans for agents to act in circumstances that the programmer did not anticipate – if this happens to be useful, and feasible, in particular applications). Plans in the plan library have labels to uniquely identify them, and such labels, like beliefs, can have annotations; meta-level information in plan labels can be used, for example, in advanced (user-defined) selection functions which could use, say, some decision-theoretic approach for choosing the best plan for an agent to use, or in scheduling intentions, etc.

set of intentions: each perceived change in the environment (as reflected in changes in beliefs) potentially create a separate intention for the agent, provided there is an applicable plan for that event. Therefore, each separate intention in the agent's set of intentions represent a particular "focus of attention" for the various tasks currently being done by the agent: they all compete in the agent's choice of intention to be further executed in a given reasoning cycle. Each intention potentially becomes a stack of partially instantiated plans. When the plan body includes a new goal to be pursued, an event is generated which when handled will lead to a new plan to be executed in order to achieve that goal, and that plan is pushed on top of the intention that required the goal to be achieved; only when that plan finishes executing (successfully) can the plan that required the goal to be achieved resume execution (hence the appropriateness of the stack structure to represent one intention). Note that the choice of plan to use in order to achieve the goals that appear in one intention is made as late as possible (and based on the plan contexts) which is an important feature for multi-agent systems, as they are typically meant to work in very dynamic environments. As we mentioned, *Jason* also has mechanisms to allow goals and intentions to be dropped or revised, but this is beyond the scope of this chapter.

There are, of course, various other structures that are required by an agent programming language interpreter (e.g., the queue of messages received from other agents that have not yet been processed by the agent). Those above are the essential ones for individual BDI agents. One unusual characteristic of AgentSpeak (and some other agent languages) as a programming language is that certain components of the interpreter are defined by the user, in particular the *selection functions*. They determine which event is to be handled next, which (applicable) plan to choose, and which intention (i.e., a focus of attention for the agent) to execute further in a given reasoning cycle. These decisions are all agent- or application-specific, hence the need to make them customizable by users. Appropriate (intelligent) scheduling of intentions in any generic way is bound to be intractable, hence the practical choice in agent languages of letting users define specific and efficient selection functions, when necessary.

The example given above was just to illustrate the basic language constructs in a way that is easy to understand; it does not cover many aspects of more advanced agent programming, which makes it difficult to illustrate some of the characteristics of BDI agents, for example the fact that agents typically have various foci of attention, and that they should reconsider intentions and drop them if they believe they are no longer achievable, or the motivation that led to the adoption of the goal no longer exists. In a later section we give a complete social simulation example; we decided to use an existing simulation in the literature to make the example more easily understandable. We chose one of the few simulations we know of where agents are not purely reactive (they are said to be "semi-cognitive"), and even then it will be clear this type of agents are extremely simple to program in *Jason*. In fact the code is much simpler than in the example above which, recall, was already too simple to demonstrate the kind of sophisticated agent behavior the language is meant to help develop. The conjecture is that by making judicious use of *Jason* features, simulations where cognitive agents display very elaborate behavior could become more popular; the simulation toolkits currently popular certainly do not lend themselves to this type of development.

Due to space restrictions, we cannot go into enough detail here to expect the reader to learn how to program in *Jason*; the interested reader should refer to [Bordini et al., 2007b] for a complete account of the current *Jason* implementation.

15.4 *Jason* Features for Simulation

The previous section described how the *agents* of a simulation can be specified using a BDI approach and that is surely the most important contribution of *Jason* for social simulation. However, to run the overall simulation we need to define other components besides the agents, one of those components is the environment where the agents are situated and that simulates the execution of their actions and provides them with the relevant perception. We also need components to control, inspect, and analyze the execution of the simulation. Some of the features provided by *Jason* to develop such components are described in the following subsections.

While the agents are programmed in AgentSpeak, the components described in this section are better programmed in Java, since the object-orientation abstractions are typically enough for these programming tasks. Some excerpts of Java code are thus shown to illustrate how those features are easily integrated into a project. To preserve readability, we will not explain all the details of the how to code those components; those details can be found in [Bordini et al., 2007b] and in the *Jason* manual.

15.4.1 Environments

Every *Jason* agent has an architecture that "binds" them to the real environment, and this architecture is responsible for the concrete execution of actions and the mapping from sensor data to symbolic (belief) representation. It is relatively easy to create a simulated environment for the agents. In that case, instead of executing the actions, the agent's architecture asks the environment component to simulate the action and to give back updated perception (when the agent reasoning cycle requires it). Simulated environments are indeed the default configuration of *Jason* projects.

The environment model should be programmed in a Java class that extends the *Jason* Environment class, where some useful functionality is provided. An environment class has three main functions:

1. Maintain a representation of the state of the environment.
2. Simulate the execution of actions required by the agents. This execution normally simply changes the state of the environment. The code that implements the simulation of the action must be written in the executeAction method. This method is called whenever an agent chooses to perform an action.
3. Provide a symbolic representation of the environment when the agents attempt to perceive the current state of the environment. A set of methods are provided to add perception to all agents or to some particular agent. The environment is a passive entity, thus it does not "send" perception to the agents when the envi-

ronment changes. Only when some agent actively senses the environment at the appropriate step of its own reasoning cycle, a representation of the environment state is made available to it.

The following code illustrates how an environment can be programmed for a simple scenario with a lamp and a switch. In this environment, the agent can do the actions `turn(on)` and `turn(off)`, and get as percepts either `light(on)` or `light(off)`.

Example 15.1

```java
public class Room extends Environment {

  // state of the environment
  boolean isOn = true;

  public void init(String[] args) {
    updatePercepts();
  }

  // update the perception of all agents with a symbolic
  // representation of the curent state of the environment
  public void updatePercepts() {
    clearPercepts();
    if (isOn)
      addPercept(Literal.parseLiteral("light(on)"));
    else
      addPercept(Literal.parseLiteral("light(off)"));
  }

  // simulates the execution of the actions
  public boolean executeAction(String agName, Structure action) {
    if (action.toString().equals("turn(on)"))
      isOn = true;
    else if (action.toString().equals("turn(off)"))
      isOn = false;
    updatePercepts();
    return true; // the action was successfully executed
} }
```

In the default implementation of the environment, the execution of actions is asynchronous and concurrent. By asynchronous we mean that it is not the agent thread that executes the action, the agent requests the execution of an action and continues its reasoning cycle without waiting until the end of the execution. Only the particular intention that is performing the action is suspended until the action is finished by the environment. By concurrent we mean that several requested actions can be executed at the same time.

Many simulations require some sort of synchronization in the environment. For instance, agent-based simulations often need the concept of simulation "step", and in one such step, each agent is is required to execute exactly one action — see the *time-stepped* approach to computer simulation as discussed as part of general simulation paradigms in Chapter 3 of this book. This kind of environment is also provided with **Jason** through the use of the SteppedEnvironment class; programmers only need to change the super-class of their environment class to start using this kind of environment (the example in Section 15.5 uses this kind of environment).

*Agent-Based Simulation Using BDI Programming in **Jason*** 461

As also discussed in Chapter 3, some simulation toolkits use the "discrete event" simulation paradigm. Although a class for this type of environment is not currently available in ***Jason***, it could be easily provided in the future. However, future direction for ***Jason*** environments is in fact to use independent approaches and tools that specialize in this, such as ELMS [Okuyama et al., 2006] and "Agents & Artifacts" [Ricci et al., 2007] (more on environments in Section 15.6). This would allow for the development of very sophisticated environments but also environments that are elegantly/declaratively specified (very much as the use of AgentSpeak for specifying autonomous agents). For example, some approaches would allow for the development of an environment simulation that runs distributedly. This could be very useful for very large scale simulations; while ***Jason*** allows agents and the environment to run in various different hosts, the environment itself can only run in one host which in some circumstances could become a distribution "bottleneck".

In the general classification of uses of multi-agent systems toolkits given in Chapter 3, the main use of ***Jason*** would fall into category "Development of Software MAS". Although we have provided changes in execution modes and extensions of environment classes to adjust ***Jason*** for traditional uses in social simulation (both other categories in that chapter), we believe that ***Jason***'s normal execution mode — where agents are completely asynchronous and the notion of "event" is individualized to each agent based on what they perceive in the environment — would provide an interesting approach for sophisticated social simulation, even though the observation and logging of simulation results might need more elaborate work than usual time-stepped simulations.

15.4.2 Execution Modes

As presented in the previous section, the environment can be used to perform some kind of synchronization regarding the actions of the agents. However, it could also be necessary to synchronize the agent's reasoning cycle regardless of the actions they are performing. To help with this, two execution modes are available with ***Jason***:

Asynchronous: all agents run asynchronously. An agent goes to its next reasoning cycle as soon as it has finished its current cycle. This is the *default* execution mode.

Synchronous: all agents perform one reasoning cycle at every "global execution step". When an agent finishes its reasoning cycle, it informs an execution controller and waits for a "carry on" signal. The controller waits until all agents have finished their reasoning cycles and then sends the "carry on" signal to them.

The controller class of the synchronous mode can be customized by the user. This feature can be used, for instance, to inspect the mind of the agents after each reasoning cycle. This is exactly what the debugging tool called "mind inspector" does. At each cycle, it asks a copy of the internal state of all agents and displays them as shown in Figure 15.2.

Users can use this feature for their own requirements. The code below illustrates how to program a controller that gets the number of intentions of the agents at each cycle. When an agent finishes its cycle, the method receiveFinishedCycle is called and an XML copy of the agent's state can be obtained.

Example 15.2

```
public class CountIntentions extends ExecutionControl {
  public void receiveFinishedCycle(String agName, boolean bp, int cycle) {
```

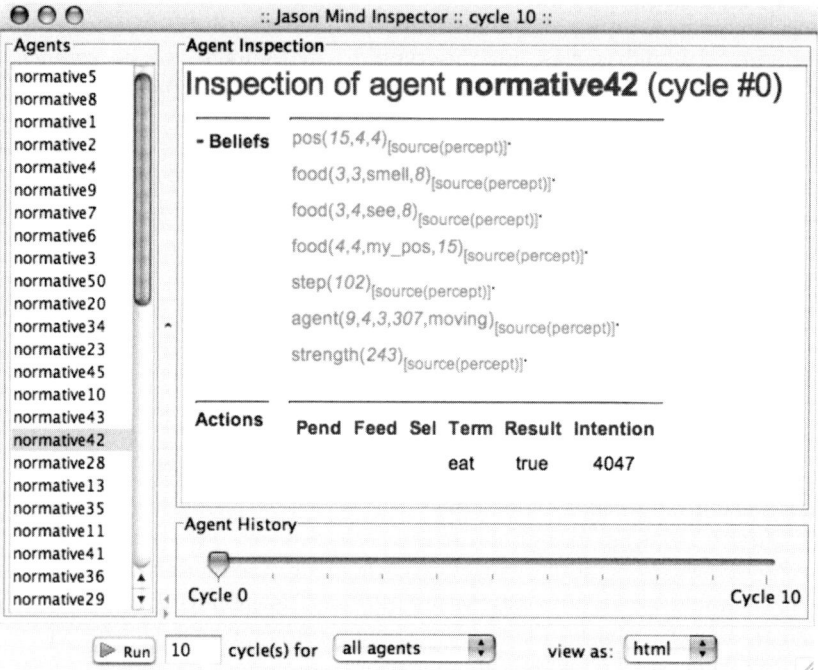

FIGURE 15.2 Screenshot of the Mind Inspector, where the beliefs, events, intentions, and actions of the agents at each step can be observed.

```
        super.receiveFinishedCycle(agName, bp, cycle);

        // get an XML description of the state of the agent
        Document doc = getExecutionControlInfraTier().getAgState(agName);

        // use XPath to count the number of intentions
        XPath xpath = XPathFactory.newInstance().newXPath();

        double ni = (Double)xpath.evaluate(
                    "count(//intention)",
                    doc, XPathConstants.NUMBER);
    } }
```

15.4.3 Internal Actions

An important construct for allowing ***Jason*** agents to remain at the right level of abstraction is that of internal actions. While actions are executed in the environment, internal actions, as the name suggests, are not related to the environment but executed "within" the agent and thus can directly manipulate the state of the agent (but only that of the agent executing it). Since internal actions are (typically) programmed in Java, they can also be used to implement any functionality not available in the AgentSpeak standard language, e.g. graphical user interface, database access, and more generally access to any legacy code.

Internal actions that start with '.' are part of ***Jason*** library of pre-defined internal actions. Internal actions defined by users should be organized in specific libraries. In the AgentSpeak code, the action is accessed by the name of the library, followed by the '.' symbol, followed by the name of the action.

For example, suppose an agent has to ask the user if they want to switch the light on when it is perceived as off by the agent, as in the AgentSpeak code below:

```
+light(off)
   <- gui.yes_no("Should I turn the light on?");
      turn(on).
```

The `gui.yes_no` is an internal action that should be programmed in Java. This internal action will receive a string as parameter and should succeed if the user's answer is "yes", otherwise it should fail. In case the internal action succeeds, the (external) action `turn(on)` will be performed. The following code implements this internal action.

Example 15.3

```
package gui;

import jason.asSemantics.*;

public class yes_no extends DefaultInternalAction {

  public Object execute(TransitionSystem ts, Unifier un,
                   Term[] args) throws Exception {

     // args[0] is the string with the message
     int answer = JOptionPane.showConfirmDialog(null, args[0].toString());

     // returns true if the user answer was YES
     return answer == JOptionPane.YES_OPTION;
} }
```

15.4.4 Customized Architectures

As mentioned in the beginning of Section 15.4.1, every agent has an architecture that is responsible for the execution of actions and the agent's perception (see Figure 15.3). This architecture can be individually customized to simulate different perceptual (dis)abilities, failures, etc. In software development, this is also useful to move from a simulated environment (e.g., used to test the system) into deployment in the real-world environment (where the architecture will interface with existing hardware/software systems to act in and perceive the environment).

The class that customizes the agent architecture can override several methods, the two most relevant for simulation are **act** and **perceive**. For example, we can simulate an agent that is able to perceive only some types of perception; another that does not see colors; and an agent with some degree of blindness (as shown in the code below). Note that the environment model determines what is actually perceptible to an agent; in this case we are interested in simulating one particular agent that will not perceive all that it could potentially perceive.

Example 15.4

```
public class BlindAgent extends AgArch {

  Random r = new Random();
```

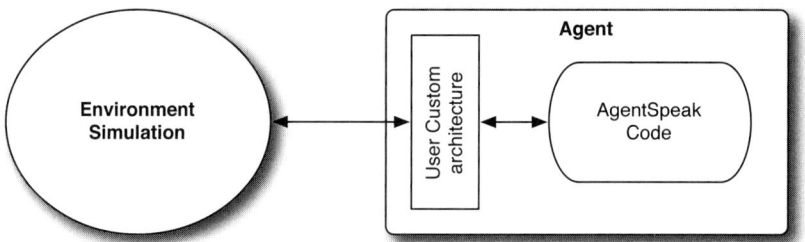

FIGURE 15.3 General view of the relation between the agent and the environment. The user sets up the simulation by defining the reasoning of the agent in the AgentSpeak code, special requirements in the architecture, and the environment simulation.

```
    public List<Literal> perceive() {
      List<Literal> percepts = super.perceive();
      if (percepts != null) {

        // randomly remove 80% of the perception
        long n = Math.round(0.8 * percepts.size());
        for (int i=0; i<n; i++)
          percepts.remove(r.nextInt(percepts.size()));
      }

      return percepts;
} }
```

15.5 Example

This section describes how a complete example can be programmed and simulated in *Jason*. In fact, the example shows how trivial the agent behavior is in the type of social simulation that some researchers have had to content themselves with, due to the limitation of agent-based simulation toolkits (and more generally, multi-agent systems technology until very recently). We have chosen the scenario used in [Castelfranchi et al., 1998]:

> "... agents as objects moving in a two-dimensional common world (a 10 x 10 grid) with randomly scattered food. An experiment consists of a set of matches, each including a fixed number of turns. At the beginning of a match, agents and food items are assigned locations at random. A location is a cell in the grid. The same cell cannot contain more than one object at a time (except when an agent is eating). The agents move through the grid in search of food, stopping to eat when they find it. The agents can be attacked only when eating: no other type of aggression is allowed. At the beginning of each turn, every agent selects an action from the six available routines: eat, move-to-food-seen, move-to-food-smelled, attack, move-random, and pause. Actions are supposed to be simultaneous and time consuming. The most convenient choice for an agent is eat. Eating begins at a given turn and may end two turns later if it isn't interrupted by aggression. To simplify matters, the eater's strength changes only when eating has been completed. Therefore, while the action of eating is gradual (to give players the chance of attacking each other), both the food's nutritional

value and the eater's strength change instantaneously. When a food item has been consumed, it is immediately restored at a randomly chosen location. This is in order to allow for all agents to survive and obtain comparable results from all simulations. The agent will then look for unoccupied food items within its "territory" (consisting of the four cells to which an agent can move in one step from its current location), choosing move-to-food-seen if any is found. If not, it will smell within its neighbourhood (extending two steps in each direction from the agents' current location), in order to choose move-to-food-smelled; the agent does not know whether this food location will be occupied or not, because agents can detect one another only within their "territory". At this point, the agent will take into consideration attacking against any other eating neighbour, weighting this option with its own norm affiliation. The outcomes of an attack are determined by the agents' respective strengths (the stronger agent is always the winner). When the competitors are equally strong, the defender is the winner. The cost of aggression is equal to the cost of receiving aggression. However, winners obtain the contested food item. Agents might be attacked by more than one agent at a time, in which case the victim's cost is multiplied by the number of aggressors. In the case of multiple aggression, only the strongest attacker carries out the attack, while the others must pass. If none of the above actions are possible, the agent is left with the sad options of move-random or pause if no close cell is free. Matches consisted of 2000 time steps, and included 50 agents with a default strength of 40, plus 25 food items with a nutritive value of 20. The costs are 0 for pausing, 1 for moving to an adjacent cell, 4 for attacking or receiving attacks".

This scenario is well known because it covers (in a simple environment) important aspects of both normative behavior as well as reputation, which are increasingly important aspects of current research and in multi-agent systems as well as, of course, fundamental issues in the social sciences.

The implementation of this simulation in *Jason* is divided into two main parts: the environment simulation and the agent behavior.

15.5.1 Environment

The environment should require one action per agent in each step, therefore its implementation extends the class **SteppedEnvironment**, as described in Section 15.4.1. The environment implements the simulation of the following actions (they are not exactly the same actions as used in the paper by Castelfranchi et al., as we chose to define more general ones):

`eat` : eat the food in the current location of the agent.
`move(X,Y)` : move the agent one step toward the location X,Y.
`random_move` : move the agent to an unoccupied random direction.
`attack(X,Y)` : attack the agent at location X,Y.
`pause` : does nothing; this action is necessary since at every step all agents have to perform one action.

The perception given to agents is as follows:

`step(N)` : the current simulation step is N.
`pos(AgId,X,Y)` : the current location of the agent doing perception, where `AgId` is the agent identification in the system, X is its column in the environment, and

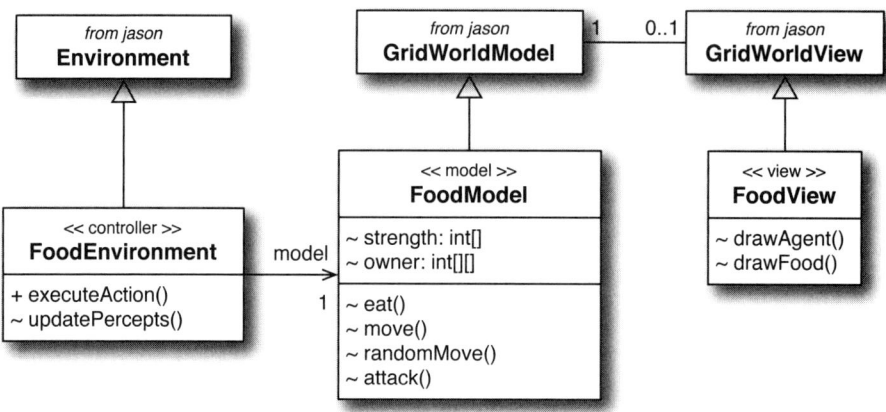

FIGURE 15.4 Class diagram with the classes used to implement the environment. The classes at the top of the figure are provided by Jason.

Y the row.

`strength(S)` : the strength of the agent doing perception is S.

`food(X,Y,P,O)` : some food is perceived at location X,Y. P indicates how far the food is from the agent, if P is `my_pos` the food is in the same location as the agent; if P is `see`, the food is one step away; and if P is `smell` the food is two steps away. As we explain later, each food item has an owner allocated to it; the variable O represents the owner of some food. If O is -1, the food has no owner.

`agent(Id, X, Y, S, A)` : the agent identified by Id is around the agent doing perception. X,Y is its location, S its strength, and A what it is doing (`moving` or `eating`).

`attacked(Id)` : the agent doing perception was attacked by the agent identified by Id.

While simple environments can be implemented in a single class as exemplified in Section 15.4.1, more complex environments demand more careful design. Most simulations require a graphical interface, and a suitable design pattern for such environments is the Model-View-Control pattern much used in object orientation. In this pattern the *Jason* environment plays the role of controller, to which a model (that maintains the state of the environment) and a view (that displays the current state of the model) should be aggregated. *Jason* provides general purpose classes for grid-based models and views, thus simplifying the development of the overall environment implementation. Figure 15.4 shows the class diagram with the relations between the classes used to implement the environment for this example. The model class contains attributes to represent the current state of the places and agents (the agents' strength and the foods' owners for example). The agents' location are maintained by the super class GridWorldModel. The model also implements useful methods for each possible action, so that the environment class in its `executeAction` method can call them when required by the agents. In a similar way, the environment uses the state of the model to provide perception to the agents in its `updatePercepts` method. Due to inheritance of several functionalities of the classes provided by *Jason*, the implementation of the environment for grid-based scenarios is quite straightforward (for users with some experience in Java programming).

15.5.2 Agents

The first type of agent described in [Castelfranchi et al., 1998] is called "blind agent". This agent attacks others whenever there are no better alternatives. The preference order of the action is defined by the following rules:

Rule-b1 If there is food in my cell, eat.
Rule-b2 If I see food in a free cell, move to there; a free cell is a cell with no agent.
Rule-b3 If I see food in an occupied cell, attack the agent at that cell.
Rule-b4 If I smell food, move to there.
Rule-b5 Otherwise, do a random movement.

Note that this first agent is very simple and its rules based only on perceptual information. The agents presented latter will be slightly more complex using communication and beliefs.

Since the agents must select one action at each simulation step, the easiest way to implement them in *Jason* is to write plans to react to the perception of a new step, following the template below for all plans (note that this is *not* how agents are normally implemented in *Jason*, this is the case here simply because these agents are purely reactive agents running in a time-stepped simulation):

```
+step(N) : <context> <- <action>.
```

The program below implements the five rules for action decision (b1–b5) using one AgentSpeak plan for each rule:

Example 15.5

```
/* -- blind agent -- */

// Rule-b1
+step(_) : food(X,Y,my_pos,_) <- eat.

// Rule-b2
+step(_) : food(X,Y,see,_) & not agent(_,X,Y,_,_) <- move(X,Y).

// Rule-b3
+step(_) : food(X,Y,see,_) & agent(_,X,Y,_,eating) <- attack(X,Y).

// Rule-b4
+step(_) : food(X,Y,smell,_) <- move(X,Y).

// Rule-b5
+step(_) <- random_move.
```

The second type of agent is called "strategic agent". This agent only attacks other eaters whose strength is not higher than their own. Its code is thus quite similar to the blind agent code, only Rule-b3 is changed to:

Example 15.6

```
/* -- strategic agent -- */
```

```
// Rule-s3
+step(_)
   :  food(X,Y,see,_) &
      agent(_,X,Y,S,eating) &
      strength(MS) & MS > S
   <- attack(X,Y).
```

The third type of agent is called "normative agent". These agents are added in the simulation to investigate the role of norms in the control of aggression. Some food units are allocated to an agent situated in its vicinity and the norm states that the agents can not attack others eating their own food.

Example 15.7

```
/* -- normative agent -- */

// Rule-n1: only eat food that is free or allocated to me
+step(_) : food(X,Y,my_pos,A) & (pos(A,_,_) | A == -1) <- eat.

// Rule-n2: if I see food allocated to me, move to there
+step(_) : food(X,Y,see,Me) & pos(Me,_,_) & not agent(_,X,Y,_,_)
       <- move(X,Y).

// Rule-n3: if I see unallocated food, move to there
+step(_) : food(X,Y,see,-1) & not agent(_,X,Y,_,_) <- move(X,Y).

// Rule-n4: if I see another agent eating food owned by others, attack it
+step(_) : food(X,Y,see,OAg) & agent(AgId,X,Y,_,eating) &
           AgId \== OAg & OAg \== -1
       <- attack(X,Y).

// Rule-n5: if I smell my food, move to there
+step(_) : food(X,Y,smell,Me) & pos(Me,_,_)  <- move(X,Y).

// Rule-n6: if I smell unallocated food, move to there
+step(_) : food(X,Y,smell,-1) <- move(X,Y).

// Rule-n7: otherwise, move randomly
+step(_) <- random_move.
```

The first experiment reported in [Castelfranchi et al., 1998] was to collect the average strength (Str), the variance of individual strengths (Var), and the number of attacks (Agg) after 2000 steps of a system formed by 50 homogeneous agents. All this information is available in the environment, so the Java class that implements the environment also collects these data. To run this simulation in *Jason*, a project file should be configured with the type and quantity of agents:

```
MAS normative_simulation {
    // the class that implements the environment is FoodEnvironment
    // parameters are: grid size, number of agents, number of food units
```

TABLE 15.1 Results of the first experiment.

Agent type	Strength	Variance	Aggressions
Blind	2793	1748	6210
Strategic	2839	1815	3170
Normative	3076	310	82

```
    environment: FoodEnvironment(10,50,25)

    agents:
        //blind     #50;
        //strategic #50;
        normative   #50;
}
```

Note that it is very easy to change the type of agents and the quantity. A screenshot of the simulation is shown in Figure 15.5. In general, the results are similar to those presented in [Castelfranchi et al., 1998] and are shown in Table 15.1: normative agents performs better (in the sense that their average strength is greater). Although the conclusion is the same, the values in Table 15.1 do not reproduce those presented in [Castelfranchi et al., 1998]. The reason for this is due to the interpretation of some parts of the scenario description that are ambiguous in the paper, so we had to make our own implementation decisions.

In the second experiment, the type of agents are mixed (e.g., 25 strategic agents with 25 normative). The result is quoted from the paper by Castelfranchi et al. below.

> "This experiment's findings can be summarized as follows: (a) in a mixed population, the normative strategy, which was the best in the standalone condition, becomes the worst. (b) The non-normative agents benefit from the existence of normative ones. (c) This requires that some penalty for transgression be introduced in the system; one such penalty can be implemented as a specific type of social knowledge, namely normative reputation".

This result has motivated the authors to define a new type of agent, called "reputation agent". Reputation agents try to adapt their behavior so that they play as normative with normative agents and as strategic against strategic agents. Therefore, they need to maintain a model of the other agents they interact with. In [Castelfranchi et al., 1998], a very simple reputation model is proposed: whenever the agent is attacked when eating its own food, the attacker is considered a cheater. In that paper, the authors also noted that the information about the reputation of others works much better if shared by several agents. Thus, reputation agents tell other agents the list of agents they think are cheaters.

The implementation of the reputation agent in *Jason* is very simple. When it perceives it has been attacked by another agent, it adds a new belief representing that that agent is a cheater (reputation based on direct experience) and broadcast that belief to the others (reputation based on collective information):

```
    +attacked(A)                       // perception event
       <- +cheater(A);                 // belief addition
          .broadcast(tell, cheater(A)). // broadcast message
```

Jason has a pre-defined implementation of the semantics of "tell" messages, so the agents do not need to be programmed to "receive" a message. The content of a tell message is automatically added to the belief base of the agents (with an annotation about the source of the belief), provided message comes from an "authorized" source*.

The belief about who are the cheaters is then used in the plans to adapt the behavior accordingly, i.e., depending on the reputation of the others. The code below contains all the rules of the reputation agent. Note that only the first and the fourth rules are slightly different from the strategic agent, showing that it is quite easy to experiment with different behaviors given the high abstraction level of the AgentSpeak language.

Example 15.8

```
/* -- reputation agent -- */

// Rule-r1: eat food that is either allocate to me, free, or
//          belongs to a cheater
+step(_) : food(X,Y,my_pos,A) &
           (pos(A,_,_) | A == -1 | cheater(A))
        <- eat.

// Rule-r2: if I see food allocated to me, move to there
+step(_) : food(X,Y,see,Me) & pos(Me,_,_) & not agent(_,X,Y,_,_)
        <- move(X,Y).

// Rule-r3: if I see unallocated food, move to there
+step(_) : food(X,Y,see,-1) & not agent(_,X,Y,_,_)
        <- move(X,Y).

// Rule-r4: attack cheaters that are weaker than me
+step(_) : food(X,Y,see,OAg) & agent(AgId,X,Y,S,eating) &
           cheater(AgId) &
           strength(MS) & MS > S
        <- attack(X,Y).

// Rule-r5: if I smell my food, move to there
+step(_) : food(X,Y,smell,Me) & pos(Me,_,_) <- move(X,Y).

// Rule-r6: if I smell unallocated food, move to there
+step(_) : food(X,Y,smell,-1) <- move(X,Y).

// Rule-r7: otherwise, move randomly
+step(_) <- random_move.
```

The result of the execution of a system with 25 strategic agents and 25 reputation agents is again similar to (but not exactly the same as) those presented in [Castelfranchi et al., 1998]: reputation agents do better than normative agents against strategic agents.

---

*Note that because the source of the information is kept in the belief base, the agent can later make judicious use of this information, e.g., if it comes not to trust a particular agent anymore. Also, by default, all agents in an agent society are allowed to communicate with each other; if, for example, security is an issue, a method can be overridden to limit the agents that are allowed to communicate with a particular agent.

Agent-Based Simulation Using BDI Programming in *Jason* 471

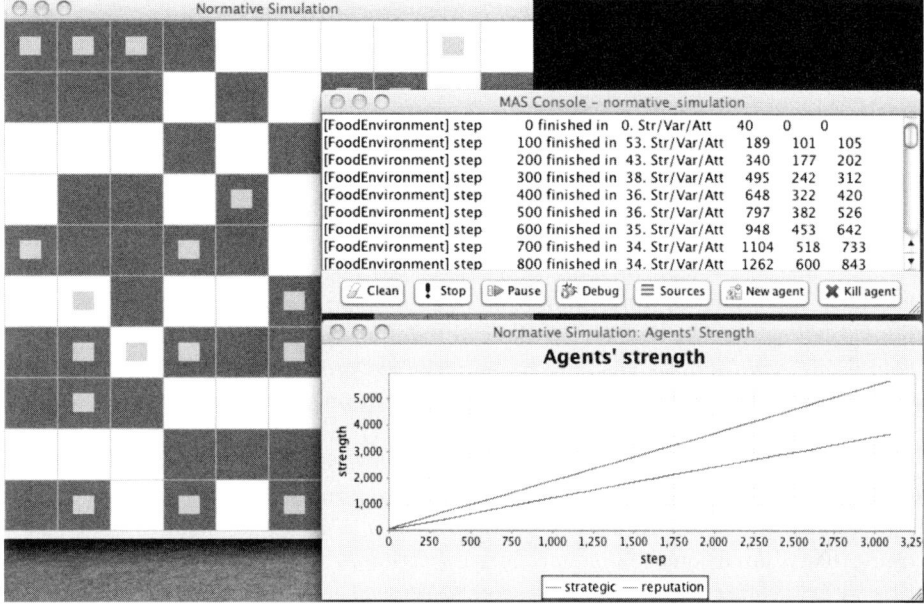

FIGURE 15.5 Screenshot of the Simulation GUI. The window to the left contains a graphical representation of the scenario: gray places are occupied by one agent and light-gray (yellow) squares represent food. The top-right window is the ***Jason*** console where statistical data (the average strength of the agents) is being displayed. We also created an extra window (bottom-right) to show a graph where we plot (using jFreechart) the sum of agents' strength (for both strategic and reputation agents) at each simulation step.

15.5.3 Results

The example has shown how this type of social simulation can be easily implemented in ***Jason***. One advantage of the approach is the readability of AgentSpeak code. Another advantage of the approach is the clear distinction between environment and agents, in particular in having a language that is specifically designed to program autonomous agents (we will give references to environment-specific languages in the next section). In this example, because the agent behavior is trivial, most of the implementation effort was in simulating the environment. As we mentioned above, these agents are simply reactive (as opposed to cognitive), they have no long-term goals and do not require any complex goal-based behavior. This is precisely the strength of ***Jason*** as a simulation platform. However, we felt we needed to use a well-known social simulation as example here. We hope that the recent availability of agent platforms such as ***Jason*** will contribute to the future development (and popularization) of much more complex forms of social simulation.

15.6 Ongoing Projects

There are a number of ongoing research projects aiming at extending or using ***Jason*** in a number of ways that will have an important impact in the sophistication of simulations that can be developed with ***Jason***. We summarize two such projects (that are more relevant for social simulation) below, for more details see [Bordini et al., 2007b, Chapter 11]. Another possible research direction would be to use the AgentSpeak interpreter of ***Jason*** in combination with existing agent-based simulation platforms (see Chapter 3 of this book). We also expect further improvement of ***Jason*** itself as a consequence of research in the area of agent-oriented programming languages (e.g., on plan selection based on reasoning about goal interactions [Thangarajah et al., 2003; Shaw and Bordini, 2007]).

Environments: As we saw in the example, *Jason* provides a very high-level language for programming agents, but still requires Java programming for simulated environments (even though a lot of support is given for that task). It would be much more convenient to also use a high-level (declarative) language created specifically for modeling environments for social simulation using cognitive agents [Bordini et al., 2005a]. This was the motivation that led to the development of the ELMS language, described in [Okuyama et al., 2005]. That work has recently been extended [Okuyama et al., 2006] to allow environment descriptions to have objects containing social norms that are to be observed only within the confines of an environment location, possibly where an institution or organization is situated (similarly to 'please refrain from smoking' or 'keep silence' signs). Another recent development [Ricci et al., 2008] is the integration of *Jason* with a well-known approach for developing multi-agent environments based on the "artifact" abstraction [Ricci et al., 2007], which could help in the development of very elaborate (distributed) environments; more on artifacts in the item below. Although this would require further investigation, it seems that other approaches to modeling environments, such as those proposed in [Bandini et al., 2002], [Weyns and Holvoet, 2004], and [Helleboogh et al., 2007], could also be potentially integrated with *Jason* for use in particular projects.

Organizations: An important part of agent-oriented software engineering is related to agent organizations, which have received much research attention in the last few years (see for example, the COIN – Coordination, Organization, Institution and Norms for agent systems – workshop series [Noriega et al., 2007]). We are currently working on allowing specifications of agent organizations (with the related notions of roles, groups, relationships between groups, social norms, etc.) to be used in combination with *Jason* for programming the individual agents. The particular organizational model we use is $\mathcal{M}\text{OISE}^+$ [Hübner et al., 2004] and an initial integration with *Jason* is discussed in [Hübner et al., 2007] (and available from http://moise.sf.net). While this latter proposal is roughly implemented by means of a library of internal actions and special agents, another approach is to place special artifacts in a virtual environment which provide *organizational actions* such as "create group" and "adopt a role" to the agents [Kitio et al., 2008]. The advantage of this second approach is that the agents can use and change the organization by performing actions in the environment in the same way they perform any other ordinary action (as proposed by the Agent & Artifacts meta-model [Ricci et al., 2007]).

A long-term goal of our work is to develop complex social simulations where emergent cognitive phenomena can be studied, associated with the well-known micro-macro link problem. In order to do that, we need to model social phenomena such as organizations as well as mental attitudes, and the combination of *Jason* with $\mathcal{M}\text{OISE}^+$ provides a suitable framework for this.

Also, we can expect further progress on each of those systems individually. There is much ongoing work on organizations, as well as on programming languages for multi-agent systems. For *Jason* in particular, various language extensions are planned, as discussed in [Bordini et al., 2007b, Section 11.2]. Some current issues in programming languages and methodologies for multi-agent systems are discussed, for example, in [Bordini et al., 2007a].

15.7 Conclusion

This chapter has shown a new approach to programming that provides powerful abstractions for the development of multi-agent systems, and we highlighted a number of features that can favor the use of multi-agent systems for (social) simulation. The approach is suitable for the implementation of much more sophisticated simulations then typically possible with existing agent-based simulation toolkits. Besides being an important abstraction for agent programming, ***Jason*** provides many features that can help with large-scale projects, such as user customizations, and the use of distribution infrastructures (such as JADE).

However, of course this does not come for free. This approach requires a lot more programming than typical agent-based simulation toolkits, and indeed it takes some effort to learn the new programming style even for those familiar with programming languages. On the other hand, future developments in agent-oriented programming, and with ***Jason*** in particular, such as libraries of architectures, actions, environments, as well as the ongoing projects discussed in the previous section, will significantly facilitate future practice of agent programming, possibly making this approach easier to use but with the potential for much more sophisticated social simulations being developed.

References

M. Alberti, F. Chesani, M. Gavanelli, E. Lamma, P. Mello, and P. Torroni. The socs computational logic approach to the specification and verification of agent societies. In C. Priami and P. Quaglia, editors, *Global Computing, IST/FET International Workshop, GC 2004, Rovereto, Italy, March 9-12, 2004, Revised Selected Papers*, volume 3267 of *LNCS*, pages 314–339. Springer, Berlin, 2005.

J. L. Austin. *How to Do Things with Words*. Harvard University Press, Cambridge, 2 edition, 1975.

S. Bandini, S. Manzoni, and C. Simone. Dealing with space in multi–agent systems: a model for situated mas. In *Proceedings of The First International Joint Conference on Autonomous Agents & Multiagent Systems, AAMAS 2002, July 15-19, 2002, Bologna, Italy*, pages 1183–1190. ACM, 2002.

F. Bellifemine, F. Bergenti, G. Caire, and A. Poggi. JADE — a java agent development framework. In Bordini et al. [2005b], chapter 5, pages 125–147.

R. H. Bordini, A. C. da Rocha Costa, J. F. Hübner, Á. F. Moreira, F. Y. Okuyama, and R. Vieira. MAS-SOC: a social simulation platform based on agent-oriented programming. *Journal of Artificial Societies and Social Simulation*, 8(3), Jun 2005a. JASSS Forum, <http://jasss.soc.surrey.ac.uk/8/3/7.html>.

R. H. Bordini, M. Dastani, J. Dix, and A. El Fallah Seghrouchni, editors. *Multi-Agent Programming: Languages, Platforms and Applications*. Number 15 in Multiagent Systems, Artificial Societies, and Simulated Organizations. Springer, New-York, 2005b.

R. H. Bordini, L. Braubach, M. Dastani, A. E. F. Seghrouchni, J. J. Gomez-Sanz, J. Leite, G. O'Hare, A. Pokahr, and A. Ricci. A survey of programming languages and platforms for multi-agent systems. *Informatica*, 30(1):33–44, 2006. Available online at `http://ai.ijs.si/informatica/vols/vol30.html\#No1`.

R. H. Bordini, M. Dastani, and M. Winikoff. Current issues in multi-agent systems development (invited paper). In G. M. P. O'Hare, A. Ricci, M. J. O'Grady, and O. Dikenelli, editors, *Proceedings of the Seventh Annual International Workshop on Engineering Societies in the Agents World (ESAW 2006), 6–8 September,*

Dublin, Ireland – Revised Selected and Invited Papers, number 4457 in LNAI, pages 38–61. Springer, Berlin, 2007a.

R. H. Bordini, J. F. Hübner, and M. Wooldridge. *Programming Multi-Agent Systems in AgentSpeak Using Jason*. Wiley Series in Agent Technology. John Wiley & Sons, Chichester, UK, 2007b. URL http://jason.sf.net/jBook.

C. Castelfranchi, R. Conte, and M. Paolucci. Normative reputation and the cost of compliance. *Journal of Artificial Societies and Social Simulation*, 1(3), 1998. URL http://jasss.soc.surrey.ac.uk/1/3/3.html. http://jasss.soc.surrey.ac.uk/1/3/3.html.

K. L. Clark and F. G. McCabe. Go! for multi-threaded deliberative agents. In J. A. Leite, A. Omicini, L. Sterling, and P. Torroni, editors, *Declarative Agent Languages and Technologies, First International Workshop, DALT 2003, Melbourne, Australia, July 15, 2003, Revised Selected and Invited Papers*, volume 2990 of *LNAI*, pages 54–75. Springer, Berlin, 2004.

M. Dastani and J. Gomez-Sanz. Programming multi-agent systems. *Knowledge Engineering Review*, 20(2):151–164, 2006.

M. Dastani, M. B. van Riemsdijk, and J.-J. C. Meyer. Programming multi-agent systems in 3APL. In Bordini et al. [2005b], chapter 2, pages 39–67.

G. de Giacomo, Y. Lespérance, and H. J. Levesque. ConGolog: A concurrent programming language based on the situation calculus. *Artificial Intelligence*, 121:109–169, 2000.

J. Dix and Y. Zhang. IMPACT: A multi-agent framework with declarative semantics. In Bordini et al. [2005b], chapter 3, pages 69–94.

A. El Fallah Seghrouchni and A. Suna. CLAIM and SyMPA: A programming environment for intelligent and mobile agents. In Bordini et al. [2005b], chapter 4, pages 95–122.

M. Fisher. Temporal development methods for agent-based systems. *Autonomous Agents and Multi-Agent Systems*, 10(1):41–66, 2004.

M. Fisher, R. H. Bordini, B. Hirsch, and P. Torroni. Computational logics and agents: a road map of current technologies and future trends. *Computational Intelligence*, 23(1):61–91, Feb. 2007.

M. P. Georgeff and A. L. Lansky. Reactive reasoning and planning. In *AAAI*, pages 677–682, 1987.

A. Helleboogh, G. Vizzari, A. Uhrmacher, and F. Michel. Modeling dynamic environments in multi-agent simulation. *Autonomous Agents and Multi-Agent Systems*, 14(1):87–116, 2007.

J. F. Hübner, J. S. Sichman, and O. Boissier. Using the MOISE+ for a cooperative framework of MAS reorganisation. In A. L. C. Bazzan and S. Labidi, editors, *Proceedings of the 17th Brazilian Symposium on Artificial Intelligence (SBIA'04)*, volume 3171 of *LNAI*, pages 506–515, Berlin, 2004. Springer. URL http://www.inf.furb.br/~jomi/pubs/2004/Hubner-sbia2004.pdf.

J. F. Hübner, R. H. Bordini, and M. Wooldridge. Programming declarative goals using plan patterns. In M. Baldoni and U. Endriss, editors, *Proceedings of the Fourth International Workshop on Declarative Agent Languages and Technologies (DALT 2006), held with AAMAS 2006, 8th May, Hakodate, Japan*, volume 4327 of *LNCS*, pages 123–140. Springer, 2006. ISBN 3-540-68959-1.

J. F. Hübner, J. S. Sichman, and O. Boissier. Developing organised multi-agent systems using the MOISE+ model: Programming issues at the system and agent levels. *International Journal of Agent-Oriented Software Engineering*, 1(3/4):370–395, 2007.

R. Kitio, O. Boissier, J. F. Hübner, and A. Ricci. Organisational artifacts and agents for open multi-agent organisations: "giving the power back to the agents". In J. Sichman, P. Noriega, J. Padget, and S. Ossowski, editors, *COIN 2007 International Workshops COIN@AAMAS 2007, Honolulu, HI, USA, May 14, 2007 COIN@MALLOW 2007, Durham, UK, September 3-4, 2007*, volume 4870 of *LNCS*, pages 171–186. Springer, New York, 2008. Revised Selected Papers.

S. Kumar, P. R. Cohen, and M. J. Huber. Direct execution of team specifications in STAPLE. In C. Castelfranchi and W. L. Johnson, editors, *Proceedings of the First International Joint Conference on Autonomous Agents and Multi-Agent Systems (AAMAS-2002), 15–19 July, Bologna, Italy*, pages 567–568. ACM Press, New York, NY, 2002.

J. A. Leite, J. J. Alferes, and L. M. Pereira. $\mathcal{MINERVA}$—a dynamic logic programming agent architecture. In J.-J. Meyer and M. Tambe, editors, *Intelligent Agents VIII – Proceedings of the Eighth International Workshop on Agent Theories, Architectures, and Languages (ATAL-2001), August 1–3, 2001, Seattle, WA*, volume 2333 of *Lecture Notes in Artificial Intelligence*, pages 141–157. Springer, Berlin, 2002.

V. Mascardi, M. Martelli, and L. Sterling. Logic-based specification languages for intelligent software agents. *Theory and Practice of Logic Programming*, 4(4):429–494, 2004.

D. Morley and K. L. Myers. The spark agent framework. In *3rd International Joint Conference on Autonomous Agents and Multiagent Systems (AAMAS 2004), 19-23 August 2004, New York, NY, USA*, pages 714–721. IEEE Computer Society, 2004.

P. Noriega, J. Vásquez-Salceda, G. Boella, O. Boissier, V. Dignum, N. Fornara, and E. Matson, editors. *AAMAS 2006 and ECAI 2006 International Workshops, COIN 2006 Hakodate, Japan, May 9, 2006 Riva del Garda, Italy, August 28, 2006*, volume 4386 of *LNAI*, 2007. Springer,Berlin. Revised Selected Papers.

F. Y. Okuyama, R. H. Bordini, and A. C. da Rocha Costa. ELMS: an environment description language for multi-agent simulations. In D. Weyns, H. van Dyke Parunak, F. Michel, T. Holvoet, and J. Ferber, editors, *Environments for Multiagent Systems, State-of-the-art and Research Challenges. Proceedings of the First International Workshop on Environments for Multiagent Systems (E4MAS), held with AAMAS-04, 19th of July*, number 3374 in Lecture Notes in Artificial Intelligence, pages 91–108. Springer-Verlag, Berlin, 2005.

F. Y. Okuyama, R. H. Bordini, and A. C. da Rocha Costa. Spatially distributed normative objects. In G. Boella, O. Boissier, E. Matson, and J. Vázquez-Salceda, editors, *Proceedings of the Workshop on Coordination, Organization, Institutions and Norms in Agent Systems (COIN), held with ECAI 2006, 28th August, Riva del Garda, Italy*, 2006.

A. Pokahr, L. Braubach, and W. Lamersdorf. Jadex: A BDI reasoning engine. In Bordini et al. [2005b], chapter 6, pages 149–174.

A. S. Rao. AgentSpeak(L): BDI agents speak out in a logical computable language. In W. Van de Velde and J. Perram, editors, *Proceedings of the Seventh Workshop on Modelling Autonomous Agents in a Multi-Agent World (MAAMAW'96), 22–25 January, Eindhoven, The Netherlands*, number 1038 in LNAI, pages 42–55. Springer,Berlin, 1996.

A. Ricci, M. Viroli, and A. Omicini. "Give agents their artifacts": The A&A approach for engineering working environments in MAS. In E. Durfee, M. Yokoo, M. Huhns, and O. Shehory, editors, *6th International Joint Conference "Autonomous Agents &*

Multi-Agent Systems" (AAMAS 2007), pages 601–603, Honolulu, Hawai'i, USA, 14–18 May 2007. IFAAMAS. ISBN 978-81-904262-7-5.

A. Ricci, M. Piunti, L. D. Acay, R. H. Bordini, J. F. Hübner, and M. Dastani. Integrating heterogeneous agent-programming platforms within artifact-based environments. In L. Padgham, D. Parkes, J. Müller, and S. Parsons, editors, *Proc. of 7th Int. Conf. on Autonomous Agents and Multiagent Systems (AAMAS 2008), 12–16 May, Estoril, Portugal*. International Foundation for Autonomous Agents and Multiagent Systems (www.ifaamas.org), 2008.

P. Shaw and R. Bordini. Towards alternative approaches to reasoning about goals. In *Proceedings of the 5th International Workshop on Declarative Agent Languages and Technologies*. Springer, New York, 2007.

M. Tambe. Agent architectures for flexible, practical teamwork. In *AAAI/IAAI*, pages 22–28. AAAI Press, Menlo Park, CA, 1997.

J. Thangarajah, L. Padgham, and M. Winikoff. Detecting and avoiding interference between goals in intelligent agents. In *proceedings of 18th International Joint Conference on Artificial Intelligence (IJCAI)*, pages 721–726. Morgan Kaufmann, San Francisco, CA, August 2003.

F. Toni. Multi-agent systems in computational logic: Challenges and outcomes of the socs project. In F. Toni and P. Torroni, editors, *Computational Logic in Multi-Agent Systems, 6th International Workshop, CLIMA VI, London, UK, June 27-29, 2005, Revised Selected and Invited Papers*, volume 3900 of *Lecture Notes in Computer Science*, pages 420–426. Springer, Berlin, 2006.

D. Weyns and T. Holvoet. A formal model for situated multi-agent systems. *Fundam. Inform.*, 63(2-3):125–158, 2004.

M. Winikoff. JACK^{TM} intelligent agents: An industrial strength platform. In Bordini et al. [2005b], chapter 7, pages 175–193.

M. Wooldridge. *An Introduction to MultiAgent Systems*. John Wiley & Sons, Chichester, UK, 2002.

16

SeSAm: Visual Programming and Participatory Simulation for Agent-Based Models

16.1 Introduction	477
16.2 Simulation Study and User Roles	478
Tasks and User Roles • User Involvement in Agent-Based Simulation Tools • Tools for Agent-Based Simulation in General	
16.3 Core SeSAm	482
Basic Model Representation • Simulation Routine and Model Interpretation • Plugin Mechanism for Extension • Problematic Details of the Language • General Aspects of Suitability	
16.4 Users, Tasks, and SeSAm	493
Principles • Developing a Conceptual Model • Visual Programming for Model Implementation • Experiment Scripting and DAVINCI for Experimenters • Online Aggregated Data Presentation and Animation • Model-Specific Interfaces • Agent Playing for Advanced Participation	
16.5 Experiences	503
Novices • Knowledgeable in Implementation, Not in Multi-Agent systems	
16.6 General Discussion and Future Work	504
Acknowledgment	506
References	506

Franziska Klügl
Universität Würzburg

16.1 Introduction

Agent-based simulation offers a lot of advantages compared to traditional (also microscopic) approaches, such as elegant representation of heterogeneous populations situated in heterogeneous environments, explicit treatment of local effects, or the possibility of formulating flexible interaction between entities based on intelligent behavior. Agent-based simulation is becoming more and more popular not only because of these properties. Additionally, it provides an intuitive and direct form of modeling. Coarsely said, entities observable in the real world can be described by entities in the virtual world. There is no abstraction step that necessarily bridges aggregation levels as in macro-simulation.

As a result, agent-based simulation seems to demand less experience in handling complex mathematics for modeling. Thus, the paradigm of agent-based modeling enables more researchers to use simulation techniques on their own. Agent-based simulation seems to be particularly attractive to people that did not use simulation techniques before, nor received any training in formalizing or programming using traditional programming languages. Whether agent-based simulation can be widely applied, depends on the availability of appropriate tools for every level of technical expertise.

Tools supporting the implementation of agent-based simulations have been developed since the early 1990's. In this connection two classes of tools can be identified: class libraries on top of a traditional programming language (such as *Swarm, www.swarm.org*, accessed June, 2008) or visual programming environments for absolute beginners (such as *AgentSheets, www.agentsheets.com*, accessed June, 2008). These early tools merely supported the implementation of a multi-agent simulation. However, there are more tasks in a simulation study than only implementing a simulation model. These other tasks, like experimentation, model analysis, etc. were hardly supported. This is still true for many existing tools, platforms and languages.

In 1995, we introduced the first version of SeSAm (She̲ll for S̲imulated A̲gent S̲ystems). By this time, SeSAm also focused on supporting implementation providing a high-level model representation framework. As the linkage to a standard programming language was different compared to class library-based tools, we had to note that the initial (Lisp-based) SeSAm was not successful. The reason was mainly that model handling was too complex. Additionally, for providing model specific add-ons for user participation, experimentation support, etc. a SeSAm user could not fall back to the basic programming language and add these tools to the given framework. In 2000/2001 we undertook a complete redesign and re-implementation of SeSAm using Java, providing more appropriate user support beyond mere implementation tasks.

In this chapter, we introduce SeSAm based on our experiences applying it as well as our experience supporting others trying to do the same. The rest of the chapter will contain three parts: at first, a short introduction to the context of human involvement into agent-based modeling and simulation based on the identification of tasks in a simulation study. The second part will contain an overview of the SeSAm system and how these different tasks are supported by it. The third part will contain a short sketch of a number of (partially interdisciplinary) projects and will discuss our future plans.

16.2 Simulation Study and User Roles

According to literature, only two or three types of people are typically involved in a simulation study: the simulation expert, the system expert, and sometimes a project manager is mentioned as a third person. We also followed this division of responsibilities in several projects and also in teaching simulation in interdisciplinary practicals[*]. In a greater context, reuse and maintenance of models have to be considered. This leads to finer considerations about persons and roles involved in the development and use of simulation models, in general and in agent-based simulation models in particular. In the following, we sketch tasks that may be assigned to specific persons or roles and may be supported appropriately by a

---

[*]Between 1997 and 2005, we executed 9 interdisciplinary practicals where computer science and biology students as teams developed agent-based simulation models of social insect behavior.

modeling and simulation platform.

16.2.1 Tasks and User Roles

In traditional industrial application domains of simulation – as mostly addressed in introductory books like [Law, 2007] – modeling and simulation experts are contracted to construct and analyze a simulation model for a particular problem or system. There are two to three different roles involved in such a study: The modeler, the SME "Subject Matter Expert" and, if necessary, a project manager. A high specialization developed for those application cases.

In scientific applications such as in social sciences, physics or biology, the system experts applied simulation for their particular research questions mostly on their own urging for a different kind of specialized tool or highly specialized (mathematical) expertise.

In contrast to the two latter cases a sustainable and efficient use, maintenance and reuse of complex simulation models becomes more and more important to involve new persons and also systematically support additional tasks. The advent of agent-based simulation has additionally changed the situation as tasks such as experimentation may become so demanding that additional persons are required for their execution. Also, the involvement of human experts without formalization training may be possible in a previously unknown way. These considerations result in the following set of tasks, in addition to the project management. We will disregard the latter in the rest of the chapter.

Design of an agent-based simulation model is the task of developing a concept model of the original system that should be analyzed or tested. This step is impossible without expertise of the system to be modeled.

Implementation of a computer simulation model means that the concept model is taken and transferred to a programming language in a way that a run-able simulation is created.

Observing and controlling is an activity that is directly connected to implementation for testing, analyzing and confirming that the implemented model sufficiently corresponds to the conceptual model and to the original system. This is denoted here as a specific task, as it should be ideally done or at least supported by people not directly involved in model design and implementation.

Observing and immersive testing denotes another form of testing and validation of the implemented model. The persons responsible for these activities should be domain experts that can directly evaluate whether the dynamics observed on the macro and on the micro level are plausible.

Calibration and experimentation is a task that involves systematic and thoughtful variation of parameter and input values for finally producing a valid model and making the simulation runs for actual deployment of the model.

Output interpretation should not be underestimated. This task is directly related to the one above. Generated output is interpreted and conclusions are drawn as the final result of the simulation study. Also, this task requires deep domain knowledge.

It is quite common in many application domains for the model developer, experimenter, observer and manager to be one and the same person. In some applications with rather industrial backgrounds characterized by long-living models, contracted model development, etc. these roles are distributed to a number of people. A consequence can be that a person, who is not familiar with the model implementation interface, is responsible for experimenting. Another person just wants to interactively observe the animation.

Before continuing with the introduction of a modeling and simulation platform, a short discussion of user involvement and existing tools is given.

16.2.2 User Involvement in Agent-Based Simulation Tools

Agent-based simulation is particularly apt for involving human users at least for two reasons.

Firstly, it forms an intuitive, direct modeling paradigm where the original system is naturally conceptualized as a collection of actors. A modeler may take the role of such an actor for specifying its behavior characterizing its rules from the Ego perspective. Therefore, for conceptualizing and formulating a model no change of the modeling perspective from micro to macro or from inside to outside the system is necessary. Thus, the direct formulation of a model is the first reason for intensive involvement of humans that are not experts in complex particular mathematics. This leads to new implementation frameworks derived from concepts of visual and end-user programming. Also domain experts can be involved in the modeling tasks more closely, as the basic model structures are more understandable than in models based on complex mathematical constructs.

The second reason is also derivable from the possible inside view. Humans may interact with agents, they may share the perspective, but also may control the model from outside. Therefore the variety of possible interactions between human user and agent-based simulation is higher than in macro or abstract simulations where only one level, namely the aggregated level of observation and parameter manipulation is possible. Additionally the richness and complexity of the agent behavior makes it interesting for a human to actually interact with the individual agents.

Therefore, user involvement is a theme that has been dealt with since the early 1990's. A. Repenning and co-workers coined the term "participatory" theater approach which combined interactive user interfaces with agent-based simulation [Ambach and Repenning, 1996]. Their ideas were implemented in the end-user programming tool AgentSheets where visually specified rules for describing agent behavior could be compiled to Java programs [Repenning et al., 2000]. The main application area of AgentSheets was in children's games or educatory simulations, therefore the practical usability for larger models was reduced. The unstructured set of rules seems to be hardly scalable. Scalability of model representation was one of the driving ideas behind SeSAm with a structured behavior representation and regarding features of the visual programming language that enhance scalability.

Kidsim [Smith et al., 1994] and its sucessors were developed for a similar audience like AgentSheets. They used programming by demonstration for implementing the behavior of agents. These simulation environments were connected to particular domains and were only used for interactive simulations for small children. SeSAm was planned as a powerful tool that allows one to model a broad variety of agent-based models. Nevertheless, learning by demonstration forms an attractive idea (see Section 16.6).

During recent years user involvement in agent-based simulation experienced a revival when role playing games and agent-based simulation were combined, see [Barreteau et al., 2001] as one of the pioneering works in this area. So-called stakeholder approaches [Moss and Edmonds, 2005] use the opportunities offered by agent-based simulation as a white-box approach for intensifying involvement of domain experts, sponsors, etc. Recently, participatory involvement of humans' playing individual agents in a multi-agent simulation were focussed on by [Guyot and Honiden, 2006]. All these works focus on particular applications, not on generic tools like SeSAm.

16.2.3 Tools for Agent-Based Simulation in General

Tools for supporting the development of agent-based simulations have become widely available since the mid of 1990's. Meanwhile a variety of tools can be found, starting from Swarm as one of the earlier class libraries for developing agent-based simulation. Newer examples are Repast (*repast.sourceforge.net*, accessed June, 2008) or MASON (*cs.gmu.edu/ eclab/ projects/mason/*, accessed June, 2008) which are basically both class libraries for efficient event-based simulation. A variety of additional tools are available like GIS-data import, class libraries for supporting the development of simulation experimentation interfaces, etc. SeSAm, in contrast to these tools, provides a proprietary language and allows one to implement simulation models without the knowledge of a traditional programming language.

Netlogo (*ccl.northwestern.edu/netlogo/*, accessed June, 2008) is a special, simple programming language initially grounded in turtle graphics that allows fast simulations. Meanwhile, the Hubnet system (*ccl.northwestern.edu/netlogo/hubnet.html*, accessed June, 2008) enhances NetLogo for participatory simulations. There are two important differences - SeSAm provides more elaborate structures for defining the agent behavior and it uses a visual programming language.

At the end of the 1990's, there were attempts to simplify modeling by providing a higher-level language built upon general simulation languages, for example MAML for Swarm [Gulyas et al., 1999], or more recently a new and promising development, Repast Simphony (*repast.sourceforge.net*, accessed June, 2008). It aims at facilitating modeling and simulation by providing appropriate user interfaces allowing the generation of Java code to GUI-based control of simulations. In contrast to this, SeSAm models are accessible using visual interfaces at any stage of use. An additional – also specification-level programmed – user simulation interface builder allows also to produce model-specific dialogs.

In addition to a survey on simulation engines for agent-based simulation [Theodoropoulos et al., 2008], this book also contains a chapter about JAMES II [Himmelspach and Röhl, 2008]. This particular simulation framework separates an explicit model specification from the executable model automatically converted to Java code.

For several years, commercial tools for agent-based simulation have also been available. An example for a general tool is AnyLogic (*www.xjtek.com*, accessed June, 2008). It provides modeling support for different modeling paradigms: agent-based modeling and event-based models as well as system dynamics models that can also be combined in one model. Although the basic agent behavior definition is based on statecharts, most items of the modeling interface connect directly to Java code. In SeSAm, Java code is completely hidden.

In particular domains specialized commercial tools for Agent-based simulations can be bought. An example is SimWalk for agent-based pedestrian simulations (*www.simwalk.ch*, accessed June, 2008). SeSAm also has been used for pedestrian simulation; the pedestrian behavior can be manipulated allowing for flexible reactions to perceptions of local congestions, etc. In SimWalk, the agent behavior is fixed, only origin and destination areas and other input value configuration can be defined. This makes it useful for standard applications, yet inhibits research.

Platforms and programming languages for developing agent systems may also be used for implementing agent-based simulation models. This can be done by adding a simulation time service to an agent platform [Braubach et al., 2006], or directly using specific agent programming languages like *Jason* [Bordini and Hübner, 2008] which is also shown in this book. The latter offers a powerful, well-structured formal language for programming agents. In addition to its focus on simulation, SeSAm is based on a more practical approach.

At the Department of Artificial Intelligence and Applied Computer Science at the University of Würzburg, we developed a modeling and simulation environment named SeSAm that

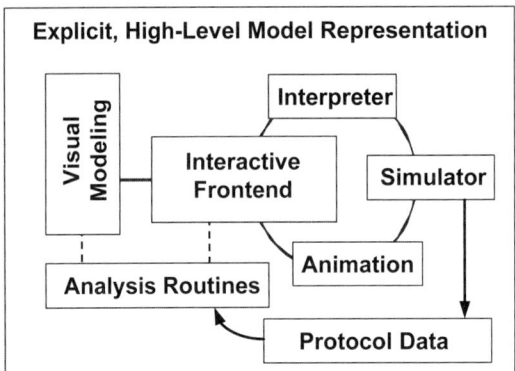

FIGURE 16.1 Overview of the SeSAm system: Based on the high-level model representation, several modules are built that allow to implement the model and its instrumentation ("Analysis Routines") conveniently using visual programming ("Visual Modeling"). The model given in the representation language is interpreted ("Interpreter") and put into a dynamic context for simulation ("Simulator"); The output data can be automatically animated ("Animation") or stored into files for later analysis ("Protocol data"). Visual programming can also be used to generate specific user interfaces for the simulation runs ("Interactive Frontend"). All these modules are based on the given model representation.

combines concepts of declarative high-level model representation and visual programming. The initial aim behind our efforts was similar to the idea of AgentSheets of providing a tool for beginners in modeling and programming. However, our intention was more ambitious as we wanted to develop a scalable platform that can be used for real-world applications. Thus, we also wanted to develop a modeling and simulation system apt for rapid-prototyping by experts. In the remainder of this chapter, we will describe the modeling and simulation system of SeSAm with a focus on the new developments concerning user tasks and their support. Finally we will discuss some anecdotic, yet symptomatic experiences we made when applying SeSAm.

16.3 Core SeSAm

In this chapter we want to describe SeSAm, a modeling and simulation environment that was constructed to support as many of the above sketched tasks as possible. SeSAm basically provides a platform for implementing and experimenting with agent-based simulation models using a higher level modeling language. Starting with some core facilities, it was stepwise enhanced to its current status. Currently SeSAm contains several components whose relations are depicted in Figure 16.1.

SeSAm is based on a specific, proprietary language for describing the elements of a multi-agent model: from the basic elements of the model, namely the structure and dynamics of agents and their environment to possible configurations, instrumentation and experiment description. Due to its declarative character there is a clear separation between model and simulator. All elements of the model description can be entered into the SeSAm system using visual programming. There are also facilities for importing data from databases, tables/spreadsheets, GIS- and CAD files, etc.

The model description is compiled into some corresponding, yet partially more efficient representation that is interpreted for actually producing model dynamics in a simulation. Due to the explicitness of the language, there are in-built facilities for animation, data gathering and visualization based on model-specific selected protocol data. Recently, the system

SeSAm: Visual Programming and Participatory Simulation for Agent-Based Models   483

has been enhanced by a tool that allows building specific user interfaces for interactive simulation, as well as a possibility to control agents from outside the simulation. All the modules depicted in Figure 16.1 will be discussed later in this chapter after the introduction of the model representation language on which they are based.

16.3.1 Basic Model Representation

The high-level modeling language of SeSAm consists of elements on different levels as depicted in Figure 16.2.

- *Primitives and data structures* form the basic language elements like in any other programming language. Primitives and data structures may be built-in or user-defined. The latter are saved together with the complete description as XML file.
- *Static structures* are the description of the structural composition of the system: entity and environment classes, state variables and their domains describing the entities' bodies, etc.
- *Configuration* of the initial situations including descriptions of a number of instances and start values for each instance, its positions, etc.
- *Description of dynamic reasoning* as the specification of agent and environmental behavior.
- *Meta-level characterization* means basically descriptions of what to do with the model: experiment scripts, model instrumentation, visualization, etc.

In the remainder of this section, we will sketch the essentials of each of these language elements.

Model Level

Primitives, User Functions and Data Types

So-called *primitives* form the basic building blocks of the model and connect the model to the basic programming language: internally, every primitive consists of two parts: a declaration part and a Java method named *execute*. The declaration part contains a structured description of input and output argument types together with some text that describes its functionality. This information is parsed by the SeSAm system for generating the basic parts of the visual interface.

There are different categories of primitives:

- *Action Primitives* are the actions that an agent may exert for manipulating its status or its environment. These primitives are connected to Java methods that modify the agent or its environment specified by the arguments given to the action. Examples are `move` or `setVariable`.
- *Sensor Primitives* collect information from the agent's environment and also from its internal state. An example is a primitive that returns a list of objects that the agent is able to perceive in some direction within some range, like `observeAllObjectsInDirection`.
- *Computational Primitives* form a quite heterogeneous set of functions that provide means for the agents to execute more or less complex computations based on the sensor information and parameters of their behavioral programm. An example is the calculation of the heading of the agent using some simple position

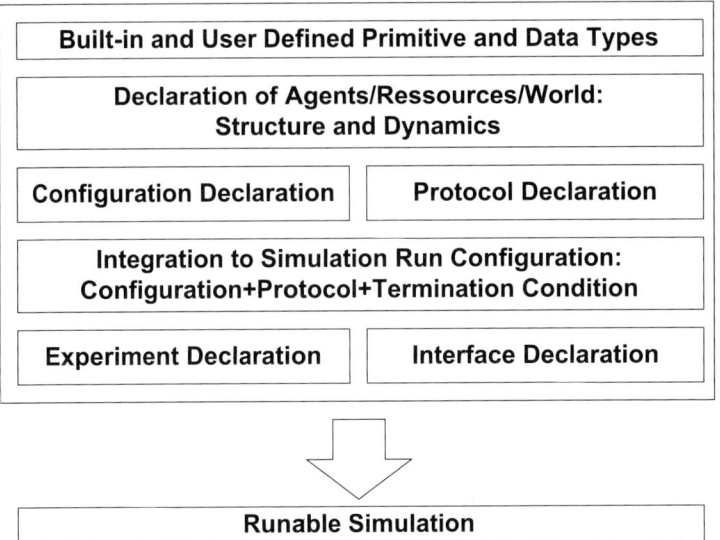

FIGURE 16.2 Building blocks of the SeSAm language: the atomic elements of the model representation are the primitives that form the basic elements of all descriptions of actions, perceptions where actual modifications happen. The concretely specified primitive calls are used for characterizing the dynamics of the agents, resources and also the world. Additionally, these entities are characterized by a structure that is able to capture their status. Based on this core model, additional information is integrated for specifying a fully functional model: Specifications of configuration (possible start situations) and of routines for collecting data to export in protocols. All this information is integrated into a full simulation run specification. The latter can be augmented by an interface declaration or several of them can be aggregated to experiments.

mathematics. Other examples are comparison predicates, list manipulation, etc.

We provided a set of primitives that can be shown to be Turing-complete [Oechslein, 2003]. A modeler may define complex nested calls of primitives as macros called *user primitives*. These macros provide abstractions in the sense of model-specific functions. The following excerpt from a SeSAm model xml file shows the definition of such a user function in a simple predator-prey model. It contains the definition of a flee action of a prey agent toward a shelter `simObject` that is computed in another user function addressed by calling `UserFunction0`. The user function possesses only one argument, namely the active agent itself. The return value is `void` denoting that it is an action.

```xml
...
<userFunction name="Flee from predator" id="UserFunction1"
    external="true" expert="false" inline="true">
    <functionCall>
        <call functionName="MoveTowardsPos">
            <call functionName="GetSpatialInfo">
                <parameterID id="FunctionArgument1"/>
            </call>
            <call functionName="GetPosition">
                <call functionName="GetSpatialInfo">
                    <call userFunctionID="UserFunction0">
                        <parameterID id="FunctionArgument1"/>
                    </call>
                </call>
            </call>
        </call>
        <call functionName="GetSpeed">
```

SeSAm: Visual Programming and Participatory Simulation for Agent-Based Models

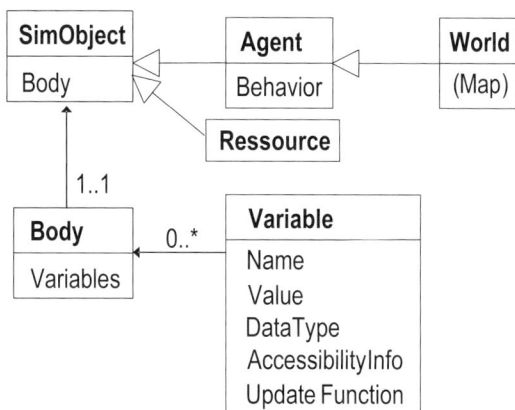

FIGURE 16.3 Basic structure for describing the structural building blocks of a SeSAm model: There are three different types of entities all derived from the `simObject` class: `Resources` that represent static objects that can be manipulated, `agents` that possess also a behavior descriptions and the world that additionally may contain other global representations such as a map. The `simObject` structure that is responsible for storing the current status of an entity. This status is stored in a collection of state variables that together form the entities "body".

```
              <call functionName="GetSpatialInfo">
                   <parameterID id="FunctionArgument2"/>
              </call>
           </call>
        </call>
    </functionCall>
    <parameter name="ich" id="FunctionArgument1">
         <simObjectType/>
    </parameter>
    <voidType/>
</userFunction>
...
```

The SeSAm modeling language provides a priori a set of atomic and abstract data types ranging from boolean, numbers, etc. to lists or hashtables. Composed data types and enumerations may be added by a modeler to provide model-specific types and abstractions in a similar way to user primitives

Structural Description

There are different structural elements and levels built upon these basic primitives. An overview can be found in Figure 16.3.

A multi-agent model consists here of a set of models of entities – in the terminology of SeSAm `simObjects`. Their structure and behavior is described on the class description level. Individual entities are characterized at the configuration or instance description level. For the simulation, actual object instances are generated from these instance descriptions. This three-level system of description allows one to separate explicitly between model and simulation run, enabling the visual programming facilities of SeSAm (see Section 16.4.3) and allowing one to add an additional compilation step for generating simulation entities from instance descriptions.

For reasons of model clarity, there are three specific different kinds of object class descriptions, derived from the generic `simObject` class description:

- Agent class descriptions form the basis for all model-specific agent classes: for example a Pedestrian Class or an Ant Class is basically an instance of an agent class description. A basic agent class description consists of a characterization of its body and its behavior.
- Resource class descriptions are used for integrating static entities populating the environment of the agents. They just contain some status that may be manipulated and accessed by the agents.
- World class descriptions are basically special agent class descriptions that contain global status variables or parameter. They enable formulation of some global level behavior. Therefore on the configuration instance level only one unique instance of a model-specific world class is allowed per situation (see below).

All `simObject` class descriptions contain declarations as to how the status of their instances is represented. In the current SeSAm version the container for all status information or individual parameter etc. is characterized as "body". The body consists of a set of variables and constants. A body description is associated with a class description. Thus, this body structure also mirrors the three levels of class description, instance and actual entity by body description, body instance description and actual body, the latter only exists during a simulation run.

Variables – also on corresponding description layers – are characterized by the following information:

- Name of the variable
- Data Type (boolean, number, string, list, hashtable, pointer to `simObject`, but also position, shape, image,...)
- Characterization: configuration parameter, state variable, output variable or supporting variable.
- Default initial value
- Interface characterization denoting whether its value is perceivable or even manipulatable for other agents

Variables may be grouped in so called "features" which can be assigned to more than one `simObject` class. Combined with user primitives that specifically work on those feature variables these language elements provide common abilities to agent, resource and the world classes. This might be useful for model elements that deal with identifiers or different types of memory, etc. Features also form the basis of an important plugin-mechanism which will be described in more detail in Section 16.3.3.

There are several predefined features that additionally provide specific data types. The most prominent feature-level plugins are spatial representations. The standard one is a continuous 2d space consisting of a feature for the environment that provides a map description at the world instance description level and a concrete map for the actual world instance, as well as variables containing the spatial information of an entity for the agent and resource classes on the different levels of declaration. Spatial information for the 2d continuous case contains not only position, but also shape in form of a polygon, speed, direction and information about visualization like color, image,... This feature also encapsulates specific primitives for spatial perception as well as several movement primitives. The treatment of spatial information as feature has the advantage of flexibility: only entities – resources or agents that possess the spatial information feature are positioned on the map. These entities may be combined with non-spatial agents like abstract, non-spatial agents, for example organizational agents. By exchanging this feature assignment, other forms of spatial representation can be used without any changes on the platform level or even concurrently.

Behavior Description

Using the above introduced primitives, dynamics can be expressed similarly to traditional programming languages. However, SeSAm provides language constructs for organizing the description of behavior of agents and of the world:

The behavior description is inspired by UML activity diagrams: consisting of activity nodes and arrows that depict transition rules between activities. Special activities are the start and the end activity. When the behavior of the agent ends up in the end node, the agent is deleted from the simulation. When an agent is generated, the interpretation of the activity diagram starts at the start node.

An agent (as well as the world entity) may have arbitrary many activity graphs or "reasoning engines" as they are called in the SeSAm terminology. This allows the modeler formulating concurrent behaviors performed by one agent – for example defining negotiation behavior in one reasoning engine and concurrent manipulation of the environment in another.

An activity encapsulates three sequences of actions that are basically sequences nested primitive calls: start actions that are performed when the activity is selected anew, standard actions that are performed once every time step as long as the agent is executing that activity and termination actions that are executed for cleaning up when the agent has decided to change its activity.

Rules are responsible for terminating one activity and selecting the next. They are linked to their predecessor activity and tested every time this activity is executed by an agent. Rules may contain arbitrarily nested function calls of primitives finally returning a boolean value as pre-conditions and an activity as post-condition. A special kind of rule has "otherwise" as condition which means that this rule fires if the condition of no other rule associated with the same activity returns true. With this special rule a sequence of activities can be enforced without detailed implementation of further conditions. Conflict resolution in the case of more than one applicable standard rule is random selection. There are two types of activity concerning their temporal aspects: "instant" or "time consuming" activities. The basic update cycle is round-based (see Section 16.3.2). This label denotes at which activity the rule-based selection and execution cycle stops waiting for the next update signal.

Activity graphs can become quite large and thus, without further structuring means, hard to conceive by a human modeler. Therefore additional language-level support is given by introducing hierarchical composition of activities. A partial graph can be replaced by a composed activity, without any change in the interpretation of the overall activity graph. Rules selecting an activity within this new graph are simply routed via the start node of the partial graph. Similarly, rules selecting activities outside of the composed activity graph are routed via the end node. Both start and end node are interpreted as instant activities without time consumption.

Declaration of Situations

As mentioned before, there is an additional description level between the agent, resource or world class and a simulation run: descriptions of possible model configuration or "situation" declarations as they are called in the SeSAm terminology. The description of a situation contains a set of instance descriptions. There may be arbitrary many agent or resource instance declarations, but only one unique instance of a world class. The reason is obvious when keeping in mind that the world instance represents the global aspects of the environment.

An instance description mainly contains – in addition to a pointer to the behavior description – a body instance description with a set of variable instance descriptions. There the important aspect is the individualization of initial values for the variables. In models with explicit spatial representations, on this level the map declaration – as an instance of the spatial feature class assigned to the world class – is available that allows for localization of entities that possess the spatial information feature.

For facilitation of situation designs with many elements, we integrated groups of instances that can be named as a group and used as a template for partial situation configurations.

Experiment Description Level

With the declaration of a situation the model itself is completely specified. However, for being able to perform useful simulation experiments additional information has to be provided.

Declaration of Model Instrumentation

The most important point is the definition of measurements that can be taken from the agent-based simulation. These can be variables defined in the world class, where data may be aggregated for protocol, but also specific output variables of prominent agents. Model instrumentation is provided from outside the model and should not be a part of the core model. Therefore, specific measurements may be explicitly defined and added to a situation description. The declaration of model instrumentation may contain primitive calls for collecting the necessary information from the multi-agent system and information on what should be done with this gathered information. Basically two options are available: writing the data to protocol files or visualization in different forms of diagrams that animate the changes throughout the simulation run.

Simulation Run Definition and Experiment Declaration

A simulation run contains not only the initial description of the situation and the instrumentation of the run. It additionally contains the description of a condition for terminating the simulation run. This condition may access the simulation time or may be a characteristic of the complete situation, for example when all agents have been deleted.

Several simulation runs can be aggregated to experiments. Instead of providing some predefined framework for describing variations of parameter values and corresponding simulation runs, we decided to provide full programming power for experiments. It allows scripting simulation runs with full flexibility. For this aim, primitives operating on situation declarations, setting the random seed, generating and executing simulation runs are provided.

16.3.2 Simulation Routine and Model Interpretation

Up to now, we briefly sketched the full model and simulation description language used by SeSAm. The semantics of the higher level elements are given by the interpreter, the semantics of the primitives by the underlying Java code. There is a compilation step between model description and simulation run for two reasons:

1. Clear separation between model and simulation
2. Possible code optimization based on techniques from compiler design, like constant folding, code in-lining, etc. These optimization steps are transparent for the user. They enable SeSAm to execute simulation runs rather fast.

SeSAm: Visual Programming and Participatory Simulation for Agent-Based Models

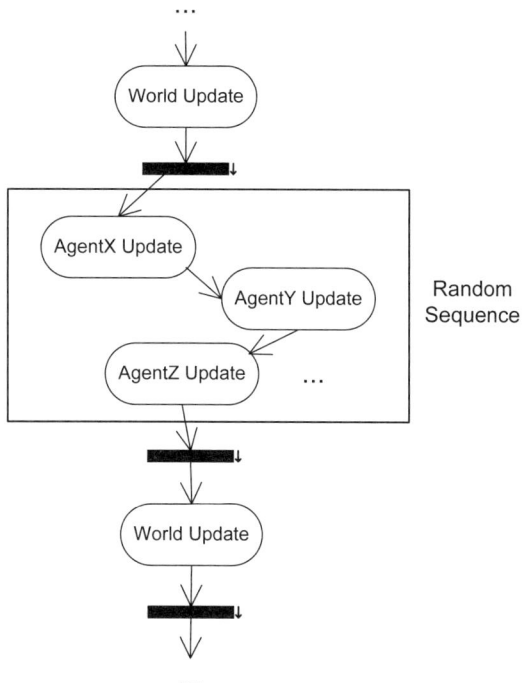

FIGURE 16.4 Interpretation cycle of SeSAm: After an update of the unique world entity, all agents are updated in a random sequence. Then, the world may again update, etc.

Basic Information about the Simulator

SeSAm provides a round-based simulation where the agents are updated one after the other in a random sequence. When all agents are treated, the world entity will be updated, thus the update of the world is the only possible synchronization point within one update round.

Each agent is fully updated – which means it senses its environment, evaluates its perceptions by going further through each of its activity graphs and finally performs the scripts associated with the current activities. The random sequence of updating the agents provides a reduced form of virtual parallelism: although each agent is fully updated before the next agent is tackled, the modeler cannot influence the actual sequence of updating the agents.

Agent Update

There are two sources of dynamics in a multi-agent model: first, the global environment which is explicitly modeled in SeSAm; second, from the agents with their behavior. As is noticeable in Figure 16.3, the world can also be seen as an agent, yet with another - more powerful - behavior repertoire as it may also access information from a global point of view.

The update sequence of an agent a is the following:

```
For every re in ReasoningEnginges of agent a in given order do
    repeat until re.currentActivity is of type time-consuming
        rules ← select rules from re.currentActivity with condition == true
        if | rules | ≥ 1
            theRule ← getRandom(rules)
```

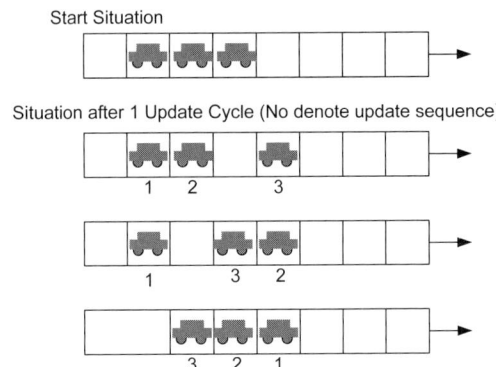

FIGURE 16.5 Different update sequences cause different outcomes. The result of the parallel case corresponds to the second line.

```
    execute termination sequence of re.currentActivity
    set re.currentActivity to next activity of theRule
    execute start sequence of new re.currentActivity
execute action sequence of re.currentActivity
```

The update cycle of the world is identical to standard agents. The main difference is that the world is allowed to use primitives affecting the complete system.

Properties of the Update Scheme

This overall update scheme has advantages and disadvantages. On the one hand its major advantage lies in its simplicity: the procedure can be easily understood by researchers who are not experts in distributed systems. No resolution in case of conflicting actions is necessary as there is no actual parallelism. On the other side, artefacts — especially when agents are learning — are prevented by the random update. It is not the case that the exact time for updating a particular agent is set; the modeler cannot determine the sequence of agent update.*

Obviously there are critical issues in the update sequence that a modeler must be aware of: with the randomization of the update sequence comes an additional, potentially hidden stochastic process that may be responsible for variations in the model output — however, if exact reproducibility of results is needed, the random seed can be explicitly set in the initialization of a simulation experiment.

Additionally the fact that one agent is updated after the other may result in different outcomes depending on the actual update sequence. This is important in applications such as traffic simulation with collision avoiding behavior when an agent may take the position that was freed by another agent immediately before. Figure 16.5 illustrates this problem by showing how the update sequence influences the result.

These basic problems can be tackled on the model level by dividing one time step into two virtual ones: one timestep in that every agent is perceiving its environment, reasoning and the selection of an action. Then the selected actions are executed in a second loop. This

---

*Unless the random seed number is fixed.

results in the fact that all agents perceive the same situation of the environment, basically as if they perceive at the same time. We used this emulation of parallelism for simulating cellular automata using SeSAm. However in our CA case, one agent was only allowed to manipulate its own status avoiding any conflicts in action performance. If concurrent modifications of the same environmental entity are possible, then conflict resolution for deciding which action actually should be taken has to be formulated on the model level ideally in the behavior definition of the world that is conceptualized as a global entity. Such a model however becomes very complex.

16.3.3 Plugin Mechanism for Extension

The language elements described above allow a clear and structured overall model representation. However it can be quite cumbersome to formulate models just using the default primitives and structures. Therefore a plugin mechanism was developed that allows enhancing the language and providing additional supporting tools. In particular a plugin may contain a set of related functionality consisting of

- Specific data types, such as a data type *path* or *schedule*
- Feature-level add-ons, like variables added to the `simObject` class to which it is assigned, like a variable that stores the personal schedule.
- Primitives using the plugin data types, interfaces to plugin variables or simply macro-like primitives as support for modeling on a higher level of abstraction
- Tools providing helpful functionality, like importing spatial information for filling situation definitions, etc.
- Plugins may also come with specific editors or panes.

As mentioned above, prominent plugins are responsible for spatial representations. There is for example a plugin for 2d worlds adding a map to the world, a spatial information data type composing position, heading and polygon-based shape representation to `simObject` classes. Alternative space representations are 3d maps, graphs connecting node- and edge-based `simObjects`, as well as raster and vector GIS-based representations. There are a variety of additional plugins from primitives that establish connections to data bases to plugins that allow generating movies from animations.

The plugin mechanism had been extensively used in SeSAmHospital which can be seen as a platform for agent-based hospital simulation with a focus on examination scheduling [Herrler, 2007]. This domain-specific simulator basically consists of SeSAm and a hierarchy of interrelated plugins providing complex data structures like clinical path representations, specific patient generation functionality, etc.

16.3.4 Problematic Details of the Language

Working with this language in a variety of simulation projects from simulation of bee behavior to large shopping behavior models or pedestrian simulations, we discovered that the language as it is now needs further improvements.

The main critical point is the weak agent interface concept. The modeler may declare appropriate variables of the agents' body as "accessible" from outside. A second tag denotes whether the value of the variable may be manipulated or not. This rudimentary representation of possible interactions may be apt for implicit interaction; however explicit, message-based communication needs additional efforts. Thus encapsulation of behavior is hardly supported by the SeSAm language, but relies on the self-restriction of the modeler.

Another missing concept is local, temporary variables that facilitate modularization of behavioral description by enabling the formulation of intermediate results in computations within an activity or a user primitive. In the current version of the language, there are some quite tricky combinations of primitives that allow us to circumvent this restriction – however such opportunities are just available to SeSAm experts.

A third drawback is the lack of a clear inheritance concept between classes that are accessible to the modeler. The modeler may just generate a list of instances of the agent description classes without specifying the relations between instances. Features – as mentioned before – remedy this problem only marginally as they currently just integrate variables and user functions. We are currently planning to enhance the features by integrating data types and partial reasoning engines moving toward a feasible model building block concept.

16.3.5 General Aspects of Suitability

Although it can be shown that the declarative language possesses the power of a general programming language, one can easily see that it is more apt for modeling a particular kind of system whereas others might be cumbersome to formulate. In [Klügl, 2008] three levels of complexity of architectures for simulated agents were identified: behavior generating that denotes mainly architectures that use planning from first principles, behavior configuring architectures, like interpretation and instantiation of skeletal plans and finally behavior describing architectures, like rule-based behavior descriptions.

In general, SeSAm is particularly apt for simulations involving rather simple agents that reside in some spatial environment and interact implicitly by manipulating the environment. Such behavior can be easily formulated using rule-based behavior descriptions. This originates in the initial applications of SeSAm in the area of social insect simulation. Related properties can also be found in traffic or pedestrian simulation where successful projects have been performed. The negotiations formulated for the hospital simulations [Herrler, 2007] show that also message-based interactions can be integrated without any problem into that frame. They involve the formulation of at least two concurrently active reasoning engines per agent: one for the actual behavior, one for accepting and interpreting incoming messages concurrently to the actual behavior. Therefore the overall agent model is slightly more complex.

In [Rindsfüser and Klügl, 2005] we formulated agent behavior as executing and manipulating daily plans stored in state variables of the agents, instead of explicitly and directly formulating the plan contents in the activity graphs. The model was sufficiently successful in reproducing travelers' daily activities, yet the representation and manipulation of such complex data structures required careful consideration. This is due to the fact that it is not appropriately supported by the overall system, in terms of providing plan structures and primitives to modify these plans.

SeSAm's range of applications has widened and experiences with more complex agent models confirm that formulating more and more complex agent behavior becomes quite cumbersome. Although a shortest path algorithm can be formulated without any problem – for example for generating a path as some kind of movement plan in a network – the integration of real behavior generating approaches is hardly possible. For planning from first principles, operator descriptions and complex situation descriptions that characterize goal states must be tackled. As SeSAm does not offer any form of logic-based description of situations nor any appropriate primitive characterization based on pre- and post-conditions, such complex architectures are basically impossible.

16.4 Users, Tasks, and SeSAm

For actually implementing a model using this SeSAm core language, several user interfaces are supporting different tasks during a simulation study.

Developing such interfaces within the same visual framework is not an easy task. Much of the advance of the SeSAm system during the last year is related to these developments. The following sections will contain a short, sketchy treatment of the different elements of the overall SeSAm model development and simulation platform offered for the different tasks mentioned above: visual modeling environment, standard animation, online aggregation and visualization of key values during simulation, interactive simulation and "agent playing" environment.

16.4.1 Principles

As mentioned in the introduction we had several, partially conflicting goals for developing SeSAm. We wanted to build a system that concurrently enables beginners to implement their model and that can be used as a fast prototyping tool for experts. Therefore the system had to fulfill the following general requirements:

- Hide potential model complexity from beginners
- Support beginners while learning to implement a model
- Make all aspects of a model accessible to experts
- Support experts while implementing a complex, large model

Thus for beginners the main issue is simplicity and consistency – for experts it is scalability of modeling and simulation. In more detail, the following ideas and principles were pursued, partially motivated by [Green and Petre, 1996] or [Kuljis, 1994].

- The main difficulty lies in selecting the appropriate primitives: therefore an expert may choose to use a different set of primitives than a beginner. The latter may use more abstract primitives that encapsulate more functionality than the former who may like to completely control every detail and search for the most effective implementation of the particular problem at hand. Good examples are movement primitives. Whereas the expert may use more basic primitives like `changeDirection`, `moveWithSpeed` or `observeObjectsInDirection`, a beginner would be happy with a primitive like `moveWhileAvoidingCollisionsWith`...
- Having learned how to deal with one editor, this experience can be used for all editors for similar elements, as GUI interaction remains the same.
- Including automatic syntax checks or missing opportunities to input something syntactically wrong.
- Functionality as provided by modern software development platforms makes it easy to change model design decisions. Examples are refactoring actions or navigation means, like searching for references or direct connection to definitions of model elements. Also support for model versioning is important for model implementation.
- Documentation facilities should be possible for any element of the model for recording why things are formulated as they are. Colors etc. can be used to mark model elements where the modeler wants to include more detail. Thus the modeler does not have to memorize a lot, but can augment the model with free text about his thoughts.

- Structuring language elements like sub-programs or modules are available in the form of user primitives, features or partial situations. Thus modeling becomes more scalable for the expert user. Predefined modules support the novice.

16.4.2 Developing a Conceptual Model

This task is not directly supported by SeSAm. One may argue that the high-level model representations – the classification in agents and resources, the way behavior is characterized – may support structured thinking about elements of a model. A basic idea of SeSAm is to bridge the gap between model specification and implementation.

16.4.3 Visual Programming for Model Implementation

In our description of the SeSAm system we will first concentrate on the task of model implementation. Providing a visual programming environment for this task was basically driven by the idea to make model logics accessible to all interested people, not only model configuration, model-produced simulation runs or data. Introductory material concerning visual programming can be found in [Chang, 1990] or more recently in [Lieberman et al., 2006].

In general one must admit that programming using a well-known textual programming language in a good environment for development may result in a more efficient modeling and simulation for the experienced programmer. Even for a person experienced as modeler using some programming language and class libraries, model implementation supported by a visual programming environment has many advantages ranging from explainability to accessibility.

Primitive Call Specification

The basic building block of the visual SeSAm programming environment is the specification of primitives and primitive calls: There is only one way to specify how a primitive is used. It is used whenever a nested primitive call has to be specified: in the basic behavior description of activities, in formulating the predicates of rules or user macros, for giving a procedure to compute initial values, ... to experiment and analysis declaration. This dialog element is shown in a screenshot in Figure 16.6.

Using such a basic element, a modeler can completely specify behavior without programming in a traditional language. The particular procedure for selecting and configuring reduces syntax error proneness as the user cannot select a primitive that returns a wrong type. Another element that deals with reducing potential errors is the way in which nested calls are to be edited. In this case the parent call is no longer accessible for manipulation which avoids inconsistencies.

In the right lower corner – mostly hidden in Figure 16.6 – three elements support fast selection by providing a short cut search item, enlarging the primitive set with "expert-mode primitives", like loops or roulette wheel functionality. A small button – the "Edgar" button in the corner right on the bottom – allows for textual search for appropriate primitive. All documentation and hints given in the primitive declaration are available there. Edgar also justifies why certain primitives are not currently available for insertion.

Forms and Tables

A typical means for specifying object instances – class declarations as well as agent instances in the model configuration – are forms where every attribute can be input using appropriate

SeSAm: Visual Programming and Participatory Simulation for Agent-Based Models 495

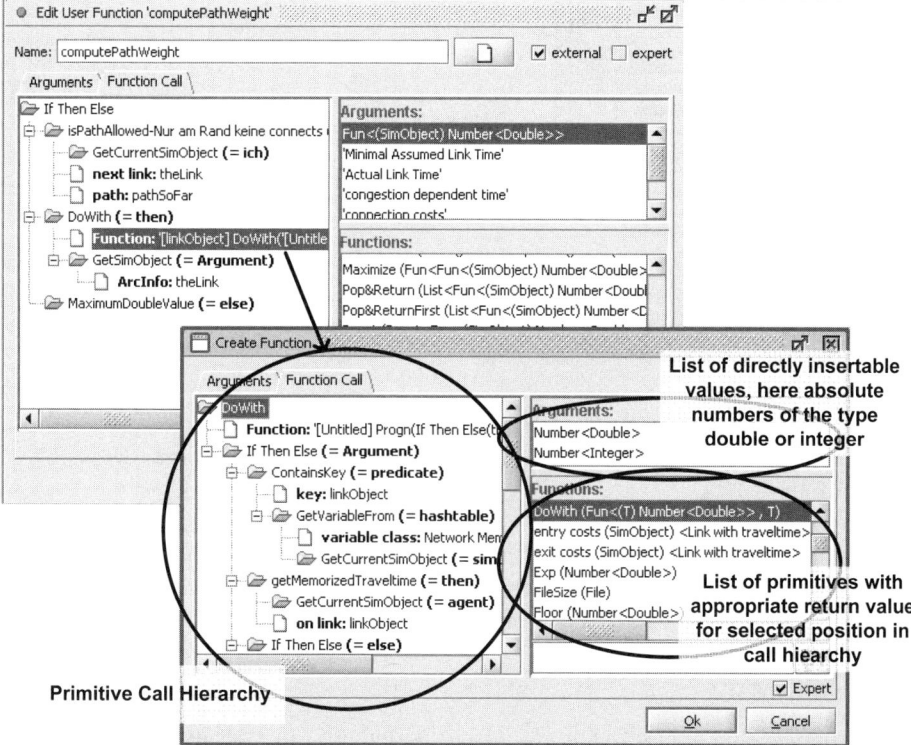

FIGURE 16.6 Screenshot of a dialog for specifying primitive calls. This basic dialog element is used everywhere in the system when this task has to be done. It consists of three parts: the current status of specification in the left half of the dialog, directly insertable values in the upper right and available primitive on the right lower half. When clicking on a primitive call, a similar dialog opens resulting in a cascade of dialogs as indicated by the arrow from one dialog to the other.

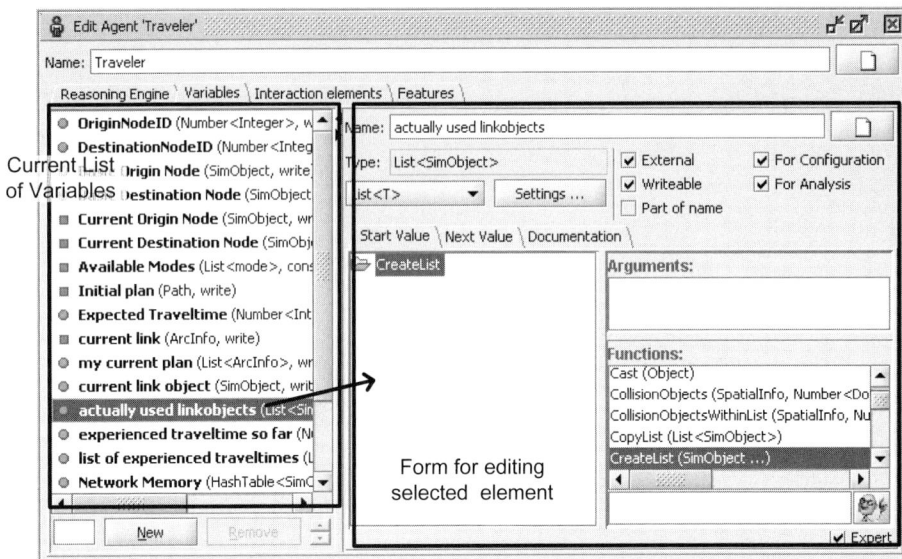

FIGURE 16.7 Screenshot of a dialog for specifying body variables. This basic dialog element is used everywhere in the system when language elements that are in a set, have to be specified. This dialog element has two parts: on the left the list is given, the pane on the right changes with selection of a list element for enabling the modeler to manipulate the selected element.

dialog items [Bamberger et al., 1997]. Forms are used in SeSAm for all fixed meta information for model elements – for example in specifying a body variable description as visible on the right side of Figure 16.7.

Tables can serve as a more compact representation of object attributes, if the data structures to represent in one cell are not too complicated. We use tables, for example, to specify the object instance declarations where the set of object variables that have to be set is not given, but depends on the class-level declaration.

Edit List Elements

Another structure that is used at many places is depicted in screenshot 16.7. On the left a list of elements is given; by selecting an item, all information about it is displayed on the right and can be edited. This dialog form is used for body variables, specification of `simObject` instances, analysis items, etc.

A direct combination of list elements and pure primitive specification can be found on several occasions. Examples are the list of user functions, the analysis items when instrumentation is specified, action sequences in activities, object lists in situations, etc. Once a modeler has learned how to deal with these modeling elements, this experience is useful all over SeSAm.

Behavior Specification

The primitive specification forms sketched above are used as low-level building blocks for defining the behavior – basically as atomic statements of the language. The structure of a reasoning engine or an activity graph is especially apt for visual constructing using some graph structure. Figure 16.8 shows an example screenshot.

There are different node shapes in addition to the previously mentioned start node, end

SeSAm: Visual Programming and Participatory Simulation for Agent-Based Models 497

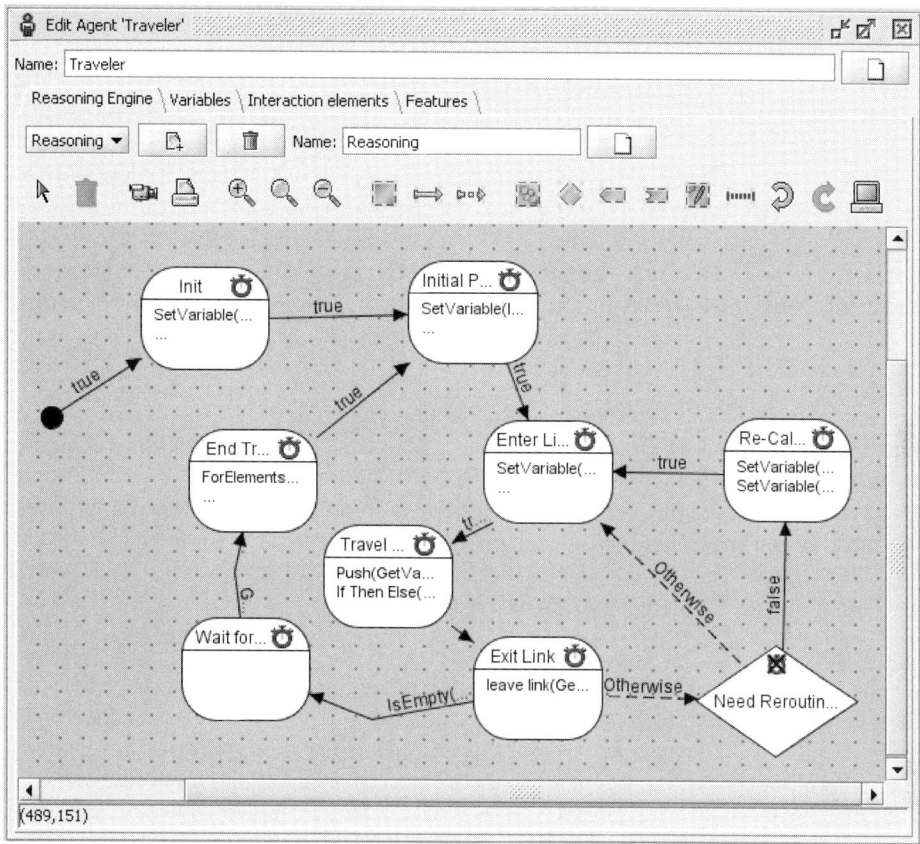

FIGURE 16.8 Screenshot of the pane for specifying the behavior description of an agent.

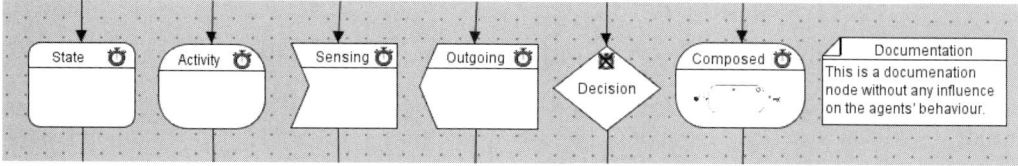

FIGURE 16.9 Different shapes of activity nodes in reasoning engine modeling. Only composed activity nodes – which hide a new activity graph – bear special semantics for interpretation. Documentation nodes (the most right ones) are ignored for behavior generation. The others only serve for enable the modeler to oversee a complex graph.

node or nested activity graphs - for depicting specific properties. These additional shapes – shown in Figure 16.9 do not result in different forms of treatment in the simulator, but merely support clarification of functionality for transparency reasons. Exceptions are the documentation nodes which are ignored. Only one look at the behavior describing graph is necessary to identify decision, sensing activities, etc. For additional overview, different colorings of activities are available. For a short look at the activity contents the beginning of the action primitive calls are also depicted using a small text size – see activities in Figure 16.8.

Edges in the activity graph editor denote rules for terminating their predecessor and initiating their successor activity. The subtitles of these edges are generated from the primitive combination that forms the condition of the rules. These generated texts can however be replaced by the modeler. Thus the modeler has several options for implementing a well-structured and clearly structured activity graph.

Configuration of Start Situations

Without additional plugins the start situation merely consists of the description of one world instance and of a set of agent and resource class instances. This is done using already known lists and tables.

Additionally editors coming with one of the spatial plugins play a particular role, as they allow positioning on maps using drag and drop functionality.

Accessibility and Documentation

Two important issues in a visual programming system are navigation and documentation. One prerequisite for useful development environments for programming is the way to find the elements that the user wants to edit. The user should not be required to memorize names, but to have directly at hand all necessary information for manipulating an element. For example, when formulating an agent's behavior, variables of the agent should be accessible with one click when they are missing or are set to the wrong data type. Also the user should not need to memorize the exact functionality of a user function, but should be able to immediately access it.

In SeSAm tool tips can be edited directly by the modeler for augmenting the documentation of the model. A right click always leads to a context menu where related elements of the definition are accessible. Every element – from user primitive to status variable to activities – can be given additional, explanatory text. These modeling GUI elements increase scalability of the implementation which is particularly important for complex multi-agent simulation models.

16.4.4 Experiment Scripting and DAVINCI for Experimenters

The second task that has to be executed during a simulation study is extensive experimentation. It is basically done for model testing and – after the necessary model quality has been guaranteed – for making the actual deployment runs generating the intended results. Whereas programming and formalization expertise can be seen as a prerequisite for users fulfilling the role of a modeler, pure experimentation contains a lot of rather simple operations with several repetitions – if the configuration to be run is known. The main intelligence in the experimentation task however has to be used for an intelligent design of experiments as well as for analyzing the output data in order to initiate additional simulation runs.

It is absolutely unacceptable to trigger all necessary simulation runs manually. Therefore the configuration of experiment scripts is or at least should be part of every modeling and simulation framework. In SeSAm there are two ways of defining and controlling simulation experimentation: the first consists of basically a script using additional primitives that operate on situation configurations, simulation run definitions. These primitives may set the random seed for ensuring exact reproducibility of results or may initiate runs based on systematic parameter variations. As the user interface is similar to all primitive call specification dialogs, the original modeler can easily input such experimentation scripts.

The second possibility was developed for supporting calibration in the PhD thesis of M. Fehler [Fehler, 2008 or 2009]. Besides other methodological advances, he developed a tool named DAVINCI that allows automatic parameter optimization for adjusting the model for maximizing a model-specific function that expresses some degree of validity of the current simulation configuration. Information about concepts and technologies behind the tool can be found in [Fehler et al., 2006] or [Fehler et al., 2005]. DAVINCI can be used more generally for systematic parameter screening as well as for all kinds of optimization based on methods ranging from tabu search to genetic algorithms. Therefore it forms a highly valuable tool for an experimenter. Due to the conceptual complexity for configuring all elements of the optimization algorithms, such as input parameter, evaluation function, parameter of the optimization, and parameter of the result presentation, this tool is implemented like a wizard that guides the user through complex configurations.

16.4.5 Online Aggregated Data Presentation and Animation

Another functionality that is primarily used for testing and analyzing the model dynamics is the animation facility and the online visualization of aggregated data during a simulation run. These techniques for observing what is happening during a simulation can be seen as a standard for simulation environments. Basically every user of the simulation models wants to observe animations.

When the animation is enabled all changes are shown immediately when they happen. If the standard 2d plugin is used, the behavior may be enriched using primitives for updating the visualization of an agent, for example by changing the image for its graphical representation, its color or intentionally draw lines or circles into the map pane for visualizing paths that the agent has followed, etc. Figure 16.10 shows a small part of the animation view of a Pedestrian simulation of the SBB Bern Railway Station [Rindsfüser and Klügl, 2007].

As the speed of a simulation run may be too slow for reasonable observation, a recording plugin has been developed that allows one to save animations as movie files for later analysis.

The results of the primitive calls for collecting output data from simulation run can be written into a file for later analysis – which is actually done during model deployment. For testing, the same data can be shown directly after generating, using either a series chart or using a block chart, depending on the volume and dynamics of the data. In Figure 16.11 the relation between analysis definition and curve is shown.

There is a small enhancement that makes the animation even more valuable: a debugger. If a simulation reaches a predefined point in its behavior definition, it stops and the agent that caused the break can be analyzed more deeply. A stepper functionality allows action-by-action advancement of the simulation execution. Also, debugger and stepper belong to the equipment of standard programming environments and thus should be also available for testing simulation models.

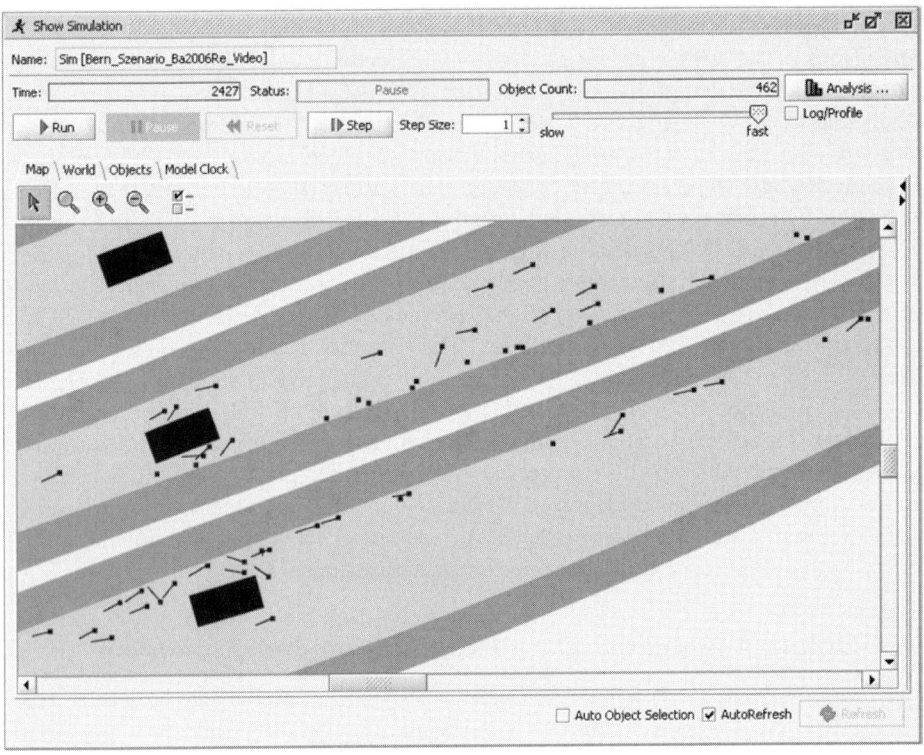

FIGURE 16.10 A part of an animation window showing agents with different attributes (colors) and their paths.

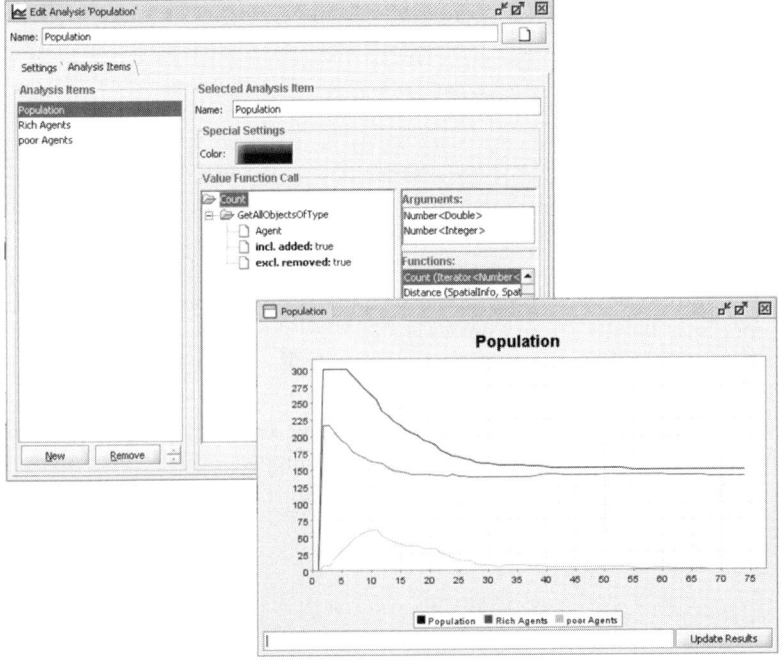

FIGURE 16.11 Based on the function specification for output function, aggregated data is shown.

16.4.6 Model-Specific Interfaces

The previous tasks required direct and extensive understanding of the implementation of the model. A user that is not involved into implementation and testing of a model, but wants to observe its dynamics, needs a model-specific user interface for "playing around" with the model. Interaction with the model may take two forms: either as an experimenter or observer from outside or alternatively from inside the running simulation. The outside view is basically standard: a human observes the dynamics of the model from outside and manipulates global parameters or sets local switches in reaction to the observation. The inside view is particular for agent-based simulation and is often referred to as participatory simulations (see [Guyot and Honiden, 2006]). In this case a human is controlling one agent, perceiving what the agent may perceive and manipulating the simulation through the agent's effectors. SeSAm supports both forms of interaction. We first tackle specific model interfaces followed by participatory simulation in the next subsection.

In order to provide functionality of interfaces for observing and controlling simulation runs in a user-adapted way, we enhanced SeSAm with a graphical GUI builder. We assume that this element of SeSAm is actually used by the modeler for providing specific interfaces for more or less experienced people using or testing the model. Examples for addressees may be stakeholders, but it is also apt for publishing demo versions of a model. Concerning its functionality, such a model-specific simulation interface corresponds to the type of model interface that a user may know from other simulation tools.

The construction of a simulation specific user interface is done in two phases. At first the modeler determines the interfaces of the model elements - for example he specifies that the variable "storage" is manipulatable by the user and then arranges the pre-defined items to an overall user interface. Figure 16.12 shows a screenshot of the visual GUI builder together with an instance of the specific user interface.

16.4.7 Agent Playing for Advanced Participation

Another development regarding SeSAm interfaces and user task is the so-called "agent playing" framework (see [Raupach, 2007]). It basically forms the logical advancement of the interactive simulation runs described in the last subsection. There a bird's eye view is used for monitoring the model dynamics from the outside. Agent-based simulation allows an inside view when a human is playing one particular agent.

This interactive element was developed not for allowing simulation games, but especially for supporting plausibility testing and validation on the agent level: we assume that for qualitative validation purposes, the modeler needs to perceive the simulated environment through the eyes of the agent, immersed into the simulation. While perceiving what the controlled agent perceives and evaluating its reactions to perceptions, as well as while observing the other agents, a human may evaluate whether the observed simulation run actually is plausible.

The "agent playing" framework consists of two parts: enhancements on the SeSAm side and a specialized piece of software that visualizes the perceptions of the agent and its actions outside of SeSAm:

- SeSAm models have to be enhanced with primitive calls sending and requesting all information marked as necessary.
- An additional program has to be developed that receives the information from SeSAm and visualizes this information appropriately. This program may also interpolate between two sets of information – for example, visualizing a smooth

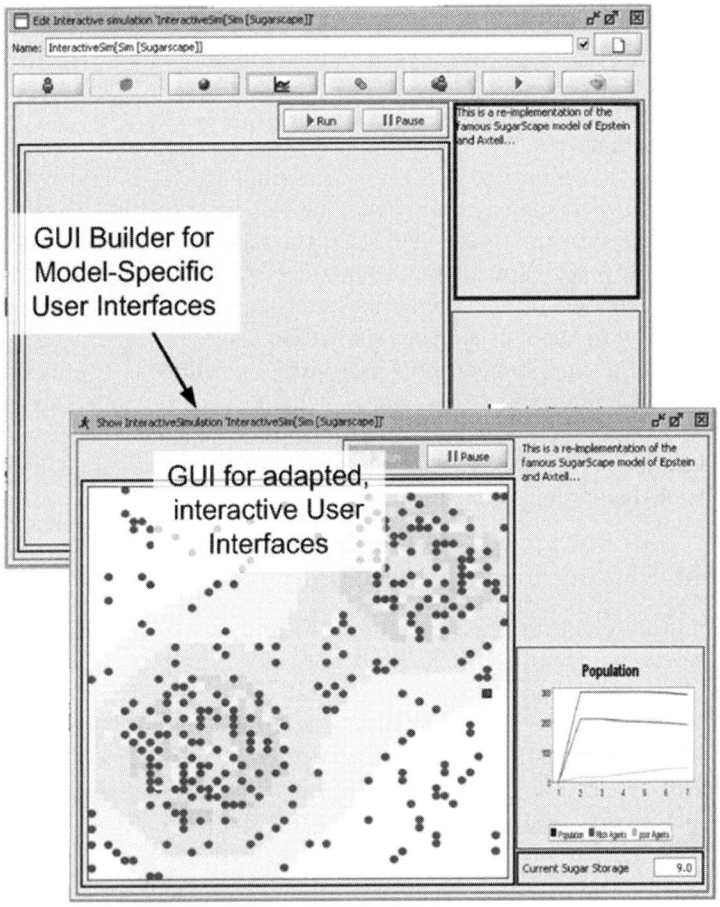

FIGURE 16.12 A model-specific interface to a simulation run is generated based on an interface configuration.

movement in a discrete world. It may be used for mere observation, but also for controlling the agents by collecting the commands from the user and thus forcing the agent to do what the human wants the agent to do.

Thus "agent playing" supports conceptual validation of an agent's behavior, as the experts of the original system can adopt the perspective of the agent. Hopefully they can see the simulated surroundings of the agent from the same perspective as in real life - resulting in more direct, immersive testing by a human expert.

16.5 Experiences

Besides use in teaching, SeSAm was and is applied in several simulation projects reaching from simulation of social insects, reproduction of shopping behavior to agent-based traffic simulation. Beyond such traditional application areas of agent-based simulation, SeSAm forms the technical base for virtual high bay warehouses that are used for software tests, requirements engineering or employee training. In several of these projects, domain experts like biologists, geographers, etc. use SeSAm independently. In other projects we use the tool ourselves. The latter allows evaluating usability directly by ourselves.

We can identify three classes of users of SeSAm who are all at least trying to execute all tasks listed in Section 16.2.1. First of all there are absolute beginners, for example domain experts who discover simulation as a scientific tool but never or hardly have used this methodology and these techniques before. Secondly one may identify a user group that has experiences in formalization of software of programs without being familiar to the multi-agent system paradigm. The third class of users are those familiar with the modeling paradigm, with programming languages as well as with the features of the SeSAm language and system. We will shortly discuss what we observed with the first two user classes, followed by some general aspects. It is quite obvious that this cannot replace a systematic evaluation, but may give indications about the feasibility of using SeSAm for doing multi-agent simulations beyond mere stability and simulation speed.

16.5.1 Novices

In particular, modeling and implementation novices were one of the premier addressees that SeSAm was developed for. In general one must admit that SeSAm is too complex for beginners, although the basic language with variables capturing the status of the entities, the activity graphs based description was accepted quite easily.

Modeling and simulation novices with a minimum of formalization training were quite successful. As experts in biology, geography or economic processes, they were not familiar with particular programming languages, but had a quite clear image of the agent-based model that they wanted to develop before starting with SeSAm. Abstraction in general was not a problem. Their minimum training consisted of programming courses in school or early university studies. Although they could not practically use the learned programming language any more, basic concepts of implementation were still present. Their major difficulty in implementation consisted of finding and selecting the appropriate primitive from the large set of atomic functions. As they knew rather exactly what they wanted to model, they had no conceptual problem and needed only episodic hints how to formulate exactly certain aspects. It was basically this user group that requested additional support for experimentation, sensitivity analysis, calibration and validation.

Another novice user group without any remarkable training in abstraction and formalization believed the promise of accessibility of agent-based simulation using SeSAm and addressed us for support. They also came with a model in mind that, however, was too vague to be implementable independent from a particular tool. Support was not restricted to how to formulate certain aspects in SeSAm, but started with a more general discussion about implementation of models that involves both abstraction and concretization. At least in three to four cases, the modeler was individually "mentored" by a computer science student who usually also implemented the model to a large extent. The advantage of SeSAm here was that the model always remained understandable for the domain modeler although he or she was not able to produce the model itself actively. Although the modeling novice was merely passive and needed a lot of support, the simulation project itself could be successfully terminated as long as the student abstained from biasing the model implementation, but thoroughly attended to the aspects that the domain modeler actually wanted to express.

It has been argued that especially the latter user group could be supported by providing a more powerful primitive set. By using abstract building blocks, a first model could be constructed. When the un-experienced modeler wonders about how certain outcomes were produced, the high-level primitives are questioned and replaced by primitive combinations that the user itself controlled. The advantage would be that the initial gap when formulating a working model is reduced. The idea of building blocks and component-based simulation is nothing new, see for example [Barros et al., 2004] or [Valentin et al., 2003], but may also be useful in the context of SeSAm.

16.5.2 Knowledgeable in Implementation, Not in Multi-Agent systems

Basically, this group of SeSAm users consisted of computer science students attending a course on multi-agent systems. They had two problems. First, lack of documentation beyond simple tutorials. This information basically confused them as they had certain expectations about the platform, but could not identify used concepts. A good example is a student that created one class for every agent to be used in the situation mixing classes and instances. Several other students had problems in understanding that activity graphs denote the complete behavior and are not passed completely once per update cycle.

Another problem was that in general too much functionality was performed by the global world entity that should have assigned to the agents. The SeSAm language does not enforce an agent-oriented implementation. Thus, it is possible to let the global entity loop through all agents and manipulate their status from this central perspective. The students in the Multi-Agent Systems course had problems abandoning the known process-oriented way of thinking and replacing it with some interaction-oriented approach, especially when they use a graph-based language that is similar to process declarations. A missing clear separation of responsibilities of global environmental model and local agent model is a drawback of SeSAm for this user group.

16.6 General Discussion and Future Work

We believe that with SeSAm an important step was taken toward the advance of simulation environments for agent-based models, despite of all its drawbacks. Coupling visual programming and simulation makes the agent-based simulation paradigm accessible for a variety of modelers that would otherwise not be able or willing to deal with agent-based simulations, as bridging the gap between a standard programming language and their model concept would be too demanding for them.

Nevertheless, there are some starting points for future improvements. Clearly, the simulation performance of SeSAm should be better. The simulator of SeSAm interprets a behavior representation that is compiled from the high-level model description given by the modeler. During this compiling step, optimization steps from compiler design are applied. A simulation run within the SeSAm framework, however, cannot be as fast as a corresponding direct implementation without the SeSAm overhead for example using Java. Clearly, one can also implement efficiently using the SeSAm high-level language based on some coarse knowledge of lower level primitive implementation which results in reasonable, yet not optimal, simulation run times. For testing alternative ways, we already experimented with a tool that generates plain Java code out of a SeSAm model [Niederle, 2005]. This tool worked for very simple models; however for applying it to more complex models, many generic possibilities of SeSAm would have to be re-implemented in that compiler tool causing a tremendous development effort with doubtable success. Thus, the starting point for increasing simulation speed should be the reduction of overhead by decoupling modeling and runtime environment in a more consequent way than done in SeSAm up to now. Additionally, the implementation of certain primitives, especially concerning spatial perception, must be improved.

There are some additional aspects that are seen as suboptimal. Modelers have to get used to the prefix notation of primitive calls that is one of the remainders of the original Lisp-based system. Another aspect that needed explanation in several cases is the separation between class and instance description. Modelers would intuitively like to start with (example) configurations, especially when spatially explicit models are to be developed. SeSAm however biasses the modeler to start with the basic structures instead of starting by arranging entities on a map.

These aspects can be remedied or avoided with sufficient training (and sufficient documentation). This alone however cannot enable a user to develop a successful simulation study as SeSAm only covers the implementation, experimentation and analysis of an agent-based simulation. The first step is model design. The step after implementation mainly consists of testing for validity. These two phases are essential. If one of them fails, the simulation study fails all together. It is completely justified that such general methodological aspects gain more and more attention, like in [Matteo et al., 2006]. There are two more visionary directions that we want to pursue in our future work. We want to investigate new ways of modeling for circumventing the design problem and secondly, provide more methodological support for all phases in an agent-based simulation study.

Up to now, when dealing with end user programming, we just considered approaches based on visual programming. Research in this direction also proposes learning by demonstration as a means for implementing agents directly by users. We want to test these forms of supervised learning and also other forms of learning and adaptive agents for supporting the development of agent-based models beyond mere implementation and analysis. We are performing first experiments with agents controlled by neural networks and by machine learning algorithms. The main problem is defining the appropriate perceptions and feedback functions that the agents may get from the environment for actually determining the direction of adaptation. In these learning agents' applications, we are not aiming at reproducing, for example, evolutionary processes, but trying to develop a tool that, for example, automatically generates the behavior of a simulated pedestrian instead of leaving the user with the cumbersome trial-and-error procedure for model design finding out which rules are the most appropriate.

Acknowledgment

Many people have contributed to create a platform like SeSAm: Christoph Oechslein was responsible for the basic implementation of Java-based version. Rainer Herrler technically supervised the developments during the last years. Manuel Fehler took care about the experimentation framework. Cornelia Triebig currently deals with reusability and Reinhard Hatko analyzes the use of adaptive agents. Generations of students have also contributed to SeSAm.

References

J. Ambach and A. Repenning. Participatory theater: Interacting with autonomous tools for creative applications. *Journal of Knowledge Based Systems*, 9(6):351–358, 1996.

S. Bamberger, U. Gappa, F. Klügl, and F. Puppe. Komplexitäsreduktion durch grafische wissensabstraktion. In *Proc. of the 4. German Conference on Expert Systems (XPS-97)*, pages 61–78. infix, 1997.

O. Barreteau, F. Bousquet, and J.-M. Attonaty. Role-playing games for opening the black box of multi-agent systems: method and lessons of its application to senegal river valley irrigated systems. *Journal of Artificial Societies and Social Simulation*, 4(2), 2001.

F. J. Barros, A. Lehmann, P. Liggesmeyer, A. Verbraeck, and B. P. Zeigler. Dagstuhl seminar on component-based modeling and simulation. http://www.dagstuhl.de/04041/, 2004.

R. H. Bordini and J. F. Hübner. Agent-based simulation using bdi programming in jason. In *this book*. 2008.

L. Braubach, A. Pokahr, W. Lamersdorf, K.-H. Krempels, and P.-O. Woelk. A generic time management service for distributed multi-agent systems. *Applied Artificial Intelligence*, 20(2):229–249, 2006.

S.-K. Chang, editor. *Principles of Visual Programming Systems*. Prentice Hall, 1990.

M. Fehler. *Kalibrierung Agentenbasierter Simulationen*. PhD thesis, Fakultät für Mathematik und Informatik, Institut für Informatik, Universität Würzburg, 2008 or 2009.

M. Fehler, F. Klügl, and F. Puppe. Calibrating agent-based simulations. In *Proc. of the EUMAS 2005, Brussels, 2005*, 2005.

M. Fehler, F. Klügl, and F. Puppe. Approaches for resolving the dilemma between model structure refinement and parameter calibration in agent-based simulations. In *Proc. of the AAMAS 2006, May 8-12, 2006, Hakodate*, pages 120–122. ACM Press, 2006.

T. R. G. Green and M. Petre. Usability analysis of visual programming environments: a cognitive dimensions framework. *Journal of Visual Languages and Computing*, 7:131–174, 1996.

L. Gulyas, T. Kozsik, and J. B. Corliss. The multi-agent modelling language and the model design interface. *Journal of Artificial Societies and Social Simulation*, 2(3), 1999.

P. Guyot and S. Honiden. Agent-based participatory simulations: Merging multi-agent systems and role-playing games. *Journal of Artificial Societies and Social Simulation*, 9(4):8, 2006. URL http://jasss.soc.surrey.ac.uk/9/4/8.html.

R. Herrler. *Agentenbasierte Simulation zur Ablaufoptimierung in Krankenhäusern und anderen verteilten, dynamischen Umgebungen*. PhD thesis, Fakultät für Mathematik und Informatik, Institut für Informatik, Universität Würzburg, 2007.

J. Himmelspach and M. Röhl. James ii - experiences and interpretations. In *this book*. 2008.

F. Klügl. Measuring complexity of multi-agent simulations – an attempt using metrics. In M. D. et al., editor, *Proc. of the LADS workshop, Durham, Sept. 2007*, number 5118 in LNAI. Springer, Berlin Heidelberg, 2008.

J. Kuljis. User interfaces and discrete event simulation models. *Simulation Practice and Theory*, 1:207–221, 1994.

A. M. Law. *Simulation Modeling and Analysis*. McGraw-Hill, New York, 4th edition, 2007.

H. Lieberman, F. Paterno, and V. Wulf, editors. *End-User Development*. Springer, Dordrecht, NL, 2006.

R. Matteo, R. Leombruni, N. J. Saam, and M. Sonnessa. A common protocol for agent-based social simulation. *Journal of Artificial Societies and Social Simulation*, 9(1):15, 2006. URL http://jasss.soc.surrey.ac.uk/9/1/15.html.

S. Moss and B. Edmonds. Towards good social science. *Journal of Artificial Societies and Social Simulation*, 8(4), 2005.

A. Niederle. Sescape – agents escaping from sesam. Master's thesis, Fakultät für Mathematik und Informatik, Institut für Informatik, Universität Würzburg, 2005.

C. Oechslein. *Vorgehensmodell mit integrierter Spezifikations- und Implementierungssprache für Multiagentensimulationen*. PhD thesis, Institute of Computer Science, University of Würzburg, 2003.

B. Raupach. Sesam als plattform für participatory simulations. Master's thesis, Fakultät für Mathematik und Informatik, Institut für Informatik, Universität Würzburg, 2007.

A. Repenning, A. Ioannidou, and J. Zola. Agentsheets: End-user programmable simulations. *Journal of Artificial Societies and Social Simulation*, 3(3):351–358, 2000.

G. Rindsfüser and F. Klügl. The scheduling agent - using sesam to implement a generator of activity programs. In H. Timmermans, editor, *Progress in Activity-Based Analysis*. Elsevier, Amsterdam, 2005.

G. Rindsfüser and F. Klügl. Agent-based pedestrian simulation. *disP*, 170(3), 2007.

D. C. Smith, A. Cypher, and J. Spohrer. Kidsim: Programming agents without a programming language. *Communications of the ACM*, 37(7):54–67, 1994.

G. Theodoropoulos, R. Minson, M. Lees, and R. Ewald. Simulation engines for multi-agent systems. In *this book*. 2008.

E. C. Valentin, A. Verbraeck, and H. G. Sol. Advantages and disadvantages of building blocks in simulation studies: A laboratory experiment with simulation experts. In *ESS'2003, Proceedings 15th European Simulation Symposium 2003 - Simulation in Industry, Delft, NL, Oct. 26-29, 2005*. SCS European Publishing House, Erlangen, 2003.

17
JAMES II - Experiences and Interpretations

17.1 Introduction	509
17.2 JAMES II	512
Using JAMES II for Multi-Agent Modeling and Simulation	
17.3 Multi-Agent Modeling and Simulation in JAMES II	514
A Modeling Formalism for the Description of Multi-Agent Systems • Simulation Algorithms • Representing Models	
17.4 Application	524
Composition Structure of the MANET Model • Equipping Nodes with Alternative User Models	
17.5 The Role of Time	528
Experiment Definition	
17.6 Experiences and Interpretations	530
17.7 Outlook	530
Acknowledgment	531
References	531

Jan Himmelspach
University of Rostock

Mathias Röhl
University of Rostock

17.1 Introduction

The need for flexible and reusable modeling and simulation environments for multi-agent systems has already been identified several times (e.g., [Gilbert, 1995; Minar et al., 1996]). Too often new modeling and simulation systems dedicated to execute a (single) concrete simulation are developed from scratch, wasting time, which might be better applied to the actual model [Minar et al., 1996].

Modeling and Simulation

Often related to the creation of new modeling and simulation (M&S) environments for single applications, M&S is prone to a number of pitfalls, which hinder the proper interpretation of the results achieved [Edmonds, 2000; Edmonds and Hales, 2003; Galan and Izquierdo, 2005]. Problems might arise anywhere in the modeling and simulation process - from theory building to the interpretation of results [Edmonds, 2000]. However, in the following we will focus on the aspects related to the usage of modeling and simulation environments.

Several issues may lead to wrong interpretations, first of all the question of which language and which level of abstraction to choose for modeling. There are arguments for using general purpose languages as well as for specialized modeling languages. If models are written in a general purpose language, modelers can easily add "not allowed" code or mix model definition and simulator code. The latter hinders the exchange of simulation algorithms and may introduce artifacts into the simulation. Advantages of general purpose languages are their widespread use and that there are nearly no restrictions with regard to the expressiveness. Specialized modeling languages ease the creation of models and prevent a developer from the integration of "not allowed" code [Gilbert, 1995]. An eXtensible Markup Language (XML)-based model description, aligned to a modeling formalism, helps to ensure that a model is well-formed and thereby that it can be executed using a variety of simulation algorithms on different hardware infrastructures. Reusing model parts is a key feature for efficient modeling. Such model components can ease the creation of single agents (e.g., by providing components of the internals), as well as the creation of complex scenarios (e.g., by providing complete agents, communication infrastructure, etc).

Whether models are created based on components or whether they are created from scratch, new models have to be validated quite carefully [Edmonds, 2000; Balci, 2003]. This might be a time intensive task because for being able to validate a model many simulation runs might be necessary and thus efficient algorithms are required. In addition, to deny any effect of a concrete simulation algorithm on the results, models should be validated using different simulation algorithms [Edmonds and Hales, 2003].

Another important issue for modeling and simulation environments is the management of experiments. For the traceability of interpretations the repeatability of experiments is a key feature of modeling and simulation software - therefore a systematic treatment of experiment intentions, parameters, and the corresponding simulation results is required.

Another pitfall might be the influence of a particular simulation algorithm on the simulation results [Edmonds and Hales, 2003]. The experimental analysis (validation + evaluation) forms an important part during the development of algorithms. If algorithms are evaluated carefully, and thus compared to alternative ones, we can judge their final performance. This is true because not every (if not none) algorithm performs best under all circumstances. I.e., the need arises to be able to select a proper algorithm out of a (huge) set of different solutions. There may even be the need to exchange an algorithm during execution time if results previously achieved indicate that the usage of the "wrong" algorithm would seriously increase the overall execution time.

Prediction of the resources required to execute a model is not easy. The efficient execution of large or computationally heavy multi agent applications may require the usage of parallel computing paradigms. For providing a seamless scalability the usage on single computers as well as on diverse sets of computing nodes should be possible: start with a model on a single computer, e.g., for model development and first runs, and continue on a distributed setup later on. This should be possible without any changes to the model.

Simulation Environments

There are several approaches and specialized modeling and simulation environments for working with multi agent models. Additionally there are many non multi agent system related developments, which have nearly no or only little impact on these, e.g., developments in massively parallelized simulations. Several new modeling and simulation systems are moving toward more general solutions. The argumentation line is always the same: software engineering can significantly reduce the effort of creating modeling and simulation applications [Minar et al., 1996; Himmelspach and Uhrmacher, 2007a] if based on reusable elements (e.g., based on a framework).

The following short overview lists several modeling and simulation environments stemming from different fields. The descriptions given here focuses on the "modeling" aspect of the environments.

Jason (see chapter 15) is a system developed for social simulations. Modeling is based on and restricted to an extended version of the AgentSpeak language. The system only provides one simulation algorithm but the simulation is maintained and controlled from an exchangeable environment, in which the agents reside and which defines the communication and order of execution of the agents.

Repast Simphony is a development in the Repast family [North et al., 2005]. Repast Simphony tries to ease modeling and experiment definition by providing graphical editors for these tasks. The system is created based on the Eclipse framework. Several additional already integrated modules provide support for different data storages, and optimization methods.

SeSAm (see chapter 16) is a multi agent modeling and simulation framework with a nice user interface. Models are created based on SeSAm UML and stored together with experiment definitions in XML-based files. A plug-in schema allows the extension of a part of the functionality originally provided. SeSAm is limited to a single execution mechanism.

Swarm has been created as a general "laboratory" for experiments with multi agent systems [Minar et al., 1996], being under development for more than 10 years now. It supports coded discrete event based models, which can be executed in a sequential or parallel manner. There is no explicit "experiment" in Swarm, experiments can be defined using a "batch mode".

CD++ [Liu and Wainer, 2007] is one of the modeling and simulation frameworks from the DEVS (Discrete Event System Specification) realm. CD++ supports experiments and DEVS based models described in well-defined text files. These can be executed either in a sequential or parallel manner. The frontend of CD++ is based on the Eclipse framework.

μ*sik* is a micro kernel approach [Perumalla, 2005] to discrete event simulation. Models have to be written in C code and there is no strict separation between model and simulator. However, due to the micro kernel approach some mechanisms are exchangeable without the need to modify the models (e.g., message passing). But μsik does not provide functionality around the pure execution of simulation runs.

The ability to provide basic functionality, which is reusable across several application domains (and modeling methodologies) is getting more and more attractive. Examples are the SSJ framework [L'Ecuyer et al., 2002] or the COLT library. Both provide a set of different functions (e.g., random number generators) reusable in many scientific applications and with varying dedication to simulations.

Most existing environments are either bound to certain modeling paradigms (e.g., the agent oriented one: ***Jason***, SeSAm, Repast), to certain modeling formalisms (e.g., ***Jason***, SeSAm, Repast, CD++, μsik), to single simulation algorithms (e.g., SeSAm), are not complete environments (SSJ, COLT), or may be less or more platform dependent (Java (e.g., SeSAm) versus C based (e.g., CD++, μsik) implementations).

In the subsequent sections we present a modeling and simulation framework which subsumes ideas from the overall modeling and simulation realm. Nevertheless the system is especially suited for the simulation of multi agent systems. The integration of these ideas is done by exploiting a concept named "Plug'n simulate"[Himmelspach and Uhrmacher, 2007a] which makes the system highly flexible and parts developed therefore reusable a cross world views, application areas and infrastructures available to users. We will illustrate some of the key features of JAMES II in combination with a model component framework integrated into the framework as a plug-in by an example application: Service trading in mobile ad-hoc networks is modeled and simulated.

17.2 JAMES II

JAMES II (a JAva-based Multipurpose Environment for Simulation, aka a Java-based Agent modeling Environment for Simulation) has been designed as an open modeling and simulation framework with strong support for experimentation [Himmelspach and Uhrmacher, 2007a; Himmelspach, 2007]. Any number of modeling formalisms can be integrated and each formalism can be supported by several sequential and parallel distributed (fine-grained) simulation algorithms*. Whereas the fine-grained approach allows the efficient execution of large and computationally intensive models, the experiment layer supports a coarse-grained execution which can be used for the parallel computation of many smaller simulations in addition.

The flexibility of JAMES II is achieved by an approach named "Plug'n simulate". The complete system is based on plug-ins and the extension points plug-ins can be installed for. The reasons for the creation of such a flexible framework are manifold:

- Modelers should not be restricted to a single modeling formalism nor a single modeling interface - the discrepancy of purposes is too large for using a single solution for everything.
- Simulations should always be executable in the best manner - thus the ability to use a single workstation, a heterogeneous network, a grid or even a high performance cluster should be there. Depending on the platform and the simulation, sequential or distributed simulation maybe of interest.
- Experiments should always be created and executed from within a well defined environment, which allows the creation of detailed experiment logs.
- Different users / user groups may require different user interfaces for modeling and simulation.
- Data collection of simulation run time data is a time consuming task. Intelligent and flexible solutions which are interchangeable may provide good solutions for any model or simulation hardware.
- The comparison of algorithms is a highly difficult task - a flexible environment containing algorithms which may be used as reference implementations is a sound base for the experimental analysis of new algorithms.

Figure 17.1 shows the basic packages of JAMES II as well as their relationships.

User Interface: The user interface is the most important part of the framework for using the framework. The user interface has to be adaptable to the needs and the knowledge of a single user or a group of users (e.g., for biologists). The user interface is split into several parts, responsible for modeling, experiment setup and result analysis / visualization.

Data sink: During a simulation run many observations have to be stored so that the run can be analyzed and visualized. The data sink interface makes no restriction to the data sink to be used - everything starting from a plain ASCII file up to a database system can be used.

Model: The base model class. JAMES II strictly separates between a model and a

---

*A simulation algorithm in JAMES II is an algorithm dedicated to the computation of a model. Examples of such algorithms are solvers for differential equations and discrete event based algorithms.

JAMES II - Experiences and Interpretations

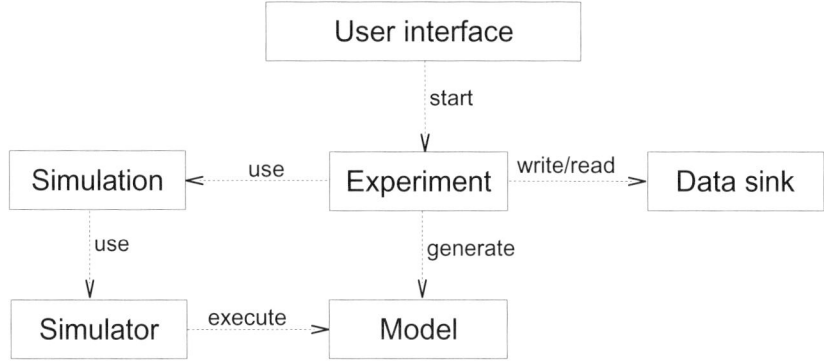

FIGURE 17.1 Sketch of packages and relationships. The lines indicate package imports, the labels indicate why this dependency is required.

simulator. All modeling formalisms supported by JAMES II must implement the base interfaces defined in this package. JAMES II itself does not provide inherent support for any modeling formalism nor any modeling paradigm.

Simulator: A simulator executes a model. In this package the base classes and interfaces for all simulators are provided and they are coming along with a set of helpers which make the creation of new simulation algorithms fairly straightforward.

Simulation: If a simulator executes a model this is called a simulation. In this package, methods and classes for creating simulations are provided. These simulations can either be distributed or not. To be executed in a distributed manner the integrated (and extensible) partitioning and load balancing subpackages come into play.

Experiment: An experiment is a set of simulation runs. Thereby the research focus can either be on simulation experiments with models or on experiments with simulation algorithms. This package provides support for creating flexible experiments with support for parameter optimization and validation.

An additional module, accessible from all other modules, is the Registry which can be considered to be the "brain" of the simulation system. During startup the plug-in types as well as the plug-ins get registered and lists with installed plug-ins or individual plug-ins can be retrieved from there.

From an use-case oriented point of view the "experiment" package is the central element in this design - an experiment is linked to the model to be simulated, it defines the number of simulation runs as well as the parameter combinations therefore and the data sink to be used. The ability to conduct different kinds of experiments is essential for all applications of modeling and simulation. A framework should not only support experiments with models but also, e.g., experiments for algorithm and model validation.

Functionality not included in the core classes, especially modeling formalisms and simulation algorithms, can be provided by using plug-ins. The plug-in mechanism allows the flexible extension of the simulation framework without the need to modify the code of the core later on. Due to the strict separation between models and simulators, simulation algorithms can be easily exchanged and thus evaluated. This makes the plug-in mechanism a base for a reliable evaluation of new simulation algorithms. The adaptation of a modeling and simulation framework for certain user groups (especially of the user interface) is crucial for its usability. This adaptation can be easily done by the plug-in mechanism or by embedding the complete JAMES II core (with all installed plug-ins) into another JAVA application [Martens and Himmelspach, 2005]. Modeling and simulation applications, formalisms, and algorithms designed for contradicting purposes (e.g., the use for demonstration purposes in teaching simulation algorithms versus highly efficient implementations) can coexist in the framework [Himmelspach and Uhrmacher, 2006].

17.2.1 Using JAMES II for Multi-Agent Modeling and Simulation

In the context of JAMES II the support for multi agent modeling and simulation is delegated to plug-ins providing a suitable formalism. One of these is the PdynDEVS (Parallel Dynamic DEVS) formalism [Himmelspach and Uhrmacher, 2004; Himmelspach, 2007] (see Section 17.3.1), another one is the MLDEVS (multi level DEVS) formalism [Uhrmacher et al., 2007]. There is freedom of choice in regards to the modeling formalism chosen for a model (or to extend the framework by new plug-ins supporting additional ways to describe multi agent systems). Both formalisms are supported by efficient simulators. The models can be described by using Java code as well as by using COMO (COmponent MOdels) a new approach using XML for describing model components and their relationships (see Section 17.3.3).

Often multi agent simulations contain stochastic parts. This leads to the requirement that simulation runs can be repeated in a controlled manner many times. This is supported by the experiment definition within JAMES II. A set of simulation runs can be defined by specifying modifications of the model's parameters. These runs are then executed either in a sequential or in a parallel manner. Using the parallel variant can drastically reduce the overall time required for the experiment. These experiments are executed without the need for any user interaction, the data can be collected in a modern database system.

JAMES II provides means for a simple analysis of the simulation run data. Due to the ability to store the data in any format or system (as long as an according plug-in exists) the analysis can be easily done with any available software.

17.3 Multi-Agent Modeling and Simulation in JAMES II

JAMES II does not provide any inherent support for any modeling formalism nor language but it can be extended by any number of those. For being able to support models with inherent dynamics we developed a new formalism, named PdynDEVS (Parallel dynamic DEVS). PdynDEVS is based on PDEVS (Parallel DEVS) [Zeigler et al., 2000] and dynDEVS (dynamic DEVS) [Uhrmacher, 2001], thus it allows the parallel processing of events in different parts of a model as well as changing the overall model structure (see Section 17.3.1). Any other modeling formalism / language suitable for multi agent modeling can be integrated as well. For each modeling formalism / language any number of simulation algorithms can be provided (see Section 17.3.2). Principally, models for JAMES II can be created by coding (in Java) or by using a model component approach (see Section 17.3.3).

17.3.1 A Modeling Formalism for the Description of Multi-Agent Systems

PdynDEVS models are being composed out of two different kinds of models. The first ones are atomic models, which encapsulate states and state transitions, the other ones are coupled models containing the network topology of a part of the model. A PdynDEVS is always a tree and thereby allows the definition of several levels of details (e.g., a concrete

JAMES II - Experiences and Interpretations 515

"agent" could be composed of several submodels).

DEFINITION 17.1 An **atomic PdynDEVS** model is defined by the tuple

$$(M, M_{init}),$$

with

- M being the least set such that $\forall\, m \in M\ \forall\, sc \in SC_M\ \exists\, n \in M : m.\rho(sc) = n$ and
- $M_{init} \in M$ the start model.

Each $M_t \in M$ is defined by a tuple

$$M_t = (X, Y, SC_M, SC_S, S, \delta_{int}, \delta_{ext}, \delta_{con}, \lambda, ta, \rho),$$

where

- $X = \{(p, v) \mid p \in InPorts \wedge v \in X_p\}$ a set of tuples, each containing an input port and an input value
- $Y = \{(p, v) \mid p \in OutPorts \wedge v \in Y_p\}$ a set of tuples, each containing an output port and an output value
- SC_M set of structural changes to be processed
- SC_S set of structural changes to be sent
- $\rho : SC_M \rightarrow M$ model transition
- S set of states
- $\delta_{int} : S \rightarrow S \times SC_S \times SC_M$ internal transition
- $\delta_{ext} : Q \times X^b \rightarrow S \times SC_S \times SC_M$ external transition
  $Q = \{(s, e) \mid s \in S \wedge 0 \leq e < ta(s)\}$ is a set of tuples ($e = elapsedtime$ in state s); X^b a bag of inputs.
- $\delta_{con} : S \times X^b \rightarrow S \times SC_S \times SC_M$ confluent transition
- $\lambda : S \rightarrow Y^b$ output function
- $ta : S \rightarrow \mathbb{R}_0^+ \cup \infty$ the *TimeAdvance*-function

The set of structural changes to be processed by ρ is communicated in analogy to external messages of classic DEVS (see Figure 17.4). Each state transition can result in structural changes in the model. Thereby each atomic model is allowed to induce structural changes anywhere in the model - if model parts are interpreted as autonomous agents a modeler has to take care of the agent's right of self-determination (if required). Structural changes are always processed from the leafs up to the top most coupled model.

Example 17.1

A PdynDEVS model will now be defined that receives and sends messages depending on its internal state. Figure 17.2 visualizes the dynamics of the model as a statechart. If active, the service model publishes and requests services in given intervals. When receiving an ActivityStatus message, the model updates the interval at which messages will be sent. Publishing and requesting services alternates until all available services have been announced. Formally, the PdynDEVS model is defined as $ServiceModel = (M, M_{init})$, with $M = \{M_1\}$, $M_{init} = M_1$, and $M_1 = (X, Y, SC_M, SC_S, S, \delta_{int}, \delta_{ext}, \delta_{con}, \lambda, ta, \rho)$, where

- $X = \{(i, v) \mid i \in InPorts, v \in X_i\}$, with $InPorts = \{\text{"act"}, \text{"response"}\}$, $X_{act} = ActivityStatus$, $X_{response} = Response$

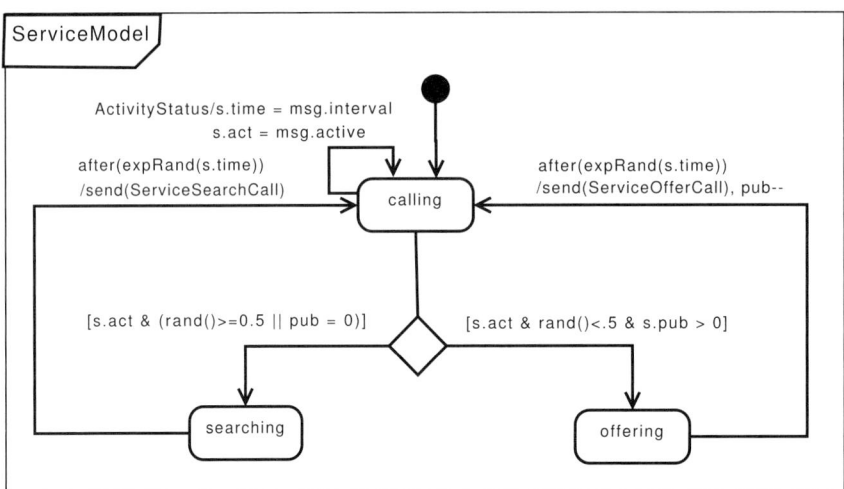

FIGURE 17.2 Dynamics of a simple service model.

- $Y = \{(i,v) \mid i \in OutPorts, v \in Y_i\}$, with $OutPorts = \{\text{"call"}\}$, $Y_{call} = Call = $ ServiceOfferCall $\cup$ ServiceSearchCall
- $S = Phase \times \mathbb{B} \times \mathbb{N} \times \mathbb{R}^{\geq 0}$, $Phase = \{\text{"calling", "searching", "offering"}\}$. In the following $(p, act, pub, time)$ is used to refer to an $s \in S$.
- $\delta_{int}((p, act, pub, time)) =$
$$\begin{cases} (\text{"offering"}, act, pub, time) & \text{if } p = \text{"calling"} \wedge act \wedge (rand() > .5 \vee pub = 0) \\ (\text{"searching"}, act, pub, time) & \text{if } p = \text{"calling"} \wedge act \wedge rand() < .5 \wedge pub > 0 \\ (\text{"calling"}, act, pub-1, time) & \text{if } p = \text{"offering"} \\ (\text{"calling"}, act, pub, time) & \text{if } p = \text{"searching"} \\ (p, act, pub, time) & \text{else} \end{cases}$$
- $\delta_{ext}(((p, act, pub), time), x) = \begin{cases} (p, x.active, pub, x.interval) & \text{if } x = ActivityStatus \\ (p, act, pub, time) & \text{else} \end{cases}$
- $\delta_{con}((s, e), x) = \delta_{ext}((\delta_{int}(s), e), x)$
- $\lambda((p, act, pub, time)) = \begin{cases} \{(\text{"call"}, ServiceSearchCall)\} & \text{if } p = \text{"searching"} \\ \{(\text{"call"}, ServiceOfferCall)\} & \text{if } p = \text{"offering"} \\ \bot & \text{else} \end{cases}$
- $ta((p, act, pub, time)) = \begin{cases} 0 & \text{if } act \wedge p = \text{"calling"} \\ uniformRand(time) & \text{if } p = \text{"offering"} \vee p = \text{"searching"} \\ \infty & \text{else} \end{cases}$
- $SC_M = \varnothing$
- $SC_S = \varnothing$
- $\rho(sc) = M_1$

This model does not use dynamic structures, i.e., M_1 is active all the time. An example for a dynamic structure model will be given for coupled models.

DEFINITION 17.2 A **coupled PdynDEVS** model is defined as a tuple

$$(CM, CM_{init}),$$

with

- CM the set of models
- $CM_{init} \in CM$ the start model

For all $CM_t \in CM$:

$$CM_t = (X, Y, SC_M, SC_S, D, \{M_d \mid d \in D\}, EIC, EOC, IC, \rho),$$

with:

- $X = \{(p, v) \mid p \in InPorts \wedge v \in X_p\}$ a set of tuples of input port and value
- $Y = \{(p, v) \mid p \in OutPorts \wedge v \in Y_p\}$ a set of tuples of output port and value
- SC_M set of structural changes to be processed or send to children
- SC_S set of structural changes to be send (to the parent)
- $\rho : SC_M \rightarrow CM$ model transition
- D set of component names
- $\forall d \in D$: $M_d = (X_d, Y_d, SC_M, SC_S, S, \delta_{int}, \delta_{ext}, \delta_{con}, \lambda, ta, \rho)$ is a Pdyn-DEVS model with $X_d = \{(p, v) \mid p \in InPorts_d \wedge v \in X_p\}$ and $Y_d = \{(p, v) \mid p \in OutPorts_d \wedge v \in Y_p\}$
- $EIC \subseteq \{((C, ip_C), (d, ip_d)) \mid ip_C \in InPorts \wedge d \in D \wedge ip_d \in InPorts_d\}$ set of external input couplings (connections between the input ports of the coupled model and the input ports of the children)
- $EOC \subseteq \{((d, op_d), (C, op_C)) \mid op_C \in OutPorts \wedge d \in D \wedge op_d \in OutPorts_d\}$ set of external output couplings (connections between the output ports of the children and the output ports of the coupled model)
- $IC \subseteq \{((a, op_a), (b, ip_b)) \mid a, b \in D \wedge op_a \in OutPorts_a \wedge ip_b \in InPorts_b\}$ the set of internal couplings (connections between the output ports of the children and the input ports of the children)

A coupled model is used to define the network structure. Similar to atomic models, coupled models differentiate between structural changes to be forwarded upwards and those to be processed by the model or forwarded downwards.

Structural changes on each level are executed one after the other. For each change (element in SC_M) ρ has to be executed resulting in subsequent transformations from the starting model to the target model $CM_t \rightarrow \cdots \rightarrow CM_{target}$ (with $CM_i.SC_M = CM_{i-1}.SC_M \backslash sc \wedge CM_{i-1}.\rho(sc) = CM_i$, with $sc \in CM_{i-1}.SC_M \wedge CM_0 = CM_t \wedge CM_{|CM_0.SC_M|} = CM_{target}$). For making the runs repeatable these changes have to be executed in a well-defined order (see [Himmelspach and Uhrmacher, 2004]). Thus, for the next structural change to be executed ($sc \in CM_i.SC_M$), the condition must hold that $\forall o \in (CM_i.SC_M \backslash sc) : priority(sc) > priority(o)$.

Having defined the new modeling formalism classes have to be realized, which allow the creation of models of this type. These classes have to be integrated into JAMES II as a plug-in.

Example 17.2

A coupled model is now used to define a second model, which may be coupled to the service

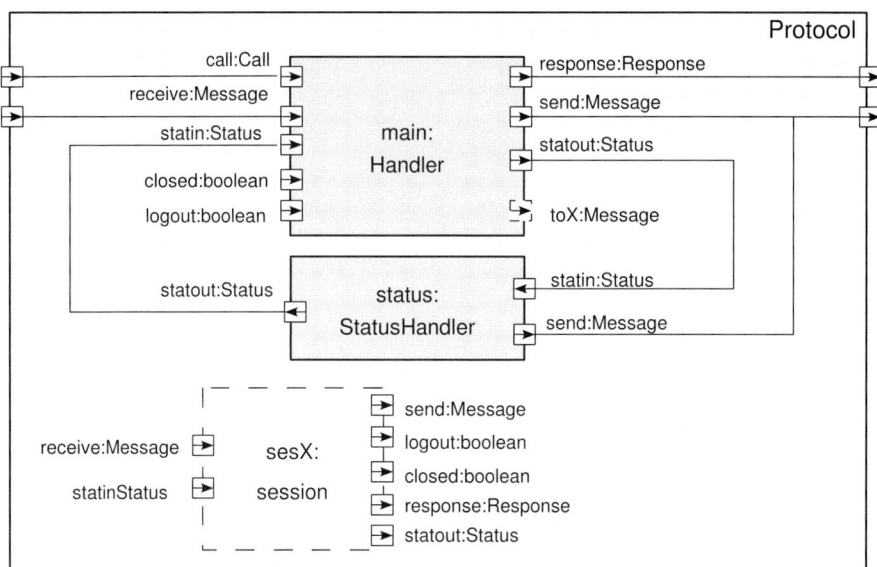

FIGURE 17.3 Permanent and temporary sub models of model Protocol.

model as introduced above. Within our application scenario* the model represents a service trading protocol. Such a protocol is responsible for processing user calls like announcements of services and requests for services. Calls are propagated transparently for the user to the network and according responses sent back to the user.

The protocol needs to be able to handle a number of calls and messages in parallel sessions. Dynamic structures are used to create and remove sessions on demand. Figure 17.3 shows the structure of the Protocol model. Whereas the Handler and Status models are permanent submodels of the protocol, incoming calls of the user and messages received from a network model initiate the creation of sessions, each being processed by a separate sub model. After having finished a session, the according sub model will be removed from the coupled model.

$CM_{init} = CM_1 = (X, Y, D, \{M_d \mid d \in D\}, EIC, EOC, IC)$, with

- $X = \{(i,v) \mid i \in InPorts, v \in X_i\}$, with $InPorts = \{\text{"call"}, \text{"receive"}\}$, $X_{receive} = Message$, and $X_{call} = Call$
- $Y = \{(i,v) \mid i \in OutPorts, v \in Y_i\}$, with $OutPorts = \{\text{"out"}\}$ and $Y_{out} = Message$
- $D = \{\text{"main"}, \text{"status"}\}$
- $\{M_d\} = \{M_{main}, M_{status}\}$
- $EIC = \{(\text{"this"}, \text{"call"}, \text{"main"}, \text{"call"}), (\text{"this"}, \text{"receive, "main"}, \text{"receive"})\}$
- $IC = \{(\text{"main, "statout"}, \text{"status"}, \text{"statin"}), (\text{"main"}, \text{"statout"}, \text{"main, "statin"})\}$
- $EOC = \{(\text{"main"}, \text{"response"}, \text{"this"}, \text{"response"}), (\text{"status"}, \text{"send"}, \text{"this"}, \text{"send"})\}$
- $SC_M = \{AddModel, AddCoupling, RemModel, RemCoupling\}$
- $SC_S = \varnothing$

---

*cf. Section 17.4 for an introduction

JAMES II - Experiences and Interpretations

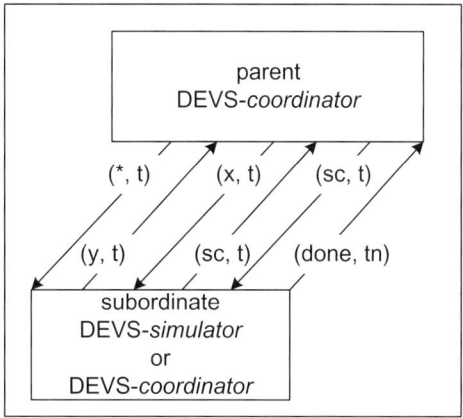

FIGURE 17.4 PdynDEVS protocol.

- $\rho(sc) = \begin{cases} CM' & \text{if } sc = AddModel(\text{"sesX"}, M_{sesX}) \\ CM & \text{else} \end{cases}$

  with all elements of CM' equal to the ones of CM except function ρ and the sets D and $\{M_d\}$, which become

  – $D' = D \cup \{\text{"sesX"}\}$

  – $\{M_d\}' = M_d \cup M_{sesX}$

  – $\rho'(sc) = \begin{cases} CM'' & \text{if } sc = AddCoupling(\text{"main, "toX", "sesX", "receive"}) \\ CM' & \text{else} \end{cases}$

  with all elements of CM'' equal to the ones of CM' except ρ'' and IC ...

The model *Protocol* permanently contains the two sub models *Handler* and *StatusHandler*. If an *AddModel* request is passed to it, ρ adds the according model to the set of sub models and thereby CM becomes CM'. The ρ function of CM' produces CM'' by adding a coupling. This ρ is applied to all structure change requests iteratively.

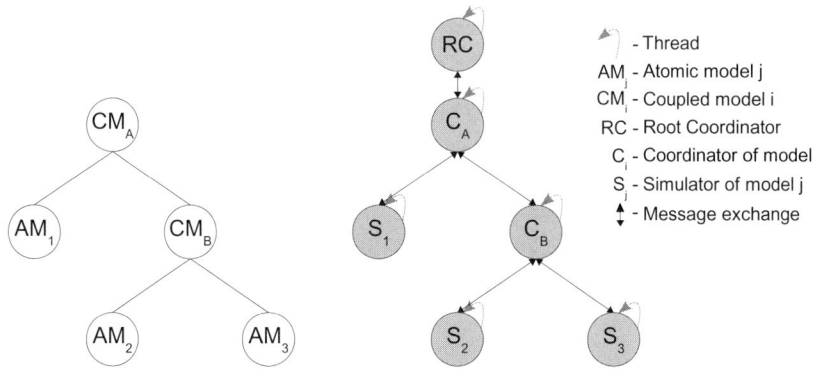

FIGURE 17.5 Abstract simulator. The classical DEVS simulator with a 1:1 mapping of models and simulators.

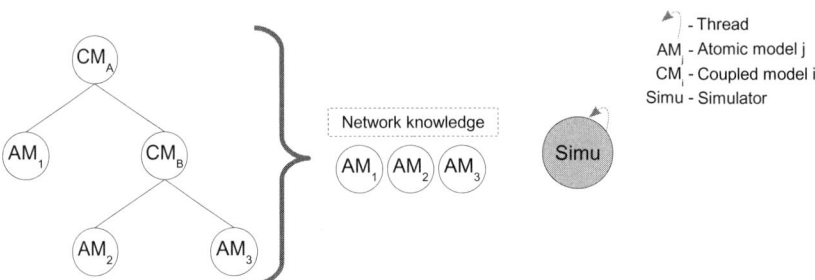

FIGURE 17.6 Flat simulator. A flat DEVS simulator – one simulator computes all models.

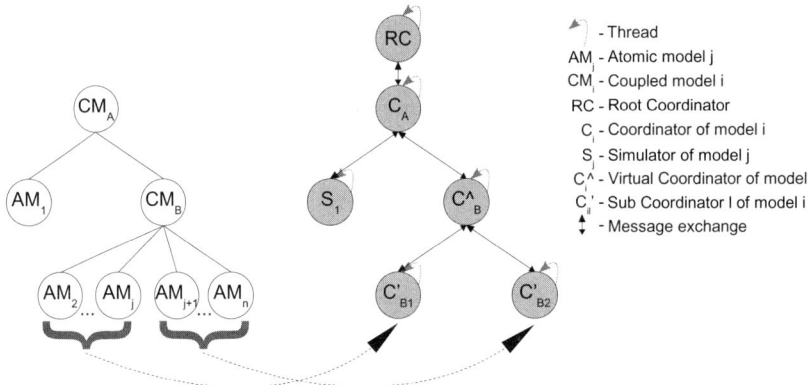

FIGURE 17.7 Combined simulator. The classical DEVS simulator in combination with sequential simulators. The sub coordinators are responsible for the computation of n of m sub models contained in a coupled model.

```
1   for each model in imminents do
2       execute model.lambda
3       influencees := union(influencees, copyOutputs(model))
4   end for
5   for each model in union(influencees, imminents) do
6       execute model.stateTransition
7       execute model.timeAdvance
8       execute structural changes
9   end for
10  bottom up execution of structural changes on the network levels
```

FIGURE 17.8 Pseudo-code of the flat sequential PdynDEVS simulator.

17.3.2 Simulation Algorithms

Simulation algorithms must be adaptable to the hardware and the model. For the formalism introduced above we have implemented three different simulation algorithms based on those presented in [Himmelspach and Uhrmacher, 2006]. The abstract threaded variant (see Figure 17.5) is very close to the abstract simulator for PDEVS as defined in [Zeigler et al., 2000]. This algorithm consists of three parts which communicate according to the protocol given in Figure 17.4 [Himmelspach and Uhrmacher, 2004]. The three parts are a coordinator for the computation of coupled models, a simulator for the computation of atomic models, and a root coordinator which is the root of the simulator tree. Out of these "components" the simulator tree is built in accordance to the model tree to be computed. This algorithm performs well if there are few computationally heavy models and if these can be computed on extra computers. The second algorithm can be used to compute a model completely on one computer (by virtually flattening the hierarchical model - see Figure 17.6 and algorithm 17.8). The third algorithm is a combination of both (see Figure 17.7) and thus can be used for the computation of heterogeneously built models. These simulation algorithms are integrated as plug-ins into JAMES II by implementing a class containing the algorithm and supporting the `IProcessor` interface of the framework [Himmelspach and Uhrmacher, 2007a; Himmelspach, 2007].

Reuse on the algorithm level comes into play here as all these algorithms require event queues for proper working. There are many different event queue implementations described in the literature - which of those to use when cannot be easily answered and mostly depends on the model and the simulation algorithm used [Himmelspach and Uhrmacher, 2007b]. JAMES II provides a set of event queue implementations which can be used alternatively.

17.3.3 Representing Models

JAMES II allows coding models directly in Java. In [Röhl and Uhrmacher, 2006] an alternative way to describe models has been proposed, which is based on XML. Data descriptions based on XML are considered to be robust, extensible, and suited to represent complex data structures [Harold, 2002]. Furthermore, XML-based storage formats integrate well with databases and facilitate the exchange of models via the world wide web.

XML documents have a tree structure consisting of nested nodes. Names of nodes give an informal hint how to interpret contents of XML documents. At the syntactical level schema languages can be used to specify occurrence and ordering constraints of nodes. Languages like XML Schema Definition (XSD) allow one to additionally constrain the content and value ranges of nodes [W3C, 2004].

XSD definitions form the basis for representing models in XML. Figure 17.9 lists the definition of the complex type *Call*. A Call may be used by a model in the context of network simulation (cf. application scenario in Section 17.4) to announce a service, to search for a service or to publish a service. These three alternatives are again defined as complex types. For example, a service search contains the network address of the requester and the name of the service to be searched. An address consists of primitive types, for which value ranges are predefined by XSD.

Figure 17.10 shows the structural part of a PdynDEVS model definition in XML, which uses XML Schema Definitions. The model contains two input ports and one output port. The state space is structured into three sub states.

In addition to an XML-based representation for model definitions, models can be equipped with explicit interface descriptions. Interface definitions orient on the Unified modeling Language 2.0 [OMG, 2005]. Most notably, published and required interfaces of each model

```xml
<xs:schema xmlns:xs="http://www.w3.org/2001/XMLSchema"
  xmlns="unihro/diane/base/service" ...>
  <xs:complexType name="Call">
    <xs:choice>
      <xs:element name="call" type="ServiceOfferCall"/>
      <xs:element name="call" type="ServiceRevokeCall"/>
      <xs:element name="call" type="ServiceSearchCall"/>
    </xs:choice>
  </xs:complexType>
  <xs:complexType name="ServiceSearchCall">
    <xs:sequence>
      <xs:element name="inquirer" type="Address"/>
      <xs:element name="serv" type="xs:string"/>
    </xs:sequence>
  </xs:complexType>
  <xs:complexType name="Address">
    <xs:attribute name="p1" type="xs:integer"/>
    <xs:attribute name="p2" type="xs:integer"/>
    <xs:attribute name="p3" type="xs:integer"/>
    <xs:attribute name="p4" type="xs:integer"/>
  </xs:complexType> ...
</xs:schema>
```

FIGURE 17.9 Definition of event types with XML Schema.

```xml
<atomic
  xmlns="http://www.informatik.uni-rostock.de/cosa/model/dynpdevs"
  xmlns:devs="http://www.informatik.uni-rostock.de/cosa/model/basicdevs">
  <devs:inport
      type="unihro/diane/base/service:Response">response</devs:inport>
  <devs:inport
      type="unihro/diane/base/activity:ActivityStatus">act</devs:inport>
  <devs:outport type="unihro/diane/base/service:Call">call</devs:outport>

  <devs:state>
    <devs:substate type="http://www.w3.org/2001/XMLSchema:int"
        init="0">id</devs:substate>
    <devs:substate type="http://www.w3.org/2001/XMLSchema:int"
        init="0">numberOfServiceTypes</devs:substate>
    <devs:substate type="http://www.w3.org/2001/XMLSchema:int"
        init="0">numberOfPublications</devs:substate>

  </devs:state>
  ...
</atomic>
```

FIGURE 17.10 Model definitions in XML based on XML Schema Definitions.

JAMES II - Experiences and Interpretations

```xml
<interface xmlns="http://www.informatik.uni-rostock.de/cosa/publici">
  <id>unihro.diane.com.service.Interface/1.0</id>
  <profile>
    <name>Service</name>
    <application_domain> MANET simulation</application_domain>
    <description> Generates service offers and service requests
       periodically. </description>
    <objective>Evaluate service trading in ad-hoc networks</objective>
    <key_abstractions>Time between calls is assumed to uniform
       distributed</key_abstractions>
    <author>Mathias Roehl</author>
  </profile>
  <param name="id" type="int" value="0"/>
  <param name="numberOfServiceTypes" type="int" value="15"
       description="how many different service types do we distinguish"/>
  <param name="numberOfPublications" type="int" value="3"
       description="how many services are going to be published"/>
  <port>
    <name>activity</name>
    <type>unihro/diane/com/service/v1/ActivityProc</type>
  </port>
  <port minMultiplicity="1">
    <name>service</name>
    <type>unihro/diane/com/service/v1/ServiceReq</type>
  </port>
  <impl>unihro.diane.com.service/1.0</impl>
</interface>
```

FIGURE 17.11 Interface Definition for a Model.

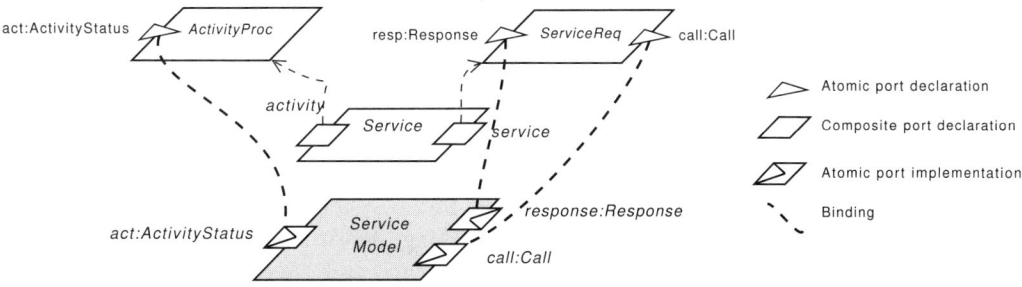

FIGURE 17.12 Relation between declared and implemented ports.

component have to be explicitly specified. Thereby internal details of a component, i.e., the implementation of model behavior, can be hidden and direct dependencies between models eliminated.

An interface definition for the model listed above is depicted in Figure 17.11. The interface contains a profile and exposes three parameters and two composite ports.

Components expose ports to indicate their provisions and requirements. Component ports may comprise a set of declarations for atomic model ports. For realizing models, port declarations need to be bound to concrete ports of model implementations. Figure 17.12 illustrates the binding of two component composite ports to three atomic model ports.

A set of components may become customized and arranged to form a *Composition* according to the aim of a simulation study. Parameters set on component instances are evaluated and dependencies between components are resolved. Composite model components support the hierarchical, modular construction of models and enable the development of larger models from smaller ones. Couplings connect ports of one model component to compatible ports of another component. Communication between models is only allowed along these couplings.

For simulation purposes composition structures are mapped automatically [Röhl, 2006] to the Parallel DEVS formalism [Zeigler et al., 2000], which forms a subset of the PdynDEVS formalism as described in Section 17.3.1.

The challenge for supporting reusable model components stems from the different characteristics of declarative and imperative representations. Declarative specifications, e.g., based on XML, ease the import and export of model definitions and are well suited to specify meta data, while declarative representations of components work well for database activities like storing, querying, and retrieving of components, they bear a significant execution overhead as models need to be interpreted at the time of execution. In contrast, imperative implementations of models are generally more efficient to execute but less eligible for database integration. Ideally one wants to combine the advantages of both "worlds", i.e., to compose efficient simulation models flexibly. To this end, the realization of the component framework is based on the strict separation between *composition* and *execution* phase. All activities that account for the composition of models work on XML data and are finished before the simulation model is executed.

17.4 Application

Mobile ad-hoc networks (MANETs) are computer networks based on wireless communication. MANETs are characterized by dynamic network topologies. Nodes induce topology changes by appearing, disappearing, and moving in a spatial environment. To provide fast and reliable connections poses a severe challenge for MANETs, because MANETs lack a central infrastructure and bandwidth as well as energy are strictly limited [Corson and Macker, 1999].

Network devices operating independently of the mains typically have limited capabilities, e.g., w.r.t. memory, battery, and performance. Complex operations require cooperation among network nodes. To make cooperation possible, resources and capabilities, in the following called services, of nodes have to be announced and need to be locatable. From the perspective of human users, services should be accessible in a transparent manner. On this account, service descriptions, service matching algorithms, incentive schemes and distributed reputation systems are developed in the project DIANE [Diane, 2007]. These mechanisms need to be thoroughly evaluated [Klein et al., 2004]. Of particular interest is how protocols scale with higher (≥ 64) number of network nodes [Röhl et al., 2007]. Concrete experiments aim at

- the comparison of different protocols for service trading in cases of different network sizes and
- the effect of different user models on the performance of a certain protocol.

Conducting experiments to evaluate service trading in MANETs requires user models to represent network nodes as autonomous actors, which move in a spatial environment and announce and request services. Developers of service trading protocols are mainly interested in the cost-benefit ratio of protocols when confronted with different kinds of user models. However, MANETs are typically simulated with special simulation systems, called network simulators [Kurkowski et al., 2005], which concentrate on the lower layers of the OSI protocol stack, e.g., routing protocols on the third layer. Within these simulation systems, models for representing movement and service behavior are simply calculated based on stochastic distributions [Tan et al., 2002]. As network traffic in MANETs is sensitive to local accu-

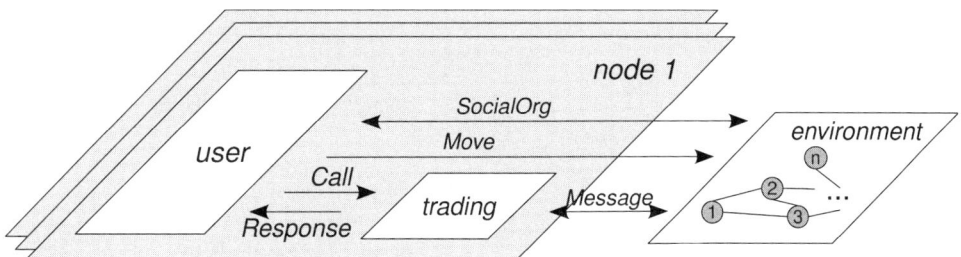

FIGURE 17.13 Conceptual model for evaluating service trading in mobile ad-hoc networks.

mulation of nodes and temporal accumulation of network usage, more complex user models are required, e.g., to allow a consistent modeling of motion and service behavior and to incorporate social aspects. Thereby users do not necessarily move individually but may form groups and move as clusters aiming at same destinations and partly synchronizing their schedules. Please note that in this setting agents are not subject to evaluation but become part of the experimental setup to evaluate service trading protocols.

Figure 17.13 shows a conceptual model for evaluating service trading in MANETs. The network consists of mobile nodes. Network connections between nodes depend on user positions. Each node comprises a user model and a service trading model such that the trading protocol mediates all network interactions and thereby provides a transparent access to services available in the network. Models are supposed to exchange the following types of events:

Call: A user initiates to offer, to revoke, or to search a certain service.
Response: The trading protocol reacts, after having performed all necessary actions, to calls of the user with according responses.
Message: Trading protocols communicate over the network with messages of arbitrary content.
Move: Movement information in a two-dimensional spatial environment.
SocialOrg: Social organization requests and responses.

User models initiate communication by sending calls to the trading protocol. Calls are passed to the protocol model, which answers user calls with *Response* events.

The concrete type of the user model and the trading protocol should be a variation point of the simulation model. Different kinds of user and protocol models should be exchangeable independently of each other within a node. For simulating service trading in MANETs with JAMES II, user and protocol models have been realized as model components.

17.4.1 Composition Structure of the MANET Model

Figure 17.14 shows the top level structure of the composite component *Manet*. The *Manet* component takes parameters to initialize the number of nodes to be simulated, the type of the user model to be used, and the type of the protocol model to be used inside each node. If the user model to be used is a social one, a further component has to be added, which represents the social environment and mediates social contacts of users.

The spatial environment represents a certain geographical area containing streets and buildings. It receives and it keeps track of all user positions. Based on positions and a constant radio range, a connectivity graph is calculated. All nodes are assumed to have an omnidirectional antenna and equal transmission power. All connections are bidirectional.

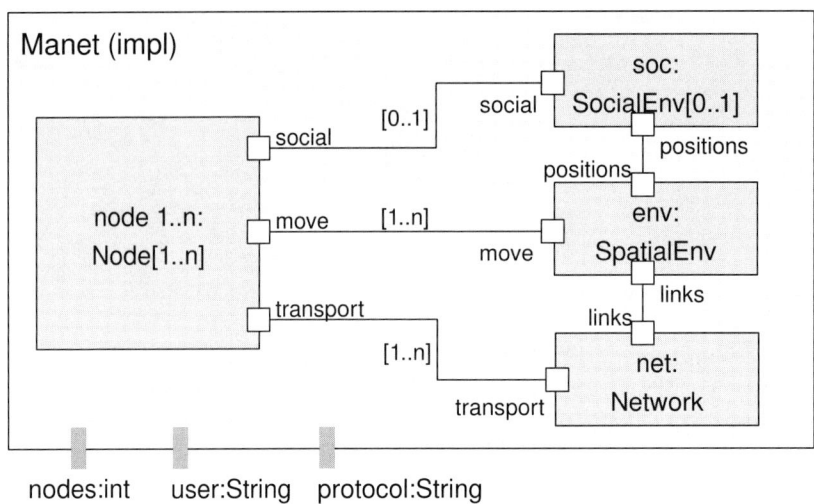

FIGURE 17.14 Top level composition structure of the MANET model.

Changes in connectivity are propagated, such that other models (e.g., the network model) can adapt their states accordingly.

The network component models the transport layer of the network. In real MANETs the whole OSI stack is part of each node. Since we are interested in the evaluation of higher level protocols, the four lower OSI layers are pooled in a centralized network component. The network component delivers messages to nodes. The connectivity of each node is calculated according to its position with respect to other nodes.

Users do not communicate directly with each other. Instead, all social interaction is mediated by the social environment. The social environment initiates group formation on the basis of spatial proximity and scheduled activities of users.

17.4.2 Equipping Nodes with Alternative User Models

Each node contains a *Protocol* component and a *User* component. For the node component, the user and protocol components are black boxes whose couplings are defined by interfaces. Each can be easily replaced by an alternative implementation that provides the same interface. We will now take a look at different implementations of user components and protocol components.

Figure 17.15 shows the composition structure of the Node component. Three different versions of the user component have been implemented up to now. They can be composed into a node alternatively, as indicated by the three dotted lines. All of these are realized as composite components.

User models represent students moving as pedestrians on a campus and performing service operations. The simple user contains an *activity* subcomponent that manages login and logout behavior. If logged in, the service component becomes notified to start publishing services and searching for services randomly according to a uniform distribution. Furthermore, the *activity* subcomponent selects destination points randomly and calculates routes to them. Routes are propagated to the *Motion* sub component, which partitions the route into small steps and executes them with a certain walking speed. After reaching the destination, a new route is requested from the *activity* component.

The second user model realizes an activity-based user behavior. Network usage and moving are not modeled independently of each other. Both depend on the activity a user is

JAMES II - Experiences and Interpretations

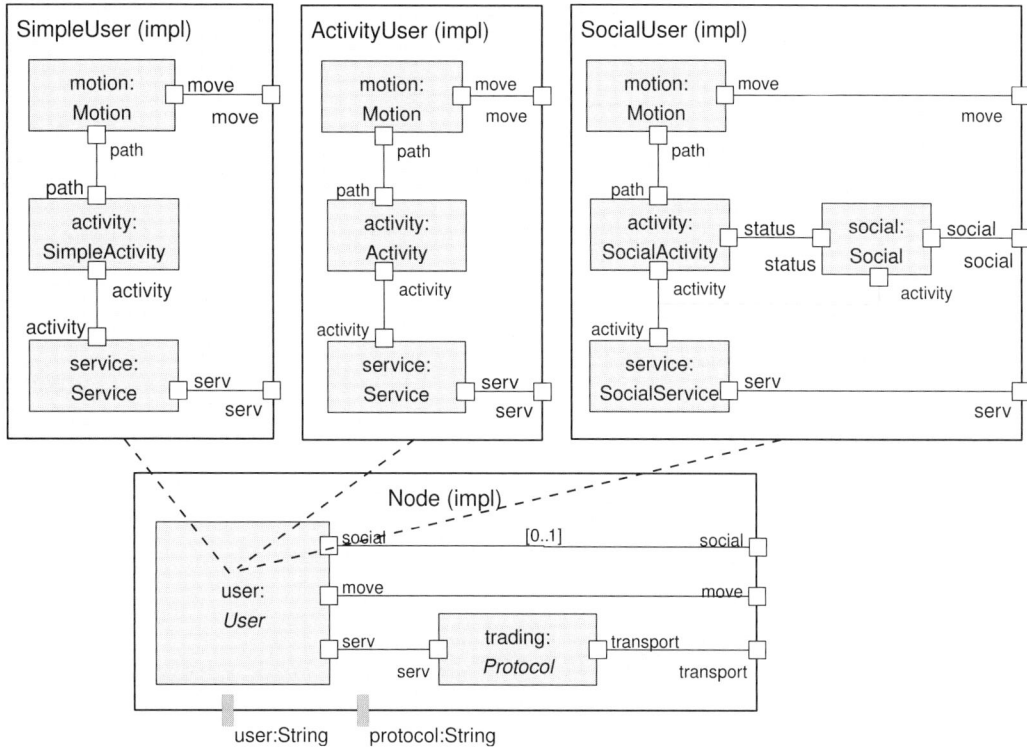

FIGURE 17.15 Structure of the node model with three different user models..

currently executing. The *Activity* model generates a schedule at the start time of the simulation. The schedule contains fixed activities, e.g., attending a lecture on the campus as well as flexible activities, such as learning, with different priorities and durations. The current activity is passed to the *Motion* and *Service* subcomponent. The service component generates service offers and requests according to the received activity. *Motion* receives path information and moves along the path until the destination is reached. If the destination is reached, an event will be produced to indicate this. Afterward, the model waits until new way points are received. Activities depend on the spatial context. Each location on the campus is only suited for a certain set of activities. The current activity constrains the choice of the next destination and thereby the motion model. Thus, motion and service behavior are both based on activities.

The third user model extends the activity-based model with social awareness. The subcomponent *Social* of each user announces planned activity to the social environment model. If users have planned similar activities and are spatially close, the social environment forms a group and selects a group leader. The group leader chooses activities, which group members may adopt. Because an activity does not uniquely define the location of performance, the group leader chooses a location out of a set of suitable ones and communicates this choice to all group members. Group members are free to choose a path to the location. Thus, social users are synchronized with respect to the next joint activity and the location where this activity will be performed.

```xml
<experiment xmlns="http://www.informatik.uni-rostock.de/cosa/experiment"
  xmlns:xsd="http://www.w3.org/2001/XMLSchema">
  <id>DianeExperiment/1.0</id>
  <model>DianeManet/1.0</model>
  <mparams>
    <param name="nodes">
      <value>50</value>
      <value>100</value>
      <value>150</value>
      <value>200</value>
      <value>250</value>
      <value>300</value>
    </param>
    <param name="protocol"> <value>Lanes</value> </param>
    <param name="user"> <value>SimpleUser</value> <value>SocialUser</value>
      </param>
  </mparams>
  <targetFormalism>http://www...de/cosa/model/execdynpdevs</targetFormalism>
  <observercfg>unihro.diane.experiments.v05.ObsNoVis</observercfg>
  <sparams>
    <startTime>0.0</startTime>
    <endTime>28800</endTime>
  </sparams>
</experiment>
```

FIGURE 17.16 Definition of a MANET experiment in XML.

17.5 The Role of Time

Modeling and simulation are both processes which require a lot of time. If a model has been created, quite often many simulation runs have to be executed - either for model validation or for executing experiments to analyze the modeled system. The three different user models and their effect on trading protocols have been evaluated in previous experiments [Röhl et al., 2007]. In case of the MANET model it turned out that the time needed for conducting the experiments increased dramatically for simulation models containing more than 200 nodes. Therefore, we conducted further experiments to evaluate the influence of node numbers on execution time in more detail.

17.5.1 Experiment Definition

Model components form the basis for experimentation. Experiments execute a certain model, derived from a model component with varying parameter values.

Figure 17.16 shows a sample experiment definition that uses the MANET model component as the root model. On the MANET component the following parameters are set:

Nodes: The number of network nodes to be simulated.
Protocol: The component Id of the trading protocol to be used inside each node.
User: The component Id of the user model to be used inside each node.

Users arrive uniformly distributed over a period of one hour, log into the network, and stay online until the end of simulation. The simulation stops one hour after all nodes have been logged in.

Figure 17.17 shows the execution times using a simple and a social user model with increasing number of nodes. For each parameter combination ten replications were made. The runs are executed by the flat sequential simulator - the currently fastest P*DEVS simulator

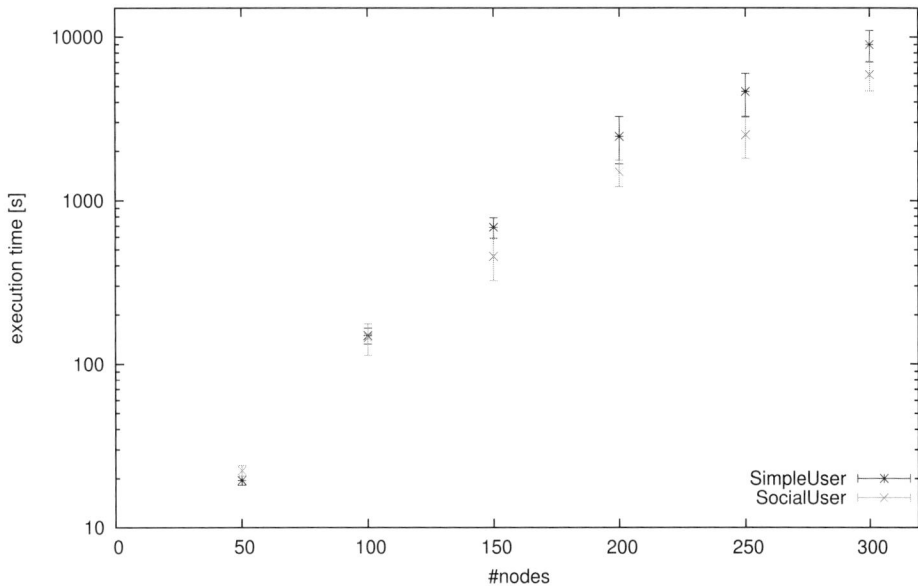

FIGURE 17.17 Execution times.

TABLE 17.1 Mean execution time per run (in seconds, from Figure 17.17) and estimated execution times (100 repetitions) in hours for alternative user models and different numbers of users using the LANES protocol.

User model/Nodes	50	100	150	200	250	300
Simple user (1) (s)	19.97	160.17	640.46	1256.81	5006.3	8970.16
Simple user (100) (h)	0.55	4.45	17.79	34.91	139.06	249.17
Social user (1) (s)	22.37	144.46	453.64	1573.07	2379.04	5869.7
Social user (100) (h)	0.62	4.01	12.6	43.7	66.08	163.05

in JAMES II. The plot shows the mean values and standard deviations of execution times on a logarithmic scale.

Surprisingly, with increasing node size, simulation runs with the more complex user model require less time than with the simple user model. Social users exchange coordination messages in addition to network messages, such that a communication overhead was expected. However, the execution times reflect the original intention for introducing the social user. Nodes clustering induced by group formation and verbal communication between group members reduce the number of messages that are actually sent over the network. Thereby, the measured execution times indirectly substantiate that adding social behavior to network users has an effect on overall network performance.

In Table 17.1 an estimation of execution times (for 100 repetitions) for the model described above is given. It can be easily seen that the simulation of the simple user with 300 nodes requires more than ≈ 249 hours (for 100 repetitions) on a single machine. The repetitions are required due to the stochastic in the model - i.e., 100 repetitions may still not be enough for getting reliable results. If the effect of slightly modified parameters (e.g. different user model, or just 50 nodes less and the same user model) shall be examined in more detail, the overall time required for all runs might be too high to execute these runs on a single computer. Here a scalable simulation framework, supporting in addition a distributed simulation of models, is essential to allow more detailed experiments with the already existing model.

The performance evaluation of the simulation model shows that the developed Manet model is suited to conduct experiments faster than real-time up to 200 nodes. However, measured execution times also indicate that robust and confidential results would benefit from advanced execution capabilities, e.g., to parallelize the execution of simulation runs with different parameter configurations.

17.6 Experiences and Interpretations

JAMES II is not only used for the simulation of completely modeled agent systems. E.g., we used JAMES II to test an agent software called Autominder [Gierke et al., 2006]. In this context the need arose to be able to integrate externally running processes into JAMES II. Two new formalisms have been created allowing such an integration: PepiDEVS (Parallel external process interface DEVS), and PepiDynDEVS (Parallel external process interface dynamic DEVS) [Himmelspach, 2007]. The first extends PDEVS, the second the PdynDEVS formalism (see 17.3.1). Models being based on these formalisms can communicate with external processes (either in paced or unpaced mode).

In addition, JAMES II is being used for nonagent models, and it is being used by third parties. By opening the framework for different user groups, developments from these can be reused easily reducing the time required to create new algorithms or models. An example for this might be a new event queue, maybe developed newly for a specialized non agent simulator, which can be automatically used for the computation of the next discrete event based agent model - this makes it possible to use basic scientific results and improvements cross research fields and application scenarios.

Flexible systems for modeling and simulation are highly required. It is not enough if modeling and simulation frameworks are restricted to single execution mechanisms. Especially multi-agent models require - due to their size and heterogeneity in regards to their computational complexity - efficient and alternatively usable algorithms for their computation. The example illustrates the time needed to compute a model and thereby the need for efficient algorithms. As we have already shown, different models may be best simulated using different algorithms / data structures [Himmelspach and Uhrmacher, 2006, 2007b; Himmelspach, 2007]. JAMES II provides the base for such a flexible system. The algorithms realized so far allow the execution of multi agent models on a diverse set of hardware (e.g., home PCs, heterogeneous LANs, grid like infrastructures). In addition, these algorithms should be adaptable in accordance to the model that shall be computed by them - e.g., by being able to integrate different event queues.

17.7 Outlook

Future research will add new formalisms and simulation algorithms to JAMES II. New formalisms will extend the usage of JAMES II for various disciplines and applications. Further advancements of model descriptions and model reuse can significantly add to gain new insights by reducing the overall time needed to create models and to experiment with these. The support of further modeling formalisms and languages for multi agent systems may add to the advancement of multi agent research due to the possibility to reuse new technologies from all fields of modeling and simulation implemented for the framework. Examples for such new technologies are new simulation algorithms which can add to the scalability of JAMES II. The experimentation with new algorithms is essential during the development of new algorithms [Tichy, 1998; Johnson, 2002; McGeoch, 2007], and in a fixed context as the one provided by JAMES II the evaluation of algorithms may additionally reveal inherent properties of these. Up to now JAMES II has been used only in heterogeneous networks and on standalone machines. In the future JAMES II shall additionally run on Grid inspired and cluster based approaches. To improve the usability of JAMES II simulation algorithms shall be automatically - based on characteristics of the model and the algorithms - selectable in the future [Ewald et al., 2008]. In addition, this mechanism could form a base to extend the framework with a dynamic adaptation mechanism of the simulation infrastructure according to structural changes of multi agent models.

Acknowledgment

This research is part of the project CoSA – *Component based framework for an effective and efficient simulation of agent systems* funded by The Deutsche Forschungsgemeinschaft (German Research Foundation) – DFG.

References

O. Balci. Verification, validation, and certification of modeling and simulation applications: verification, validation, and certification of modeling and simulation applications. In *WSC '03: Proceedings of the 35th conference on Winter simulation*, pages 150–158. Winter Simulation Conference, 2003. ISBN 0-7803-8132-7.

S. Corson and J. Macker. Mobile ad hoc networking (MANET): Routing protocol performance issues and evaluation considerations. http://www.ietf.org/rfc/rfc2501.txt, Jan. 1999. Network Working Group Memo (Request for Comments: 2501).

Diane. DIANE-Projekt: Services in ad hoc networks (Dienste in Ad-Hoc-Netzen). http://www.ipd.uni-karlsruhe.de/DIANE (zugegriffen am 19. September 2007), 2007.

B. Edmonds. The use of models - making MABS more informative. In S. Moss and P. Davidsson, editors, *MABS*, volume 1979 of *Lecture Notes in Computer Science*, pages 15–32, Berlin, Heidelberg, 2000. Springer. ISBN 3-540-41522-X.

B. Edmonds and D. Hales. Replication, replication and replication: Some hard lessons from model alignment. *J. Artificial Societies and Social Simulation*, 6(4), 2003.

R. Ewald, J. Himmelspach, and A. M. Uhrmacher. An algorithm selection approach for simulation systems. In *Proceedings of the PADS*, pages 91–98, Rome, Italy, June 2008. IEEE Computer Society.

J. M. Galan and L. R. Izquierdo. Appearances can be deceiving: Lessons learned reimplementing axelrod's 'evolutionary approach to norms'. *Journal of Artificial Societies and Social Simulation*, 8(3), 2005. ISSN 1460-7425. URL http://jasss.soc.surrey.ac.uk/8/3/2.html.

M. Gierke, J. Himmelspach, M. Röhl, and A. Uhrmacher. Modeling and simulation of tests for agents. In *Proceedings of MATES*, 2006.

N. Gilbert. Environments and languages to support social simulation. In *Social Science Microsimulation*, pages 457–458, 1995.

E. R. Harold. *Processing XML with Java*. Pearson Education, Boston, 2002.

J. Himmelspach. *Konzeption, Realisierung und Verwendung eines allgemeinen Modellierungs-, Simulations und Experimentiersystems - Entwicklung und Evaluation effizienter Simulationsalgorithmen*. Dissertation, Universität Rostock, 2007.

J. Himmelspach and A. M. Uhrmacher. Processing dynamic PDEVS models. In D. DeGroot and P. Harrison, editors, *Proceedings of the 12th IEEE International Sym-*

posium on MASCOTS, pages 329–336, Volendam, The Netherlands, Oct. 2004. IEEE Computer Society. ISBN 0-7695-2251-3. URL http://csdl.computer.org/comp/proceedings/mascots/2004/2251/00/22510329abs.htm.

J. Himmelspach and A. M. Uhrmacher. Sequential processing of PDEVS models. In A. G. Bruzzone, A. Guasch, M. A. Piera, and J. Rozenblit, editors, *Proceedings of the 3rd EMSS*, pages 239–244, Barcelona, Spain, Oct 2006.

J. Himmelspach and A. M. Uhrmacher. Plug'n simulate. In *Proceedings of the Spring Simulation Multiconference*, pages 137–143. IEEE Computer Society, Mar. 2007a.

J. Himmelspach and A. M. Uhrmacher. The event queue problem and PDEVS. In *Proceedings of the DEVS Integrated M&S Symposium*, pages 257–264. SCS, Mar. 2007b.

D. Johnson. A theoretician's guide to the experimental analysis of algorithms. In *Fifth and Sixth DIMACS Implentation Challenges*, 2002. URL citeseer.ifi.unizh.ch/johnson96theoreticians.html.

M. Klein, M. Hoffman, D. Matheis, and M. Müssig. Comparison of overlay mechanisms for service trading in ad hoc networks. Technical Report TR 2004-2, University of Karlsruhe, Oct. 2004.

S. Kurkowski, T. Camp, and M. Colagrosso. MANET simulation studies: The incredibles. *ACM's Mobile Computing and Communications Review*, 9(4):50–61, 2005.

P. L. L'Ecuyer, L. Meliani, and J. Vaucher. SSJ: a framework for stochastic simulation in java. In J. L. Snowdon and J. M. Charnes, editors, *Proceedings of the 34th Winter Simulation Conference: Exploring New Frontiers, San Diego, CA, Dec. 8-11, 2002*, pages 234–242. ACM, 2002.

Q. Liu and G. Wainer. Improving CD++ parallel simulation engine. In *Proceedings of the 39th IEEE/SCS Annual Simulation Symposium*, Norfolk, VA, USA, Mar. 2007. IEEE.

A. Martens and J. Himmelspach. Combining intelligent tutoring and simulation systems. In P. Fishwick and B. Lok, editors, *Proceedings of the International Conference on Human-Computer Interface Advances for Modeling and Simulation (SIMCHI'05)*, pages 65–70, New Orleans, USA, Jan. 2005. SCS, The Society for Modeling and Simulation International. ISBN 1-56555-288-1.

C. C. McGeoch. Experimental algorithmics. *Commun. ACM*, 50(11):27–31, 2007. ISSN 0001-0782. doi: http://doi.acm.org/10.1145/1297797.1297818.

N. Minar, R. Burkhart, C. Langton, and M. Askenazi. The SWARM simulation system: a toolkit for building multi-agent simulations. Technical report, Santa Fe Institute, June 1996.

M. J. North, T. Howe, N. T. Collier, and J. R. Vos. The repast simphony runtime system. In C. M. Macal, M. N. North, and D. Sallach, editors, *Proceedings of the Agent 2005 Conference on Generative Social Processes, Models, and Mechanisms*, pages 151–158. Argonne National Laboratory and The University of Chicago, USA, 2005.

OMG. UML superstructure specification version 2.0 (document formal/05-07-04). Available online via http://www.omg.org/cgi-bin/doc?formal/05-07-04, July 2005.

K. Perumalla. μsik: A micro-kernel for parallel/distributed simulation systems. In *ACM/IEEE/SCS Workshop on Parallel and Distributed Simulation (PADS)*, page to appear, Monterey, CA, Juni 2005. IEEE Computer Society Press.

M. Röhl. Platform independent specification of simulation model components. In *20th European Conference on Modeling and Simulation, ECMS 2006*, pages 220–225, Bonn, Sankt Augustin, Germany, May 28th-31th 2006.

M. Röhl and A. M. Uhrmacher. Composing simulations from XML-specified model components. In *Proceedings of the Winter Simulation Conference, Monterey, CA, Dec. 3-6*, pages 1083–1090. WSC, 2006.

M. Röhl, B. König-Ries, and A. M. Uhrmacher. An experimental frame for evaluating service trading in mobile ad-hoc networks. In *Mobilität und Mobile Informationssysteme (MMS 2007)*, volume 104 of *Lect. Notes Inform.*, pages 37–48, 2007.

D. S. Tan, S. Zhou, J.-M. Ho, J. S. Mehta, and H. Tanabe. Design and evaluation of an individually simulated mobility model in wireless ad hoc networks. In *Communication Networks and Distributed Systems Modeling and Simulation Conference 2002*, San Antonio, TX, 2002. URL `citeseer.ist.psu.edu/tan02design.html`.

W. F. Tichy. Should computer scientists experiment more? *Computer*, 31(5):32–40, May 1998. URL `http://csdl.computer.org/comp/mags/co/1998/05/r5032abs.htm`.

A. M. Uhrmacher. Dynamic structures in modeling and simulation - a reflective approach. *ACM Transactions on Modeling and Simulation*, 11(2):206–232, Apr. 2001.

A. M. Uhrmacher, R. Ewald, M. John, C. Maus, M. Jeschke, and S. Biermann. Combining micro and macro-modeling in DEVS for computational biology. In *Proceedings of the Winter Simulation Conference, Washington, DC, Dec. 9-12*. WSC, 2007.

W3C. XML Schema part 0: Primer second edition. `http://www.w3.org/TR/2004/REC-xmlschema-0-20041028/`, Oct. 2004. W3C Recommendation 28 October 2004.

B. P. Zeigler, H. Praehofer, and T. G. Kim. *Theory of Modeling and Simulation*. Academic Press, London, 2nd edition, 2000.

Glossary

AGV transportation system: A system that consists of a set of automated guided vehicles (AGVs), controlled by a multi-agent system, that transport loads in a warehouse.

Architectural view: A representation of a coherent set of architectural elements and the relations among them. A view presents a particular perspective on the architecture or a part thereof.

Artifact: According to the A&A Metamodel, artifacts are passive, reactive computational entities aimed at the agent's use, working as agent's instruments and tools. Artifacts typically implement those behaviors that cannot or do not require to be modeled as goal-oriented, and provide agents with the services and functions they require to ease the achievement of the agent's goals.

Automated guided vehicle (AGV): An unmanned, computer-controlled vehicle that transports loads in a warehouse and that uses a battery as its energy source.

Avatar: A software agent in the polyagent modeling construct that manages the correspondence with a domain entity. It persists as long as its entity, generates ghosts to explore alternative possible behaviors, and may employ complex agent logic such as BDI reasoning.

Behavioral implicit communication: It is a form of communication that, to be effective, it is sufficient that the receiver reads the act or structure as a cue and not as a signal. In particular the receiver is able to use a practical behavior or a trace of a practical behavior.

Car-following model: A car-following or follow-the-leader model describes the longitudinal movement of a driver-vehicle agent. It assumes that drivers react to the stimulus from neighboring vehicles with the dominant influence from the direct leader. Together with lane-changing models, car-following models constitute the basis of microscopic traffic simulations.

Cellular automata crowd model: In this approach pedestrians are viewed as specific states for the cells, units in which the environment is discretized. The movement is generated by the uniform CA transition rule. The environment has an explicit and discrete representation.

Communication: The production of any act or structure which alters the behavior of other agents, which evolved (or is learnt or deliberated) because of that effect, and which is effective because the receiver's response has also evolved (or is learnt or designed). Defined in this way, communication is a form of coordination by hetero-adjustment.

Continuous strategy space: Instead of a finite (or discrete) set of strategies a continuum of strategies is considered.

Coordination: It is a form of individual social action in which an agent adapts to another agent's behavior in order to increase its own success. More precisely it is adaptation by reading cues of interference. Coordination is direct when the cues of interference are the other agent's behavior themselves. Coordination is indirect when, by acting on the basis of some cue other than the behavior of some other agent, the agent is also adapting its own behavior to those of the other agent.

Crowd (of pedestrians): There is no precise and shared definition of the term; a qualitative but effective definition is "(Too) many people in (too) little space".

Cue: Any animate or inanimate feature of the environment that an agent is adapted (thanks to evolution, learning or deliberation) to use as a guide for future actions.

Decision model: Represents a model of possible decision making options in the physical system. Different what-if scenarios are created by using the decision model. It is used only in the context of an symbiotic simulation control system (SSCS) and a symbiotic simulation decision support system (SSDSS).

Delegate MAS: A multi-agent system that is assigned a task to perform on behalf of a higher-level agent; in the polyagent construct, corresponds to a collection of ghosts supporting an avatar.

Differentiated cells: Cells which have a specialized function such as blood or heart cells and which have (at least in general) lost the potential to "dedifferentiate" and become less specialized. Fully differentiated cells have lost the ability to divide.

Differentiation and dedifferentiation: The first is the process of becoming more specialised which may or may not be independent of division, and the second is any process where a cell becomes more "stem-like".

DIME: DIME stands for Diplomatic, Intelligence, Military, and Economic actions.

Discrete strategy set: A finite set of strategies.

Driver-vehicle agent: In a microscopic traffic simulation, the driver-vehicle agent is the atomic entity which represents the internal characteristics of a driver (driving style) as well as properties of the vehicle such as acceleration and braking capabilities. The operational tasks are the longitudinal movement as well as changing lanes.

Dynamic environment: An environment in which the operating conditions of a multi-agent control system are continuously changing. In a dynamic environment, an agent cannot determine the outcome of its actions a priori, as other activities happening in the environment can have a significant impact on the outcome of actions.

Dynamic uncertainty: Uncertainty about a system variable that is due, not to lack of knowledge (e.g., sensor noise), but to the repeated iteration of nonlinear processes.

Emergence: A phenomenon observed in a system is said to be emergent if and only if it cannot be explained in the same terms (same structures, same laws) describing the states and dynamics of the generating system.

Evolutionary game theory: The evolutionary version of Game Theory arose from the application of traditional Game Theory to biology. Its central concepts are the replicator dynamics and evolutionary stable strategies.

External influence: Represents influence factors which have their origins in the environment of the physical system. They affect the behaviour of the physical system but cannot be controlled by it or the simulation system.

FactionSim: FactionSim is an environment that captures a globally recurring sociocultural "game" that focuses upon inter-group competition for control of resources (Security, Economic, and Political Tanks).

Functional components: Realise various functionalities which are required for typical activities in a symbiotic simulation system. They are implemented by means of capabilities in the agent-based framework.

Ghost: A software agent in the polyagent modeling construct that explores one possible behavior for its avatar. It is transient to make room for later ghosts from its avatar. Ghosts interact with each other and with avatars, not directly, but through digital pheromones that they deposit in the environment. This lightweight reasoning allows the concurrent execution of multiple ghosts.

GSP: The Goals, Standards, and Preference tree is a multi-attribute value structure. It is designed to capture an agent's short term needs and motivations (Goals), its preferred means to achieving its end (Standards), and its long-term desires (Preferences).

HBM: HBM stands for human behavior modeling.

Hematopoeitic stem cells: The hematopoietic stem cells (HSCs), which can be found mostly in the bone marrow, are responsible for the constant replacement of blood cells lost to normal turnover processes as well as to illness or trauma.

Hematopoietis: The entire process of production and maintenance of all types of blood cells which exhibit very different functions, such as transport of oxygen, production of antibodies to fight infection and blood clotting.

Hetero-adjustment: It is a strategy of coordination in which an agent coordinates with the obstacles or the opportunities created by another agent's actions by changing its own behavior in order to influence the behavior of the other agent. When the interference is negative, hetero-adjustment takes the form of impediment. When the interference is positive, hetero-adjustment takes the form of inducement.

High Level Architecture (HLA): HLA is general purpose architecture for distributed computer simulation systems. IEEE 1516 is a standardized process for developing interoperable HLA based federations. With HLA, computer simulations can communicate to other computer simulations.

IDE: IDE stands for an Integrated Development Environment.

Influence/Reaction model: This model, which has been proposed in the mid nineties by Ferber, defines an action representation providing a solution to the representation of simultaneous actions in multi-agent systems.

Inter-driver variability: Inter-driver variability comprises the heterogeneity between driver-vehicle agents with respect to driving style and vehicle properties. In the agent-based approach, inter-driver variability is implemented by different sets of parameter values for each agent.

Interference: The process of reciprocal causal influence of agents' living in a common environment. Interference is the most basic social relation between the agents since each individual success is also a function of the obstacles (negative interference) or opportunities (positive interference) created by the actions of other agents.

Intra-driver variability: Intra-driver variability describes the observation that even identical driver-vehicle agents may respond differently to a given traffic situation depending on the agent's history and experience. In an agent-based approach intra-driver variability can be modeled, e.g., by means of time-dependent model parameters.

Key indicator: A quantitative measure used by the simulation system in order to compare the simulation results of different what-if scenarios with each other, or with the physical system.

Macroscopic traffic simulation: In contrast to an agent-based simulation, the macroscopic simulation is based on aggregated quantities such as density or average velocity. Macroscopic models make use of the picture of traffic flow as a physical flow of a fluid.

Microscopic traffic simulation: The microscopic simulation approach describes the motion of the vehicles on an individual basis including the interactions with other vehicle-driver agents and with the road infrastructure. It can be considered as an example of a multi-agent simulation.

Microsimulation: Proposed by Orcutt in 1957, microsimulation was the first modeling approach which tried to directly represent the characteristics and behaviors of individual entities in system.

Modeling relation: Defined in the scope of the Zeigler's framework for M&S, the modeling relation concerns the study of the relation between a simulation model and its corresponding source system (i.e. the phenomenon to be modeled). This study notably addresses validity issues in computer simulation experiments.

OODA: OODA refers to the Observe, Orient, Decide, and Act loop that is involved in an agent's decision-making process.

PageRank: PageRank is an algorithm that has been used in Google. It assigns a numerical weighting to each element of a hyperlinked set of the World Wide Web. The algorithm may be applied to any collection of entities with quotations and references.

Participatory simulation: Participatory Simulation is a concept that basically stands for human involvement in a running simulation. The human does not only control the time advance by stepping or stopping the simulation, but may manipulate various facets of the simulated situation. Sometimes, the notion focusses on a human taking over the control of a particular agent similar to role-playing experiments. A related notion is serious gaming.

Particle-based crowd model: In this approach, pedestrians are viewed as particles subject to forces, representing geometrical features of the environment, goals of pedestrians, interactions among them and the environment. The environment is implicitly represented in the "laws of motion".

Pedestrian: A person traveling on foot, whether walking or running. It is considered the basic analytical unit of pedestrian and crowd models.

Pheromonal communication: A pheromone is an explicit signal whose production is evolved (or learnt or deliberated) for influencing another agent, and that is effective because the receiver's response has also evolved (or learnt or designed). Though it is a form of indirect communication, it is not an instance of stigmergic behavior since the receiver's response depends on the fact that the traces are considered signals and not only cues.

Pheromone: In biology, a chemical marker deposited in the environment to enable stigmergic coordination. In polyagent systems, a scalar variable, indexed to a location in the environment, that agents can augment and read.

PMESII: PMESII stands for Political, Military, Economic, Social, Informational, and Infrastructure systems of a given country of interest.

PMF: A Performance Moderator Function (PMF) is a micro-model covering how human performance (e.g., perception, memory, or decision-making) might vary as a function of a single factor (e.g., sleep, temperature, boredom, grievance, and so on).

PMFserv: PMFserv is a unifying architecture for agent cognition built from best available PMF theories.

Polyagent: A modeling construct consisting of a single avatar and a collection of ghosts that explore alternative behaviors for the avatar.

Prediction horizon: The distance in the future at which prediction becomes unreliable due to dynamic uncertainty.

Progenitor cells: Are somewhere between stem cells and differentiated cells. Their definition is vague in the literature but essentially they have less flexibility in their range of differentiation possibilities but still retain some stem cell like properties.

Reference model: A model (including corresponding values for its parameters) that describes the current behaviour of the physical system with sufficient accuracy. The reference model is essential for symbiotic simulation model validation system (SSMVS) and symbiotic simulation anomaly detection system (SSADS).

Reinforcement learning: Is a sub area of machine learning which is concerned with learning from experience (trial and error) instead of learning from examples (or a teacher).

Replicator dynamics: The replicator dynamics are a system of differential equations which describe how a population of strategies evolves through time.

RoboCup: RoboCup is an international research and education initiative. Its ultimate goal is to develop a team of fully autonomous humanoid robots that can win against the human world champion team in soccer by 2050.

RoboCupRescue: RoboCupRescue project is one of RoboCup leagues. The intention of the league is to promote research and development in the rescue domain by involving software/physical robotic agents for search and decision support systems, and evaluation benchmarks that are all integrated into a system in future.

Self-adjustment: It is a strategy of coordination in which an agent coordinates with the obstacles or the opportunities created by another agent's actions by changing its own behavior. When the interference is negative, self-adjustment takes the form of avoidance. When the interference is positive, self-adjustment takes the form of exploitation.

Self-organizing system: A system able to re-organize itself by managing the relations between components (either topological, structural or functional) upon environment perturbations, solely via the interactions of its components, with no need of external forces super-imposing organization.

Simulation relation: Defined in the scope of the Zeigler's framework for M&S, the simulation relation concerns the verification of a simulator with respect to a simulation model. This study notably should ensure that the simulator does generate the behavior of the model correctly.

Situated agents crowd model: In this approach pedestrians are reified as autonomous entities situated in an environment, characterized by an explicit representation. Agents perceive their local context, decide on their movements according to (possibly heterogeneous) behavioural specifications, interact with the environment and with other agents.

Socionics: An approach to socially inspired computing which tries to use knowledge from social science theories about adaptive social systems to learn from them in order to improve the adaptivity of computer systems.

Software architecture: The structure or structures of a system, which comprise software elements, the externally visible properties of those elements, and the relationships among them.

Software-in-the-loop simulation: A simulation in which the software of a real control system is embedded in the simulation loop.

Stakeholder approach: In multi-agent social simulation, the term Stakeholder Approach is used for characterizing a strong involvement of domain experts, clients or other people that are knowledgeable about the structure and dynamics of the original system when developing an agent-based simulation. It basically corresponds to the guideline in traditional simulation domains to involve contracting entities into the development of a simulation model.

Stem cells: Cells which have the ability to maintain their own population and to re-generate the population after disease or injury through cell division which is called proliferation. They can also produce specialised functional cells through the process of differentiation.

Stem cell niche: Specific areas in the body where stem cells are abundant. With hematopoietic stem cells (HSCs) for example this is in the bone marrow. The niche acts as a micro-environment for regulating the behavior of stem cells.

Stigmergy (Gardelli, Viroli, Omicini): First observed in social insects, stigmergy is today generally regarded as a set of mechanisms promoting environment-based coordination, where coordination is achieved by using specific markers—pheromone in social insects, or their digital equivalent in computational systems.

Stigmergy (Parunak, Brueckner): Coordination of multiple agents by making and sensing changes to a shared environment.

Stigmergy (Tummolini, Mirolli, Castelfranchi): Is a form of indirect coordination in which an agent adapts its behavior to a specific set of cues: i.e. the practical behavioral traces of other agents. Stigmergic self-adjustment is adaptation by the detection of practical behavioral traces. Stigmergic communication is influence by an evolved (or learned or deliberated) production of this form of cues.

Stochastic model checking: Model checking is a formal technique for automatically verifying the properties of a target systems against its model. When applied to stochastic systems, model checking makes it possible to compute the actual likelihood for the properties under test.

Symbiotic simulation: A simulation paradigm in which a simulation system is closely coupled with a physical system. Both symbionts, the physical system and the simulation system, may benefit from this relationship.

Visual programming: Visual Programming refers to specifying a program in an at least two-dimensional fashion. Basic strategies found in Visual Programming systems are direct manipulation of concrete objects and their graphical notations, immediate feedback, and explicitness.

What-if analysis: An essential concept in symbiotic simulation which is concerned with the evaluation of a number of what-if scenarios by means of simulation. The purpose of the what-if analysis process is to determine the optimal what-if scenario in the given application context.

What-if scenario: A what-if scenario consists of a particular model, an initial state for the simulation, and information regarding external influence. Its purpose depends on the type of symbiotic simulation.

Index

A

A&A meta-model, 136
    architectural pattern, 138
    overview, 137
    and role of environment in self-organizing systems, 136–137
Acceleration strategy, 331
Accessible environments, 18
Accommodation relation, 37
ACL language, 20
Action, 16, 17
    temporal granularity problem, 25, 26
Activation list, rank in, 25
Actual behaviors, in theoretical environments, 33
Adaptation
    by reading cues of interference, 249
    in self-organizing MAS, 134
Adaptive cruise control, 352
Adaptive MAS, 139
Adaptor capability (A-C), 379, 380
Advance planning script, 122
Agent architectures, 14
    cognitive architectures, 15–16
    hybrid architectures, 16
    reactive architectures, 14–15
Agent-based computational demography, 56
Agent-based framework
    actuation layer, 376
    architectural requirements, 372–373
    capability-centric solution, 375–376
    existing architectures, 373–374
    layers and associated capabilities, 376–377
    perception layer, 376
    process layer, 376
    for symbiotic simulation, 357–359
    web services *vs.* agent-based approach, 374–375
Agent-based modeling (ABM), 9–11, 112, 407
    environment in, 10
    interactions, 10
    micro-level entries, 10
Agent-based simulation, 109
    tools for, 481–482
    using BDI programming in Jason, 451–452
Agent-based social simulations (ABSS), 11–12, 424
    analysis based on probability model, 441–447
    architecture analysis, 447–448
    validity issues, 438–441
Agent-based stem cell modeling, 389–390, 392
    agent actions, 399
    agent modeling framework, 395–400
    agent modeling tools, 407–412
    agent world state in, 396
    agentifying existing approaches, 400–407
    and behavioral potential, 393
    biological domain, 390–392
    biological plausibility, 394, 407
    CA approach, 401–403
    cell agent behavior, 399–400
    cell agent-environment interface, 398
    cell agent external requests, 400
    cell agent internal actions, 399
    cell agent signals, 399–400
    cell agent-simulation engine interface, 398
    cell agent specification, 404
    cell agents, 395, 397
    conceptual framework, 414
    determined cell nodes, 402
    empty nodes, 402
    environment, 395, 397
    existing model drawbacks, 394–395
    experimental limitations, 392–393
    extension to agent modeling framework, 400
    FIPA-compliant tools, 407
    framework component behavior, 399–400
    framework component interfaces, 398
    framework components, 396–398
    HSC biology, 390–392
    internal state information, 398
    MASON toolkit, 408–409

model rules, 402–403
model uses, 393
multilevel model, 394
node state, 401
perceive actions, 399
performance planning, 399
probability function, 413
rationale for agents, 394–395
re-formulation using agent-based approach, 403–405
Roeder-Loeffler model of self-organization, 406–407
role of graph nodes, 403–404
safety properties, 405
simulation engine, 395, 397–398
simulation engine behavior, 400
simulation engine-environment interfaces, 398
StarLogo, 407
state information, 396
stem cell agent definition, 404
stem cell division representation, 403
Sugarscape tool, 407
Swarm multi-agent simulation platform, 407–408
system state, 491
theoretical simplifications, 393
unique cell identifiers, 404
visualization, 413, 414
Agent behavior, modeling, 16–17
Agent class descriptions, SeSAm, 486
Agent communication, 69–70
Agent communication language, 244
Agent decision-making
  constraint-based methods, 110
  derivational methods, 110
  dynamic uncertainty and, 110
  game theoretic methods, 110
  polyagent model, 111–118
  polyagents and, 109–111
  process irreducibility limitations, 110
  spatial irreducibility limitations, 110
Agent logical processes (ALP), 97
Agent memory, 61
  and cultural transmission, 62
Agent modeling tools, for stem cell modeling, 407
Agent motion, 22
Agent-oriented software engineering (AOSE), 137

Agent perception, 275
Agent programming languages, 452–453
Agent roles, in different environments, 65–67
Agent specification, 77
Agent thinking time, 94
Agent update, in SeSAm toolkit, 489–490
Agents, v
  in A&A meta-model, 137
  autonomy of, 10
  continuous approach to perception, 20
  coupling with environment, 24
  defined, 13–14
  different roles in different environments, 65–67
  inputs and outputs, 18
  Jason example, 467–471
  modeling behavior time, 23
  situatedness, 10
  sociability of, 10
  social system models, 281
  as three-phase process, 16
  for traffic simulation, 325–327
AgentSpeak, 458, 470, 471
  Jason variant, 454
AGV agent software
  control interface of AGV agents, 184
  execution time of AGV agents, 185
  simulation model, 183
AGV simulation platform architecture, 185
  activity repository, 191–192, 193
  activity transformer repository, 192, 196
  API translator repository, 192
  architectural rationale, 189–190, 193–194, 196, 197–199, 201
  aspect-oriented approach to embed control software, 197–199
  battery law, 202
  block interface, 201
  clock manager component, 199
  collision detection, 202
  component and connector view of execution tracker, 199–201
  component and connector view of simulated environment, 190–194
  component and connector view of simulation engine, 194–196
  control API interface, 189
  controller software module, 187

INDEX

customizable presentation of environment state, 194
customizable state updating strategy, 194
decoupling synchronization from simulated environment, 194
duration mapping repository, 200, protocol-based transport assignment 205
elements and their properties, 187–189, 190–193, 194–196, 199–200
environment inspector module, 192, 193
environment source component, 195–196
environment source repository, 192–193
evaluating, 201–205
execution blocker module, 200
execution tracker module, 188, 196
experiment setup, 202–203
explicit support for as-fast-as-possible simulations, 190
field-based transport assignment, 205
flexibility of AGV simulator, 202
flexibility of embedding MAS, 198
functional requirements, 186
hull-based collision avoidance, 204–205
influence repository, 192, 193
interaction law repository, 192
interface descriptions, 189, 193, 196, 201
low coupling between controller and simulation platform, 189
low coupling between simulated environment and simulation engine, 189–190
low coupling due to data repositories, 193–194
MAS development supported by, 204–205
measurements, 203
notify interface, 196
performance measurements, 204
quality requirements, 186–187
reaction law repository, 192, 193
read interface, 201
requirements, 186–187
reusable infrastructure, 194, 201
separating monitoring from synchronization, 201
separating simulation from application concerns, 199
simulated environment module, 188
simulation engine encapsulating all synchronization, 196
simulation engine module, 188
simulation platform element, 187
simulator measurements, 202–203
state repository, 191, 193
state updater module, 192, 196
support for controller execution tracing, 198
sync interface, 196
top-level module decomposition view, 187–190
trace interface, 189, 201
virtual environment based routing, 204
AGV simulator, requirements, 176
AGV steering system, 175–176
AGV transportation system, 172, 535
   3-D view, 174
   AGV agent software integration, 183–185
   cheese factory example, 173
   collision avoidance, 174
   control system, 173
   deadlock avoidance, 174
   functionalities in control system, 173–175
   maintenance, battery charging, and resource saving, 175
   manipulation of dynamism in warehouse environment, 183
   physical setup, 173
   routing, 174
   simulation model, 181–185
   simulator requirements, 176
   steering system, 175–176
   transport assignment, 174
   warehouse environment dynamism, 182–183
   warehouse environment simulation model, 181–182
   warehouse environment sources of dynamism, 183
   warehouse environment structure, 181–182
AIMSUN, 342
Alternative threat management scenarios, 362

Altruistic genes, 257
Annotations, in Jason, 454
Anomaly detection, in symbiotic simulation, 364
Ant agents, 21, 133, 141
    pheromonic diffusion in, 145
Ant-colony optimization, 117
Ant colony simulation, 13
AOSE, 137
    integrating formal approach with, 140
3APL, 452
Apoptosis, 391
Applicability requirement, for symbiotic simulation, 372
Applications
    agent communication, 69–70
    cellular automata, 58–62
    complex systems simulation, 53–56
    discrete event simulation, 62–63, 63
    dynamic microsimulation predicting demographic processes, 57–58
    early agent models, 63
    microsimulation, 63
    polyagents, 118–124
    role of environment, 67–68
    role of interactions, 67
    1960s voting behavior, 57
    system dynamics, 63
Architectural pattern, for self-organizing MAS, 138–139
Architectural view, 535
ARCHON, 116
Artifacts, 135, 535
    in A&A meta-model, 137
    maintenance by environmental agents, 138
Artificial animals, 12–13
Artificial laboratories, 4, 5, 7
Artificial Life (AL), 5
Aspect-oriented programming, 197
Aspect weaving, 198
Atomic agentified entities (AE), 38
Atrocities Event Data (AED), 283, 285
Augmented experiment, 36, 37
Automap, 286
Automated data extraction technology, 272, 284–289
    challenges, 288
    summary, 287
    theme codes, relevance, and reliability, 291
Automated guided vehicle (AGV), 535
Automated semantic analysis tools, 286
Automated web-scraping tools, 287
Autonomic Computing proposal, 139
Autonomous agents, pedestrians, 305–306
Autonomy
    of agents, 10
    modeling, 38
    in self-organizing MAS, 134
Avatars, 116, 535
    in BEE process, 123
    in polyagent model, 114
    work-piece, 118
Avoidance behaviors, 22

B

Banana game, 223–224, 237
    continuous replicator dynamics (CRD), 238
    equilibrium solution, 238, 239
Battle of the sexes game
    direction field plot, 229, 232
    Q-learning plots, 232
    strategic form, 219
BDI agents, 456, 458
BDI architecture, 15
BDI programming. *See also* Belief-Desire-Intention (BDI) architecture
    in Jason, 451–453
Behavior-based robotics, 260
Behavior Evolution and Extrapolation (BEE) process, 123
Behavior module, 26
    and participatory design, 35–36
Behavior time, of agents, 23
Behavioral communication, and implicit signals, 250
Behavioral emulation and extrapolation, 124
Behavioral implicit communication, 535
Behavioral messages, 251
Behavioral traces, 248, 251
    as cues, 248
Behaviors, 27
Belief-Desire-Intention (BDI) architecture, 15, 16, 78, 261, 375
Beliefs, in Jason, 453
Binary Decision Diagrams (BDD), 143

Biological modeling, 269
    plain diffusion in, 145
    and self-organization, 133
Blackboard systems, 243, 244
Boids simulation, 82
Braking strategy, 331, 332

C

Capabilities, in symbiotic simulation framework, 375–377
Car-following model, 326, 331, 535
Causation, upward and downward, 54
CD++, 511
Cell agents
    in agent-based stem cell modeling, 395, 397
    behavior, 399–400
    external requests, 400
    internal actions, 399
    signals, 399–400
Cell growth factors, 392
Cell intrinsic regulatory processes, 391
Cellular automata, 58–62, 63, 268, 303, 304, 315
    in crowd behavior modeling, 301
    crowd model, 535
    operation of reformulated approach, 405–406
    pedestrians as states of, 304–305
    re-formulation using agent-based approach, 403–405
    stem cell modeling approach, 401–403
    for traffic simulation, 329
Centralized event list, 95
Changing environments, simulation robustness to, 435–436
CHARON toolkit, 84, 85, 95–96
Checkpointing, in MASON toolkit, 408
Chemical master equations (CME), 394
Chronic myeloid leukemia, modeling, 406
CLAIM, 452
CNET simulator, 5–6
Coercion, *vs.* communication, 249
Coffeemeter, as driving comfort measure, 345–346
Cognitive agent environment, 273
Cognitive agent modeling, 274
    major PMF models, 275–278
Cognitive appraisal, 275
Cognitive architectures, 15–16, 276

Cognitively detailed models, 271
Coherency, 37
Coleman boat, 54
Collaborative stigmergic coordination, 252
Collective pattern emergence
    speed limit impacts, 347–349
    stop-and-go waves, 346–347
    store-and-forward strategy for inter-vehicle communication, 349–351
    in traffic simulations, 346
Collision avoidance, in AGV transportation system, 174
Comfortable deceleration, 332, 336, 341
Commensalism, 361
Commercial Off The Shelf (COTS) software, 274
Communication, 535
    agent, 69–70
    between agents, 20
    behavioral, 250
    as coordination mechanism, 250
    by message passing, 20–21
    and signals, 249–250
    stigmergic, 250–251
    vs. coercion, 249
    vs. stigmergy, 247
Comparative Study of Electoral Systems, 283
Competition
    inter-group, 278
    stigmergy in, 251–252
Competitive interference, 251
Competitive task architecture, 15
Complex social systems, 272
Complex systems, 55, 56, 64–68
    agent-based stem cell modeling, 393
    global perspective for, 8
    MAS for studying, 78
    simulation, 53–56
Computational agents, 35
Computer simulation, 3
Confidence Index, 293
Conflict, 251
Conflict resolution, shared state and, 98
Conflictual stigmergic coordination, 252
Connectionist architectures, 15
Constraint-based decision-making approaches, 110
Continuous environments, 18, 19

Continuous replicator dynamics (CRD), 237
- banana game, 238
- equilibrium solution, 238, 239

Continuous replicator equations, resulting dynamics, 237

Continuous Stochastic Logic (CSL), 145

Continuous strategy spaces, 535
- draining of pay-off streams in, 234–236
- evolutionary dynamics in, 232–236
- mutation as engine for diffusion in, 232–233
- non-isotropic mutations and evolutionary flows in, 234
- replicator dynamics in, 215–217, 226–228
- in robot soccer, 216
- strategic games with, 222–224

Continuous time, 22

Continuous-Time Markov Chains, 142

Control software, embedding, 197, 207

Control systems, in symbiotic simulation, 362

Cooperation, stigmergy in, 251–252

Cooperative interference, 251

Coordination, 535
- as adaptation to other agents' behaviors, 248
- collaborative and conflictual stigmergic, 252
- and cues of interference, 248–249
- evolution of indirect, 257–258
- stigmergic cues in, 243–245, 248–251

Correlates of War (COW) database, 282, 285

Countries, as complex social systems, 273

Country data, 272

Country databases, 272, 273
- limitations, 283–284

Country modeling
- and automated data extraction technology, 284–289
- catalog building issues, 288
- challenges, 271–274
- and cognitive agent modeling, 274–278
- coverage concerns, 288
- error rate testing, 288
- evidence reliability, 288
- existing, 282–284, 285
- and factions, 278–282

FactionSim testbed, 278–282
- and integrative knowledge engineering, 290–296
- keyword synonym issues, 288
- seamless integration issues, 288
- and social agents, 278–282
- and subject matter expert studies, 289–290
- and surveys, 289–290

Coupling, agents and environment, 24–26

Create scenario activity, 370–371

Crowd behavior modeling, 301–303, 319, 536
- future research directions, 321
- guidelines with situated cellular agents approach, 306–310
- pedestrian dynamics context, 303–306
- pedestrian modeling scenario, 310–315
- phases diagram, 307
- from SCA model to implementation, 315–319
- spatial infrastructure and active elements of the environment, 308–309

Crowd simulation, 268

Cues, 536
- of interference, 248–249

Cultural Simulation Model (CSM), 286

Cultural transmission, and memory, 62

Cumulative variables, 98

D

DARPA JFACC program, 120

DARPA RAID program, 122

Data collection, 272

Data distribution, 97–98

Data extraction, 272

Databases, country modeling with, 271–274

DAVINCI, 498–499

DDDAS, 361
- architecture for autonomous agents, 373, 374
- in housing scenario example, 381
- *vs.* symbiotic simulation, 359

Deadlock avoidance, in AGV transportation system, 174

Decision-making. *See also* Agent decision-making
- challenges for simulation-based, 111–112

nonlinearities in, 111
polyagents for, 109
in symbiotic simulation, 360, 361
Decision management capability (DECM-C), 379, 380
Decision models, 536
Decision support, 362
in disaster response, 425
Declarative goals, in Jason, 456
DECM-C, 379, 380, 381
Dedifferentiation, 536
Defection strategy, in Prisoner's dilemma game, 220
Delegate MAS, 115, 116, 536
Deliberation, 16, 17
Demographic processes, predicting by microsimulation, 57–58
Derivational decision-making approaches, 110
Desired acceleration, 336
Desired speed, 335
in ICM, 331
Desired time gap, 332, 336
Determined cell nodes, 402
Deterministic environments, 18
DEVS simulation, 62
flat, 520
DGensim, 87–89, 96
execution time measurement, 207
Differential diagnosis, 292, 295
Crusade example, 294
example, 296
in integrated knowledge engineering, 295
Differential equation system (DES), 8
Differentiated cells, 536
Diffusion, mutation as engine for, 232–233
Digital pheromone field, 114
Diplomatic, Informational Military, and Economic (DIME) actions, 273, 274, 279, 289, 297, 536
Direct communication, 250, 261, 262
limitations, 244
and MAS concept, 243
Direct payoff-matrix, 222, 223
Direction field plots
Battle of the sexes game, 229, 232
Matching pennies game, 229
Prisoner's dilemma game, 228
and Q-learning experience, 231

Disaster classification, 426
Disaster management systems, 426
Disaster needs, 425–426
Disaster prevention plans, simulation use in, 438
Disaster simulations, time sequence analysis, 445
Disaster-struck areas, effect of shape, 440
Disconfirming evidence, weighting, 293
Discrete decision modeling
lane change modeling, 338–339
traffic light approaches, 339–341
traffic simulation models, 338
Discrete event-based time, 22
Discrete event simulation, 62–63, 63
Discrete event simulation (DES) paradigm, 79
Discrete event simulation engines, 79
CHARON, 84
DES paradigm, 79
JAMES II, 80–81, 82
LEE toolkit, 80
MAML, 84
RePast, 81, 83
SeSAm, 81, 82
SIM_AGENT, 84
Simulation toolkit taxonomy, 85
standard operation, 80
Swarm, 83
toolkit survey, 79–84
Discrete strategy set, 536
Discrete strategy spaces
discrete-time replicator dynamics, 224–225
evolutionary dynamics and reinforcement learning, 230
evolutionary dynamics in, 228–232
game categorization and examples, 218–220
multi-population replicator dynamics in, 225–226
and normal form games, 217–218
and Q-learning, 230–231
replicator dynamics in, 215–217, 224–226
replicator dynamics with mutation in, 226
strategic games with, 217–222
Discrete time, 22, 24
Discrete time replicator dynamics, 224–225

Discretized environments, 18, 19
Distributed Artificial Intelligence (DAI), 4
    challenges for, 34–35
Distributed intelligence, MAS for studying, 78
Distributed Monitoring Vehicle Testbed (DVMT) project, 6. *See also* DVMT project
Distributed problem solving systems, 7
Distributed sensing system (DSS), 5
Distributed simulation, 97
    Interest Management (IM) in, 97–98
Distributed virtual environments (DVEs), 97
Domain related simulators, 39
Downward causation, 54
Driver-vehicle agents, 326, 328, 329–330, 536
    characteristics, 330
    IDM simulation, 333
Driving, as cognitive process, 330
Driving comfort measure, 345, 346
Driving styles, 335–336
Driving task models, 331
    inter-driver variability, 335–336
    intra-driver variability, 337–338
DVMT project, 6–7
Dynamic Data Driven Application Systems (DDDAS), 99, 358. *See also* DDDAS
Dynamic environments, 18, 536
    AGV transportation system simulation model, 181–185
    complexity of simulations, 168
    controller, 179
    dynamism in environment, 178–179
    environment structure, 177–178, 178
    extending modeling framework for, 209
    extending software architecture for, 210
    interaction law, 179
    manipulation of dynamism, 179
    MAS software modeling, 179–181
    modeling challenges, 177
    modeling constructs, 180
    modeling framework overview, 177
    modeling MAS in, 177
    reaction law, 179
    real-world execution time representation, 180
    simulation model, 181–185
    software-environment interactions, 180–181
    sources in warehouse environment, 183
    sources of dynamism, 179
    supporting perception in, 209
    supporting sources of dynamism in, 209–210
    in warehouse simulation, 182–183
Dynamic uncertainty, 110, 536
Dynamics, polyagents, 125
Dynamism, in self-organizing MAS, 134
DYNAMOD model, 63

E

Early agent models, 63
Earthquake management, 423, 424
Ecological models, 68
Embedding, in MASON toolkit, 408
Emergence, 536
Emergent properties, 38, 134, 135
    in ant colonies, 13
    in traffic simulations, 346
EMIL project, 69–70
Empirical target matter, 32
Engineering divergence phenomenon, 40
Entropy, and mutation, 231
Environment, 27
    in ABM, 10
    active elements, 308–309
    in agent-based stem cell modeling, 395, 397
    coupling agents with, 24
    discretized, 19
    as essential MAS component, 17
    extending distribution of simulated, 210
    in fitness models, 255
    in Jason toolkit, 459–461
    lattice model, 302
    modeling, 18–19
    modeling temporal evolution, 23
    pedestrian modeling scenario, 312
    in polyagent model, 115–116
    registration of traces by, 257
    role in MAS, 67–68
    role in self-organizing systems, 136–137
    in software systems, 169–170
Environment module, 26

and need for VR, 33–35
Environmental agents
　characterization of, 141
　defined in plain diffusion system, 148
　in legacy systems, 138
　for self-organizing MAS, 138
Environmental impacts, transportation systems, 325
Environmentally-mediated interactions, 112–113
EOS testbed, 62
Episodic environments, 18
Equation-Based Modeling (EBM), 9
Equidistant time steps, in social science simulations, 64
Equilibrium distance, 334
Equilibrium gap, 333
Equilibrium properties, in IDM, 333
Equilibrium traffic, 333
Equilibrium velocity, 334
Equilibrium velocity-density relations, 335
Ethnic Conflict and Civil Life (ECC), 283, 285
Evaporation service, in pheromone-bases systems, 137
Event-based time, 22, 26
Event databases, 272, 282
Event handlers, 79
Evidence tables, 292
　positive and negative, 287
Evolution, understanding stigmergy through, 253–260
Evolutionary computation, 116
Evolutionary dynamics
　analysis in game categorization, 228–229
　of continuous replicator equations, 237
　in continuous strategy spaces, 232–236
　in discrete strategy spaces, 228
　and draining of pay-off streams in strategy space, 234–236
　of multiple agents playing strategic games, 214
　and Q-learning, 230
　and Q-learning experiments, 231
　relationship to reinforcement learning, 230
Evolutionary flows, with non-isotropic mutations, 234

Evolutionary game theory (EGT), 215, 216, 536
Evolutionary stable strategies, 221–222
Execution analysis, 28
Execution time
　measurement, 207–208
　specification, 208–209
Execution tracker, 199–201
Experimental frame
　model accommodation to, 37–38
　in Zeigler's framework, 29
Experimentation in the large, 35
Experimentation in the small, 34
Expert surveys, 272
　country modeling with, 271–274
Explicit signals, 252
Exponential distribution, 142
Extensibility requirement, for symbiotic simulation, 372
External influences, 536
　in symbiotic simulation, 360
External triggering, in symbiotic simulation, 370
EyeSim simulation, 208–209

F
Fact-based agents, 53
Factions, 278–282
　identification, 283
　profiles, 279–280
FactionSim testbed, 278–282, 297, 536
　models and components, 279
Factory scheduling, with polyagents, 118–119
Federation
　in HLA, 87
　node structure in HLA_RePast, 92
Federation object model (FOM), 87
Feedback loop, in A&A meta-model, 138
Field plots. *See* Direction field plots
Fields, signaling with, 21
FIPA compatible middleware, 92
FIPA-compliant tools, 407
Firefighter situation assessment, 362
Fitness models
　baseline individual model, 254–255
　hungriness simulation, 259, 260
　social simulation, 257–258
Fitzhugh-Nagumo model, 237
Fokker-Planck equation, 233

Food neophobia, 254
Forecasting systems, in symbiotic simulation, 363
Formal tools, 142
    for developing self-organizing MAS, 133–136
Forms, in SeSAm, 494, 496
Frame problem, 25
Framework for M&S, 28. *See also* Zeigler's Framework for modeling and simulation
FREESIM traffic simulator, 342
Frustration effect, in drivers, 338
Fully actual systems, 33
Fully theoretical systems, 33
Functional components, 536
Functionally accurate, cooperative distributed systems (FA/C), 6

G

Gaia methodology, 139, 161
Game categorization, evolutionary dynamics of, 228–229
Game theoretic decision-making approaches, 110
Game theory concepts, 214, 217, 278
    evolutionary stable strategies, 221–222
    Nash equilibrium, 220
    normal form games, 217–218
    Pareto optimality, 220–221
    Q-learning in games, 230–231
    strategic games with discrete strategy sets, 217–222
Gateway federates, 92
Gauntlet routing problem, 121
General Inquirer, 286
General purpose frameworks, 315
General sum games, 218
Generic simulators, 39
Genetic networks, evolution of, 234
Gensim, 87–89
Ghosts, 113, 537
    in BEE process, 123
    emission by avatars, 118
    integration of information from multiple, 121
    in polyagent model, 114–115
Global Barometer Surveys (GBS), 283, 285
Global Terrorism Database, 285
Goal adoption, 251

Goal driven behaviors, 305
Goal trees, 275, 294
Goals, in Jason, 453
Groups, parameters for social system models, 281
GSP, 537

H

Hanshin-Awaji earthquake, 423, 425
HBM, 537
Hematopoiesis control mechanisms, 391–392
Hematopoietic stem cell, 389, 391, 537. *See also* HSC biology
Hematopoietic system, 391
Hetero-adjustment, 537
High Level Architecture, 87, 537
    Ambassador abstraction, 88
HLA Jade, 02–93, 92–93
HLA_Agent, 89–90, 96, 98
    running SIM-TILEWORLD testbed simulation, 91
HLA_Jade, 96
HLA_RePast, 90–92, 96, 98
Housing scenario predictions, SSADS example, 381
HSC biology, 390–391
    hematopoiesis control mechanisms, 391–392
    hematopoietic system, 391
Human behavior modeling (HBM), 537
Human-behavior-modeling (HBM), 271
Human-inspired behaviors
    and movement in social field, 55
    simulating, 11–12
Human terrain systems, 272
Human Terrain Tool (HTT), 273
Hungriness simulation, 259, 260
Hybrid symbiotic simulation systems, 364
Hybrid symbolic-subsymbolic reasoning, 125

I

Implicit signals, 250, 258
Inaccessible environments, 18
Indirect communication, 245, 250
Indirect coordination, 243, 244
    evolution, 257–258
Indirect interactions, in self-organizing MAS, 137

Individual-based modeling, 7–9
    microsimulation approach as forerunner, 9
Influence/Reaction model, 38, 42, 537
Influence theory, 249
Information fusion, with polyagents, 126
Insect pheromone fields, 112
Institutions, parameters for social system models, 281
Instrumentation, for MAS, 98–99
Integrated Development Environment (IDE), 276, 537
Integrative knowledge engineering, 290–296
Intelligent Driver Model (IDM), 331–335
    equilibrium distance, 334
    equilibrium velocity-density relations, 335
Intelligent transportation systems (ITS), 325
Intention ants, 120
Inter-driver variability, 326, 335–336, 537
Inter-vehicle communication, 327, 350, 352
    store-and-forward strategy, 349–351
Interactions, 2
    between agents, 20–22
    communication by message passing, 20–21
    complexity of micro-level, 8
    environmentally-mediated, 112–113
    pheromone-based, 115
    between predator and prey, 8
    role in MAS, 67
    using signals, 21–22
Interactive decision making, 215
Interest management (IM), 97–98
Interference, 240, 537
    cooperative and competitive, 251
    cues of, 248–249
Internal actions, in Jason, 455, 462–463
Internal triggering, in symbiotic simulation, 370
Interoperability, and HLA, 87
Interpreter, in Jason, 456–459
Interviews, subject matter expert difficulties, 289
Intra-driver variability, 326, 337–338, 537
Intra-state conflict and Interventions (ICI), 283, 285
Intra-state conflict datasets, 282
Iterated coupled maps, 329

Iterative approaches, self-organizing MAS, 140

J
JACK, 452
JADE agent toolkit, 373, 384, 452, 453
Jadex, 452
JAMES II toolkit, 80–81, 85, 93–94, 420–421
    abstract simulator, 519
    combined simulator, 520
    data sink, 512
    declared *vs.* implemented ports, 523
    equipping nodes with alternative user models, 526–528
    event types definition with XML Schema, 522
    example application, 524–528
    excavation times, 529
    experiences and interpretations, 530
    experiment definition, 513, 528–530
    flat simulator, 520
    interface model definition, 523
    MANET experiment in XML, 528
    model class, 512–513
    model definitions in XML, 522
    modeling and simulation, 509–510, 514–524
    for multi-agent modeling and simulation, 514
    outlook, 530–531
    packages and relationships, 513
    PdynDEVS models, 514
    PdynDEVS protocol, 519
    PdynDEVS simulator pseudo-code, 520
    permanent and temporary sub models, 518
    Registry module, 513
    representing models, 521, 523–524
    role of time, 528–530
    service model, 516
    service trading model for mobile ad-hoc networks, 525
    simulation algorithms, 521
    simulator, 513
    user interface, 512
JAMES simulation platform, 23
Jason toolkit, 420, 511
    agent-environment relationship, 464
    agent organizations in, 472

annotations, 454
asynchronous execution mode, 461
belief base in, 457
class diagram for environment implementation, 466
customized architectures, 463–464
declarative goals in, 456
environments in, 459–461, 472
example agents, 467–471
example application, 464–471
example environment programming, 460, 465–466
execution modes, 461–462
features for simulation, 459–464
internal actions in, 455, 462–463
interpreter, 456–459
language constructs, 453–456
Mind Inspector, 462
ongoing projects, 471–472
plan library in, 457–458
programming MAS using, 453–459
reputation agent, 469
set of events in, 457
set of intentions in, 458
Java-based Agent Modeling Environment for Simulation (JAMES) II, 80–81. *See also* JAMES II toolkit
Java Virtual Machine (JVM), 92

K
Kansas Event Data Systems (KEDS), 286
Key indicators, 537
    in symbiotic simulation, 369
KQML, 20

L
Lane-changing behaviors, 327
    incentive criterion, 339
    velocity differences, 339
Language evolution, 69
Latent Energy Environment (LEE) toolkit, 80. *See also* LEE toolkit
Layers, in symbiotic simulation framework, 376–377
Leaders, and country modeling, 296
LEE toolkit, 80, 85
Legacy systems, 138
Levels concept, 64
Local causality constraint, 86
Logical processes (LPs), in PDES, 86

Low-dimensional paradigm, 78

M
MABS, 36–37
MACE, 7
Macro level, linking with micro level, 38
Macroscopic traffic modeling, 327–329, 538
MAML toolkit, 84, 85
Man-made disasters, 426
MANET model
    alternative user models for nodes, 526–528
    composition structure, 525–526
MANTA project, 13, 59
Marks, 21
MAS principles, 13
    agent architectures, 14–16
    agent behavior modeling, 16–17
    agents, 13–14
    environment, 17–19
    incompleteness of, 28
    interactions, 20–22
    MAS as three correlated modeling activities, 26–28
    time modeling, 22–26
MAS simulations
    accommodations to experimental frame, 37–38
    behavior module issues, 35–36
    environment module issues, 33–35
    issues using Zeigler's framework, 41
    merging participatory design and VR, 36–37
    model representation of source system, 32–37
    simulator correctness, 39–40
    using Zeigler framework, 32–40
MAS toolkits, 77
MASON toolkit, 408–409
    advantages, 403
    cell class UML diagram, 411
    cell type UML diagram, 411
    implementation of CELL, 409–410
    implementation of Roeder-Loeffler model, 410–414
    world state UML diagram, 412
Matching pennies game
    direction field plot, 229
    strategic form, 219–220
MATSim, 326

INDEX 555

Maximum acceleration, in IDM, 331–332
Memory, agent-based, 61
Memory effect, and intra-driver variability, 337
Memoryless property, 142
MESS simulation, 208
Message passing, 20–21
Metabolic pathways modeling, 394
Methodological issues, self-organizing systems, 139–140
Micro level
    linking with macro level, 38
    modeling with ABM, 9
Microscopic traffic simulation, 327–329, 538
Microscopic traffic simulation software, 341–342
    numerical integration, 343–344
    simulator design, 343
    visualization, 344–346
Microsimulation approach, 9, 63, 538
    demographic processes prediction, 57–58
Minimizing Overall Braking Induced by Lane Changes (MOBIL), 339. *See also* MOBIL meta-model
Minimum distance, 336
    in IDM, 332
Minorities at Risk (MAR), 283, 285
MOBIL meta-model, 327, 339
Mobile ad-hoc networks (MANETs), 524
    model composition structure, 525–526
Mobility, desire for personal, 325
Mode split, 326
Model, in Zeigler's framework, 29–30
Model calibration, 363
Model checking, 135, 143, 160
    with MTBDD, Sparse, and Hybrid, 145
    with PRISM tool, 144
Model classification, 363
Model design, 28
Model execution, 28
Model management capability (MODM-C), 377. *See also* MODM-C
Model validation
    in SSFS workflow, 367
    in symbiotic simulation, 363–364
Modeling and simulation (M&S)
    agent-based modeling approach, 9–11

    agent-based social simulation (ABSS), 11–12
    ant colonies, 13
    artificial animals, 12–13
    CNET simulator, 5–6
    DVMT project, 6–7
    flocks and ants, 12–13
    MACE, 7
    for MAS, 4, 5–7
    MAS for, 7–13
    microsimulation approach, 9
    need for individual-based, 7–9
    plain diffusion, 147–148
    Reynolds's boids, 12
    Zeigler's framework, 29–31
Modeling phase, 140
    plain diffusion, 147–148
    PRISM tool in, 144
    self-organizing MAS, 141–142
Modeling relation, 31, 538
Modeling specifications, sharing efficiently, 40
Models, unfolding, 64–68
MODM-C, 377, 383, 384
Modularization, in symbiotic simulation, 375–376
Multi-agent based simulation (MABS), 407
Multi-Agent Computing Environment (MACE), 7. *See also* MACE; MAML toolkit
Multi-agent control systems, 170–171. *See also* Multi-agent systems
Multi-agent embryogeny, 33
Multi-agent evolution and learning, 216
Multi-Agent Modeling Language (MAML) toolkit, 84
Multi-Agent System architectures, 78–79
Multi-Agent System Simulator (MASS), 23
Multi-agent systems, v, 407
    approach to complex systems simulation, 302–303
    characteristics, 170–171
    correlation with EGT, 216
    and crowd behavior modeling, 301
    developing self-organizing, 133–136
    development of software MAS, 78–79
    discrete event simulation engines for, 79–85
    flexibility of embedding, 198–199
    instrumentation and visualization for, 98–99

as metaphor, 268
non-discrete nature of, 226
programming languages for, 452–453
programming with Jason, 453–459
role of software architecture for simulating, 167–169
schematic view, 171
for simulation, 267–269
simulation for, 107–108
software integration challenges, 168
software modeling, 179–181
symbiotic simulation for, 99
vehicular traffic, 327
Multi-agent systems (MAS)
application perspective, 53
basic principles, 13–28
and CNET simulator, 5–6
and direct communication, 243
execution problems, 2
modeling by MABS, 37–38
for M&S, 4, 7–13
M&S for, 5–7
one-shot simulator problem, 39
predecessors and alternatives, 56–64
as programming paradigm, 4
replication problem, 39–40
role of environment, 67–68
role of interactions, 67
scheduling, 24–26
simulating as three modeling modules, 26–28
and simulation, 3–5
simulation engines for, 77–78
for studying complexity, 78
for studying distributed intelligence, 78
Multi-agent world, 14
Multi-population replicator dynamics, 225–226
Multi-state games, 240
Multilayered environment specification, 320
Mutation
continuum limit, 232
as diffusion engine in continuous strategy spaces, 232–233
and entropy, 231
and evolution of genetic networks, 234
non-isotropic, 234
replicator dynamics with, 226
Mutualism, 358, 361

N

Nash equilibrium, 220, 228
  of prisoner's dilemma, 221–222
Natural disasters, 426
Negative evidence, 287
Negative interference, 249
Nesting, 64–68
Netlogo simulation applet, 315
Network delays, with DGensim, 88
Network metamanagement agents, 78
Neumann boundary condition, 233
Neural networks, 255, 256
  in LEE toolkit, 80
  and stigmergic communication, 256
NEW TIES project, 69
Newsfeeds, 273
  country modeling with, 271–274
  webscraping, 272
Non-associative reinforcement learning tasks, 230
Non-isotropic mutations, 234
Nondeterministic environments, 18
Nonlinear dynamical system, 113, 237
Normal form games, 217–218

O

OASYS, 286
Observe, orient, decide, and act (OODA) loop, 275, 538
Observe physical system activity, 369
On-line simulations, 358
  vs. symbiotic simulations, 359
One-shot simulators, 39
Online adaptation, 97
OODA, 538
Opinion polls, 272
Ordinary differential equations (ODE), in stem cell modeling, 394
Organizational Risk Analyzer, 286
Outsourcing decisions, symbiotic simulation example, 380–381

P

Paced mode, 80–81
Packet-World simulation, 209
PageRank, 538
Paradigmatic validity, 37
Parallel discrete event simulation (PDEVS), 86–87, 93
Parallel engines, scalability issues, 96

INDEX

Parallel simulation engines, 86
- Charon, 95–96
- DGensim, 87–89
- Gensim, 87–89
- High Level Architecture, 87
- HLA JADE, 92–93
- HLA_Agent, 89–90
- HLA_RePast, 90–92
- parallel discrete event simulation, 86–87
- SPADES, 94–95
- taxonomy, 96
- toolkit survey, 87–96

Parameter analysis, 125
PARAMICS, 342
Parasitism, 361
Pareto dominance, 221
Pareto optimality, 220–221
Partial differential equation (PDE), 227
- diffusion type, 234
- in stem cell modeling, 394
- in strategy space and time, 217

Participatory design, 35–36, 538
- merging with VR, 36–37
- with SeSAm, 477–478

Particle-based crowd model, 538
Particles, pedestrians as, 303–304
Passive abstractions, 315
Pay-off streams, 221
- draining in strategy space, 234–236

PDEVS simulator tree, 93
Pedestrian dynamics, 301, 303, 319, 538
- pedestrian movement, 304
- pedestrians as autonomous agents, 305–306
- pedestrians as particles, 303–304
- pedestrians as states of cellular automata, 304–305

Pedestrian modeling scenario, 310–311
- agent attitudes, 312
- data acquisition, 321
- environment, 312
- higher level linguistic support, 316
- immobile active elements, 311
- methodological issues, 321
- methods, 318
- modeling assumptions, 311
- passengers, 312–313
- psycho/sociological considerations, 321
- relevance of perceived signals, 313
- screenshots, 314
- simulation results, 313–315
- validation techniques, 321

Pedestrian movement, 304
- sample effect, 305

Pedestrians, 309–310
- as autonomous agents, 305–306
- as particles, 303–304

Perception agents, 88, 89
Perception-deliberation-action, 16, 17
Performance moderator function (PMF), 274
Personality traits, output from automated web-scraping tools, 287
Pheromonal communication, 246, 538
- as explicit signals, 252
- non-stigmergic nature of, 252–253
- role of stigmergy in evolution, 253

Pheromone-based systems, 145
- evaporation service in, 137

Pheromones, 538
- ant agents, 21
- evaporation of, 114
- insect construction of, 112
- as probabilities, 125
- propagation of, 115
- use by ants, 13
- vs. stigmergic cues, 252–253

Physical system observation, 361
- in symbiotic simulation systems, 369

Plain diffusion, 136
- average value probability, 155
- base rate parameter, 148
- instance evolution, 151–153
- instantaneous global rate, 156
- modeling, 147–148
- node probability values, 159–160
- performance evaluation, 155
- preliminary scalability analysis, 158
- probability distribution, 151, 154
- probability values, 156, 157
- problem statement, 146–147
- reference network topology, 146
- self-organizing MAS case study, 145–146
- tuning, 154–157
- verifying, 150–154

Plans
- applicable, 454
- in Jason, 453

Player/Stage/Gazebo simulation platform, 206
    execution time measurement, 207
Plugins, SeSAm toolkit, 491
PMESII, 538
PMF models, 275–278, 538
PMFserv, 273, 276, 297, 539
    deployment, 276
    implemented theories, 277
Political, Military, Social, Economic, Informational, and Infrastructure (PMESII) effects, 273, 274, 279, 289, 297, 538
Political Instability Task Force (PIT), 283, 285
Politico-economic indicators, 272
Polyagent architecture, 113–114, 116
    avatars in, 114
    and delegate MAS, 115
    environment in, 115–116
    ghosts in, 114–115
Polyagent model, 111
    architecture, 113–115
    and continuous short-range prediction, 113
    and environmentally-mediated interactions, 112–113
    related work, 116–118
    and simulation-based decision making challenges, 111–112
Polyagents, 108, 539
    for agent decision-making, 109–111
    applications, 118–124
    dynamics, 125
    factory scheduling and control applications, 118–119
    future research, 124–126
    hybrid symbolic-subsymbolic reasoning, 125
    information fusion applications, 126
    and multi-agent paradigms, 116–117
    new applications, 126
    parameter analysis, 125
    pheromones as probabilities, 125
    prediction applications, 122–124
    and simulation, 117–118
    swarming over task networks, 126
    terminology, 118
    theoretical opportunities, 124–125
    universality, 125
    vehicle routing applications, 119–122
PolynDEVS protocol, 519
Portability, in MASON toolkit, 408
Positive evidence, 287
Positive interference, 249
Practical behavior
    baseline individual model, 254–255
    evolution of, 256–257
    stigmergic cues and, 247–248
    traces, 257, 262
    work and traces of, 247–248
Prediction
    continuous short-range, 113
    polyagent applications, 122–124
    in symbiotic simulation, 370–371
Prediction horizon, 539
Predictive validity, 30
PRISM Probabilistic Symbolic Model Checker, 144. *See also* PRISM tool
    higher-level language atop, 161
PRISM tool, 144
    modeling phase, 144
    simulation phase, 144–145
    strategy specification, 149
    tuning phase, 145
    verification phase, 145
Prisoner's dilemma game
    defecting as dominant strategy, 220
    direction field plot, 228
    Nash equilibrium of, 221–222
    strategic form, 219
Probabilistic Computational Tree Logic (PCTL), 145
Probability distribution, plain diffusion case study, 154, 157
Process irreducibility, 110
Profiler+, 286
Progenitor cells, 539
Programming languages, for MAS, 452–453
Protest and Coercion (PCD), 283, 285
Pseudo code
    PolynDEVS simulator, 520
    traffic simulator design, 343
Public events management, 302

Q

Q-learning
    Battle of the sexes plots, 232
    dynamics in games, 230–231
    and evolutionary dynamics, 230

multi-agent, 216
Q-learning experiments, 231

R

Reactive architectures, 14–15
ReadMe, 286
Real-map experiments, 438–441
Real-time script, 122
Real-time traffic visualization, 344–345
Real-world execution time, modeling in simulation model, 180
Real-world living entities, 35
    difficulty of representing, 42
Recursive Porus Agent Simulation Toolkit (RePast), 81, 83. *See also* RePast toolkit
Recursive simulation, 117
Reference models, 539
Reinforcement learning, 539
    relating evolutionary dynamics in discrete strategy spaces with, 230
Reliability
    of automatically extracted evidence, 288
    and differential diagnosis, 292–293
    subject matter experts, 289
RePast toolkit, 12, 81, 83, 85, 316, 511
    GUI, 83
    simulation model screenshot, 315
Replication problem, 39–40
Replicative validity, 30, 36
Replicator dynamics, 108, 224, 539
    in continuous strategy spaces, 226–228
    in discrete and continuous strategy spaces, 215–217
    in discrete strategy spaces, 224–226
    discrete-time, 224–225
    and evolutionary dynamics in continuous strategy spaces, 232–236
    and evolutionary dynamics in discrete strategy spaces, 228–232
    and game theory concepts, 217–223
    multi-population, 225–226
    with mutation, 226
    and Q-learning, 230
    resulting dynamics of continuous replicator equations, 237
Request time notify, 94
Rescue management needs, 425–426
Rescue scenarios, 427

Resource class descriptions, SeSAm toolkit, 486
Resource description Framework Extractor (T-Rex), 286
Resource distribution, with MAS, 108
Results analysis, in symbiotic simulation, 371–372
Reynolds's boids, 12
Rich media export, in MASON toolkit, 408–409
Ritualization, 253
RoboCupRescue project, 420, 423–425, 539
    2004 Challenge Game, 436, 444
    ABSS analysis based on probability model, 441–447
    ABSS architecture, 447–448
    ABSS assumptions, 442
    administrative boundaries map, 440
    agent ability and map dependency, 436–437
    agent behavior formulation and presentation, 441–444
    agent competitions lessons, 434–437
    agents, 429
    agents' behavior, 428
    architecture, 427–433, 428–430
    burning rate changes, 437, 439, 447
    changes in rescue performance, 436
    competition logs analysis results, 444–447
    component simulators, 429
    consistency with other methods, 439–440
    disaster and rescue management needs, 425–426
    disaster simulations, 428
    effects of shape of disaster-struck areas, 440
    eigenvalue comparisons, 445
    fire brigade agent action patterns and state transition, 443
    GIS module, 429
    header and use of commands between kernel and agents, 433
    ignition points relative to estimated earthquake, 441
    inter-module communication, 431–432
    kernel module, 430
    Kobe simulation snapshots, 437
    lessons learned, 434–438

normalized team scores, 435
number of agents and parameters of disaster situations, 435
properties analysis, 443
protocol communication during simulations, 431
real-map experiments, 438–441
and real-world applications, 437–438
rescue activities simulations, 428
rescue agent behavior model, 442
rescue scenarios, 427
rescue simulation and soccer simulation, 447
robustness to changing environments, 435–436
search and rescue operations in unknown areas, 434–435
simulation period, 427
simulation progress, 430–431
simulation results assessment, 448
target sizes, 427
TCP/UDP protocol, 432–433
time sequence analysis of disaster simulations, 445, 446
time sequence of burning rate, 447
and validity of agent-based social simulations, 438–441
viewer modules, 429
world model and representation, 431–432
RoboCupRescue simulation (RCRS), 423
RoboCupSoccer simulation, 420
Robot soccer, 216–217
Robotic vehicles, path planning for, 119
Robustness, in ant colonies, 13
Roeder-Loeffler model of self organization, 406–407
implementation in MASON, 410–414
Role assignment, to team members, 66
Route planning, 122
Routing, in AGV transportation system, 174
Rule-based agents, 53
modeling language relying on, 12
Rules of action, 142
Runtime infrastructure (RTI), 87

S

Safe distance strategy, 331
Safety time gap, in IDM, 334

SCA models
    AbstractAgent class methods, 318
    implementing, 315–316
    screenshots, 319
    simplified class diagram, 317
    supporting and executing, 316–319
Scalability, 97
    of parallel engines, 96
    plain diffusion case study, 158
    in symbiotic simulation, 372–373
SCEM-C, 377, 380, 381, 382, 383
Scenario creation, 384
    in symbiotic simulation, 370–371
Scenario management capability (SCEM-C), 377. See also SCEM-C
Scheduling, 27
Scheduling module, 27
Scope and resolution, 55
Search and rescue operations, in unknown areas, 434–435
Self-adaptation, in ant colonies, 13
Self-adjustment, 539
    stigmergic, 250–251
Self-organization, 38, 108, 133
    adaptive characteristics, 134
    in ant colonies, 13
    autonomy in, 134
    dynamic characteristics, 134
    historical definition, 134
    organization in, 134
    Roeder-Loeffler model, 406–407
    and stigmergic communication, 246
    in traffic dynamics, 328
Self-organizing MAS
    A&A meta-model, 136–139
    architectural pattern, 138–139
    methodological approach for engineering, 140–144
    methodological issues, 139–140
    model checking techniques, 160
    modeling phase, 140, 141–142
    non-linear dynamics, 135
    plain diffusion case study, 145–158
    PRISM tool for, 144–145
    role of environment in, 136–137
    simulation and formal tools for developing, 133–136
    simulation phase, 140, 142
    time to dynamic equilibrium, 160
    tuning phase, 140, 143–144

INDEX

verification phase, 140, 143
Self-organizing systems, 539
Semantic Web agents, 78
Sematectonic communication, 247
Semiconductor manufacturing
    SSCS in, 362, 380–381
    symbiotic simulation in, 358
Sensor capabilities, 379, 381, 382, 383
    in symbiotic simulation, 377
SeSAm toolkit, 81, 85, 420, 511
    abstract simulator, 419
    accessibility and documentation, 498
    action primitives, 483
    activity node shapes, 497
    agent-based simulation model design, 479
    agent class descriptions, 486
    agent playing for advanced participation, 501, 503
    agent update, 489–490
    aggregated data, 500
    animation window example, 500
    basic model representation, 483–488
    behavior description, 487, 497
    behavior specification, 496, 498
    body variables specification, 496
    boids simulation/animation, 82
    calibration and experimentation, 479
    class descriptions, 486
    computational primitives, 483–484
    computer simulation model implementation, 479
    conceptual model development, 494
    core modeling and simulation environment, 482–492
    data types, 483–485
    DAVINCI for experimenters, 498–499
    declaration of model instrumentation, 488
    declaration of situations, 487–488
    dynamic reasoning, 483
    edit list elements, 496
    experiences with, 503–504
    experiment declaration, 488
    experiment description level, 488
    experiment scripting, 498–499
    flat simulator, 520
    forms and tables, 494, 496
    future work, 504–505
    GUI, 82
    interpretation cycle, 489
    language building blocks, 484
    meta-level characterization, 483
    model level, 483–488
    model-specific interfaces, 501, 502
    node model structure, 427
    novice experiences, 503–504
    observing and controlling activity, 479
    observing and immersive testing, 479
    online aggregated data presentation and animation, 499
    outcomes and update sequences, 490
    output interpretation, 479
    plugin mechanism for extension, 491
    predefined features, 486
    primitive call specification, 494, 495
    primitives, 483–485
    principles, 493–494
    problematic language details, 491–492
    resource class descriptions, 486
    sensor primitives, 483
    simulation GUI, 471
    simulation routine and model interpretation, 488
    simulation run definition, 488
    simulation study and user roles, 478–482
    simulator information, 489
    start situation configurations, 498
    static structures, 483
    structural building blocks, 485
    structural description, 485–487
    suitability aspects, 492
    synchronous execution mode, 461
    tasks and user roles, 479–480
    tell messages semantics, 469
    tools for agent-based simulation, 481–482
    triggering events, 454
    update scheme properties, 490–491
    user functions, 483–485
    user involvement, 480
    users and tasks, 493
    users knowledgeable in implementation only, 504
    variables grouped as features, 486
    visual programming and participatory simulation with, 477–478, 494
    world class descriptions, 486
Shared state, 97–98

and conflict resolution, 98
Shell for Simulated Agent Systems (SeSAm), 81. *See also* SeSAm toolkit
Signals, 21–22, 249
    by cell agents in agent-based stem cell modeling, 399–400
    and communication, 249–250
    implicit, 250, 258
    relevance of perceived, 313
    traces as, 260
SIM-AGENT toolkit, 84, 85, 89
SIM-TILEWORLD testbed simulation, 90, 91
SIMA-C, 377, 380, 382, 383
SIMM-C, 377, 383
Simulated escalation scenarios, 362
Simulation, v, 2, 107–108
    cellular automata, 58–62
    combining with formal tools, 133–136
    complex systems, 53–56
    as computational tool, 4
    discrete events, 62–63
    of dynamic environments, 168
    MAS for, 267–269
    and multi-agent systems (MAS), 3–5
    plain diffusion, 148–150
    role of software architecture, 167–169
    self-organizing MAS, 133–136
    voting behavior, 1960s, 57
Simulation analysis, 371
Simulation analysis (SIMA-C) capability, 377. *See also* SIMA-C
Simulation-based decision-making, challenges, 111–112
Simulation engines, 77–78
    in agent-based stem cell modeling, 395, 396, 397–398
    behavior in agent-based stem cell modeling, 400
    discrete event type, 79–85
    issues for future research, 96–99
    MAS architectures for, 78–79
    parallel simulation engines, 86–96
Simulation interoperability, 87
Simulation management capability (SIMM-C), 377. *See also* SIMM-C
Simulation phase, 140
    plain diffusion, 148–150
    PRISM tool in, 144–145
    self-organizing MAS, 142

Simulation relation, 30–31, 31, 539
Simulation updates, 371
Simulator, in Zeigler's framework, 30
Simulator correctness, 39–40
    in Zeigler's framework, 30–31
Simultaneous actions, modeling, 38
SITSIM, start and final states, 60
Situated agents crowd model, 539
Situated cellular agents, 306, 307. *See also* SCA models
    crowds modeling with, 306–310
    interaction mechanisms, 307
    in stem cell modeling, 394
Situatedness, 268
    of agents, 10
Situation awareness, decreases in, 261–262
Sociability, of agents, 10
Social agent environment, 273
Social agents, 278–282
Social dynamics, 59
Social force model, 304
Social insects, 244
Social models, 78
    parameter types, 281
Social science simulations, equidistant time steps in, 64
Social simulations, 2, 55, 78
    fitness model, 257–258
    in Jason, 471
    results, 258
    and stigmergic communication, 258
    validity of agent-based, 438–441
Socio-cognitive agents, 278
Socio-cultural games, 278
Socionics, 539–540
SODA methodology, 139
Software architecture, 540
Software engineering
    and environments, 169–170
    simulation in, 135
Software-environment interactions, 180–181
Software execution time, modeling in simulation model, 180
Software-in-the-loop simulation mode, 172, 208, 540
Software MAS, 78, 108
Software MAS simulations, 167–169
    AGV simulator evaluation, 201–205
    AGV transportation system, 172–176

background, 169–172
embedding control software, 207–209
future research directions, 209–210
and MAS characteristics, 170–171
modeling in dynamic environments, 177–185
related work, 205–209
simulation platform architecture, 185–201
and software-in-the-loop simulation mode, 172
special-purpose simulation platforms, 205–207
system and environment, 169–170
Source system
accurate representation of, 32–37
model variations, 33
in Zeigler's framework, 29
Space abstractions, for SCA model, 308
SPADES toolkit, 94–95
architecture, 94
execution time measurement, 207–208
SPARK simulation platform, 205–206, 452
Spatial infrastructure, in crowds modeling, 308
Spatial irreducibility, 110
Special-purpose simulation platforms, 205–207
Speech act theory, 20, 244
Speed limit impacts, 327, 347–349
SSADS, 361, 364, 365
scenario creation in, 370
simulation example, 381–382
trigger conditions, 369
workflow, 368, 369
SSCS, 361, 362, 365
scenario creation in, 370
simulation example, 380–381
trigger conditions, 369
workflow, 366–367, 367
SSDSS, 361, 362, 365
example simulation, 379–380
scenario creation in, 370
trigger conditions, 369
workflow, 366, 367
SSFS, 361, 363, 365
simulation example, 382–383
in WIPER system, 382
workflow, 367, 368
SSMVS, 361, 363–364, 365

scenario creation in, 370
simulation example, 382–383
trigger conditions, 369
in WIPER system, 382
workflow, 368
Stakeholder approach, 540
Standards Tree, 275
STAPLE, 452
StarLogo, 407
State space explosion problem, 161
and limitations of model checking, 135, 143
State updates, in RePast, 91
Stateless games, 240
Static environments, 18
STEAM, 452
Stem cell biology, 390–392, 540
Stem cell division, 403
Stem cell fate, 390, 393
Stem cell modeling. *See also* Agent-based stem cell modeling
agent-based, 389–390
Stem cell niche, 392, 401, 413, 540
Stigmergic communication, 245, 250–251
baseline individual model, 254–255
evolution, 258–260
and evolution of practical behavior, 256–257
and evolution of stigmergic self-adjustment, 257–258
and indirect coordination, 257–258
neural network, 256
results of baseline simulation, 256
results of hungriness simulation, 260
social simulation results, 258
vs. pheromonal communication, 252–253
vs. stigmergic self-adjustment, 259
Stigmergic coordination, collaborative and conflictual, 252
Stigmergic cues, 250, 258
derivation of pheromones from, 253
evolutionary approach, 243–245
operational model, 245
as practical behavioral traces, 247–248
uses in coordination, 243–245
vs. pheromones, 252–253
Stigmergic self-adjustment, 245, 250–251
evolution of, 257–258
vs. stigmergic communication, 259

Stigmergy, 38, 108, 115, 136, 540
    as communication, 244
    in cooperation and competition, 251–252
    false definitions, 245–247
    future research, 260–261
    role in eliciting additional work, 247
    role in evolution of pheromonal communication, 253
    understanding through evolution, 253–260
    uses in coordination, 248–251
    vs. communication, 247
Stochastic differential equations (SDE), in stem cell modeling, 394
Stochastic model checking, 540
Stop-and-go waves, 327, 346–347
Store-and-forward strategy, inter-vehicle communication, 349–351
STORY, 286
Strategic games
    Battle of the sexes, 219
    categorization and examples, 218–220
    with continuous strategy spaces, 222–224
    with discrete strategy sets, 217–222
    and evolutionary stable strategies, 221–222
    Matching pennies, 219, 220
    and Nash equilibrium, 220
    normal form games, 217–218
    and Pareto optimality, 220–221
    Prisoner's dilemma, 219
Structural validity, 30
    multi-agent embryogeny, 33
Subject matter expert studies, 273, 289–290
Sugarscape, 61, 407
SUMO traffic simulator, 342
Support networks, evolution of, 59
Surveys, and country modeling, 289–290
Swarm toolkit, 83, 85, 407–408
Swarming, over task networks, 126
Switching dynamics, 240
Symbionts, 357
Symbiotic simulation, 99, 268, 540
    activities, 369–372
    actuation layer, 376, 378
    agent-based approach, 374–375
    agent-based generic framework, 357–359, 372–378
    analyze results activity, 371–372
    analyze simulation activity, 371
    applicability requirement, 372
    applications, 358, 378–383
    capability-centric solution for framework architecture, 375–376
    classification overview, 365
    concepts, 359–361
    conceptual showcases, 379–383
    create scenario activity, 370–371
    decision making in, 360, 361
    evaluate trigger conditions activity, 369–370
    execution workflows, 366
    existing architectures, 373–374
    extensibility requirement, 372
    external influences, 360
    future research, 384
    hybrid systems, 364
    key indicators, 360
    model function, 359
    modularization, 375
    observe physical system activity, 369
    perception layer, 376, 377
    physical system in, 361
    process layer, 376, 377
    proof of concept showcase, 378–379
    reference model, 359
    scalability requirement, 372–373
    SSADS example, 381–382
    SSCS example, 380–381
    SSDSS example, 379–380
    SSMVS and SSFS example, 382–383
    symbiotic simulation anomaly detection systems (SSADS), 364
    symbiotic simulation control systems (SSCS), 362
    symbiotic simulation decision support systems (SSDSS), 362
    symbiotic simulation forecasting systems (SSFS), 363
    symbiotic simulation model validation systems (SSMVS), 363–364
    system activities, 366
    systems classification, 361–365
    update simulation activity, 371
    vs. DDDAS, 359
    vs. online simulation, 359
    web services approach, 374–375
    what-if scenarios, 360

WIPER emergency response system, 373
workflows and activities, 365–372
Symbiotic simulation anomaly detection systems (SSADS), 364. *See also* SSADS
Symbiotic simulation control systems (SSCS), 362. *See also* SSCS
Symbiotic simulation decision support systems (SSDSS), 362. *See also* SSDSS
Symbiotic simulation forecasting systems (SSFS), 363. *See also* SSFS
Symbiotic simulation model validation systems (SSMVS), 363–364. *See also* SSMVS
Symbiotic simulation systems workflows, 365–366
 SSADS workflow, 368
 SSCS workflow, 366–367
 SSDSS workflow, 366
 SSFS workflow, 367
 SSMVS workflow, 368
System dynamics, 56, 63
System state space, 113
System trajectory, over time, 113
Systems biology, 269

T

Tables, in SeSAm, 494–496
Task networks, swarming over, 126
Temporal evolution, of environments, 23
Temporal granularity, of actions, 25, 26
Termite construction behavior, 245, 246
Test-beds, 2, 5, 77, 420
Text extraction tools, 290
Theoretical behaviors, in virtual reality, 33, 34
Theoretical environments, moving to actual from, 35
Theoretical source system, 32, 33
Thick agents, 271
TileWorld MAS testbed, 98
Time-continuous traffic models, 328–329
Time modeling, 22–23
 agent behavior time, 23
 equidistant time steps in social science simulations, 64
 scheduling, 24–26
 temporal evolution of environment, 23
Time-stepped approach, 84, 89
 in DES, 79
 in Jason, 460
Tools, 419–421
Traces. *See also* Behavioral traces
 pheromones and stigmergic cues as, 252
 of practical behavior, 247–248
 practical behavioral, 247–248
 registration by environment, 257
 as signals, 260
 social functions, 259, 260
 stigmergic cues as, 247–248
 vs. cues, 260
Traffic assignment, 326
Traffic dynamics, 326
Traffic flow propagation, 326
Traffic light approaches, 339–341
Traffic safety simulation, 329
Traffic simulation agents, 325–327
 acceleration strategy, 331
 adaptive cruise control, 352
 behavioral attributes, 329
 braking deceleration, 341
 braking strategy, 331, 332
 cellular automata (CA), 329
 coffeemeter, 345–346
 collective pattern emergence, 346–351
 comfortable deceleration, 332, 336, 341
 3D animation, 345
 desired acceleration, 336
 desired speed, 331, 335
 desired time gap, 332, 336
 discrete decision modeling, 338–341
 driver-vehicle agents, 329–330
 driving task models, 331–338
 equilibrium distance, 334
 equilibrium gap, 333
 equilibrium properties, 333
 equilibrium traffic, 333
 equilibrium velocity, 334
 frustration effect, 338
 Intelligent Driver Model (IDM), 331–335
 inter-driver variability, 335–336
 intra-driver variability, 337–338
 iterated coupled map approach, 329
 lane change modeling, 338–339
 macroscopic *vs.* microscopic approaches, 327–329

maximum acceleration, 331, 332
memory effect, 337
microscopic traffic simulation software, 341–346
minimum distance, 332, 333, 336
prediction of future congestion levels, 326
relaxation time, 337
rush hour traffic and speed limits, 348
safe distance strategy, 331
safety time gap, 334
simulation controllers, 341, 342
speed limit impacts, 347–349
stop-and-go waves, 346–347
stopping criterion, 340
time-continuous models, 328–329
traffic light approaches, 339–341
transversal hopping, 349
two-lane freeway visualization, 345
Transitions, in plain diffusion case study, 148
Transport assignment
in AGV transportation system, 174
field-based, 205
protocol-based, 205
Transport planning, 326
Transportation systems
efficiency, 325
traffic dynamics, 326
transport planning, 326
vehicle dynamics, 326
Transversal hopping, 349
Traveler decisions, 326
Trigger conditions, in symbiotic simulation systems, 369–370
Triggering events, in Jason, 454
Trip distribution, 326
Trip generation, 326
Tuning phase, 140
plain diffusion, 154–157
PRISM tool in, 145
self-organizing MAS, 143–144

U

UAV path planning, 362, 373–374
as mobile application, 375
SSDSS example, 379–380
symbiotic simulation in, 358
Ubersim simulation platform, 206
Uhrmacher, A.M., xi

Ultimatum game, 223
Uncertainty estimation, 296
Underground station
pedestrian modeling scenario, 310–311
screenshots, 314
Universality, polyagents, 125
Unmanned aerial vehicles (UAVs), 358. *See also* UAV path planning
Uppsala Conflict Database (UCD), 282, 283, 285
Upward causation, 54
User agents, *vs.* environmental agents, 138
User involvement, in ABS tools, 480
User roles, with SeSAm, 478–482

V

Validation and verification (V&V), 31
Validity
paradigmatic, 38
in Zeigler's framework, 30
Value tree models, 292
Vehicle dynamics, 326
Vehicle routing, with polyagents, 119–122
Vehicular traffic MAS, 268
Verification phase, 140
plain diffusion, 140–154
PRISM tool in, 145
self-organizing MAS, 143
Virtual reality (VR), 41
as environment model issue, 33–35
merging with participatory design, 36–37
theoretical behaviors in, 33, 34
VISSIM, 342
Visual programming, 494, 540
with SeSAm, 477–478
Visualization
for MAS, 98–99
in MASON toolkit, 408
separation from model in MASON toolkit, 408
Voting behavior simulation, 57

W

Warehouse simulation. *See* AGV transportation system
Web crawlers, 286
Web services, in symbiotic simulation, 374–375
Webots simulation platform, 206

execution time specification, 209
Webscraping, newsfeeds, 272
Wet bench tool set (WBTS), 378, 379
Weta Digital MASSIVE software, 12
Weyns, D., xi
What-if scenarios, 540
  and decision support, 362
  in SSCS workflow, 366
  in SSDSS workflow, 366
  in stem cell modeling, 395
  in symbiotic simulation, 360
WIPER emergency response system, 373, 374
  SSMVS and SSFS examples, 382
WORC-C, 377, 380, 381, 383
Work
  and practical behavioral traces, 247–248
  role in stimulation agent reactions, 245
Work-piece avatars, 118
Workflow controller capability (WORC-C), 377. *See also* WORC-C
World class descriptions, SeSAm toolkit, 486
World Values Survey (WVS), 283, 285

X
XML Schema, definition of event types with, 522
XRaptor simulation platform, 205

Z
Zeigler's Framework for modeling and simulation, 29, 41
  environment model issues, 33–35
  experimental frame in, 29
  model in, 29–30
  modeling relation, 30, 31
  simulation relation, 30–31, 31
  simulator correctness, 30–31
  simulator in, 30
  source system in, 29
  studying MAS simulations using, 32–40
  three fundamental questions, 31
  validity modeling relation, 30
Zero sum games, 218